THE FRESH-WATER ALGAE OF THE UNITED STATES

McGRAW-HILL PUBLICATIONS IN
THE BOTANICAL SCIENCES

Edmund W. Sinnott, *Consulting Editor*

ARNOLD An Introduction to Paleobotany
CURTIS AND CLARK An Introduction to Plant Physiology
EAMES Morphology of the Angiosperms
EAMES Morphology of Vascular Plants: Lower Groups
EAMES AND MACDANIELS An Introduction to Plant Anatomy
HAUPT An Introduction to Botany
HAUPT Laboratory Manual of Elementary Botany
HAUPT Plant Morphology
HILL Economic Botany
HILL, OVERHOLTS, POPP, AND GROVE Botany
JOHANSEN Plant Microtechnique
KRAMER Plant and Soil Water Relationships
KRAMER AND KOZLOWSKI Physiology of Trees
LILLY AND BARNETT Physiology of the Fungi
MAHESHWARI An Introduction to the Embryology of the Angiosperms
MILLER Plant Physiology
POOL Flowers and Flowering Plants
SHARP Fundamentals of Cytology
SINNOTT Plant Morphogenesis
SINNOTT, DUNN, AND DOBZHANSKY Principles of Genetics
SINNOTT AND WILSON Botany: Principles and Problems
SMITH Cryptogamic Botany
  Vol. I. Algae and Fungi
  Vol. II. Bryophytes and Pteridophytes
SMITH The Fresh-water Algae of the United States
SWINGLE Textbook of Systematic Botany
WEAVER AND CLEMENTS Plant Ecology

There are also the related series of McGraw-Hill Publications in the Zoological Sciences, of which E. J. Boell is Consulting Editor, and in the Agricultural Sciences, of which R. A. Brink is Consulting Editor.

THE FRESH-WATER ALGAE OF THE UNITED STATES

BY

GILBERT M. SMITH

Stanford University

SECOND EDITION

McGRAW-HILL BOOK COMPANY

New York Toronto London

1950

THE FRESH-WATER ALGAE OF THE UNITED STATES

12 13 14 - MP - 1

ISBN 07-058809-0

PREFACE

The years since the appearance of the first edition have seen the addition of many genera and species to the known fresh-water algal flora of this country. For certain groups (Xanthophyceae, Chrysophyceae, Dinophyceae) the number of genera known to occur in the United States has been more than doubled. In addition, many genera known from but one or two localities in 1933 are now known to have a much wider distribution. The morphology and mode of reproduction of many of the genera described in the first edition are also more fully known than they were seventeen years ago, and this has necessitated a change in the systematic position of certain genera.

This edition follows the general plan of the first edition, except for treatment of the species in the various genera. In the first edition, where there were less than ten species in a genus, each was named and briefly characterized. In this edition the species of a genus are listed and references given to sources where a complete description of each of them can be found. Another change is the addition of a survey of the Charophyceae, Cryptophyceae, and Chloromonadales, groups that were not treated in the first edition.

The completeness of this edition has been greatly enhanced by the helpful cooperation of many phycologists. For supplying information or furnishing specimens I am indebted to Dr. Mary A. Pocock and to Professors L. H. Flint, D. L. Jacobs, C. E. Taft, E. N. Transeau, L. A. Whitford, and R. D. Wood. Special thanks are due Professors G. W. Prescott and R. H. Thompson for furnishing numerous original drawings and for permitting me to incorporate in this book their as yet unpublished additions to the algal flora of this country.

GILBERT M. SMITH

STANFORD, CALIF.
October, 1950

CONTENTS

CHAPTER 1

NATURE, EVOLUTION, AND CLASSIFICATION OF THE ALGAE

Position of the Algae in a Natural System. The conception of the nature and systematic position of the group of plants known as the algae has been, and still is, continually changing. Although Linnaeus[1] gave the name *Algae* to one of his orders of plants, this cannot be considered the first recognition of them as a definite part of the plant kingdom, since the group that he called the algae consisted very largely of *Hepaticae*. The first delimitation of the Algae as we now interpret the term is that of A. L. de Jussieu,[2] but his characterization of the group is practically worthless since it is based entirely upon macroscopic features. Even when specialists began to devote their attention to this group of plants and had described many genera and species, there was for a long time but little knowledge of the cell structure and the methods of reproduction. Thus, we find C. A. Agardh,[3] the great pioneer in the study of algae, describing them as follows:

Plantae aquaticae acotyledoneae & agamae; gelatinosae, membranaceae vel coriaceae; filamentosae, laminosae vel tandem foliosae; colore virides, purpureae vel olivaceae; articulatae vel continuae; sporidia aut pericarpiis inclusa aut superficiei inspersa foventes.

Endlicher's[4] inclusion of algae, lichens, and fungi in an assemblage (kingdom) called the *Thallophyta* marks the recognition of a morphological distinction that is still followed today, especially in the widely used classification of Eichler,[5] that divides plants into *Thallophyta*, *Bryophyta*, *Pteridophyta*, and *Spermatophyta*. Endlicher separated the Thallophyta from other plants because of their lack of differentiation into stems and leaves. It is quite impossible to draw a sharp distinction between Thallophyta and Bryophyta on such a basis: certain algae, as the Laminariales, have a clear differentiation into stems and leaves; many liverworts have a plant body that is a simple thallus. It is possible, however, to base a clear-cut distinction between Thallophyta and other plants, on the structure of the gamete- and the spore-containing organs. In the Thallophyta the sex organs are one-celled; or when multicellular, as in certain Phaeophyceae,

[1] LINNAEUS, 1754. [2] DE JUSSIEU, A. L., 1789. [3] AGARDH, 1824.
[4] ENDLICHER, 1836. [5] EICHLER, 1886.

do not have the gametes surrounded by a layer of sterile cells. Bryophytes, and plants immediately above them in the evolutionary scale, have multicellular sex organs, in which there is an outer layer of sterile cells. Sporangia of Thallophyta are always one-celled: those of higher plants are always many-celled. Another distinction between the Thallophyta and other plants is the fact that the zygotes of Thallophyta never develop into multicellular embryos while still within the female sex organs.

Granting the distinctiveness of the assemblage of plants called the Thallophyta, there then arises the question: Is this a natural division of the plant kingdom that may, in turn, be divided into *Algae* and *Fungi?* To accept the Thallophyta as a natural division of the plant kingdom implies acceptance of the view that all algae are more or less closely related to one another. The question of the phylogenetic relationships between the algae rests, in turn, upon the discovery of an adequate basis for classifying them. The test of time has shown the inadequacy of taxonomic classifications of algae based either upon the organization of the plant body or upon the method of reproduction.

It has become increasingly clear during the past quarter century that the morphology and physiology of the individual cells are the fundamental basis upon which the algae must be classified. This evidence shows that there are several series among the algae, each of which has cells with certain distinctive morphological and physiological traits. Chief among the morphological characteristics is the structure of the motile cell, and for most of the series among the algae there is a striking constancy in its organization, especially with respect to the number, arrangement, and relative length of the flagella. On the physiological side there is, throughout each series, a constancy in the pigments present in the plastids (Table I), and a constancy in the chemical nature of the food reserves accumulating through photosynthetic activity. For example, the Chlorophyceae always have flagella of equal length, a predominance of green pigments in their plastids, and usually store photosynthetic reserves as starch. The Xanthophyceae on the other hand, always have flagella of unequal length, a predominance of yellow pigments in their plastids, no formation of starch, and usually store photosynthetic reserves as oils. This constancy with which the morphological and physiological cellular characteristics obtain in each series, and the marked differences between the various series (Chlorophyceae, Myxophyceae, Chrysophyceae, etc.) suggest very strongly that they have originated from pigmented ancestors quite different from one another.

Acceptance of the view that the various series among the algae are more or less independent of one another means that the Thallophyta cannot be considered a natural division of the plant kingdom. The *Algae*, likewise,

TABLE I. PRINCIPAL PIGMENTS OF THE DIFFERENT CLASSES OF ALGAE (BASED ON STRAIN, IN PRESS)

	Myxophyceae	Rhodophyceae	Xanthophyceae	Chrysophyceae	Bacillariophyceae	Phaeophyceae	Dinophyceae	Chlorophyceae	Euglenophyceae
Chlorophylls:									
Chlorophyll *a*	+++	+++	+++	+++	+++	+++	+++	+++	+++
Chlorophyll *b*	0	0	0	0	0	0	0	++	+
Chlorophyll *c*	0	0	0	...	+	+	+	0	0
Chlorophyll *d*	0	+	0	...	0	0	0	0	0
Chlorophyll *e*	0	0	+	...	0	0	0	0	0
Carotenes:									
α-Carotene	...	+	...	...	0	0	0	+	
β-Carotene	+++	+++	+++	+++	+++	+++	+++	+++	+++
ε-Carotene	...	...	...	...	+	0	...	0	
Flavicin	+	...	...	...	0	0	...	0	
Xanthophylls:									
Lutein	?	++	0	+	0	0	0	+++	?
Zeaxanthin	?	...	0	...	0	0	0	+	
Violaxanthin	...	...	...	...	0	+	0	+	
Flavoxanthin	...	...	...	...	0	+	...	?	
Neoxanthin	...	...	0	...	0	+	0	+	
Fucoxanthin	...	?	0	+	++	++	0	0	0
Neofucoxanthin A	...	...	0	...	+	+	0	0	0
Neofucoxanthin B	...	...	0	...	+	+	0	0	0
Diatoxanthin	...	...	0	...	+	?	0	0	0
Diadinoxanthin	...	...	0	...	+	?	+	0	0
Dinoxanthin	...	...	0	...	0	?	+	0	0
Neodinoxanthin	...	...	0	...	0	0	+	0	0
Peridinin	...	...	0	...	0	0	++	0	0
Myxoxanthin	++	...	0	...	0	0	0	0	0
Myxoxanthophyll	++	...	0	...	0	0	0	0	0
Unnamed	?	?	++	?		+			+
Phycobilins:									
r-Phycoerythrin	0	+++	0	?	0	0	0	0	0
r-Phycocyanin	0	+	0	?	0	0	0	0	0
c-Phycoerythrin	+	0	0	?	0	0	0	0	0
c-Phycocyanin	+++	0	0	?	0	0	0	0	0

+++ indicates the principal pigment in each of the four groups of pigments.
++ indicates a pigment comprising less than half of the total pigments of the group.
+ indicates a pigment comprising a small fraction of the total pigments of the group.
? indicates small quantities of a pigment whose source or identification is uncertain.
0 indicates known absence of a pigment.
... indicates lack of knowledge concerning the presence of certain pigments in some classes of algae.

cannot be regarded as a particular subdivision of the plant kingdom. This does not mean that the word "alga" must be abandoned, since it is still of great service as a descriptive term for designating simple plants with an autotrophic mode of nutrition.

Organisms to Be Placed among the Algae. Until the beginning of the twentieth century, it was customary to recognize the following four classes of algae: Chlorophyceae, Phaeophyceae, Rhodophyceae, and Myxophyceae (Cyanophyceae). Diatoms were universally included among the algae and placed either in the Phaeophyceae or in a class distinct from other classes.

During this time botanists rarely questioned the practice of protozoologists who placed all motile unicellular and colonial flagellated organisms in the class Mastigophora of the phylum Protozoa. An exception must be made in the case of the volvocine (*Chlamydomonas-Volvox*) series. Here, beginning nearly a century ago,[1] botanists began calling certain members of this series algae but made no attempt to assign them a definite place among the algae. This was first done by Rabenhorst (1863) who placed the *Chlamydomonas-Volvox* series in the group of grass-green algae to which he gave the name Chlorophyllaceae.

When, at the turn of the century, the Xanthophyceae (Heterokontae) were segregated[2] from the grass-green algae (Chlorophyceae), certain pigmented flagellates were included in the series. Later the chrysomonads and the dinoflagellates each were shown to be related to organisms of an unquestionable algal nature. The euglenoids and cryptomonads are also related to organisms of an algal type, but types that are not so highly developed as in the case of the algal types related to the chrysomonad and the dinoflagellate series.

Thus, with the possible exception of the chloromonads, all the various groups (orders) of flagellates which protozoologists place in the subclass Phytomastigina of the class Mastigophora are phylogenetically connected to organisms of a truly algal nature. Disregarding, for the present, the interrelationships between them, these phylogenetic series (classes) may be briefly characterized as follows:

1. *Chlorophyceae*, in which the photosynthetic pigments are localized in chromatophores that are grass-green because of the predominance of chlorophylls *a* and *b* over the carotenes and xanthophylls. Photosynthetic reserves are usually stored as starch, and its formation is intimately associated with pyrenoids. Motile cells have flagella (generally two or four) of equal length and borne at the anterior end. Most members of the class reproduce sexually.

2. *Euglenophyceae*, in which the photosynthetic pigments are approximately the same as in Chlorophyceae and are localized in grass-green

[1] Braun, 1851; Cohn, 1853. [2] Luther, 1899.

chromatophores. Paramylum, an insoluble carbohydrate, is the chief food reserve. Motile cells have either one or two flagella and have them inserted in a gullet at the anterior end of the cell.

3. *Xanthophyceae* (*Heterokontae*), in which the photosynthetic pigments are localized in chromatophores that are yellowish green because of a predominance of beta-carotene over chlorophylls *a* and *e*. Pyrenoids are usually lacking, and the reserve foods are stored as fats or as leucosin. The cell wall frequently consists of two overlapping halves and contains little, if any, cellulose. Motile cells have two flagella of unequal length at the anterior end.

4. *Chrysophyceae*, in which the photosynthetic pigments are localized in chromatophores that are usually golden brown because of a predominance of carotenes and xanthophylls over the chlorophyll. Oils are formed in abundance, and sometimes there is also a formation of leucosin. Motile cells may have a single anterior flagellum or two anterior flagella of unequal or of equal length. Members of this series have an endogenous formation of cysts surrounded by a two-parted silicified wall with a terminal pore. Sexual reproduction is of rare occurrence in this class.

5. *Bacillariophyceae* (*Bacillarieae*), in which the photosynthetic pigments are localized in chromatophores that are usually a deep golden brown because of a predominance of carotenes and xanthophylls (especially, fucoxanthin) over chlorophylls *a* and *c*. The cell wall regularly consists of two overlapping halves which are highly silicified. Reproduction by flagellated swarmers has been found in certain genera, but the precise nature of these motile bodies is unknown. Sexual reproduction is of widespread occurrence and is immediately preceded by meiosis.

6. *Phaeophyceae*, in which the photosynthetic pigments are localized in chromatophores that are olive yellow to deep brown because of a predominance of carotenes and a series of xanthophylls (notably, fucoxanthin) over chlorophylls *a* and *c*. The most abundant reserve product of photosynthesis is a polysaccharide, laminarin. All members of the class, of which there are numerous marine members, have a filamentous or a more elaborate organization. Motile reproductive cells are pyriform with two laterally inserted flagella. Sexual reproduction is found in most of the genera.

7. *Dinophyceae*, in which the photosynthetic pigments are localized in chromatophores that are yellowish green to deep golden brown because of a predominance of carotenes and a unique series of xanthophylls over chlorophylls *a* and *c*. Food reserves are stored as starch or as oil. Motile cells have a transverse furrow in which the two flagella are inserted. One flagellum encircles the cell transversely, the other extends vertically backward. Sexual reproduction is of rare occurrence in this class.

8. *Myxophyceae* (*Cyanophyceae*), in which the photosynthetic pigments

are not localized in chromatophores. In addition to chlorophyll *a*, beta-carotene, and two unique xanthophylls, the cells contain two phycobilin pigments c-phycocyanin and c-phycoerythrin. The cells do not have a definitely organized nucleus. The chief food reserve is a carbohydrate, cyanophycean starch. Sexual reproduction is unknown for this series and there is never a formation of flagellated reproductive cells.

9. *Rhodophyceae*, in which the photosynthetic pigments are localized in chromatophores that are usually reddish in color because of a predominance of r-phycoerythrin over the other pigments (r-phycocyanin, chlorophylls *a* and *d*, alpha- and beta-carotene, and the xanthophyll lutein). The chief food reserve is floridean starch, a carbohydrate intermediate between starch and dextrin. Motile reproductive cells are never found within this series. The sexual organs are of a unique type, and practically all members of the class reproduce sexually.

10. *Cryptophyceae*, in which the photosynthetic pigments are localized in variously colored chromatophores. Reserve foods are usually stored as starch or as starch-like compounds. Motile cells are compressed, biflagellate, and with a superficial curved furrow extending back from the terminal or lateral insertion of the flagella.

11. *Chloromonadales*, in which the photosynthetic pigments are localized in chromatophores of a distinctive green color. The chief food reserve is oil. Motile cells are biflagellate and with the flagella inserted in a reservoir at the anterior end. A majority of the genera have trichocysts within the cells.

Evolution of Plant-body Types among Algae. The modern conception of the nature of evolution of the thallus among algae originated from observations[1] which showed that there was a marked parallelism between the Chlorophyceae and the Xanthophyceae, and that practically all types of cellular or colonial organization among the Chlorophyceae have their counterparts among the Xanthophyceae. This was accepted as an interesting though not significant fact until it was shown that the types of plant-body construction found in the Chrysophyceae, Dinophyceae, and certain other series can also be homologized with those of the Xanthophyceae and Chlorophyceae.[2]

The explanation of this parallelism is based upon the hypothesis that only four basic types of body (thallus) construction can be evolved from a primitive motile unicellular ancestral form. This hypothesis also holds that the types of plant body evolved from the motile unicellular ancestor of one series are essentially like those evolved from the motile unicellular forms of other series. The idea of evolution in different directions from a motile unicellular ancestor was originally proposed to account for the

[1] Pascher, 1913. [2] Pascher, 1914, 1925, 1927.

various types of body construction found in the Chlorophyceae.[1] Here it was postulated that there are three main evolutionary lines or tendencies from the motile unicell: (1) the *volvocine tendency*, in which the individual cells become organized into a colony but retain their vegetative motility; (2) the *tetrasporine tendency*, in which there is a loss of motility, except in reproductive stages, but a retention of the capacity for vegetative division; (3) the *chlorococcine* (*siphonaceous*) *tendency*, in which there is a loss of motility, except at the time of reproduction, and a loss of the ability to divide vegetatively. To these should be added (4) the *rhizopodal tendency*, in which there is evolution toward an amoeboid type of organization.

In its wider application[2] it is held that these tendencies are also found in phylogenetic series other than the Chlorophyceae, and that each of the algal series has repeated the same experiments in evolution of body types. Certain of these experiments were foredoomed to failure, since the potentialities are extremely limited. Thus, the volvocine tendency, or evolution of a volvocine colony, cannot develop a colony of any appreciable size and have the individual cells retain their motility. The siphonaceous tendency, with its abolition of vegetative cell division but retention of nuclear division, is also an experiment that was not particularly successful.

The tetrasporine experiment is along a line that results in a nonmotile holophytic organism capable of infinite variation. The beginnings of evolution along the tetrasporine line are found in those algae in which motile cells are imprisoned within a gelatinous sheath. Many of the Volvocales have temporary colonies of this nature, and certain of the lower Tetrasporales, as the Chlorangiaceae, have colonies in which motile cells are more or less permanently imprisoned within gelatinous sheaths. Colonies of the *Palmella* type are a step in advance of this. Here, the cells within the gelatinous matrix are without flagella but may develop them at any time and so return directly to a motile unicellular condition. Cells in colonies of the *Palmella* type have the capacity to divide vegetatively, and cell division may result in amorphous colonies of indefinite size or in colonies with a definite shape. The next step in the tetrasporine evolution is the loss of the cell's capacity to return directly to a motile condition. However, this is not accompanied by a loss of the capacity to divide vegetatively. When such cells divide vegetatively, the daughter cells may separate from each other or may remain united. Separation of daughter cells after cell division results in an immobile unicellular organism. Cohesion of the daughter cells, followed by further divisions in the same plane, results in an unbranched filament; the simplest of the filamentous types and the forerunner of the various branched types.

[1] Blackman, 1900; West, G. S., 1904.

[2] Fritsch, 1929, 1935; Fritsch and West, 1927; Pascher, 1914, 1925, 1927.

Evolution along the chlorococcine line (in which there is a loss of motility of the vegetative cells and a loss of the capacity to divide vegetatively) may begin with the primitive unicellular flagellate. It may also arise from unicellular members of the tetrasporine line by a loss of the capacity for vegetative division. The siphonaceous type is a modification of the chlorococcine type in which the nuclei retain their ability to divide and the cells develop the capacity to elongate indefinitely.

Evolution along the tetrasporine and the chlorococcine lines results in immobile organisms that are distinctly plant-like in organization. Evolution of the motile unicell along the rhizopodal line results in an organism with an amoeboid method of movement and nourishment. In certain series, as the Chlorophyceae, amoeboid stages are but temporary. These have been found in motile unicellular forms, such as *Chlamydomonas*,[1] and in gametes or zoospores of specialized filamentous forms, such as *Stigeoclonium*.[2] In other phylogenetic series, as the Chrysophyceae, Heterokontae, and the Dinophyceae, there has been a rhizopodal evolution to a state where the cells are in a more or less permanent amoeboid condition.[3]

The parallelism in the evolution of plant-body types among the various algal series with motile unicellular forms comes out most strikingly when the data are arranged in tabular form (Table II). Analysis of the data immediately brings out the fact that progressive evolution within the various series has not reached the same level: certain series have progressed but little beyond the motile unicell, others have advanced to the point where they have a complex plant body. The point which Table II does not bring out is the relative abundance of the different types in the various phylogenetic series. Some of them, as the Chrysophyceae and Dinophyceae, have but few advanced types; others, as the Chlorophyceae and Phaeophyceae, have many advanced types.

The preceding paragraphs have stressed the evolution of certain algal series from motile pigmented ancestors markedly different in their basic morphology and physiology. This postulation of the origin of certain series from motile unicells does not mean that all algal series have arisen in this fashion. The complete lack of flagellated vegetative and reproductive cells in the Myxophyceae suggests very strongly that motile cells have never been present in the myxophycean series. If the Myxophyceae have developed from a flagellated ancestor, one would expect to find flagellated reproductive cells somewhere among the many genera of this series.

The problem of determining the origin of Rhodophyceae and of Phaeophyceae is extremely difficult, since the simplest of them have a rather

[1] PASCHER, 1918. [2] PASCHER, 1915. [3] PASCHER, 1917.

TABLE II. THE PARALLELISM IN EVOLUTION OF PLANT-BODY TYPES AMONG THE CLASSES OF ALGAE WITH KNOWN FLAGELLATED UNICELLS

Unless otherwise noted, the examples cited are found in the fresh-water flora of the United States

	Type of plant-body construction	Chlorophyceae	Xanthophyceae	Chrysophyceae	Dinophyceae	Euglenophyceae
	Motile unicell	Chlamydomonadaceae, Sphaerellaceae, etc.	Heterochloridales	*Mallomonas*, *Chromulina*, etc.	*Glenodinium*, *Peridinium*, etc.	Euglenaceae, etc.
Tetrasporine tendency	Palmelloid colony with imprisoned motile cells	Chlorangiaceae	....	....	....	*Colacium*
	Palmelloid colony with nonflagellated vegetative cells	Palmellaceae, Tetrasporaceae, etc.	Chlorosaccaceae	*Chrysocapsa*, *Hydrurus*, etc.	*Gloeodinium*	
	Simple filaments	Ulotrichaceae, etc.	Tribonemataceae	*Nematochrysis* (Pascher, 1925)		
	Branched filaments	Chaetophoraceae, etc.	*Monocilia*	*Phaeothamnion*	*Dinothrix* (Pascher, 1914)	
Chlorococcin tendency	Chlorococcoid cells	Chlorococcales	Heterococcales	*Chrysosphaera* (Pascher, 1925)	*Dinastridium* (Pascher, 1927)	
	Siphonaceous cells	Siphonales	*Botrydium*			
Volvocine tendency	Dendroid colonies	....	....	*Dinobryon*		
	Globose colonies	Volvocaceae, etc.	....	*Chrysosphaerella*, *Synura*, etc.		
Rhizopodal tendency	Unicellular	Transitory stages only	*Myxochloris* (Pascher, 1930*B*)	*Chrysamoeba*	*Dinamoeba* (Pascher, 1915*C*)	
	Plasmodial colonies	....	*Chlorarachnion* (Geitler, 1930)	*Chrysidiastrum*		

complex organization, and all connecting links with hypothetical unicellular ancestors are unknown. One explanation for this absence of primitive Rhodophyceae and Phaeophyceae is that they developed in the ocean at a time when it was much less saline than at present, and that there was a dying off of the more primitive forms as the salinity of the ocean increased. There was, however, a survival of certain advanced types among the Rhodophyceae and Phaeophyceae, and these constituted a fresh starting point for the even more complex red and brown algae found in the present-day marine flora. The universal presence of motile zoospores and gametes throughout the phaeophycean series indicates that this series originated in a motile unicellular ancestor; the lack of motile reproductive cells in the rhodophycean series seems to show that, similar to the Myxophyceae, these algae have come from a nonflagellated unicellular ancestor.

Classification of the Algae. If the Thallophyta cannot be considered as constituting a natural division of the plant kingdom, how can the algae be brought into harmony with the International Botanical Code which says (Article 10), "every individual plant belongs to a species, every species to a genus, every genus to a family, every family to an order, every order to a class, and every class to a division"? It is clear that the algae must be separated into a number of divisions coordinate in rank with the Bryophyta, Pteridophyta, and Spermatophyta. Modern discussions on the phylogeny and classification of algae hold that certain of the classes noted above (see page 4), are sufficiently distinct to warrant recognition as divisions of the plant kingdom.[1] However, other classes have so many features in common that they are evidently related to one another. Thus, the number of divisions necessary for a complete classification of the algae is less than the number of classes. The first recognition of an affinity between classes was that which showed a relationship between the Xanthophyceae, Chrysophyceae, and Bacillariophyceae.[2] Features in common to these three classes include: cell walls composed of two overlapping halves, silicified cell walls, motile cells with similarities in flagellation, a distinct type of resting cell (cyst), and similarities in the nature of food reserves. Despite differences in the chlorophylls and xanthophylls (see Table I), there seems to be good ground for placing the three in a single division, the *Chrysophyta*. The golden-brown chromatophores of the Phaeophyceae resemble those of Chrysophyta, but there are some differences in the pigments causing the brown color. Since there are striking differences in the food reserves and in structure of motile reproductive cells, the Phaeophyceae should be placed in a separate division, the *Phaeophyta*. The Myxophyceae and Rhodophyceae are the only algae in which there are

[1] PASCHER, 1914, 1921, 1931. [2] PASCHER, 1914.

phycobilin pigments, but these pigments are not identical in the two,[1] and the two classes differ in their chlorophylls and xanthophylls (see Table I). The differences in nuclear organization, localization or nonlocalization of pigments in chromatophores, and presence or absence of sexual reproduction are so striking that there does not seem to be a phylogenetic connection between the two classes. Thus the Rhodophyceae are to be placed in one division, the *Rhodophyta*, and the Myxophyceae in another, the *Cyanophyta*. The chlorophycean series is also so distinctive that it should be placed in a separate division, the *Chlorophyta*. Similarities in pigmentation and food reserves of Euglenophyceae and Chlorophyceae tempt one to place the Euglenophyceae in the Chlorophyta, but it seems better to place them in a separate division, the *Euglenophyta*. The Dinophyceae have sufficient distinctiveness to be placed in another division, the *Pyrrophyta*. Some phycologists[2] think that the Cryptophyceae should be included in the Pyrrophyta; others[3] think that they should not. For the present it seems better to consider the Cryptophyceae a class of uncertain systematic position and not to place them in any of the divisions mentioned above. The question of the proper disposition of the chloromonads is even more difficult, and in their case, also, it seems best to group them among algae of uncertain systematic position.

Relationship of Algae to Other Plants. According to the International Rules of Botanical Nomenclature the primary step in a classification of the plant kingdom is the establishment of *divisions*. The multicellular plant body in certain of the algal divisions, as the Cyanophyta, is merely an aggregation of individuals into a colony of simple construction. Other phylogenetic lines, as the Rhodophyta and Phaeophyta, include algae of large size, external complexity of form, and with some differentiation of tissues within the plant body. In reality, the six divisions listed above represent six kingdoms, all plant-like in nature. In five of these kingdoms there has not been an evolution of anything more complex than an algal type of organization. Thus these five kingdoms would have but one division each. The kingdom of the grass-green plants consists of a number of divisions of which the grass-green algae (Chlorophyta) are the most primitive and lead successively to the Bryophyta, Pteridophyta, and Spermatophyta.

[1] STRAIN (in press). [2] PASCHER, 1914, 1927. [3] GRAHAM (in press).

CHAPTER 2

THE DISTRIBUTION AND OCCURRENCE OF FRESH-WATER ALGAE

One ordinarily thinks of the fresh-water algae as plants largely restricted to standing and running waters and occasionally growing in other habitats. This is far from the truth. Algae are of widespread occurrence in moist situations (as tree trunks, walls, woodwork, rocks, and damp soil) where they frequently occur as an extended stratum consisting of either a single species or a mixture of species. There is, in addition, a rather long list of what may be called "algae of unusual habitats." These algae include those growing endophytic in other plants, leaf epiphytes, perforating algae of molluscs and calcareous rocks, snow algae, thermal algae, epizoic algae, and certain others.

Geographical Distribution of Algae. One of the striking features of the fresh-water algal flora is its cosmopolitanism. Many species are found in all parts of the world, from the tropics to the polar regions, and in a variety of habitats. Other species are restricted to particular habitats, but even these may be found at stations thousands of miles apart.

Some of the fresh-water algae, such as *Trentepohlia* and *Pithophora*, are more abundant in the tropics than in temperate regions, and a few of them appear to be limited by temperature. The best known of these are *Cephaleuros*, a green alga parasitic upon the leaves of several Angiosperms, and *Compsopogon*, a red alga found in the southern states of this country, the West Indies, and in Central America.

The only group of fresh-water algae in which there is any evidence of endemism is that of the Desmidiaceae. A specialist shown a collection rich in species of desmids, but not told the source of the collection, would be able to tell whether it came from Europe, Australia, the Indo-Malay region, or the Americas, but even in this family the majority of species are cosmopolitan.

Dispersal of Algae. The cosmopolitanism of most species, and the localism of others, are dependent upon the methods by which algae become distributed from one locality to another. All discussions of the means by which alga are dispersed[1] have been based upon general observations rather than detailed study, and it is not definitely known whether algae are trans-

[1] Borge, 1897; Ström, 1926; Wille, 1897.

ported in a vegetative or in a resting stage. It is generally assumed that vegetative cells cannot withstand desiccation and would perish while being transported through the air. Studies[1] on the viability of algae in dry soils show that this is far from the case and that several algae without zygotes or spore-like stages withstand desiccation for more than a quarter century. The importance of zygotes and resting cells has been greatly overemphasized in discussions on modes of dispersal,[2] and it is probable that dissemination of vegetative cells is of greater importance than that of resting cells.

Streams assist in the dispersal of algae, but the two major agencies transporting algae from one locality to another are birds and the wind. Those who argue for transportation by birds[3] hold that most of the algae are carried in half-dried mud adhering to aquatic birds' feet, but lodging of algae among their feathers may be fully as important a factor. Transfer of plankton algae is brought about by migratory aquatic birds moving from one body of water to another. The almost complete lack of resting stages among plankton algae shows that this transfer must take place in a vegetative condition. The effectiveness of transportation of plankton algae in a vegetative condition is to be seen in the widespread distribution of *Stylosphaeridium stipitatum* (Bachm.) Geitler and Gimesi, a chlorophycean epiphyte restricted to *Coelosphaerium Naegelianum* Unger, a distribution that can be accounted for only by assuming that colonies of *Coelosphaerium* have been carried from place to place.

Transportation by wind is rarely recognized as an important factor, but studies on the movement of dust clouds demonstrate the importance of wind as an agency of dispersal. Dust clouds originating in Arizona and New Mexico, or in the western portion of the Great Plains, have traveled east of the Mississippi before falling to the earth.[4] The finding of diatoms among these dust particles that have been carried for more than a thousand miles shows that wind transportation of algae is not a mere theoretical possibility. It is true that there is no evidence at hand to show that the diatoms just mentioned were alive but, since we know that diatoms of dried soil remain viable for long periods,[5] it is not unreasonable to assume that some of them deposited from the dust clouds were alive.

The best proof of this transportation by wind is to be seen in the widespread distribution of certain algae that are restricted to special habitats. For example, *Dunaliella* is a chlamydomonad of world-wide distribution but known only from brine lakes, salterns, and oceanside pools, with a sodium chloride content of 12 to 17 per cent. Another example is to be found in the waters of hot springs in this country and elsewhere. Hot

[1] Bristol-Roach, 1919, 1920. [2] Ström, 1924*A*. [3] Beger, 1927.
[4] Winchell and Miller, 1918, 1922. [5] Bristol-Roach, 1920.

springs in all parts of the world contain certain Myxophyceae that are not found elsewhere than in thermal waters. The almost universal occurrence of these saline and thermal algae in small habitats, remote from one another, shows that wind is a very effective agency in dispersing algae. Dispersal of thermal algae cannot possibly be ascribed to birds since it is inconceivable that a bird would fly directly from one hot spring region to another.

ENVIRONMENTAL FACTORS AFFECTING THE GROWTH OF ALGAE

The successful introduction of algae into new localities by the methods just mentioned depends upon their being transferred to a suitable habitat. Water is essential for growth of the alga in the new location, but there are many other factors, any one of which may prevent its growth. Chief among these are light, temperature, and the chemical composition and pH of the water. Attempts to evaluate the effects of these various factors have been based upon two general methods of investigation: (1) growth of algae in pure culture and (2) chemical and physical studies of various types of habitat. The results obtained by either method are so inconclusive and contradictory that one can make only very broad generalizations concerning the effects of the various factors.

Light. Light is an essential for photosynthesis, but algae differ markedly in respect to their tolerance of light intensity. Some of the aerial algae, as terrestrial species of *Vaucheria*, are indifferent to the intensity of light; other aerial algae, particularly many Myxophyceae, grow only in shaded habitats. Aquatic algae of unshaded pools are usually considered as growing in full sunlight, but the more deeply submerged ones are not so strongly illuminated as those at the surface, because there is a geometrical decrease in intensity of illumination with an arithmetic increase in depth. Despite this, many algae show a greater rate of photosynthesis at a certain depth than they do at the surface.[1]

As is well known, there is a qualitative penetration of light in water and a greater absorption at the red end of the spectrum (see Table III). However, the extent to which the various rays penetrate the water is affected by the color of the water, its turbidity, and the amount of dissolved salts.[2] The vertical distribution evident in the attached algae of deep-water lakes[3] is directly correlated with the intensity of illumination at different levels; and there are certain algae, especially Rhodophyceae and Myxophyceae, that are found only many meters below the surface.

According to the theory of complementary chromatic adaptation (see page 546), light also affects the coloration of algae containing phycoerythrin and phycocyanin.

[1] MANNING, JUDAY, and WOLF, 1939. [2] PIETENPOL, 1918.
[3] OBERDORFER, 1927; ZIMMERMANN, 1927.

Temperature. Temperature rarely plays a direct role in the acclimatization of algae in new localities, but it has a very important effect in its acceleration or retardation of growth and reproduction. Under exceptional conditions, as in thermal algae and the cryovegetation, the temperature of the habitat restricts the algal population to certain species. A few algae, as *Hydrurus*, are found in very cold waters, but in most cases temperature is not a factor determining the nature of the algal flora, since most species are able to develop if other conditions are favorable.

Temperature is probably the limiting factor for the algae restricted to tropical and subtropical regions. This is shown by the known cases where tropical algae have been introduced into colder regions[1] and where they developed luxuriantly the first summer but were unable to over-winter in the colder climate.

TABLE III. THE AMOUNT OF SURFACE ILLUMINATION PENETRATING TO VARIOUS DEPTHS IN A FRESH-WATER LAKE (THE LAKE OF CONSTANCE*)

Depth, m.	Intensity of illumination compared with that of the surface, per cent		
	Red	Blue	Violet
1	61.8	66.3	72.5
5	9.0	12.8	20.0
10	0.81	1.6	4.0
20	0.0066	0.026	0.16
30	0.000053	0.00044	0.0064

*OBERDORFER, 1927.

Inorganic Compounds. The elements essential for growth of the algae are the same as those necessary for the growth of higher plants. Calcium is not an essential element for many algae,[2] but certain of them are unable to develop in a medium lacking it. Silicon is an additional element necessary for the growth of diatoms.

The marked differences in the algal flora of various regions are directly correlated with the amount of calcium. The algae of soft-water regions (*calciphobes*) are largely desmids and certain species of Chrysophyceae, Myxophyceae, and Chlorophyceae. These algae are rarely found in the *calciphilic* flora of hard-water regions. The distinction between calciphilic and calciphobic algae rests more upon the pH value of the water than upon the amount of calcium,[3] since it has been shown that the bottom and littoral flora of lakes in limestone regions may have a rich calciphobic element[4] if suitable conditions of acidity prevail. The pH value of the

[1] COLLINS, 1916. [2] PRINGSHEIM, E. G., 1946. [3] STRÖM, 1926.
[4] STRÖM, 1924*A*.

water has been shown to be a limiting factor for many algae,[1] especially species of desmids, but these observations must be repeated for many localities before one can make generalizations upon the distribution of particular species.

Calcium and magnesium are also of importance in their influence upon the total number of algae present, because their bicarbonates furnish a supplemental supply of carbon dioxide for photosynthesis. The greater abundance of algae in hard-water lakes, as compared with their abundance in soft-water lakes, is traceable directly to a utilization of dissolved bicarbonates in photosynthesis.[2]

One would suppose that the total mineral content of the water would be an important factor governing the occurrence of algae, but this does not seem to be the case. Thus, the extensive algal flora of Devils Lake, North Dakota, a brackish-water lake with a salinity of about 1 per cent, is quite similar to that of other lakes in the region that have a much smaller proportion of dissolved salts.[3] In the case of "brine lakes" (see page 22), the nature of the algal flora is directly traceable to the mineral content of the water.

Laboratory experiments have shown that small quantities of many elements are toxic to algae, but iron is the only one of these that is of importance in nature. Most algae grow best when the Fe_2O_3 content of the water is 0.2 to 2.0 mg. per liter, and there is a distinct toxic effect when the amount of available iron is over 5 mg. Many natural waters have a total iron content of more than 5 mg. per liter, but these waters are not toxic because of the buffer action of organic compounds or of calcium salts.[4]

Many of the higher plants show a decided preference for nitrates as the source of nitrogen. A majority of the algae grow equally well when their nitrogen is obtained from nitrates, nitrites, or ammonium compounds. Algae may also utilize more complex nitrogenous compounds as the source of nitrogen. Under certain outdoor conditions the nature and amount of available nitrogenous compounds have a direct influence upon the algal flora. This is especially true of barnyard waters, where the abundance of chlamydomonads and euglenoids is traceable to the richness in ammonia. Another example of an abundant nitrogen supply determining the nature of the algal flora is to be seen in the predominance of *Prasiola* on rocks covered with the droppings of sea birds.

Organic compounds dissolved in the water affect the nature of the algal flora, but the problem is so complex that it is impossible to determine their

[1] WEHRLE, 1927. [2] BIRGE and JUDAY, 1911.
[3] MOORE, 1917; MOORE and CARTER, 1923. [4] USPENSKI, 1927.

effect upon the flora. This is especially true of peaty moors and swampy areas where the characteristic flora is undoubtedly due in considerable part to the dissolved organic materials.

THE ALGAE OF DIFFERENT TYPES OF HABITAT

The striking differences between the algae of various habitats tempt one to undertake an elaborate ecological classification with or without special ecological terms, but one soon runs into difficulties. For example, it is difficult to draw precise limits between aerial algae that obtain all their water from moisture in the air, and terrestrial algae that obtain their moisture partly from the air and partly from the ground water. It is also difficult to draw sharp distinctions on the basis of the nature of the habitat, since there are no precise limits between pools and lakes, or between swiftly running and more slowly moving waters.

For convenience in discussion, the habitats of algae will be divided into three groups: (1) aerial habitats, (2) aquatic habitats, and (3) unusual habitats.

Aerial Habitats. Aerial algae have been defined[1] as algae that obtain their water wholly or in large part from moisture in the air. They are also able to endure drought without entering upon special resting stages. Strictly aerial algae are found on the bark and leaves of trees, on woodwork, on stones, and on rocky cliffs. Most of these algae belong to the Chlorophyceae and *Protococcus*, *Trentepohlia*, and *Prasiola* are conspicuous members of the aerial flora. A moisture-laden atmosphere favors the development of aerial algae. This is usually due to abundant rainfall, as in northern Europe and in the tropics. However, aerial algae may develop in considerable abundance in relatively arid regions where there is a localized area of moisture-laden air. An excellent example of this is seen in the profuse development of *Trentepohlia* on trees next to the shore of the Monterey peninsula, California. Aerial algae usually grow on the shaded side of the substratum, but protection from prevailing winds is probably of greater importance than shade.

Adaptation to an aerial existence depends more upon the water-retaining than upon the water-absorbing capacity of the cells.[2] The ability to withstand desiccation seems to rest upon the high osmotic concentration within the cell and the highly viscous state of the protoplasm.[3]

Terrestrial algae are more nearly aerial than aquatic, but it is impossible to differentiate between aerial and terrestrial on the basis of the source of water. The number of soil-inhabiting algae is far greater than was for-

[1] PETERSEN, 1915, 1928. [2] HOWLAND, 1929; SCHMID, 1927.

[3] FRITSCH, 1922; FRITSCH and HAINES, 1923.

merly supposed and includes many species of Chlorophyceae, Myxophyceae, Xanthophyceae, and Bacillariophyceae.[1] Certain of the soil-inhabiting genera (*Botrydium*, *Protosiphon*, *Botrydiopsis*, *Zygogonium*) are strictly terrestrial, but most of the genera are also known from aquatic habitats.

In the Middle West development of soil algae into an extensive green coating on the soil occurs only during especially rainy years; in California, there is regularly a development of such coatings during the rainy winter months. Sometimes the algae grow in large patches, an acre or more in extent, that consist largely of a single genus. *Vaucheria* is a good example of this. More often, terrestrial algae grow in small patches, a few inches in diameter, that contain a number of species. The particular species present often depend upon the texture and chemical composition of the soil. The effect of soil texture upon the nature of the algal flora comes out quite clearly when one compares the algae growing on a well-beaten path with those growing on bare loose soil next to the path. Examples of the influence of chemical environment are found in the restriction of *Zygogonium* to acid soils and in the restriction of *Prasiola* to soil rich in nitrogenous matter.

Terrestrial algae also grow beneath the surface of the soil and even at a depth of a meter or more. These algae are rarely evident when one makes a microscopical examination of samples of subterranean soil, but they soon appear in cultures started from such samples.[2]

Soil-dwelling algae, like strictly aerial algae, are able to withstand prolonged desiccation. This is especially evident in regions with a long rainless season, as California, where a coating of algae appears upon the soil a few days after the rainy season begins. Many of these algae pass through the dry season in a vegetative condition. The length of time that soil algae can withstand desiccation is remarkable, and some of them are able to resume growth after drying for 50 years.[3]

Moist and inundated rocks offer a substratum intermediate between subaerial and strictly aquatic habitats. Moist rocks have been interpreted[4] as subaerial habitats with a maximum of aeration. Dampening and flooding of such rocks may be due to continual seepage from faults in exposed rocky cliffs, or to the spray from waterfalls. Algae growing on damp rocks are mostly filamentous Myxophyceae (especially *Stigonema* and *Scytonema*), but sometimes there are large gelatinous masses of desmids, especially *Mesotaenium* and certain species of *Cosmarium*. Rocks moist-

[1] Bristol-Roach, 1919, 1920; Fritsch and John, 1942; Petersen, 1928, 1935.

[2] Bristol-Roach, 1919, 1926; Fritsch and John, 1942; Moore and Carter, 1926; Petersen, 1935; Singh, 1939.

[3] Bristol-Roach, 1919, 1920. [4] Ström, 1926.

ened by the spray from waterfalls are frequently covered with such filamentous Chlorophyceae as *Ulothrix*, *Stichoccus*, and *Cladophora*.

Aquatic Algae. The habitats of strictly aquatic algae are of four general types:[1] (*a*) flowing waters, (*b*) ponds and lakes, (*c*) pools and ditches, (*d*) bogs and swamps.

The algae of swiftly running waters are more distinctive than those of any other type of aquatic habitat and include a larger percentage of species restricted to the particular habitat. Algae found in rapids and waterfalls include representatives of the Myxophyceae, Rhodophyceae, Chrysophyceae, Chlorophyceae, and Bacillariophyceae. Rhodophyceae have an especial tendency to be restricted to rapid waters and many of them [*Hildenbrandia*, *Audouinella* (*Chantransia*), and *Lemanea*] are known only from the turbulent portions of streams. Sometimes the temperature of the water is also of importance in determining the nature of the flora, and there are certain algae, notably *Hydrurus* and certain species of *Prasiola*, that are found only in very cold and very rapid waters.

The algae of swiftly flowing waters are of two distinct morphological types: encrusting algae and those in which the greater part of the thallus projects into or trails in the current. Algae of the encrusting type include *Hildenbrandia*, Chamaesiphonaceae, and various diatoms.[2] To these should be added *Gongrosira*, *Fridaea*, *Rivularia*, and other filamentous algae in which the thallus is heavily impregnated with lime. *Lemanea*, *Audouinella*, *Batrachospermum*, and *Hydrurus* are the most frequent of cataract algae with a trailing thallus.

Slower flowing portions of streams have an algal flora markedly different from that of rapids. Streams with clear water often have the stones and boulders covered with trailing strands of *Ulothrix*, *Stigeoclonium*, *Draparnaldia*, *Cladophora*, and filamentous Myxophyceae. *Vaucheria*, Zygnemataceae, and Oedogoniaceae often occur in abundance in gently flowing portions of streams but rarely in a fruiting condition. Except for the rapidly flowing portions, streams have a monotonous algal flora and one far less rich than that of lakes and pools.

Quiet stretches of streams and backwaters have an algal flora composed in part of the same algae found in portions where the current is moving slowly. In addition to these there are also algae characteristic of pools and other standing waters.

There are two distinct elements in the flora of lakes and ponds: the *benthos*, or shore and bottom algae; and the *plankton*, or free-floating algae. The plankton of lakes and ponds may contain algae that have drifted from the benthos, but for the most part it consists of algae not found in the benthos. Plankton algae may be present in sufficient quantity to

[1] West, G. S., 1916. [2] Fritsch, 1929*A*.

color the water, but more often they are so scanty that special means must be employed (see page 31) to obtain a quantity sufficient for microscopical examination. A majority of the species found in the plankton are unicellular rather than colonial, but conditions may be reversed as far as the total number of individuals is concerned. This is especially true of "water blooms," where the algae developing in quantity are usually colonial Myxophyceae.[1]

Some hundreds of species are known only from the plankton, and the great majority of these are Desmidiaceae and Chlorococcales.[2] The latter are especially striking and include *Micractinium*, *Golenkinia*, *Chodatella*, and many species of *Tetraëdron*. Species of Myxophyceae, Bacillarieae, and Chrysophyceae are less numerous, but they often compose the bulk of the plankton. Many of the plankton algae are morphologically adapted to a pelagic life, because of their long gelatinous bristles, wide gelatinous sheaths, flattened colonies, or coiled filaments.[3] All the features just mentioned tend to give these algae greater buoyancy.

Size and depth are not important factors in the development of a true plankton flora by a lake or pond. Permanent pools less than a meter deep and a hectare in extent may have a plankton flora as distinctive as that of a body of water the size and depth of Lake Superior. However, the plankton community (*heleoplankton*) of small and shallow pools is often quite different from the *limnoplankton* of larger lakes.[4] This is especially noticeable in regions where small ponds adjoin deep lakes, as in northwestern Iowa,[5] and where the heleoplankton is much richer in *Chodatella*, *Micractinium*, and *Golenkinia* than is the limnoplankton.

On the basis of the species present, the plankton flora may also be divided into two distinct types: the *Caledonian* and the *Baltic*. The Caledonian type, so-called[6] because it was first discovered in Scotland,[7] is rich in species but small in bulk. Desmids are the predominating algae in this plankton flora, and there are relatively few Chroococcales and Hormogonales. Lakes of the Baltic type have relatively few species in the plankton, but these are usually present in considerable quantity. The Baltic phytoplankton has a predominance of Chlorococcales and Myxophyceae, and a conspicuous poverty in species of Desmidiaceae. Desmid-rich lakes are usually found in geological formations older than the Mississippian, and it has been held[8] that the antiquity of the region is the primary cause of a plankton flora of the Caledonian type. However, mere antiquity of a region is not the cause since most lakes are of very recent formation, and the entire plankton population has been introduced since the glacial age.

[1] Smith, G. M., 1924. [2] Smith, G. M., 1920, 1924, 1924*A*, 1926.
[3] Lemmermann, 1904*A*; Smith, G. M., 1924; West, W. and G. S., 1909.
[4] Zacharias, O., 1898. [5] Smith, G. M., 1926. [6] Teiling, 1916*A*.
[7] West, W. and G. S., 1903*A*. [8] West, W. and G. S., 1906, 1909.

Lakes of the Caledonian type have also been found in areas more recent than the Mississippian.[1] The factor determining the nature of the plankton flora is the hardness of the water, and a Caledonian flora is found only in waters poor in calcium.[2]

Studies on the plankton algae in artificial lakes of known age show[3] that it takes considerable time before a lake develops a true plankton flora, and that the plankton organisms are not introduced from the surrounding drainage area. However, a true plankton flora may develop in artificial pools within a very short time if the waters are inoculated with the proper algae. This is well illustrated by the true plankton flora of 5-year-old pools at a goldfish hatchery near Palo Alto, California.

Many of the species in the benthic flora of ponds and lakes are also found in small pools and ditches. Differences in composition of the benthos from lake to lake are in part dependent upon hardness of the water, and in part upon the marshy, sandy, or rocky nature of the shore. The only distinctive portion of the benthic flora is that at the bottom of deep lakes, where the algae grow in greatly diminished light in which the intensity of violet rays is more than a hundred times that of red and blue rays. The quality of the light profoundly affects the nature of the flora, and at a depth of 15 to 20 m. there are more Myxophyceae and Rhodophyceae than there are Chlorophyceae.[4] The few Chlorophyceae which grow at deep levels, as *Dichotomosiphon tuberosus* (A.Br.) Ernst and *Cladophora profunda* Brand, are rarely found in the upper portion of the benthos. Deep-water Chlorophyceae are of the same color as those growing near the surface; depth-inhabiting Rhodophyceae and Myxophyceae are almost always of a bright red color.

The algal flora of permanent and semipermanent pools and ditches is richer and more varied than that of any other type of habitat. Floating filamentous Chlorophyceae, especially Zygnemataceae and Oedogoniaceae, occur in abundance. Bacillariophyceae, Myxophyceae, Xanthophyceae, and Chlorococcales grow intermingled with the filamentous Chlorophyceae but occur in even greater abundance among or upon the submerged macrophytes.

The semiepiphytic flora growing in gelatinous masses on submerged plants in bogs and swamps is even richer than that growing on plants in pools and ditches. This is especially the case in swampy areas of soft-water regions, where there is a predominance of desmids in the algal flora.

Algae of Unusual Habitats. There is a surprisingly long list of algae that have become restricted to unusual environmental conditions or to particular substrata.

[1] Smith, G. M., 1920. [2] Smith, G. M., 1924; Ström, 1926.

[3] Smith, G. M., 1924.

[4] Geitler, 1928*C*; Oberdorfer, 1927; Pascher, 1923; Zimmermann, 1927.

Certain algae have become adapted to extraordinary conditions of temperature. At one extreme are those growing on snow and ice; at the other extreme are those growing in water whose temperature is near the boiling point.

Algae of snow and ice, the *cryovegetation*, are usually regarded as being restricted to regions of perpetual snow. This is not always the case since, as in Southern California, they are also found at the top of mountains where there is a complete melting of the snow every summer. Snow algae are often present in sufficient abundance to color the snow red, yellowish green, green, or brown. Red snow is often due to *Chlamydomonas nivalis* (Bauer) Wille and to a development of hematochrome in the large immobile akinetes. Green snow on European mountains is usually due to *Raphidonema*, but in this country it may be due to other algae. Thus, *Chlamydomonas yellowstonensis* Kol was found to be the predominant organism in green snow in Yellowstone National Park,[1] and an undetermined species of *Euglena* the predominant organism of green snow in Nebraska.[2] A predominance of the desmid *Ancylonema Nordenskioldii* Berggr. colors the snow brownish to purplish.[3] The total list of algae in the cryovegetation of any region may be twenty or more. The majority are members of the Chlorophyceae.

Thermal algae are present wherever there are hot springs. In this country the most extensive development of thermal algae is in Yellowstone National Park. Here[4] some of them can grow and multiply at temperatures as high as 85°C. Practically all the strictly thermal algae are Myxophyceae.

The "brine lakes" found in arid regions of this and other countries represent an unusual environment with a distinctive algal flora. Truly halophytic algae have been defined[5] as those able to develop in solutions with a sodium chloride concentration of more than 3 molar (17:55 per cent). The best known halophytic Chlorophyceae are *Dunaliella* and *Stephanoptera.*

There are many algae found only in association with specific plants or animals. Most of the animals bearing particular algae live in or about the water, but there are a few land animals regularly harboring algae. The most interesting of the latter is the three-toed sloth where a Protococcus-like alga growing upon the hairs gives the animal a distinctly greenish coloration.[6] Algae associated with aquatic animals may be epizoic or endozoic. Epizoic algae may grow upon animals of large size, as *Basicladia* and *Dermatophyton* upon turtles; or on animals of small size, as *Characium* upon plankton crustacea. Endozoic algae may have a para-

[1] Kol, 1941. [2] Kiener, 1944. [3] Kol, 1944. [4] Copeland, 1936.
[5] Hof and Frémy, 1932. [6] Sonntag, 1922.

sitic or a symbiotic relationship with the host. In some cases, as the colorless Myxophyceae[1] and Dinophyceae,[2] there is a true parasitism in which the alga is nourished by the host. Other endozoic algae are space parasites. The "perforating algae" found in shells of molluscs are the best known space parasites and include representatives of the Myxophyceae and Chlorophyceae. Symbiotic algae contribute far more than they receive from the animal. In a majority of cases, the symbiotic algae are species of *Chlorella* or of related genera, and these are found[3] in a wide variety of uni- and multicellular animals.

The list of algae associated with particular plants is also extensive. Most of the algae epiphytic upon other algae are not restricted to a particular host. The host is usually one of the filamentous Chlorophyceae. Some phycologists[4] hold that cessation of elongation of the host is the primary factor resulting in an accumulation of epiphytes upon filamentous algae, but others[5] think that chemical composition of cell walls of the host is the major factor. Young, vigorously growing Zygnemataceae are poor in epiphytes, because the outer pectose layer is continually dissolving away; old filaments are often covered with epiphytes after the pectose layer is dissolved away.

Lichens are the best known case of symbiosis of algae with other plants, but algae are also symbionts with plants other than fungi. Several species of Myxophyceae have been found within the protoplasts of colorless Tetrasporales and Chlorococcales.[6] In this symbiosis the photosynthetic activity of the Myxophyceae supplies the needs of both plants. Some of the algae within the tissues of higher plants (as the *Anabaena* within *Azolla*, or the *Nostoc* within *Anthoceros*) are examples of symbiosis in which the two plants have no mutual effect on each other. Other algae within higher plants, as the *Anabaena* in roots of *Cycas*,[7] represent a true symbiosis in which each member of the pair contributes to the mutual support.

There are also several algae that are parasitic upon the leaves and twigs of land plants. All these algae belong to the Chlorophyceae, and the best known are *Phyllosiphon*, *Cephaleuros*, *Chlorochytrium*, and *Rhodochytrium*. These parasites are found upon a wide variety of unrelated hosts and may or may not cause injury to the plant in which they are growing.

Periodicity of Fresh-water Algae. The seasonal succession of algae is a problem that has long attracted the attention of algologists, but many of the studies on periodicity have been restricted to too limited areas to warrant generalizations. Practically all studies on periodicity of the phytoplankton show that there is a spring maximum of diatoms, sometimes fol-

[1] Langeron, 1924; Petit, A., 1926. [2] Chatton, 1920. [3] Buchner, 1921.
[4] Cholnoky, 1927*B*. [5] Tiffany, 1924. [6] Pascher, 1929. [7] Spratt, 1911.

lowed by a second maximum in the fall, an early summer maximum of Chlorophyceae, and a late summer and early fall maximum of Myxophyceae.[1] In lakes of the Caledonian type, these seasonal changes are not so pronounced, chiefly because the Myxophyceae do not increase in the autumn.[2]

There is a similar periodicity in the benthos of lakes and ponds, and in the algae of pools and ditches, but this is more difficult to record, because the relative abundance and the time of fruiting of many species must be taken into consideration. Weekly observations on pools in central Illinois, continued through several years, show that the algae fall into six natural groups:[3]

1. *Winter annuals*, that begin their vegetative activity in the autumn and fruit in early spring (March and April). This group includes species of *Tribonema*, *Draparnaldia*, *Tetraspora*, and one or two of *Spirogyra* and *Oedogonium*.
2. *Spring annuals*, that begin their vegetative activity in late autumn or early spring and attain their maximum development and reproduce during May. This group includes many Zygnemataceae, Oedogoniaceae, and other filamentous Chlorophyceae.
3. *Summer annuals*, that germinate in the spring and fruit most abundantly during July and August. Species of *Oedogonium* and Zygnemataceae are the most prominent in this series.
4. *Autumnal annuals*, that begin vegetative development late in the spring and are most abundant in the autumn. This includes still other species of *Oedogonium* and certain Myxophyceae, as *Gloeotrichia natans* (Hedw.) Rab.
5. *Perennials*, whose vegetative cycle may be, or is, continuous from year to year. Cladophoraceae and certain Desmidiaceae belong to this series.
6. *Ephemerals*, that may appear in numbers for a short time at any season.

Fluctuating changes in the temperature, dissolved salts, light intensity, and dissolved gases undoubtedly have a marked effect upon periodicity, but the interaction of these various factors is so complex that one can only hazard guesses concerning the specific action of each. The early development of certain algae in the spring (*Tribonema*, Diatomaceae, Volvocales) is correlated with their capacity to grow vigorously at low temperatures. Failure of the water to attain higher temperatures may also check the usual seasonal appearance of certain algae. An example of this is seen in the failure of certain plankton Myxophyceae to develop in abundance during especially cold summers.[4] Temperature may also affect periodicity by accelerating or retarding physiological processes. Thus, temperature is one of the factors modifying the time of fruiting of various species of *Spirogyra*[5] (Fig. 1).

[1] WESENBERG-LUND, 1904.
[2] SMITH, G. M., 1920; WESENBERG-LUND, 1905; WEST, W. and G. S., 1912
[3] TRANSEAU, 1913, 1916. [4] WESENBERG-LUND, 1904. [5] TRANSEAU, 1916.

The annual variation in the intensity and duration of sunlight is another agency affecting the periodicity of algae. Some idea of the amount of this variation may be gained from measurements of light at the Lake of Constance (Switzerland). These measurements show that the total illumination under midsummer conditions is about ten times that at midwinter.[1] There are also midsummer and midwinter differences in the amount of light penetrating various levels below the surface of the water. This seasonal fluctuation in illumination has been correlated with the maximal development of certain green algae, as *Ulothrix zonata* (Weber and Mohr) Kütz. and *Spirogyra adnata* Kütz., at different levels below the surface.

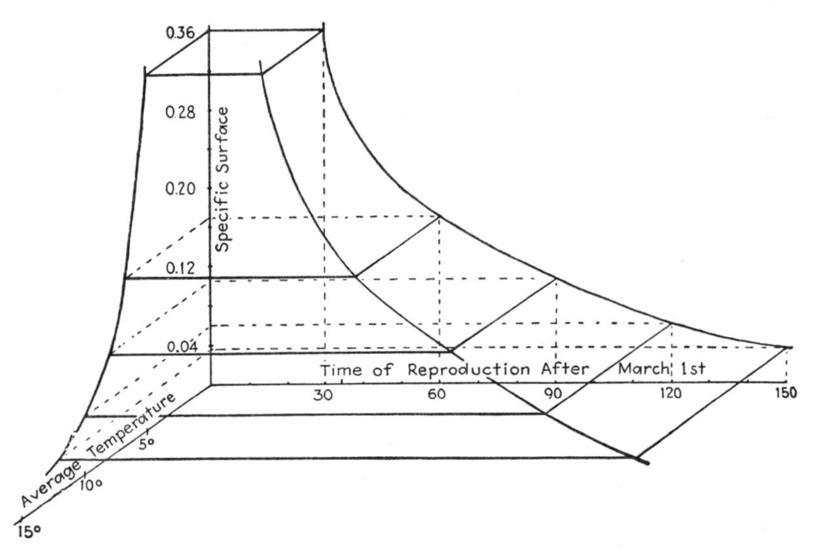

Fig. 1. Curve showing the interrelation between temperature, specific surface (ratio between surface and volume), and the time of fruiting in *Spirogyra*. (*From Transeau*, 1916.)

The late spring and early summer fruiting of many algae has been ascribed to a concentration of the dissolved salts due to a general lowering of the water level. There is little, if any, causal relationship between concentration of dissolved salts and fruiting.[2] Seasonal fluctuations in dissolved salts, caused by evaporation, are so slight that they have but little effect upon periodicity. Pools may have a sudden increase in the amount of dissolved salts following heavy rains, because of runoff from the surrounding drainage basin. The sudden addition of waters rich in silicon and nitrogen may disturb the regular periodicity and cause certain algae, especially diatoms, to develop in greater number.[3] Periodicity

[1] Oberdorfer, 1927. [2] Transeau, 1913, 1916. [3] Pearsall, 1923.

may also be affected by the increase in the amount of dissolved organic compounds as the season progresses,[1] but the precise effect of the organic compounds is unknown. The marked seasonal change in the amount of carbon dioxide and oxygen dissolved in the water[2] is also a factor to be taken into consideration, but it is impossible to correlate such changes with the appearance and disappearance of the various components in the algal flora.

[1] Hodgetts, 1921–1922. [2] Birge and Juday, 1911.

CHAPTER 3

COLLECTION, PRESERVATION, AND METHODS OF STUDYING THE FRESH-WATER ALGAE

Equipment for Collecting. The first requisite for the collection of algae is an adequate supply of containers. One-ounce wide-mouthed bottles are very convenient, and a simple method of transporting the bottles on a field trip is to place them in the divided cartons in which eggs are sold by the dozen. One should also take along a few bottles of larger size, for samples of water and for the collection of Volvocales and other motile organisms. Bottles made of ordinary glass are not satisfactory for the collection of Chrysophyceae and other algae of very acid waters, because, even within a few hours, sufficient alkali goes into solution to damage the algae. Containers of resistant glass (Pyrex) should, therefore, be used when collecting the algae from soft-water localities. Some of these algae inhabiting acid waters are so sensitive to alkalies that they go to pieces within a very short time when ordinary commercial slides and cover glasses are used in making water mounts for microscopical examination.

The collector should carry along also a supply of sheets of wax paper for transporting collections of terrestrial algae without undue drying. Wax paper is also useful when collecting mats of filamentous algae that do not have to be placed in water for transportation to the laboratory.

Equipment for collecting plankton algae is discussed on page 30.

Many phycologists recommend the use of special rakes or dipnets for gathering algae growing so deeply in the water that one cannot reach them with the hand. The plant grapple devised by Pieters[1] is as effective as any rake and has the great advantage of being small enough to be carried in the pocket. Another useful piece of collecing apparatus is a giant pipette, made by slipping an atomizer bulb over a 1-ft. length of glass tubing ¼ or ⅜ in. in diameter. One should also carry a stout pocketknife for scraping encrusting algae from stones, hacking off fragments of soft rock, and cutting off surface slivers from wood covered with algae.

Methods of Collection. It is much better to make a thorough sampling of a few pools or other habitats, when on a collecting trip, than to attempt to gather one or two samples from a large number of stations. Nothing

[1] Ward and Whipple, 1918.

is more annoying when one finds a collection rich in some rare alga, or containing a form too immature for determination, than to be unable to state precisely where the collection was made. For this reason, it is highly desirable to make precise notes at the time each collection is made, especially if several collections are made from the same body of water, and to make the notes sufficiently detailed to enable one to return to the exact spot where each collection was made. This may be done by numbering the bottles and making the notes in a record book. A simpler method is to write the necessary information on a small slip of paper and place the paper in the bottle. If the collection proves one worthy of preservation, it is then advisable to rewrite the data on a label and affix it to the bottle.

When making collections, the bottles should be filled with algae to not more than a quarter of their capacity and water added nearly to fill the bottle. If the bottles are stuffed with algae to their full capacity, deterioration is apt to set in, and material is often unfit for examination, even after a few hours' storage in stoppered bottles. Living algae show great variability in their ability to withstand storage in corked bottles containing an adequate volume of water. Some, as *Vaucheria*, show signs of deterioration within 2 or 3 hr.; others, such as *Cladophora* or *Oscillatoria*, may be kept in stoppered bottles for several hours without injuring them. In any case, it is advisable to uncork all bottles as soon as one returns to the laboratory.

One rarely collects all the algae present if collecting is restricted to the shores of pools. The best procedure in collecting from pools is to wade in, with or without rubber boots, to where the water is 1 or 2 ft. deep, and then collect samples from the surface and from the bottom. Samples from deeper portions of the pool, or from areas that one cannot reach by wading, may be gathered by means of the plant grapple described above. The filamentous algae one finds floating on the surface of the water are usually Zygnemataceae or Oedogoniaceae. Specific differences in these two families are based in large part upon the structure of their mature zygotes, and ripe fruiting material is essential for taxonomic studies. Microscopical examination at the time of collection will ensure the gathering of suitable material. One of the best procedures for this examination in the field is Taylor's method[1] of mounting a small portable microscope on a tripod designed for use with a hand camera and setting up the tripod in the pond. A hand lens with a magnification of 15 or 20 diameters is, however, of sufficient magnification to enable one to distinguish between sterile and fruiting material in the field.

In addition to collecting the free-floating filamentous forms, one should examine and collect submerged twigs, dead culms of aquatics, and stones,

[1] TAYLOR, 1928*A*.

since these often harbor algae that do not grow free-floating. The best method of gathering the algal sludge at the bottom of pools and other waters is to use the giant pipette described on a previous page. The flocculent algal masses accumulating around stems of aquatics may be collected by pinching the stem between the fingers, with the hand palm upward, and then gently drawing the hand upward through the water. It takes time to develop the knack of lifting flocculent material in the open hand and through the water, but the richness of the material repays one for time spent in learning the method. The viscous brownish or greenish liquid obtained by squeezing submerged aquatics, from which the superfluous water has been allowed to drain, contains many algae. Such squeezings collected from *Myriophyllum*, *Utricularia*, and *Sphagnum* are usually rich in Chlorococcales, Desmidiaceae, and Bacillarieae.

Permanent or semipermanent pools usually yield the greatest variety and abundance of algae. This does not mean that temporary pools should be neglected, since they often contain quantities of Myxophyceae and Volvocales. Collections made from standing waters in barnyards, and other places where the water is rich in nitrogenous compounds, frequently contain unicellular or colonial Volvocales in abundance.

As a general rule, streams are a poorer source of algae than standing waters, but streams should not be ignored since certain of the Ulotrichales and Rhodophyceae are restricted to clear, swiftly running water. These swift-water algae are apt to be abundant around dam sites, especially if the dam has a wooden spillway. Rocks continually moistened by the spray from dams, as well as those dampened by the spray from waterfalls, support a type of algal vegetation not found elsewhere. Algae also develop in extensive felt-like or gelatinous masses upon rocky cliffs where there is a continuous trickle of water. Myxophyceae, especially *Scytonema* and *Stigonema*, are the most frequently encountered algae on dripping cliffs.

The collector should also bear in mind that damp soil, damp brickwork, and wet wood harbor algae not found elsewhere. In regions with a definite rainy season, as California, where the ground is continuously soaked for 2 or 3 months, practically every bare spot of soil is rich in such algae as *Vaucheria*, *Botrydium*, *Cylindrospermum*, *Oscillatoria*, or *Microcoleus*. In the eastern part of the United States terrestrial algae are conspicuous only during especially wet seasons. The muddy shores of retreating pools and slowly drying streams are often densely covered with these terrestrial algae. When gathering terrestrial algae, it is best to take up a portion of the substratum on which they are growing, and to wrap the material in wax paper for transportation to the laboratory.

Collections from what appear to be the least promising places, as a shallow rainwater puddle in a path, may yield algae of interest. This justifies the statement that no place should be passed by too carelessly, since surprising discoveries may be missed by such lack of attention. One should also keep in mind the algae associated with animals, the perforating algae growing in shells of molluscs, the distinctive flora in the backs of turtles, and the epiphytes on the smaller crustacea. There are also algae that are associated with specific aquatic plants, as the *Chlorochytrium* in *Lemna*, and the usual epiphytism of *Coleochaete* on old culms of *Typha*. European phycologists have described a number of epiphytes and endophytes from leaves of *Sphagnum*, and many of these doubtless occur in this country also. The collector should also remember that certain algae, as *Rhodochytrium* and *Phyllosiphon*, are found within leaves of phanerogams that are not aquatic in habit.

Plankton algae can be obtained, in quantity sufficient for study, only by means of special nets. Sometimes, when a "water bloom" prevails, one may gather a sample by filling a bottle directly with green water, but such collections ordinarily contain but one or two species and give no idea of all the plankton algae present in the lake or pond. Plankton nets may be purchased from dealers in biological supplies, or one may make his own nets by following the directions given by Ward and Whipple (1918). However, even the finest of bolting silk nets catch but few of the smaller algae. It is, therefore, advisable to supplement the net catches with collections obtained by filtration of water through cotton disks. This may be done by using the apparatus originally designed for testing the amount of sediment in milk.[1] Plankton collections uncontaminated by shore algae can be obtained only by towing the net from a boat. Such surface collections usually contain specimens of all the phytoplankton in the water but, if one wishes collections from the lower levels, these may be obtained by means of a plankton pump.

Laboratory Study of Living Algae. To prevent undue deterioration, one should uncork all bottles containing algae immediately upon return from any collecting trip. On an overnight collecting trip, it is helpful to remove the stoppers from the bottles when one returns to his lodgings and to recork them only for transportation back to the laboratory. If one is on a trip where it will be several days before the collections can be examined, the material should be preserved at once, since the collections would otherwise deteriorate before the laboratory is reached.

If one wishes to obtain a complete record of all the algae present, it is necessary to examine the collections while they are still alive. There is no satisfactory method for preserving motile Chlorophyceae and Chrysophyceae so that the flagella and other delicate structures remain intact.

[1] WHIPPLE, 1927.

Many other algae, especially Chlorococcales, preserve in an indifferent fashion, and it is practically impossible to make out such cytological structures as the chromatophores from preserved material. Myxophyceae, Vaucheriaceae, Zygnematales, and Oedogoniaceae can be studied almost as well from preserved material as when in a living condition.

Since the motile forms are usually the first to disintegrate, it is best to begin study of a collection by looking for them. Many motile algae are positively or negatively phototactic and tend to accumulate at one side of the surface of the water when collecting bottles are placed so that they receive one-sided illumination. Algae accumulating at the surface of the water should be removed by means of a pipette and examined directly in water mounts. Samples should also be taken from both the shaded and illuminated sides at the bottom of the bottle, since some of the motile forms may accumulate there, instead of at the surface.

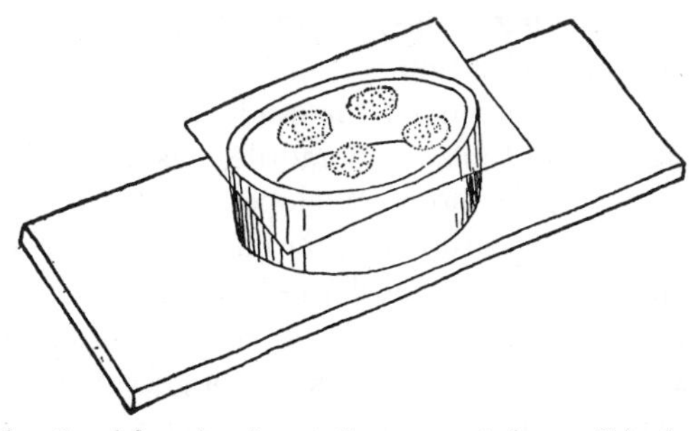

FIG. 2. A hanging-drop culture mounted on a slide ring.

The rapid disappearance of oxygen from the water and the intense illumination necessary for microscopical examination cause most swarming algae to come to rest and to show signs of disintegration within a few minutes after a water mount is put on the stage of a microscope. These deleterious effects may be overcome by placing motile algae in hanging-drop mounts, where they often remain alive for days. Hanging drops on "slide rings," obtainable from any dealer in microscope supplies, are to be preferred to slides with a shallow well. The most convenient size of slide ring is one 18 by 5 mm., and it should be firmly cemented to an ordinary slide by means of vaseline, or a mixture of beeswax and vaseline, and the free edge of the slide ring coated with vaseline. One to four droplets of the material to be studied are placed upon a cover slip; the cover slip is then inverted and pressed firmly against the vaseline-coated margin of the slide ring (Fig. 2). Hanging drops in properly prepared mounts do not evaporate for days and so enable one to follow the history of a particular cell and to make drawings of it at various stages of its development.

Another advantage of the hanging drop is the tendency of motile cells to gather at one side of the drop and to remain immobile without retracting their flagella. It is usually difficult to study the flagella of algae in ordinary water mounts, because the motile cells are moving too swiftly, and those that have come to rest generally do so with the flagellated pole downward. All too often in ordinary mounts, individuals that have become immobile start to move just after one has adjusted the camera lucida and started to draw them. This difficulty may be overcome by placing a drop of material on a slide and inverting it for 15 to 30 sec. over a vial containing a 1 per cent solution of osmic acid before placing a cover slip on the slide. Flagella also remain intact when cells are killed with the iodine potassium iodide solution used in testing for starch, or when cells are treated with Noland's combined fixing and staining mixture.[1] Noland's mixture is excellent for flagella but has the disadvantage of obscuring the cell contents. Motile forms killed by any of the foregoing methods should be compared with living cells in hanging drops, because flagella sometimes assume an abnormal position during fixation.

Intra-vitam stains recommended for study of protozoa[2] are sometimes useful in studying the protoplasts of motile algae. A simple method for staining cell walls and gelatinous envelopes of motile and other algae is that of placing the alga in a drop of water on a slide, stirring the water with the point of an indelible pencil until the water becomes purple, and then examining microscopically. An addition of a minute drop of india ink to the mount makes visible the most watery of gelatinous sheaths. Microchemical tests for determining the nature of food reserves and the chemical composition of cell walls will be found in any handbook on microchemistry.

Determination of species always involves accurate measurements. These may be computed from camera lucida drawings, but it is easier to use an ocular micrometer when measuring algae. Descriptions of the various types of ocular micrometer and of the method of calibrating them are to be found in most textbooks on microscopy. If the study of an alga involves anything more than a determination of the species, nothing is more valuable than a detailed accurate camera lucida drawing. In fact, even when a determination of the species is the object in view, accurate drawings are helpful in recording critical characters. The drawings should be made with a pencil of sufficient hardness (3H or 4H) to permit considerable handling of the drawings without smudging. A hard smooth-surface drawing paper should be used, because one often wishes to ink in the drawings. Reynold's three-ply Bristol board is a very satisfactory paper. The drawings should be affixed to sheets of uniform size (11- by 8½-in. sheets of typewriter paper of good quality are very convenient), each sheet being devoted

[1] NOLAND, 1928. [2] WENRICH, 1937.

to one species. The drawings may be lightly glued to the sheet, but the gummed corners used for mounting photographs are more satisfactory, since they permit the removal of the drawings from the sheet without injuring them. The iconography of original drawings that one gradually accumulates is of invaluable service in affording a record of past observations and as a means of comparing a specimen under observation with one collected some years ago.

After observing the motile forms in a collection and making hanging-drop cultures of those desired for further study, one should next take up the Chlorococcales. These algae can be studied quite satisfactorily in water mounts, if they are in a healthy living condition. One difficulty frequently encountered in temporary water mounts is a drifting of the object from the field, just as the camera lucida has been set or as the drawing is half completed. The drifting is due to evaporation of water from the edges of the cover glass and may be overcome by smearing all edges of the cover glass with oil before making the mount. Mounts sealed in this manner remain intact for days, and the cells show no abnormalities for the first few hours.

Many of the Chlorophyceae reproduce by means of zoospores or zoogametes, and one should always be on the lookout for these motile stages. The shock of change from the natural environment to the laboratory often induces a formation of zoospores, or of gametes, the day after collections are made. Frequently the swarming induced by the changed environment takes place at daybreak and is completed within a couple of hours. However, the student need not be at the laboratory by daybreak if he has placed the collection in a darkroom the night before. Algae brought from the darkroom to the laboratory at any time before noon often swarm freely, shortly after the transfer. Sometimes the swarming can be delayed until the afternoon by placing the collection in a dark icebox and then bringing it into the light. Algae which do not swarm when first brought into the laboratory may sometimes be induced to do so by keeping them in darkness or at low temperatures for two or three days and then bringing them into the laboratory, or by transferring them from the sunny to the shaded side of the laboratory, or vice versa. Aerial and terrestrial algae often swarm freely when flooded with water. This may take place within a very few minutes, as is the case with *Trentepohlia*, but more often it occurs the day after flooding. The failure of algae to produce swarmers in the laboratory is sometimes due to traces of illuminating gas in the air. Collections placed in greenhouses, or other buildings not supplied with gas, may show swarming stages that do not appear in duplicate collections kept in the laboratory.

There is no certain method for obtaining reproductive stages of algae; and when one is fortunate enough to encounter swarming, all other studies should be neglected. Hanging-drop mounts of swarming algae give more accurate information concerning the duration of swarming than do ordinary water mounts, because the swarming period in the latter is almost always prematurely shortened. The use of hanging drops also enables one to follow the development of germlings or the ripening of zygotes for several successive days and to make a series of drawings of a particular specimen.

There are no short cuts for studies on cell division, or studies on the nuclear behavior preceding swarming. In a majority of the fresh-water algae these stages occur during the night. It is often necessary to follow the life history of an alga from sunset to daybreak to obtain such stages. Fixation at hourly intervals will generally yield a sufficiently close series for cytological study.

Germinating stages of aplanospores, akinetes, and of zygotes are much more infrequent than are swarming stages of vegetative cells. Such cases of germination as have been found in the past have been largely a matter of good fortune. Methods developed for inducing a germination of zygotes of Volvocales[1] may prove useful in studying the resting cells of other algae.

The general procedure just described for the Chlorophyceae is equally applicable to the Xanthophyceae, Dinophyceae, and Chrysophyceae. The same is also true of Myxophyceae and Rhodophyceae, except for the portion dealing with swarming stages. The systematics of the Bacillariophyceae hinge so largely upon structure and ornamentation of the cell wall that living material is a hindrance rather than a help in taxonomic studies. Methods used in studying diatoms are briefly mentioned on page 35.

Preservation of Fresh-water Algae. The methods used for the preservation of algae for future examination depend upon the purposes for which the algae are desired and upon the algae to be preserved. Material for taxonomic studies may be preserved in fluids. The simplest procedure is to add sufficient commercial formalin to the collection to make a 2 to 4 per cent solution, recork the bottles, and store for future reference. Another simple method of preservation is to replace the water in the collection with a mixture of formalin, acetic acid, and alcohol,* in which the algae may be stored indefinitely. Numerous attempts have been made to devise solutions that will preserve the color of chromatophores, but these are not particularly successful except with certain Chlorophyceae and Xanthophyceae. The most satisfactory of these is the mixture proposed by Keefe.[2]

Some of the collections preserved in liquids are certain to go to dryness through failure to cork firmly, faulty corks, or deterioration of the corks. Dried-out collections cannot be be restored by adding more liquid. It is,

[1] Strehlow, 1929.

* Glacial acetic acid, 30 cc.; commercial formalin, 65 cc.; 50 per cent alcohol, 1,000 cc.

[2] Keefe, 1926

therefore, advisable to replace the original preserving fluid with a 5 per cent solution of glycerin in 50 per cent alcohol before permanent storage of collections. The glycerin prevents complete drying out of the collection and enables one to add more liquid and thus save the material.

Many phycologists build up a series of permanent mounts, for rapid reference to algae previously collected, and to avoid a continual opening of bottles containing materials stored in liquids. These preparations are usually mounted in glycerin, or in glycerin jelly. Directions for making mounts of this sort may be found in any of the general handbooks on microtechnique. Since a certain percentage of the mounts deteriorate through evaporation, it is advisable to make duplicate mounts of the rarer algae.

Permanent mounting of desmids involves a special technique since empty cells must be selected, and the cells must be so oriented that they may be viewed from the front, top, and side. Bullard's method[1] for picking out and orienting individual desmids is invaluable to one making taxonomic studies of this group. Preparations made in this manner enable one to use the "type slide" method of the diatomist, a method that permits rapid and repeated examination and comparison of specimens.

Preparation of diatoms for study involves a "cleaning" of the cells (a destruction of all organic matter in the cell with suitable reagents) and a mounting of the siliceous remains in a medium with a high refractive index to bring out the markings on the wall.* Diatomists make permanent mounts of cleaned material as "strewn slides," or pick out individual frustules with a mechanical finger and mount them as type slides. Diatoms are treated in the same manner as other algae when it comes to the study of their protoplasts.

Earlier workers in the field of taxonomy made herbarium specimens of all algae. The custom of building up herbariums of fresh-water algae fell into disrepute about the turn of the century, but today several phycologists are reviving the practice. The value of a herbarium specimen depends in part upon the manner in which it is prepared. Drouet[2] recommends drying the material as quickly as possible and without the aid of heat. He finds that the best preparations are secured when the material is reduced to a dried condition in open air (in front of a shaded open window or an electric fan) within an hour after removal from the habitat. When Myxophyceae are dried in this manner, the shape of the cells and even protoplasmic characters of color and granulation are preserved. The coarser forms may be dried

[1] Bullard, 1921.

* General descriptions of the techniques of diatomists are found in Bellido, 1927; Edwards, Johnson, and Smith, 1877; Hanna and Driver, 1924; Hustedt, 1927; Mann, 1922; and Van Heurck, 1880–1885.

[2] Drouet, 1938.

on small squares of herbarium paper; the smaller filamentous forms and all unicellular forms are more easily examined when dried on sheets of mica. Myxophyceae and the zygotes of Oedogoniaceae and Zygnemataceae return to approximately the original condition when soaked in water. The cells of other algae are usually too badly shriveled to be of much service, but sometimes they may be restored to a condition suitable for microscopical examination by placing them in a drop of lactic acid and gently warming.

Cytological Methods. Studies on the details of cellular structure and on nuclear phenomena involve a killing, fixing, and staining of the material. The reader is referred to Johansen's treatise[1] for a comprehensive discussion of these subjects.

Cultivation of Algae. There is no general method by which collections of algae may be kept alive in the laboratory for indefinite periods. Aquatic algae can usually be maintained for some time if a small quantity of material is put in a glass container with a large volume of water and not exposed to direct sunlight. Water from the habitat where the alga is growing is often the best culture solution, and in this day of automobile transportation it is not a difficult matter to carry an adequate supply of water back to the laboratory. The various nutritive solutions devised for the cultivation of algae are successful with certain algae but not with others. In attempting to cultivate any particular species in nutrient solutions, it would be well to consult Bold's[2] extensive review to see whether or not previous investigators have found a suitable nutrient solution for growing the alga. Sometimes, failure of algae to grow in the laboratory is due to leakage of gas. The high temperature to which laboratories are heated in this country is also detrimental to the growth of algae and not infrequently the cause of a deterioration of cultures.

Cultures of aquatic algae in aquariums or in hanging drops can often be maintained long enough to permit the ripening of immature fruiting material or the development of germination stages. However, algae maintained under laboratory conditions so frequently develop abnormal thalli that one should compare the laboratory-grown algae with freshly gathered specimens to avoid misleading conclusions. The erroneous ideas concerning the polymorphism of algae, prevailing a half century ago, were based in large part upon monstrosities appearing in laboratory cultures. Attention should be called to the fact that many algae maintained in laboratory cultures for long periods and not developing cells of monstrous shape may be abnormal in so far as the cell contents are concerned. This is particularly true of reserve foods, and one almost always finds algae in long-standing cultures so densely packed with starch or fats that the chromatophores and other cell organs are obscured.

[1] JOHANSEN, 1940. [2] BOLD, 1942.

Terrestrial and aerial algae may be maintained for a considerable time by placing them on thoroughly soaked filter paper in a dish covered with a sheet of glass. The paper should remain moist at all times, but at no time should enough water be added to flood the specimens.

When a mixture of aquatic or of terrestrial algae is placed in Beijerinck's, Knop's, or any of the various culture solutions, some of the algae present in small numbers will eventually be present in abundance. This is due to the fact that the new environment favors the growth of certain species that were at a disadvantage as compared with other species in the original environment. For a number of years, Pringsheim has made use of such selective (enrichment) cultures and thereby obtained cultures of species ordinarily unobtainable in laboratory culture. For details concerning soil-water, putrefactive, and other types of selective culture, the reader is referred to discussions of the topic by Pringsheim.[1]

Pure Cultures of Algae. There is more or less confusion in the use of the term "pure culture." According to the usage of some authors, a pure culture is one that contains one species of alga; others understand it to be a culture of a single species of alga that is also free from other organisms including bacteria and fungi. To differentiate between the two, the term *unialgal culture* has been proposed[2] to designate one which contains but a single species of alga but which may contain other organisms. The term *pure culture* is reserved for one which contains a single species of alga and is absolutely free from other organisms.

A large number of algae have been grown in unialgal or in pure culture. Unialgal cultures of filamentous algae can often be obtained by transferring a single filament to a suitable nutrient solution. Pure cultures of filamentous algae have been obtained[3] by shaking material in repeated changes of water, or by the agar plate method.

The only method of obtaining many Chlorococcales and other small algae in quantity is to isolate them in unialgal or in pure culture. Many of the unicellular algae may be isolated by the agar plate method of the bacteriologist. Beijerinck's solution in 1½ per cent agar is excellent, but agar may be added to any of the nutrient solutions listed by Bold[4] as suitable for growing algae. The tube containing the nutrient agar is melted in the usual manner, and a drop of water containing the algae to be isolated is added. The agar is then poured in Petri dishes in the customary manner. In order to be certain that colonies of algae developing in the agar will not be too close together, it is best to plate successive dilutions of the inoculum. After the agar hardens, the Petri dishes may be placed in diffuse daylight, but never in direct sunlight. Even better results are obtained if the light

[1] Pringsheim, E. G., 1946, 1946A. [2] Smith, G. M., 1914.
[3] Czurda, 1926, 1926A. [4] Bold, 1942.

source is a fluorescent lamp and the distance of the cultures from the lamp is such that intensity of illumination is 50 to 75 foot-candles. Under ordinary laboratory conditions, minute green dot-like algal colonies become evident in the agar in from 10 to 15 days. All algal colonies do not develop at the same rate, and the first to appear are usually those of *Chlorella* and other simple forms. For this reason, it is best to let the plate cultures stand 20 days or more before attempting to isolate the different species. If the colonies are not too close to one another, they may be cut from the agar in the Petri dish and streaked on slanted tubes of agar. An alga growing on an agar slant may then be used as an inoculum for starting a culture in a nutrient liquid.

Unicellular algae do not form colonies of characteristic appearance in agar plates, and so the kind of alga cannot be determined by macroscopical examination. A series of isolations, by the method just described, would, therefore, contain many duplicates. This duplication may be avoided by determining the genus or species at the time a colony is cut from the assemblage of colonies in the original Petri dish.[1] A colony cut from the agar is placed on a sterile slide, covered with a sterile cover glass, crushed slightly, and examined microscopically. If the alga is one desired for study, the cover is removed from the slide and the colony transferred to a sterile agar slant. In this manner 200 to 300 colonies may be examined in a single day and perhaps only 20 of them retained for study, thereby saving much needless inoculation on agar slants before determining the species or genus.

Nutrient agar is a selective medium, and certain species in the inoculum do not develop into colonies when plated in Petri dishes. Because of this, some phycologists prefer to make all isolations from the mixed inoculum by means of micropipettes. Special micropipettes have been devised[2] for isolating unicellular algae, and when one has mastered the technique of manipulating them, a large number of isolations can be made in a relatively short time. The problem is not so much a matter of picking out the individual organisms as it is one of finding a suitable nutrient medium once they have been picked out.

[1] SMITH G. M., 1914. [2] PRINGSHEIM, E. G., 1946, 1946*A*.

CHAPTER 4

DIVISION CHLOROPHYTA

The Chlorophyta, or grass-green algae, have their photosynthetic pigments localized in chromatophores which are grass-green because of the predominance of chlorophylls *a* and *b* over the carotenes and xanthophylls. There are several xanthophylls not found in other algae, and of these lutein is the most abundant (see Table I, page 3). Photosynthetic reserves are usually stored as starch, and its formation is intimately associated with an organ of the chromatophore, the pyrenoid. Motile stages have flagella of equal length and, with a few exceptions, the zoospores and motile gametes have two or four flagella. Although sexual reproduction is not a feature distinguishing Chlorophyta from other algae, it is a phenomenon of wide occurrence within the group and in the various orders it ranges all the way from isogamy to oögamy.

Occurrence. Since the fresh-water Chlorophyta of this country outnumber the combined species of all other algae, it is not surprising to find that practically every collection examined contains at least a few representatives of the division. Samples gathered from permanent or semipermanent pools usually contain some Chlorophyta, and many samples consist almost wholly of them. The filamentous species are the most conspicuous of those in quiet waters, but there are often many unicellular and colonial species intermingled with them. In most cases, also, the pools contain small sessile species, either upon the coarser algae, upon submerged stems and leaves of phanerogams, or upon stones or woodwork in the water. Temporary pools and puddles are usually not so rich in green algae as are pools of a more permanent nature, and one is more apt to find Myxophyceae than Chlorophyta in such habitats. On the other hand, temporary waters rich in nitrogenous compounds, as the puddles in barnyards, usually contain green algae, especially members of the Volvocales.

The green algae that one may expect to find in quiet portions of streams are mostly the same as one finds in pools, but in slowly flowing waters there are often genera not found in standing waters. Just as the green algae of slowly moving waters differ somewhat from those of quiet water, so does one find that certain green algae are restricted to the swiftly moving waters of cataracts, waterfalls, and the spillways of dams. *Chlorotylium*, *Fridaea*, and certain *Ulothrix* species are found only in such stations.

Although Chlorophyta are rarely the predominating organisms in the phytoplankton of ponds and lakes, the number of species in the fresh-water plankton is very large. The number of genera known only from the plankton is greater in this division than in any other. As a general rule, plankton Chlorophyta are most abundant during late spring and early autumn, but this periodicity is not sharply marked, and they are to be found at all times when plankton hauls can be made from ponds and lakes.

The Chlorophyta also comprise an important part of the subaerial algal flora. *Zygogonium* and *Hormidium* may form felt-like masses several square meters in extent on bare soil. Usually, however, green algae growing on the surface of soil are not in conspicuous patches. Terrestrial green algae may grow below, as well as upon, the surface of the soil and at a depth of 50 cm. or more. Damp rocks and cliffs have more blue-green than grass-green algae, but one sometimes encounters gelatinous masses of green algae in such habitats. The Chlorophyta within such masses are usually species of *Mesotaenium* or of *Cosmarium*. The strictly aerial algae, which would include those found on tree trunks, damp woodwork, and moist brickwork, are usually unicellular forms and are often found in pure stands. *Trentepohlia* is an important aerial alga and, when present, occurs in abundance. On the Monterey peninsula, California, for example, *T. aurea* var. *polycarpa* (Nees. and Mont.) Hariot is so abundant on the Monterey cypress (*Cupressus macrocarpa* Hartw.) that the foliage of these trees has a distinct orange color.

In California, and in certain other Western states, there are several lakes whose content of dissolved salts is from two to fifteen times that found in the ocean. The most striking of these is Searles Lake, California, with 33 per cent dissolved salts in the water, and where the salt forms a solid surface crust strong enough to bear the weight of an automobile. Even under such rigorous physiological conditions, one finds that there is a growth and development of *Dunaliella* and *Stephanoptera* in sufficient abundance to color the salt crust a bright green.

There are several published records of the occurrence of red snow in the Sierra Nevada and the Rocky Mountains. In most cases, the coloration was assumed to be due to *Chlamydomonas nivalis* (Bauer) Wille, but this has been established by microscopical examination in only two cases.[1] Green snow has been found in the Yellowstone National Park and shown to be due to an association of *Chlamydomonas yellowstonensis* Kol with several other Chlorophyta.[2]

A number of Chlorophyta known for this country grow in communal relationship with other organisms. The communal relationship may be one of "space parasitism," as *Gomontia* growing in shells of *Unio*. When

[1] Kiener, 1946; Smith, G. M., 1933. [2] Kol, 1941.

green algae grow within the leaves of phanerogams, the relationship is not so clear. Possibly certain species of *Chlorochytrium*, as *C. Lemnae* Cohn, are simple space parasites. The relationship is certainly one of parasitism in the case of *Rhodochytrium spilanthidis* Lagerh., which grows in the leaves of *Ambrosia atremisiaefolia* L., since the algal member of the association lacks chlorophyll. Judging by the effect on leaves in which they grow, the siphonaceous *Phyllosiphon Arisari* Kühn and the ulotrichaceous *Cephaleuros virescens* Kunze are also true parasites. Green algae may also be symbionts, and the association may be with an animal, as the symbiosis of *Chlorella* and *Hydra*; or the symbiosis may be with another plant, as is the case with the *Trebouxia* found in many lichens.

Cell Structure. A few of the more primitive Chlorophyta have naked protoplasts, but in the great majority of cases the protoplast lies within a definite wall which is a secretion product of the protoplast. Even where there is no definite wall the exterior portion of the protoplasm is rigid and, as in *Pyramimonas*, the cell has a characteristic shape. The characteristic shape in genera with a cell wall is, therefore, probably due to the protoplast itself rather than to the enclosing wall. All cells which are surrounded by a wall have the wall composed of at least two concentric layers. The innermost layer is composed wholly or in large part of cellulose.[1] In the great majority of cases the cellulose portion is homogeneous in structure, but it may be composed of concentric cellulose layers as in *Cylindrocapsa*. External to the cellulose layer is a layer of pectose. It is not clear whether this is a conversion product derived directly from the cellulose or a direct secretion of the protoplast that filters through the cellulose wall. The latter appears to be the case with the desmids in which definite pores are present in the wall. In most algal cells, the outermost portion of the pectose is converted into a water-soluble pectin that dissolves in the surrounding medium. It is very probable, also, that the formation of pectose is a continuous process throughout the vegetative life of the cell. There are many algae, including the Zygnemataceae, where the amount of pectose secreted just about balances that dissolved away. The result is a sort of equilibrium in which the thickness of the pectose layer remains practically constant.[2] If the formation of pectose ceases, as it does during conjugation in Zygnemataceae, there comes a time when the pectic layer becomes dissolved away and where there is nothing external to the cellulose layer. Confirmation of this view is seen in the lack of epiphytes on actively growing vegetative cells of *Zygnema* and *Spirogyra* (because the surface upon which the epiphytes would grow is continually dissolving away) and the frequent presence of epiphytes upon old or conjugating filaments in these genera. Not all algae establish this balance in the gelatinous portion of the

[1] TIFFANY, 1924; WURDACK, 1923. [2] TIFFANY, 1924.

wall and certain of them, as *Gloeocystis*, have the gelatinous portion increasing indefinitely in thickness. There are also genera where the external portion of the pectose becomes impregnated with chitin and where the wall is therefore composed of three distinct layers.[1] This is especially the case with *Cladophora* and *Oedogonium*, genera which apparently secrete but small amounts of pectose. Even if one agrees with those who deny the presence of chitin in the algae,[2] it is clear that these genera have an outermost wall layer that inhibits, or greatly reduces, the dissolving away of the pectose portion. Proof of this is seen in the frequency with which epiphytes are found at all stages of development in these genera with impermeable superficial wall layers.

The most conspicuous organ of the protoplast is the chloroplast. The amount of pigments present is extremely variable and ranges all the way from a quantity sufficient to color the plastid a brilliant green to an amount so small that there is only a tinge of color. Certain Chlorophyta, as *Polytoma*, completely lack photosynthetic pigments. Vegetative cells of certain genera and zygotes of many genera have the green color masked by a red pigment called *hematochrome*. In vegetative cells of *Trentepohlia* the so-called hematochrome is beta-carotene, and in vegetative cells of *Haematococcus* the hematochrome is a ketonic carotinoid—euglenorhodone.[3] Anthocyan pigments have been found in the cell sap of a few Zygnematales, including *Mesotaenium*, *Pleurodiscus*, and *Mougeotia*.

Chloroplasts of green algae always have a shape characteristic for the particular genus or species. Old cells of *Scenedesmus*, *Hydrodictyon*, and several other algae seem to have the pigments diffused throughout the cytoplasm, but young cells of all of them have definite chloroplasts. The shape of the chloroplast from genus to genus is extremely varied. The massive cup-shaped chloroplast characteristic of so many species of *Chlamydomonas* is also found in many other Volvocales and in Tetrasporales. The widespread occurrence of cup-shaped chloroplasts among these lower Chlorophyta gives good reason for supposing that this is the primitive type. However, even in *Chlamydomonas* there are species where the chloroplast is stellate or is H-shaped in optical section. More advanced green algae, as the Ulotrichales, usually have cells with a chloroplast that is parietal in position, laminate, and entire or perforate. A few fresh-water algae, as *Eremosphaera*, and many marine Siphonales have numerous small discoid chloroplasts in the peripheral portion of the cytoplasm. The most striking of all chloroplasts are those found in the Desmidiaceae, and the range from species to species and genus to genus is almost infinite.[4]

Most chloroplasts contain a special organ, the *pyrenoid*. Structurally it

[1] Wurdack, 1923. [2] Wettstein, F. v., 1921. [3] Strain (in press).
[4] Carter, N., 1919*A*, 1919*B*, 1920, 1920*A*.

consists of a central proteinaceous core, which in turn is ensheathed by minute plates of starch. Strictly speaking, the term should be applied only to the proteinaceous core, but in common usage it is often applied to both the core and the surrounding starch plates. Formerly, opinion was divided as to whether the pyrenoid proper is intimately concerned with starch formation or merely a reserve protein which has nothing to do with starch formation. Practically, all present-day phycologists adhere to the first view. The first study of pyrenoids with the aid of modern cytological technique was in *Hydrodictyon* where it was shown[1] that the core of a pyrenoid becomes differentiated into two parts: one destined to become impreg-

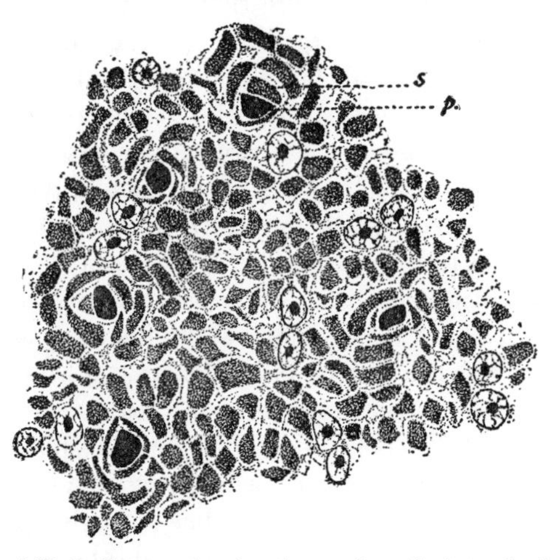

FIG. 3. Pyrenoids of *Hydrodictyon*, showing the cutting off of starch plates. (*From Timberlake*, 1901.)

nated with starch, the other to remain unchanged (Fig. 3). The part undergoing change gradually gives more and more of a starch reaction, eventually moves away from the unchanged part, and becomes one of the starch plates surrounding the central core. A somewhat similar relationship between starch and the central core has been found in *Closterium*,[2] *Tetraspora*,[3] and *Chlorococcum*.[4]

Pure cultures of green algae furnish considerable evidence on the role of the pyrenoid in starch formation. When the cells are grown in total darkness but in the presence of glucose or some other simple carbohydrate, there is an accumulation of starch around the protein core. This seems to show that there is a division of labor in the plastids of green algae. The chlorophyll-containing portion of the plastid synthesizes CO_2 and water into some

[1] TIMBERLAKE, 1901. [2] LUTMAN, 1910. [3] MCALLISTER, 1913.
[4] BOLD, 1931.

simple carbohydrate, and the pyrenoid core converts these simple compounds into starch.

In vigorously growing cultures with rapidly dividing cells, as in cultures of *Chlamydomonas*, the only starch plates present are those encircling the pyrenoid core. As growth slows down in the culture, starch plates are to be found in portions of the chloroplast away from the pyrenoid core. It seems very probable that these latter starch plates were formed in association with the central core and that attempts[1] to differentiate between stroma starch and pyrenoid starch are without adequate foundation.

Cells of small size and with a correspondingly small chloroplast usually have but one pyrenoid, but even here there may be species, as in *Chlorogonium* and *Chlamydomonas*, where there are regularly several pyrenoids within a chloroplast. Cells with large chloroplasts have many pyrenoids; irregularly distributed as in *Oedogonium* and *Mougeotia*, or in a linear series as in *Spirogyra* and *Spirotaenia*. Mature cells whose chloroplast contains numerous pyrenoids may have but a single pyrenoid when young and the number increases as the cell grows in size (*Hydrodictyon*), or there may be several pyrenoids even in the youngest cell (*Oedogonium*, *Spirogyra*). For *Closterium*[2] and *Kentrosphaera*[3] it has been shown that increase in number of pyrenoids is brought about by division of pyrenoids already present in the cell, but whether or not this is true of all genera with more than one pyrenoid in a chloroplast is as yet unsettled. As far as chloroplasts with one pyrenoid are concerned, it has been definitely established that pyrenoids in daughter cells may be formed *de novo* or by a division of that in the parent cell. A formation of pyrenoids *de novo* has been found in *Pediastrum*,[4] *Tetraedron*,[5] *Characium*,[6] *Tetraspora*,[7] and several other genera. Formation of pyrenoids by division of preexisting has been found in *Chlorococcum*[8] and in certain Zygnematales, including *Hyalotheca*.[9]

Chlorophyta also store reserve foods as oils. The oil droplets so frequently found in old vegetative cells and in zygotes are undoubtedly conversion products from starch. A formation of oil, instead of starch, is a regular phenomenon in vegetative cells of *Schizochlamys* and *Mesotaenium*.

All the green algae have a definitely organized nucleus with a distinct nuclear membrane, one or more nucleoli, and a chromatic network. The amount of chromatic material in a nucleus is often so scanty that the space between nucleolus and the membrane is almost colorless, but many nuclei, as in *Oedogonium*, *Cladophora*, and *Spirogyra*, have considerable chromatic material. Nuclear division, except in certain cells of Charales, is mitotic and similar to that found in higher plants. Vegetative cells of all Volvo-

[1] Klebs, 1891. [2] Lutman, 1910. [3] Reichart, 1927.
[4] Smith, G. M., 1916*B*. [5] Smith, G. M., 1916*A*. [6] Smith, G. M., 1916*A*.
[7] McAllister, 1913. [8] Bold, 1931. [9] Potthoff, 1927.

cales are uninucleate, and the same is true for many more advanced green algae. A coenocytic condition is, however, known both for Chlorophyta which have no vegetative cell division and for those whose cells divide vegetatively. In genera without vegetative cell division, as in *Pediastrum*, the number of nuclei may remain small until just before the formation of zoospores and gametes and then suddenly increase; or the number of nuclei may continually increase during vegetative life of a cell (*Characium*, *Dichotomosiphon*, *Hydrodictyon*). Coenocytic cells of the type found in *Cladophora* and *Rhizoclonium* have a gradual increase in number of nuclei as the cells increase in size and not a sudden increase in number just prior to cell division.

Flagella of a sufficient number of Volvocales and of swarmers of other Chlorophyta have been investigated[1] to warrant the assumption that there is but one type of flagellum throughout the entire series. This is the whiplash type[2] and one in which an axial filament is surrounded by a cytoplasmic sheath for a greater part of its length. The cytoplasmic sheath usually ends abruptly, the naked portion of the axial filament extending beyond it being known as the *end piece*.

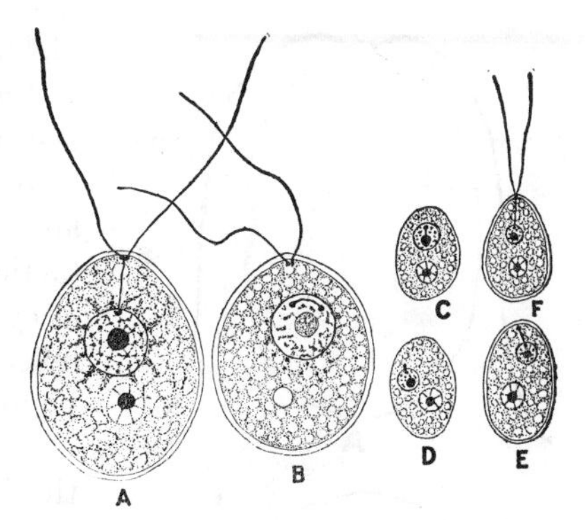

FIG. 4. Neuromotor apparatus of *Chlamydomonas nasuta* Korshikov. *A*, neuromotor apparatus of a vegetative cell. *B*, at the beginning of cell division. *C–F*, development of the neuromotor apparatus in a young cell. (*All after Kater*, 1929.) (× 1284.)

Flagella of vegetative cells of Volvocales are intimately connected with a neuromotor apparatus. In *Polytoma uvella* Ehr.[3] and *Chlamydomonas nasuta* Korshikov[4] the neuromotor apparatus is of an elaborate type (Fig. 4). There is a granule (*blepharoplast*) at the base of each flagellum, and the two are connected by a transverse fiber (the *paradesmose*) which is connected with a descending fiber (the *rhizoplast*). The rhizoplast runs down to and connects with an intranuclear centrosome. The available evidence indicates that *Volvox*[5] and *Eudorina*[6] also have a blepharoplast-rhizoplast-centriole neuromotor apparatus in which the centriole is intranuclear. A similar system has been shown for *Polytomella*[7] and *Haematococcus*,[8] but here the centrosome is extranuclear.

The *eyespot* or *stigma* is a photoreceptive organ intimately concerned with directing the movement of flagella. It may be a part of the neuromotor

[1] VLK, 1938. [2] PETERSEN, 1929. [3] ENTZ, 1918. [4] KATER, 1929.
[5] ZIMMERMANN, 1921. [6] HARTMANN, 1921.
[7] KATER, 1925. [8] ELLIOT, 1934.

apparatus, as has been claimed,[1] but as yet there has been no cytological demonstration of the fact. Characteristically there is a single eyespot in vegetative cells of Volvocales, and in gametes and zoospores of other green algae. The color of an eyespot varies from orange-red to reddish brown, and it may be circular, oval, or sublinear in outline. It is usually located near the base of the flagella, but it may lie in the equatorial or posterior portion of a cell.

In *Chlamydomonas* (Fig. 5*A*) and possibly other unicellular Volvocales, the eyespot consists of two portions: a biconvex hyaline portion, which is the photosensitive portion, and a curved pigmented plate.[2] *Gonium*, *Eudorina*, and *Volvox* have a more complicated type of eyespot (Fig. 5*B*). Here there is a definite biconvex lens, exterior to the pigmented cup and the photosynthetic portion.[3] Phototactic responses in organisms with this type of eyespot are thought to be due to selective reflection from the concave surface of the pigmented portion.

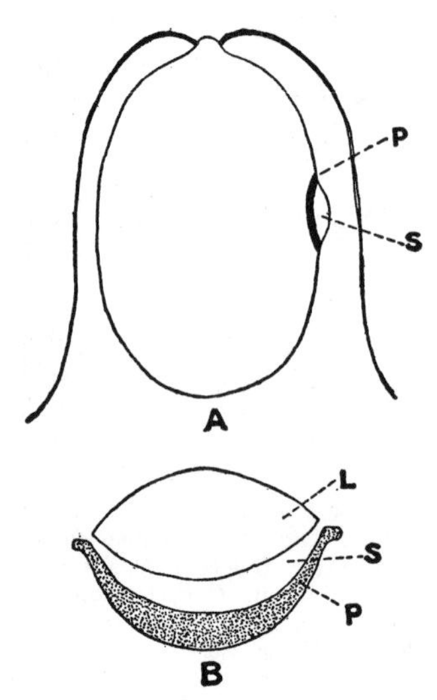

FIG. 5. *A*, diagram of an eyespot of *Chlamydomonas*, showing the pigment *P* and the photosensitive substance *S*. *B*, diagram of a cross section of the eyespot of *Volvox*, through the lens *L*, photosensitive substance *S*, and the pigment cup *P*. (*Modified from Mast*, 1928.)

Most of the Volvocales and certain of the Tetrasporales, as *Stylosphaeridium* and *Asterococcus*, have contractile vacuoles. In most biflagellated genera, there are two contractile vacuoles near the base of the flagella. Some biflagellate genera have more than two vacuoles, and these, as in *Haematococcus*, may lie beneath any part of the protoplast's surface. When two vacuoles are present, these usually contract alternately. The contraction is sudden, and the distension is slow. Because of this, they are sometimes called *pulsating vacuoles*. It is thought that their function is excretory and that the liquid discharged from the vacuole is expelled from the cell. Contractile vacuoles in cells of Chlorophyta probably function independently of one another and are not interconnected to form the complex systems such as are found in Chrysophyceae and certain Protozoa.

The vacuoles present in cells of genera more advanced than the Tetrasporales are of the familiar type found in most plant cells. Immature cells developing from zoospores, as those of *Hydrodictyon*, have innumerable minute vacuoles scattered throughout the cytoplasm, many of which grad-

[1] MAST, 1928. [2] MAST, 1928. [3] MAST, 1916, 1928.

ually increase in size. Now and then two or more of the enlarging vacuoles coalesce. Gradually, therefore, the number of conspicuous vacuoles in the cell become smaller, and eventually they unite to form a single large central vacuole. Sometimes, as in *Sphaeroplea*, there are several large vacuoles within the mature cell, or, as in *Spirogyra*, the central vacuole is incompletely divided by strands of cytoplasm. Species of the higher green algae which do not have a conspicuous central vacuole are chiefly those that have become adapted to a subaerial existence. It is thought[1] that this lack of large vacuoles is the chief reason why *Protococcus*, *Trentepohlia*, *Prasiola*, and other algae can live where there is a greatly reduced water supply.

Cell Division. One of the characteristics of the Chlorococcales and of the Siphonales is the inability of their cells to divide vegetatively. Vegetative cell division is, however, a phenomenon found in all other orders of the Chlorophyceae and occurs in both uninucleate and multinucleate cells. Division is intercalary in most unbranched filamentous species, and, except for the basal cell, any cell in the filament may divide. Genera in which the thallus is a branching filament may have intercalary divisions, but more often the divisions are restricted to the terminal portions of the branches though not necessarily to the apical cells.

Cytokinesis of uninucleate cells is always preceded by a mitotic division of the nucleus; coenocytic cells may or may not have nuclear divisions preceding cell division. It is usually held that cytokinesis in cells of Chlorophyta is not due to the formation of a cell plate (phragmoplast). The demonstration of a phragmoplast for *Spirogyra*,[2] *Tetraspora*,[3] and possibly for *Eremosphaera*,[4] shows that cell division of Chlorophyta may be by means of a cell plate. The usual method of cell division is by a furrowing of the plasma membrane midway between the cell ends. This linear furrow deepens until it has cut entirely through the cell and so produced two daughter protoplasts. Although division by means of a transverse furrow is almost universal, there is great variability in the time at which the new transverse walls are formed and in the method of their formation. Most cells dividing by means of a transverse furrow secrete wall material within the furrow as it deepens. In fact, cross-wall formation follows so closely upon the furrowing, that cell division is often thought to be caused by the inward growth of a transverse septum. The complete division of the protoplast before the beginning of wall formation is found in *Microspora* and *Oedogonium*. *Microspora* has an intercalation of an H-shaped piece of wall material between the daughter protoplasts; *Oedogonium* has the secretion of a transverse wall after elongation of the unique lateral ring of wall material (see page 197). Cell division in constricted desmids, to which the

[1] de Puymaly, 1924. [2] McAllister, 1931. [3] McAllister, 1913.
[4] Mainx, 1927.

majority of species belong, begins with a division of the nucleus and an elongation of the isthmus. There then follows a transverse division in the region of the isthmus. Each daughter cell is at first asymmetrical and consists of one semicell and half of the isthmus portion from the mother cell. The old semicell remains unchanged, but the isthmus portion enlarges until it is identical in size and ornamentation with the old semicell (Fig. 222, page 312).

Cytokinesis may be accompanied by a division of the chloroplast and pyrenoid. Cells with the chloroplast a longitudinal strip (as in *Spirogyra* and *Mougeotia*), or a transverse girdle (as in *Ulothrix* and *Stigeoclonium*), have cytokinesis dividing the chloroplast transversely. Cells with longitudinally symmetrical chloroplasts, as *Chlamydomonas*, have a longitudinal division of the chloroplast. In the more specialized uninucleate cells which have a chloroplast axial to both poles of the nucleus (*Zygnema*,[1] *Hyalotheca*,[2] and *Cylindrocystis*[3]), there is, after nuclear division, a transverse division of each chloroplast with migration of a daughter nucleus to a point midway between each pair of daughter chloroplasts. *Cladophora*, *Sphaeroplea*, and other genera with numerous chloroplasts have no division of the chloroplasts accompanying cell division.

Many of the Chlorophyta show a very marked diurnal periodicity in the time at which nuclear and cell division take place. In the great majority of cases, these divisions occur during the night. Nuclear division usually begins within an hour or two after sundown and is often completed by the early morning hours. The occurrence of division at night rather than during the daytime may possibly be correlated with the greater accumulation of reserve foods following the photosynthetic activity of the daytime.

Vegetative Multiplication. In colonial genera, cell division increases the size of the colony but does not bring about a formation of new plants. Accidental breaking of the colony, especially in the case of filamentous genera, may result from external causes, as aquatic animals feeding upon the alga, or the action of water currents. The formation of zoospores or aplanospores in certain parts of the filament often weakens the walls and tends to bring on a breakage in these portions. In some cases, as *Microspora* and *Oedogonium*, the cell wall breaks transversely when zoospores are liberated and so severs the filament into two or more portions.

Some filamentous species, as those of *Stichococcus* and small-celled species of *Spirogyra*, have a strong tendency to separate into individual cells or short series of a few cells each. Such fragments may then grow into long filaments. Fragmentation is especially frequent in species of *Spirogyra* that have an annular infolding of the transverse walls (replicate end walls), the fragmentation resulting from the eversion of the replicate walls induced

[1] Merriam, 1906. [2] Potthoff, 1927. [3] Kauffmann, 1914.

through changes in the cell's turgidity (Fig. 202, page 287). It has been held that many of the transverse walls of filamentous algae have a middle lamella of pectic material.[1] If this be true of *Stichococcus*, changes in the composition of the middle lamellae might be one of the causes of the dissociation found in this genus.

Asexual Reproduction. The commonest method of asexual reproduction is by the formation of zoospores. From the phylogenetic standpoint, zoospores may be looked upon as a temporary reversion to the primitive ancestral flagellated condition. Zoospore formation, like cell division, frequently takes place at night, and the spores are usually liberated at daybreak. Sudden changes in the environment often stimulate profuse sporulation, and it is no unusual experience for the collector to find, the day after collection, *Stigeoclonium* and *Draparnaldia* producing zoospores so abundantly that but few traces of the original filaments remain. Change from light to darkness, a transfer from an aerial to an aquatic environment, as in *Trentepohlia* and *Vaucheria*, and a change from running to quiet water also stimulate zoospore formation. However, change of external factors does not bring about the formation of zoospores with the regularity that some claim,[2] and one cannot always be certain of obtaining zoospores at the desired time by modifying the environment.

Zoospores are formed in vegetative cells morphologically similar to others in the colony, except for the Trentepohliaceae which form them in special sporangia. All vegetative cells in a colony may be able to produce zoospores, as in the Hydrodictyaceae, or spore formation may be restricted to certain cells. The restriction of zoospore formation to certain cells is especially prominent in genera with branching filaments, where it usually takes place only in young, vigorously growing portions of the thallus. In most unbranched filamentous genera, as *Ulothrix* and *Oedogonium*, spore formation may take place in all cells but those functioning as holdfasts. Colonies with all cells potentially capable of producing zoospores rarely have all of them doing so simultaneously; spore formation is usually restricted to isolated cells (*Pediastrum*) or to short series of cells.

Zoospores may be formed singly or in numbers within a cell. The production of a single swarm spore within a vegetative cell (as in *Coleochaete*, Oedogoniales, *Microspora*, and the finer branches of Chaetophoraceae) is a very simple process and merely a rejuvenescence of the protoplast and a reversion to a primitive type of organization where it has flagella and an eyespot. The formation of more than one zoospore within a cell is a more complicated process and may take place in a variety of ways. Uninucleate cells may have a division of the nucleus followed by a cytokinesis into two

[1] Tiffany, 1924. [2] Klebs, 1896.

uninucleate protoplasts (*Ulothrix*,[1] *Ulva*,[2] *Sorastrum*[3]). This is followed by a simultaneous division of the two protoplasts, and these, in turn, may divide two or three times in succession. The number of angular protoplasts (immature zoospores) thus formed is always a multiple of two, and may be 2, 4, 8, 16, 32, or even more. Instead of having cytokinesis after the first mitosis, a uninucleate cell may, as in certain individuals of *Pediastrum*,[4] have two or three simultaneous divisions of the daughter nuclei, to form four or eight nuclei, and then a division of the protoplast into a like number of zoospores. This multinucleate condition preceding zoospore formation may be the result of simultaneous nuclear divisions during the vegetative growth of the cell (*Characium*)[5] or to more or less irregular divisions during growth (*Hydrodictyon*).[6] Coenocytic cells may also have a series of simultaneous nuclear divisions just before swarm-spore formation (*Pediastrum*).[4]

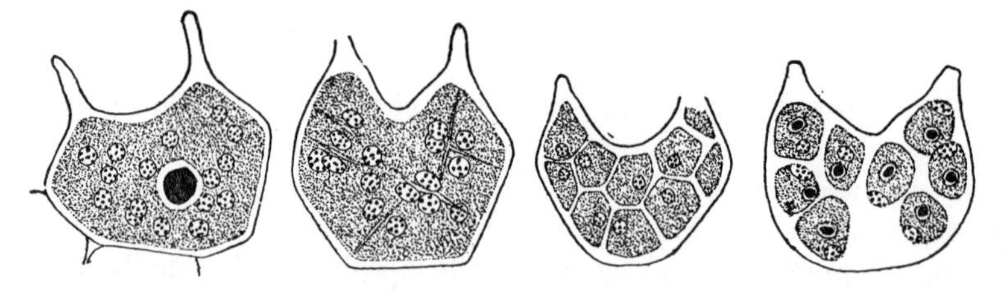

FIG. 6. Progressive cleavage of the protoplast of *Pediastrum Boryanum* (Turp.) Menegh. to form zoospores. (× 1330.)

No matter what the origin of the multinucleate condition, zoospore formation is almost always due to the type of cytokinesis known as "progressive cleavage" (Fig. 6); that is, a division into large multinucleate masses and a redivision of these until there are uninucleate protoplasts. Progressive cleavage in the algae is by furrowing, and the furrowing may begin next the plasma membrane and the vacuolar membranes (*Hydrodictyon*), or the cleavage furrows may first appear in the interior of the cytoplasm (*Characium*, *Pediastrum*). The uninucleate protoplasts formed by progressive cleavage are at first angular, but before their liberation they become rounded unless they are too densely crowded within the old mother-cell wall. Blepharoplast granules have been noted in the zoospore formation of a few Chlorophyta (*Hydrodictyon*,[6] *Oedogonium*[7]), and it is very probable that development of flagella by zoospores of other green algae is due to a neuromotor apparatus.

Cleavage stages leading to the formation of zoospores frequently take

[1] GROSSE, 1931. [2] CARTER, 1926. [3] GEITLER, 1924*B*.
[4] SMITH, G. M., 1916*B*. [5] SMITH, G. M., 1916*A*. [6] TIMBERLAKE, 1902.
[7] GUSSEWA, 1930.

place during the very early morning hours, and swarming occurs shortly after sunrise. The close correlation between swarming and illumination can be demonstrated by keeping fertile algae in darkness until late in the forenoon or until after midday. Here swarming does not take place until they have been brought into light.

Zoospores may be liberated through a definite lateral or terminal pore in the wall of the parent cell (Fig. 7). In *Cladophora*, pore formation is obviously due to a local gelatinization of the cell wall, and it is probable that the same is true for many other green algae. Sometimes, as in *Microspora*, the entire wall gelatinizes. Liberation of zoospores may also be by a transverse separation of the wall into two equal parts (*Microspora*), or into two unequal parts (*Oedogonium*). Many zoospores are enclosed within a delicate gelatinous vesicle when first discharged. In rare cases (*Pediastrum*, *Sorastrum*), the vesicles are persistent structures, and the zoospores do not escape from them. Usually the vesicles are transitory structures and dissolve within a few minutes after the spore mass is discharged. For some algae, including *Pediastrum*,[1] it is evident that the vesicle is the innermost wall layer of the parent cell. The mechanism by which the spore mass and the surrounding vesicle are expelled from the old parent-cell wall is as yet not definitely established. The discharge does not appear to be due to amoeboid movement of the, as yet, inactive zoospores; probably it is due to an excretion of water similar to that found in *Prasinocladus*.[2]

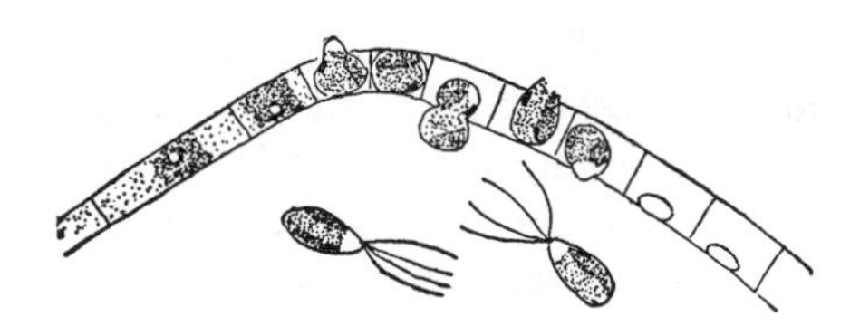

FIG. 7. Zoospores of *Draparnaldia glomerata* (Vauch.) Ag. (× 800.)

Zoospores of Chlorophyta are always without a cell wall and always have the flagella at the anterior end. Most zoospores have two or four flagella, but in Oedogoniaceae and Derbesiaceae there is a whorl of many flagella. Within the protoplast are one or two contractile vacuoles, one or more chloroplasts, and a single nucleus. There is usually a single eyespot, but in exceptional cases there may be more than one.

The length of time that zoospores remain motile is dependent upon both the particular species and the environmental conditions. For example, zoospores of *Pediastrum*[3] move actively for only 3 or 4 min., those of *Coleochaete*,[4] for an hour or so, but those of *Ulothrix*[5] may continue movement for 2 or 3 days. The marked effect of environmental conditions upon the duration of swarming has long been known. The environmental factors of greatest importance are light and temperature. For example, in *Ulo-*

[1] HARPER, 1918. [2] LAMBERT, 1930.
[3] HARPER, 1918. [4] WESLEY, 1928. [5] KLEBS, 1896.

thrix zonata (Weber and Mohr) Kütz., too intense illumination tends to reduce the period of swarming, and darkness tends to extend it.[1] In this same species, many zoospores remain actively motile for 48 hr. if the temperature is 3 to 4°C., but the majority swim about for only 7 to 9 hr. if the temperature is 12 to 15°C.

Many zoospores show phototactic responses during the swimming period, responding positively or negatively according to the light intensity. Toward the end of the swarming period the movement becomes more and more sluggish, and a slow lashing of the flagella often continues after zoospores have come to rest. Zoospores that have ceased moving are usually resting upon some solid object in the water, but in a few cases, as in *Acanthosphaera* and *Sphaeroplea,* the quiescent spore is free-floating. Soon after coming to rest, the flagella disappear and the spore secretes a wall. The cell thus formed may have the same shape as the free-swimming spore or, as in *Pediastrum,* there may be marked changes in the protoplast's shape before a wall is secreted. In a majority of cases, as in *Oedogonium,* they come to rest with the anterior end downward. In rare cases, as in *Characium,*[2] the anterior end may be upward. Cases are also known[3] where zoospores come to rest with one side against the substratum. The one-celled plants resulting from metamorphosis of zoospores are attached to the substratum by means of pectic substances in the newly formed walls. Sometimes, as in *Characium* and *Cylindrocapsa,* a definite gelatinous stalk is secreted. The firmness of adherence to the substratum may also be enhanced by a development of rhizoidal processes in the lower part of the germling. In *Oedogonium*[4] the development of these processes is a direct response to contact stimuli, and it has been shown that processes are lacking or greatly reduced in germlings growing on a smooth surface.

The remarkable amoeboid stages found in a few green algae, chiefly Ulotrichales, are to be looked upon as modified zoospores. Not only do these amoeboid stages move in the same fashion as *Amoeba,* but they may also ingest solid foods and divide in the same manner. Such amoeboid stages may be formed from zoospores which have been swarming a short time (*Aphanochaete*[5]), from protoplasts of vegetative cells (*Stigeoclonium*[6]), or from germinating aplanospores (*Tetraspora*[6]).

Not infrequently the angular protoplasts formed by cleavage within a cell do not develop into zoospores. Instead, each protoplast becomes rounded and secretes a wall distinct from the parent-cell wall. Such *aplanospores,*[7] which may also be formed singly within a vegetative cell,

[1] STRASBURGER, 1878. [2] KOSTRUN, 1944; SMITH, G. M., 1916*A*.
[3] KOSTRUN, 1944. [4] PEIRCE and RANDOLPH, 1905. [5] PASCHER, 1909.
[6] PASCHER, 1915. [7] WILLE, 1883.

are to be regarded as abortive zoospores in which the motile phase has been omitted. Aplanospores are regularly formed by certain genera of Chlorophyceae, as *Microspora* (Fig. 8), and only occasionally in others, as *Ulothrix*. Aplanospores may be liberated from the old parent-cell wall before germination takes place, or they may grow into a new filament while still within the parent-cell wall. Aplanospores with greatly thickened walls are usually called *hypnospores*.

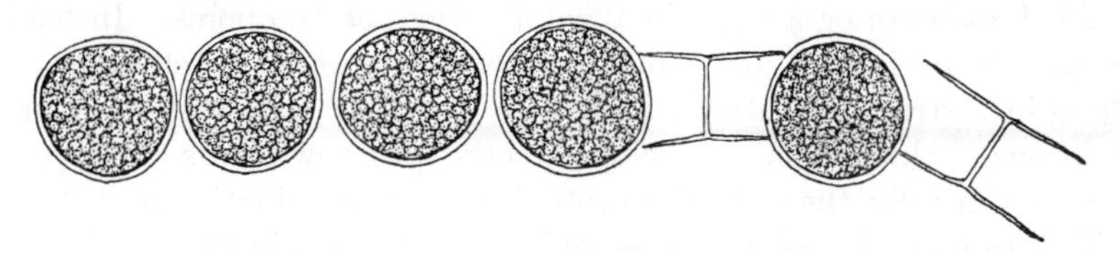

Fig. 8. Aplanospores of *Microspora Willeana* Lagerh. (× 800.)

Aplanospores that have the same distinctive shape as the parent cell are called *autospores*. Autospore formation is the only known method of reproduction in certain families of the Chlorococcales. They may also be formed occasionally by zoosporic genera, as is the case in *Desmatractum*.[1] In genera where zoospore formation has been completely suppressed, as those belong-

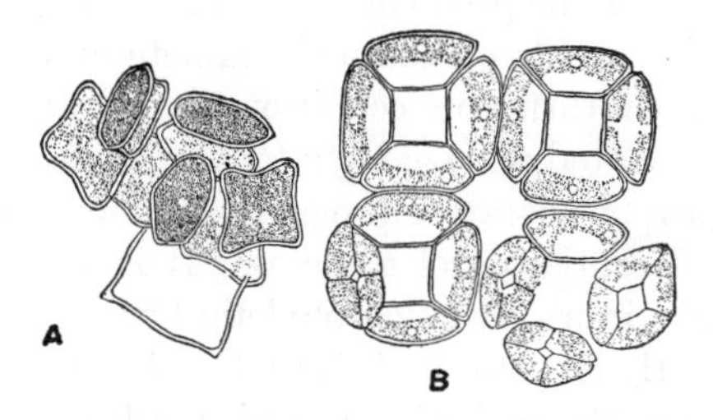

Fig. 9. Autospores. *A*, *Tetraëdron minimum* (A. Br.) Hansg., a species in which the autospores separate from one another after liberation. *B*, *Crucigenia fenestrata* Schmidle, a species in which the autospores remain united in an autocolony. (× 1000.)

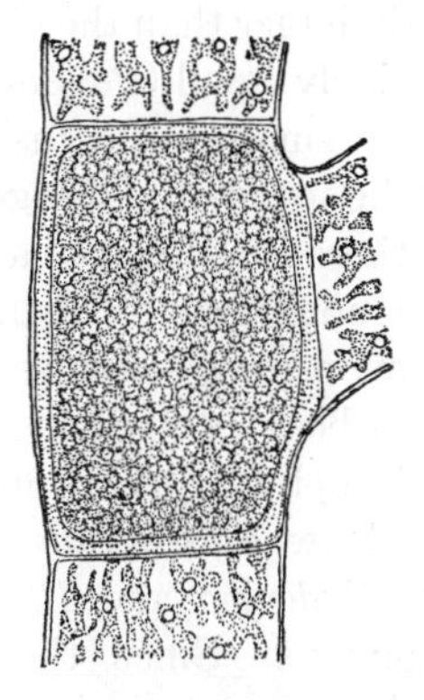

Fig. 10. Akinete of *Pithophora Oedogonia* (Mont.) Wittr. (× 300.)

ing to the Oöcystaceae and Scenedesmaceae, the autospores may become separated from one another when liberated from the parent-cell wall (*Chodatella*, *Tetraëdron*) (Fig. 9*A*), or they may remain permanently united in an autocolony whose cells are arranged in a specific manner (*Scenedesmus*, *Crucigenia*) (Fig. 9*B*).

[1] Pascher, 1930*A*.

Vegetative cells may also develop into spore-like resting stages with much thicker walls and more abundant food reserves. These bodies, *akinetes*,[1] may always be distinguished from aplanospores by the fact that the additional wall layers around the protoplast are fused with the wall of the parent cell. A formation of akinetes is a regular phenomenon in certain genera (*Pithophora*) (Fig. 10); and occasional in others (*Tetraspora, Microspora, Pediastrum*). An akinete should not be regarded as a modified zoospore or a stage in the formation of zoospores. Instead, it should be considered a direct modification of a vegetative cell and one resulting in a structure better adapted to tide the alga over unfavorable conditions. Akinetes may develop directly into new plants (*Pithophora*[2]) or, as is generally the case, the protoplast of a germinating akinete may divide into a number of zoospores which are liberated in the usual fashion.

Sexual Reproduction. The Chlorophyta are an evolutionary series in which gametic union is of widespread occurrence. In the simplest and most primitive case there is a fusion of a flagellated gamete (*zoogamete*) with another zoogamete of identical size and structure. Among algae with this *isogamous* gametic union, it is impossible to make morphological distinctions between male and female gametes. Isogamy leads to a condition of *anisogamy* where both gametes are flagellated, but where one of a fusing pair is regularly larger than the other. The difference in size between the two may be relatively small (*Pandorina*), or it may be pronounced. In anisogamous algae the smaller of a fusing pair is considered the male gamete and the larger the female. Anisogamy leads, in turn, to a condition of *oögamy* and where there is a union of a small flagellated male gamete (*antherozoid*) with a large nonflagellated gamete (*egg*). Isogamy, anisogamy, and oögamy represent a progressive series in differentiation of gametes. It is a series that has been evolved independently in at least five phyletic lines among the Chlorophyta. The most frequently cited example is that of the colonial Volvocales where *Gonium* is isogamous, *Pandorina* is inconspicuously anisogamous, *Eudorina* and *Pleodorina* are markedly anisogamous, and *Volvox* is oögamous. Since the *Gonium-Pandorina-Eudorina-Pleodorina-Volvox* series shows a progressive increase in number of cells in a colony and a progressive differentiation into somatic (vegetative) and gametic cells, it is frequently held that evolution from isogamy to oögamy is correlated with increase in complexity of vegetative structure. The unicellular Volvocales show that this is not necessarily the case since an evolution from isogamy to oögamy is to be seen in both *Chlamydomonas* and *Chlorogonium*. Thus, *Chlamydomonas Snowiae* Printz is isogamous, *C. Braunii* Gorosch. is anisogamous, and *C. coccifera* Gorosch. is oögamous. In *Chlorogonium*, *C. euchlorum* Ehr. is isogamous and *C. oögamum* Pascher is oögamous. Among

[1] WILLE, 1883. [2] WITTROCK, 1877.

the Ulotrichales, *Ulothrix* is isogamous, *Aphanochaete* is anisogamous, and *Chaetonema* and *Coleochaete* are oögamous. A similar series occurs among siphonaceous algae where *Protosiphon* is isogamous, *Bryopsis* and *Codium* are anisogamous, and *Dichotomosiphon* is oögamous.

Study of green algae in unialgal and pure culture has shown that certain species are *homothallic* (monoecious) and can have a union of gametes derived from a single parent cell. Other species are *heterothallic* (dioecious) and have the gametes fusing in pairs only when the two come from cells of different parentage.

Among a few primitive one-celled motile forms, two vegetative cells may function directly as gametes and fuse with each other. This occurs among both homo- and heterothallic species of *Chlamydomonas*. Most green algae do not have this fusion of vegetative cells, and there is only a fusion of swarmers (gametes) whose special function is to fuse in pairs. When this occurs among one-celled motile green algae, the stages leading to a formation of gametes are not the same as those in which cells divide vegetatively to form daughter cells (which are the morphological equivalent of zoospores). Thus, the protoplast of a vegetative cell of *Chlorogonium elongatum* Dang. divides to form 8, 16, or 32 gametes but only 2, 4, or 8 daughter cells when it divides vegetatively. In many algae, as *Ulothrix* and *Enteromorpha*, gametes are biflagellate and zoospores are quadriflagellate.

Strictly speaking, any cell producing gametes is a *gametangium*, but this term is ordinarily used only when cells producing gametes are morphologically different from vegetative cells. A few isogamous Chlorophyta, as the Trentepohliaceae, have morphologically distinct gametangia. Practically all the oögamous green algae have gametangia of distinctive shapes and ones in which the male gametangium (*antheridium*) is morphologically different from the female gametangium (*oögonium*).

Isogamous and anisogamous Chlorophyta almost invariably discharge their gametes from the parent cell, formation and ripening of the zygote occurring external to the thallus. With a few exceptions, as *Chaetonema*,[1] eggs of oögamous genera are retained within the oögonium. Fertilization takes place within the oögonium, and the resultant zygote is liberated only when the oögonial wall decays.

A union of gametes both of which are without flagella (*aplanogamy*) is a feature which immediately separates the Zygnematales from all other Chlorophyta. There are minor variations in the manner in which aplanogametes are brought in contact and in their behavior prior to union. In Zygnemataceae there is an establishment of a tubular connection between two cells functioning as gametangia. Some Zygnemataceae are truly isogamous and have the two gametes meeting and fusing in the conjugation tube (*Mougeotia* and certain species of *Zygnema*). Other Zygnemataceae

[1] Meyer, K., 1930.

have a morphological isogamy accompanied by a physiological anisogamy in which one gamete is actively amoeboid and the other passive (*Spirogyra* and certain species of *Zygnema*). Most Desmidiaceae do not form a conjugation tube but have an escape of the aplanogametes from the parent-cell walls before they fuse with each other.

Phycological literature contains numerous records of cases where gametes that have not united with gametes of opposite sex develop into new plants or into resting stages identical in appearance with zygotes. The evaluation of these records of *parthenogenesis* is a difficult matter, since they have been observed in mixed cultures in aquariums or in mixed collections of material brought in from the field. Pure and unialgal cultures started from single cells definitely establish that a parthenogenetic germination of gametes may take place. *Chlorogonium euchlorum* Ehr.,[1] *Cladophora Suhriana* Kütz.,[2] and *Ulva Lactuca* L.[3] may be cited as examples of this. The parthenogenetic development of gametes of Zygnemataceae into zygote-like bodies (*parthenospores*) is a well-established fact.

Clear proof of parthenogenesis in certain species does not mean that gametes of all species of a genus may be parthenogenetic. For example, *Ulva Lactuca* L. is parthenogenetic, but studies by the writer show that this is not true for *U. lobata* (Kütz.) S. and G.

Certain Chlorophyta, especially *Oedogonium*[4] and *Spirogyra*,[5] show a marked seasonal periodicity in the occurrence of sexual reproduction. This usually takes place in the spring or early summer, but certain species of these genera fruit only in the autumn. In the case of *Spirogyra*, sexual reproduction seems to take place only when a certain amount of reserve food has accumulated, and the accumulation of reserve food is dependent directly upon the ratio between cell surface, cell volume, and temperature (see Fig. 1, page 25). Environmental changes other than temperature may also influence fruiting, and it is a matter of common knowledge that *Spirogyra* usually remains sterile when growing in flowing water. The prevalent idea that a diminution in the water supply induces fruiting is erroneous, since it has been shown[6] that there is an even more abundant fruiting if weather conditions are such that the water level of the pool does not fall. The reason for the apparent increase in fruiting of algae when water levels are falling is due to the fact that many algae fruit in late spring and at a time when rainfall is usually decreasing. Although environmental conditions do affect sexual reproduction, they do not affect it to the extent that sexual reproduction can be induced at any time by the experimenter. Most studies which claim complete control of sexual reproduction by altering the environment[7] have not been substantiated by other workers.

[1] SCHULZE, 1927. [2] FÖYN, 1934. [3] FÖYN, 1934A.
[4] TIFFANY and TRANSEAU, 1927. [5] TRANSEAU, 1916.
[6] TRANSEAU, 1913. [7] KLEBS, 1896.

The Zygote and Its Germination. There is usually not a disappearance of flagella when isogamous and anisogamous gametes unite, and the resultant zygote may swarm in much the same fashion as a zoospore before it comes to rest, loses its flagella, and secretes a wall. All zygotes of oögamous Chlorophyta are, of course, immobile from the beginning, and they all secrete a wall within a fairly short time. When first formed, the zygote wall is a thin homogeneous structure, but as ripening proceeds, it often becomes differentiated into three layers, the two outer of which contain cellulose. The ornamentation and coloration characteristic of many ripe zygotes are developed chiefly in the median wall layer. Protoplasts of young resting zygotes are of a bright green color and contain such food reserves as were present in the gametes. Photosynthesis by the ripening zygote results in an accumulation of still more reserve food. In young zygotes these food reserves consist almost wholly of starch; later on, there is often a conversion of the starch into oil. There is often, also, a development of hematochrome in sufficient abundance to color the protoplast a bright red or orange-red.

Sooner or later, the two nuclei contributed by the two gametes fuse with each other. The fate of the chloroplast or chloroplasts contributed by each gamete is harder to follow. In the case of *Zygnema*[1] and certain desmids,[2] it seems fairly certain that chloroplasts derived from the female gamete persist and those contributed by the male gamete degenerate.

Fusion of gamete nuclei (syngamy) is followed by meiosis at some later stage in the life cycle. Meiosis among the green algae was first demonstrated in *Coleochaete*[3] and shown to take place when the zygote nucleus divides. Subsequently a meiotic division of the zygote nucleus was discovered in several other Chlorophyta. Among these were members of the Volvocales,[4] Ulotrichales,[5] Oedogoniales,[6] Chlorococcales,[7] and Zygnematales.[8] The demonstration of a meiotic division of the zygote nucleus among so many orders of Chlorophyta seemed to warrant the generalization that this was to be expected throughout the entire division. However, this generalization became invalid when it was shown[9] that division of a zygote may be equational. This is now known to hold for *Chaetomorpha*,[10] *Cladophora*,[11] *Ulva*,[12] *Enteromorpha*,[13] and *Draparnaldiopsis*.[14] The significance of this variation in the time of meiosis will be noted on page 59.

Among genera with an equational division of the zygote nucleus, the germination of a zygote usually occurs within a day or two after gametic

[1] Kurssanow, 1911. [2] Potthoff, 1927. [3] Allen, C. E., 1905.
[4] Zimmermann, 1921. [5] Grosse, 1931. [6] Gussewa, 1930.
[7] Mainx, 1931*A*.
[8] Kauffmann, 1914; Kurssanow, 1911; Potthoff, 1927; Tröndle, 1911.
[9] Föyn, 1929; Hartmann, 1929. [10] Hartmann, 1929. [11] Föyn, 1929.
[12] Föyn. 1934*A*. [13] Ramanathan, 1939*A*. [14] Singh, 1945.

union. Among genera with a meiotic division of the zygote nucleus, the zygote undergoes a period of ripening before meiosis takes place.

The time interval between gametic union and the time thick-walled resting zygotes are capable of germination is not the same for all green algae. In *Chlamydomonas Reinhardi* Dang. the writer has found that zygotes may be induced to germinate 20 days after they are formed; in *Ulothrix* zygotes germinate 5 to 9 months after their formation,[1] and in *Oedogonium* germination takes place after a resting period of 12 to 14 months.[2]

Genera whose vegetative cells do not produce zoospores may form them when the zygote germinates (*Sphaeroplea*), but more commonly, as in the Zygnematales, no zoospores are formed by a germinating zygote if vegetative cells do not produce them. Many genera have the germinating zygote producing four zoospores. For certain genera (*Ulothrix*, *Oedogonium*, *Hydrodictyon*) this has been shown to be due to a meiosis in which all four nuclei are functional. There is also evidence for a similar condition in certain other genera for which meiosis has not been demonstrated. For example, in heterothallic species of *Gonium*[3] and *Chlorogonium*[4] two of the zoospores are "plus" in nature and two are "minus."

Germinating zygotes may produce more than four zoospores. The writer has found that, at times, germinating zygotes of *Chlamydomonas Reinhardi* Dang. form 8 instead of 4 zoospores, and that zygotes of *C. intermedia* Chod. regularly produce 16 or 32 zoospores. Germinating zygotes of *Coleochaete* produce 8 to 32 zoospores, and this has been shown[5] to be due to equational nuclear division following the meiotic divisions.

There may also be a formation of less than four zoospores when a zygote germinates. For *Eudorina* this has been shown[6] to be due to a degeneration of certain of the nuclei. A production of but two zoospores may be due to a single division of the zygote nucleus (*Oedogonium*[7]). Here it has been shown that the zoospores have a diploid nucleus and that filaments developing from them are double the size of those coming from zoospores with haploid nuclei. Degeneration of certain nuclei in a germinating zygote is common among the Zygnematales, an order where there is a meiotic division of the zygote nucleus. *Zygnema*[8] and *Spirogyra*[9] have a degeneration of three of the four nuclei resulting from meiosis and so have a production of but one germling when a zygote germinates. *Closterium*,[10] *Cosmarium*,[10] and *Netrium*[11] have two of the nuclei from meiosis large and persistent, and two small and degenerate. When zygotes of these desmids germinate, there is a formation of two germlings, each of which contains a large and a degenerate nucleus. A zygote of *Hyalotheca*[12] contains two

[1] GROSSE, 1931. [2] MAINX, 1931. [3] SCHREIBER, 1925.
[4] SCHULZE, 1927. [5] ALLEN, C. E., 1905. [6] SCHREIBER, 1925.
[7] MAINX, 1931. [8] KURSSANOW, 1911. [9] TRÖNDLE, 1911.
[10] KLEBAHN, 1891. [11] POTTHOFF, 1928. [12] POTTHOFF, 1927.

large and two small nuclei after meiosis, but when the zygote germinates there is a degeneration of one of the large as well as both of the small nuclei. There is, therefore, but one germling from a zygote in this genus.

Life Cycles. The simplest possible type of life cycle is that found in *Chlamydomonas*. Here division of a vegetative cell results in the formation of two, four, or eight motile daughter cells which may function as gametes. Zygotes resulting from gametic union have a meiotic division of the zygote nucleus and produce zoospores which function directly as one-celled vegetative plants. The life cycle of this primitive green alga consists, therefore, of an alternation of a one-celled haploid phase with a one-celled diploid phase. Such an alternation is not obligatory in the sense that the haploid phase must always give rise to the diploid phase, and there may be a succession of haploid phases before production of the diploid phase. The alternation is obligatory in the sense that the diploid phase cannot give rise to further diploid phases but must always form the haploid phase. Among most unicellular algae with alternating haploid and diploid phases, the vegetative functions, especially photosynthesis, are centered in the haploid phase. *Chlorochytrium* may be cited as a unicellular green alga where the opposite condition obtains and where the vegetative functions center in the diploid phase. This alga must be considered as having an obligatory alternation of haploid and diploid one-celled phases, if the very temporary coenocytic condition during gametogenesis is not taken into consideration.

Beginning with the primitive condition of an alternation of unicellular haploid and diploid phases, there may be an evolution of a multicellular generation on either the haploid or the diploid side of the life cycle. The great majority of Chlorophyta have had this on the haploid side. This has resulted (as in *Spirogyra*, *Oedogonium*, or *Coleochaete*) in a life cycle in which a multicellular haploid generation alternates with a unicellular diploid phase. Many such algae have a reduplication of the haploid generation by means of zoospores or other asexual reproductive bodies. *Codium*[1] represents a case where there has been an interpolation of equational division of the diploid nucleus between syngamy and meiosis. Here we have what is essentially a multicellular diploid generation alternating with a unicellular haploid phase.

Particular interest attaches to the similar alternating multicellular generations found in *Enteromorpha*, *Ulva*, *Cladophora*, *Chaetomorpha*, and *Draparnaldiopsis*. Here the alternation may be strictly obligate, the diploid generation always producing zoospores germinating into the haploid generation, and the haploid generation always forming gametes which fuse to form a zygote growing into the diploid generation. In all these genera, the two generations are morphologically alike and cannot be dis-

[1] WILLIAMS, 1925.

tinguished from each other until the time of reproduction. The morphological similarity of the two generations is significant; the fact that all these generations with identical haploid and diploid generations are heterothallic and isogamous or feebly anisogamous may or may not be significant.

It should be noted that this alternation of identical generations has appeared independently in three families not closely related to one another (Cladophoraceae, Ulvaceae, Chaetophoraceae). In each of these families there are genera with an alternation of a multicellular haploid generation with a unicellular diploid phase. Many Chaetophoraceae have such a life cycle: *Urospora*[1] and possibly *Spongomorpha*[2] are examples of this among the Cladophoraceae, and *Monostroma*[3] is an example among the Ulvaceae.

As to the origin of a life cycle with identical multicellular generations, it is not improbable that this came from an ancestor with a multicellular haploid generation and a one-celled diploid phase. It may have originated by suppression of meiosis in the germinating zygote. The contribution by each haploid gamete of an identical set of genes for size and shape of thallus resulted in the zygote growing into a diploid multicellular thallus identical with the haploid multicellular thallus. The capacity for meiosis, originally present in the zygote, is transmitted through each cell generation in development of the diploid thallus and usually becomes effective when the diploid plants are fully mature, although it may become evident in juvenile plants. Whether or not every cell in a thallus exhibits this capacity for undergoing meiosis is largely one of vegetative activity; all cells may do so, as in *Ulva* and *Enteromorpha*, or only young actively growing cells may undergo meiosis, as those at the ends of branches in *Cladophora*. *Cladophora glomerata* (L.) Kütz. is unusual in that there is no meiosis immediately prior to zoospore formation.[4] Early stages in nuclear division prior to zoospore formation are suggestive of meiosis, but later stages are mitotic. As a result, the zoospores give rise to diploid plants which do not undergo meiosis until just before gametogenesis.

Evolution among the Chlorophyta. The various types of thallus organization postulated in the theory of plant body types (see page 6) are represented in much greater abundance among the Chlorophyta than among other algae. There are numerous examples of the motile unicellular type among the Volvocales, but no one would go so far as to hold that any of these is the particular genus from which the chlorophytan series has developed. The volvocine tendency in which the vegetative cells become organized into definite colonies but still retain their motility is represented by several genera. The culminating member of the series, *Volvox*, probably represents the limit of colonial development where the vegetative cells retain their motility.

[1] Jorde, 1933. [2] Smith, G. M., 1946. [3] Yamada and Saito, 1938.
[4] List, 1930.

The tetrasporine tendency in which vegetative cells are immobile but retain their capacity to return directly to a motile condition is found in temporary palmella stages of many unicellular Volvocales (including *Chlamydomonas*, *Carteria*, and *Haematococcus*). From such organisms it is but a small step to truly palmelloid genera, as *Tetraspora* and *Apiocystis*, where the immobile vegetative cell is the dominant phase and the vegetative cells are only temporarily motile. These tetrasporine genera also have a vegetative division of their cells. Cell division within the palmelloid mass may take place indiscriminately and so produce an amorphous colony (*Palmella*); or cell division may predominate in one plane and result in an elongate cylindrical colony [*Tetraspora cylindrica* (Wahlb.) Ag.] or a cylinder that is profusely branched [*Palmodictyon varium* (Näg.) Lemm.].

Palmelloid algae are characterized by a strong tendency on the part of the cells to divide into fours or eights and for the daughter cells thus formed to become separated from one another because of a secretion of gelatinous materials. Restriction of cell division to a transverse bipartition and the restriction of all divisions to the same plane result in a simple filament of cells. The most primitive of the filamentous forms derived from palmelloid colonies would probably have the seriately arranged cells separated from one another by gelatinous material and have much the same organization as the filaments of *Geminella*. Filaments of the *Ulothrix* type, where the cells abut on one another, represent a more advanced condition. Transition to the filamentous condition is accompanied by another feature, a loss of the protoplast's ability to return directly to the motile condition. There are no filamentous Chlorophyceae with this feature so characteristic of Tetrasporales. Many of the filamentous algae do, however, retain the capacity of forming motile reproductive cells, and it is interesting to note that when zoospores or gametes are formed there is usually a division of the protoplast into four, eight, or more motile cells.

In contrast with the Xanthophyceae and the Chrysophyceae, the Chlorophyta have a multitude of filamentous genera. Many of these have simple filaments; there are even more in which the filament is branched. Branches are usually developed by a cell sending out a lateral projection and then forming a cross wall between the projection and the parent cell. Frequently, these initials are formed only toward the distal end of the thallus, but the formation of branches is not definitely restricted to the apical cell as is the case with Rhodophyceae. Green algae with a branching filamentous thallus are always sessile, and such free-floating individuals as one encounters are plants which have become accidentally detached from the substratum. The thallus may have no difference in size between the various branches (*Trentepohlia*, *Microthamnion*), or there may be a distinct main axis which bears lateral branches (*Draparnaldia*, *Stigeoclonium*).

Organization of the thallus into a branched filamentous system may also be accompanied by a differentiation into a prostrate and an erect portion, as in *Trentepohlia* and *Gongrosira*, but nowhere among the branching fresh-water Chlorophyta is there the development of a complex thallus comparable to that of Rhodophyceae and Phaeophyceae. The genera thus far cited all have laxly branched thalli. There are also genera and species in which the branching is compact; either a discoid layer flattened against the substrate (*Protoderma* and *Coleochaete scutata* Bréb.), or a pseudoparenchymatous tissue several cells in thickness (*Gongrosira*). There are several genera in which the branching thallus consists of but a few cells (*Protococcus*, *Thamniochaete*, *Chaetonema*). Possibly these are primitive forms, but it is much more likely that they are reduced. In fact, such Chlorophyta as *Dicranochaete* and *Chaetosphaeridium* are to be interpreted as algae in which the branching thallus has been reduced to a single cell, rather than as examples of the chlorococcine type.

Another potentiality of the tetrasporine evolutionary line is found in the Ulvaceae and Prasiolaceae. Here the thallus may be an erect monostromatic layer (*Monostroma*), a distromatic layer (*Ulva*), a hollow sac (*Enteromorpha*), or a solid cylinder (*Schizomeris*). The origin of such thalli is not to be sought in a modification of a branching thallus. Probably it originated by cells of a simple filament developing the capacity to divide in more than one plane; or by a unicellular form developing this capacity.

There are also numerous examples of the *chlorococcine tendency* (siphonaceous tendency) among the green algae. Most genera of this type belong to the Chlorococcales and are derivatives from motile unicellular ancestors that have lost their vegetative motility and their ability to divide vegetatively, except for the production of reproductive cells. Comparatively few of those who have concerned themselves with the evolution among the Chlorophyceae have recognized that the lack of vegetative division of the chlorococcine algae automatically excludes them from the evolutionary line leading from the Volvocales to the Ulotrichales.[1]

Chlorococcum and *Trebouxia*, with their uninucleate solitary cells reproducing by means of zoospores, are among the simplest known members of this phylogenetic series. Evolution has proceeded in several directions from this simple type. In some cases, as in the Oöcystaceae, the zoospores have lost their power of locomotion and the only means of reproduction is by a formation of autospores. On the other hand, the swarming zoospores may be retained within the parent cell wall (*Hydrodictyon*) or within a vesicle (*Pediastrum*, *Sorastrum*) and develop into cells that are united with one another to form colonies of definite shape. There are also many genera, all referable to the Scenedesmaceae, in which autospores, instead of zoospores, become united to form colonies of definite shape.

[1] Blackman, 1900; West, 1916.

In some of the foregoing chlorococcine types, the cell remains uninucleate until the time of reproduction; in others (*Tetraëderon*, *Pediastrum*) there are one or two nuclear divisions during the vegetative life and two or more additional mitoses just before reproduction. In at least one case (*Hydrodictyon*), there is a continual increase in the number of nuclei throughout the life of the cell. This retention of the capacity for nuclear division, even though the cell itself cannot divide, is also found in *Characium*, a genus in which the cell shape and cell size are definitely limited. Continued growth of such a coenocyte, instead of limited growth, would lead to a simple siphonaceous form such as *Protosiphon*. From this it is not a great step to the more elaborate coenocytic Siphonales.

The rhizopodal tendency is not present among Chlorophyta except as a temporary stage of zoospores (see page 52).

The foregoing discussion of evolution among green algae has ignored progressive changes in gametic union and the question of an alternation of generations. As already pointed out (see page 54), there has been an evolution from isogamy to oögamy in several distantly related groups among green algae. This shows that those systems which make sexuality the fundamental starting point for classifying green algae are founded on a wholly artificial basis. The postponement of meiosis and the resultant appearance of two alternating generations are extremely interesting facts but have not been important factors in evolution within the Chlorophyta. This type of life cycle has arisen independently in at least three families, but in none of them has it become sufficiently established to give rise to a long phylogenetic series.

Classification. Phycologists are in general agreement concerning the natural affinities of many groups of genera (variously included in the Volvocales, Hydrodictyaceae, Ulvaceae, Cladophoraceae, Oedogoniaceae, Zygnemataceae, Desmidiaceae, and Characeae). But when it comes to limits of groups larger than the family, there is great diversity of opinion. For example, some phycologists place the Ulotrichaceae and Chaetophoraceae in separate orders; others include them in the same order. Another example is seen in the inclusion of the Ulvaceae among the Ulotrichales by some and their segregation in a separate order by others.

All phycologists have long recognized that the Oedogoniaceae and the conjugating algae (Zygnemataceae and Desmidiaceae) are more or less remote from other green algae. Formerly these differences were considered sufficient to warrant placing the conjugates in one subclass, the Akontae, and the Oedogoniaceae in another, the Stephanokontae. Today most phycologists agree that the magnitude of difference is not greater than that of an order.

Finally there is the question of the proper distribution of the charas. Some think that their difference is not greater than that of ordinal rank and so include all green algae in a single class, the Chlorophyceae. Others consider them so different that they merit recognition as a separate division, the Charophyta. Still others place them in a class coordinate in rank with the Chlorophyceae. As among these three alternatives, the writer prefers the third in which the Chlorophyta are divided into two classes the Chlorophyceae and the Charophyceae.

CLASS 1. CHLOROPHYCEAE

The Chlorophyceae include all the green algae except the few genera assigned to the Charophyceae. Members of the Chlorophyceae may be unicellular or multicellular. Multicellular Chlorophyceae never have the verticillate vegetative organization or the ensheathing structures around the sex organs that are found in members of the Charophyceae.

Opinion varies as to the number of orders that should be recognized among the Chlorophyceae. The reader interested in the differences between the system of classification followed in this book and other recent systems is referred to the tabular summaries given by Fritsch.[1]

ORDER 1. VOLVOCALES

Genera assigned to this order include those Chlorophyceae in which the vegetative cells are flagellated and in which the cells are actively motile during vegetative phases of the life cycle. Many of the genera belonging to the order are unicellular; others have a fixed number of motile cells united in colonies and the cells arranged in a manner characteristic for the genus.

Most genera have ovoid, cordiform, pyriform, or fusiform cells, but some have cells that are compressed or with an irregular outline. All genera have the flagella borne at the anterior end. The great majority of genera are biflagellate, but some have quadriflagellate or octoflagellate cells. Many of the unicellular Volvocales may retract their flagella (or fail to develop them) and enter upon a temporary immobile phase in which the cell is surrounded by a copious gelatinous envelope. Continued cell division, when in an immobile state may result in a large amorphous colony with many cells irregularly distributed throughout a gelatinous matrix. Such *Palmella* stages are most frequently encountered among individuals living under subaerial conditions. Any cell in a palmelloid colony may, at any time, develop flagella and return directly to the motile condition.

The typical volvocaceous cell is enclosed by a definite wall and one in which there is always a cellulose layer next the protoplast. Frequently there is a layer of pectic material external to the cellulose layer. Gelatinous envelopes around the cells are especially common among the colonial

[1] FRITSCH, 1944.

genera, and in many of these the individual envelopes are completely fused with one another to form a homogeneous colonial matrix. There are a few genera (*Phacotus*, *Pteromonas*) in which the wall consists of two overlapping halves, sometimes impregnated with calcium carbonate, and in which the halves separate from each other at the time of liberation of daughter cells.

Volvocales lacking a definite wall, but in which the peripheral portion of the protoplasm is rigid, are usually placed in a family by themselves, the Polyblephardiaceae. Although it is generally agreed this naked condition is the more primitive, it does not necessarily follow that all naked-celled Volvocales are of a primitive type. It is quite possible that certain genera now placed in the Polyblepharidaceae are derived from ancestors that had a wall and that this primitive type of organization is secondarily acquired.

Cells of the Volvocales have protoplasts without a central vacuole, but they do have contractile vacuoles. In most genera there are two contractile vacuoles at the base of the flagella but, as in *Haematococcus*, there may be more than two vacuoles and these may lie just within any part of the plasma membrane. The eyespot in the volvocaceous cell is always solitary and generally located in the anterior half. As already noted (page 46), it may have a simple or a complex organization. The chlorophyll-containing portion of the cytoplasm, the chloroplast, is usually a massive cup-shaped structure occupying the greater portion of the space within the plasma membrane. Many of these cup-shaped chloroplasts contain a single pyrenoid, but some typically have more than one pyrenoid. It is becoming increasingly evident that there are innumerable modifications of the cup-shaped chloroplast, ranging all the way from structures that are U-shaped in optical section to those that are H-shaped. Other Volvocales have chloroplasts consisting of a central mass with numerous broad radial projections to the plasma membrane, or they may have the central mass greatly reduced or even lacking. The last-named case results in a number of separate discoid parietal chloroplasts. Since species with the foregoing types of chloroplast have been described for a single genus,[1] it does not seem as if there were two or three basic types of chloroplast structure among the Volvocales which, in turn, gave rise to distinct phylogenetic series among the higher Chlorophyceae. The amount of pigmentation in the chloroplast is extremely variable, and there are genera in which there is but a trace of chlorophyll (*Pseudofurcilia*) or in which there is no chlorophyll (*Polytoma*). A lack of chlorophyll does not exclude such genera from the Volvocales, since other features of cell structure are typical, including the presence of a pyrenoid and a regular formation of starch.

[1] Pascher, 1930.

These colorless genera cannot be placed in a family by themselves but must be ranged alongside genera with chlorophyll and put in the appropriate family.

Vegetative cells of Volvocales are uninucleate. Division of the nucleus is mitotic and may be parallel to the long axis of a cell (*Polytoma*,[1] *Chlorogonium*[2]) or at right angles to the long axis (*Eudorina*,[3] *Carteria*[1]).

Asexual reproduction of unicellular genera is by division to form two, four, or eight daughter cells. Genera without a cell wall always have a longitudinal division into two daughter cells, and cleavage may begin at the posterior or the anterior end. Genera with a cell wall may have a longitudinal or a transverse division of the protoplast. In the formation of more than two daughter protoplasts within a wall there is first a bipartition into two uninucleate protoplasts and then a division of each of these. The final series of divisions is followed by formation of a wall around each daughter protoplast. The daughter cells are usually liberated by a gelatinization of the parent-cell wall.

Asexual reproduction of colonial genera is by repeated division of a cell to form a daughter colony. In the more primitive genera (*Gonium*, *Pandorina*) every cell in a colony produces a daughter colony. Among advanced genera certain cells are always vegetative and others reproductive. For example, all cells in the anterior half of a colony of *Pleodorina californica* Shaw are vegetative. Reduction in number of reproductive cells is carried still further in *Volvox* where only 4 to 20 cells in the posterior half of a colony may be capable of dividing to form daughter colonies. Genera belonging to one of the colonial families, the Volvocaceae, have the divisions forming daughter colonies taking place in a very regular sequence (see page 94).

Gametic union is found in most members of the order and, according to the particular species, may be isogamous, anisogamous, or oögamous. The zygotes develop a thick wall and enter upon a period of rest before they are capable of germination. Division of the zygote nucleus has been shown to be meiotic in *Chlamydomonas*[4] and in *Volvox*.[5] The regular production of four zoospores by germinating zygotes of certain other species, including the demonstration that two of these are "plus" and two are "minus" in heterothallic species,[6] would seem to show that vegetative cells of all Volvocales are haploid.

Six families may be recognized among the Volvocales, and all of them have representatives in this country. The first three families listed on succeeding pages are somewhat artificial, but most of the genera assigned

[1] Entz, 1918. [2] Dangeard, 1898. [3] Hartmann, 1921.
[4] Moewus, 1936. [5] Zimmermann, 1921.
[6] Moewus, 1935; Schreiber, 1925; Schulze, 1927.

to each of them seem closely related. Some phycologists group all colonial genera, except *Stephanosphaera*, in a single family, but differences in arrangement of cells and in mode of formation of daughter colonies justify recognition of the Volvocaceae and the Spondylomoraceae as separate families.

Family 1. Polyblepharidaceae

Genera belonging to this family have a protoplast which is not enclosed by a cellulose wall. The cells may change slightly in shape as they move through the water, but the peripheral portion of the protoplast is always sufficiently rigid to give the cell a characteristic shape. Cells may be uni-, bi-, quadri-, or octoflagellate, and with the flagella longer or shorter than the cell. All genera have an eyespot, but there may or may not be contractile vacuoles at the base of the flagella. The chloroplast is usually cup-shaped and with a single pyrenoid. There are genera without chlorophyll, as *Polytomella* but in which the saprophytic nutrition results in an accumulation of starch.

Reproduction usually takes place while the cells are actively motile and by a longitudinal cleavage which may begin at the anterior or posterior ends. Some biflagellate genera, as *Phyllocardium* and *Dunaliella*, have one flagellum from a parent cell going to each daughter cell and a formation of one new flagellum by each daughter cell.

Resting stages of a cyst-like nature are formed by most genera of the family. When these cysts germinate, they develop directly into motile cells.

Sexual reproduction is isogamous.[1] Zygote germination has been seen in but one genus (*Dunaliella*), and here there is usually a formation of four zoospores.

The genera found in this country* differ as follows:

1. Cells with one flagellum .... 1. **Pedinomonas**
1. Cells with more than one flagellum 2
 2. With two or four flagella 3
 2. With six to eight flagella .... 8. **Polyblepharides**
3. Cells biflagellate 4

[1] Korshikov, 1927; Lerche, 1937.

* *Chloraster* has been reported (Kofoid, 1910) from the United States but is a genus of very questionable validity.

Some phycologists (*e.g.*, Pascher, 1927*A*) place *Collodictyon* among the Polyblepharidaceae; but, unlike other colorless Volvocales, it does not contain starch. This indicates that its affinities are with the Protomastiginae of the Protozoa rather than with the Volvocales. If *Collodictyon* is a member of the Polyblepharidaceae, it should be included in the algal flora of this country since it has been found in California, Ohio, and Tennessee.

3. Cells quadriflagellate 6
 4. Cells compressed .... 4. **Heteromastix**
 4. Cells not compressed 5
5. Cells with longitudinal ridges in anterior portion .... 2. **Stephanoptera**
5. Cells without longitudinal ridges .... 3. **Dunaliella**
 6. Cells spindle-shaped .... 6. **Spermatozoopsis**
 6. Cells not spindle-shaped 7
7. With chlorophyll .... 5. **Pyramimonas**
7. Without chlorophyll .... 7. **Polytomella**

1. **Pedinomonas** Korshikov, 1923. *Pedinomonas* has compressed cells that are ellipsoidal to subcircular in front view. The cells are naked and without an envelope or a differentiated periplast. There is but one flagellum, and it is directed backward as a cell swims through the water. A single contractile vacuole lies near the base of the flagellum. The chloroplast is sickle-shaped and contains a single conspicuous pyrenoid surrounded by starch granules.[1]

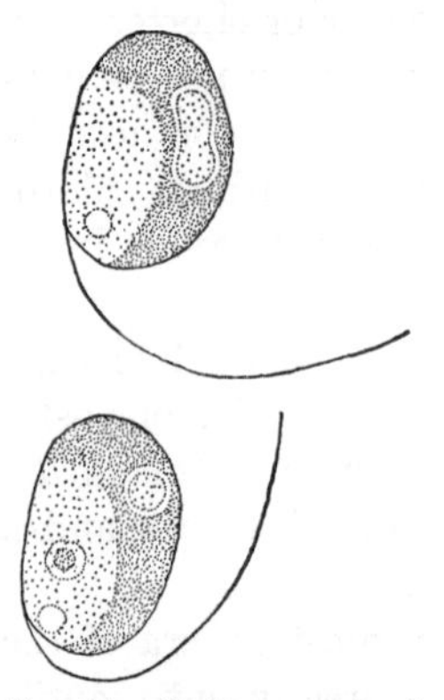

Fig. 11. *Pedinomonas minor* Korshikov. (*After Korshikov*, 1923.)

Asexual reproduction is by bipartition and takes place while a cell is motile. There may be a development of palmella stages.

Sexual reproduction is by fusion of two vegetative cells, and the newly formed zygote may be uni- or biflagellate. Eventually the zygote loses its flagella, assumes a spherical shape, and becomes surrounded by an envelope.

P. minor Korshikov (Fig. 11) has been found both in Ohio and Tennessee,[2] and *P. rotunda* Korshikov has been found in Ohio.[3] For descriptions of these two species, see Pascher (1927*A*).

2. **Stephanoptera** Dangeard, 1910 (*Asteromonas* Artari, 1913). The solitary cells of this alga are pyriform and with the broad anterior end four-lobed to six-lobed. When viewed from above, the cells are stellate in outline and with four to six broadly rounded projections. Cells of *Stephanoptera* are without a wall, and the two flagella are longer than the cell. There are two contractile vacuoles at the base of the flagella. The chloroplast is cup-shaped, and its anterior end has as many laminate projections as there are lobes to the cell. The elongate eyespot usually lies toward the anterior end of a cell. The nucleus is fairly conspicuous.

Reproduction is by a longitudinal division that begins at the anterior end.[4] The vertical division halves the chloroplast and pyrenoid and

[1] Korshikov, 1923. [2] Lackey, 1939, 1942.
[3] Lackey, 1939. [4] Dangeard, 1912.

distributes one flagellum to each daughter cell. Each daughter cell becomes biflagellate by formation of a new flagellum during later stages of division.

At times the cells retract their flagella, assume a spherical shape, and secrete an enclosing membrane. Such "cysts" are thought to germinate directly.

Sexual reproduction has not been noted in this alga.

S. gracilis (Artari) G. M. Smith (Fig. 12) has been found in several salterns and brine lakes in the western part of the United States. For a description of it as *Asteromonas gracilis*, see Artari (1913).

3. **Dunaliella** Teodoresco, 1905. The cells of *Dunaliella* are ovoid to narrowly pyriform and with two long flagella at the anterior end. In

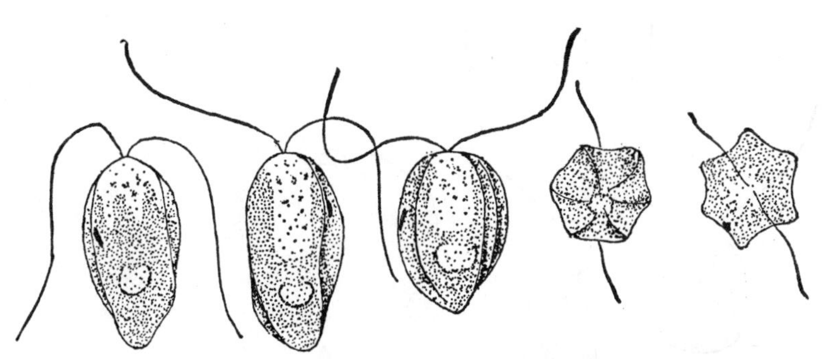

FIG. 12. *Stephanoptera gracilis* (Artari) G. M. Smith. (× 1300.)

the posterior end is a more or less cup-shaped chloroplast containing a single pyrenoid. According to the species, there is or is not an eyespot in the anterior end. Under unfavorable conditions[1] the green of the chloroplast and even the colorless portion of the protoplast are masked with hematochrome.

Asexual reproduction takes place by longitudinal division and while cells are in motion. A cell may develop into a cyst by rounding up and secreting a wall. The finding of many empty cyst walls[2] would seem to indicate that germination of cysts is direct.

Sexual reproduction is isogamous and results in a spherical zygote with a smooth wall.[3] Germination of a zygote may result in a production of as many as 16 zoospores. Genetic analysis of the products from germinating zygotes indicates that division of the zygote nucleus is meiotic.[4]

[1] LERCHE, 1937. [2] HAMBURGER, 1905.
[3] LERCHE, 1937; TEODORESCO, 1905.
[4] LERCHE, 1937.

Dunaliella is of world-wide distribution and almost invariably present in salterns and brine lakes. The optimum NaCl concentration for most species is between 4 and 8 per cent, but they can tolerate much higher concentrations.[1] *Dunaliella* has been collected from brine lakes in Utah, Nevada, and California. Two species, *D. salina* (Dunal) Teodor. (Fig. 13) and *D. Peircei* Nicolai have been found in the United States. For descriptions of them, see Lerche (1937).

4. **Heteromastix** Korshikov, 1923. The naked cells of this polyblepharid are conspicuously flattened and broadly ellipsoidal, subhexagonal, or reniform in outline. One of the two flagella is somewhat longer than the other. The chromatophore, which lies toward the posterior end of a cell, may have a conspicuous cup-shaped depression or a very slight depression. It contains a single spherical pyrenoid surrounded by several starch plates or a somewhat compressed pyrenoid surrounded by two large starch plates.

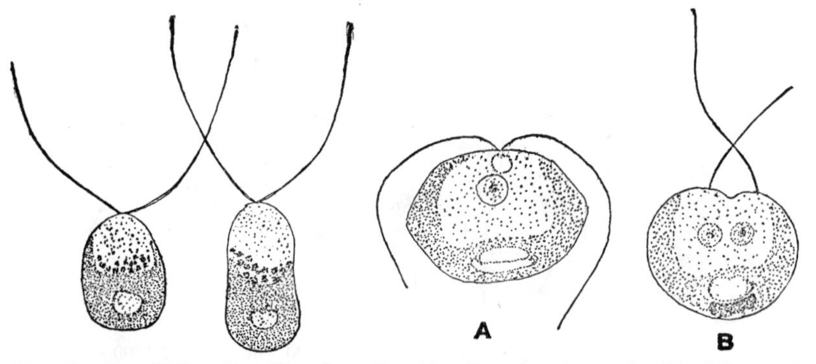

FIG. 13. *Dunaliella salina* (Dunal) Teodor. (× 860.)

FIG. 14. *Heteromastix angulata* Korshikov. *A*, vegetative cell. *B*, dividing cell. (× 2600.)

Sometimes there is a minute eyespot at the anterior end of a cell. There is a single contractile vacuole beneath the flagella. The nucleus is conspicuous and anterior. In addition to the distinctive morphological features, *Heteromastix* has a characteristic irregular motion as it moves through the water. This irregular motion has been ascribed[2] to one flagellum acting as a trailing flagellum while a cell is swimming about.

Reproduction takes place while cells are motile and is by a longitudinal cleavage that begins at the anterior end.

A fusion of cells of equal size has been observed,[2] and the quadriflagellate zygote remains motile for a considerable time before it becomes immobile, assumes a spherical shape, and secretes a smooth wall.

In this country the type species, *H. angulata* Korshikov (Fig. 14) is known only from California.[3] For a description of it, see Pascher (1927*A*).

[1] LERCHE, 1937. [2] KORSHIKOV, 1923. [3] SMITH, G. M., 1933.

5. **Pyramimonas** Schmarda, 1850. The naked solitary cells of this genus are sometimes hemispherical but more often pyriform. The posterior end of a cell is always broadly rounded; the anterior end is always retuse and four-lobed. The cells have four flagella that are inserted close together in a depression of varying depth at the anterior end of a cell. Each flagellum has a minute blepharoplast at its base, and the blepharoplasts are connected with the nuclear membrane by a fairly thick rhizoplast.[1] There are two contractile vacuoles near the base of the flagella. The chloroplast is typically cup-shaped and with the anterior end deeply incised to form four lobes, each of which may, in turn, have a median longitudinal incision.[2] There is a single pyrenoid. The eyespot is con-

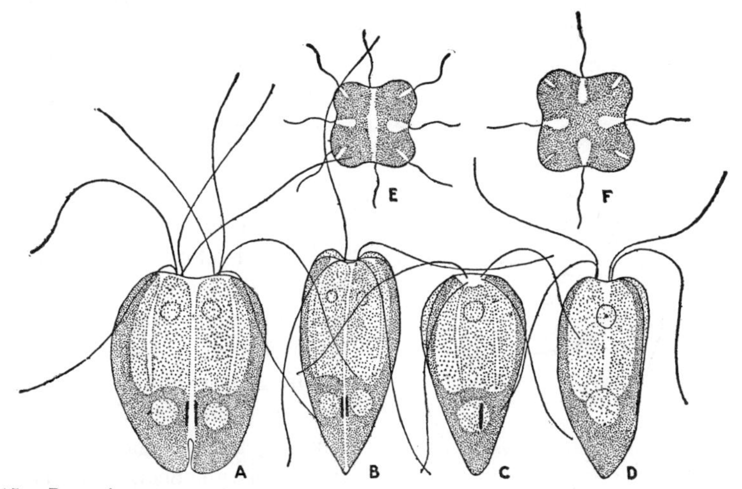

Fig. 15. *Pyramimonas tetrarhynchos* Schmarda. *A–B*, dividing cells. *E–F*, vertical view from the posterior pole. (× 1000.)

spicuous and anterior to submedian in position. Although there is no definite wall around a cell, the peripheral portion of the cytoplasm (periplast) seems to be distinct from the rest of the cytoplasm and may even be caused to separate off in large blisters by the application of suitable reagents.[3]

Reproduction is by longitudinal division and takes place while cells are in motion. Following nuclear division, there is a formation of two sets of flagella at the anterior end, and the two become remote from each other before division is completed (Fig. 15*A*).

Subglobose cysts with a thick hyaline gelatinous covering are formed at

[1] Bretschneider, 1925.

[2] Geitler, 1925*D*; Griffiths, 1909.

[3] Griffiths, 1909.

times, and these cysts may be aggregated within a common gelatinous matrix to produce a *Palmella*-like stage.[1]

A union of approximately similar gametes has been recorded for *P. reticulata* Korshikov.[2] Germinating zygotes of this species give rise to four zoospores.

Four species, *P. inconstans* Hodgetts, *P. montana* Geitler, *P. reticulata* Korshikov, and *P. tetrarhynchos* Schmarda (Fig. 15) have been found in this country. For descriptions of them, see Pascher (1927*A*).

6. **Spermatozoopsis** Korshikov, 1913. The naked cells of this alga are spindle-shaped and usually bent in an arc. The cells are quadriflagellate, with two contractile vacuoles near the base of the flagella, and with a linear chloroplast that lies along the convex side of a cell. There are no pyrenoids. The colorless cytoplasm on the concave side of a cell contains several granules of food reserves.[3]

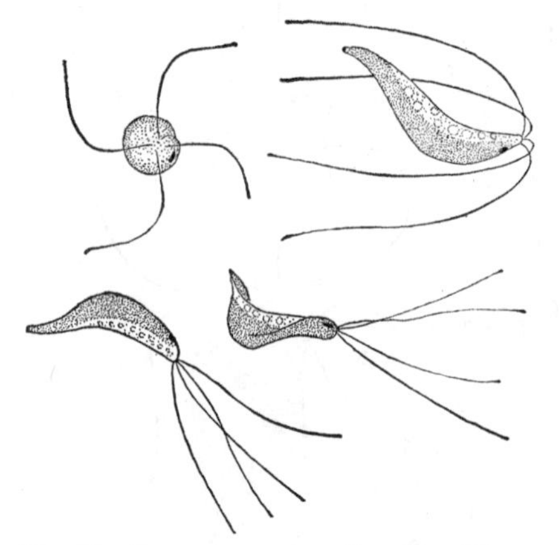

FIG. 16. *Spermatozoopsis exultans* Korshikov. (*After Korshikov*, 1923.)

Asexual reproduction is by longitudinal division into two daughter cells and takes place while the cells are motile.

Biflagellate individuals have also been recorded for this genus.[4] Presumably they are gametes, but their union has not been observed.

S. exultans Korshikov (Fig. 16) has been found in Indiana, Ohio, and Kentucky.[5] It was present in abundance at two stations in Ohio, and all individuals were quadriflagellate.[6] For a description of this species, see Pascher (1927*A*).

7. **Polytomella** Aragão, 1910. Thus far this is the only known member of the family with colorless cells. The cells are characteristically pyriform to ovoid but often show considerable change in shape as they move through the water. Occasionally there are transitory conical processes at the posterior end of a cell.[7] The cells are without a wall, but, as is the case with *Pyramimonas*, the periplast may be separated from the rest of the cytoplasm by the use of suitable reagents. The four flagella, which are

[1] DANGEARD, 1889; HODGETTS, 1920. [2] KORSHIKOV in PRINTZ, 1927.
[3] KORSHIKOV, 1913. [4] KORSHIKOV, 1913; PASCHER, 1927*A*.
[5] BRINLEY and KATZIN, 1942; LACKEY, 1939. [6] LACKEY, 1939.
[7] KATER, 1925.

somewhat shorter than the cell, are connected with the nucleus by a blepharoplast-rhizoplast neuromotor apparatus.[1] There are two contractile vacuoles near the base of the flagella, and an orange-red eyespot may at times be present[2] in the anterior end of a cell. No pyrenoids have been found in association with the starch grains, which are often present in such abundance as to obscure the structure of the protoplast. The nucleus is not usually evident in living cells, but stained preparations show that it lies in the central portion of a cell.

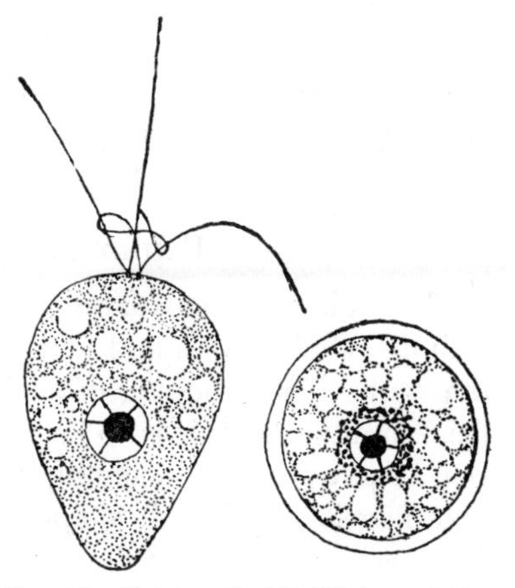

Fig. 17. *Polytomella Citri* Kater. (*After Kater*, 1925.)

Asexual reproduction is by longitudinal division and while the cells are in motion. The mitotic spindle lies at right angles to a cell's long axis,[3] and cytoplasmic cleavage follows very shortly after mitosis. Each daughter cell has two flagella derived from the parent cell and two which are newly formed. The four flagella of each daughter cell are completely developed before the daughter cells separate.[4] Encystment is of frequent occurrence and takes place by a cell retracting its flagella, assuming a spherical shape, and secreting a cellulose wall.

P. Citri Kater (Fig. 17) and *P. agilis* Aragão have been found in this country. For a description of the former, see Kater (1925); for the latter, see Pascher (1927*A*).

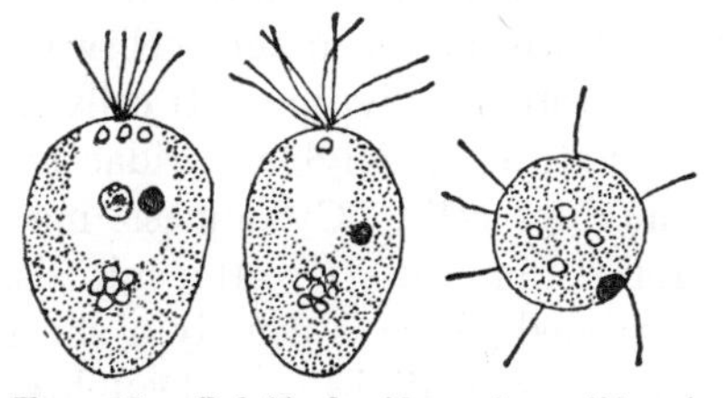

Fig. 18. *Polyblepharides fragariiformis* Hazen.

8. **Polyblepharides** Dangeard, 1888. The solitary naked cells of this alga are ovoid, cylindrical, or subpyriform and with six or eight flagella. At the base of the flagella are one (?), two, or four contractile vacuoles, and a short distance beneath them is a nucleus with a conspicuous nucleolus. The cup-shaped chloroplast is massive. It may contain a pyrenoid or lack one and merely have a cluster of starch grains. An eyespot may be present or lacking; if present, it lies at the level either of the nucleus or of the pyrenoid.

[1] Kater, 1925. [2] Doflein, 1916*A*. [3] Aragão, 1910; Kater, 1925.
[4] Aragão, 1910.

Asexual reproduction is by longitudinal division while the cells are motile. There may be a formation of spherical cysts with a smooth gelatinous envelope. A germinating cyst has an amoeboid escape of the protoplast from the cyst wall and a development of flagella after escapement.

P. fragariiformis Hazen (Fig. 18) is known from Vermont and *P. singularis* Dang. from North Carolina. For a description of the former, see G. M. Smith (1933); for the latter, see Whitford (1943).

Family 2. Chlamydomonadaceae

The Chlamydomonadaceae include all unicellular Volvocales with a definite cellulose wall and with all or a portion of the protoplast adjoining the wall. The grouping of such genera in one family results in an artificial family, but this is to be preferred to the even more artificial recognition of two families including, respectively, the genera with two and with four flagella. The cell shape is extremely varied from genus to genus. Most genera are biflagellate, a few are quadriflagellate, but none of them has, as in the Polyblepharidaceae, more than four flagella. In the majority of cases, the chloroplast is cup-shaped, but chloroplasts of some species are laminate, stellate, or even divided into a number of small discoid bodies. According to the species, the chloroplasts contain one, two, or several pyrenoids; or lack them. There are also genera without chlorophyll, but in which the cells accumulate starch through a saprophytic mode of nutrition. Most genera have an eyespot, and practically all of them have contractile vacuoles at the base of the flagella. All genera are uninucleate.

Asexual reproduction is by division into two, four, or eight daughter protoplasts which form cell walls while still within the parent-cell wall. There is usually a disappearance of flagella before cell division. If cells do divide while in a motile condition, there is not, as in Polyblepharidaceae, a contribution of persistent flagella to daughter cells. Cytokinesis may be longitudinal or transverse and always follows immediately after mitosis. If four or eight daughter protoplasts are formed, there is usually a short period of rest before the second or third series of divisions. Daughter cells are liberated by a rupture or by a gelatinization of the parent-cell wall. If there is a gelatinization of the wall, the daughter cells may not escape before they produce a new cell generation or several cell generations. Such palmella stages are known for a number of genera, and in each of them the cells may become motile at any time. Akinetes have been reported for a number of genera, but little is known concerning their mode of formation or their mode of germination.

Gametic union may be isogamous, anisogamous, or oögamous, and study in pure and unialgal culture has shown that some species are homothallic and others heterothallic. Zygotes form a thick wall and usually undergo a period of rest before germinating to form four or more zoospores.

The genera of Chlamydomonadaceae found in this country* may be distinguished as follows:

1. Cells with chlorophyll.. 2
1. Cells without chlorophyll.. **3. Polytoma**
   2. Motile cells biflagellate... 3
   2. Motile cells quadriflagellate... 10
3. Vertical view circular.. 4
3. Vertical view not circular... 7
   4. Flagella inserted close together.. 5
   4. Flagella remote from each other............................ **8. Gloeomonas**
5. Cells fusiform.. **7. Chlorogonium**
5. Cells not fusiform.. 6
   6. Protoplast same shape as envelope...................... **1. Chlamydomonas**
   6. Protoplast not same shape as envelope.................. **2. Sphaerellopsis**
7. Cells compressed.. 8
7. Cells not compressed... 9
   8. Front view ovate.. **4. Platychloris**
   8. Front view circular... **5. Mesostigma**
9. Posterior end with several conical projections................ **9. Brachiomonas**
9. Posterior end without conical projections........................ **6. Lobomonas**
   10. Cells compressed.. 11
   10. Cells not compressed... **10. Carteria**
11. Posterior end pointed.. 12
11. Posterior end broadly rounded.................................. **11. Platymonas**
   12. Lateral margins of cell wall wing-like...................... **12. Scherffelia**
   12. Lateral margins of wall not wing-like..................... **13. Chlorobrachis**

1. **Chlamydomonas** Ehrenberg, 1833. The solitary free-swimming cells of *Chlamydomonas* (Fig. 19) may be spherical, ellipsoidal, subcylindrical, or pyriform. The two flagella are inserted fairly close together, and the contour of a cell may or may not be distinctly papillate in the region bearing the flagella. There is always a definite cellulose wall layer, and some species have a distinct gelatinous layer external to the cellulose layer. Most species have a delicate to massive cup-shaped chloroplast; other species have chloroplasts that are laminate, stellate, or H-shaped in optical section. There is usually but one pyrenoid within a chloroplast, but some species regularly have more than one pyrenoid. Typically there are two contractile vacuoles near the base of the flagella, but the number and position of the vacuoles are not constant for the genus. The shape and posi-

* *Furcilia*, a genus found in the United States, has been included among the Volvocales (Pascher, 1927*A*; G. M. Smith, 1933) but Jane (1944) has given good reasons for referring it to the Protomastigineae of the Protozoa.

tion of the eyespot are fairly constant for any given species, but taking the genus as a whole the eyespot may lie anywhere between insertion of the flagella and the base of a cell. It may be circular, oval, or sublinear in outline. In the only species critically examined,[1] there is a neuromotor apparatus of the blepharoplast-rhizoplast-centriole type.

Asexual reproduction is by longitudinal division of the protoplast into two, four, or eight daughter protoplasts which lie within the parent-cell wall. Each daughter protoplast secretes a wall and develops a neuromotor apparatus before it is liberated.[1] Liberation is usually by a gelatinization of the parent-cell wall. Failure of daughter cells to escape from the gelatinized wall results in a palmella stage in which cell division may continue until there are hundreds of cells. *Chlamydomonas* may also form aplanospores and akinetes.

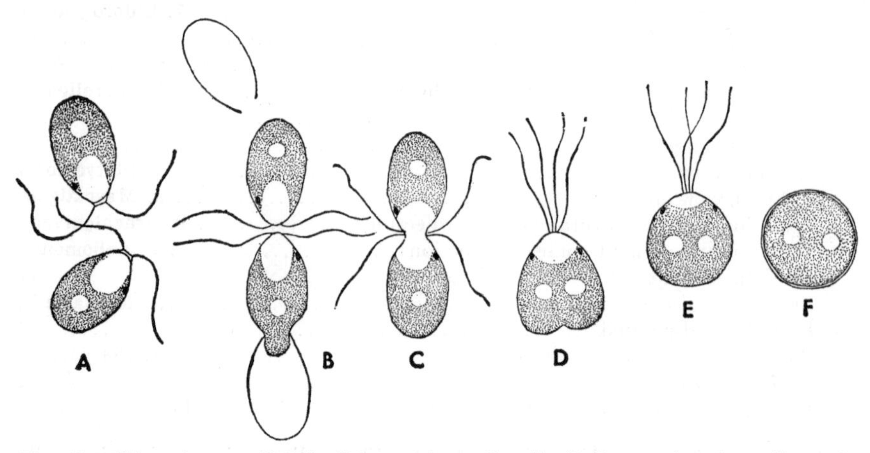

FIG. 19. *Chlamydomonas Snowiae* Printz. *A*, motile cells, *B–D*, gametic union. *E*, motile zygote. *F*, resting zygote. (× 1300.)

Some species are homothallic, others are heterothallic, and gametic union may be isogamous, anisogamous, or oögamous. Isogamous species usually have the cells capable of functioning as gametes when conditions are favorable. When such cells unite, there is frequently an escape of the protoplasts from the enclosing walls before the two unite (Fig. 19). In the anisogamous heterothallic *C. Braunii* Gorosch. a cell of a male clone divides to form 8 or 16 daughter cells and one of a female clone forms 2 or 4 daughter cells.[2] In the two oögamous species, *C. coccifera* Gorosch.[3] and *C. oögamum* Moewus,[2] a male cell divides to form 8, 16, or 32 small cells, one of which unites with a large female cell that has become immobile and lost its flagella.

The quadriflagellate zygote of isogamous species may remain motile for

[1] KATER, 1929. [2] MOEWUS, 1938. [3] GOROSCHANKIN, 1905.

a few hours or may swim about for as many as 15 days[1] before coming to rest and secreting a wall. Zygotes of a few species remain green, but in most species the chlorophyll eventually becomes masked by hematochrome and the reserve starch converted into oil. When a zygote germinates, its protoplast usually divides to form four zoospores, but in some species there may be a formation of eight zoospores (*C. Reinhardi* Dang.) or 16 to 32 zoospores (*C. intermedia* Chod.). A meiotic division of the zygote nucleus has been observed in a hybrid zygote formed by crossing two species.[2] Genetic analysis of zoospores from germinating zygotes of several species also shows that division of the zygote nucleus is meiotic.[3]

Nearly 30 species have been recorded from the United States. The most noteworthy of these are the snow algae, *C. nivalis* (Bauer) Wille and *C. yellowstonensis* Kol. For monographic treatments of the genus, see Pascher (1927*A*), and Gerloff (1940).

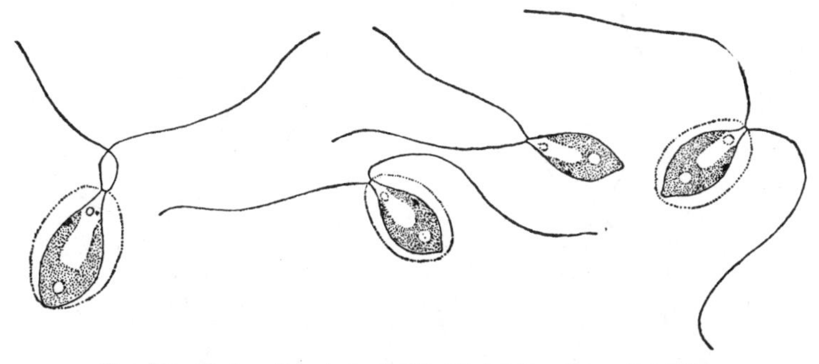

FIG. 20. *Sphaerellopsis fluviatilis* (Stein) Pascher. (× 900.)

2. **Sphaerellopsis** Korshikov, 1925. The chief difference between *Sphaerellopsis* and species of *Chlamydomonas* surrounded by a broad gelatinous envelope is that in *Sphaerellopsis* the protoplast and the enclosing sheath are of quite different shape.[4] *Sphaerellopsis* also has the daughter cells developing the characteristic gelatinous sheath before they are liberated from the old parent-cell wall. The broadly fusiform protoplast of this alga has a fairly massive chloroplast with a single pyrenoid at its posterior end. There is an eyespot toward the anterior end of a cell, and two contractile vacuoles lie just beneath the insertion of two body-long flagella. Asexual reproduction is by division into four or eight daughter cells. Sexual reproduction is isogamous.

The only record for *Sphaerellopsis* in this country is the finding of *S. fluviatilis* (Stein) Pascher (Fig. 20) in California.[5] For a description of it, see Pascher (1927*A*).

[1] STREHLOW, 1929. [2] MOEWUS, 1936. [3] MOEWUS, 1938, 1939, 1940*A*.
[4] KORSHIKOV, 1925. [5] SMITH, G. M., 1933.

3. **Polytoma** Ehrenberg, 1838. The free-swimming cells of this genus are similar in shape to those of certain *Chlamydomonas* species, but they always lack chlorophyll. The two flagella at the anterior end are usually as long as the cell and inserted close to each other in a small anterior papilla. There is a definite cellulose wall around the protoplast, and this may or may not have an external layer of gelatinous material. There are two contractile vacuoles at the base of the flagella. Sometimes there is an eyespot in the anterior end of the cell, but this cell organ is often lacking. The single nucleus lies in the long axis of the cell and usually about a third of the distance from the anterior to the posterior end. A centriole-rhizoplast-blepharoplast neuromotor apparatus connects the flagella with the nucleus.[1] The colorless cytoplasm of the cell may be densely packed with starch, or the starch may be restricted to the anterior

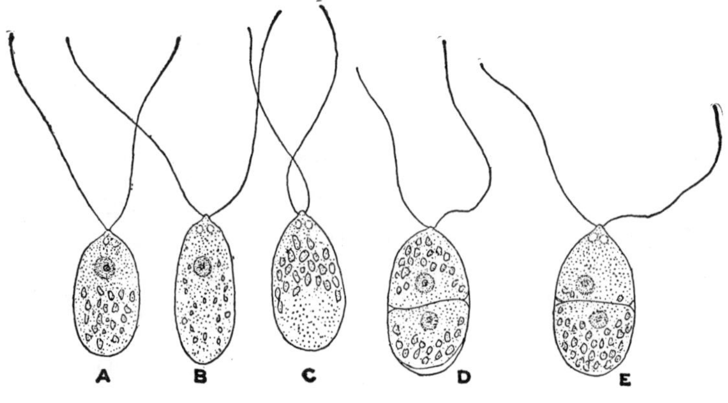

FIG. 21. *Polytoma uvella* Ehr. (× 900.)

or to the posterior portions of the cytoplasm.[2] As is the case with most other colorless Volvocales, there is no pyrenoid within the cell. Sometimes there are droplets of a yellowish-brown oil in addition to the starch.

The physiology of *Polytoma* is better known than that of other colorless Chlorophyceae.[3] It has been shown that its nutrition is strictly saprophytic and that the organism grows as well in darkness as in light. The foods which *Polytoma* can utilize in starch manufacture include a wide range of carbohydrates and salts of fatty acids.

Asexual reproduction is by transverse division (Fig. 21*D* and *E*). If there are further divisions, these are longitudinal.[1] The newly formed daughter protoplasts elongate to assume the characteristic cell shape,

[1] DANGEARD, 1912; ENTZ, 1918. [2] JIROVEC, 1926; PRINGSHEIM, E., 1927.
[3] JACOBSEN, 1910; PRINGSHEIM, E., 1921; PRINGSHEIM and MAINX, 1926.

develop a neuromotor apparatus, and secrete a cell wall while they are still within the wall of the parent cell. Aplanospores have been found by a number of investigators.[1] These are generally formed singly within a cell, and there is a marked shrinking away and rounding up of the protoplast before it secretes its special wall.

Gametic union is isogamous, and both heterothallism and homothallism are found within the genus.[2] The zygote becomes spherical, forms a smooth wall, and sometimes the protoplast becomes deeply colored with hematochrome. Genetic analysis shows[2] that division of the zygote nucleus is meiotic. When a zygote germinates,[3] its protoplast divides into two, four, or eight parts all of which may become zoospores, but it is not unusual to have one or more protoplasts aborting and so to have only one to six zoospores coming from a germinating zygote.

P. uvella Ehr. (Fig. 21) has been reported from various parts of this country, and *P. granulifera* Lackey has been found in Ohio. For a description of the former, see Pascher (1927*A*); for the latter, see Lackey (1939).

4. **Platychloris** Pascher, 1927. Cells of this alga are smaller than those of most other unicellular Volvocales and are markedly compressed. In front view, they are oval in outline; in side view, they are narrowly elliptical and with a length about six times the breadth. The cells are biflagellate, and the length of the flagella is more than four times that of a cell. There is a single contractile vacuole in the anterior end of a cell. The chloroplast is a flattened plate and lies in the posterior portion of a cell. It lacks a pyrenoid.[4]

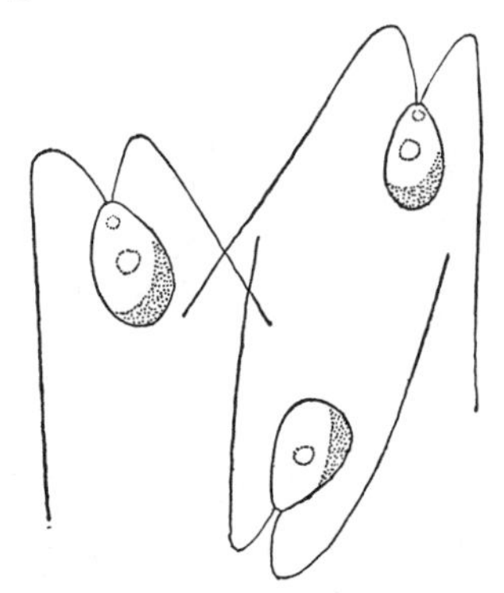

FIG. 22. *Platychloris minima* Pascher. (*After Pascher*, 1927*A*.)

Asexual reproduction is by division of the protoplast into four daughter cells.

P. minima Pascher (Fig. 22), the only known species, has been found in Ohio.[5] For a description of it, see Pascher (1927*A*).

5. **Mesostigma** Lauterborn, 1899. *Mesostigma* has markedly compressed disk-shaped cells, but unlike other compressed unicellular Volvocales the compression is in the polar instead of the equatorial axis. As

[1] DANGEARD, 1901; FRANZÉ, 1894; KRASSILSTSCHIK, 1882; PRINGSHEIM, E., 1921.
[2] MOEWUS, 1935*A*. [3] MOEWUS, 1935; STREHLOW, 1929.
[4] PASCHER, 1927*A*. [5] LACKEY, 1939.

a result of compression in the polar axis, the two flagella seem to be borne on the middle of a cell instead of at the anterior end. The cells are without a well-defined wall, but there is an evident gelatinous envelope external to the protoplast. There are two to seven contractile vacuoles near the base of the flagella. The chloroplast is cup-shaped,[1] but the base of the cup is so thin that the chloroplast has been erroneously described as ring-shaped.[2] According to the species, a chloroplast lacks pyrenoids or has two of them.

Asexual reproduction is by division into two daughter cells and takes place when the organism is at rest.[3]

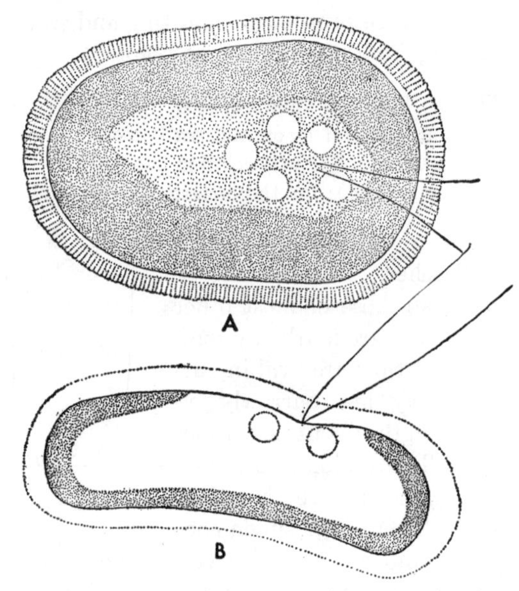

Fig. 23. *Mesostigma grande* Korshikov. *A*, top view. *B*, side view. (*After Korshikov,* 1938.) (× 1065.)

M. viridis Lauterb. and *M. grande* Korshikov (Fig. 23) have been reported from Ohio, Indiana, and Kentucky.[4] For a description of *M. viridis* see Lund (1937); for *M. grande*, see Korshikov (1938).

6. **Lobomonas** Dangeard, 1898. *Lobomonas* has an organization quite similar to that of *Chlamydomonas*, except for the irregularly distributed blunt processes on the cell wall. The cells are solitary, biflagellate, and with the length of the flagella about equal to that of a cell. There are

[1] Korshikov, 1938. [2] Lauterborn, 1899; Lund, 1937. [3] Lund, 1937.
[4] Brinley and Katzin, 1942; Lackey, 1939.

two contractile vacuoles at the base of the flagella, and a sublinear eyespot lies near the anterior end of a cell. The chloroplast is cup-shaped and with a single excentrically disposed pyrenoid in its posterior portion.[1]

Asexual reproduction is by division of a cell into four or eight daughter cells. The first cleavage appears to be transverse but really is longitudinal since the protoplast revolves 90 deg. from its original position within the wall. Liberation of daughter cells is by a gelatinization of the parent-cell wall.[2]

Sexual reproduction has not been recorded for the genus, if *L. pentagonia* Hazen is placed among the species of *Diplostauron*.

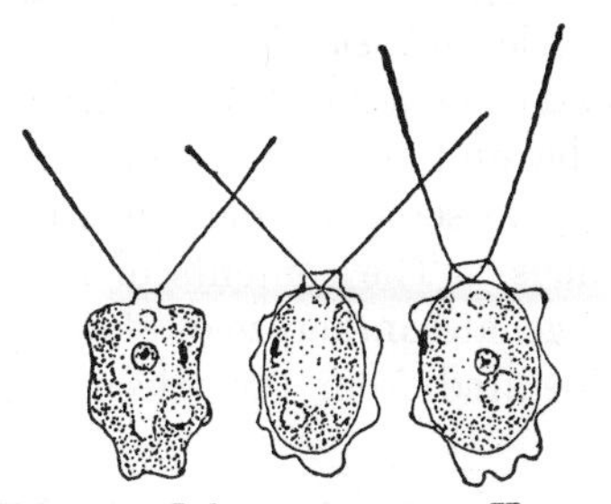

FIG. 24. *Lobomonas rostrata* Hazen. (*After Hazen*, 1922*A*.) (× 1375.)

L. rostrata Hazen (Fig. 24) has been collected[2] in rain-water pools in New Jersey and Vermont. For a description of it, see Hazen (1922*A*).

7. **Chlorogonium** Ehrenberg, 1830. Members of this distinctive genus are immediately distinguishable from other pigmented unicellular Volvocales by their fusiform cells, in which the posterior end is pointed and the anterior end is narrowly rostrate. The two flagella at the anterior pole are about half as long as the cell. There is a delicate wall around the protoplast, and two contractile vacuoles can generally be seen at the anterior end of the cell. The chloroplast may be massive and constituting most of the protoplast, or it may be distinctly laminate. According to the species the chloroplast contains one, two, or several pyrenoids or lacks them entirely. Chloroplasts with two pyrenoids have them axial to the centrally located nucleus (Fig. 25*D*); those with several pyrenoids have them irregularly scattered (Fig. 25*A*). Most species have a sublinear eyespot toward the anterior end of the cell.

Asexual reproduction begins with a transverse division of the protoplast, and this may be followed by one or two more series of transverse divisions.[3] The newly formed protoplasts then elongate in the long axis of the parent cell and each secretes a wall of its own (Fig. 25*G* and *H*). Liberation of these zoospores is by a gelatinization of the parent-cell wall.

Sexual reproduction of most species is by the fusion in pairs of equal-sized biflagellated gametes[4] (Fig. 25*I*). Thirty-two or sixty-four iso-

[1] DANGEARD, 1898; HAZEN, 1922*A*. [2] HAZEN, 1922*A*.

[3] DANGEARD, 1898; HARTMANN, 1919; JACOBSEN, 1910.

[4] DANGEARD, 1898; KLEBS, 1883; KRASSILSTSCHIK, 1882*A*; SCHULZE, 1927; STREHLOW, 1929.

gametes are usually formed by a single cell, and gamete formation begins with a series of simultaneous nuclear divisions.[1] After 32 or 64 nuclei have been formed, there is a cleavage of the cytoplasm into uninucleate gametes. Isogametes are naked or enclosed by a wall[2] and with the flagella as long as the body of the gamete or only half as long. Four isogamous species [*C. euchlorum* Ehr., *C. elongatum* Dang., *C. leiostracum* Streh., and *C. neglectum* (Korshik.) Pascher[2]] have been shown to be heterothallic.[3] The uniting gametes lose their flagella during fusion, and the resultant zygote soon becomes spherical and secretes a wall. One species is oögamous.[4] The contents of male cells divide to form 64 or 128 acicular biflagellate antherozoids that are surrounded by a delicate vesicle when first liberated from the parent-cell wall. The entire protoplast of a female

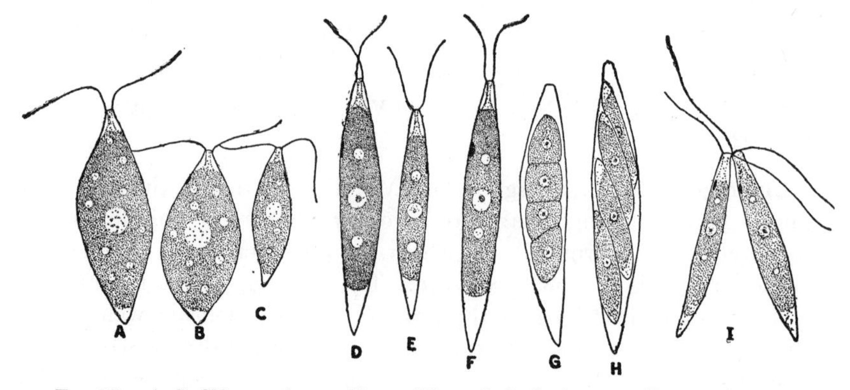

FIG. 25. *A–C, Chlorogonium euchlorum* Ehr. *D–I, C. elongatum* Dang. (× 650.)

cell develops into a globose egg, which escapes in an amoeboid fashion from the parent-cell wall.

Germinating zygotes have a division of their contents into four zoospores, and genetic analysis shows[3] that division of the zygote nucleus is meiotic.

C. euchlorum Ehr. (Fig. 25*A–C*), *C. elongatum* (Dang.) Franzé (Fig. 25*D–I*), and *C. spirale* Scherffel and Pascher are known to occur in this country. For descriptions of them, see Pascher (1927*A*).

8. **Gloeomonas** Klebs, 1886. *Gloeomonas* differs from other members of the family in having the two flagella remote from each other and in having a gelatinous layer of uniform thickness external to the cell wall. The cell contains numerous discoid chloroplasts which are without pyre-

[1] DANGEARD, 1898. [2] STREHLOW, 1929.
[3] SCHULZE, 1927; STREHLOW, 1929. [4] PASCHER, 1931*A*.

noids.[1] There is a conspicuous eyespot in the anterior end of a cell, and at the base of the flagella are two contractile vacuoles.

Cell division takes place while the cells are immobile.

Sexual reproduction has not been observed in this genus.

G. ovalis Klebs (Fig. 26) has been found[2] in Maine and Massachusetts. For a description of it, see Pascher (1927*A*).

9. **Brachiomonas** Bohlin, 1897. The solitary free-swimming cells of this alga have an anterior half that is subhemispherical, and a posterior half that is distinctly conical. The equatorial region may have four stout, recurved, conical processes, equidistant from one another; or the equatorial processes may be reduced to inconspicuous bumps.[3] The cell has a delicate wall, and the two flagella are generally longer than a cell. Contractile vacuoles appear to be lacking in this genus,[4] but there is a conspicuous sublinear eyespot in the equatorial region. The single massive chloroplast may fill most of the protoplast, or the lateral and posterior projections may be colorless. A chloroplast contains a single pyrenoid.

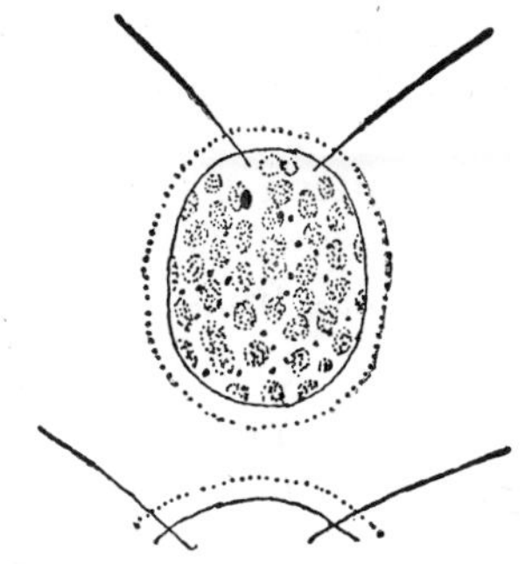

FIG. 26. *Gloeomonas ovalis* Klebs. (*After Pascher*, 1927*A*.)

Asexual reproduction by division into four or eight (rarely two) daughter cells takes place while a cell is motile. The first and second cleavages are longitudinal; if a third series of cleavages occur, they are transverse. Daughter cells assume their characteristic shape before liberation from the parent-cell wall. A formation of aplanospores has been noted, and this has been ascribed[5] to increased salinity of the water.

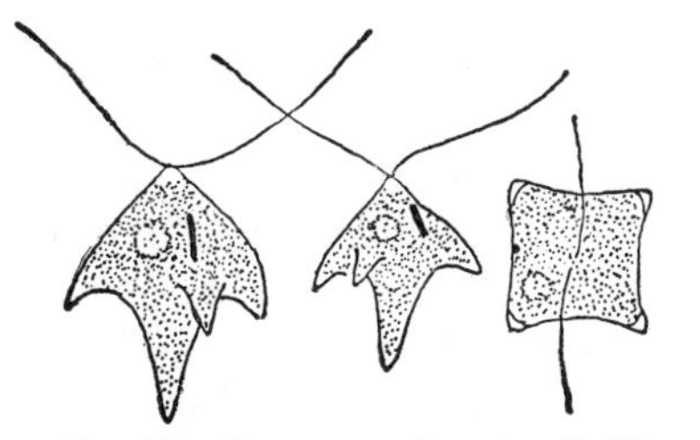

FIG. 27. *Brachiomonas submarina* Bohlin. (*After Hazen*, 1922.) (× 1440.)

Sexual reproduction is isogamous. The protoplast of a cell divides to form 16 or 32 gametes which are somewhat different in shape from the cell producing them.[6] The fusing gametes, which appear to be naked,

[1] KLEBS, 1886; PASCHER, 1927*A* [2] SMITH, G. M., 1933.
[3] BOHLIN, 1897*A*; HAZEN, 1922; WEST, G. S., 1908. [4] HAZEN, 1922.
[5] GABRIEL, 1924. [6] HAZEN, 1922; WEST, G. S., 1908.

unite laterally. The quadriflagellate zygotes may remain motile for some time before they assume a spherical shape and secrete a smooth thick wall. Old zygotes usually have the green color masked by a reddish-orange hematochrome.

B. submarina Bohlin (Fig. 27) has been found along both the Atlantic and Pacific coasts of this country in littoral rock pools so high above the ordinary tide limits of the ocean that they are only occasionally dashed with salt spray. European phycologists have found this species in strictly fresh-water habitats. For a description of *B. submarina*, see Pascher (1927*A*).

10. **Carteria** Diesing, 1866. The free-swimming cells of *Carteria* differ from those of *Chlamydomonas* only in that they are quadriflagellate. *Carteria* has, from species to species, much the same range in cell shape as does *Chlamydomonas*, and the anterior pole of a cell may be smooth or papillate. The chloroplast is usually cup-shaped, but there are species with chloroplasts that may be either laminate, or H-shaped in optical section. A chloroplast may contain a single pyrenoid or several pyrenoids; or pyrenoids may be lacking. An eyespot may or may not be present.

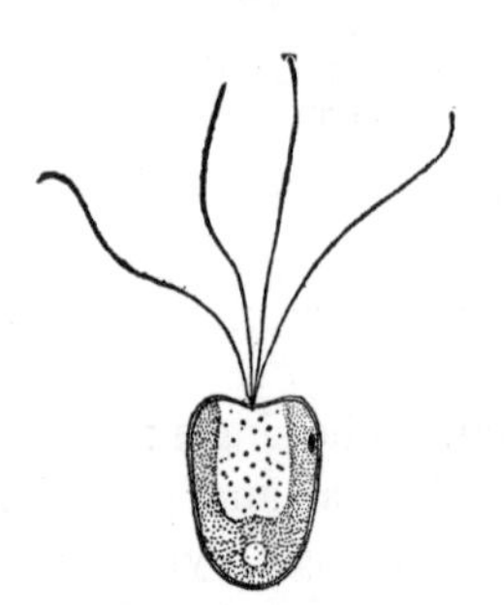

FIG. 28. *Carteria cordiformis* (Carter) Dill. (× 1000.)

Asexual reproduction is similar to that of *Chlamydomonas* and, as in the case of that genus, there may be a retention of daughter cells within the gelatinized parent-cell wall and a formation of a palmella stage.

Most species are isogamous or slightly anisogamous and with a shedding of the cell walls as cells fuse in pairs.[1] The resultant octoflagellate zygote may swim for a considerable time before it loses its flagella, assumes a spherical shape, and develops a thick wall. One species is oögamous. The contents of male cells divide to form a number of small quadriflagellate antherozoids.[2] The entire protoplast of a female cell becomes a single large nonflagellated egg which escapes from the parent-cell wall before an antherozoid swims to and unites with it.

Eight species have been found in the United States. For descriptions of six of them, *C. cordiformis* (Carter) Dill (Fig. 28), *C. crucifera* Korshikov, *C. globosa* Korshikov, *C. globulosa* Pascher, *C. Klebsii* (Dang.) France, and C. *multifilis* (Fresen.) Dill, see Pascher (1927*A*). For descriptions of the other two, see Tiffany (1934), for *C. dissecta* Tiffany, and see Bold (1938), for *C. ellipsoidalis* Bold.

[1] DANGEARD, 1898; DILL, 1895.

[2] RAMANATHAN, 1942.

11. **Platymonas** G. S. West, 1916. This genus is like the foregoing in having quadriflagellate cells but differs in having compressed cells that are subrectangular to broadly ellipitical when seen in vertical view. The four flagella lie in a depression at the anterior end of a cell and are inserted very close together. Vertical views of cells show that the flagella are in two distinct pairs (Fig. 29*E–F*) and are not quadrately arranged as in *Carteria*. There is a delicate wall around the protoplast, but this is not usually evident except in reproducing individuals. Some species have two contractile vacuoles at the base of the flagella; contractile vacuoles are not evident in other species. An eyespot is always present and, according to the species, near the anterior end or in the posterior third of a cell. The chloroplast is cup-shaped, massive or delicate, and entire, or with the an-

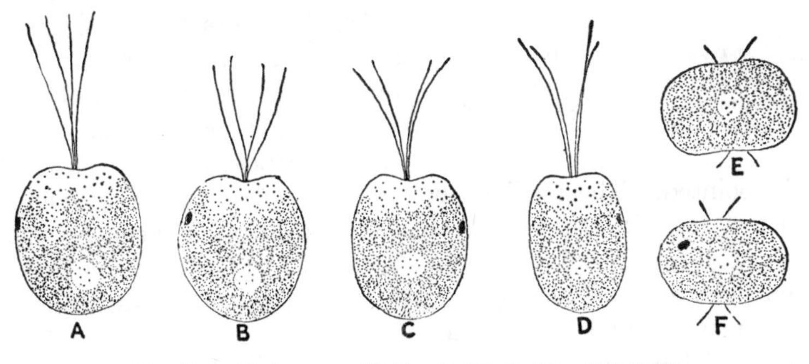

Fig. 29. *Platymonas elliptica* G. M. Smith. (× 975.)

terior end incised into four lobes. There is a single pyrenoid which may be spherical or cup-shaped.

Asexual reproduction is by longitudinal division to form two or four daughter cells.

Sexual reproduction has not been observed in this alga.

Two species, *P. subcordiformis* (Wille) Hazen and *P. elliptica* G. M. Smith (Fig. 29), have been found in this country. *P. subcordiformis* has been found on both the Atlantic and Pacific coasts in brackish-water pools rich in droppings from gulls. It is a distinctly brackish-water species but will continue to grow and multiply when transferred to fresh water. For a description of *P. elliptica,* see G. M. Smith (1933); for *P. subcordiformis,* see Wille (1903*B*) under *Carteria subcordiformis* Wille.

12. **Scherffelia** Pascher, 1911. Cells of *Scherffelia* are quadriflagellate and markedly compressed. They differ from those of other Chlamydomonadaceae with compressed cells in that the lateral margins of the cell wall are flattened and wing-like. The wing-like portion of a wall may

extend the whole length of a cell or be restricted to the anterior portion of a cell.[1] The cells contain two or more contractile vacuoles and a conspicuous eyespot. There may be two chloroplasts extending longitudinally from base to apex of a cell; or the two may be connected by a bridge in the basal portion of a cell. The chromatophores do not contain pyrenoids. Asexual reproduction is by longitudinal division into four daughter cells.[2]

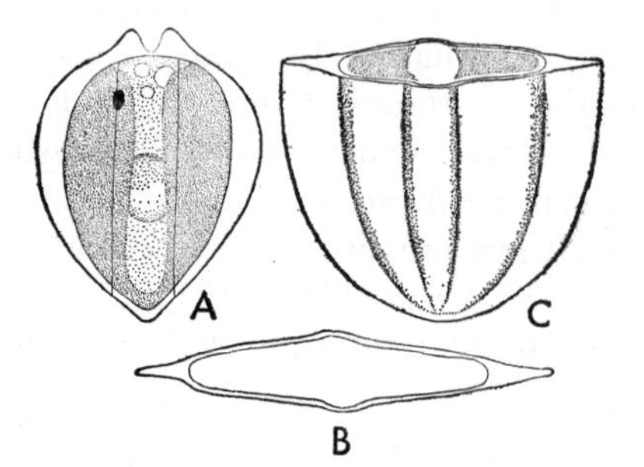

FIG. 30. *Scherffelia phacus* Pascher. *A*, front view. *B*, transverse section. *C*, stereodiagram of basal half of a cell. (*After Pascher*, 1927*A*.)

S. phacus Pascher (Fig. 30) has been found in Ohio and Tennessee.[3] For a description of it, see Pascher (1927*A*).

13. **Chlorobrachis** Korshikov, 1925. *Chlorobrachis* is unicellular and also quadriflagellate. The anterior pole of a cell is cylindrical and truncate; the posterior pole, conical and acutely pointed. The median portion of a cell is markedly compressed and, when seen in front view, has four quadrately arranged truncate processes. The protoplast does not extend to the cell wall in the acutely pointed posterior pole and in the truncate lateral processes. The chloroplast is massive and without a pyrenoid. There is a conspicuous contractile vacuole near the anterior pole and a large eyespot submedian in position.[4]

The only reproductive stage thus far observed is a rounding up of the protoplast into a cyst-like sphere.[4]

The only known species, *C. gracillima* Korshikov (Fig. 31), has been found in abundance in a small stream in Indiana polluted by waste from a distillery.[5] For a description of this species, see Pascher (1927*A*).

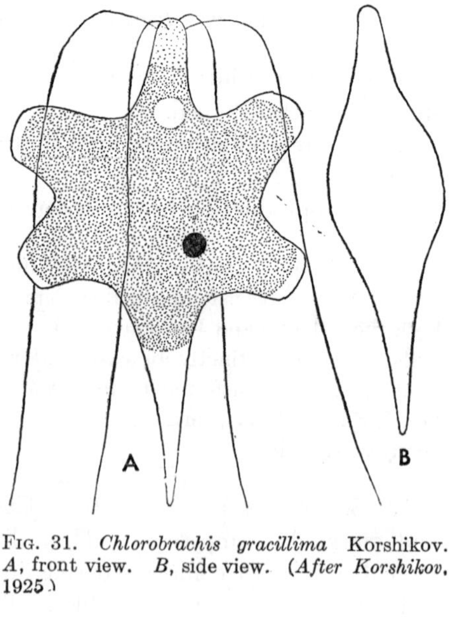

FIG. 31. *Chlorobrachis gracillima* Korshikov. *A*, front view. *B*, side view. (*After Korshikov*, 1925.)

[1] CONRAD, 1928*B*; PASCHER, 1927*A*.
[2] PASCHER, 1927*A*.
[3] LACKEY, 1942; LACKEY *et al.*, 1943.
[4] KORSHIKOV, 1925.
[5] LACKEY, 1942*A*.

Family 3. Phacotaceae

Cells of members of this family of unicellular Volvocales are enclosed within a wall-like structure, the *lorica*, which does not contain cellulose and which is frequently impregnated with calcium compounds or ferric compounds. In most genera, the cell has not the same shape as the lorica, and there is a water-filled space between the two. Some genera have a lorica composed of two overlapping halves; other genera have an entire lorica. Biflagellate genera, when seen in polar view (Figs. 34*E*, 35*C*), have the two flagella axial to each other and at an angle to the major axis of a cell. The protoplast has the usual chlamydomonad structure of a cup-shaped chloroplast with one or more pyrenoids, contractile vacuoles at the base of the flagella, and an eyespot.

Asexual reproduction is by division to form two, four, or eight daughter cells. According to its structure the lorica separates into two halves (valves) or breaks irregularly at the time of liberation of daughter cells.

In all instances thus far noted sexual reproduction is isogamous.

The genera found in this country differ as follows.

1. Cells biflagellate 2
1. Cells quadriflagellate .... 8. **Pedinopera**
    2. Lorica compressed 3
    2. Lorica not compressed 7
3. Compressed face with projections .... 3. **Wislouchiella**
3. Compressed face without projections 4
    4. Halves of lorica evident in vegative cells .... 1. **Phacotus**
    4. Halves of lorica not evident in vegetative cells 5
5. Lorica smooth 6
5. Lorica verrucose .... 2. **Thoracomonas**
    6. Lorica much narrower at posterior pole .... 5. **Cephalomonas**
    6. Lorica not narrower at posterior pole .... 4. **Pteromonas**
7. Lorica without pores .... 6. **Coccomonas**
7. Lorica with pores .... 7. **Dysmorphococcus**

1. **Phacotus** Perty, 1852. *Phacotus* has free-swimming biflagellate cells in which the protoplast lies within a compressed lorica composed of two overlapping halves. The lorica is usually dark colored, impregnated with lime, and with the outer face sculptured. Except at its anterior end, the protoplast is not in contact with the lorica, and the intervening region is filled with a watery gelatinous substance. The outline of the protoplast is ovate or subovate in front and side views. There are usually two contractile vacuoles at the base of the flagella, and the eyespot may be anterior or posterior in a cell. The chloroplast is massive, cup-shaped, and may contain one or several pyrenoids.

Asexual reproduction is by longitudinal division into four or eight daughter cells.[1] During cell division, there is an increase in the amount of gelatinous material next to the protoplast, and the accumulation of this material forces apart the overlapping halves of the lorica. The flagellated daughter cells may escape from the gelatinous matrix, or they may remain embedded within it to form a palmella stage. If further cell divisions take place and there is no escape of daughter cells, the palmella stage contains the half-loricas of several cell generations.

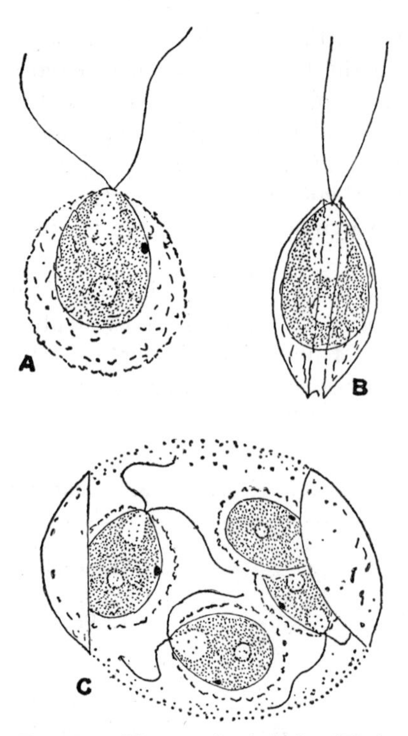

FIG. 32. *Phacotus lenticularis* (Ehr.) Stein. *A–B*, front and side views of a vegetative cell. *C*, liberation of four daughter cells by separation of halves of parent-cell wall. (× 975.)

Sexual reproduction is unknown for *Phacotus*.

P. angustus Pascher, *P. glaber* Playf., and *P. lenticularis* (Ehr.) Stein (Fig. 32) have been found in this country. For descriptions of them, see Pascher (1927*A*).

2. **Thoracomonas** Korshikov, 1925. The cells of this alga are somewhat compressed. The lorica is of a firm gelatinous texture, irregularly verrucose, and with the verrucae a deep brown because of impregnation with ferric hydroxide. The protoplast may nearly fill the lorica, or there may be a considerable interval between lorica and protoplast. There are two flagella, and these have the diagonal insertion characteristic of Phacotaceae. Two contractile vacuoles may or may not be evident at the base of the flagella. There is a cup-shaped chloroplast, with a shallow or a deep depression, that may have one to three pyrenoids or lack them. The eyespot is median in a cell.

Asexual reproduction is by division into two, four, or eight daughter cells that may be liberated by an irregular breaking of the lorica or by the lorica separating into two halves.

Thus far, the only species known for the United States is *T. Phacotoides* G. M. Smith (Fig. 33). For a description of it, see G. M. Smith (1933).

[1] DANGEARD, 1898.

3. **Wislouchiella** Skvortzow, 1925. This genus has the protoplast surrounded by a strongly compressed lorica with a broad wing-like expansion. Front and side views show that each compressed face has two blunt cylindrical projections; one projecting upward from its insertion near the lorica's apex, the other projecting backward from its insertion near the level of the base of the protoplast.[1] Vertical views of the lorica show that the processes on its opposite faces are axial to each other. However, the axial plane of the superior pair lies at an angle with the plane of the inferior pair. The surface of the lorica and its processes is minutely verrucose, and the verrucae may be brownish or colorless. The protoplast is ovate in side view and rhomboidal in vertical view. There are two long flagella, and these have the diagonal insertion characteristic of the family. At the base of the flagella are two contractile vacuoles. The chloroplast is massive, cup-shaped, and with a single submedian pyrenoid. The eyespot lies a short distance back from the anterior end.

W. planctonica Skvortzow (Fig. 34) has been found in California and in several states of the Mississippi Valley. For a description of it, see Skvortzow (1925).

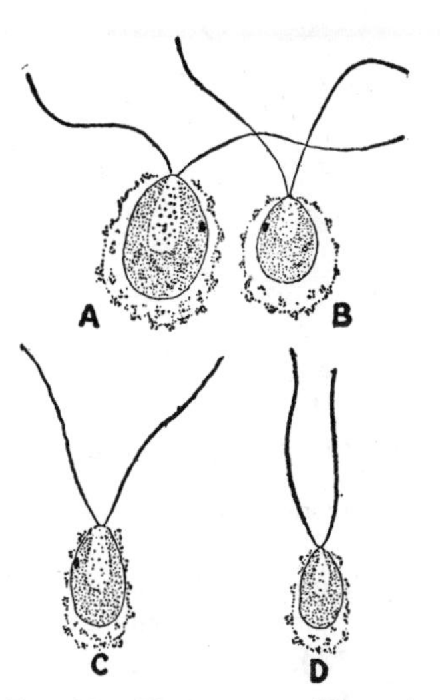

FIG. 33. *Thoracomonas Phacotoides* G. M. Smith. *A–B*, front views. *C–D*, side views. (× 975.)

4. **Pteromonas** Seligo, 1887. *Pteromonas* differs from other Phacotaceae with a bivalved lorica in that the lorica surface is smooth. When seen in front view, the outline of the lorica may be circular or oval and with the anterior end cut off abruptly; or the outline may be subrectangular and with or without papillate projections at the angles. When seen in vertical view, the contents fit snugly against the lorica which has a linear or sigmoid projection extending beyond the poles of the cell. The front view of a protoplast is pyriform and with the two flagella borne at the narrow anterior end. There are two contractile vacuoles at the base of the flagella. The chloroplast is massive and cup-shaped. It may contain one or several pyrenoids. The eyespot is anterior and ellipitical or sublinear in outline.[2]

Asexual reproduction is by longitudinal division into two or four daughter

[1] SKVORTZOW, 1925.

[2] LEMMERMANN, 1900*C*; WEST, G. S., 1916*B*.

cells which are liberated by a separation of the two halves of the lorica.[1] The springing apart of the two halves seems to result from the accumulation of gelatinous material as in *Phacotus*.

When sexual reproduction takes place the protoplast divides to form 8, 16, or 32 naked biflagellate gametes. Gametic union is isogamous. The zygote is spherical and with a smooth wall. Germinating zygotes give rise to four or eight zoospores.[2]

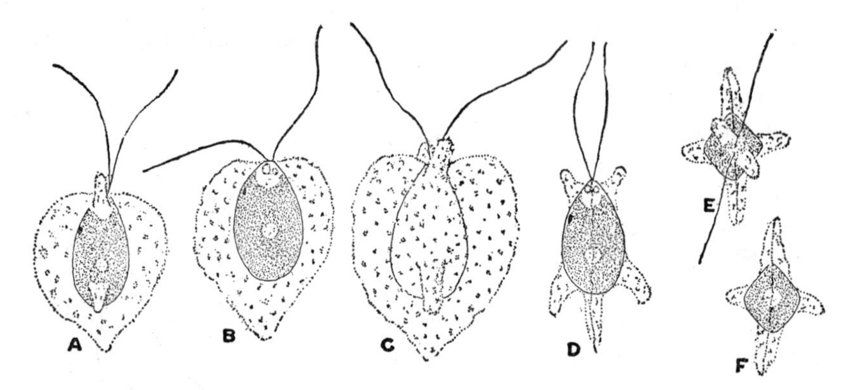

FIG. 34. *Wislouchiella planctonica* Skvortzow. *A*, surface view from the front. *B*, optical section. *C*, front view with the protoplast omitted. *D*, side view. *E*, vertical view from the anterior end. *F*, vertical view from the posterior end. (× 975.)

P. aculeata Lemm. (Fig. 35), *P. angulosa* (Carter) Lemm., and *P. cruciata* Playf. have been found in the United States. For descriptions of them, see Pascher (1927*A*).

5. **Cephalomonas** Higinbotham, 1942. Cells of this genus have a rigid, brittle, colorless to yellowish, compressed lorica. As seen in front view, a cell is sharply differentiated into anterior and posterior halves, the former with a breadth more than double that of the latter. The cells are biflagellate, and in polar view the flagella lie diagonal to each other. The protoplast usually occupies all the space within a lorica. It contains a massive cup-shaped chloroplast with one pyrenoid, an eyespot, and a nucleus.[3]

Asexual reproduction is by division into two or four naked ellipsoid daughter cells which are liberated by a breaking of the lorica into several irregularly shaped fragments. Soon after liberation, a cell assumes the characteristic shape and secretes a lorica.[3]

Sexual reproduction is isogamous and by a fusion of naked biflagellate

[1] GOLENKIN, 1892; SELIGO, 1887.
[2] DANGEARD, 1889; GOLENKIN, 1892.
[3] HIGINBOTHAM, 1942.

ellipsoid gametes. The naked quadriflagellate zygote swarms for 1 to 3 days before it comes to rest, becomes spherical, and secretes a granular wall.

The single species, *C. granulata* Higinbotham (Fig. 36) is known only from the original description by Higinbotham (1932) of material collected in Maryland.

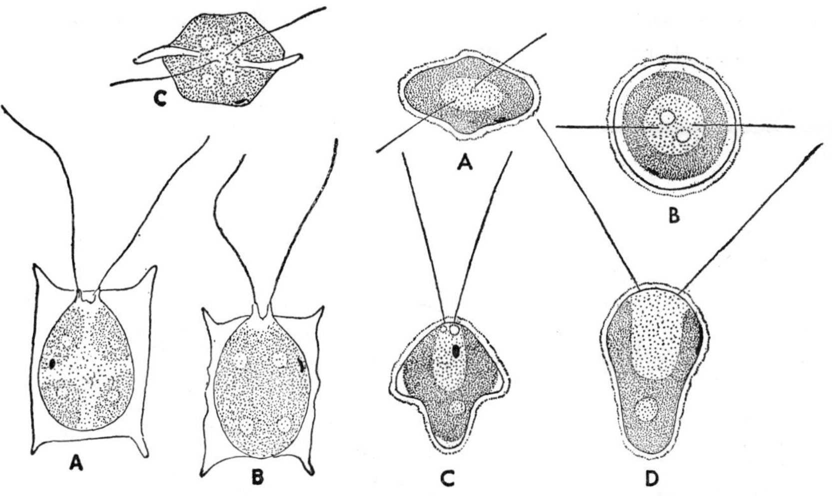

FIG. 35. *Pteromonas aculeata* var. *Lemmermanni* Skuja. *A–B*, front views. *C*, vertical view. (× 975.)

FIG. 36. *Cephalomonas granulata* Higenb. *A–B* vertical views. *C*, front view. *D*, side view. (*After Higinbotham*, 1942.) (× 1650.)

6. **Coccomonas** Stein, 1878. The protoplast of this alga lies within a globose to cardiform homogeneous lorica with a circular pore at the an-

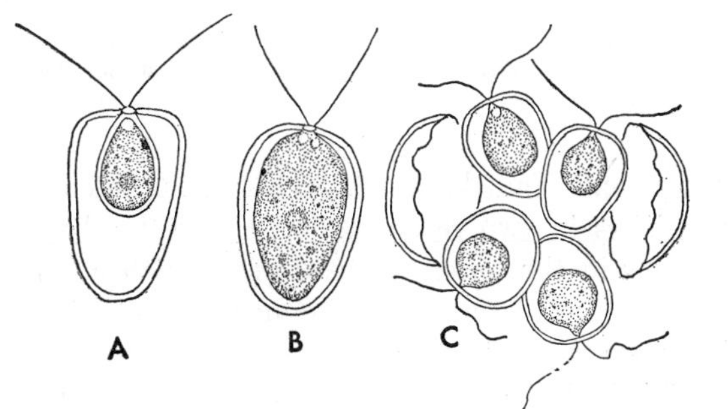

FIG. 37. *Coccomonas orbicularis* Stein. *A–B*, vegetative cells. *C*, liberation of daughter cells by irregular breaking of lorica. (*After Stein*, 1878.) (× 650.)

terior end. The lorica is impregnated with lime and iron compounds, the latter at times being present in such quantity as to color it a dark brown and obscure the protoplast. The protoplast, which partially fills the lorica, is biflagellate and with the flagella projecting through a common pore in the lorica. It contains two contractile vacuoles, an eyespot, and a cup-shaped chloroplast with one pyrenoid.[1]

Asexual reproduction is by division of a protoplast into four daughter protoplasts each of which develops a lorica before liberation by an irregular fragmentation of the parent lorica into two parts.

C. orbicularis Stein (Fig. 37) has been recorded from several states of the Ohio River Valley. For a description of it, see Pascher (1927*A*).

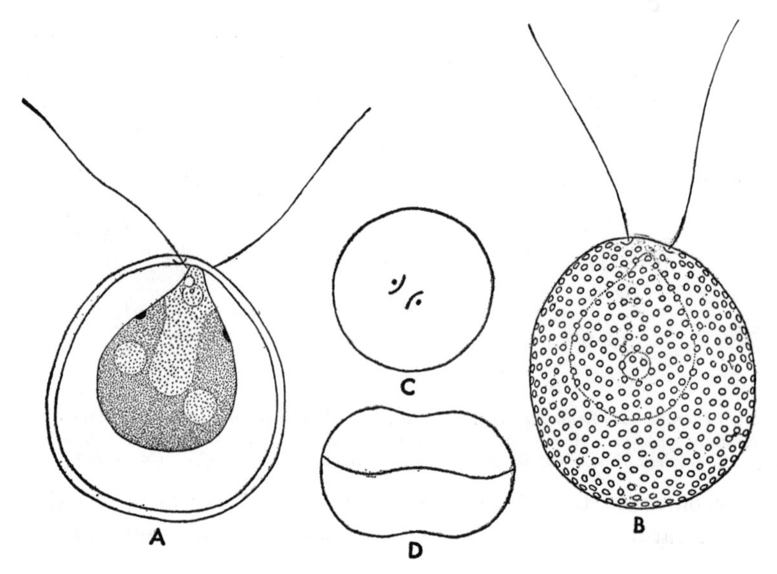

Fig. 38. *Dysmorphococcus variabilis* Takeda. *A*, optical section of a cell in an early stage of division. *B*, surface view of lorica. *C–D*, anterior and posterior vertical views. (*After Bold*, 1938.) (× 1875.)

7. **Dysmorphococcus** Takeda, 1916. This genus resembles *Coccomonas*, but the lorica differs in two respects. The two flagella project through two small openings in the lorica, instead of through a common opening;[2] and the lorica has been described as being granulate, but this granulate appearance has been shown[3] to be due to minute pores. The protoplast, which does not fill the lorica, may have two or several contractile vacuoles. The chloroplast is cup-shaped and with one to several pyrenoids.

Asexual reproduction is by division of the protoplast into two daughter protoplasts which are liberated by an irregular fragmentation of the lorica

[1] Pascher, 1927*A*.

[2] Takeda, 1916. [3] Bold, 1938; Korshikov, 1925.

into two parts.[1] Daughter cells do not have an evident lorica at the time they are liberated.

D. variabilis Takeda (Fig. 38) has been found in Maryland.[2] In his description of this species, Pascher (1927*A*) mistakenly calls it "*D. Fritschii*," and it has been recorded under this erroneous name from Ohio, West Virginia, and Tennessee.[3]

8. **Pedinopera** Pascher, 1925. This genus is immediately distinguishable from other Phacotaceae found in this country by its quadriflagellate cells. The lorica is compressed, has a granulate surface, and is with or without longitudinal ridges in addition to the granules. As in many other members of the family, the lorica is brownish in color. The protoplast, which only partially fills the lorica, has a cup-shaped chloroplast. An eyespot may or may not be present.[4]

Reproductive stages have not been noted for this alga.

P. granulosa (Playf.) Pascher (Fig. 39) has been found in Tennessee.[5] For a description of it, see Pascher (1927*A*).

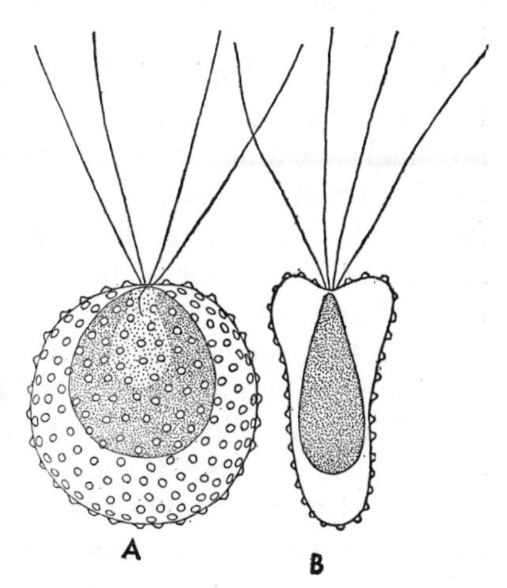

Fig. 39. *Pedinopera granulosa* (Playf.) Pascher. *A*, front view. *B*, optical section of side view. (*After Playfair*, 1918.) (× 660.)

Family 4. Volvocaceae

This family includes those colonial Volvocales in which there is a formation of a flat plate (*plakea*) during early development of a colony. The number of cells in a colony is a multiple of two, and each of the cells is surrounded by a gelatinous sheath. The sheaths may be distinct from or confluent with one another. Some species of *Volvox* have evident cytoplasmic strands connecting the cells one to another. Other genera have no visible cytoplasmic strands, but the behavior of severed colonies furnishes indirect evidence[6] that there are ultramicroscopic connections between the cells. Most of the genera have the cells arranged in a hollow spherical layer one cell in thickness.

Individual cells of a colony may be spherical, ovoid, pyriform, or hemispherical. They are biflagellate and with a typical chlamydomonas structure. Most of them have two contractile vacuoles, an eyespot, and one

[1] Korshikov, 1925. [2] Bold, 1938.
[3] Brinley and Katzin, 1942; Lackey, 1942.
[4] Pascher, 1925*B*; Playfair, 1918. [5] Lackey, 1942. [6] Bock, 1926.

to several pyrenoids. All cells of a colony may be identical in size and structure, or they may differ sufficiently to give the colony a definite polarity. This difference may be merely in size of eyespots at the two poles of a colony (*Pandorina*, *Eudorina*). In the most advanced genera (*Pleodorina*, *Volvox*), the differentiation reached a point where there are large reproductive cells (*gonidia*) and small cells purely vegetative in function.

Asexual reproduction is by repeated division of a cell or of a gonidium to form an autocolony. All cell divisions in formation of an autocolony are longitudinal and in a very regular sequence. The first three series of division result in a curved eight-celled rectangular plate (plakea) in which four of the cells are cruciately arranged and in mutual contact with one another at the center of the plate (Fig. 47*B*). By the time the young colony has become 16-celled, the curving of the plakea is often so pronounced that the colony is a hollow sphere, but one in which there is an opening (*phialopore*) at one pole (Fig. 47*C*). Depending upon the extent to which longitudinal cell division continues, the colony eventually has 4, 8, 16, 32, 64, 128, 256, or more cells. In the case of *Gonium*, development rarely progresses beyond the 16-celled stage, and there is but little curving of the plakea. Among genera with the plakea developing into a hollow sphere, and until fairly late in development of a colony, the nucleus in each cell lies toward the cell pole facing the interior of a colony. Subsequently each cell has its nucleus at the pole facing the exterior of the colony. This change in position of nuclei results from the colony turning itself inside out (inverting) through the phialopore during later stages of development.

Palmella stages are of rare occurrence among Volvocaceae and are known only for *Gonium*.[1] Akinetes are also uncommon and, when they are formed as in *Gonium*, they are produced by isolated cells.[2]

Sexual reproduction is isogamous in *Gonium*, the most primitive member of the family. It is anisogamous in most of the genera, and in *Volvox* it is oögamous. Zygotes of Volvocaceae have a thick wall, which may be smooth or ornamented. Division of the zygote nucleus has been shown[3] to be meiotic in *Volvox*, and study of sexuality of colonies derived from germinating zygotes of certain other species[4] indicates that their nuclei divide meiotically. When a zygote germinates, the protoplast may divide into four zoospores which escape singly (*Eudorina*) or united as a four-celled colony (*Gonium*). In *Volvox* there is regularly a formation of a single zoospore.

[1] Dangeard, 1916; Hartmann, 1924. [2] Crow, 1927*A*; Migula, 1890.
[3] Zimmermann, 1921. [4] Schreiber, 1925; Schulze, 1927.

The genera of Volvocaceae found in this country differ as follows:

1. Colonial envelope flattened 2
1. Colonial envelope globose 3
  2. Envelope with an anterior-posterior differentiation .... 6. **Platydorina**
  2. Envelope without anterior-posterior differentiation .... 1. **Gonium**
3. Not over 256 cells in a colony 4
3. Over 500 cells in a colony .... 8. **Volvox**
  4. All cells of a colony alike in size 5
  4. With cells of two distinct sizes .... 7. **Pleodorina**
5. Cells forming a sphere 6
5. Cells not forming a sphere .... 4. **Stephanoon**
  6. Cells close together .... 2. **Pandorina**
  6. Cells not close together 7
7. Cells hemispherical .... 3. **Volvulina**
7. Cells spherical .... 5. **Eudorina**

1. **Gonium** Mueller, 1773. The colonies of *Gonium* typically have 4, 16, or 32 biflagellate cells arranged in a flat quadrangular plate. The cells are embedded in a common gelatinous matrix and are connected to one another by gelatinous strands of a tougher consistency. Sixteen-celled colonies have 12 peripheral cells (three on each side and with their long axes in the plane of the colony) and four quadrately disposed central cells with their long axes radial to the plane of the colony. Four-celled colonies have the cells so disposed that their long axes are parallel or slightly divergent. The cells in a *Gonium* colony, which are joined to one another by very delicate cytoplasmic threads,[1] are ovoid to pyriform in shape and with two flagella at the anterior end. There are two contractile vacuoles at the base of the flagella, and the single eyespot lies in the anterior end of the cell. The eyespot is of the simple chlamydomonad type.[2] The chloroplast is cup-shaped and contains a single pyrenoid.

Because of the nonradial arrangement of the four central cells, the 16-celled colonies swim through the water with a somersault-like revolution. Four- and eight-celled colonies have the anterior ends of the cells directed forward as they move through the water.

Asexual reproduction takes place by the simultaneous formation of autocolonies by all cells in a colony. In the case of 16-celled colonies, an accidental breaking of the colony leads immediately to asexual reproduction,[3] but single cells breaking away from colonies of the four-celled species do not reproduce immediately.[4] Each daughter colony has a colonial envelope of its own at the time when it is liberated by a gelatinization of the envelope of the parent colony.

[1] Harper, 1912. [2] Mast, 1928. [3] Bock, 1926.
[4] Smith, G. M., 1931*C*.

Solitary cells which have broken away from a colony may develop into akinetes[1] or into palmella stages.[2]

Division stages in formation of gametes are identical with those in daughter colony formation, but the gametes separate from one another when liberated from the parent colony. Gametic union is isogamous, and certain of the species are known[3] to be heterothallic. The quadriflagellate zygote soon comes to rest, becomes spherical, and secretes a smooth thick wall. A germinating zygote forms four zoospores which are usually joined in a four-celled colony when liberated. Sometimes the zoospores escape singly. The distribution of sex in colonies derived from the four zoospores indicates that division of the zygote nucleus is meiotic.[3]

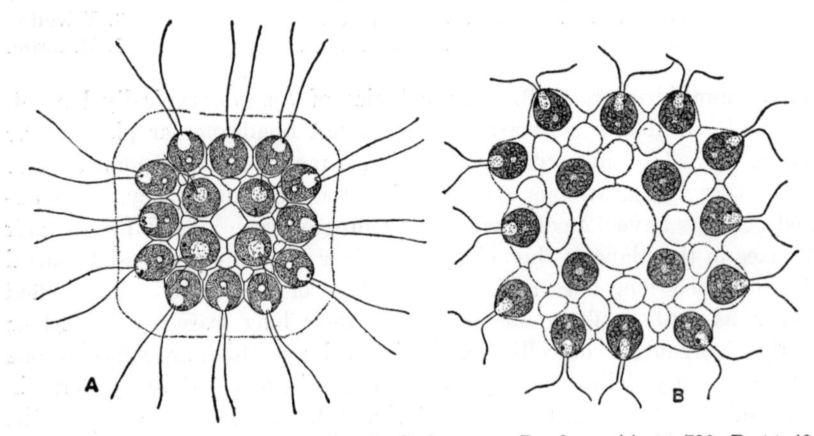

Fig. 40. *A*, *Gonium pectorale* Muell. *B*, *G. formosum* Pascher. (*A*, × 780; B, × 400.)

Gonium is usually found sparingly intermingled with other algae of pools and ditches. Occasionally it occurs in almost pure culture in temporary puddles rich in organic matter. *G. discoideum* Prescott, *G. formosum* Pascher (Fig. 40*B*), *G. pectorale* Muell. (Fig. 40*A*), and *G. sociale* (Duj.) Warming are the species found in this country. For a description of *G. discoideum*, see Prescott (1942); for descriptions of the others, see Pascher (1927*A*).

2. **Pandorina** Bory, 1824. The colonies of this alga are subspherical to ellipsoidal and with 4, 8, 16 or 32 biflagellate cells embedded within a homogeneous colonial envelope. There may also be an outer gelatinous sheath of a more watery consistency (Fig. 41*A*). The cells are arranged in a hollow sphere within the colonial envelope and are generally so close together that they are somewhat flattened by mutual pressure. They are

[1] Chodat, 1894; Migula, 1890; Schussnig, 1911. [2] Dangeard, 1916; Migula, 1890.

[3] Schreiber, 1925.

usually pyriform but may be oblate spheres. Pyriform cells have the two flagella borne at the broad anterior end. The eyespot is on the face of the cell toward the exterior of a colony. All eyespots in a colony may be of the same size, or there may be marked differences in size between those on opposite poles of a colony. There are two contractile vacuoles at the base of the flagella. The chloroplast is cup-shaped and massive; it may have a smooth outer face and contain one pyrenoid, or its outer face may be ridged and contain several pyrenoids.

Asexual reproduction is by a simultaneous formation of daughter colonies by all cells of a colony. Prior to reproduction, a colony ceases moving actively, sinks to the bottom of the pool, and the colonial envelope be-

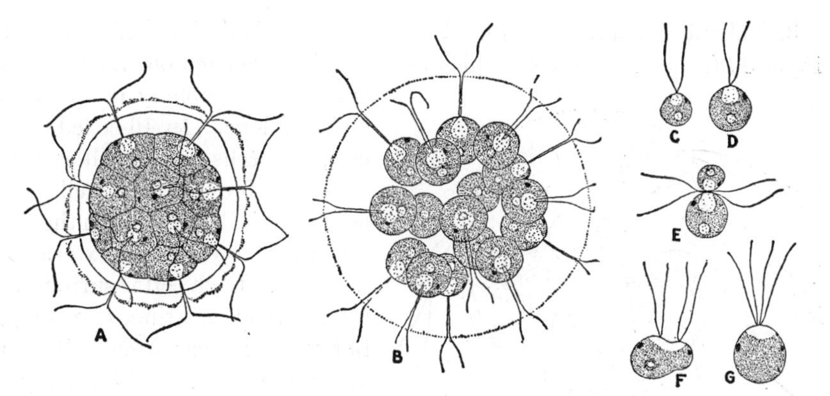

FIG. 41. *Pandorina morum* Bory. *A*, vegetative colony. *B*, colony of female gametes. *C*, male gamete. *D*, female gamete. *E–F*, fusing gametes. *G*, zygote. (× 1300.)

comes more watery and swollen. Each cell divides into a typical plakea which is bowl-shaped instead of a hollow sphere. There is an inversion of a young bowl-shaped colony to form a sphere in which the phialopore is closed.[1] After inversion, each cell develops a pair of flagella, and the daughter colony swims away from the greatly gelatinized envelope of the parent colony.

Pandorina is heterothallic, and gametic union is anisogamous. Divisions leading to the formation of gametes are identical with those in asexual reproduction. However, colonies composed of cells destined to function as gametes are *Eudorina*-like and with a watery gelatinous envelope. These colonies may swim through the water, but sooner or later the individual cells (gametes) escape from the colonial matrix and move about singly.[2] Male gametes are somewhat smaller and swim more actively than female gametes. Fusion may be terminal or lateral (Fig. 41*E–F*). The quadriflagellate zygote remains motile for a short time before it loses

[1] MORSE, 1943; TAFT, 1941. [2] PRINGSHEIM, N., 1870.

its flagella and secretes a wall. Old zygotes have a smooth wall and a protoplast colored red by hematochrome. When a zygote germinates,[1] there is a liberation of a single biflagellate zoospore. After swarming for a time, it loses its flagella, secretes a wide gelatinous envelope, and divides and redivides to form a typical colony.

Although rarely found in abundance, *Pandorina* is more frequently encountered than any other member of the family. It is also a regular constituent of the plankton of lakes with fairly hard waters. Of the two species found in this country, *P. morum* Bory (Fig. 41) is known to be widespread, and *P. charkowiensis* Korshikov is known only from California and North Carolina. For descriptions of them, see G. M. Smith (1920, 1931*B*).

3. **Volvulina** Playfair, 1915. *Volvulina* has spherical colonies of 4, 8, 16, or 32 (generally 16) cells which lie some distance from one another just within the periphery of the colonial envelope. In 16-celled colonies, the cells are in alternating tiers of four.[2] The cells are hemispherical and lie with the flattened end toward the exterior of a colony. Each cell is surrounded by a broad gelatinous envelope, but these appear fused with one another unless stained in an appropriate manner. Each cell is biflagellate and has a pale cup-shaped chloroplast without pyrenoids, an eyespot, and two to several contractile vacuoles. Eyespots in the anterior tier of cells are larger than those of other cells and may be lacking in cells in the posterior half of a colony.[2]

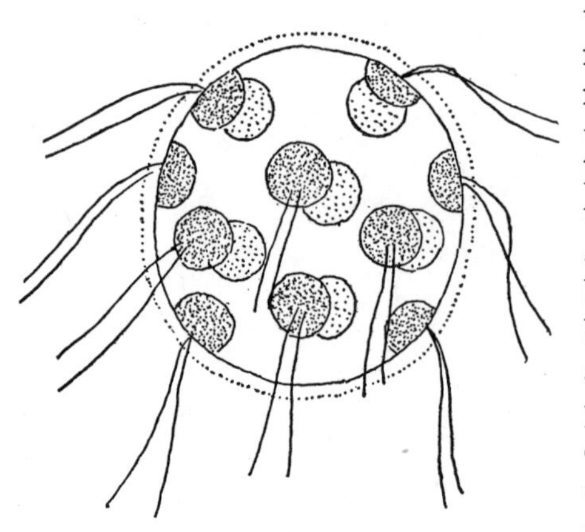

FIG. 42. *Volvulina Steinii* Playfair. (*After Korshikov*, 1938*A*.) (× 630.)

Asexual reproduction is by division and redivision of each cell in a colony to form an autocolony.[3]

Gametic union has not been observed, but, to judge from the appearance of recently formed zygotes,[2] it is similar to that in *Pandorina*.

There is but one species, *V. Steinii* Playfair (Fig. 42). The only published record of its occurrence in the United States is the report of it from Minnesota.[4] Miss Mary A. Pocock informs me that she has found this alga in a culture inocu-

[1] KORSHIKOV, 1923; PRINGSHEIM, N., 1870. [2] KORSHIKOV, 1938*A*.
[3] GESSNER, 1931; KORSHIKOV, 1938*A*; PLAYFAIR, 1915*A*. [4] TILDEN, 1935.

lated with soil from Nebraska. For a description of *V. Steinii,* see Playfair (1915*A*).

4. **Stephanoon** Schewiakoff, 1893. Colonies of this alga have a spherical to oblately spherical gelatinous envelope. There are either 8 or 16 cells, and they lie in two transverse tiers near the equator and just within the periphery of the colonial envelope. They are so arranged that the four or eight cells of one tier alternate with the four or eight in the other tier. The cells are biflagellate and with a cup-shaped chloroplast containing a pyrenoid, and an eyespot.

Asexual reproduction is by division of each cell of a colony into a daughter colony.[1] The arrangement of cells in a young colony suggests that there

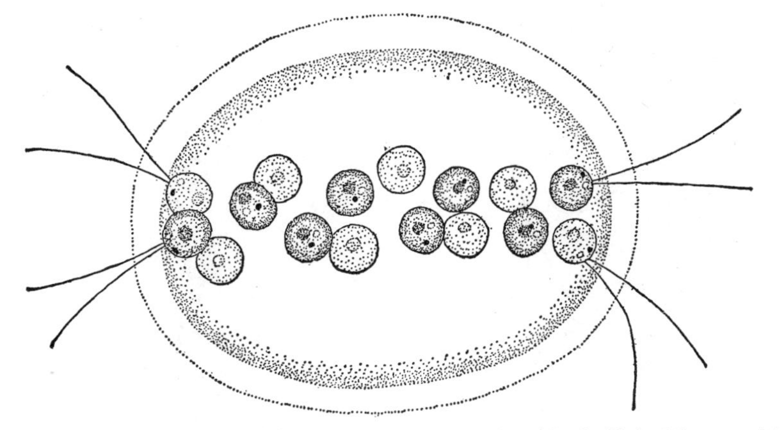

FIG. 43. *Stephanoon Askenasii* Schewiakoff, a species not found in the United States. (*After Schewiakoff in Printz,* 1927.) (× 800.)

is a formation of a plakea as in other Volvocaceae, but this has not been demonstrated beyond all doubt.

S. Wallichii (Turn.) Wille (Fig. 43) has been found in Ohio.[2] For a description of it, see Pascher (1927*A*).

5. **Eudorina** Ehrenberg, 1932. *Eudorina* has spherical to obovoid colonies with a homogeneous envelope that may have mamillate projections at the posterior pole.[3] There are 16, 32, or 64 cells in a colony, and these lie some distance from one another in a single layer toward the periphery of the colonial envelope. Frequently the cells are in distinct transverse tiers. If a colony is 32-celled, the anterior and posterior tiers contain four cells each and the three median tiers eight cells each.[3] The cells are

[1] FRITSCH, 1918. [2] TIFFANY, 1921.

[3] CHODAT, 1902; CONRAD, 1913; HARTMANN, 1921; SMITH, G. M., 1931.

spherical, and all of a colony are approximately the same size. The length of the two flagella is two to four times the diameter of a cell, and the cells may be with or without a conical elevation where the flagella are inserted. There are two contractile vacuoles at the base of the flagella, and the single eyespot lies at the anterior end of a cell. Certain species have a progressive diminution in size of eyespots from the anterior to posterior poles, and eyespots may even be lacking in the lowermost tier of cells. The chloroplast is cup-shaped, massive, and, according to the species,[1] contains one or several pyrenoids.

Asexual reproduction is by means of autocolonies, each cell developing into a plakea in the manner characteristic of the family. All cells in a

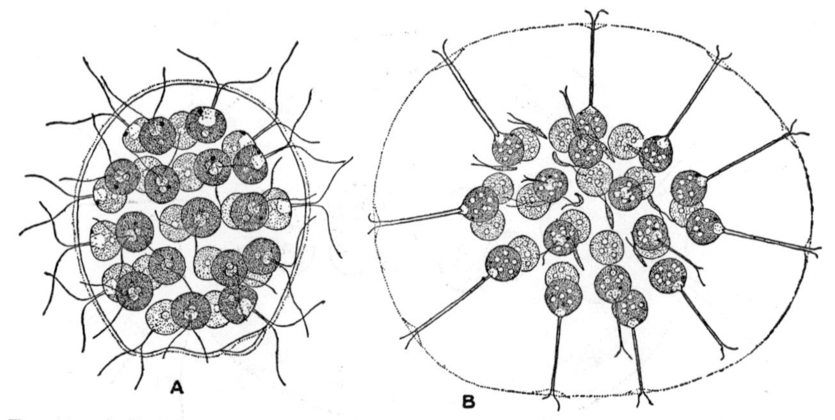

Fig. 44. *A, Eudorina unicocca* G. M. Smith. *B*, female colony of *E. elegans* Ehr., in which antherozoids are swimming within the colonial envelope. (*A*, × 780; *B*, × 320.)

colony usually produce daughter colonies, but sometimes certain cells in a colony are sterile.[2]

According to the species, *Eudorina* is homothallic or heterothallic.[3] Gametic union is an advanced type of anisogamy. A vegetative cell divides to form a packet of spindle-shaped male gametes which swim as a unit to a female colony and there dissociate into individual gametes which penetrate and swim slowly within the gelatinous envelope of a female colony. The biflagellate female gametes are globose and are never liberated from the gelatinous envelope of a colony. The male gamete has been variously described as having its anterior end,[4] its posterior end,[5] and its side[3] fuse with the female gamete. The zygotes remain within the colonial envelope and soon lose their flagella and secrete a smooth wall. The zygotes eventually become free by a decay of the colonial envelope, and sooner or later their protoplasts become deeply colored with hematochrome.

[1] Smith, G. M., 1931. [2] West, G. S., 1916; West, W. and G. S., 1912. [3] Meyer, K. I., 1935. [4] Pocock, 1937. [5] Iyengar, 1937.

The first step in germination is a swelling of the zygote wall and a formation of a sac-like extrusion at one side.[1] Usually the extrusion contains one reddish biflagellate zoospore and two or three small hyaline bodies which are probably degenerate zoospores. After liberation from the vesicle, a zoospore swims about for a time and then, in the same manner as a vegetative cell, divides and redivides to form a colony.

Eudorina is another colonial genus of frequent occurrence in pools and ditches, It is also common in the plankton of soft-water lakes. Two species, *E. elegans*. Ehr. (Fig. 44*B*) and *E. unicocca* G. M. Smith (Fig. 44*A*), have been found in this country. For descriptions of them, see G. M. Smith (1931).

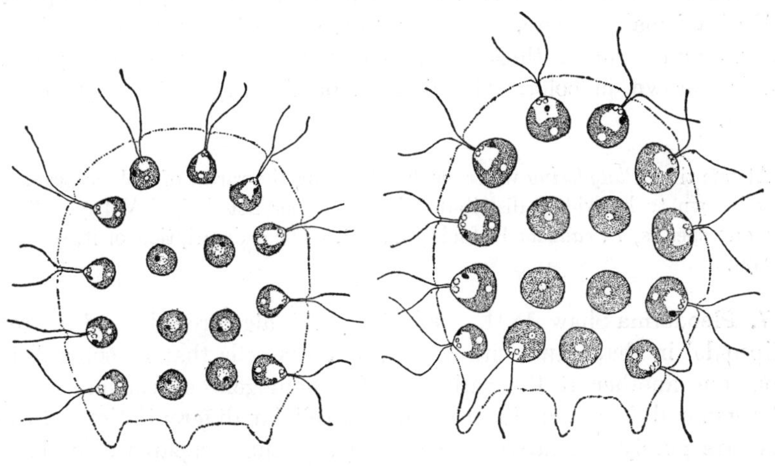

FIG. 45. *Platydorina caudata* Kofoid. (× 530.)

6. **Platydorina** Kofoid, 1899. *Platydorina* has a flattened horseshoe-shaped colony, which may be somewhat twisted from left to right. The anterior end of the colonial envelope is semicircular, and the posterior end has three to five conical projections. Colonies may be 16- or 32-celled, but in either case the cells are in a single layer. Sixteen-celled colonies have 10 marginal and 6 interior cells; 32-celled ones have 12 marginal and 20 interior ones. Marginal cells have their long axes parallel to the plane of the colony. Interior cells have their long axes vertical to the colony and are so arranged that they alternately face in opposite directions. Each cell is biflagellate and either oblately spherical or pyriform. A cell has two contractile vacuoles, an eyespot, a cup-shaped chloroplast with one pyrenoid, and a single nucleus internal to the cup of the chloroplast. All cells in a colony are of the same size, and there is but little difference in size of eyespots at the anterior and posterior poles of a colony.[2]

[1] OTROKOV, 1875; SCHREIBER, 1925.

[2] KOFOID, 1899; SMITH, G. M., 1926; TAFT, 1940.

Asexual reproduction is by a simultaneous division of all cells in a colony to form autocolonies. There is the usual plakea and inversion after cell division ceases. Immediately after inversion, a young colony is globose, but it soon assumes the flattened shape characteristic of the genus.[1]

Platydorina is heterothallic and anisogamous. Male gametes are formed by cell division into a plakea that inverts to form a globose mass which escapes from the parent colony and does not dissociate into individual gametes until it swims near a female colony.[1] Female colonies have a direct functioning of cells as female gametes. Each gamete escapes through a pore in the portion of the colonial envelope immediately external to it.[1] Gametic union is effected by a male gamete boring into the posterior half of a female gamete. The resultant biflagellate zygote soon loses its flagella and forms a thick wall. Later its contents become yellow or reddish brown in color. Germination of the zygote has not been observed.

At one time *Platydorina* was considered among the rarest of Volvocaceae. It is now known to be widely distributed in the upper Mississippi Valley. There is but one species, *P. caudata* Kofoid (Fig. 45). For a description of it, see Kofoid (1899).

7. **Pleodorina** Shaw, 1894. Colonies of this alga are spherical to broadly ellipsoidal in shape and with 32, 64, and 128 cells that lie some distance from one another at the periphery of a homogeneous colonial envelope. The genus differs from *Eudorina* in the cellular differentiation into those that are purely vegetative in character and those capable of dividing to form daughter colonies. All but four anterior cells, or all but the two anterior tiers of cells, in a colony may be reproductive, or those in the anterior half may be vegetative and those in the posterior half reproductive. It is impossible to distinguish between the two types of cells in young colonies, but as colonies grow older the reproductive cells become two or three times the diameter of vegetative cells and have several instead of a single pyrenoid.[2] The cells are spherical to ovoid in shape, with an anterior eyespot, and with two contractile vacuoles at the base of the flagella. The chloroplast is cup-shaped, thin-walled, and with a single pyrenoid if the cell is vegetative. Reproductive cells (gonidia), which often lose their flagella and eyespot, have a massive chloroplast with several pyrenoids.

Asexual reproduction is by simultaneous division of all gonidia into autocolonies. There is the plakeal sequence and inversion characteristic of the family.[3]

[1] TAFT, 1940.
[2] KOFOID, 1898; SHAW, 1894; SMITH, G. M., 1920. [3] DORAISWAMI, 1940.

Sexual reproduction may be anisogamous or oögamous. Most colonies produce only male or only female gametes, but occasionally a colony may produce both kinds of gametes.[1] Development of male gametes is similar to that in *Eudorina*, and a mass of gametes behaves in the same manner.[2] According to the species, cells functioning as female gametes retain[3] or lose[4] their flagella. In either case, male gametes enter the gelatinous

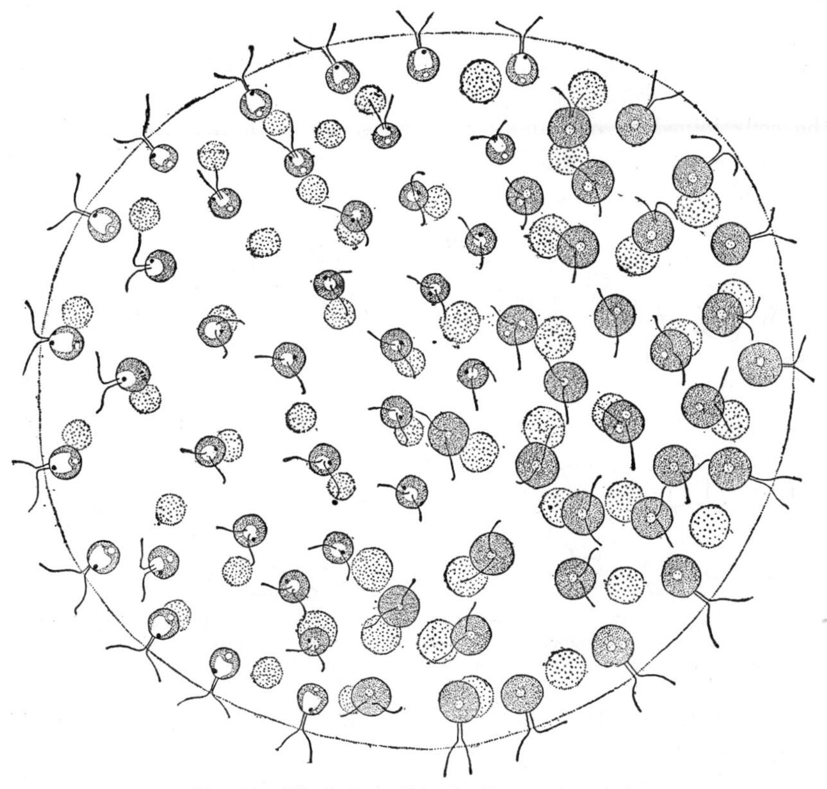

FIG. 46. *Pleodorina californica* Shaw. (× 400.)

matrix of female colonies and there unite with the female gametes. A ripe zygote has a smooth or a finely granulate wall and a protoplast deeply tinged with hematochrome. Its method of germination is unknown.

Two species, *P. californica* Shaw (Fig. 46) and *P. illinoisensis* Kofoid, are found in this country. For descriptions of them, see Collins (1909). Some phycologists think that *P. illinoisensis* belongs in *Eudorina* rather than in *Pleodorina*.

[1] DORAISWAMI, 1940; TIFFANY, 1935.
[2] CHATTON, 1911; DORAISWAMI, 1940; MERTON, 1908; TIFFANY, 1935.
[3] DORAISWAMI, 1940. [4] TIFFANY, 1935.

8. **Volvox** Linnaeus, 1758. *Volvox* may be distinguished from all other members of the family by the large number of cells in a colony. The number may be as low as 500 (512 ?) or as high as approximately 50,000. The colonies are spherical to ovoid and have the biflagellate cells in a single layer just within the periphery of the gelatinous colonial matrix. Each cell is surrounded by a gelatinous sheath of its own, and the sheaths may be confluent with or distinct from one another. In the latter case, they are angular by mutual compression and usually hexagonal. There is gelatinous material of a more watery consistency internal to the gelatinous sheaths of the cells. Most species have ovoid cells. Some species have the cells joined to one another by conspicuous or delicate cytoplasmic strands, a connection which becomes established early in development of colonies.[1]

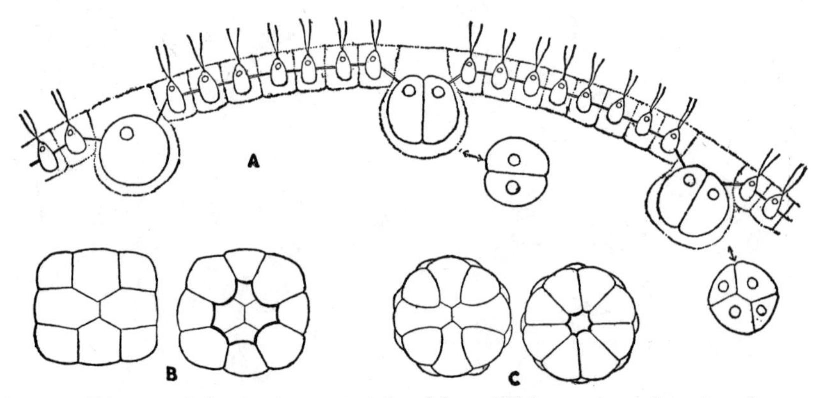

FIG. 47. Diagram of the development of the plakea of *Volvox*. *A*, portion of a colony containing 1-, 2-, and 4-celled stages of plakeal development. *B*, 8-celled plakea as seen from above and from below. *C*, similar views of a 16-celled plakea.

Most of the cells in a colony are vegetative in nature and are incapable of giving rise to new colonies. Each vegetative cell is biflagellate and with two contractile vacuoles near the base of the flagella or with two to five contractile vacuoles irregularly distributed in the anterior end of a cell. There is a cup-shaped to laminate chloroplast toward the posterior pole of a cell, and it usually contains but one pyrenoid. The nucleus is centrally located and is connected with the flagella by a neuromotor apparatus of the blepharoplast-rhizoplast-centriole type.[2] Each vegetative cell has a single anteriorly located eyespot, those of cells toward the anterior end of a colony being somewhat larger than those in cells at the posterior end.

Young colonies have all cells alike in size. As a colony grows older, 2 to 50 asexual reproductive cells (gonidia) may be differentiated in the

[1] JANET, 1912; MEYER, 1896. [2] ZIMMERMANN, 1921.

posterior half of a colony. Each gonidium lies within a globular gelatinous sac projecting toward the interior of a colony. Gonidia usually have a diameter ten or more times that of vegetative cells, have several pyrenoids within the chloroplast, and lack both an eyespot and flagella. Each gonidium divides in a plakeal sequence (Fig. 47), and during colony development there is the usual inversion.[1] Each daughter colony remains within the gelatinous sac originally containing the gonidium until it escapes by moving through a pore-like opening at the free face of the sac.

Sexual reproduction is oögamous, and according to the species the colonies are homothallic or heterothallic. Antherozoids are developed from enlarged cells resembling gonidia. Usually there are relatively few

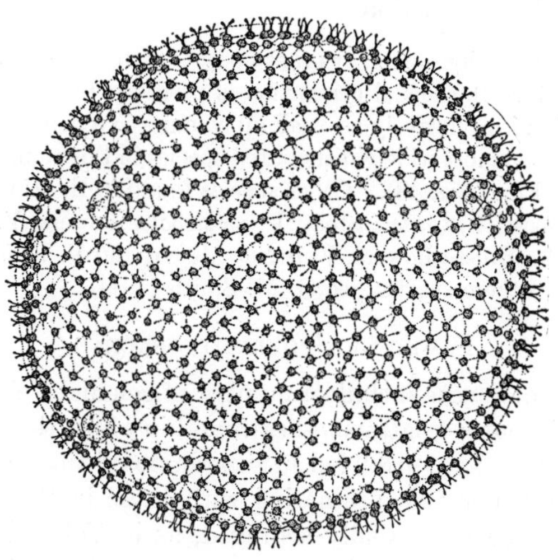

FIG. 48. *Volvox aureus* Ehr. (× 200.)

of these cells, but in certain species a majority of or all cells of a colony may produce gametes.[2] According to the species, a cell divides to form 16, 32, 64, 128, 256, or 512 fusiform biflagellate gametes. Cell division is in a plakeal sequence and with an inversion to form a disk-shaped or globose mass of antherozoids.[3] The colony-like mass of antherozoids is liberated as a unit and does not break up into individual free-swimming antherozoids until it approaches the vicinity of an egg. Only a small percentage of cells in a colony functions as eggs. An egg resembles a young gonidium. When fertilization takes place, the individual antherozoids swim slowly through the gelatinous sheath around an egg and prob-

[1] KUSCHAKEWITSCH, 1931; POCOCK, 1933, 1938; POWERS, 1908; ZIMMERMANN, 1925.

[2] POCOCK, 1938; POWERS, 1908; SMITH, G. M., 1944.

[3] POCOCK, 1933, 1933*A*, 1938.

ably enter it from the side.[1] There also may be a development of unfertilized eggs into parthenospores.[2] After fertilization, the zygote forms a thick, smooth, or stellate wall and develops sufficient hematochrome to color the protoplast an orange red. Zygotes do not germinate until some time after they are liberated by death and decay of the colony in which they were produced. Prior to germination, there is a meiotic division of the nucleus.[3] In germination there is a splitting of the zygote's outer wall layer (exospore) and an extrusion of the inner wall layer (endospore) as a vesicle which surrounds the reddish protoplast. The protoplast may become a biflagellate zoospore, but it rarely escapes from the vesicle.[4] Development into a colony is by the same sequence of plakeal stages as is found in asexual reproduction.

When found in any ditch, pool, or other body of water, *Volvox* is usually present in abundance. Since the sexual generation is preceded by several generations in which reproduction is exclusively asexual, all colonies in a collection may be sexual or asexual. The two most widely distributed species in this country are *V. globator* L. and *V. aureus* Ehr. (Fig. 48). Other species found in the United States are *V. africanus* G. S. West, *V. Carteri* Stein, *V. perglobator* Powers, *V. Powersi* (Shaw) Printz, *V. spermatosphaera* Powers, *V. tertius* Meyer, and *V. Weissmannia* Powers. For descriptions of these species, see G. M. Smith (1944).

Family 5. Spondylomoraceae

The Spondylomoraceae include a number of colonial Volvocales in which the colony is without a gelatinous sheath and in which the cells, when more than four in number, are arranged in superimposed tiers of four cells each and with those in one tier alternating with those in the next. The long axes of the cells are not radially arranged, as in Volvocaceae, but all lie parallel to the long axis of a colony. Most genera have chloroplasts without pyrenoids, but this cannot be considered a family character since *Pascheriella* has pyrenoids. The most important difference between this family and the Volvocaceae is a lack of a plakeal sequence in divisions leading to formation of a daughter colony.

Cells of Spondylomoraceae are usually napiform and with two or four flagella at the anterior end. There are two contractile vacuoles at the base of the flagella, and the eyespot may be anterior or posterior. The chloroplast is generally massive, cup-shaped, and without a pyrenoid.

When asexual reproduction takes place, there may be a simultaneous division of all cells into daughter colonies, or there may be a division of certain cells only. The first two divisions of a protoplast are longitudinal; if further divisions occur, these may be in any plane. Daughter colonies

[1] Lander, 1910. [2] Mainx, 1929. [3] Zimmermann, 1921.
[4] Kirchner, 1883; Metzner, 1945; Pocock, 1933*A*, 1938.

are liberated by a gelatinization of the surrounding parent-cell wall. Akinetes have been found in one genus[1] and aplanospores in another.[2]

Sexual reproduction is isogamous. So far as known, members of the family are homothallic and either with a union of gametes formed by the same cell (*Pascheriella*[3]) or by different cells in the same colony (*Pyrobotrys*[4]). The zygote may retain the flagella derived from the gametes for some time before losing them, assuming a spherical shape, and secreting a wall. Germination of the zygote has not been observed in members of this family.

The three genera found in this country differ as follows:

1. Colonies with two or four cells .... 1. **Pascheriella**
1. Colonies with more than four cells 2
    2. Cells biflagellate .... 3. **Pyrobotrys**
    2. Cells quadriflagellate .... 2. **Spondylomorum**

1. **Pascheriella** Korshikov, 1928. Colonies of *Pascheriella* are usually four-celled but sometimes two-celled. There are never 8 or 16 cells in a colony. The colonies lack a gelatinous envelope, and the adhesion between the cells is so slight that individual cells often break away and swim about by themselves. Four-celled colonies have alternate tiers of two cells each, so arranged that the transverse axis of one tier is perpendicular to the transverse axis of the other. All cells in a colony have their longitudinal axes approximately parallel. The cells are ovoid, slightly compressed, and at times with a small papilla at the anterior pole. They are biflagellate and with the flagella about as long as the cell. There are two contractile vacuoles at the base of the flagella, and the eyespot is anterior. The protoplast contains a laminate longitudinal chloroplast, nearly as long as the cell, that lies next the free face of the cell. It contains a single pyrenoid. The four chloroplasts of a four-celled colony are, therefore, cruciately arranged with respect to one another.[5]

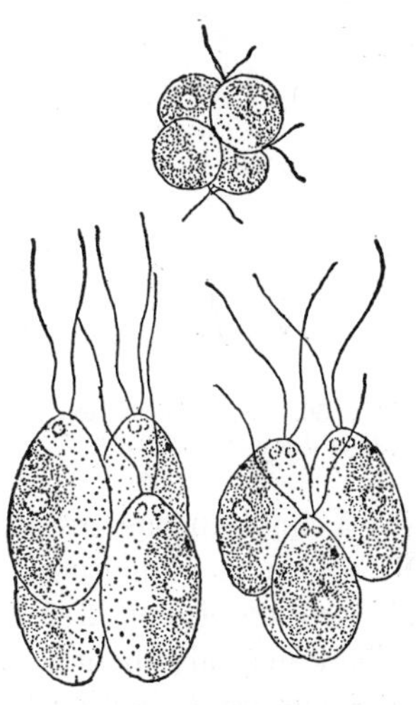

FIG. 49. *Pascheriella tetras* Korshikov. (× 975.)

All cells of a colony produce daughter colonies, though not necessarily simultaneously. Daughters are formed by two successive longitudinal

[1] SCHILLER, 1927. [2] SCHULZE, 1927.
[3] KORSHIKOV, 1928A. [4] STREHLOW, 1929.
[5] KORSHIKOV, 1928A.

divisions of the protoplast and are liberated by a gelatinization of the parent-cell wall.

Sexual reproduction is isogamous, and each cell of a colony produces two biflagellate gametes.[1]

There is but one species *P. tetras* Korshikov (Fig. 49). Thus far for the United States it has only been found in California.[2] For a description of it, see Pascher (1927*A*).

2. **Spondylomorum** Ehrenberg, 1848. The colonies of this alga are usually 8- or 16-celled, and with the cells superimposed in alternating tiers of four each. All cells have their flagellated ends toward the anterior pole of a colony. At the anterior end of a cell is a conical protuberance which bears four flagella. A cell is enclosed by a definite wall, and the protoplast is nearly filled with a massive protoplast that is without a pyrenoid. There are two alternately contracting vacuoles at the base of the flagella, and the nucleus lies a short distance back from them. The eyespot is usually linear and toward the posterior end of a cell.[3]

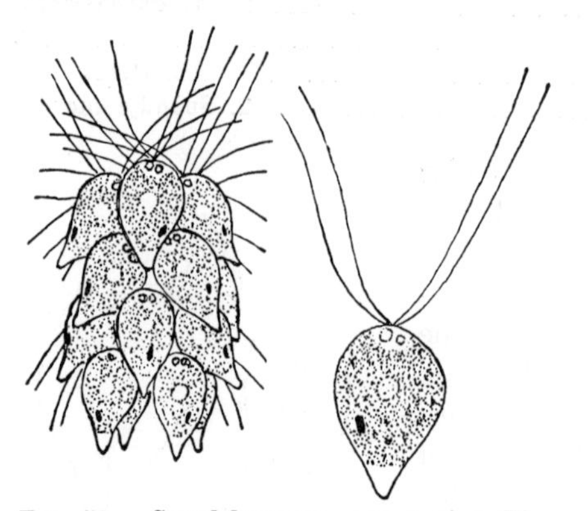

FIG. 50. *Spondylomorum quaternarium* Ehr. (*After Jacobsen*, 1911.)

Asexual reproduction takes place while a colony is in motion and is by a simultaneous division of all cells into daughter colonies. The first two divisions are longitudinal,[4] but further divisions are at various angles. Liberation of daughter colonies is by gelatinization of the parent-cell wall. There may also be a formation of akinetes.[5]

Sexual reproduction has not been reported for this genus.

There are several authentic records for the occurrence of *S. quaternarium* Ehr. (Fig. 50) in the United States. For a description of it, see Pascher (1927*A*).

3. **Pyrobotrys** Arnoldi, 1914 (*Chlamydobotrys* Korshikov, 1924). *Pyrobotrys* has the same mulberry-shaped colonies as *Spondylomorum* and has the cells alternately arranged in tiers of four cells each. It differs in having biflagellate cells. As is the case with the preceding genus, the

[1] KORSHIKOV, 1928*A*. [2] SMITH, G. M., 1933.
[3] KORSHIKOV, 1928*A*; SCHILLER, 1927; SCHULZE, 1927.
[4] STICKNEY, 1909. [5] SCHILLER, 1927.

cells have massive cup-shaped chloroplasts without pyrenoids. In *Pyrobotrys* the eyespot may lie at the anterior end of a cell.

Division to form daughter colonies usually takes place simultaneously in all cells of a colony.[1] The first two divisions appear to be longitudinal; further divisions are without definite orientation. Liberation of daughter colonies is as in *Spondylomorum*. Aplanospores may also be formed.[1]

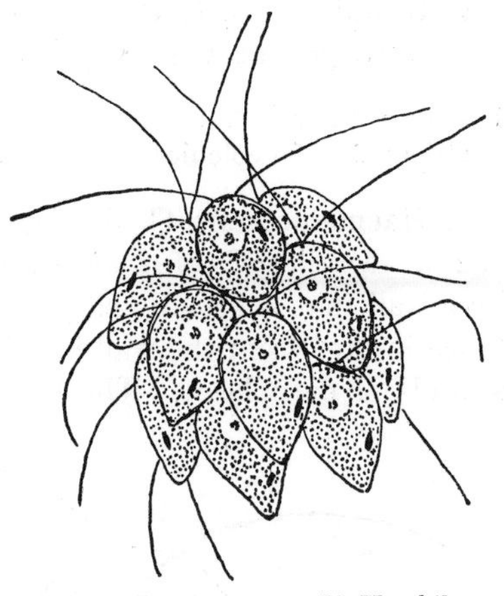

FIG. 51. *Pyrobotrys gracilis* Korshikov.

Gametic union is isogamous, and the colonies are homothallic, but apparently with a fusion of gametes from different cells.[2] Four or eight gametes are formed within a cell. A pair of fusing gametes become apposed at the anterior ends, and the flagella persist after fusion is completed. The quadriflagellate zygote, which is more or less napiform, remains motile for some time and secretes a wall during the period of motility. After the zygote ceases swarming, the protoplast assumes a spherical shape and secretes a thick wall distinct from the first-formed wall.[3]

Korshikov (1924) described the genus *Chlamydobotrys* which he later (1938) placed as a synonym of *Pyrobotrys*. Both of the two species found in this country, *P. gracilis* Korshikov (Fig. 51) and *P. stellata* Korshikov, have been found in several localities. For a description of them as species of *Chlamydobotrys*, see Pascher (1927*A*).

FAMILY 6. HAEMATOCOCCACEAE

Genera belonging to this family have cells with protoplasts that are connected to the cell wall by numerous cytoplasmic processes. The cells are biflagellate, and either solitary or united in colonies with a definite organization. The number of contractile vacuoles is greater than two, and they may be in other parts of the protoplast than just beneath the flagella. The chloroplasts may contain several pyrenoids.

Asexual reproduction is by division into a definite number of daughter cells. In the case of the colonial genus, these remain united to one another

[1] SCHULZE, 1927. [2] KORSHIKOV, 1938; STREHLOW, 1929.
[3] STREHLOW, 1929.

to form an autocolony. Palmella stages, aplanospores, and akinetes are also known.

Sexual reproduction is isogamous and with the formation of a spherical smooth-walled zygote that germinates to form four to eight zoospores.

The two genera found in this country may be distinguished as follows:

1. Cells solitary.. 1. **Haematococcus**
1. Cells united in colonies.................................... 2. **Stephanosphaera**

1. **Haematococcus** C. A. Agardh, 1828 (not *Sphaerella* Sommerfelt,* 1824). Motile cells of *Haematococcus* are solitary, biflagellate, and enclosed by a wall that is broadly ellipsoid to ovoid. The protoplast lies some distance inward from the wall and is connected with it by numerous delicate strands of cytoplasm. The intervening space between the wall and the

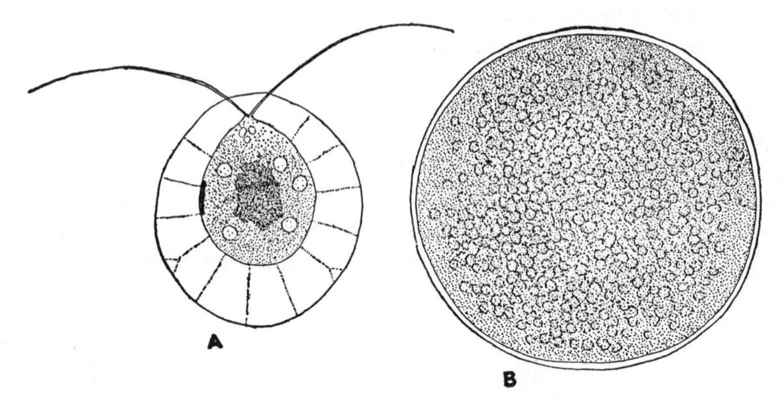

Fig. 52. *Haematococcus lacustris* (Girod.) Rostaf. *A*, vegetative cell. *B*, akinete. (× 800.)

protoplast is filled with a watery gelatinous substance. There are two flagella at the anterior end of a cell, and the portion of each flagellum between the wall and protoplast lies within a gelatinous canal. It is very unusual to find motile cells in which the chloroplast is not more or less obscured by a hematochrome that is a ketonic carotinoid—euglenorhodone.[1] If the red pigment is not too abundant, one can see that there is an eyespot in the equatorial region, contractile vacuoles here and there beneath the plasma membrane, and several pyrenoids within the chloroplast.[2]

Asexual reproduction may take place by division of free-swimming cells into two or four daughter cells (macrozoospores). More frequently a cell

* The type of the genus *Sphaerella* is the red-snow alga, *S. nivalis* (Bauer.) Sommerf. Since this is now known to be a *Chlamydomonas*, the generic name *Sphaerella* is invalid.

[1] Strain (in press). [2] Hazen, 1899; Wollenweber, 1909.

develops into an akinete (Fig. 52*B*) which attains a diameter three or four times that of a vegetative cell. Protoplasts of akinetes have their structure completely obscured by the red pigment. When an akinete germinates, its protoplast divides into 4, 8, or 16 zoospores that are liberated by a sac-like swelling of the akinete wall. Division of an akinete's protoplast may result in a formation of aplanospores instead of zoospores. Aplanospores are liberated in the same manner as zoospores.[1]

Akinetes which have been subjected to adverse conditions, such as cold, rapid drying, or starvation, usually have the protoplast dividing into 32 or 64 biflagellate swarmers. These are sometimes called "microzoospores,"[2] but it has been shown[3] that they are gametes which cannot develop parthenogenetically into vegetative cells. Gametes may also be formed by repeated division of the protoplast of a vegetative cell. When within the wall of a parent cell, the gametes are arranged in the form of a hollow cup and are in cytoplasmic connection with one another.[4] Gametic union is isogamous and the quadriflagellate zygote may remain motile for some time before losing its flagella, rounding up, and secreting a thick wall.

The only species thus far found in this country is *H. lacustris* (Girod.) Rostaf. (Fig. 52) (*H. pluvialis* Flotow). It frequently occurs in almost pure stands in hollows of rocky ledges temporarily filled with rain water. Another place where it has been repeatedly found is in ornamental urns in cemeteries. It is of less frequent occurrence in concrete basins. For a description of this species as *H. pluvialis*, see Pascher (1927*A*).

2. **Stephanosphaera** Cohn, 1852. Colonies of *Stephanosphaera* are usually 8-celled, but they may be 16-, 4-, or 2-celled. The cells lie with their long axes parallel and embedded within a globose gelatinous envelope. The cells are naked, truncately spindle-shaped, biflagellate, and with elongate cytoplasmic processes both from the poles and from the medial portion. A cell has several contractile vacuoles that lie irregularly distributed beneath the surface of the protoplast.[5] The chloroplast is a reticulate sheet. The number of pyrenoids varies from one to five, but usually there are two and they lie axial to each other.

Asexual reproduction is by a simultaneous formation of autocolonies by all cells of a colony.[6]

In sexual reproduction, each cell of a colony divides and redivides to form a globose clump of 16, 32, or 64 biflagellate fusiform gametes.[7] The clumps dissociate into individual gametes that fuse in pairs as they swim about within the gelatinous envelope of the parent colony. The zygote

[1] Hazen, 1899; Peebles, 1909. [2] Hazen, 1899.
[3] Peebles, 1909; Schulze, 1927. [4] Pocock, 1937. [5] Wollenweber, 1909.
[6] Hieronymus, 1884. [7] Hieronymus, 1884; Moewus, 1933.

is spherical and with a smooth wall. When it germinates, there is usually a formation of four zoospores but the number may range from five to eight.[1] At first after liberation, the zoospores lie within a vesicle extruded from the zygote wall. Each zoospore develops into a vegetative colony, either after escape from the vesicle or while still lying within it.

There is but one species, *S. pluvialis* Cohn (Fig. 53). It is an exceedingly rare organism and is found only in rain pools in granitic rocks. In this country, the only published record for it is from Massachusetts.[2] The late Prof. J. C. McKee informed me he had found it in Virginia, and Dr. M. A. Pocock informs me that she has found it in Echo Lake, near Lake Tahoe, California. For a description of *S. pluvialis*, see Pascher (1927*A*).

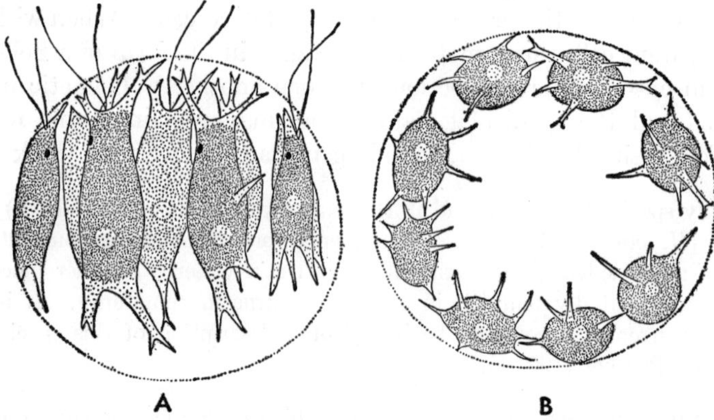

FIG. 53. *Stephanosphaera pluvialis* Cohn. *A*, front view. *B*, polar view. (*After Prescott and Croasdale*, 1937.) (× 480.)

ORDER 2. TETRASPORALES

This order includes a heterogeneous assemblage of Chlorophyceae whose vegetative cells are immobile but have the ability to divide vegetatively. Several of the Tetrasporales are so closely allied to the Volvocales that their cells have many characteristics of the volvocaceous cell. Among these characteristics are a palmelloid organization of the thallus the ability of immobile cells to return directly to a motile condition, and, occasionally, the presence of an eyespot or contractile vacuoles.

A few members of the family are unicellular and with the cell surrounded by a gelatinous sheath; the majority have the cells united to one another in nonfilamentous gelatinous colonies. Colonial genera have the cells irregularly distributed throughout, or at the periphery of, microscopic or macroscopic colonies. Gelatinous sheaths around the individual cells may

[1] STREHLOW, 1929. [2] PRESCOTT and CROASDALE, 1937.

be distinct or confluent; in rare cases the sheaths are so joined to one another that the colony is dendroid.

Most Tetrasporales have spherical or ovoid cells, and a few have fusiform ones. Typical members of the order have the protoplasts enclosed by a wall in which there is a cellulose layer next the protoplast and a wide pectic layer external to the cellulose. The gelatinous portion of the wall may be homogeneous or stratified. It is usually secreted in equal amounts on all sides of the cell, but sometimes, as in *Hormotila*, the secretion of pectic material is restricted to one side of the cell. In such cases, the cells are borne on branched gelatinous stalks. The absence of a cellulose layer around the protoplast, which obtains in *Prasinocladus*,[1] may be characteristic of all the Chlorangiaceae.

Algae belonging to the Tetrasporales always have uninucleate cells and usually but a single chloroplast in a cell. The majority of genera have cup-shaped chloroplasts, but there are cases where it is stellate and central (*Asterococcus*) or disciform and parietal (*Schizochlamys*). Chloroplasts usually contain a pyrenoid surrounded by a sheath of starch grains, but pyrenoids may be lacking and there may be a formation of fats instead of starch (*Schizochlamys*). Genera belonging to the Tetrasporaceae have two (sometimes four or more) long, delicate, immobile cytoplasmic processes (*pseudocilia*) at the anterior end of the cell. Some of the genera with pseudocilia, and some without them, have two contractile vacuoles, anterior in location, as in a chlamydomonad cell. Eyespots are also known for certain genera; these may be of regular occurrence in the vegetative cell (*Prasinocladus*) or only in young cells (*Malleochloris*).

Most members of the order have a vegetative division of their cells into two, four, or eight daughter cells, which are permanently retained within the colony. As the colony increases in size, it may become broken into two or more portions, each of which continues growth as an independent colony. Certain genera, especially those with individual gelatinous envelopes around the cells, have a very strong tendency to fragment as a colony becomes older and thus have a regular propagation by vegetative multiplication. Reproduction of a colony may also take place by the direct metamorphosis of vegetative cells into zoospores, which escape from the colonial envelope, swarm for a time, then come to rest, withdraw their flagella, secrete a gelatinous envelope, and then grow into new colonies by vegetative cell division. Since these motile cells are formed by a direct metamorphosis of vegetative cells, they are homologous with vegetative cells of Chlamydomonadaceae and not with zoospores.

True zoospores are also found in the Tetrasporales and are formed by a division of an immobile cell's protoplast into two, four, or eight zoospores

[1] ZIMMERMANN, 1925*A*.

which are liberated by rupture or by gelatinization of the parent-cell wall. Aplanospores and akinetes are of rather common occurrence among members of the order and are often so thick-walled as to be hypnospores. Germinating hypnospores usually grow directly into new colonies, but in *Tetraspora* the protoplast escapes in an amoeboid fashion and continues to live in an amoeboid manner for some time.

The genera in which sexual reproduction has been observed are isogamous and with both gametes flagellated.

Some phycologists think that the genera whose immobile vegetative cells have contractile vacuoles, eyespots, and pseudocilia are too closely related to the chlamydomonad type to warrant separation from the Volvocales.

Of the four families described on succeeding pages, the Palmellaceae are probably in the line leading from the Volvocales to the more advanced Chlorophyceae, but the Chlorangiaceae are the most primitive as far as organization of the vegetative cells is concerned.

Family 1. Palmellaceae

Members of the Palmellaceae have their cells united in small gelatinous colonies which are generally amorphous, but which may be of definite shape. The gelatinous sheaths around the cells may be distinct from one another and homogeneous, or concentrically stratified; or the sheaths may be confluent into a homogeneous colonial matrix. A few genera with gelatinous sheaths evident around the individual cells have secretion of gelatinous material restricted to one side of a cell, thus causing a development of branched tubular colonies.

Cells of Palmellaceae are usually spherical or ellipsoidal, and with a thin cellulose layer next to the protoplast. According to the genus, the chloroplast is cup-shaped, stellate, or discoid. There is usually a pyrenoid within a chloroplast, but certain genera regularly lack pyrenoids. At times, contractile vacuoles and eyespots are present in vegetative cells of some members of the family. These structures are found more frequently in young cells than in old ones.

Vegetative multiplication by a fragmentation of colonies is of frequent occurrence. New colonies may also result from nonflagellated cells developing flagella and behaving as zoospores.

Zoospores are formed by division of the cell contents into 2, 4, 8, or 16 zoospores. These rarely become changed into aplanospores, but a development of vegetative cells into akinetes is known for several genera.

All cases of sexual reproduction thus far recorded are isogamous.

The genera found in this country differ as follows:

1. Chloroplast stellate .... **7. Asterococcus**
1. Chloroplast not stellate **2**

2. Cells solitary .... 3. **Gloeocystis**
2. Cells in colonies 3
3. Colonies simple or branched tubes 4
3. Colonies not tubular 5
4. Cells uniseriate .... 4. **Hormotila**
4. Cells multiseriate .... 5. **Palmodictyon**
5. Sheaths of individual cells distinct .... 3. **Gloeocystis**
5. Sheaths of individual cells indistinct 6
6. Cells pyriform .... 6. **Askenasyella**
6. Cells globose 7
7. Colonies spherical, with few cells .... 2. **Sphaerocystis**
7. Colonies amorphous, with many cells .... 1. **Palmella**

1. **Palmella** Lyngbye, 1819; emend., Chodat, 1902. Colonies of this alga are amorphous and of microscopic or macroscopic size. The cells are spherical to broadly ellipsoidal, and the gelatinous sheaths around

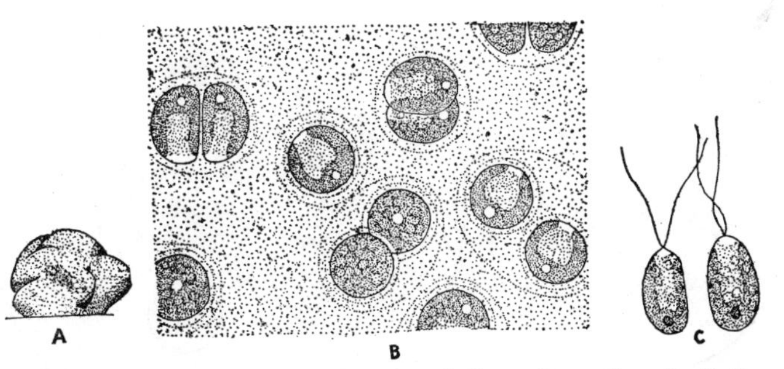

FIG. 54. *Palmella miniata* var. *aequalis* Näg. *A*, thallus. *B*, portion of a thallus. *C*, zoospores. (*A*, × ½: *B–C*, × 600.)

them are wholly or partially confluent with one another to form a gelatinous matrix of indefinite extent. The chloroplast is cup-shaped and with a single pyrenoid at its base. Some species have hematochrome masking the chlorophyll or coloring the whole protoplast.

Growth of a colony takes place by cell division in all planes. Colony reproduction may result from direct metamorphosis of vegetative cells into biflagellate motile cells. This may take place in recently formed or in old cells. It is of frequent occurrence when individuals growing on damp soil become flooded with water. Some of these motile cells have an eyespot, others lack them.

Asexual reproduction is by division of the cell contents into 4, 8, or 16 zoospores. Akinetes are also known for the genus.

Gametes are formed by division of a protoplast into 32 or 64 biflagellate gametes.[1] Gametic union is isogamous.

[1] CHODAT, 1902.

Formerly, phycologists described many species which they referred to this genus. Many of these "species" have been shown to be growth phases (palmella stages) in the life histories of other algae, especially Volvocales. When these palmella stages are excluded from the genus, there remain three or four good species of which *P. miniata* Liebl. (Fig. 54) and *P. mucosa* Kütz. have been found in the United States. For a description of these two species, see Lemmermann (1915).

2. **Sphaerocystis** Chodat, 1897. The colonies of this alga are always free-floating and usually with a perfectly spherical, homogeneous, colonial envelope. Colonies contain 4, 8, 16, or 32 spherical cells which lie equidistant from one another and toward the periphery of the colonial envelope. Sometimes certain cells in a colony have divided into four or eight small daughter cells which are surrounded by a fairly distinct, spherical, gelatinous envelope. In rare cases, the colonies are irregular in shape and with a large number of cells. Young cells have a massive, cup-shaped chloroplast containing a single pyrenoid. Older cells have the chloroplast completely filling the cell, but the single pyrenoid is still evident.

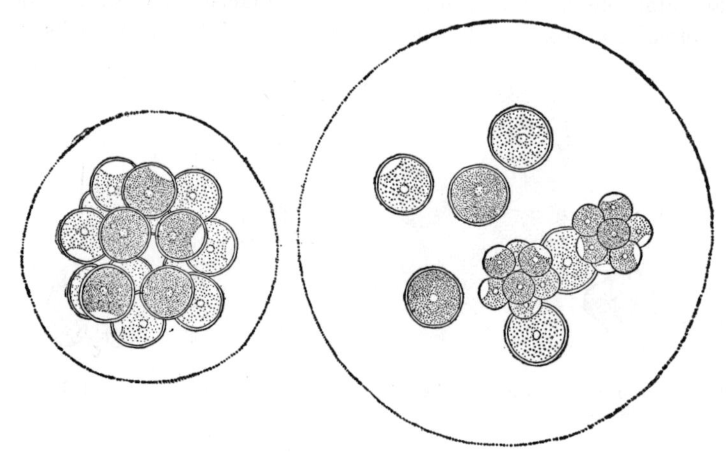

FIG. 55. *Sphaerocystis Schroeteri* Chod. (× 1000.)

There is a regular multiplication of colonies by a softening of the colonial matrix and an escape of young colonies surrounded by a spherical colonial envelope of their own. The direct metamorphosis of vegetative cells into motile cells is rare but may take place in young or in old cells.[1]

Sphaerocystis is a strictly planktonic genus and one of common occurrence in lakes everywhere in this country. There is but one species, *S. Schroeteri* Chod. (Fig. 55). For a description of it, see G. M. Smith (1920).

3. **Gloeocystis** Nägeli, 1849. The cells of *Gloeocystis* are sometimes solitary but more often are united in colonies of small size. The gelatinous

[1] CHODAT, 1897.

sheaths around the cells may be lamellated or without stratification, but in either case they are never confluent with one another. The cells are spherical to ovoid and with a thin cellulose wall. Young cells contain a single, massive, cup-shaped chloroplast with one pyrenoid; older cells often have a diffuse chloroplast completely filling the cell and one containing numerous starch granules.

Because of the lack of confluence of the cell sheaths, there is a very strong tendency for colonies to fragment before they become of any size. New colonies may also result from direct metamorphosis of immobile cells into motile cells. Akinetes have been recorded[1] but are rather uncommon.

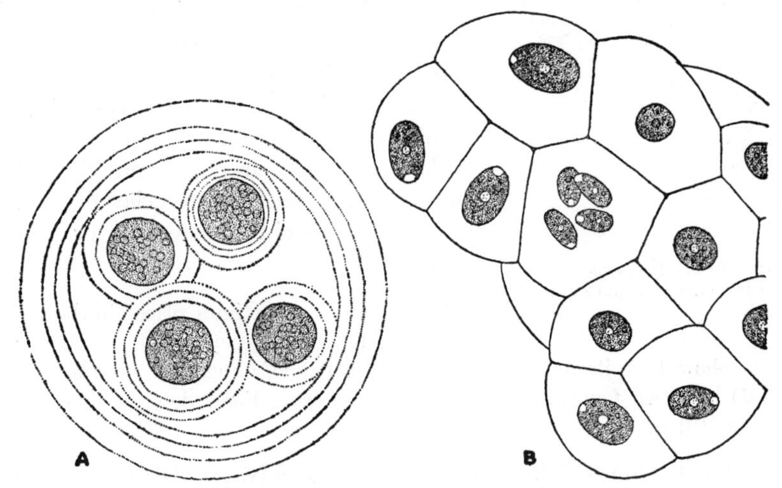

Fig. 56. *A*, *Gloeocystis gigas* (Kütz.) Lagerh. *B*, *G. ampla* Kütz. (*A*, × 400; *B*, × 600.)

Asexual reproduction is by division of the cell contents into four or eight biflagellate zoospores.

Sexual reproduction has not been observed.

Some species of *Gloeocystis* are aquatic, others are terrestrial. The following have been found in the United States: *G. ampla* Kütz. (Fig. 56*B*), *G. fenestralis* (Kütz.) A. Br., *G. gigas* (Kütz.) Lagerh. (Fig. 56*A*), *G. Paroliniana* (Menegh.) Näg., *G. planctonica* (W. and G. S. West) Lemm., *G. rupestris* (Lyngb.) Rab., and *G. vesiculosa* Näg. For descriptions of *G. ampla* and *G. planctonica*, see Lemmermann (1915); for the others, see Collins (1909).

4. **Hormotila** Borzi, 1883. Young colonies of this alga have much the same appearance as *Gloeocystis*, and each of the spherical cells is enclosed by a thick many-layered gelatinous envelope.[2] These juvenile colonies may have the same fragmentation into daughter colonies as is found in *Gloeocystis*. As a colony grows older and if there is no fragmentation, the

[1] Hansgirg, 1886. [2] Borzi, 1883.

secretion of gelatinous material becomes restricted to one side of a cell and there is a gradual development of lamellated tubes at one side of the cells. Cell division, followed by excentric secretion of gelatinous material by each daughter cell, may result in a dichotomously branching gelatinous tube, or the cells may become seriately arranged, some distance from one another within an unbranched tube. The chloroplast is diffuse and fills the entire protoplast. Usually it is so densely packed with starch that its structure cannot be determined, but sometimes one or two pyrenoids can be seen.[1]

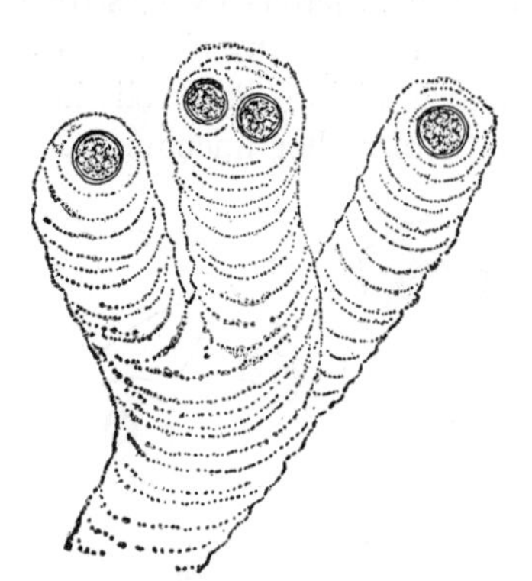

FIG. 57. *Hormotila mucigena* Borzi. Drawn from a herbarium specimen. (× 650.)

Old cells may develop into ovoid to subcylindrical akinetes which become two to five times the size of vegetative cells. When an akinete germinates, its protoplast divides into 8, 16, 32, or 64 daughter protoplasts[2]. These may become biflagellate zoospores liberated by a gelatinization of one side of the old akinete wall, or aplanospores.

This genus is sometimes placed in the Chlorangiaceae but the *Gloeocystis*-like juvenile stages seem to show that its affinities are with the Palmellaceae. *H. mucigena* Borzi (Fig. 57) has been found in California and in Iowa. For a description see Lemmermann (1915).

5. **Palmodictyon** Kützing, 1845 (*Palmodactylon* Nägeli, 1849). The chief feature distinguishing this genus from others in the family is its organization into a simple, anastomosing or branched, tubular thallus in which the cellular arrangement is multiseriate. There may be a broad homogeneous sheath around each cell or group of two or four cells, or the cellular sheaths may be completely fused with one another to form a homogeneous colonial matrix. The cells are spherical, and there are usually two or three curved, parietal, laminate chloroplasts within a cell. Pyrenoids are lacking. Cell division is largely in one plane, resulting in a growth of a colony in one direction.

Biflagellate swarm spores have been recorded.[3] There may also be a formation of thick-walled akinetes which germinate directly into new plants.[4]

The only distinction between *Palmodactylon* and *Palmodictyon* is the lack of sheaths around individual cells or groups of cells in *Palmodactylon*. Since both genera have the same type of laminate parietal chloroplasts without pyrenoids, it seems best to follow those who combine the two rather than those who consider them distinct. The two species found in this country are *P. viride* Kütz. (Fig.

[1] WEST, W. and G. S., 1907. [2] BORZI, 1883. [3] LEMMERMANN, 1915.
[4] WEST, G. S., 1904.

58*A–B*) and *P. varium* (Näg.) Lemm. (Fig. 58*C–D*), the latter including *Palmodactylon simplex* Näg. and *Palmodactylon subramosum* Näg. For descriptions of *P viride* and *P. varium*, see Lemmermann (1915).

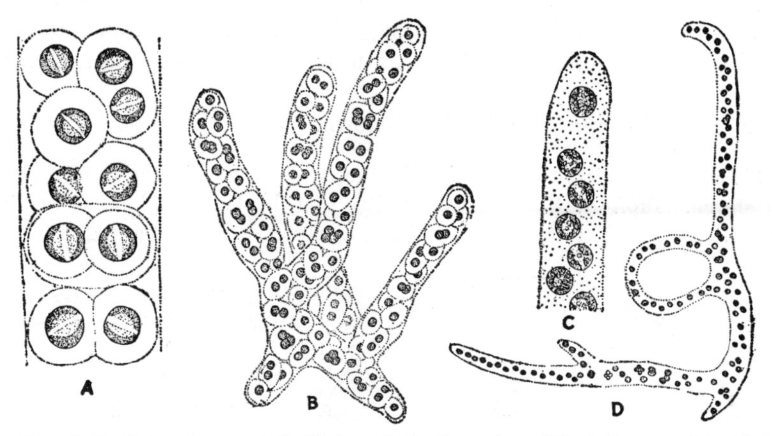

FIG. 58. *A–B*, *Palmodictyon viride* Kütz. *C–D*, *P. varium* (Näg.) Lemm. (*A*, × 1000; *B*, *D*, × 200; *C*, × 500.)

6. **Askenasyella** Schmidle, 1902. This colonial alga has a broad, globose, or irregularly shaped, colorless, gelatinous matrix within which the

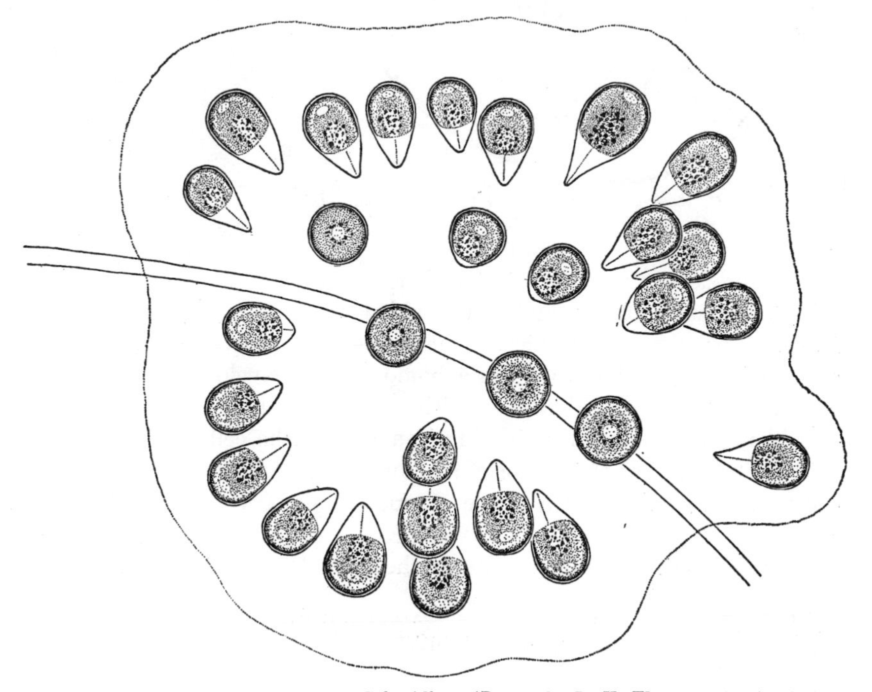

FIG. 59. *Askenasyella chlamydopus* Schmidle. (*Drawn by R. H. Thompson.*) (× 810.)

cells are radially arranged. The cells are markedly pyriform, and all of them lie with the broad posterior end toward the exterior of a colony. There is a single cup-shaped chloroplast, containing a single pyrenoid, at the posterior end of a cell. The hyaline anterior portion of a cell contains a delicate thread which runs from the cell apex to the chloroplast. A cell contains granules of starch which lie near the inner face of the chloroplast.

The method of reproduction is unknown.

This genus was placed among the Xanthophyceae[1] until it was shown[2] to contain starch. Prof. R. H. Thompson writes that he has found *A. chlamydopus* Schmidle (Fig. 59) in Kansas. For a description of it, see Pascher (1927*A*).

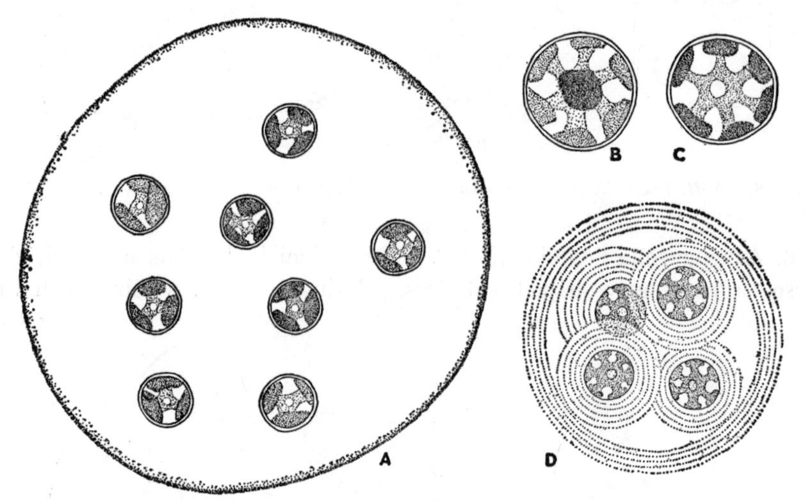

FIG. 60. *A–C*, *Asterococcus limneticus* G. M. Smith. *D*, *A. superbus* (Cienk.) Scherffel. (*A*, × 500; *B–C*, × 1000; *D*, × 300.)

7. **Asterococcus** Scherffel, 1908. The spherical to subspherical cells of this alga may be solitary and with a wide gelatinous envelope, or united in 2-, 4-, 8-, or 16-celled colonies. One species has an envelope with several concentric strata; the other has a homogeneous envelope. The cell contains a single central chloroplast with a variable number of rays from the central mass, each ray terminating in a disk at the cell wall. The pyrenoid is single and at the center of a chloroplast. Sometimes there are two contractile vacuoles and a single eyespot in the peripheral portion of a cell.[3]

Vegetative cell division results in the formation of two, four, or eight

[1] PASCHER, 1925*A*; PRINTZ, 1927. [2] PASCHER, 1937.
[3] SCHERFFEL, 1908; SMITH, G. M., 1916*C*.

daughter cells, and these may be liberated by a gradual dissolution at one side of the stratified envelope which enclosed the parent cell. Vegetative cells may develop directly into motile cells which escape from the gelatinous envelope,[1] or the protoplast may escape before developing flagella.[2]

Two species, *A. limneticus* G. M. Smith (Fig. 60*A–C*) and *A. superbus* (Cienk.) Scherffel (Fig. 60*D*), have been found in this country. For a description of *A. limneticus*, see G. M. Smith (1920), for *A. superbus*, see Lemmermann (1915).

Family 2. Tetrasporaceae

This family differs sharply from others of the order by the presence of two or more long rigid cytoplasmic processes (*pseudocilia*) at the anterior end of a cell. Practically always the cells are united in microscopic or macroscopic colonies that may be pyriform, subspherical, cylindrical, or amorphous in shape. Whether solitary or colonial, the cells are always enclosed by a sheath or colonial matrix that is homogeneous in structure. Colonial species have the cells toward the periphery of the colonial matrix and often in twos, fours, or rings of eight. The anterior pole of a cell is toward the outer face of a colony, and the pseudocilia often project beyond the colonial envelope.

Cells of most species have massive cup-shaped chloroplasts containing a single pyrenoid, but some have parietal chloroplasts without pyrenoids. There may also be such primitive chlamydomonad features as contractile vacuoles within immobile vegetative cells.

Growth of a colony is by division of its cells into two, four, or eight daughter cells. The parent-cell wall enclosing daughter cells may become completely gelatinized and fuse with the colonial matrix (*Tetraspora*, *Apiocystis*), or it may be cast off in one, two, or four pieces which remain unchanged within the colonial matrix (*Schizochlamys*). When first formed, daughter cells are close together, but as they become older they become more and more remote from one another because of a secretion of gelatinous material. Colony fragmentation, resulting from weak adhesion of gelatinous envelopes, is very rare in this family, and vegetative multiplication of a colony is brought about only by an accidental breaking. The direct metamorphosis of vegetative cells into a motile condition is known for all genera, and these are liberated by a gelatinization of the colonial envelope. Thick-walled akinetes (hypnospores) may also be formed from vegetative cells.

Asexual reproduction is by division of a cell's protoplast into two, four, or eight bi- or quadriflagellate zoospores.

[1] Cienkowski, 1865. [2] Chodat, 1895.

Sexual reproduction is isogamous and by a union of biflagellate gametes. The quadriflagellate zygote usually remains motile for some time before it becomes immobile and secretes a wall. Germinating zygotes of *Tetraspora* have the protoplast dividing to form four or eight aplanospores which remain embedded within a common gelatinous matrix where they divide to form vegetative cells.

The four genera found in this country differ as follows:

1. Colonial envelope containing fragments of old-cell walls...... 3. **Schizochlamys**
1. Colonial envelope without evident remains of old-cell walls.................. 2
   2. Mature colonies of microscopic size.. 3
   2. Mature colonies of macroscopic size.......................... 1. **Tetraspora**
3. Colonies pyriform.. 2. **Apiocystis**
3. Colonies crustose.. 4. **Chaetopeltis**

1. **Tetraspora** Link, 1820. Colonies of *Tetraspora* are macroscopic or microscopic, attached or free floating, and spherical, saccate, vermiform, or irregularly expanded. Most species have a colonial matrix of a watery texture, but sometimes, as in *T. cylindrica* (Wahlb.) Ag., the matrix is tough and cartilaginous. The colonial matrix is structureless, except for an individual gelatinous sheath surrounding each group of recently formed cells. The cells are spherical in shape and lie toward the periphery of the colonial envelope. There is a certain tendency for them to be in twos or in fours, but in many cases they are irregularly scattered. The face of the cell toward the exterior of the colony bears two long cytoplasmic processes (pseudocilia) which may extend only to the surface of the colonial envelope or may extend beyond it.[1] Although pseudocilia are characteristic of all species, they are not always evident in all individuals. Within the cell are a nucleus and a massive, cup-shaped chloroplast containing a single pyrenoid. Cells of old colonies often have chloroplasts so densely packed with starch that the chloroplast's shape is obscured.

Growth of the colony is by the cells dividing into two or four daughter cells. Cytokinesis in cells of *Tetraspora* is by means of a cell plate developed on the mitotic spindle.[2] The gelatinous envelope around each group of daughter cells, resulting from the gelatinization of the old parent-cell wall, is quite distinct when the cells are first formed but gradually merges with the colonial envelope as the daughter cells increase in size and become more remote from one another. Colonies may grow to over 1 m. in length or diameter, but they usually become broken into smaller pieces before they attain such a size. At any time in the development of a colony, all or certain of the cells may be metamorphosed into biflagellate

[1] Klyver, 1920; Schröder, 1902. [2] McAllister, 1913.

zoospores. These escape from the colonial matrix, swim about for a time, then withdraw their flagella, secrete a gelatinous envelope, and develop into new colonies by vegetative cell divisions. Vegetative cells may also develop into thick-walled akinetes (hypnospores) with brown, sculptured walls. Germinating hypnospores[1] have an amoeboid liberation of the protoplast, and this may remain amoeboid through several cell generations before assuming the usual shape and structure of vegetative cells.

Sexual reproduction is by the division of the cells into four or eight

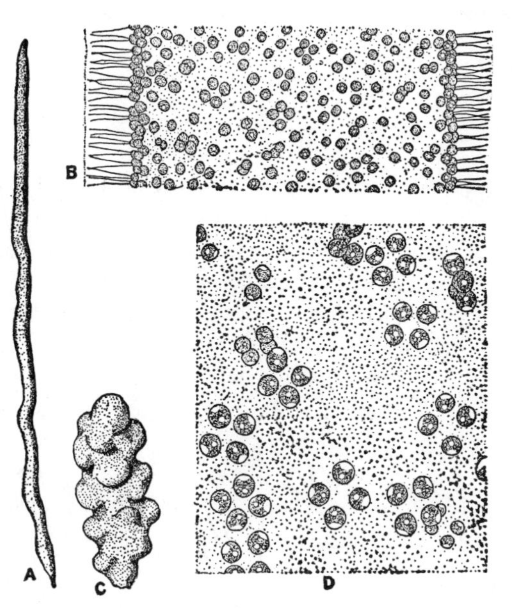

FIG. 61. *A–B, Tetraspora cylindrica* (Wahlb.) Ag. *C–D, T. lubrica* (Roth) Ag. (*A, C,* × ½; *B,* × 155; *D,* × 500.)

biflagellate gametes.[2] Gametes have a more pronounced pyriform shape than do zoospores, a more distinctly cup-shaped chloroplast, and an eyespot at the anterior end. Certain species are heterothallic.[3] The gametes become apposed in pairs at their anterior ends and fuse laterally. The zygote swarms for some time after its formation but eventually comes to rest, secretes a wall, and grows to twice its original size. When it germinates, the protoplast divides to form four or eight aplanospores which lie within a common matrix formed by the gelatinization of the old zygote wall. The vegetative cells resulting from the germination of these aplano-

[1] PASCHER, 1915. [2] KLYVER, 1929. [3] GEITLER, 1931*A*.

spores remain within a common matrix and so develop into a compound colony.[1]

Tetraspora is often found in abundance in quiet waters. Numerous species have been described, but most of these are merely growth forms. The following species have been found in this country: *T. cylindrica* (Wahlb.) Ag. (Fig. 61*A–B*), *T. gelatinosa* (Vauch.) Desv., *T. lacustris* Lemm., *T. lamellosa* Prescott, *T. limneticus* W. and G. S. West, *T. lubrica* (Roth) Ag. (Fig. 61*C–D*). For a description of *T. lamellosa*, see Prescott, 1944; for descriptions of the others, see Lemmermann, 1915.

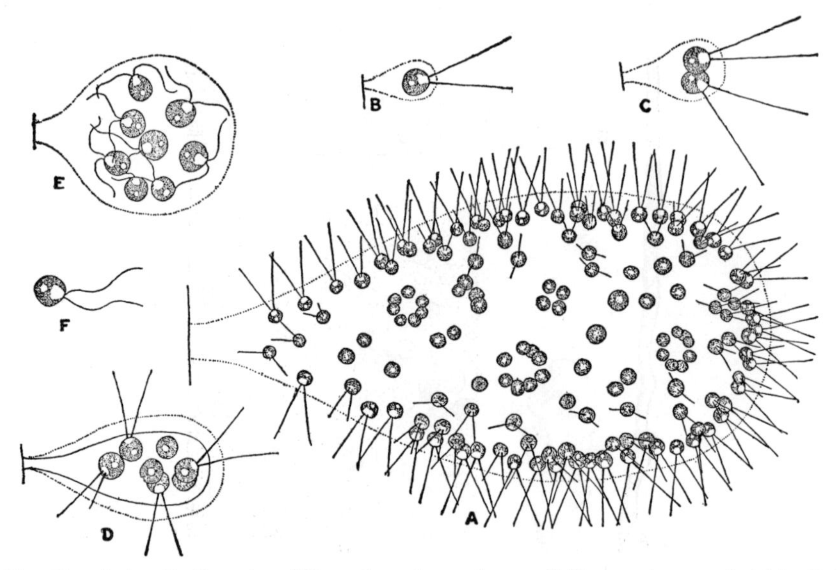

FIG. 62. *Apiocystis Brauniana* Näg. *A*, mature colony. *B–D*, one-, two-, and eight-celled colonies. *E*, metamorphosis of vegetative cells into zoospores at the eight-celled stage. *F*, zoospore. (*A*, × 200; *B–F*, × 400.)

2. **Apiocystis** Nägeli, 1849. Colonies of *Apiocystis* are always microscopic, pyriform to clavate, and attached by a stipe-like base. They never contain more than a few hundred spherical cells and usually less than 50. The colonial matrix often has an outer sheath of a more watery nature. The cells lie toward the periphery of the matrix and are irregularly distributed more frequently than in groups of four or rings of eight. Each cell bears two long pseudocilia on the face toward the exterior of a colony, and the pseudocilia extend for a considerable distance beyond the colonial

[1] KLYVER, 1929.

matrix. A cell contains a well-defined, massive, cup-shaped chloroplast with a single pyrenoid. Cell division is usually into four daughter cells whose gelatinous sheaths soon become completely fused with the colonial matrix. *Apiocystis* does not have the fragmentation of a colony found in many other Tetrasporales, and there is rarely an accidental breaking of a colony. A direct metamorphosis of cells into a motile condition is of frequent occurrence. When this does occur, it is not unusual to find all cells of a colony becoming motile and swimming about within the colonial matrix for some time (Fig. 62*E*). Motile cells that have been liberated and have ceased swarming come to rest upon a firm substratum, usually one of the coarser filamentous Chlorophyceae, and secrete a gelatinous envelope which adheres firmly to the substratum. The first few cell divisions in development of this unicellular stage into a colony are simultaneous, and the number of cells in a young colony is always a multiple of two (Fig. 62*B–D*); later divisions are not simultaneous.

Sexual reproduction is isogamous, four or eight biflagellate gametes being produced by a cell.[1]

Apiocystis is usually found in quiet water and upon old filaments of Cladophoraceae, *Oedogonium*, or *Vaucheria*. When found at any station, it is usually present in abundance. The single species, *A. Brauniana* Näg. (Fig. 62) is widely distributed in this country. For a description of it, see Collins (1909).

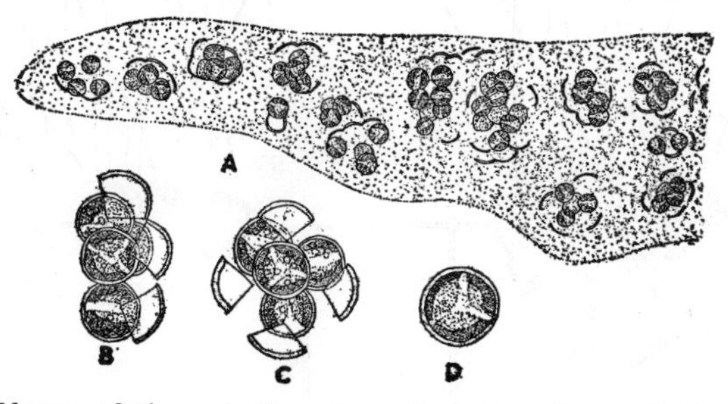

FIG. 63. *Schizochlamys gelatinosa* A. Br. *A*, portion of a colony. *B–C*, fragmentation of parent-cell wall. *D*, vegetative cell. (*A*, × 300; *B-D*, × 1000.)

3. **Schizochlamys** A. Braun, 1849. The spherical cells of *Schizochlamys* are usually united in amorphous colonies. One species has solitary cells.[2] Colonial species have the cells irregularly distributed within a structureless matrix which contains fragments of walls of the various cell generations. Sometimes a cell contains a massive cup-shaped chloroplast with one pyrenoid, but more often the chloroplasts are parietal, laminate, with or with-

[1] CORRENS, 1893; MOORE, S. L., 1890. [2] SMITH, G. M., 1922*A*

out a pyrenoid, and with oil instead of starch. A cell may bear several pseudocilia and have two contractile vacuoles at their base,[1] but these structures are not usually evident unless brought out by staining. When cell division takes place, there is a bi- or quadripartition of the protoplast and a splitting of the wall into two or four parts. The parts of the wall remain unchanged within the colonial matrix for a considerable time.

Asexual reproduction is by division of a cell's protoplast into two, four, or eight bi- or quadriflagellate zoospores.

This genus is easily distinguished from other Tetrasporales by the persisting fragments of old cell walls within the colonial matrix. Three species, *S. compacta* Prescott, *S. delicatula* W. West, and *S. gelatinosa* A. Br. (Fig. 63) have been found in the United States. For a description of *S. compacta*, see Prescott (1944); for descriptions of the other two, see Lemmermann (1915).

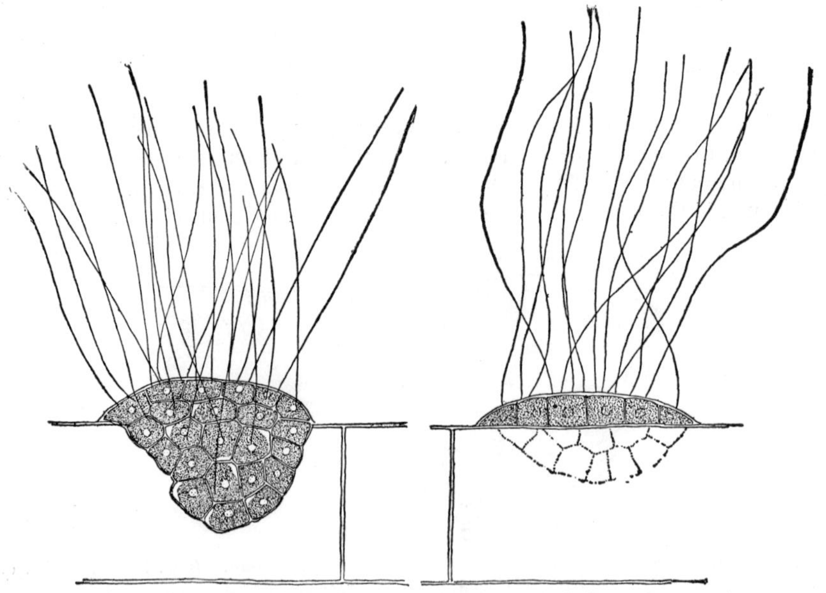

FIG. 64. *Chaetopeltis orbicularis* Berth. *A*, surface view of thallus. *B*, optical section of thallus. (× 485.)

4. **Chaetopeltis** Berthold, 1878. *Chaetopeltis* is an epiphytic alga with a discoid thallus but one cell in thickness. The cells are angular by mutual compression but with a tendency toward a radial arrangement. The thallus has a thin gelatinous sheath from the surface of which arise numerous long, exceedingly delicate, gelatinous setae (pseudocilia). The protoplast

[1] SCHERFFEL, 1908*A*.

has a single, laminate, entire, or perforate chloroplast covering most of the side toward the free face of the cell. There is one pyrenoid in a chloroplast.[1] On the side of the protoplast next to the substratum are two contractile vacuoles.[2]

Asexual reproduction is by division of the protoplast into two, four, or eight quadriflagellate zoospores. When first liberated, the zoospores are surrounded by a vesicle derived from the innermost layer of the parent-cell wall.[3]

A union of isogamous biflagellate gametes has also been recorded.[4]

C. orbicularis Berth. (Fig. 64) occurs in this country and usually as an epiphyte on coarse filamentous green algae. For a description of it, see Heering (1914).

Family 3. Chlorangiaceae

The Chlorangiaceae include certain sessile algae whose protoplasts are enclosed by narrow tough gelatinous sheaths and which are attached to the substratum by a stipe-like prolongation of the anterior portion of the sheath. Some members of the family are unicellular; others are colonial and with the cells united in dendroid colonies.

The cellular organization is more consistently primitive than in other families of the order. The cells usually have contractile vacuoles at their anterior end and often contain an eyespot in the anterior or in the equatorial region. The chloroplast is cup-shaped and lies at the posterior pole of a cell; it is usually massive and entire but it may be lobed and reticulate.

True palmella stages are rarely formed by members of this family. When cell division takes place, the protoplast divides into two, four, or eight daughter protoplasts which develop flagella. The flagellated cells may be liberated from the old cell sheath (*Stylosphaeridium*, *Malleochloris*) or they may remain permanently surrounded by a gelatinous sheath (*Prasinocladus*).

Sexual reproduction, so far as known, is isogamous and by a union of biflagellate gametes.

Genera belonging to this family are closer to the Volvocales than others in the order and have even been included in the Volvocales by those[5] who recognize the Tetrasporales as a distinct order. In spite of the fact that the flagella are not evident, the vegetative cells may be interpreted as a volvocaceous type permanently imprisoned within a gelatinous sheath. The apparent lack of cellulose in the sheath around protoplasts would seem

[1] Geitler, 1944; Bohlin; 1890; Huber, 1892; Möbius, 1890. [2] Korshikov, 1935.

[3] Korshikov, 1935; Huber, 1892; Möbius, 1888. [4] Möbius, 1888.

[5] Pascher, 1927*A*.

to indicate that the Chlorangiaceae are derived from Polyblepharidaceae rather than from Chlamydomonadaceae.

The four genera in the local flora may be differentiated as follows:

1. Cells solitary.. 2
1. Cells in dendroid colonies.. 3
   2. Stipe much narrower than cell.......................... 1. **Stylosphaeridium**
   2. Stipe nearly as broad as cell............................. 2. **Malleochloris**
3. Stipe much narrower than cell................................... 3. **Chlorangium**
3. Stipe nearly as broad as cell.................................... 4. **Prasinocladus**

1. **Stylosphaeridium** Geitler and Gimesi, 1925. The cells of this alga are solitary, though often gregarious, and almost always growing within the gelatinous envelope of *Coelosphaerium* or *Anabaena*. They grow directly upon cells of the host plant and are attached by a delicate stipe-like prolongation of the anterior end of the cellular sheath. The protoplast has a distinct polarity and always has the anterior end downward. The chloroplast is massive, contains one pyrenoid, and lies at the posterior pole of a cell. The colorless cytoplasm at the anterior end contains a nucleus and two contractile vacuoles. Vegetative cells do not have eyespots.

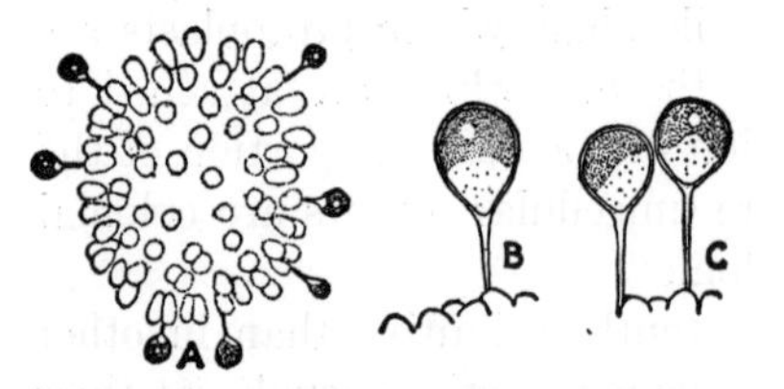

FIG. 65. *Stylosphaeridium stipitatum* (Bachm.) Geitler and Gimesi. *A*, cells epiphytic upon a colony of *Coelosphaerium*. *B–C*, individual cells. (*A*, × 400; *B–C*, × 1000.)

Reproduction is by division of the protoplast into two, four, or eight broadly ellipsoidal biflagellate zoospores. These lack an eyespot but have two contractile vacuoles at the anterior end and have a chloroplast with or without a pyrenoid.[1]

In sexual reproduction the protoplast divides to form 16 biflagellate gametes. Gametic union is isogamous.[2]

S. stipitatum (Bachm.) Geitler and Gimesi (Fig. 65) is widely distributed in the United States as an epiphyte upon planktonic thalli of *Coelosphaerium Naegelianum* Unger. For a description of it as *Characium stipitatum* (Bachm.) Wille, see G. M. Smith (1920).

2. **Malleochloris** Pascher, 1927. This unicellular alga has subspherical sessile cells which grow epiphytic upon Cladophoraceae. A cell is surrounded by a delicate sheath, sometimes of a reddish tinge, whose anterior end forms a broad stipe attaching the alga to the substratum. A stipe is about a third as long as a cell and is often impregnated with ferric hydroxide. A cup-shaped chloroplast containing a single pyrenoid lies toward the pos-

[1] BACHMANN, 1907; GEITLER, 1925*C*. [2] PASCHER, 1927*A*.

terior pole of a cell. The nucleus is central in position, and young cells often have an eyespot at the level of the nucleus. Old cells always lack an eyespot.

Asexual reproduction is by division of the protoplast into two or four quadriflagellate zoospores. Liberated zoospores swarm for a short time and then come to rest upon the substratum with the anterior end downward. Soon after coming to rest, they retract their flagella and secrete a gelatinous envelope.

In sexual reproduction the protoplast divides to form 8 or 16 quadriflagellate gametes whose union is isogamous. The octoflagellate zygote remains motile for a considerable time before losing the flagella, assuming a spherical shape, and secreting a sparsely verrucose wall.[1]

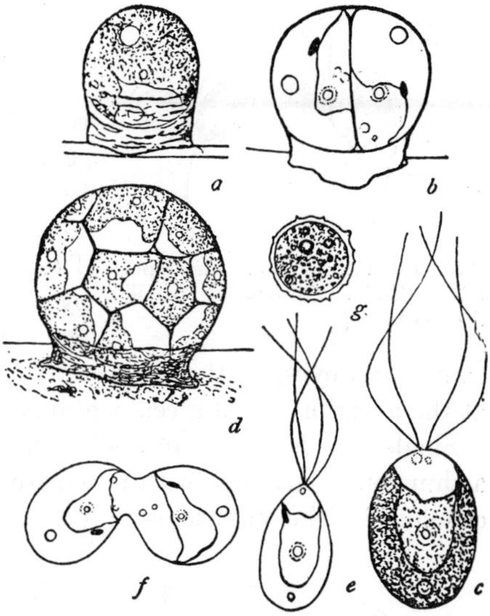

In this country *M. sessilis* Pascher (Fig. 66) is known only from Massachusetts.[2] For a description of it, see Pascher (1927*A*).

FIG. 66. *Malleochloris sessilis* Pascher. (*From Pascher*, 1927*A*.)

3. **Chlorangium** Stein, 1878. *Chlorangium* has dendroid colonies in which the cells are only at the tips of branches of a branching gelatinous stalk. The gelatinous branches are much narrower in diameter than are the cells. The cells are more or less spindle-shaped, with a distinct polarity, and one in which the anterior end is downward. The protoplast contains one or two laminate chloroplasts without pyrenoids, a median nucleus, and two contractile vacuoles in the end next to the stipe.[3]

Reproduction is by metamorphosis of the protoplast into a biflagellate motile cell which, after escaping from the enclosing sheath and swimming about for a time, comes to rest with the anterior end downward and develops a gelatinous stipe at this end.

Chlorangium is usually found epizoic upon crustaceans. Prof. G. W. Prescott writes that he has found *C. stentorium* Stein (Fig. 67) in Wisconsin. For a description of it, see Lemmermann (1915).

[1] PASCHER, 1927*A* [2] SMITH, G. M., 1933. [3] STEIN, 1876.

4. **Prasinocladus** Kuckuck, 1894. *Prasinocladus* is a dendroid colonial alga in which the cells are found only at the extremities of a branching tubular colonial envelope. There are two types of colony: the *lubricus* type in which the tubular envelope is solid and with transverse layers of gelatinous material; and the *subsalsus* type in which the envelope is hollow and with a succession of water-filled cylindrical compartments. The protoplasts are ovoid and with the inverted polarity characteristic of the family. The cup-shaped chloroplast lies toward the posterior pole of a cell. In vigorously growing cells it has four longitudinal lobes which are joined to one another by irregular anastomoses. There is a single large pyrenoid surrounded by starch plates except for the area facing the nucleus. Both young and old cells contain an eyespot at the level of the nucleus. The colorless cytoplasm at the anterior end of a cell contains a large noncontractile vacuole.[1]

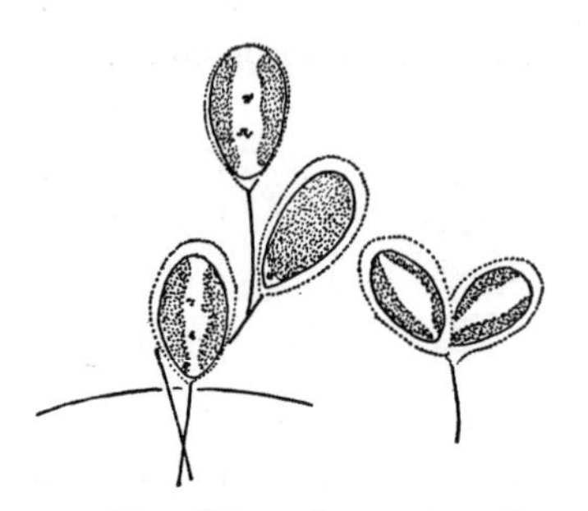

FIG. 67. *Chlorangium stentorium* (Ehr.) Stein. (*Drawn by G. W. Prescott.*)

Single-celled plants are sessile and with the protoplast surrounded by a thin tough envelope with two distinct layers,[1] neither of which contains cellulose.[2] The outer layer may break at its apex, and the protoplast, still

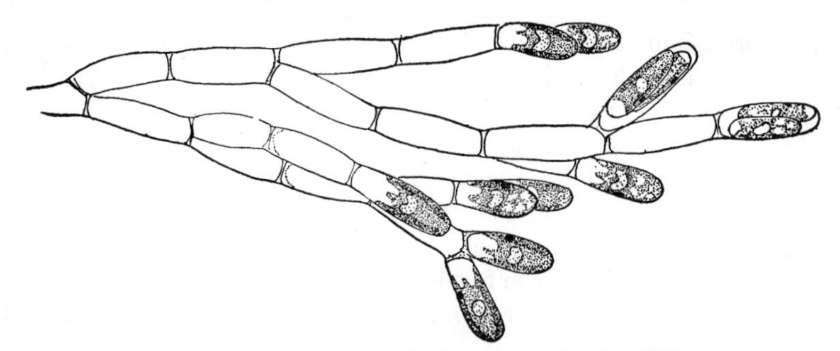

FIG. 68. *Prasinocladus lubricus* Kuckuck. (× 480.)

surrounded by the inner layer, migrate to its summit. Here, the anterior portion of the protoplast retracts somewhat from the enveloping layer and then secretes an additional layer of enveloping material. Repetition of the partial escape from the enclosing sheath, and the formation of new sheaths, result in formation of the transversely septate envelope characteristic of the *subsalsus* type.[2] From time to time, the cell at the end of

[1] LAMBERT, 1930; ZIMMERMANN, 1925*A*.

[2] ZIMMERMANN, 1925*A*.

the tube divides longitudinally into two daughter protoplasts each of which becomes the initial of a new branch. A pushing of the protoplasts to the top of the envelope results in formation of a colony of the *lubricus* type.[1]

Colonies of the *subsalsus* type have the protoplasts bearing flagella at the time when they slip to the apex of the enclosing sheath. If the protoplast remains surrounded by the inner layer of the sheath, there is soon a disappearance of the flagella. Sometimes there is a gelatinization of the inner layer of the sheath and an escape of the quadriflagellate protoplast. It swarms for a time and then settles down on some firm object, retracts or discards its flagella, and secretes a gelatinous envelope. *Prasinocladus* may also have the protoplast developing into a spherical aplanospore which remains enclosed within the old cell sheath.[2]

P. lubricus Kuck. (Fig. 68) is a marine organism which has been found on both the Atlantic and the Pacific coasts of the United States. It is generally found in brackish water, but it can live under practically fresh-water conditions. For a description of it, see Collins (1909).

Family 4. Coccomyxaceae

This family includes a number of genera that are not closely related to one another and have but little in common aside from the nonfilamentous organization of the colonies and the ability of their cells to divide vegetatively. Some of the genera, as *Elakatothrix* and *Ourococcus*, have been thought to be related to certain Chlorococcales, but their multiplication by vegetative cell division excludes them from the Chlorococcales. The Coccomyxaceae cannot be considered typical Tetrasporales since their cells do not have the capacity to metamorphose directly into a motile condition. In fact, most genera in the Coccomyxaceae are not even known to reproduce by means of zoospores. There is the alternative of placing them in a group called "unicellular Chlorophyceae of uncertain systematic position,"[3] but this does not show that there are distinct, though somewhat remote, affinities with the algae ordinarily placed in the Tetrasporales.

Most genera belonging to the Coccomyxaceae have ellipsoidal or fusiform cells that divide at right angles to the long axis. The cells are usually surrounded by a broad gelatinous envelope and may be joined to one another in colonies of limited or indefinite extent. There is generally a single parietal chloroplast within a cell, with or without a pyrenoid.

Cell division is the only method of reproduction in some of the genera. The daughter cells may separate immediately after cell division, or they

[1] Kuckuck, 1894; Zimmermann, 1925*A*.
[2] Dangeard, 1912; Zimmermann, 1925*A*. [3] Pascher, 1915*D*.

may remain embedded in the colonial envelope. Certain genera also reproduce by means of biflagellate zoospores or by means of aplanospores (autospores).

Sexual reproduction has not been observed in any of the genera belonging to the family.

The genera belonging to the family differ as follows:

1. Cells solitary, without a gelatinous envelope 2
1. Cells in colonies, usually with a gelatinous envelope 3
    2. Cells fusiform .... 1. **Ourococcus**
    2. Cells subcylindrical, with broadly rounded poles .... 4. **Nannochloris**
3. Colony a flat plate .... 6. **Dispora**
3. Colony not a flat plate 4
    4. Cells spherical to subspherical .... 7. **Chlorosarcina**
    4. Cells elongate 5
5. Cells fusiform .... 5. **Elakatothrix**
5. Cells ellipsoidal to cylindrical 6
    6. Colonies two- to eight-celled .... 3. **Dactylothece**
    6. Colonies many-celled and cells irregularly arranged .... 2. **Coccomyxa**

1. **Ourococcus** Grobéty, 1909; emend., Chodat, 1913 (*Keratococcus* Pascher, 1915). The cells of this alga are usually solitary. They are fusiform and straight, sigmoid, lunate, or irregularly bent. Both poles of a cell may be acutely pointed, or one pole may be pointed and the other broadly rounded. A cell contains a single parietal laminate chloroplast that usually has a single pyrenoid.

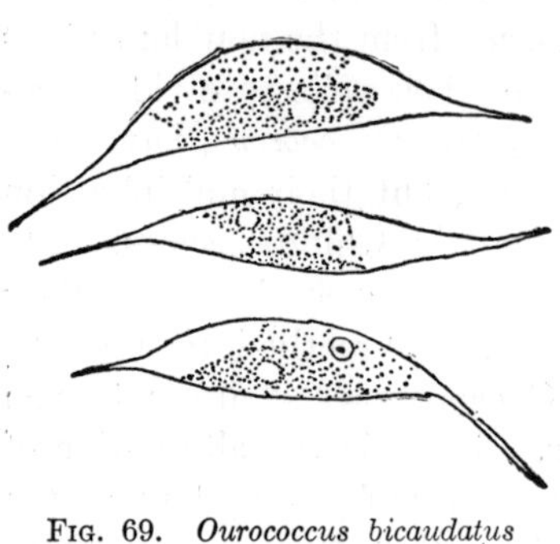

Fig. 69. *Ourococcus bicaudatus* Grobéty.

The usual method of multiplication is by transverse division. There may also be a division of the protoplast into two or four autospores that are liberated by a rupture of the parent-cell wall.[1]

O. bicaudatus Grobéty (Fig. 69) has been recorded from Massachusetts[2] and Kansas.[3] For a description of it, see Thompson (1938).

2. **Coccomyxa** Schmidle, 1901. The cells of *Coccomyxa* are ellipsoidal to cylindrical. They are surrounded by a gelatinous sheath and are usually united in colonies of indefinite extent by a confluence of cellular sheaths. The cells are irregularly distributed within the colonial matrix, and there is only a very slight tendency for them to have their long axes parallel with

[1] Chodat, 1913; Pascher, 1915*D*.
[2] Lewis, 1924. [3] Thompson, 1938.

one another. The chloroplast is a longitudinal plate partially encircling a cell. It lacks a pyrenoid.[1]

Cell division is transverse and usually in a plane diagonal to the long axis. There may also be a division of the cell contents into two or four aplanospores (autospores).[2]

Coccomyxa may grow as a free-living aerial alga, as an epiphyte on lichens, or as an endophyte within lichens. *C. dispar* Schmidle (Fig. 70) has been recorded from several localities in this country. For a description of it, see Pascher (1915*D*).

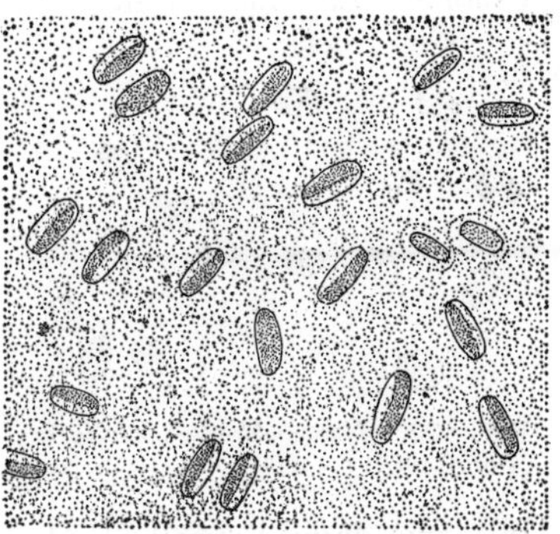

FIG. 70. *Coccomyxa dispar* Schmidle. (× 650.)

3. **Dactylothece** Lagerheim, 1883. The cells of *Dactylothece* are oblong-ellipsoidal and each is enclosed by a broad lamellated sheath. They may be solitary or united end to end in colonies of two or four that are surrounded by the sheath of the old parent cell. There is a parietal laminate chloroplast, about two-thirds as long as a cell, that is with or without a pyrenoid.

Cell division is always transverse and the two daughter cells become remote from each other through secretion of a gelatinous envelope.[3]

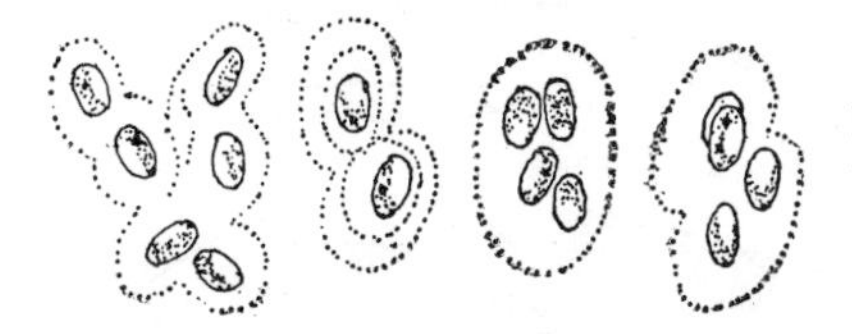

FIG. 71. *Dactylothece Braunii* Lagerh., a species not found in the United States. (*After G. S. West*, 1904.) (× 420.)

D. confluens (Kütz.) Lagerh. (Fig. 71) is listed[4] as occurring in Massachusetts. For a description of it, see Pascher (1915*D*).

4. **Nannochloris** Naumann, 1919. The cells of *Nannochloris* are solitary and without a gelatinous envelope. They are subspherical to subcylindrical and with a small discoid chloroplast near one pole. The chloroplast is without a pyrenoid.

Reproduction is by transverse division at right angles to the long axis.[5]

This unicellular alga resembles *Stichococcus* but never has the cells united in short filaments. *N. bacillaris* Naumann (Fig. 72) has been isolated in pure culture in this country. For a description of it, see Naumann (1919).

[1] ACTON, 1909; CHODAT, 1909, 1913; JAAG, 1933; SCHMIDLE, 1901.

[2] CHODAT, 1909, 1913.

[3] LAGERHEIM, 1883; PASCHER, 1915*D*; WEST, G. S., 1904. [4] COLLINS, 1909.

[5] NAUMANN, 1919

5. **Elakatothrix** Wille, 1898 (*Fusola* Snow, 1903). Mature cells of *Elakatothrix* are fusiform. They are surrounded by a broad homogeneous gelatinous envelope and are usually in colonies of two, four, or many cells. All cells in a colony have their long axes more or less parallel to one another. Colonies may be free-floating at all times, or epiphytic when young and free-floating when mature.[1] A colony increases in size as a result of cell division, and the two daughter cells formed by transverse division of a parent cell lie axial to each other and have one pole broadly rounded for some time following division. The chloroplast is laminate and partially encircles a cell. It may contain one or two pyrenoids.

Fig. 72. *Nannochloris bacillaris* Naumann. (× 800.)

Reproduction is by transverse division into two daughter cells that may remain within the gelatinous envelope or may escape from it. There is also a formation of brownish akinetes.[2]

Some species of *Elakatothrix* are planktonic, others are nonplanktonic. *E. americana* Wille, *E. gelatinosa* Wille (Fig. 73*A–B*), and *E. viridis* (Snow) Printz (Fig. 73*C–F*) have been found in this country. For a description of *E. americana*, see Wille (1899); for the other two, see G. M. Smith (1920).

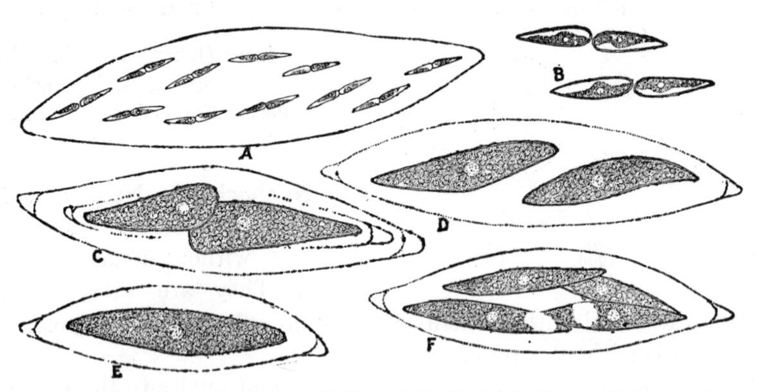

Fig. 73. *A–B*, *Elakatothrix gelatinosa* Wille. *C–F*, *E. viridis* (Snow) Printz. (*A*, × 333; *B*, × 666; *C–F*, × 866.)

6. **Dispora** Printz, 1914. *Dispora* has free-floating plate-like colonies, one cell in thickness, in which the cells tend to lie in groups of four within a colorless gelatinous matrix. The individual cells, when lying free from one another, are broadly ellipsoidal and with a thin wall. The protoplast contains a single parietal broadly disciform chloroplast without pyrenoids. Growth of a colony is by cell division in two planes.[3]

[1] Wille, 1898, 1899. [2] Wille, 1909. [3] Printz, 1914.

The only method of reproduction is by fragmentation of colonies.

As far as vegetative organization is concerned, this genus is similar to *Crucigenia,* but the two differ in method of growth of a colony. In *Dispora* this is due to vegetative cell division; in *Crucigenia* it is due to formation of four autospores by a cell and their development into vegetative cells without escaping from a colony. Prof. G. W. Prescott writes that he has found *D. crucigenioides* Printz (Fig. 74) in Wisconsin. For a description of it, see Printz (1914).

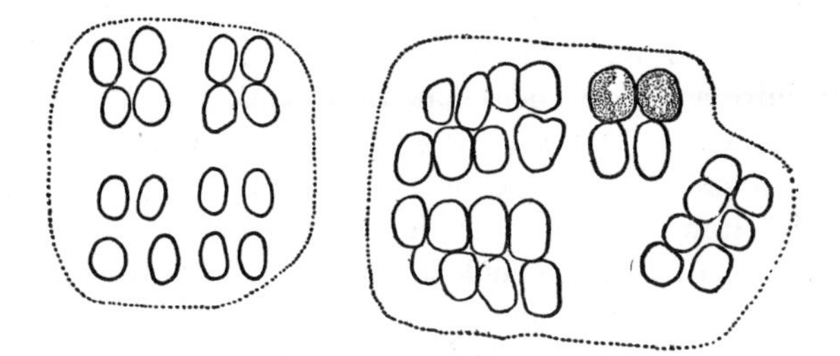

FIG. 74. *Dispora crucigenioides* Printz. (*Drawn by G. W. Prescott.*)

7. **Chlorosarcina** Gerneck, 1907. The cells of *Chlorosarcina* are free-living or are endophytic within various aquatic phanerogams. They are spherical to angular by mutual compression and are joined to one another in small packets with or without a narrow gelatinous sheath. Colonies increase in size by cell division, but they dissociate into fragments or into individual cells before they attain any appreciable size. The cell walls are relatively thin. Each cell contains a single parietal cup-shaped chloroplast that may be solid or reticulate. One or more pyrenoids may be present in a chloroplast, or pyrenoids may be lacking.[1]

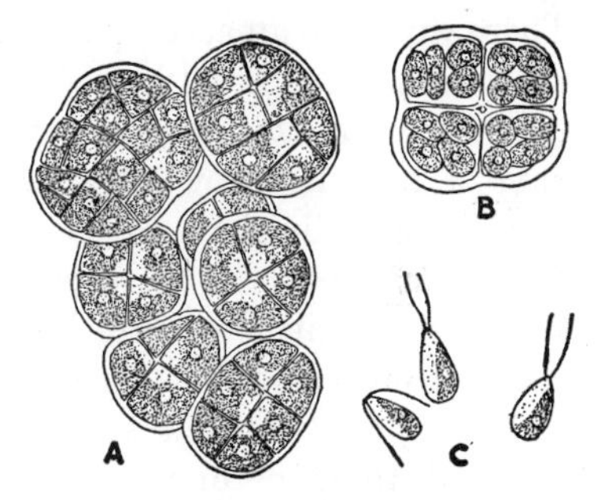

FIG. 75. *Chlorosarcina consociata* (Klebs) G. M. Smith. (× 975.)

Vegetative multiplication by fragmentation is of frequent occurrence. Asexual reproduction is by means of biflagellate zoospores. Four or eight zoospores are formed within a cell and are liberated by a softening at one side of the parent-cell wall.

The three species found in the United States are *C. consociata* (Klebs) G. M. Smith (Fig. 75), *C. lacustris* (Snow) Lemm., and *C. parvula* (Snow) Lemm. For a description of *C. consociata* as *Chlorosphaera consociata* Klebs and for the other two as species of *Chlorosarcina*, see Lemmermann (1915*A*).

[1] ARTARI, 1892; GERNECK, 1907; SNOW, 1903.

ORDER 3. ULOTRICHALES

The Ulotrichales are filamentous algae in which the cells are usually uninucleate and with a single laminate parietal chloroplast. The usual method of asexual reproduction is by means of zoospores, which may have two or four flagella; but aplanospores and akinetes are not at all uncommon. Gametic union is found in many genera of the order and ranges from isogamy to oögamy.

The thallus may be a simple filament, but more often it is a branched filament. Branching filaments of certain genera have the various branches aggregated into a subparenchymatous tissue. Some genera have the filament reduced to a few cells or even a single cell.

The usual taxonomic arrangement of the Chlorophyceae is to place the Chlorococcales immediately after the Tetrasporales and before the Ulotrichales. Such an arrangement implies that the Ulotrichales are not immediately derived from the Tetrasporales but have come from the Tetrasporales through the Chlorococcales. The arrangement followed in this book is designed to emphasize the fact that the Ulotrichales have been evolved directly from the Tetrasporales. Development of the tetrasporaceous type of plant body into a ulotrichaceous filamentous type would result from a limiting of cell division to one plane only, accompanied by a failure to develop gelatinous material between the two daughter cells. There would be an accompanying loss from the protoplast of such primitive characters as contractile vacuoles and eyespots and the ability to return directly to a motile condition. Certain Tetrasporales, as *Palmodictyon*, have a tendency toward a filamentous organization in their restriction of most cell divisions to one plane. Such truly ulotrichaceous algae as *Geminella* appear to be so close to the Tetrasporales that they have not lost the capacity for secreting gelatinous material between the daughter cells. However, it is much more probable that *Geminella* and other Chlorophyceae with a similar organization are Ulotrichales that have reverted to a permanently filamentous palmella stage. The frequent occurrence of palmella stages among the primitive families of the order (Ulotrichaceae, Chaetophoraceae), as contrasted with their rare occurrence among advanced families (Coleochaetaceae, Trentepohliaceae, Cladophoraceae), is good evidence that the Ulotrichales have been derived from ancestors with a palmelloid organization.

The branching so characteristic of a majority of genera in the order represents a more advanced condition than the simple filament. Ulotrichales with a branching thallus often have the plant body divided into a prostrate and an erect portion. Because of this, they have been thought to have an organization fundamentally different from those with unbranched filaments[1] and have been placed as a distinct order. Such differences are, however, not of sufficient magnitude to warrant the establishment of separate orders for the branched and unbranched genera. Germlings of genera with a branched thallus are usually unbranched during the very early stages of their development. None of these branching genera have the formation of branches restricted to an apical cell, and the great majority of branches are developed from intercalary and not from terminal cells of the filament. Mature thalli of branching forms show great diversity from genus to genus. Many genera have a freely branched erect portion in which all branches are of the same size (*Stigeoclonium*, *Microthamnion*), or in which there is a differentiation into large primary branches and small lateral ones (*Draparnaldia*). Other branching genera have the thallus assuming the form of a flattened disk (*Protoderma*, *Ulvella*) and one in which the cells are so densely crowded that the branches are quite obscure. Genera epiphytic upon other algae, or upon water plants, often have the thallus reduced to a few irregular branches (*Aphanochaete*, *Chaetonema*) or to an irregularly shaped few-celled mass. *Chaetosphaeridium* and *Dicranochaete* are epiphytic Ulotrichales in which the thallus has been reduced to a unicellular structure.

Cells of the Ulotrichales have the protoplasts surrounded by a wall in which the innermost layer is composed chiefly of cellulose. The layer immediately outside of the cellulose is of a pectic nature; either a thin tough layer, or broad and watery (*Draparnaldia*, *Geminella*). In certain cases, as *Cylindrocapsa*, the cellulose portion of the cell wall is stratified. The successive layers of wall material around the protoplast may be continuous from one end of the cell to the other, or the wall may consist of two overlapping halves. In the first type of cell wall, characteristic of most Ulotrichales, there is a longitudinal stretching of the whole wall when daughter cells elongate. In the second type of cell wall, found in *Microspora*, *Radiofilum*, and *Binuclearia*, elongation of the two daughter protoplasts causes the overlapping halves of the wall to pull away from each other. This is generally followed by the interpolation of an H-shaped piece of wall material between the two daughter protoplasts. Cell walls of Ulotrichales may also bear setae, quite different in structure from the hair-like attenuation at the ends of branches in *Chaetophora* or the modified terminal cells in branches of *Fridaea* or *Chaetonema*. The setae borne by cells of *Coleochaete* are very delicate cytoplasmic processes which project through a pore in the wall and are surrounded in the basal portion by a sheath of gelatinous material. The setae of *Chaetosphaeridium* appear to be of a similar nature. Those of *Dicranochaete* also have a cyto-

[1] Fritsch, 1916, 1935; West and Fritsch, 1927.

plasmic thread but one which is ensheathed with gelatinous material throughout its entire length.

The primitive type of cell in the Ulotrichales is undoubtedly one containing a single nucleus and a single laminate parietal chloroplast. If the cells are large, the chloroplast is often a median girdle, with entire or lobed margins, completely or partially encircling the protoplast. It may contain one to several pyrenoids or lack them entirely. Sometimes, however, the chloroplasts lie next the end and side walls and are entire or perforate. Small cells, especially those at the ends of branches, usually have the chloroplast covering all faces of the cell. Most Ulotrichales have green thalli; a few have a regular development of hematochrome in vegetative cells (*Trentepohlia*) or in cells containing gametes (*Cylindrocapsa*). Advanced members of the order may have multinucleate instead of uninucleate protoplasts (Trentepohliaceae).

Vegetative multiplication of the plant may result from an accidental breaking of the filament. This is usually due to some external agency, as water currents or aquatic animals, rather than to an insecure adhesion between the cells (*Stichococcus*) or to a breaking because of the formation of zoospores (*Microspora*). If the accidentally broken plant is one with unbranched filaments, the portions into which it is broken may grow indefinitely. Ulotrichales with a branching thallus often have but little growth of the severed portion.

Asexual reproduction by means of bi- or quadriflagellate zoospores is of regular occurrence in most members of the order. Zoospores may be formed in sporangial cells of different shape from vegetative cells (Trentepohliaceae) or directly within vegetative cells. Unbranched genera usually have all cells in the filament, except those comprising the holdfast, potentially capable of producing zoospores: branching genera, as *Stigeoclonium* and *Draparnaldia*, often have spore production limited to the smaller branches. Typically, there is a division of the protoplast to form 2, 4, 8, or 16 zoospores, but if the cell is small there may be the production of but a single zoospore. The production of one zoospore only is because of the cell's small size and is not to be compared with the direct metamorphosis of vegetative cells into zoospores found in Tetrasporales. Some genera, as *Ulothrix*, form zoospores of two types, macrozoospores and microzoospores. These differ not only in size but may also differ in such morphological characters as the number of flagella and the position of the eyespot and in such physiological characters as the length of the swarming period or phototactic responses. Zoospores of the Ulotrichales are always naked and usually with an anterior eyespot and a laminate chloroplast at the posterior end. Even in those species that seem to be permanently free-floating (*Cylindrocapsa*, *Microspora*), the zoospore comes to rest upon some firm object before it develops into a filament. Most of the genera reproducing by means of zoospores may also have a development of aplanospores instead of zoospores. Aplanospores may be liberated by the decay of the old mother-cell wall surrounding them, or they may develop into filaments *in situ*. Some genera, as *Microspora*, have a rounding up of the entire protoplast and the development of a single aplanospore within a cell. Such aplanospores do not seem to be abortive zoospores, and they often are hypnospores. A number of genera are also known to form akinetes.

The usual method of sexual reproduction is by the union of like-sized motile gametes. These gametes are ordinarily formed within cells similar in shape and structure to vegetative cells, but the Trentepohliaceae form gametes with special gametangia. Although the majority of Ulotrichales are isogamous, there has been a development of oögamy in at least three independent phylogenetic lines within the order. Beginning with the isogamy of the more primitive Chaetophoraceae, there has been a development of a pronounced type of anisogamy in *Aphanochaete* and this has led, in turn, to a complete loss of motility in the large female gamete (egg) discharged from the oögonium of *Chaetonema*. *Coleochaete* represents a higher type of oögamy where the egg is retained within an oögonium. *Cylindrocapsa* and *Sphaeroplea* are the only unbranched Ulotrichales whose sexual reproduction is oögamous, but they differ so much from each other that their oögamy must have been developed independently.

Several Ulotrichales are said to have a parthenogenetic development of gametes into thalli. Although this is doubtless true in certain cases, such assertions should be accepted with caution, especially in the case of genera where zoospores and gametes have the same number of flagella and might be mistaken for each other.

Zygotes are naked when first formed but soon secrete a wall. Oögamous genera have zygotes in which there is considerable accumulation of reserve food and in which the protoplast often becomes colored with hematochrome. As is the case with many other Chlorophyceae, there is usually a considerable interval between union of gametes and germination of the zygote. Meiosis of the zygote nucleus has been found in *Coleochaete*[1] and on *Ulothrix*,[2] and the production of four zoospores by germinating zygotes of several other genera would seem to indicate that division of their zygote nuclei is also meiotic. This means that these algae have a life cycle in which a haploid multicellular gametophyte alternates with a one-celled diploid phase. On the other hand, at least two genera are known to have an alternation of a multicellular haploid generation and a multicellular diploid generation. In one of these genera (*Draparnaldiopsis*[3]), the alternation is isomorphic; in the other (*Stigeoclonium*[4]), it is heteromorphic.

[1] ALLEN, 1905. [2] GROSSE, 1931. [3] SINGH, 1945. [4] JULLER, 1937.

The Ulotrichales may be divided into two suborders; the Ulotrichi neae and the Sphaeropleineae.

SUBORDER 1. ULOTRICHINEAE

The Ulotrichales in which the cells are uninucleate and with a parietal chloroplast constitute a natural group, the Ulotrichineae, but one in which there is great diversity in structure of the thallus. The Ulvaceae are often included in this assemblage, because of similarities in cell structure, but the thallus structure is so fundamentally different that they are better placed in a separate order.

The Ulotrichineae are usually divided into families distinguished from one another on the basis of thallus structure and on method of sexual reproduction. In the families recognized on pages to follow, isogamy and oögamy are held to be characteristics insufficient for establishment of separate families unless they are supplemented by features in the structure of the thallus. This is in accordance with the universally accepted treatment of the Volvocaceae that places both isogamous and oögamous genera in the same family.

FAMILY 1. ULOTRICHACEAE

Genera belonging to this family have their cells seriately united in unbranched filaments. The unbranched filament is thought to be the primitive type of thallus among Ulotrichales, but certain genera with unbranched filaments may be retrogressive from a branching type. All genera referred to the Ulotrichaceae have a cell wall that is not composed of two overlapping halves. Most genera have cylindrical cells whose flattened ends abut on one another. More rarely the cells have broadly rounded poles and a length greater than or less than the breadth. Genera with the latter type of cell usually have a broad gelatinous sheath within which the cells are in contact with one another (*Radiofilum*) or lie some distance from one another (*Geminella*). Young filaments of aquatic species are usually sessile and affixed by a discoid holdfast developed by the lowermost cell. The filament may remain sessile throughout its further development (*Uronema*), or the terminal portion may break away and continue growth as a free-floating filament. Repeated vegetative multiplication in such free-floating individuals eventually gives a tangled mass which appears to have been free-floating from the beginning.

Cells of Ulotrichaceae are uninucleate and with a girdle-shaped chloroplast partially encircling the protoplast. Most genera have chloroplasts with pyrenoids, and species with large cells often have more than one pyrenoid.

Increase in the number of filaments may be due to an accidental breaking and a continuation of growth by the various fragments. This is the usual method of reproduction in genera with a loose adhesion of the cells and is found both in terrestrial (*Stichococcus*) and strictly aquatic (*Geminella*) members of the family. Palmella stages are of rather rare occurrence among the Ulotrichaceae.

Asexual reproduction by means of zoospores is known for most genera in the family. There may be a regular production of a single zoospore by a vegetative cell (*Hormidium*), or the protoplast may divide to form 2, 4, 8, or 16 zoospores (*Ulothrix, Uronema*). The zoospores are bi- or quadriflagellate. All zoospores may be of one type, or there may be a formation of two types differing from each other in size and in other morphological and physiological characters. Zoospores generally escape through a round pore in the parent-cell wall. They may be enclosed by a delicate vesicle when first liberated or may be free from the instant they are discharged. The swarming period rarely lasts for more than a day, after which the zoospore comes to rest and secretes a cell wall. The germling is usually affixed to a substratum by a holdfast, but there may not be a development of a holdfast (*Hormidium*). Aplanospores are of common occurrence in the Ulotrichaceae and may be formed singly within a cell or in multiples of two. The aplanospore may germinate while it is still enclosed by the parent-cell wall, but usually there is a disintegration of the parent-cell wall before germination takes place. Akinetes have also been recorded for a majority of genera in the family, but these are of much rarer occurrence than are aplanospores. An akinete may germinate directly into a new filament, or its protoplast may divide to form a number of zoospores.

Sexual reproduction is known for but few genera, and all cases thus far observed have a fusion of motile gametes in pairs. The uniting gametes, which may be isogamous or slightly anisogamous, are always biflagellate and naked. The zygote formed by their fusion may retain the flagella for a time, but eventually it assumes a spherical shape and secretes a wall. Germinating zygotes may produce either zoospores or aplanospores. Four spores are usually formed at the time of germination, but the number may be greater.

Generic differences among Ulotrichaceae are based upon structure of the cells and structure of the filament. Since the structure of filaments is subject to considerable variation in certain genera, there is more or less disagreement among phycologists as to the precise generic limits in certain cases.

The genera found in this country differ as follows:

1. Filaments without a gelatinous sheath.. 2
1. Filaments with a broad gelatinous sheath.. 7
   2. Filaments never with more than 10 cells..................................... 3
   2. Filaments usually with more than 10 cells................................... 4

3. Ends of filaments pointed.................................... 3. **Raphidionema**
3. Ends of filaments not pointed....... .... 4. **Stichococcus**
   4. With cells in pairs... 7. **Binuclearia**
   4. Not with cells in pairs... 5
5. Filaments without a basal cell............................... 6. **Hormidium**
5. Filaments with a basal cell... 6
   6. Free end of filament pointed.................................... 2. **Uronema**
   6. Free end of filament not pointed................................ 1. **Ulothrix**
7. Cells cylindrical, often remote from one another and in pairs..... 5. **Geminella**
7. Cells subspherical, always touching one another................. 8. **Radiofilum**

1. **Ulothrix** Kützing, 1833. *Ulothrix* has unbranched filaments of indefinite length which, at least when young, are sessile and affixed to the substratum by a special basal cell. All cells of a filament, except the basal cell, are capable of division and of forming zoospores or gametes. The cell walls may be thick or thin, and homogeneous or stratified. The cells are always uninucleate and with a single girdle-shaped chloroplast that partly or completely encircles the protoplast. According to the species, a chloroplast extends the whole length of a cell, or only a part of its length, and contains one or several pyrenoids.

Vegetative multiplication may take place by an accidental breaking of a filament, but there is never the regular dissociation into few-celled fragments that is found in *Hormidium*. All cells but the holdfast cell are capable of producing zoospores, but those in the distal portion often produce them in advance of those in the proximal portion. Species with narrow cells produce one, two, or four zoospores per cell and all zoospores from a filament are alike. Species with broad cells, as *U. zonata* (Weber and Mohr) Kütz., produce two types of zoospores;[1] two, four, or eight large quadriflagellate zoospores with an anterior eyespot or 4, 8, 16, or 32 small quadriflagellate zoospores with a median eyespot. The macrozoospores swarm for 24 hr. or so, and temperature has but little effect on the swarming; microzoospores swarm 2 to 6 days if the temperature is below 10° C., otherwise they disintegrate. The reported occurrence of biflagellated zoospores has recently been denied.[2] The zoospores are usually liberated through a pore in the lateral wall and are surrounded by a delicate vesicle when first liberated. Zoospores which are not discharged from the parent-cell wall may each secrete a wall and so become thin-walled aplanospores. These aplanospores usually germinate within the parent-cell wall.[3] There may also be a rounding up of the entire protoplast within a cell to form a large thick-walled aplanospore.[4]

Gametes are formed by the same repeated bipartition of the protoplast as in zoospore formation, but the number of daughter protoplasts produced

[1] Klebs, 1896; Pascher, 1907. [2] Grosse, 1931. [3] Dodel, 1876.
[4] West, G. S., 1916.

is 8, 16, 32, or 64.[1] As is the case with zoospores, the mass of gametes is surrounded by a delicate vesicle when first discharged from the parent cell, but the vesicle soon disappears. The gametes are biflagellate, pyriform, and with an eyespot. They fuse in pairs with one another, but fusion takes place only between gametes coming from different filaments.[2] There is no parthenogenetic development of gametes into vegetative filaments.[2] The zygote remains motile for a short time after its formation, but it soon comes to rest, secretes a thick lamellated wall, and enters upon a resting period during which there is considerable accumulation of reserve food. Division of a zygote nucleus is meiotic. A germinating zygote has a division of the protoplast into 4 to 14 (16?) uninucleate protoplasts which may develop into aplanospores[3] or into zoospores.[4]

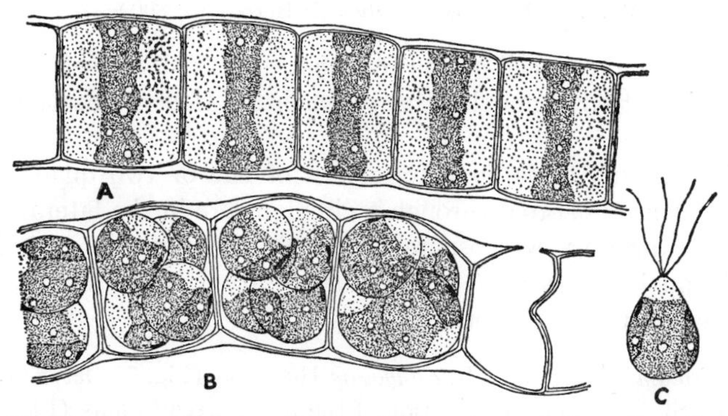

FIG. 76. *Ulothrix zonata* (Weber and Mohr) Kütz. *A*, vegetative cells. *B*, formation of macrozoospores. *C*, macrozoospore. (× 1000.)

The life history of *Ulothrix* consists of an alternation of a haploid multicellular generation with a one-celled diploid phase. In the assemblage of haploid individuals there are certain ones that produce gametes only, others that produce both gametes and zoospores, and still others that produce zoospores only.[5]

Ulothrix is usually found in quiet or running water, but it also grows on cliffs moistened by the spray from waterfalls. Some species, especially *U. zonata* (Weber and Mohr) Kütz., are distinctly cold-water plants, appearing in early spring, disappearing during the summer, and then reappearing in the autumn. At times it is difficult to distinguish between *Hormidium* and those species of *Ulothrix* in which the filaments break loose and grow in free-floating masses. However, these *Ulothrix* species do not have the filaments breaking up into short filaments of a few cells as is the case with *Hormidium*. The following species have been found in

[1] DODEL, 1876; KLEBS, 1896; PASCHER, 1907; WEST, G. S., 1916.
[2] GROSSE, 1931.
[3] JÖRSTAD, 1919; KLEBS, 1896. [4] DODEL, 1876. [5] GROSSE, 1931.

this country: *U. aequalis* Kütz., *U. cylindrica* Prescott, *U. moniliformis* Kütz., *U. oscillarina* Kütz., *U. subconstricta* G. S. West, *U. tenerrima* Kütz., *U. tenuissima* Kütz., *U. variabilis* Kütz., and *U zonata* (Weber and Mohr) Kütz. (Fig. 76). For a description of *U. cylindrica*, see Prescott (1944); for *U. subconstricta*, see G. M. Smith (1920); for the other species, see Collins (1909).

2. **Uronema** Lagerheim, 1887. *Uronema* is quite like *Ulothrix* in general appearance but has filaments that are always sessile, are of limited growth,

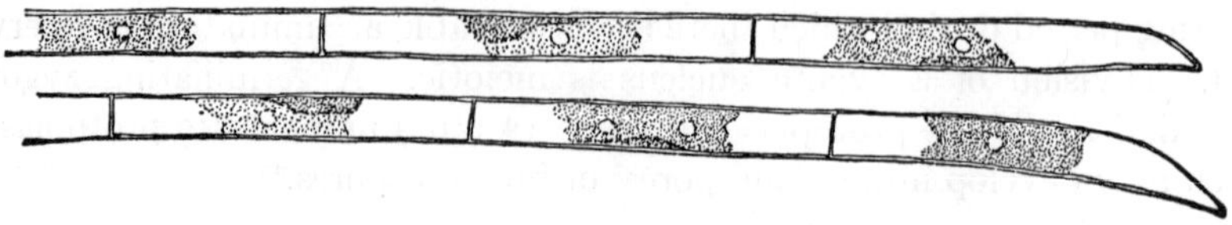

FIG. 77. *Uronema elongatum* Hodgetts. (× 800.)

and have an acuminate cell at the free end. The chloroplast is laminate, parietal, and partially encircles the protoplast. It contains one to three pyrenoids.

Asexual reproduction is by a formation of one or two quadriflagellate zoospores that are liberated through a circular pore in the lateral wall of a cell. After swarming, a zoospore comes to rest with its anterior pole downward and becomes affixed to the substratum by a discoid holdfast. Aplanospores may also be formed singly within a cell.[1]

U. confervicolum Lagerh. and *U. elongatum* Hodgetts (Fig. 77) have been found in the United States. For a description of the former, see Collins (1909); for the latter, see Hodgetts (1918).

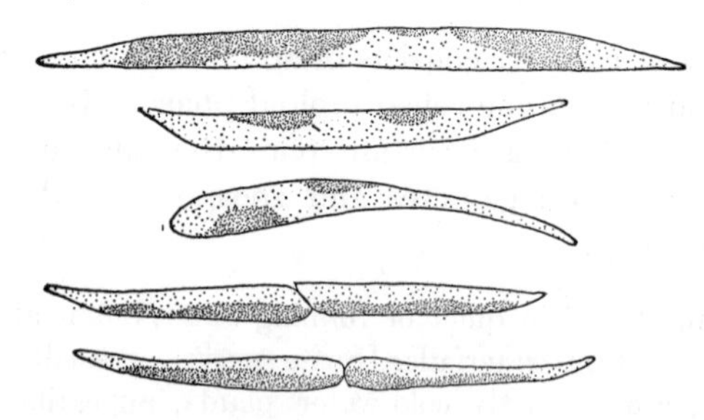

FIG. 78. *Raphidonema nivale* Lagerh. (*After Chodat*, 1896.)

3. **Raphidonema** Lagerheim, 1892. Filaments of this alga are unbranched and rarely more than a half-dozen cells in length. They frequently dissociate into solitary cells. The cells are cylindrical and with the ends not abutting on other cells gradually tapering and acutely pointed. The chloroplast is parietal, laminate, more or less lobed, and without pyrenoids.[2]

[1] HODGETTS, 1918; LAGERHEIM, 1887. [2] LAGERHEIM, 1892*A*; CHODAT, 1896.

Reproduction is exclusively vegetative and by fragmentation of filaments.

Most species grow in permanent snow fields on high mountain peaks, and they may be present in sufficient abundance to color the snow green. *R. nivale* Lagerh. (Fig. 78) and *R. tatrae* var. *yellowstonensis* Kol have been found in Yellowstone National Park.[1] For a description of the former, see Heering (1914); for the latter, see Kol (1941).

4. **Stichococcus** Nägeli, 1849. Cells of this alga are cylindrical and with rounded poles. According to the species, the length equals the breadth or is more than eight times the breadth. The cells are united end to end in filaments without gelatinous sheaths but, because of the tendency of the cells to break away from one another, the filaments are rarely more than two or three cells in length. Frequently no filamentous organization is evident because of dissociation into individual cells. The chloroplast is laminate, parietal, and encircles less than half a cell. There are no pyrenoids.

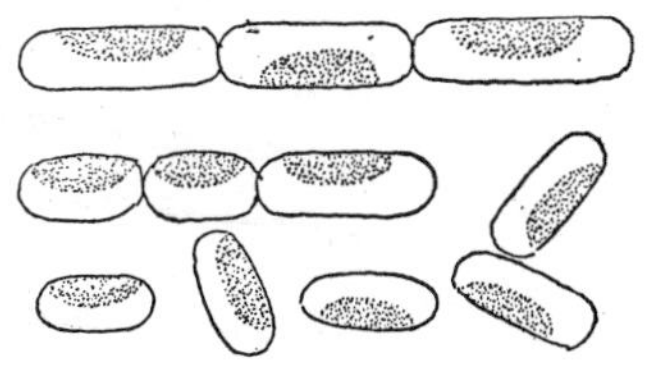

FIG. 79. *Stichococcus bacillaris* Näg. (× 1300.)

Reproduction is solely by fragmentation, and there is never a formation of zoospores, aplanospores, or zoogametes.[2]

Some phycologists[3] expand the generic concept to include most species of *Hormidium*, but this inclusion of zoosporic forms whose chloroplasts have pyrenoids is illogical.

Stichococcus grows on damp soil. *S. bacillaris* Näg. (Fig. 79) has been recorded from numerous localities in this country, but it is very probable that in the majority of cases the material thus identified was a *Stichococcus* stage of *Hormidium*. For a description of *S. bacillaris*, see Grintzesco and Péterfi (1932).

5. **Geminella** Turpin, 1828; emend., Lagerheim, 1883. *Geminella* has free-floating or sessile filaments surrounded by a tubular envelope of homogeneous structure. The individual cells are usually cylindrical and with broadly rounded poles. They may be remote from one another and equidistant or in pairs, or they may be so close together that their poles touch. The equatorial region of a cell contains a laminate chloroplast, usually with one pyrenoid, that partially encircles the protoplast.[4]

The usual method of reproduction is by a fragmentation of the filaments. Zoospores are not definitely known for this genus. Akinetes with brownish walls have been reported for one species.[5]

[1] KOL, 1941. [2] GRINTZESCO and PÉTERFI, 1932. [3] GAY, 1891; HAZEN, 1902.
[4] LAGERHEIM, 1883. [5] WEST, G. S., 1904.

Ulothrix may, at times, have *Hormospora* stages in which the filaments have broad gelatinous sheaths, but *Hormospora* stages of *Ulothrix* do not have cells with broadly rounded poles or the cells remote from one another. The species found in the United States are *G. crenulatocollis* Prescott, *G. ellipsoidea* (Prescott) comb. nov., *G. flavescens* (G. S. West) Wille, *G. interrupta* Turp. (Fig. 80*D*), *G. minor* (Näg.) Hansg. (Fig. 80*C*), *G. mutabilis* (Näg.) Wille, *G. ordinata* (W. and G. S. West) Heering, and *G. spiralis* (Chod.) G. M. Smith (Fig. 80*A–B*). For a description of *G. crenulatocollis*, see Prescott (1944); for *G. ellipsoidea*, see Prescott

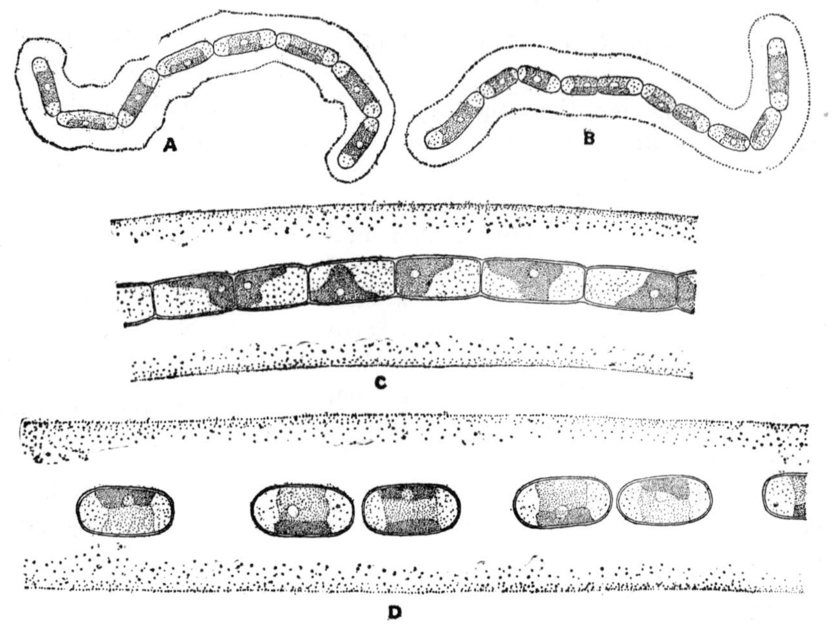

FIG. 80. *A–B*, *Geminella spiralis* (Chod.) G. M. Smith. *C*, *G. minor* (Näg.) Hansg. *D*, *G. interrupta* Turp. (*A–B*, × 650; *C–D*, × 1000.)

(1944), under *Hormidiopsis ellipsoideum* Prescott; for *G. flavescens*, see Prescott and Croasdale (1942); for *G. spiralis*, see Heering (1914), under *Gloeotila contorta;* for the remaining species, see Heering (1944), under *Geminella.*

6. **Hormidium** Kützing, 1843; emend., Klebs, 1896. The cells of this alga are cylindrical and united end to end in unbranched filaments that are without the basal cell differentiated into a holdfast. A filament is not surrounded by a gelatinous sheath. The cells are uninucleate and with a single laminate parietal chloroplast that does not encircle more than half of a protoplast. There is one pyrenoid within a chloroplast. In certain species, there is a marked tendency to dissociate into short *Stichococcus*-like filaments of two to six cells.

Asexual reproduction is by means of biflagellate zoospores formed singly within a cell and liberated through a pore in the lateral cell wall.[1] Aplanospores may be formed singly within a cell.

An isogamous fusion of biflagellate gametes has been reported for one species.[2]

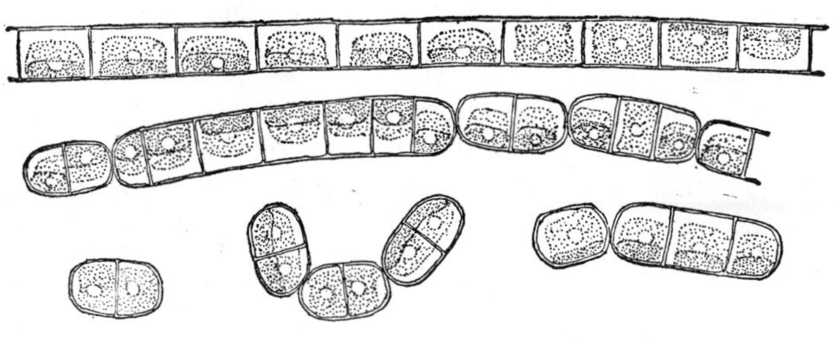

FIG. 81. *Hormidium subtile* (Kütz.) Heering. (× 975.)

Hormidium is closely related to *Ulothrix* but differs in that the filaments are not attached by a basal cell, and by the fact that zoospores are biflagellate. Some species are terrestrial or aerial; other species are aquatic and grow in free-floating masses. The species found in the United States are *H. flaccidum* (Kütz.) A. Br., *H fluitans* (Gay) Heering; *H. Klebsii* G. M. Smith, *H. rivulare* Kütz., *H. scopulinus* (Hazen) comb. nov. (*Stichococcus scopulinus* Hazen), and *H. subtile* (Kütz.) Heering (Fig. 81). For a description of *H. Klebsii*, see Heering (1914), under *H. nitens;* for descriptions of the others as species of *Stichococcus*, see Hazen (1902).

7. **Binuclearia** Wittrock, 1886. The cells of *Binuclearia* are cylindrical, with flattened end walls, and serially united in unbranched filaments that are without a gelatinous envelope. The protoplasts are cylindrical, with

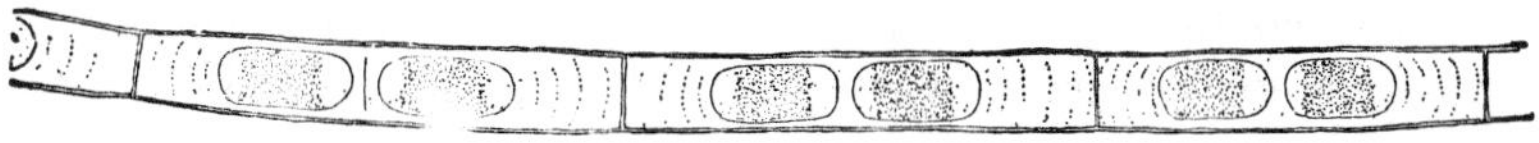

FIG. 82. *Binucleria tatrana* Wittrock. (× 830.)

broadly rounded ends, and much shorter than the cell wall. The intervening space between protoplast and end walls of a cell is filled with stratified layers of a gelatinous substance. When a cell divides, the two daughter protoplasts are separated by a thin wall for some time following division; consequently, cells frequently appear as if there is a pair of protoplasts within a cell. Eventually there is an accumulation of gelatinous material next to the new cross wall that causes the protoplasts to lie equidistant from one another. A protoplast contains a single laminate chloroplast, without a pyrenoid, that completely encircles the cell contents.[3]

[1] HAZEN, 1902; KLEBS, 1896; PASCHER, 1907; WILLE, 1912.
[2] WILLE, 1912. [3] SCHRÖDER, 1898*A*, WEST, G. S., 1904; WITTROCK, 1886.

The discovery of sessile germlings affixed by a discoid holdfast[1] indicates that zoospores are formed by this alga, but as yet they have not been observed. Spherical thick-walled aplanospores are known for *Binuclearia*.[2]

The two species found in the United States are *B. eriensis* Tiffany and *B. tatrana* Wittrock (Fig. 82). For descriptions of them, see Tiffany (1937).

8. **Radiofilum** Schmidle, 1894. This alga has spherical to sublenticular cells that are serially joined in unbranched filaments. The filaments are enclosed by a broad gelatinous sheath that frequently has radial fibrils of a denser gelatinous substance. In at least one species (*R. conjunctivum* Schmidle), the cell wall is composed of two helmet-shaped halves, in which the line of juncture between the two halves is generally evident. After

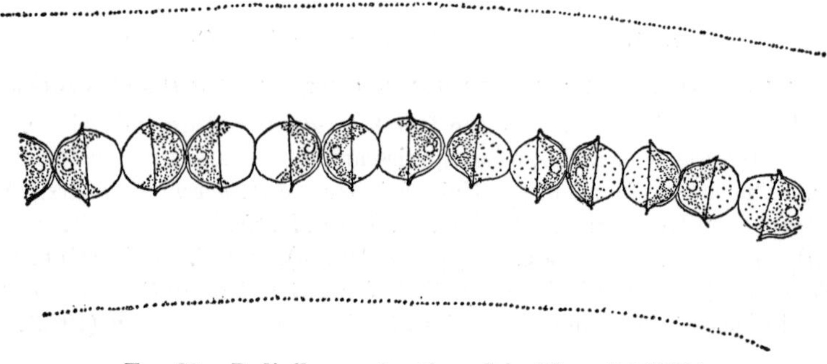

FIG. 83. *Radiofilum conjunctivum* Schmidle. (× 1200.)

cell division in this species, there is an interpolation of two new half-walls between the two old ones. *R. conjunctivum* has a single cup-shaped chloroplast, containing one or two pyrenoids, within each cell. The chloroplast is so arranged that the base of the cup lies next to one pole of a cell. During cell division, the chloroplast is so divided that the concavities of the chloroplasts in the two daughter cells face each other.

The filaments of *Radiofilum* are so fragile that there is often a vegetative multiplication by fragmentation. Thus far, zoospores, aplanospores, and zoogametes have not been observed.[3]

The two species found in this country are *R. conjunctivum* Schmidle (Fig. 83) and *R. irregulare* (Brunnth.) Wille. For descriptions of them, see Collins (1918).

FAMILY 2. MICROSPORACEAE

Microspora Thuret, 1850; emend., Lagerheim, 1888. This genus, the only one in the family, has the cells united in unbranched filaments that

[1] SCHRÖDER, 1898*A*. [2] WEST, G. S., 1904; WITTROCK, 1886.
[3] BRUNNTHALER, 1913; SCHMIDLE, 1894; WEST, G. S., 1904.

are usually sessile when young but are sometimes free-floating from the beginning. The cell walls in a filament are H-shaped pieces so articulated that each protoplast is enclosed by the conjoined halves of two successive H-pieces. Internal to the H-pieces is a very thin layer of cellulose that lies immediately next to the protoplast. The H-pieces are heavily impregnated with cellulose[1] and sometimes are distinctly stratified. The outermost layer of the filament is pectic in nature. In cell division there is the formation of a very thin cellulose layer around each daughter protoplast and then the interpolation of an H-shaped piece of wall material between the two as the two H-pieces that formerly enclosed the mother cell pull apart from each other. The cells are uninucleate, with the nucleus generally lying in a bridge of cytoplasm across the middle of the central vacuole. There is often so much reserve starch that but little can be made

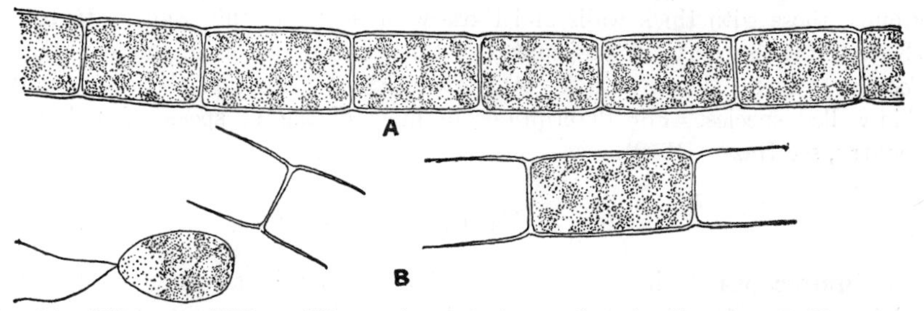

Fig. 84. *Microspora Willeana* Wittr. *A*, vegetative portion of a filament. *B*, liberation of zoospore. (× 900.)

out concerning the structure of the chloroplast. In young vigorously growing cells (Fig. 84*A*), the chloroplast is an irregularly expanded, perforate, and reticulate sheet covering both the sides and ends of the protoplast. Pyrenoids are lacking in chloroplasts of *Microspora*.[2]

Asexual reproduction is by the formation of zoospores; these may be formed singly within a cell, or the protoplast may divide to form 2, 4, 8, or 16.[3] When more than one zoospore is formed, there is a bipartition of the protoplasts after each mitosis. The zoospores may be liberated by a disarticulation of the H-pieces in the wall of the mother cell (Fig. 84*B*), or there may be a gelatinization of the sides of the H-pieces and a swimming of the zoospores through these gelatinized walls. Both bi- and quadriflagellate zoospores have been reported for *Microspora*.[4] After swarming for a short time, the zoospore comes to rest and secretes a thin wall. It may remain free-floating after it ceases swarming, but usually the swarming spore develops into a germling affixed by a discoid holdfast. Aplano-

[1] Tiffany 1924. [2] Lagerheim, 1889; West, G. S., 1916.
[3] Meyer, K., 1889. [4] Hazen, 1902; Skuja, 1934; West, G. S., 1916.

spores (Fig. 8, page 53) are of common occurrence in *Microspora*. They are usually spherical and formed singly within a cell. Akinetes are of somewhat rarer occurrence and have very thick walls. When akinete formation or aplanospore formation does take place, it generally occurs in several successive cells in a filament. Germinating akinetes[1] have a division of the protoplast into four daughter protoplasts, either before or after escape from the akinete wall, and a direct development of the protoplasts into filaments.

Sexual reproduction has not been recorded for *Microspora*.

Microspora is of common occurrence in pools and ditches, especially during cooler months of the year. The wall structure is so like that of *Tribonema* that the two are sometimes confused; but the presence of starch in *Microspora* affords an easy method of differentiating between the two. The species fall into two groups: those with thick walls and those with relatively thin walls. *M. amoena* (Kütz.) Rab. is the most widespread of thick-walled species in this country; *M. floccosa* (Vauch.) Thur. and *M. stagnorum* (Kütz) Thur. are the commonest of thin-walled species. For descriptions of most of the 11 species found in this country, see Hazen (1902).

FAMILY 3. CYLINDROCAPSACEAE

Cylindrocapsa Reinsch, 1867. *Cylindrocapsa* is another of the Ulotrichales so distinctive that it must be placed in a separate family. The filaments are unbranched and usually with the cells uniseriately arranged within a tough tubular envelope. In rare cases, the cells are irregularly arranged or side by side. Under certain conditions of growth, there is a complete loss of the filamentous organization and a development of a palmella stage quite similar in appearance to *Gloeocapsa*.[2] Each cell in a filament is surrounded by a cellulose wall[3] which is laid down in concentric strata. The tubular sheath enveloping a filament is pectic in nature. In most specimens, it is impossible to make out the structure of the chloroplast, but in favorable material it can be seen that there is a single stellate chloroplast with a single pyrenoid.[4]

Vegetative multiplication may take place by fragmentation of filaments. Asexual reproduction is either by means of biflagellate zoospores with two contractile vacuoles and an eyespot[5] or by means of quadriflagellate zoospores.[6] Zoospores are formed singly or in twos or fours, within a cell. Germination stages of zoospores are sessile and affixed to the substratum by a gelatinous holdfast (Fig. 85*B–C*). Cell division in young filaments is

[1] WEST, G. S., 1916. [2] CIENKOWSKI, 1876.
[3] TIFFANY, 1924. [4] IYENGAR, 1939.
[5] CIENKOWSKI, 1876. [6] IYENGAR, 1939.

transverse, and the two daughter cells each develop a wall with several concentric layers (Fig. 85*E*).

Sexual reproduction is oögamous.[1] Cells developing into sex organs may be distinguished from vegetative cells by their reddish color.[2] In the formation of antheridia, certain vegetative cells divide and redivide to form a double file of red-colored cells, each of which gives rise to two short biflagellate fusiform antherozoids. Cells developing into oögonia have a considerable increase in size of their protoplasts and a swelling of the wall layers. Just before fertilization, there is a development of a pore at one side of the oögonial wall. One species has been reported[3] which resembles nannandrous species of *Oedogonium* in that there are one-celled sexual plants epiphytic upon the vegetative filaments. The female plant

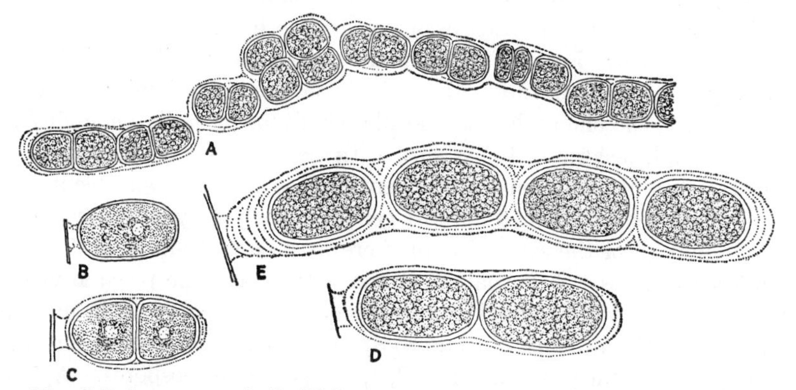

FIG. 85. *Cylindrocapsa geminella* Wolle. *A*, portion of a mature filament. *B–D*, early stages in development of a filament. (*A*, × 325; *B–D*, × 650.)

produces a single egg, and the male plant produces four quadriflagellate antherozoids. Fertilization takes place within an oögonium. Mature zygotes are spherical and with a smooth thick wall enclosing a bright red protoplast.

Cylindrocapsa is frequently encountered in collections from pools and ditches though rarely in abundance. Fruiting material is of rare occurrence and is usually found only in late spring.[4] For descriptions of the two species found in this country, *C. conferta* W. West and *C. geminella* Wolle (Fig. 85), see Heering (1914).

FAMILY 4. CHAETOPHORACEAE

The Chaetophoraceae are immediately distinguished from the foregoing families by their branching thalli. They differ from other branching

[1] IYENGAR, 1939; CIENKOWSKI, 1876. [2] CIENKOWSKI, 1876. [3] IYENGAR, 1939.
[4] TRANSEAU, 1913*A*.

Ulotrichineae in lacking the sporangia found in Trentepohliaceae and in lacking the seta characteristic of Coleochaetaceae. Most Chaetophoraceae have a plant body differentiated into a prostrate portion and an erect freely branched portion. The erect portion may have all branches the same size, or there may be a differentiation into large primary and small lateral branches. In either case, the ultimate branches are usually attenuated and often with the terminal cell or cells drawn out into a long hyaline seta. Genera in which the plant body is freely branched but with relatively few cells (*Aphanochaete*, *Chaetonema*) do not have a differentiation into prostrate and erect portions. There are some genera in which the branches are closely applied to one another, in a regular or irregular fashion, to form a discoid or irregularly expanded subparenchymatous thallus. Palmella stages are of common occurrence in certain genera, as *Stigeoclonium*, and of rare occurrence in others.

The cell structure is quite uniform throughout the family, and the cells are always uninucleate and with a single laminate parietal chloroplast. In most cases, the chloroplast is a girdle wholly or partially encircling the protoplast and, according to its size, with one to several pyrenoids.

Vegetative multiplication by fragmentation is of infrequent occurrence. Zoospores are readily formed in most genera. They may be bi- or quadriflagellate and all of the same size, or there may be a production of macro- and microzoospores. A formation of zoospores may take place anywhere in a thallus except in terminal and rhizoidal cells. Small cells in ultimate branchlets usually produce but one zoospore; larger cells in a filament may have a division of the protoplast into 2, 4, 8, 16 or more zoospores. Aplanospores and akinetes are also known for many Chaetophoraceae.

Many of the genera also reproduce sexually. In most cases, sexual reproduction is isogamous and with biflagellate gametes formed in considerable number within a cell. There are also anisogamous genera. The only case of oögamy thus far found in the family is in *Chaetonema*.

The family includes numerous genera, many of which are strictly freshwater in habit. Genera* found in the United States may be distinguished from one another as follows:

1. Plant body freely branched.. 2
1. Plant body pseudoparenchymatous.. 11
   2. Filaments bearing long one-celled hairs.................................. 3

* *Saprochaete*, a colorless filamentous aquatic plant discovered in North Carolina by Coker and Shanor (1929), was interpreted by them as a colorless alga related to *Stigeoclonium*. However, unlike other colorless Chlorophyceae, it does not store reserve carbohydrates as starch. There is also no formation of zoospores, a feature of frequent occurrence in Chaetophoraceae. These characters indicate that *Saprochaete* is a fungus and not a colorless alga.

Since *Pseudochaete* is a genus of such questionable validity, it is best to follow the practice of Tiffany (1937) who places the single species as a species of *Stigeoclonium*.

2. Filaments lacking long one-celled hairs 5
3. Thallus erect on substratum .... 11. **Thamniochaete**
3. Thallus prostrate on substratum 4
4. Thallus with short erect branches .... 13. **Chaetonema**
4. Thallus wholly prostrate .... 12. **Aphanochaete**
5. Ends of branches tapering to a sharp point 6
5. Ends of branches not tapering to a sharp point 9
6. Main branches broader than lateral branches 7
6. All branches approximately the same breadth 8
7. Main branches with alternate long and short cells .... 4. **Draparnaldiopsis**
7. Main branches with all cells the same length .... 3. **Draparnaldia**
8. Plant body of definite macroscopic shape .... 2. **Chaetophora**
8. Plant body not of definite macroscopic shape .... 1. **Stigeoclonium**
9. Thallus endophytic in walls of other algae .... 7. **Entocladia**
9. Thallus not endophytic in walls of the algae 10
10. Thallus encrusted with lime, cells of different length .... 6. **Chlorotylium**
10. Thallus not encrusted with lime, cells of approximately the same length .... 5. **Microthamnion**
11. Thallus one cell in thickness .... 8. **Protoderma**
11. Thallus more than one cell in thickness 12
12. Restricted to carapaces of turtles .... 9. **Dermatophyton**
12. Not restricted to carapaces of turtles .... 10. **Pseudoulvella**

1. **Stigeoclonium** Kützing, 1843. This alga has a plant body differentiated into an erect and a prostrate portion. The prostrate portion, which attaches the alga to the substratum, is either pseudoparenchymatous or irregularly branched and gives rise to numerous erect filaments. The erect filaments are somewhat sparsely branched, with an alternate or opposite branching, and have the obscure major branches and lateral branches attenuate to long multicellular hairs. The filaments are often enclosed by a broad gelatinous sheath, but this is of a very watery consistency and not usually visible unless demonstrated by special methods. *Stigeoclonium* has pseudoparenchymatous palmella stages, and it has been shown[1] that these may be induced in various ways.

Cells of both filamentous and palmella stages are uninucleate and with a single chloroplast. Larger cells in a filament have a transverse zonate chloroplast with several pyrenoids; smaller cells have the chloroplast girdling the whole length of a cell and usually contain but a single pyrenoid.

Vegetative multiplication may take place by fragmentation, but fragments from the erect portion do not grow so vigorously when severed from the prostrate portion. Zoospore formation takes place freely in *Stigeoclonium,* and one often finds all cells in the smaller branches sporulating a day after collections are brought into the laboratory. Zoospores are generally formed singly within a cell, and development of flagella is due to a blepharoplast-rhizoplast-centriole type of neuromotor apparatus.[2] The

[1] Livingston, 1900, 1901, 1905, 1905*A*.
[2] Reich, 1926.

zoospores are quadriflagellate. In some species, there is a formation of quadriflagellate swarmers of two sizes. In certain of these, the smaller of the two are zoosporic in nature[1] and hence quite properly called micro-zoospores. In other cases, the smaller of the two are gametes.[2] After swarming for a time, a zoospore comes to rest with its anterior pole downward and secretes a wall after disappearance of the flagella. According to the species, the one-celled germling grows directly into a vertical filament that later develops the procumbent portion of a thallus from its lowermost cells; or the one-celled germling first grows into a prostrate branching system from which erect branches arise.[3] Cases are also known where the zoospore metamorphoses into a rhizopodial stage that remains amoeboid for some time before developing into a filament.[4] Aplanospores are generally formed singly within a cell and in several successive cells. Akinetes are sometimes formed in abundance.[5]

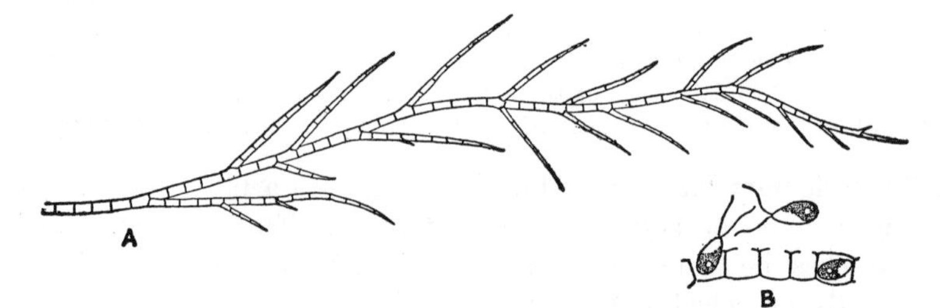

Fig. 86. *Stigeoclonium tenue* (Ag.) Kütz. *A*, portion of a filament. *B*, liberation of zoospores. (*A*, × 125; *B*, × 400.)

Sexual reproduction is isogamous. Some species have a fusion of biflagellate gametes to form a quadriflagellate zygote,[1] others have a fusion of quadriflagellate gametes to form an octoflagellate zygote.[2] After swarming for a time, the motile zygote rounds up, loses its flagella, and secretes a wall. The zygote may enter upon a period of rest or germinate within a day or two. When resting zygotes germinate, there is a formation of four zoospores.[2] If the zygote germinates immediately, there is an equational division of the zygote nucleus and a formation of a short filament of diploid cells.[1] The diploid generation forms quadriflagellate zoospores, and it is presumed, though not definitely established, that zoospore formation is immediately preceded by meiosis.[1]

Stigeoclonium (Fig. 86) is a common alga in slowly flowing clear-water brooks, in springs, and in the overflow from fountains and watering troughs. It is most abundant during late spring and early autumn. Approximately 20 species are known to occur in this country. For a description of a majority of them, see Collins (1909).

[1] Juller, 1937. [2] Godward, 1942.
[3] Fritsch, 1903; Godward, 1942; Ström, 1921.
[4] Pascher, 1915. [5] Tilden, 1896.

2. Chaetophora, Schrank, 1789. This alga has macroscopic thalli that are of an exceedingly tough consistency and hemispherical, spherical, or elongate, and irregularly tuberculate. Globular colonies have the branches radiating from a palmelloid base and repeatedly branched throughout or with a fasciculate branching at their apices. The branches are invested with a tough gelatinous envelope, and the envelopes of adjoining branches may be distinct or fused with one another. The basal portion of a colony is often impregnated with calcium carbonate, and in rare cases the entire colony may be strongly calcified. Species with elongate colonies have the filaments intertwined to form an axial strand that bears numerous short

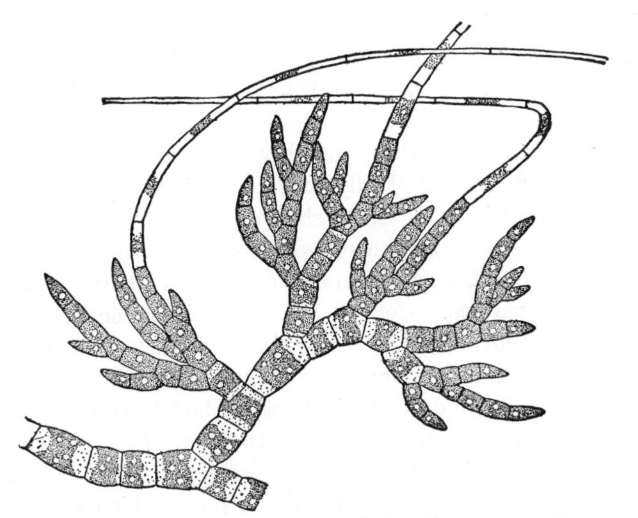

FIG. 87. *Chaetophora incrassata* (Huds.) Hazen. (× 400.)

lateral branches throughout its length. In both types of colony, the ultimate branchlets are often prolonged into long multicellular hairs that gradually taper to a fine point.

There is but a single chloroplast in a cell. In larger cells, it is a transversely zonate band containing several pyrenoids; in smaller cells, it covers the entire side wall, overlaps the end walls, and contains but a single pyrenoid.

Thalli of *Chaetophora* are of so tough a consistency that there is rarely vegetative multiplication by fragmentation. Asexual reproduction is by the formation of quadriflagellate zoospores. Akinetes, when present, are usually formed by cells of the ultimate branchlets, but sometimes they are formed in any cell of a thallus.[1]

Sexual reproduction is isogamous and by a union of biflagellate gametes.

[1] HAZEN, 1902.

Chaetophora does not require so well-aerated water as does *Stigeoclonium* and is often found in abundance in standing water and epiphytic upon submerged vegetation or attached to stones or woodwork. Specific differences are based upon macroscopic appearance of a thallus and the method of branching of the filaments. The species found in this country are *C. attenuata* Hazen, *C. elegans* (Roth) Ag., *C. incrassata* (Huds.) Hazen (Fig. 87), and *C. pisiformis* (Roth) Ag. For descriptions of them, see Hazen (1902).

3. **Draparnaldia** Bory, 1808. The plant body of *Draparnaldia* is a macroscopic, amorphous, pale-green mass of a very watery consistency. The prostrate portion of the plant body is usually but little developed. The erect portion, which may be of indefinite extent, has a conspicuous differentiation into primary branches and short fasciculate lateral branchlets, whose terminal filaments are drawn out into long hyaline setae. The fasciculate branches may be borne anywhere on the primary branches, except the basal portion, where they are replaced by rhizoid-like branches. The primary branches and the branchlets are embedded in a very broad gelatinous matrix composed chiefly of pectic acid.[1] Cells of the primary branches are cylindrical to barrel-shaped, all about the same length, and with a transversely zonate chloroplast. Chloroplasts of axial cells may be entire or reticulate, with smooth or toothed edges, and they always contain several pyrenoids. Cells of the lateral branchlets are cylindrical, with a chloroplast covering the entire side wall and generally containing but a single pyrenoid. Palmella stages are rarely formed by *Draparnaldia*, but under certain conditions, as an increase in the CO_2 content of the water, or an increase in the amount of nitrates, the organization of the thallus is quite like that of a *Stigeoclonium*.[2]

Asexual reproduction is by means of zoospores which are formed only in the cells of the lateral branchlets. They are usually formed singly within a cell, but sometimes the protoplast divides to form two or four of them.[3] Zoospores are quadriflagellate and swarm but a few minutes before coming to rest with the anterior end downward. The germling developing from a zoospore has a long seta at the distal end. By the time a germling is four- or five-celled, there is a modification of the basal cell to form a rhizoid-like holdfast.[4] Instead of producing zoospores, the cells of branchlets may develop into thick-walled akinetes (hypnospores) whose protoplasts are colored a deep orange.[5]

Sexual reproduction is isogamous and by a fusion of quadriflagellate gametes.[6] These become amoeboid before they fuse in pairs. Gametes which have not fused with one another round up and become thick-walled

[1] Wurdack, 1923. [2] Uspenskaja, 1930.
[3] Johnson, 1893; Klebs, 1896; Pascher, 1907. [4] Johnson, 1893.
[5] Gay, 1891; Klebs, 1896. [6] Klebs, 1896; Pascher, 1907.

parthenospores. When a zygote germinates, it gives rise to two or four germlings.[1]

Draparnaldia is usually found only in clear cool running water. The thalli look somewhat like those of *Tetraspora*, but the two can be distinguished from each other in the field with a good hand lens. The following species are found in this country: *D. acuta* (Ag.) Kütz., *D. glomerata* (Vauch.) Ag. (Fig. 88), *D. Judayi* Prescott, *D. platyzonata* Hazen, *D. plumosa* (Vauch.) Ag., and *D. Ravenelii* Wolle. For a description of *D. Judayi*, see Prescott (1944); for *D. Ravenelii*, see Collins (1912), for the others, see Hazen (1902).

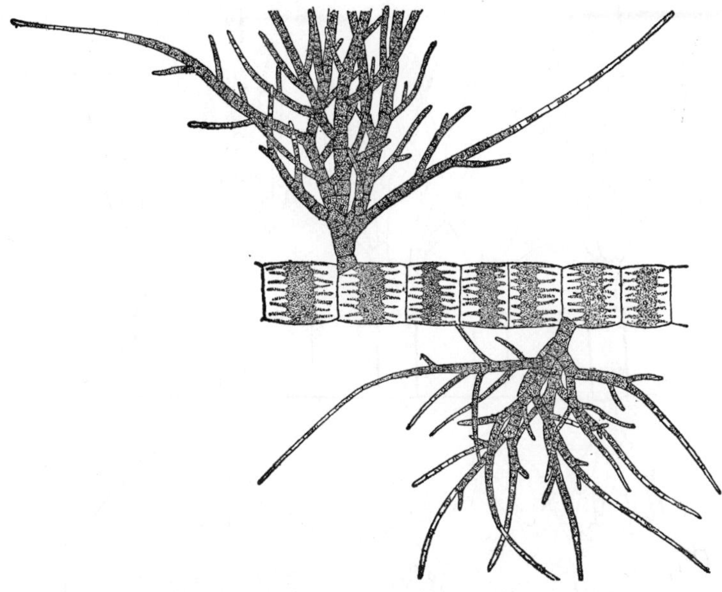

FIG. 88. *Draparnaldia glomerata* (Vauch.) Ag. (× 185.)

4. **Draparnaldiopsis** Smith and Klyver, 1929. The thallus of this alga has the same differentiation into long and short branches as *Draparnaldia*, but in *Draparnaldiopsis* the long branches have long and short cells alternating with one another in regular succession.[2] Both types of cell in the long branches have zonate chloroplasts with several pyrenoids. The lateral branchlets are borne only by short cells of long branches. The basal cell of a branchlet is cuneiform and di- or trichotomously forked at its apex. Cells borne by the basal cell are, in turn, forked at their apices, and there may be a succession of two to four of these forked cells. Most of the ultimate branches of a branchlet terminate in one or more elongate cells without chloroplasts. Other cells in a branchlet have a laminate

[1] PASCHER, 1907. [2] BHARADWAJA, 1933; SMITH and KLYVER, 1929.

parietal chloroplast containing one or two pyrenoids. The whole thallus is invested with a broad gelatinous envelope, and the thallus is attached to the substratum by rhizoidal branches.

Reproduction is known for one species only, *D. indica* Bharadwaja. Here the life cycle consists of an alternation of morphologically identical haploid gametophytes and diploid sporophytes. The chromosome number is doubled at the time of gametic union and is halved at the time zoospores are formed.[1] The sporophyte produces quadriflagellate zoospores of two

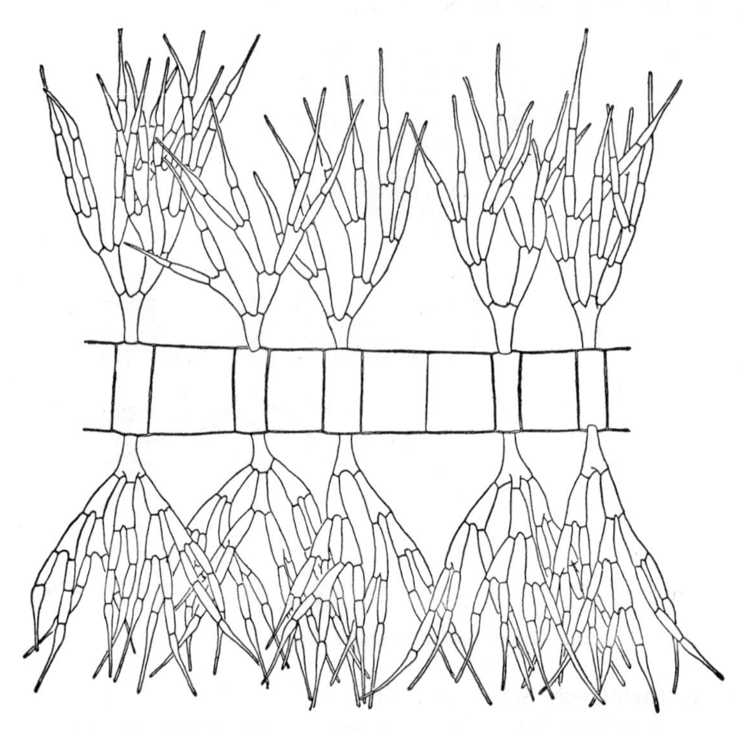

FIG. 89. *Draparnaldiopsis alpina* Smith and Klyver. (× 480.)

sizes, macrozoospores and microzoospores. Microzoospores are formed in cells of the lateral branchlets. The zoospores develop into gametophytes when they germinate.[2] There may also be a formation of aplanospores and akinetes.[3] Gametophytes form their gametes in cells of lateral branchlets. The gametes are biflagellate, gametic union is isogamous, and there is rarely a union of gametes from the same thallus.[4] There is an immediate germination of the zygote, and division of the zygote nucleus is mitotic.

[1] SINGH, 1945. [2] SINGH, 1942, 1945. [3] MITRA, 1943. [4] SINGH, 1942.

The type species, *D. alpina* Smith and Klyver (Fig. 89) is known only from California. For a description of it, see Smith and Klyver (1929).

5. **Microthamnion** Nägeli, 1849. This alga has a densely branched thallus which rarely grows to more than 1 cm. in height and is always attached to some firm object by means of a bulbous basal cell. All branches of a thallus are of the same diameter, and there is but little attenuation toward the apices of branches. Branches originate as lateral outgrowths from the upper end of a cell, and there is a marked tendency for branching to be unilateral. Cell division in an outgrowth does not take place until it has attained a considerable length, and the transverse wall is laid at some distance from the base of an outgrowth. The thallus is without a gelatinous envelope. There is a single parietal laminate chloroplast extending nearly the whole length of a protoplast and partially encircling it. Chloroplasts are without pyrenoids, and reserve foods accumulate mainly as fats. Asexual reproduction is by biflagellate zoospores.

Four or eight zoospores are generally formed within a cell, and these are liberated by a rupturing of the apex of the lateral cell wall. Zoospores form no rhizoidal outgrowths when they germinate.[1] Akinetes have been recorded but are of rare occurrence.[2]

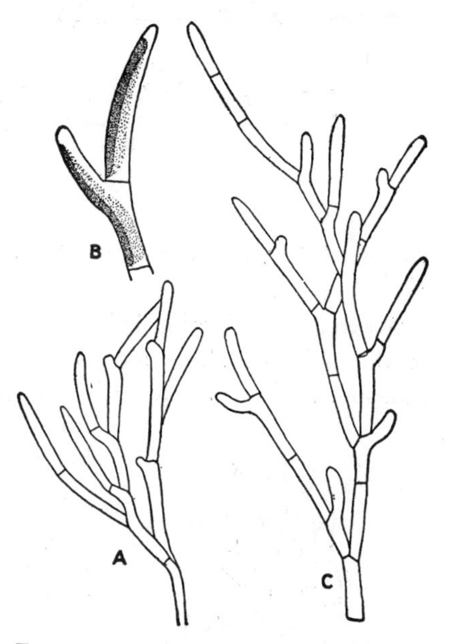

FIG. 90. *A–B*, *Microthamnion Kuetzingianum* Näg. *C*, *M. strictissimum* Rab. (*A*, *C*, × 375; *B*, × 1000.)

Both of the two species found in this country, *M. Kuetzingianum* Näg. (Fig. 90*A–B*) and *M. strictissimum* Rab. (Fig. 90*C*), are widely distributed. For descriptions of them, see Hazen (1902).

6. **Chlorotylium** Kützing, 1843. The thallus of this alga is sessile, hemispherical to irregularly pulvinate, and generally strongly incrusted with lime. The interior of a thallus may be zonate or homogeneous. The cells are cylindrical and are joined end to end in filaments in which branching is almost wholly unilateral. The cells are usually of different length, one or two elongate ones with pale chloroplasts alternating with a succes-

[1] GREGER, 1915; HAZEN, 1902; RAYSS, 1929. [2] RAYSS, 1929.

sion of three to seven short ones with conspicuous chloroplasts. Chloroplasts are parietal, laminate, and with a pyrenoid.[1] Palmella stages may be formed at times, and these have *Gloeocystis*-like cells joined together in an irregular mass.[2]

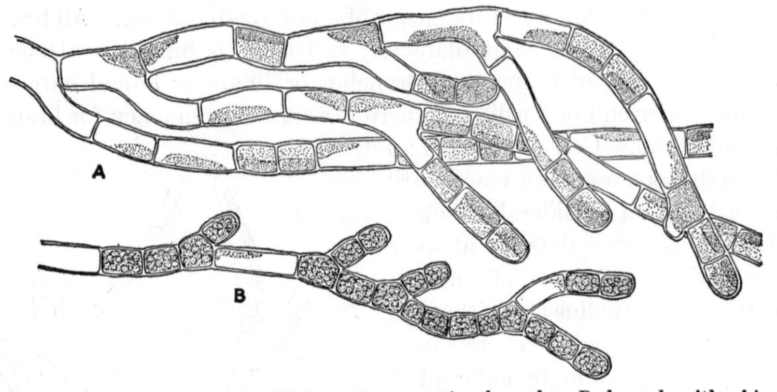

FIG. 91. *Chlorotylium cataractum* Kütz. *A*, vegetative branch. *B*, branch with akinetes. (× 485.)

Asexual reproduction is by a formation of biflagellate zoospores.[3] Cells of the palmella stage have a formation of 4 to 16 quadriflagellate (?) zoospores. Akinetes are of frequent occurrence in *Chlorotylium* and are formed by short cells in a filament. They have quite thick walls, and the protoplast is often of a reddish color.

C. cataractum Kütz. (Fig. 91) has been recorded[4] from New York and Kentucky. The specific name implies that it is found only in swiftly running water, but in certain European localities it is quite as abundant in standing water.[5] For a description of *C. cataractum*, see Collins (1909).

7. Entocladia Reinke, 1879. This microscopic alga has an irregularly branched to subparenchymatous thallus in which none of the cells bears long hyaline hairs. *Entocladia* usually grows endophytic within the wall layers of other algae or aquatic phanerogams (Fig. 92). Growth of a thallus is mostly by division of the terminal cells. The cells are

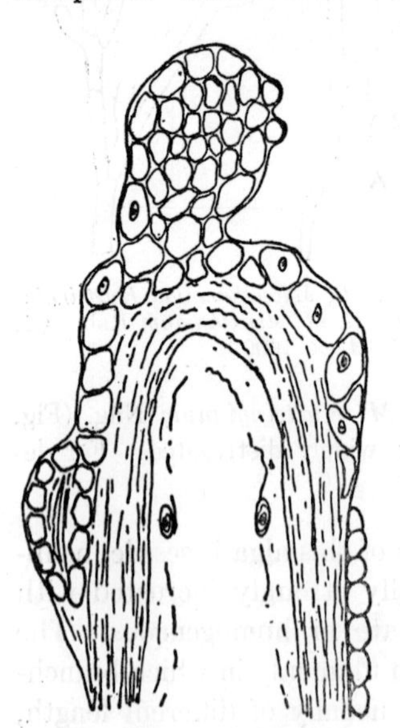

FIG. 92. *Entocladia Cladophora* (Hornby) G. M. Smith, a species not found in the United States. (*From Hornby*, 1918.)

[1] PRINTZ, 1927. [2] REINKE, 1879.
[3] HANSGIRG, 1886. [4] COLLINS, 1909.
[5] HANSGIRG, 1886, 1892.

uninucleate and with a single laminate chloroplast that has one or more pyrenoids.

Asexual reproduction is by division of the cell contents into eight or more quadriflagellate zoospores with a conspicuous eyespot.[1] When a zoospore ceases swarming, it comes to rest upon the host and soon sends out a short rhizoidal protuberance that penetrates the host.

Sexual reproduction is isogamous and by a union of biflagellate gametes, of which 8 to 16 are formed within a cell.

Entocladia is a genus whose species are largely marine but which also has them in fresh water. The two species reported from fresh water in this country, *E. Pithophorae* (G. S. West) G. M. Smith and *E. polymorphum* (G. S. West) G. M. Smith, both grow endophytically within cell walls of Cladophoraceae. For descriptions of them as species of *Endoderma*, see Collins (1909).

8. **Protoderma** Kützing, 1843; emend., Borzi, 1895. The thallus of this alga is microscopic, epiphytic, and a discoid layer one cell in thickness. The central portion of a thallus is pseudoparenchymatous; the marginal portion consists of short and irregularly branched radiating filaments, more or less parallel to one another. There are no erect filaments or setae. Within each cell is a typical parietal ulotrichaceous chloroplast containing one pyrenoid. Under certain conditions the thallus develops into a *Gloeocystis*-like palmella stage.[2]

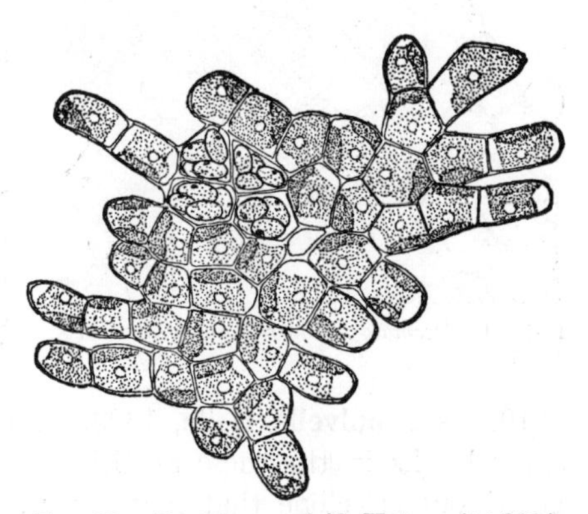

FIG. 93. *Protoderma viride* Kütz. (× 800.)

Zoospores may be formed by any cell in the plant body, but they are usually produced by the pseudoparenchymatous cells at the center of a thallus. Four or eight biflagellate zoospores are usually formed within a cell. Aplanospores may also be formed.[1]

P. viride Kütz. (Fig. 93), the only species found in this country, is more frequently found on submerged phanerogams than on other algae. This alga is so similar to palmella stages or juvenile stages of *Stigeoclonium* that its presence is extremely dubious if *Stigeoclonium* is present at the collecting station. For a description of *P. viride*, see Collins (1909).

9. **Dermatophyton** Peter, 1886. The thallus of *Dermatophyton* is a sessile orbicular disk, a few millimeters in diameter, that grows attached

[1] BORZI, 1895; HUBER, J., 1892; KYLIN, 1935. [2] BORZI, 1895.

to the dorsal carapace of turtles.[1] Juvenile thalli have irregularly arranged cells, but older thalli frequently have radially arranged cells with a radial axis two to three times longer than the tangential axis. Young thalli are one cell in thickness, older ones are four to five cells in thickness in the central portion. The cells are multinucleate and with several pyrenoids.[2]

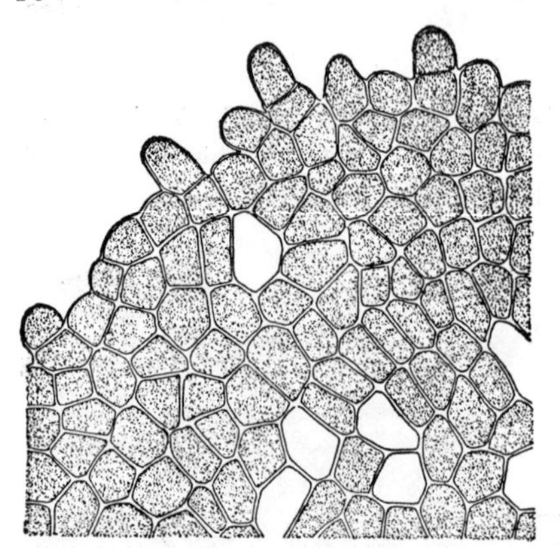

Fig. 94. *Dermatophyton radians* Peter. (× 485.)

Asexual reproduction is by means of zoospores. Cells producing zoospores lie in the central portion of a thallus, become globose, and have a considerable increase in the number of nuclei prior to zoospore formation.[2]

There is but one species, *D. radians* Peter (Fig. 94). It has been found in Massachusetts[3] and Kansas.[4] Feldmann (1938) has shown that the practice[5] of referring this species to *Ulvella* is incorrect. For a description of *D. radians*, see Collins (1909).

10. **Pseudoulvella** Wille, 1909. The thallus of this alga is discoid and several cells in thickness at the center. The thallus is invested with a gelatinous envelope that may bear a few evanescent gelatinous bristles. Under certain conditions, the plant body is an irregularly shaped mass of cells. The cells are uninucleate and with a laminate, more or less reticulate, chloroplast containing a single pyrenoid.[6]

Zoospore formation is restricted to the portion of the thallus more than one cell in thickness. Four, eight, or sixteen pyriform quadriflagellate zoospores are formed within a cell.[6] When a zoospore ceases swarming, it secretes a wall and grows directly into a new thallus.

P. americana (Snow) Wille (Fig. 95) has been found in Michigan[7] and Iowa.[8] For a description of it as *Ulvella americana*, see Snow, 1899.

11. **Thamniochaete** Gay, 1893. The thallus of *Thamniochaete* is microscopic and rarely contains more than a half-dozen cells. It is epiphytic upon other algae and grows erect upon the host plant. The lowermost cell may be like other cells of the thallus, or it may be modified to form a rhizoidal holdfast. The plant body is distinctly filamentous and simple,

[1] Peter, 1886. [2] Feldmann, 1939. [3] Collins, 1909. [4] Thompson, 1938.
[5] Fritsch, 1935; Printz, 1927; Smith, G. M., 1933.
[6] Philipose, 1946. [7] Snow, 1899. [8] Prescott, 1931.

or with rudimentary branches. The terminal cells, and sometimes the intercalary cells also, bear one or more long unicellular setae that gradually taper from a swollen base to an acute apex. The remaining cells in a thallus are irregularly cylindrical, with a single parietal laminate chloroplast that contains one pyrenoid.[1]

Asexual reproduction is by zoospores formed singly within a cell.[2] There may also be a formation of brownish akinetes.[3]

T. Huberi Gay (Fig. 96) has been recorded from California.[4] For a description of it, see Heering (1914).

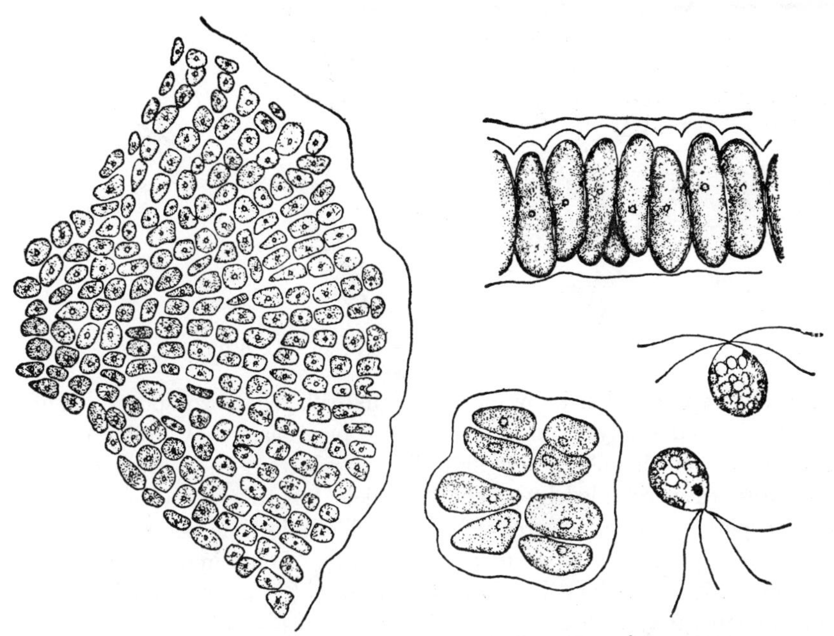

FIG. 95. *Pseudoulvella americana* (Snow) Wille. (*From Snow*, 1899.)

12. **Aphanochaete** A. Braun, 1851. This microscopic alga is epiphytic, prostrate upon the host, and with simple or irregularly branched filaments; tapering somewhat at their extremities but not terminating in long unicellular setae. The median cells of the thallus are globose to globose-cylindrical; those toward the extremities are cylindrical to barrel-shaped; the terminal cells are conical. The cells may bear one or more long, hyaline, unicellular setae on their dorsal side. The setae, which taper from a swollen base to an acute apex, are probably modified erect branches.

[1] GAY, 1893; WEST, G. S., 1904. [2] CHADEFAUD, 1932. [3] GAY, 1893.
[4] SMITH, G. M., 1933.

Under certain cultural conditions, the setae are multicellular.[1] Median cells of the filament have a single parietal laminate chloroplast, usually with several pyrenoids; cells toward the ends of the filaments have chloroplasts with one pyrenoid.

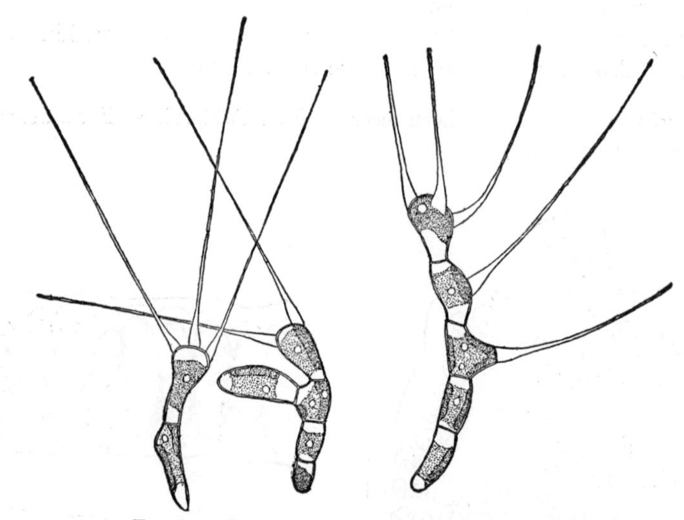

FIG. 96. *Thamniochaete Huberi* Gay. (× 600.)

Asexual reproduction is by means of quadriflagellate zoospores. These may be formed singly within a cell, or the protoplast may divide to form two or four of them.[2] When first discharged from the parent cell, the zoospores are surrounded by a delicate vesicle, but this soon disappears and the zoospores swim about freely for a considerable time before they come to rest and germinate to form new filaments. Sometimes the swarming zoospores withdraw their flagella and move about in an amoeboid manner.[3] Aplanospores may also be formed singly within a cell.[4]

Sexual reproduction is anisogamous and with a union of large quadriflagellate female gametes and small quadriflagellate male gametes.[5] *Aphanochaete* is homothallic, and the female gametes are formed singly within enlarged cells in the central portion of the thallus. Male gametes are formed singly or in pairs within small colorless cells resulting from repeated cell divisions at the tips of branches. Both male and female gametes are surrounded by a gelatinous vesicle when first liberated, but the vesicles soon disappear. Male gametes swim actively, female gametes move quite sluggishly; and the two fuse with their colorless anterior end apposed. The zygote secretes a thick wall soon after gametic union and

[1] FRITSCH, 1902; CHODAT, 1902. [2] HUBER, J., 1892, 1894.
[3] PASCHER, 1909. [4] WEST, G. S., 1904. [5] HUBER, J., 1894.

enters upon a period of rest during which the protoplast becomes filled with a reddish oil.

Aphanochaete is of common occurrence upon old filaments of *Oedogonium*, *Vaucheria*, and Cladophoraceae. *A. polychaete* (Hansg.) Fritsch, *A. repens* A. Br. (Fig. 97), and *A. vermiculoides* Wolle have been found in this country. For descriptions of them, see Heering (1914).

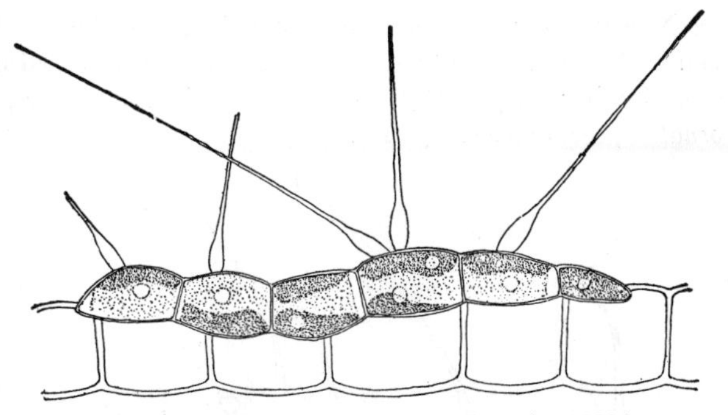

FIG. 97. *Aphanochaete repens* A. Br. (× 1000.)

13. **Chaetonema** Nowakowski, 1876. This alga grows epiphytic upon *Batrachospermum*, *Tetraspora*, *Chaetophora*, and other algae with a conspicuous gelatinous envelope. It is microscopic and consists of irregularly twisted main branches which bear short lateral branches terminating in long hyaline unicellular setae. Lateral branches arise as cylindrical outgrowth from the middle of cells in a main branch, and the first cross wall is formed at some distance from the point of origin (Fig. 98*A*). Palmella stages are at times formed by *Chaetonema*, and giant cells of the palmella stage divide by a yeast-like budding.[1] The cells of filaments contain a single laminate chloroplast with one or more pyrenoids. The chloroplast is usually zonate and restricted to the median part of a cell.

Asexual reproduction is by means of quadriflagellate zoospores. Two zoospores are generally formed within a cell.[2] During the period of swarming, they often remain within the gelatinous envelope of the host plant. When the zoospore comes to rest, it becomes spherical, secretes a wall, and then sends out a germ tube at its anterior end. The first cell cut off at the distal end of the germ tube may be rhizoidal in nature.

Sexual reproduction is oögamous.[3] Antheridia are usually formed in short series and are developed from cells in short lateral branches or from cells in the main axis. Cells developing into antheridia become barrel-shaped and have pale chloroplasts that are without pyrenoids (Fig. 98*D*). The protoplast of each antheridium divides to form eight biciliate anthero-

[1] HUBER, J., 1894. [2] HUBER, J., 1892. [3] MEYER, K., 1930.

zoids that are liberated through a pore in the side wall. Oögonia are usually developed from the terminal cells of lateral branches, though sometimes from cells in the main axis. Cells functioning as oögonia become much larger than vegetative cells. The entire protoplast of an oögonium develops into an egg which is partially extruded through an apical pore in the oögonial wall (Fig. 98*B* and *C*).

C. irregulare Nowak. (Fig. 98) has been found in Massachusetts,[1] and *C. ornatum* Transeau in Alabama.[2] Both species were found growing within the gelatinous envelopes of other algae. For a description of *C. irregulare*, see Collins, 1918; for *C. ornatum*, see Transeau (1943).

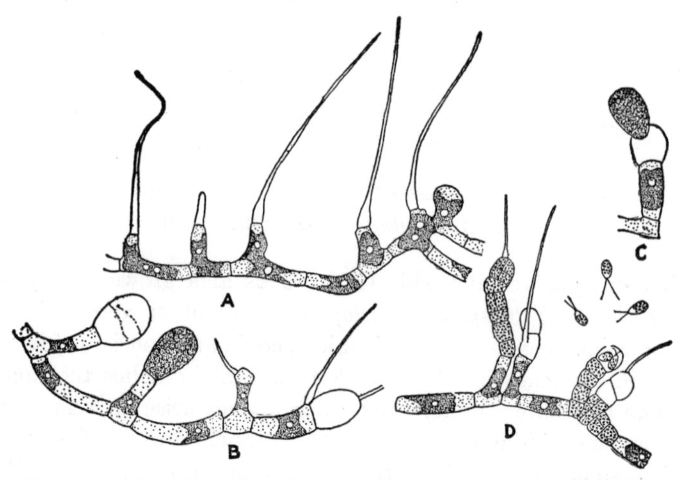

FIG. 98. *Chaetonema irregulare* Nowak. *A*, vegetative filament. *B–C*, filaments with oögonia. *D*, a filament with an antheridium. (*After K. Meyer*, 1930.)

FAMILY 5. PROTOCOCCACEAE

Protococcus Agardh, 1824 (*Pleurococcus* Meneghini, 1837; *Pseudopleurococcus* Snow, 1899). This genus, the only one in the family, has cells that may be solitary or may divide to form packets of two, three, four, or more cells. Solitary cells are spherical to broadly ellipsoidal and with a fairly thick wall that is without a gelatinous envelope. Packets of two or more cells are formed by division of solitary cells. The first division is transverse; if further divisions take place, the plane of division is at right angles to the preceding one. When the alga grows submerged in water, cell division may continue until there are 50 or more cells in a colony and there may be a development of a profusely and irregularly branched condition.[3] However, such *Pseudopleurococcus* stages are not always produced when the

[1] COLLINS, 1918; PRESCOTT and CROASDALE, 1942. [2] TRANSEAU, 1943.
[3] SNOW, 1899.

alga grows submerged in water. Cells of *Protococcus* have a single parietal laminate chloroplast that is more or less lobed at the margins. The chloroplasts are usually without pyrenoids, but in some instances they have them.

There is no formation of zoospores, and reproduction is exclusively by means of cell division. The two daughter cells may remain united after cell division, or they may separate from each other and become globose.

Protococcus is almost universally interpreted as a reduced form from a branching ulotrichaceous ancestor. This interpretation is based upon the fact that it may grow into an irregularly branched filament when growing in an aquatic instead of an aerial environment.

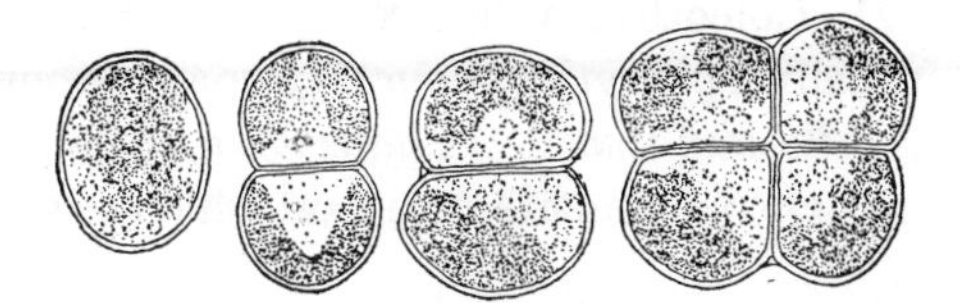

FIG. 99. *Protococcus viridis* Ag. (× 1300.)

Protococcus is, perhaps, the commonest green alga in the world. It is generally found forming a green coating on stone walls, woodwork, or the trunks of trees. According to popular tradition, such green coatings are most abundant on the shaded north side of trees, but this is often not the case,[1] since the effect of prevailing winds is even more important than light intensity in determining the position. The importance of prevailing winds in governing the position of *Protococcus* is due to the fact that the alga has a great capacity for absorbing water from humid air[2] and therefore grows best on the side of a tree where the air is dampest.

There are four or five good species in the genus, but there has been no critical study of species found in this country. *P. viridis* Ag. (Fig. 99) is world-wide in distribution.

FAMILY 6. COLEOCHAETACEAE

Vegetative cells of algae belonging to this family bear long cytoplasmic setae which are partly or wholly surrounded by a gelatinous envelope. All cells in a thallus may be setiferous, or certain cells only may bear setae. Members of this family are always sessile and usually grow on other plants. Some Coleochaetaceae are unicellular and solitary or gregarious; others are multicellular and with cells united in filaments that are erect, prostrate, or laterally united to form a disk one to three cells in thickness. The cells are uninucleate and with a single parietal laminate chloroplast covering most of the protoplast's surface. Usually there is a single pyrenoid within a chloroplast.

Asexual reproduction is by means of biflagellate zoospores, that are formed singly or 2, 4, 8, or 16 within a cell.

Sexual reproduction is isogamous or oögamous.

[1] KRAEMER, 1901. [2] SCHMID, 1927.

The genera found in this country differ as follows:

1. Thallus multicellular, filamentous, or pseudoparenchymatous.. 1. **Coleochaete**
1. Thallus unicellular, though sometimes gregarious 2
   2. Cells with more than one seta 3
   2. Cells with a single seta 4
3. With a gelatinous sheath at the base of each seta .... 4. **Conochaete**
3. Without a gelatinous sheath at the base of each seta .... 5. **Oligochaetophora**
   4. Seta unbranched .... 2. **Chaetosphaeridium**
   4. Seta branched .... 3. **Dicranochaete**

1. Coleochaete de Brébisson, 1844. *Coleochaete* is a sessile alga that is usually epiphytic, though it may be endophytic within the cell walls of *Nitella* and *Chara*. The cells may be joined end to end in dichotomously branched filaments some of which are prostrate and others erect; or all

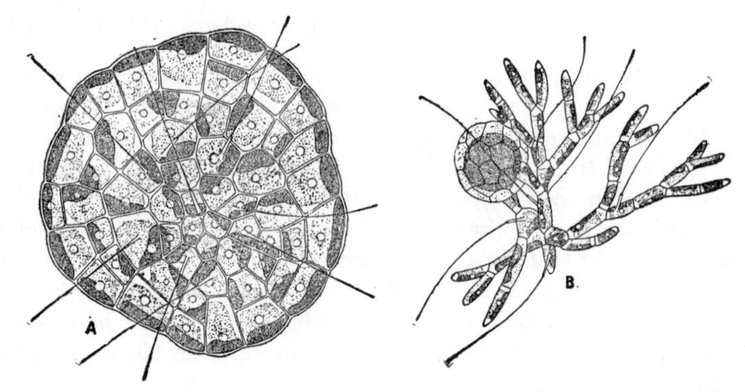

FIG. 100. *A, Coleochaete scutata* Bréb. *B, C. pulvinata* A. Br. (*A*, × 250; *B*, × 75.)

the filaments may be prostrate and with the branches distinct from one another or laterally apposed to form a parenchymatous disk that may be more than one cell in thickness in the central portion. In all these types of plant body, certain cells bear a single, long, unbranched cytoplasmic seta whose base is ensheathed by a cylinder of gelatinous material. In the formation of a seta there is first the formation of a small pore in the cell wall. The development of the seta is due to a blepharoplast that lies immediately beneath the pore.[1] Cells of *Coleochaete* are uninucleate and with a single laminate chloroplast that partially or wholly encircles the protoplast. There is usually one large pyrenoid within the chloroplast.

Asexual reproduction is by means of biflagellate zoospores that are formed singly within a cell. Isolated cells in a thallus may produce zoospores at any time of the year, but in the spring there is frequently a

[1] WESLEY, 1928.

production of zoospores by every cell in plants living over from the previous summer. The zoospore escapes by moving in an amoeboid manner through a small pore in the parent-cell wall and then swarms for an hour or so before coming to rest.[1] Soon after it becomes quiescent, there is the formation of a cell wall, and this is shortly followed by cell division. Germlings containing but a few cells have the cellular arrangement characteristic of the species, and at least one of these early formed cells bears a seta. Aplanospores with fairly thick walls may also develop singly within a cell.[2]

Most species of *Coleochaete* also reproduce sexually, although there are dwarf species that form zoospores only.[3] Sexual reproduction is oögamous and, according to the species, the plants are heterothallic or homothallic. In *C. pulvinata* A. Br.[4] and *C. Nitellarum* Jost,[5] the antheridia are bluntly conical and usually borne at the tips of branches. In *C. scutata* Bréb., the antheridia develop from cells midway between center and periphery of the discoid thallus.[6] A vegetative cell divides into two antheridial mother cells, by whose division antheridia are formed. Antheridia of *Coleochaete* each produce a single antherozoid, which may be green or colorless. The oögonium of *C. pulvinata*[7] is a flask-shaped structure with a long colorless neck, the *trichogyne*. It results from the metamorphosis of a one-celled lateral branch. Oögonia of other species do not have a long trichogyne. After these oögonia have been differentiated from cells at the tips of branches, there is a continuation of thallus growth beyond and around the oögonia. These species have, therefore, a zone of ripe zygotes some distance in from the periphery of a thallus. Fertilization takes place by an antherozoid swimming to and fusing with the egg within an oögonium. After fertilization, the zygote secretes a heavy wall and increases greatly in size. During enlargement of a zygote, there is a growth of branches from the cell below the oögonium and from neighboring cells to form a parenchymatous sheath more or less completely enclosing the oögonium. The oögonium with its enclosing sheath of cells, which soon become reddish brown, is termed a *spermocarp*. Spermocarps remain dormant over winter.[8]

A meiotic division of the zygote nucleus takes place at the end of the winter rest period.[9] A cleavage of the cytoplasm follows after each of the two nuclear divisions in meiosis, and the four protoplasts may continue division until there are 8 to 32 of them. Each protoplast of the final series of divisions becomes a biflagellate zoospore. The zoospores are liberated

[1] Lambert, 1910*A*; Pringsheim, N., 1860; Wesley, 1930. [2] Wesley, 1928.
[3] Lambert, 1910*A*. [4] Oltmanns, 1898. [5] Lewis, I. F., 1907.
[6] Wesley, 1930. [7] Oltmanns, 1898; Pringsheim, N. 1860. [8] Wesley, 1930.
[9] Allen, C. E., 1905.

by a breaking of the spermocarp wall into two halves and, after they cease swarming, they develop into thalli.[1]

Coleochaete is widely distributed in this country and is usually found on dead culms of *Typha* and *Sagittaria* that are submerged throughout the year. Small thalli are sometimes epiphytic on *Vaucheria* and *Oedogonium*. The species found in the United States are *C. divergens* Pringsh., *C. irregularis* Pringsh., *C. Nitellarum* Jost, *C. pulvinata* A. Br. (Fig. 100*B*), *C. Sampsonii* Transeau, *C. scutata* Bréb. (Fig. 100*A*), and *C. soluta* (Bréb.) Pringsh. For a description of *C. Sampsonii* see Transeau, 1943; for the others, see Collins, (1909).

2. **Chaetosphaeridium** Klebahn, 1892. This microscopic epiphytic alga has spherical to ovoid cells, each of which bears a single long seta at its distal end. Usually the alga is unicellular, though often growing in dense clusters and with or without a common gelatinous envelope around the cluster. Sometimes it is multicellular and with the cells connected to one another by fairly long empty gelatinous tubes. The seta has an axial cytoplasmic filament whose basal portion is ensheathed by a short cylinder of gelatinous material. The cells are uninucleate and have one or two laminate parietal chloroplasts, each usually containing a single pyrenoid.[2]

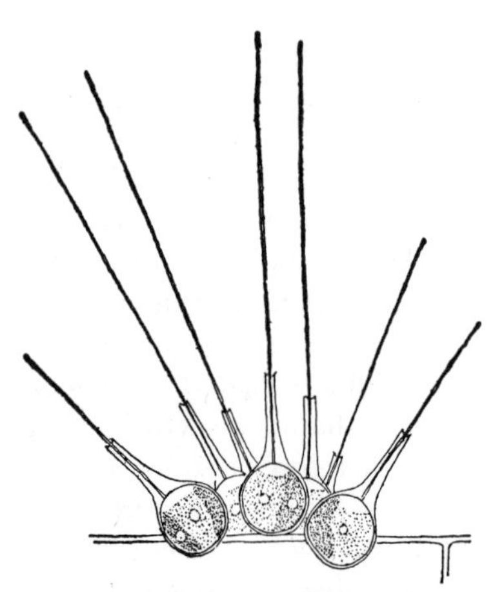

Fig. 101. *Chaetosphaeridium globosum* (Nordst.) Klebahn. (× 500.)

Cells of *Chaetosphaeridium* divide vegetatively. Division is usually transverse, with the lower daughter protoplast migrating to the apex of a long tubular outgrowth from the base of the parent-cell wall and then secreting a wall of its own and forming a seta.[3] Sometimes division is vertical and without a formation of a tubular outgrowth.

Asexual reproduction is by means of zoospores, four of which are formed within a cell.[2]

Chaetosphaeridium is a rather common epiphyte on filaments of the coarser Chlorophyceae, especially in regions where the water is soft. *C. globosum* (Nordst.) Klebahn (Fig. 101), *C. ovale* G. M. Smith, and *C. Pringsheimii* have been found in this country. For a description of *C. ovale*, see G. M. Smith (1916*C*); for the other two, see Hazen (1902).

[1] Chodat, 1898; Oltmanns, 1898; Pringsheim, N., 1860.
[2] Hazen, 1902; Klebahn, 1892*A*, 1893. [3] Klebahn, 1892*A*.

3. **Dicranochaete** Hieronymus, 1887. Thalli of this alga are epiphytic, unicellular, and distinguishable from other unicellular green algae by the presence of an erect dichotomously branched seta which arises from the side of the cell next to the substratum. The cells are usually solitary, but they may be gregarious and so densely crowded in short series of three or four that the plant body appears to be a true filament. When viewed from the side, the cells are semicircular in outline; when seen from above, the outline is reniform. A cell is enclosed by a homogeneous or lamellated wall, sharply differentiated into a cap-like portion covering the cell apex and a much thicker portion which encloses the rest of the cell. The cap portion consists largely of cellulose, the lower portion does not.[1] Setae of young cells are repeatedly forked cytoplasmic filaments surrounded for

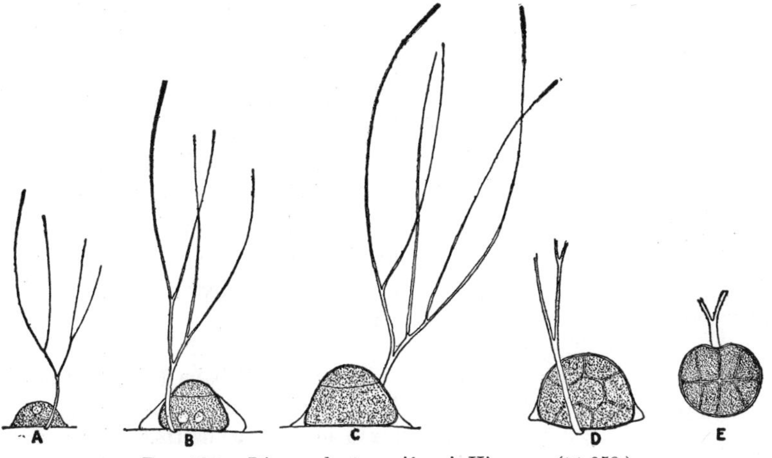

FIG. 102. *Dicranochaete reniformis* Hieron. (× 650.)

their entire length by a sheath of gelatinous material; those of old cells lack the cytoplasmic filament within the gelatinous sheath. The cells are uninucleate and with a single inverted cup-shaped chloroplast next to the upper side. A chloroplast usually contains two or three pyrenoids.[2]

Cells of *Dicranochaete* do not divide vegetatively. Asexual reproduction is by formation of 4 to 32 biflagellate zoospores which lie within a gelatinous vesicle when first liberated by abscission of the cap-like portion of the cell wall. After the swarming period, which lasts for but a few minutes, the zoospores settle down on some substratum, with their anterior ends downward, and retract their flagella. The seta is developed from what was the anterior end of a cell soon after retraction of the flagella.[2]

[1] HIERONYMUS, 1892; HODGETTS, 1916; WEST, G. S., 1912*A*.

[2] HIERONYMUS, 1892; HODGETTS, 1916.

Biflagellate gametes have been observed,[1] and it is thought that these fuse to form a quadriflagellate zygote that swarms for some time before it comes to rest and forms a wall.

Dicranochaete is a rare alga, which is more often epiphytic on submerged phanerogams and bryophytes than epiphytic upon other algae. Thus far in this country *D. reniformis* Hieron (Fig. 102) has been found only in Maine[2] and California.[3] For a description of it, see Collins (1918).

4. **Conochaete** Klebahn, 1893. *Conochaete* is epiphytic and with a plant body consisting of a cluster of cells. The number of cells in a cluster increases by cell division in a plane perpendicular to the substratum.[4] Each cell is surrounded by a fairly broad gelatinous envelope and bears two to several setae. Each seta consists of an axial filament whose base is ensheathed by a short cylinder of gelatinous material. The cells are uninucleate and generally contain a single parietal chloroplast with one pyrenoid.

Asexual reproduction is by a formation of four or eight zoospores which are liberated by a rupture of the parent-cell wall.[5]

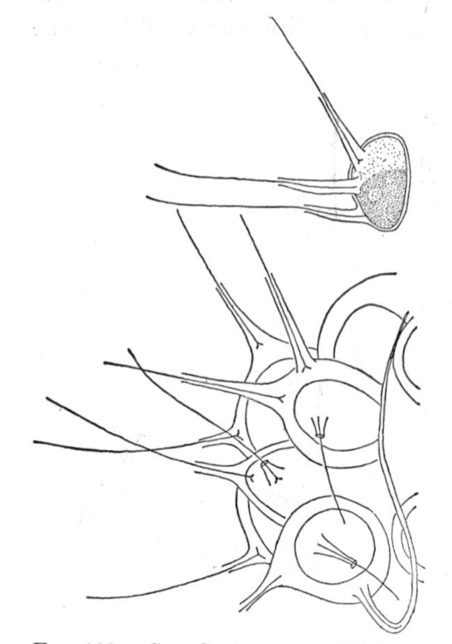

FIG. 103. *Conochaete comosa* Klebahn. (*After Prescott and Croasdale*, 1942.)

C. comosa Klebahn (Fig. 103) has been found epiphytic on coarse filamentous algae in Massachusetts.[6] For a description of it, see Heering (1914).

5. **Oligochaetophora** G. S. West, 1911. The cells of this alga are epiphytic and solitary, or in clusters of two to six. A cluster of cells may or may not be surrounded by a common gelatinous envelope.[7] The cells are globose to ellipsoidal and bear two to four long simple setae without a gelatinous sheath at their base. Within a cell is a single parietal chloroplast, with or without a pyrenoid, and a minute nucleus.

The method of reproduction is unknown.

The conventional practice of considering this genus closely related to

[1] HODGETTS, 1916. [2] COLLINS, 1918. [3] SMITH, G. M., 1933.
[4] KLEBAHN, 1893; SCHMIDLE, 1899*B*. [5] SCHMIDLE, 1899*B*.
[6] PRESCOTT and CROASDALE, 1942.
[7] WEST, G. S., 1911; PRESCOTT and CROASDALE, 1942.

Chaetosphaeridium and *Conochaete* is open to question because the structure of setae of *Oligochaetophora* is not the same as in these two genera.

O. simplex G. S. West (Fig. 104), the only species, has been found in Massachusetts.[1] The American material differed from the original description (G. S. West, 1908) in that the cells had a common gelatinous envelope and the chloroplasts had a pyrenoid.

FAMILY 7. TRENTEPOHLIACEAE

Algae belonging to this family differ from other Ulotrichineae in that their zoospores and gametes are formed within special cells differing in shape and structure from vegetative cells. Sporangia and gametangia may be terminal or intercalary in position and solitary or in series. Typical members of the family (*Trentepohlia, Cephaleuros*) have these reproductive cells markedly different from vegetative cells, but some genera do not have marked differences between the two (*Gongrosira, Leptosira*). In fact, it is hard to draw a definite line between these simpler Trentepohliaceae and the Chaetophoraceae. Many phycologists place *Gongrosira, Leptosira*, and other genera without strongly differentiated gametangial and sporangial cells in the Chaetophoraceae. Genera belonging to the Trentepohliaceae have a branching thallus, but one in which there is often a sharp differentiation into a prostrate and into an erect portion. The branches may be wholly free from one another (*Trentepohlia, Fridaea*), united except for the free ends of certain erect branches (*Gongrosira*), or wholly united to form a flattened discoid thallus (*Cephaleuros, Phycopeltis*). Sometimes the thallus has long unicellular setae (*Fridaea*) or multicellular setae (*Cephaleuros*), but setiferous branches are of much rarer occurrence than in the Chaetophoraceae.

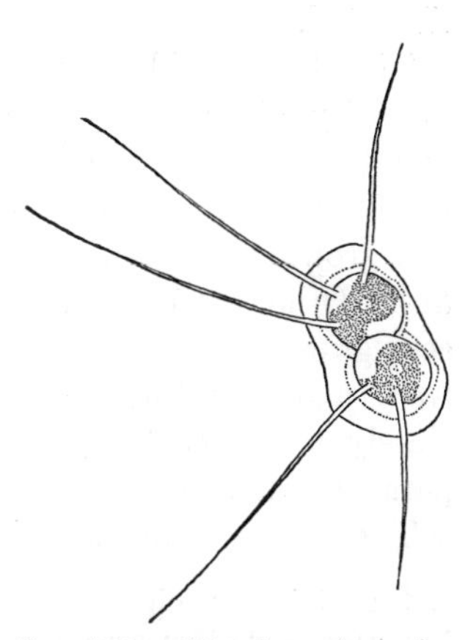

FIG. 104. *Oligochaetophora simplex* G. S. West. (*After Prescott and Croasdale*, 1942.)

The cells are characteristically uninucleate, but old filaments that have ceased to grow may have the cells containing a half-dozen or more nuclei. The chloroplasts are quite variable in this family: they may be the parietal bands characteristic of the Ulotrichineae, or they may be reticulate or broken up into a number of separate pieces. Pyrenoids may or may not be present, but starch is usually present in abundance. In special cases (*Trentepohlia*), the protoplast becomes deeply tinged with hematochrome.

[1] PRESCOTT and CROASDALE, 1942.

All members of the family reproduce asexually by means of zoospores, which are formed in sporangial cells. The liberation of the zoospores may take place while the sporangia are attached to the plant, or the sporangia may become detached from the plant and be dispersed by the wind. Such wind-borne sporangia (*Trentepohlia, Cephaleuros*) immediately produce zoospores if they chance to fall where there is sufficient moisture. Aplanospores and akinetes are also formed by many genera of the family.

Biflagellated gametes are known for several genera. These are borne in cells of much the same appearance as sporangia, but the gametangia are usually attached to the thallus at the time when their gametes are liberated.

The Trentepohliaceae are remarkable among the Ulotrichales for the diversity of habitats in which they grow. Some of them are strictly aquatic and grow in quiet or running water. These aquatic members of the family may be free-living or definitely associated with other organisms and growing upon the surface of, or within, the host (*Gongrosira, Gomontia*). Others of the family are aerial. Some of these aerial Trentepohliaceae grow on damp rocks and stones. Others are epiphytic on the bark of trees and sometimes, as in certain species of *Trentepohlia*, are rather closely restricted to specific genera. Still others of the family are true parasites and restricted to specific phanerogams.

The genera found in this country differ as follows:

1. Free living or epiphytic 2
1. Perforating other organisms, limestone, or dead wood 7
   2. Protoplasts of vegetative cells with hematochrome .... 6. **Trentepohlia**
   2. Protoplasts of vegetative cells without hematochrome 3
3. Cells moniliform .... 7. **Physolinum**
3. Cells not moniliform 4
   4. Filaments without long setae 5
   4. Filaments with long unicellular setae .... 5. **Fridaea**
5. Branching mostly unilateral .... 3. **Ctenocladus**
5. Branching not markedly unilateral 6
   6. Sporangia strictly terminal .... 4. **Gongrosira**
   6. Sporangia intercalary .... 1. **Leptosira**
7. Thallus without setae .... 2. **Gomontia**
7. Thallus with multicellular setae .... 8. **Cephaleuros**

1. **Leptosira** Borzi, 1883. The thallus of *Leptosira* consists of a dense tuft of yellowish-green subdichotomously branched filaments that gradually taper toward their apices but do not end in long hairs. Some of the branches are prostrate, but the majority are erect and densely radiate. The cells are barrel-shaped, and those lower in a filament have greatly thickened walls. The cells are uninucleate and with a single peripheral chloroplast without a pyrenoid.[1]

[1] Borzi, 1883; Steil, 1944.

Asexual reproduction is by means of zoospores, and any thick-walled cell in the older part of a thallus may become a sporangium. There is a repeated bipartition of the sporangial contents to form ovoid biflagellate zoospores which lie within a vesicle when liberated through a circular pore in the sporangial wall. Zoospores may germinate directly into new filaments,[1] or they may develop into *Characium*-like stages whose protoplasts divide to form four aplanospores.[2] These aplanospores are liberated by a gelatinization of the parent-cell wall and develop directly into multicellular thalli.

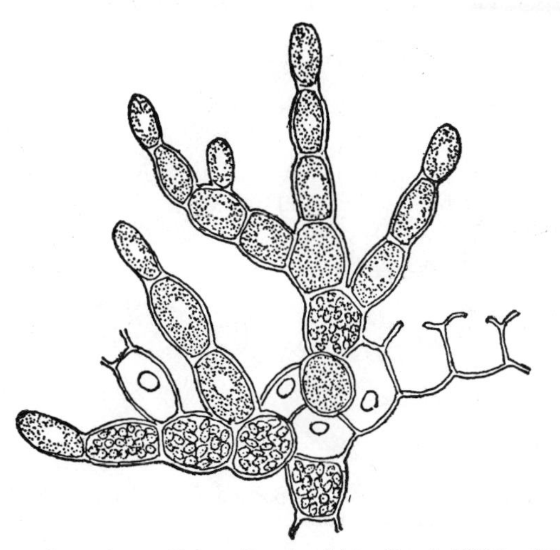

FIG. 105. *Leptosira mediciana* Borzi. (*After Borzi*, 1883.) (× 630.)

A production of biflagellate gametes which fuse posteriorly when they unite in pairs has been described[2] for *Leptosira*. The zygote is spherical and with a thick wall.

The type species, *L. Mediciana* Borzi (Fig. 105) has been found in Massachusetts[3] and Wisconsin.[4] For a description of it, see Collins (1912).

2. **Gomontia** Bornet and Flahault, 1888. Most species of this perforating alga grow in the calcareous shells of molluscs, but some of them perforate limestone rocks,[5] submerged woodwork,[6] or the dense mats of *Cladophora* balls.[7] Thalli of species perforating shells of molluscs usually have a greatly and irregularly branched mass of filaments immediately below the surface of the shell. The cells in this portion of the thallus are quite irregular in shape and often so densely crowded as to form a pseudo-

[1] STEIL, 1944; VISCHER, 1933. [2] BORZI, 1883.
[3] COLLINS, 1912. [4] STEIL, 1944.
[5] CHODAT, 1898. [6] MOORE, G. T., 1918. [7] ACTON, 1916.

parenchymatous mass. Numerous branched filaments, composed of long cylindrical cells, grow downward from the under side of the irregularly branched portion and penetrate deep into the substratum.[1] The cells have fairly thick stratified cellulose walls. Cells of some species are uninucleate;[2] those of other species are multinucleate.[3] The chloroplast is laminate, perforate, and with several pyrenoids.

Vegetative cells in the superficial portion of a thallus enlarge to form sporangia whose walls frequently have one or more thick rhizoid-like processes at one pole. Quadriflagellate zoospores are formed in large numbers within a sporangium.[4] Sporangia may also enter upon a long

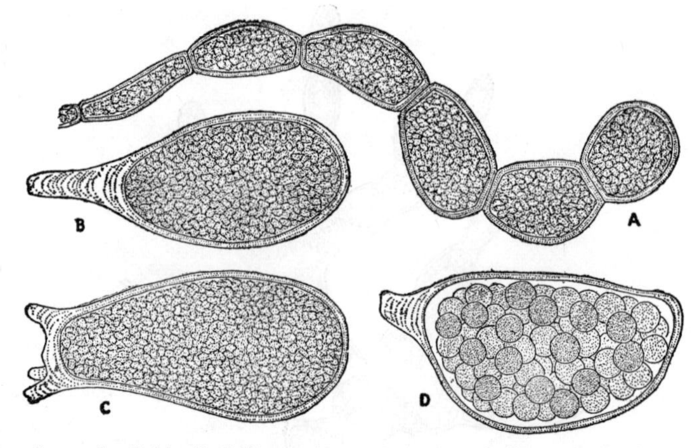

FIG. 106. *Gomontia Holdenii* Collins. *A*, portion of a thallus. *B–C*, akinetes. *D*, an akinete containing aplanospores. (× 485.)

rest period before giving rise to zoospores; such resting sporangia are quite like akinetes in their structure and behavior. Cultural studies of germination of zoospores show[2] that they develop into irregularly shaped cells resembling sporangia. Zoospores which have failed to escape from sporangia may develop into such cells *in situ*.[5] On this account it has been held[2] that the sporangium is the only vegetative cell in the life cycle and that the filaments found in association with it belong to some other alga. However, the fact that filaments are always found in association with the sporangia indicates that they are an integral part of the alga.

Most species of *Gomontia* are marine, but two fresh-water species, *G. Holdenii* Collins (Fig. 106) and *G. lignicola* G. T. Moore, have been found in this country. For a description of *G. Holdenii*, see Collins (1909); for *G. lignicola*, see G. T. Moore (1918).

[1] BORNET and FLAHAULT, 1888*A*, 1889; CHODAT, 1898. [2] KYLIN, 1935.
[3] ACTION, 1916. [4] KYLIN, 1935; WILLE, 1906. [5] LAGERHEIM, 1885.

3. **Ctenocladus** Borzi, 1883 (*Lochmiopsis* Woronichin and Popova, 1929). The thallus of this alga is gelatinous, spherical to hemispherical, and 5 mm. or more in diameter. It is of a moderately firm texture and not impregnated with lime. The filaments within the gelatinous matrix are distinctly radiate in arrangement and with a tendency toward unilateral branching. Lateral branches arise as outgrowths from the upper end of a cell, and the first cross wall is laid down some distance from the point of origin of a branch. There may be a formation[1] of palmella stages whose cells give rise to zoospores. Vegetative cells of filaments are cylindrical, with a length several times the breadth, and all cells in a branch are of the same length. A cell contains a singly parietal laminate chloroplast with one to several pyrenoids.

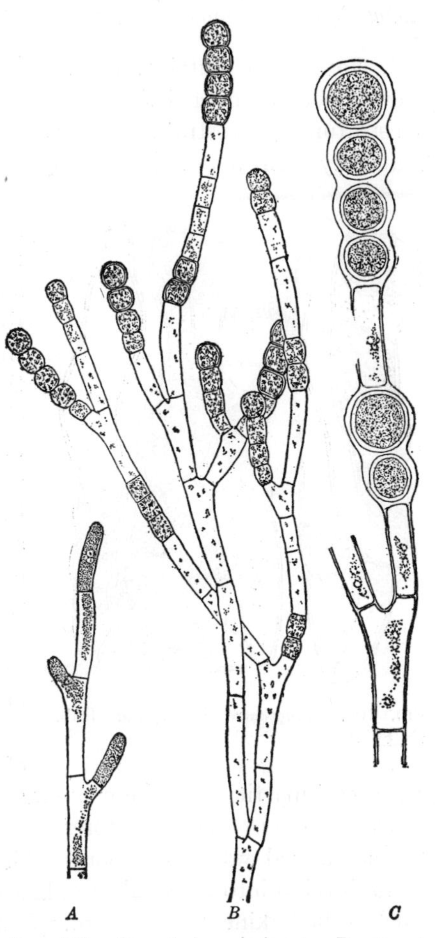

FIG. 107. *Ctenocladus circinnatus* Borzi. *A* portion of a young vegetative filament. *B–C* portions of old filaments with akinetes (*A–B*, × 325; *C*, × 650.)

Asexual reproduction is by a formation of zoospores within cylindrical sporangia borne terminally at the tips of branches. The sporangial contents divide to form 4, 8, 16, or 32 biflagellate zoospores that are liberated through a terminal pore in the sporangial wall.[2] Fully grown thalli usually have numerous akinetes which are borne in short catenate series that may be terminal or intercalary. When first formed, akinetes of each series abut on one another; as they grow older, they become somewhat separated from one another through gelatinization of the outer wall layer.

Sexual reproduction is isogamous and by a union of biflagellate gametes of which 30 to 60 are formed within a gametangium.

[1] BORZI, 1883. [2] BORZI, 1883; WORONICHIN and POPOVA, 1929.

Ctenocladus is a brine organism in which growth is active up to a NaCl concentration of 1.5 mol. and in which the resting stages remain viable in a saturated solution of NaCl.[1] *C. circinnatus* Borzi (Fig. 107) has been found in both inland and coastal brine pools in California. For a description of it as *Gongrosira circinnata* (Borzi) Schmidle, see Heering (1914).

4. **Gongrosira** Kützing, 1843. The thallus of *Gongrosira* is sessile and grows upon submerged timbers, stones, and shells. It is attached to the substratum by prostrate branches so densely interwoven with one another that they form a pseudoparenchymatous layer several cells in thickness. The upper surface of the pseudoparenchymatous portion gives rise to numerous erect filaments, sometimes sparingly branched, in which cells at the distal end are somewhat broader than those in the basal portion. The thallus is often encrusted with calcium carbonate; sometimes the encrustation is so heavy that the alga forms a hard green coating on the substratum. Cell walls are fairly thick and at times distinctly stratified. The chloroplast is laminate, parietal, and with one or more pyrenoids. In many cases, the cells are so densely filled with starch granules that the chloroplasts are quite indistinct.

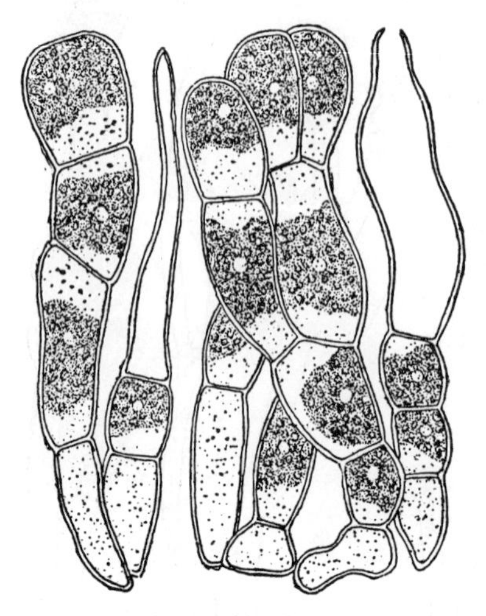

FIG. 108. *Gongrosira Debaryana* Rab. (× 800.)

Asexual reproduction is by means of biflagellate zoospores formed by repeated bipartition of the contents of sporangia borne terminally on the erect filaments.[2] The zoospores are liberated through a terminal pore in the sporangial wall. Zoospores which are not liberated from a sporangium develop into thin-walled aplanospores. There is also a formation of thick-walled akinetes whose protoplasts are reddish in color.[3]

The two species known to occur in this country are *G. Debaryana* Rab. (Fig. 108) and *G. lacustris* Brand. For descriptions of them, see Collins (1909, 1918).

5. **Fridaea** Schmidle, 1905. The thallus of this alga is a sessile, hemispherical to irregularly expanded, greenish mass, rarely over 1 cm. in diameter. The plant mass is more or less impregnated with lime. It

[1] RUINEN, 1933.
[2] BRISTOL-ROACH, 1920; SCHAARSCHMIDT, 1883; WILLE, 1883*A*, 1887*B*.
[3] SCHAARSCHMIDT, 1883.

is composed of repeatedly branched filaments, in which growth is largely terminal and in which all but two or three cells at the tips of branches are usually without chloroplasts.[1] Branches arise as lateral outgrowths from the distal end of a cell. Many of the cells have their upper ends drawn out into a long hollow seta with a swollen base. The chloroplasts of cells in rapidly growing branches are laminate, parietal, with one to four pyrenoids, and usually lie at the distal end of a cell. Slowly growing plants usually have the cell so densely packed with starch that but little of the cell structure is evident.

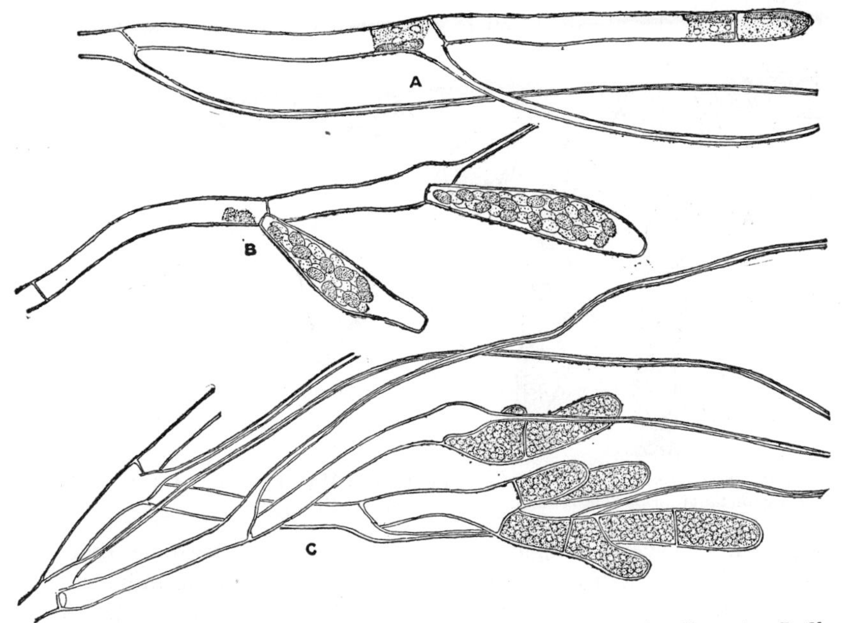

Fig. 109. *Fridaea torrenticola* Schmidle. *A*, apex of a young vegetative filament. *B*, filament with sporangia. *C*, apex of an old filament. (× 485.)

Asexual reproduction is probably by means of zoospores, many of which are produced within bottle-shaped sporangia borne laterally on the branches (Fig. 109*B*).

This genus differs from other Trentepohliaceae in that certain cells have their distal ends bearing long, hollow, *Bulbochaete*-like setae. In this country *F. torrenticola* Schmidle (Fig. 109) is known only from a single station in California.[2] For a description of it, see Heering (1914).

6. **Trentepohlia** Martius, 1817 (*Chroolepus* Agardh, 1824; *Nylandera* Hariot, 1890). *Trentepohlia* is a strictly aerial alga in which the thallus

[1] Schmidle, 1905. [2] Smith, G. M., 1933.

is filamentous and branched. The major portion of the plant body may be prostrate and with very few short erect branches; or the erect portion may be more extensive than the prostrate portion. The color of the plant mass is typically a brownish to yellowish red. Branching of erect filaments may be predominantly alternate, opposite, or unilateral. The cells are cylindrical to slightly moniliform and rarely of a length more than twice the breadth. Cell walls are lamellate and composed almost wholly of cellulose.[1] Some species have the wall layers parallel to one another and encircling the protoplast; other species have the wall layers outwardly and upwardly divergent. In the latter type of wall, there is usually a cap of pectose on the terminal cell of each branch. These caps, which have

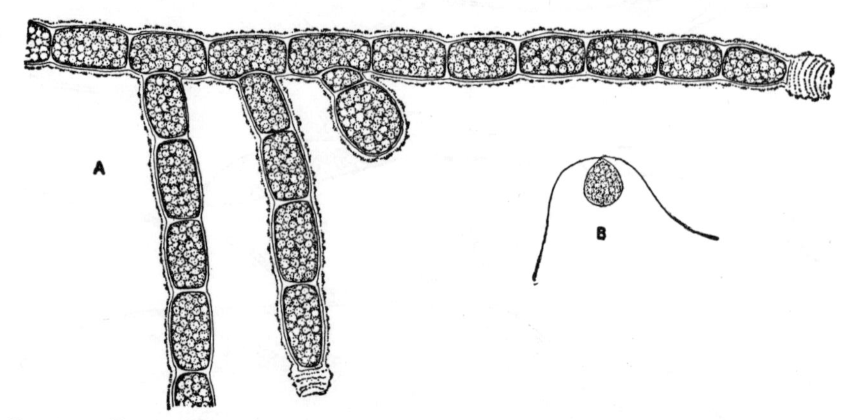

FIG. 110. *Trentepohlia aurea* var. *polycarpa* (Nees and Mont.) Hariot. *A*, portion of thallus with a gametangium on one branch. *B*, gamete. (× 975.)

successive transverse lamellae, may become so cumbersome that they impede apical growth of a filament.[1] The protoplasts are uninucleate in young cells but may be multinucleate in older ones. There are several parietal chloroplasts in each cell that, according to the species, are discoid, spiral bands, or combinations of the two.[2] Usually the number and structure of chloroplasts are completely obscured by development of beta carotene in such quantity as to color the entire protoplast a deep orange-red.

Many species have flagellated swarmers formed within two kinds of special cells: sessile intercalary or intercalary or terminal cells, and stalked cells which are always borne terminally and shed when mature. Both kinds of cell are often called sporangia, but it is very probable that sessile "sporangia" are gametangial in nature. The detachable reproductive cells are sporangial in nature and in several,[3] if not all, species the

[1] WEST and HOOD, 1911. [2] GEITLER, 1923.
[3] MEYER, K., 1909, 1936, 1936*A*, 1937.

zoospores formed within are quadriflagellate. Detachment of a sporangium is due to a development of special wall layers between a sporangium and the cell below it. Sporangia are dispersed as wind-borne spore-like bodies that immediately produce zoospores when moistened.[1] Under certain conditions, the contents of a sporangium become divided into aplanospores instead of into zoospores. Akinetes of *Trentepohlia* are generally produced in several successive cells in the prostrate portion of a thallus. They have quite thick walls and germinate directly into filaments.[2]

Gametangia may be intercalary or terminal, but in neither case are they shed from the thallus. Gametangia and sporangia of the same species are different in shape.[3] The gametes are biflagellate, and a fusion in pairs has been observed in certain species,[4] but has not been found among liberated gametes of other species.[5] In some species, gametes may develop parthenogenetically into filaments,[6] in other species, they do not.[7]

Trentepohlia is an aerial alga which is especially abundant in the tropics but which at times is found in temperate and arctic regions. It grows as a felted layer on rocks and on leaves and bark of trees. Certain species are the algal component of lichens. Printz (1939) has written the most recent monograph of the genus, and the following species that he recognizes have been recorded from the United States: *T. abietina* (Flotow) Hansg., *T. aurea* (L.) Martius (Fig. 110), *T. effusa* (Kremp.) Hariot, *T. Iolithus* (L.) Wallr., *T. lagenifera* (Hildebr.) Wille, and *T. odorata* (Wiggers) Wittr.

7. **Physolinum** Printz, 1921. The thallus of this alga is an irregularly branched filament that is not differentiated into prostrate and erect portions. The cells are markedly moniliform, a feature immediately distinguishing *Physolinum* from other branched filamentous green algae. Cell division is also distinctive. A cell about to divide sends out a papillate outgrowth whose tip eventually becomes the same size and shape as the cell. After this, a cross wall is formed in the narrow isthmus connecting the cell and the enlarged outgrowth. Within each cell is a parietal laminate chloroplast, with or without band-shaped projections. At times the chromatophore fragments into several disk-shaped chromatophores. There are no pyrenoids.[8]

Reproduction is by a formation of several small spherical aplanospores within certain intercalary cells. The aplanospores are liberated by a rupture of the parent-cell wall.[8] Vegetative multiplication by fragmentation of filaments is of frequent occurrence.

[1] Gobi, 1871; Karsten, 1891. [2] Meyer, K., 1909. [3] Meyer, K., 1909, 1936, 1936*A*, 1937. [4] Karsten, 1891; Lagerheim, 1883; Wille, 1887. [5] Meyer, K., 1936, 1936*A*, 1937. [6] Meyer, K., 1936*B*. [7] Meyer, K., 1938.
[8] Printz, 1921*A*.

P. monilia (de Wildm.) Printz (Fig. 111) has been found by Robert Runyon on trunks of trees near Brownsville, Texas. For a description of it, see Printz (1939).

8. **Cephaleuros** Kunze, 1829 (*Mycoidea* Cunningham, 1879). *Cephaleuros* grows as a subcuticular or as an intercellular parasite of the leaves and twigs of *Magnolia, Thea, Rhododendron, Piper*, and certain other plants of tropical and subtropical regions. The portion of the host containing the alga is a circular to irregular slightly elevated spot whose diameter is

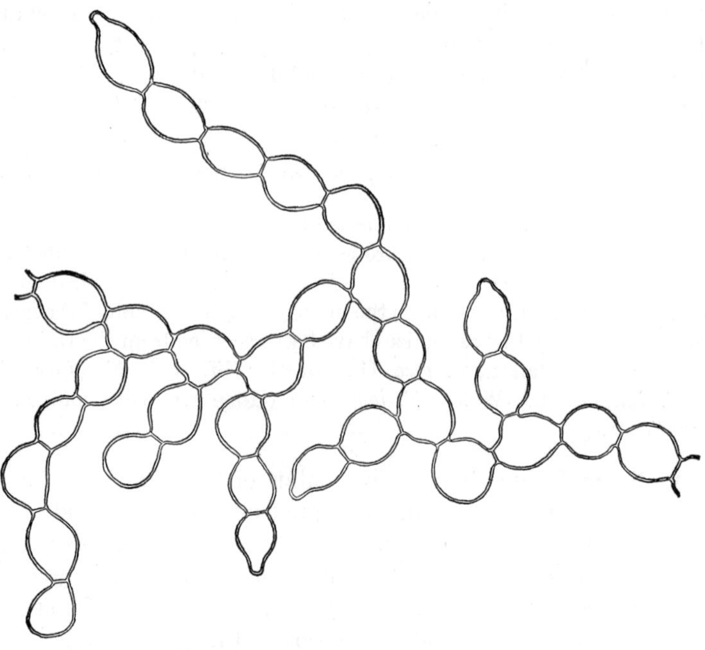

FIG. 111. *Physolinum monilia* (de Wildm.) Printz. (× 490.)

usually less than 2 mm. but may be as much as 1 cm. (Fig. 112*A*). The spots are greenish gray in color, velvety in texture, and often with the surface bearing reddish-brown hair-like structures. The thallus of the alga is a pseudoparenchymatous tissue, one to several cells in thickness, in which the cells are radiately arranged in much the same manner as in the discoid species of *Coleochaete*. The discoid mass, which grows just beneath the cuticular layer of epidermal cells, often has irregular branches on its under side, which grow down between the cells of the epidermis and deeper lying tissues. The upper surface of the algal mass bears numerous unbranched filaments which project vertically through the cuticle. Some of these erect filaments are sterile hairs; others bear a cluster of sporangia or gametangia at their apex. Cells of both the sterile and the fertile hairs are usually reddish brown, because of hematochrome in their protoplasts; cells in other parts of the thallus lack hematochrome. Cells of the hairs contain numerous discoid to irregularly shaped parietal chloroplasts which may be free from one another or united to form a green reticulum. The chloroplasts are without pyrenoids.[1]

Asexual reproduction is by means of zoospores which are formed in sporangia produced at the extremities of certain erect hairs. The sporangia are borne in clusters, and each lies at the end of a short stalk cell (Fig. 112*D*). When mature, the sporangia break away and are carried in all directions by the wind. They produce biflagellate zoospores as soon as they are moistened; if the sporangium is one that has fallen on a leaf or twig of a suitable host, the zoospore may germinate to form a new thallus after it ceases swarming.[2]

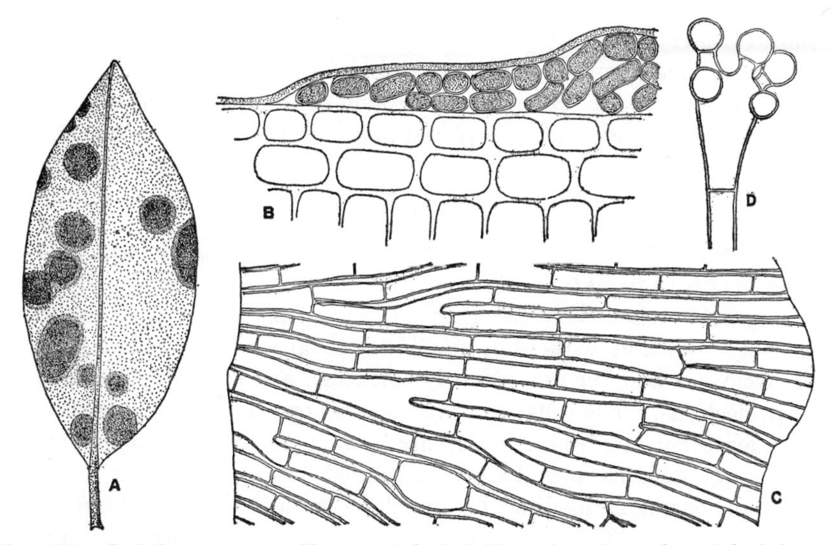

FIG. 112. *Cephaleuros virescens* Kunze. *A*, leaf of *Magnolia* with patches of *Cephaleuros*. *B–C*, vertical section and surface view of thallus. *D*, fertile filament with sporangia. (*A*, × ½; *B–D*, × 375.)

There is also a formation of biflagellate gametes within gametangia resulting from enlargement of certain cells in the pseudoparenchymatous portion of a thallus.[3]

[1] CUNNINGHAM, D. D., 1877; KARSTEN, 1891; THOMAS, 1913; WARD, 1883.

[2] CUNNINGHAM, 1877; KARSTEN, 1891; MANN and HUTCHINSON, 1907; WARD, 1883; WOLF, 1930.

[3] KARSTEN, 1891

Cephaleuros is a parasitic alga of wide distribution in tropical and subtropical regions and occurs on a number of hosts. It is a serious parasite when growing on certain cultivated plants, notably tea and pepper, and it is a parasite of some economic importance on *Citrus* growing in Florida.[1] *C. virescens* Kunze (Fig. 112) is widely distributed in the South Atlantic and Gulf Coast states of this country and is most frequently encountered on leaves of *Magnolia*. For a description of it, see Printz (1939).

SUBORDER 2. SPHAEROPLEINEAE

Because of their multinucleate cells the Sphaeropleaceae, with the single genus *Sphaeroplea*, are generally ranged alongside the Cladophoraceae. Since other characters are more like those of Ulotrichales, it is best placed there. However, the family differs so markedly from families assigned to the Ulotrichineae, that it is placed in a separate suborder, the Sphaeropleineae.

FAMILY 1. SPHAEROPLEACEAE

1. **Sphaeroplea** Agardh, 1824. This genus has long cylindrical cells, with a length 15 to 60 times the breadth, that are united end to end in free-floating unbranched filaments (Fig. 114*A*). The side walls of a cell are relatively thin and without a gelatinous sheath. End walls separating cells from one another are unevenly thickened and sometimes with knob-like projections. The cells contain numerous transverse cytoplasmic septa which are separated from one another by large vacuoles. Each cytoplasmic septum contains several nuclei, a band-shaped chloroplast with several pyrenoids, or numerous discoid chloroplasts certain of which contain pyrenoids. There is a thin layer of cytoplasm between the vacuoles and the side wall of a cell, but chloroplasts are lacking in this portion of the cytoplasm. Here and there in a cell, a septum develops a vacuole, the enlargement of which divides the septum into two equal parts that become more and more remote from each other as the vacuole increases in size. The resulting increase in length of a cell does not continue indefinitely since, sooner or later, there is a formation of a transverse wall. Growth in length of a filament may continue indefinitely, but usually the filaments become accidentally severed before attaining a length of a few centimeters. Vegetative multiplication by an accidental breaking of filaments is the only method of asexual reproduction in *Sphaeroplea*.

Sexual reproduction is usually oögamous, but it may be by a very advanced type of anisogamy. Eggs and antherozoids are usually produced in separate filaments, but sometimes they are formed in alternate cells of the same filament[2] (Fig. 113). *Sphaeroplea* is unique among oögamous Chlorophyceae in that the sex organs are not of distinctive shape. The

[1] WINSTON, 1938. [2] RAUWENHOFF, 1888.

first step in formation of an antheridial cell is an increase in number of nuclei, a division of the chloroplasts, and a disappearance of the pyrenoids.[1] There is then a progressive cleavage into angular uninucleate protoplasts each of which metamorphoses into a naked, spindle-shaped, biflagellate antherozoid[2] (Fig. 113). The antherozoids escape through several small pores in the lateral wall of a cell forming antherozoids. Oögonial cells do not have an increase in number of nuclei prior to cleavage of the cytoplasm into eggs. When first formed, the eggs are multinucleate, but later there is a degeneration of all nuclei but one. This is not accompanied by a disappearance of chloroplasts or pyrenoids.[3] The number and size of eggs within an oögonial cell are extremely variable, even in the same species. In most species, the diameter of eggs is more than half that of the

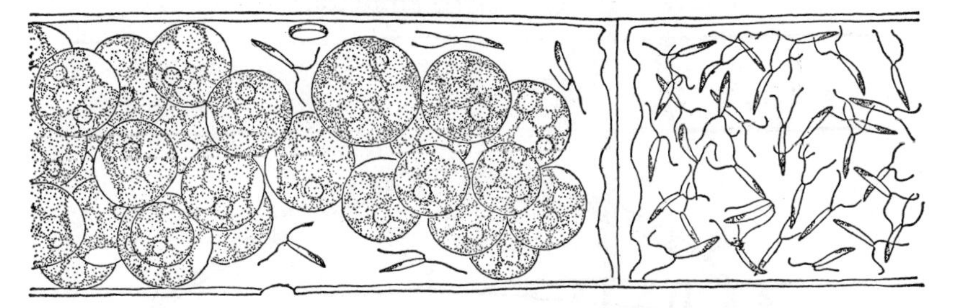

FIG. 113. *Sphaeroplea cambrica* Fritsch. Portion of an oögonial cell in which the eggs are being fertilized, and a portion of an antheridial cell. (× 650.)

oögonial cell, and they lie in a single to double longitudinal series within the cell. More rarely the eggs have a diameter less than a quarter that of the oögonial cell and lie in multiple longitudinal series within it. Oögonial cells containing mature eggs have small pores in the lateral walls (Fig. 113); and the antherozoids enter through these pores, swim about between the eggs, and eventually unite with them. In *S. cambrica* Fritsch there are at times large biflagellate female gametes with contractile vacuoles at the anterior end.[4] They move about sluggishly within the oögonial cell for only a short time and become immobile and lose their flagella before antherozoids unite with them. In *S. tenuis* Fritsch, fertilization probably takes place outside the oögonial cell, and there are good reasons for thinking that both male and female gametes of this species are motile.[5] Soon after fertilization, the zygote secretes a thick wall, with an ornamentation typical for the species, and the protoplast becomes bright red. The zygotes are eventually liberated by decay of the oögonial cell wall, and they may remain dormant for several years before germinating.

[1] KLEBAHN, 1899. [2] GOLENKIN, 1899; KLEBAHN, 1899.
[3] GILBERT, 1915; KLEBAHN, 1899. [4] PASCHER, 1939*A*. [5] FRITSCH, 1929.

When germination takes place,[1] there is usually a division of its protoplast to form four biflagellate zoospores; but one, two, or eight zoospores are sometimes formed. The zoospores are ovoid when first liberated, but shortly before or after they cease swarming they become spindle-shaped and with greatly attenuate poles. After coming to rest, the protoplast secretes a cell wall, but there is no formation of a holdfast. This free-floating cell (Fig. 114*D–E*) increases to many times its original length before it divides transversely.

Sphaeroplea is quite sporadic in occurrence, one year appearing in abundance at a given station and often not reappearing in succeeding years. Its vegetative period lasts but a few weeks. It is most often found in periodically flooded gravels,

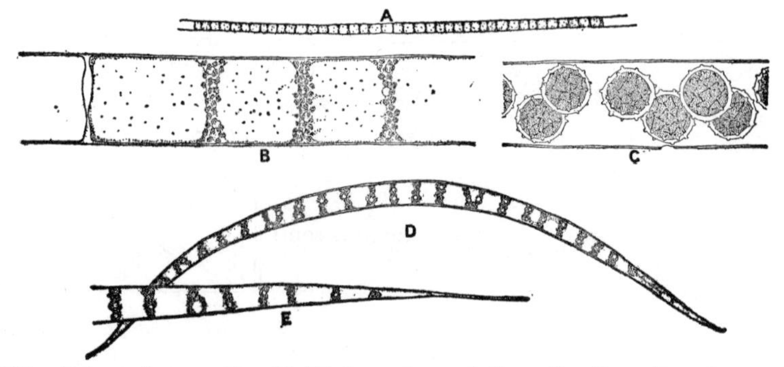

Fig. 114. *Sphaeroplea annulina* (Roth) Ag. *A*, vegetative cell. *B*, portion of a vegetative cell. *C*, portion of an oögonial cell containing ripe zygotes. *D–E*, one-celled germlings. (*A*, × 60; *B–C*, × 650; *D–E*, × 325.)

as pools in gravel pits. Sometimes it appears in flooded meadows. Two species, *S. annulina* (Roth) Ag. (Fig. 114) and *S. cambrica* Fritsch, are known for the United States. For descriptions of them, see Fritsch (1929).

ORDER 4. ULVALES

The Ulvales have uninucleate, more or less cubical, cells laterally united to form a thallus that is either an expanded mono- or distromatic sheet, a hollow tube, or a solid cylinder. Each cell contains a single cup-shaped or laminate chloroplast that usually contains a single pyrenoid.

Vegetative multiplication may take place by an accidental breaking of the thallus or by an abscission of proliferous shoots of the thallus. Asexual reproduction is by means of quadriflagellate zoospores, 4, 8, 16, or 32 of which are formed by repeated bipartition of the protoplast of a cell. Aplanospores and akinetes are not known to occur within the order.

Sexual reproduction is isogamous or anisogamous and by a fusion of biflagellate gametes. Most members of the order are heterothallic. The

[1] Cohn, 1856; Heinricher, 1883; Meyer, K., 1906; Rauwenhoff, 1888.

zygote may enter upon a period of rest or may germinate immediately. Species with an immediate germination of the zygote have an isomorphic alternation of generations and with cells of the diploid sporophytic generation dividing meiotically at the time of zoospore formation.

A large majority of the species are marine, but fresh-water representatives of both families are found in this country.

Family 1. Ulvaceae

Thalli of Ulvaceae have the cell divisions in two planes. A thallus may begin development as a hollow tube, with a wall one cell in thickness, and remain tubular throughout its entire development (*Enteromorpha*) or the tube may split and become a sheet one cell in thickness (*Monostroma*). In other cases, the thallus is never tubular. According to whether the cell divisions are largely restricted to one plane or take place freely in both planes, the thallus is a narrow ribbon never more than a few cells broad (*Percusaria*) or is an expanded sheet (*Ulva*). Most members of the family have cells that are angular by mutual compression and separated from one another by walls of medium thickness. In some species of *Monostroma*, the cells are rounded in outline and tend to lie in groups of four and at some distance from one another within a more or less homogeneous matrix. *Capsosiphon* also has gloeocapsoid cells, each surrounded by a gelatinous sheath, which are arranged in vertical series.

Sessile Ulvaceae are attached to the substratum by discoid holdfasts. In *Ulva*,[1] where the holdfast has been most thoroughly studied, the holdfast consists of tubular prolongations developed by cells in the lower part of a thallus. Near the point of attachment to the substratum, they emerge from the thallus and become closely appressed to one another to form a pseudoparenchymatous mass. Ends of the prolongations are somewhat swollen and multinucleate. There may be a formation of cross walls in the multinucleate portion. Other cells in a thallus are uninucleate and with a single chloroplast that usually contains one pyrenoid. *Monostroma* and *Enteromorpha* may multiply vegetatively by abscission of small proliferous outgrowths from the thallus. In *Ulva* and possibly other members of the family, there is a regenerative formation of new blades from old holdfast cells that have lived through the winter. Asexual reproduction is by repeated bipartition of the protoplast into 4, 8, 16, or 32 daughter protoplasts which are metamorphosed into quadriflagellate zoospores and liberated through a pore in the old parent-cell wall. The germling developed from a zoospore may be filamentous and with a rhizoidal basal cell, until it is several cells in length; or the germinating zoospore

[1] Delf, 1912.

may develop into an irregular mass of cells from which develops a thallus with the structure characteristic of the genus.

Sexual reproduction is by a fusion of biflagellate gametes and almost invariably by a fusion of gametes from separate thalli. Most species are isogamous, but anisogamous species have been found in *Enteromorpha*,[1] *Monostroma*,[2] and *Ulva*.[3] The quadriflagellate zygote remains motile for a short time and then retracts its flagella and secretes a wall. According to the genus, there is an immediate germination of the zygote without any enlargement, or it increases to several times its original diameter and does not germinate for several weeks. Cultural studies of gametes show that those of certain species germinate parthenogenetically, whereas those of others do not.[4] Cultural studies have also shown that species with an immediate germination of the zygote have an isomorphic alternation of generations.[5] Cytological study[6] and genetic analysis[7] show that the asexual plants are diploid and that meiosis occurs at the time of zoospore formation. Species have also been found where there is no sexual generation and where there is a succession of generations producing quadriflagellate zoospores.[8] Although not investigated cytologically, it is thought that the zoospores are diploid because of a lack of meiosis just prior to their formation. *Monostroma* has no alternation of generations, the greatly enlarged zygote producing zoospores which develop into sexual plants.

The Ulvaceae are primarily marine in habit, but certain of them grow equally well in salt, brackish, and fresh waters. This indifference to salinity is best exemplified by species of *Enteromorpha* growing on river boats traveling from salt to fresh water on alternate days.[9] Certain Ulvaceae also grow in brine lakes where salinity of the water is much higher than in the ocean.

The two genera found in fresh waters in this country differ as follows:

1. Mature thallus a hollow tube .... 1. **Enteromorpha**
1. Mature thallus a monostromatic sheet .... 2. **Monostroma**

1. **Enteromorpha** Link, 1820. Mature thalli of *Enteromorpha* are always hollow tubes with a wall one cell in thickness. Tubes of species growing in salt water are usually simple, but those growing in fresh water are usually branched and with numerous lateral proliferations. Young plants are sessile and attached by a basal rhizoidal cell or by basal multicellular

[1] KYLIN, 1930; MOEWUS, 1938*A*. [2] SUNESON, 1947; YAMADA, 1932. [3] SMITH, G. M., 1947.

[4] BLIDING, 1933; CARTER, N., 1926; FÖYN, 1934*A*; MOEWUS, 1938*A*; SUNESON, 1947; YAMADA and SAITO, 1938.

[5] BLIDING, 1933; FÖYN, 1929, 1934*A*; HARTMANN, 1929; MOEWUS, 1938*A*.

[6] FÖYN, 1929, 1934*A*; HARTMANN, 1929; RAMANATHAN, 1939.

[7] MOEWUS, 1938*A*. [8] BLIDING, 1933, 1938. [9] OSTERHOUT, 1906.

rhizoids. Older plants may be free-floating. The cells are angular by mutual compression, and irregularly disposed or with a tendency to lie in vertical series. Each cell is uninucleate, and the single parietal chloroplast usually contains but one pyrenoid, though some species[1] may have more than one.

Vegetative multiplication by an abscission of proliferous shoots is not uncommon in *Enteromorpha*. Asexual reproduction is by means of quadriflagellate zoospores, and they may be formed in any cell of a thallus except the lowermost ones. A cell may produce 4, 8, 16, or 32 zoospores,

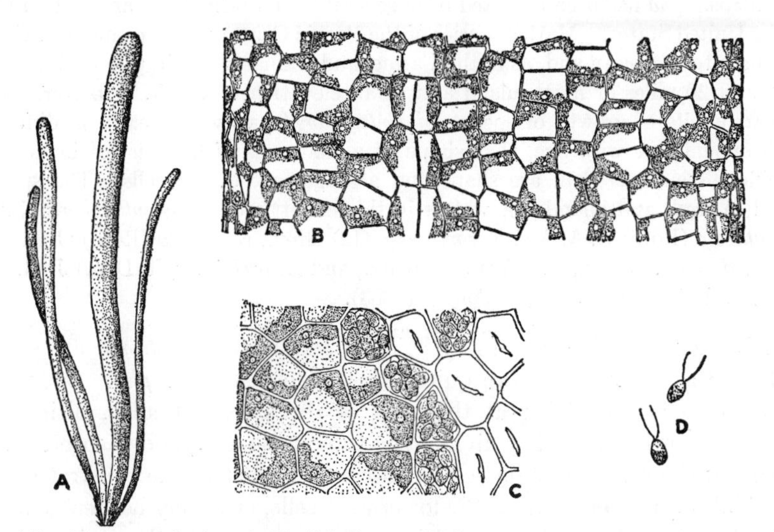

FIG. 115. *Enteromorpha intestinalis* (L.) Grev. *A*, cluster of young plants. *B*, vegetative portion of thallus. *C*, fertile portion of gametophyte. *D*, gametes. (*A*, natural size; *B*, × 485; *C–D*, × 650.)

and they escape through a pore or through a slit-like opening in the cell wall. At the end of the swarming period, a zoospore settles down on some firm object and secretes a wall. The first cell division in growth of a germling is transverse, the lower cell developing into a rhizoidal holdfast. Transverse division continues until there is a filament of a few cells, after which cell division is both vertical and transverse.

Sexual reproduction is by means of biflagellate gametes (Fig. 115), and all species thus far investigated have been found[2] to be heterothallic. Gametic union is isogamous in most of these species, but certain of them[3] are anisogamous. Germination of a zygote takes place within a day or

[1] BLIDING, 1938.
[2] BLIDING, 1933, 1938; HARTMANN, 1929; MOEWUS, 1938*A*; RAMANATHAN, 1939*A*.
[3] KYLIN, 1930; MOEWUS, 1938*A*.

two after it is formed, and division of the zygote nucleus is equational. A parthenogenetic germination of gametes has been recorded[1] for certain species, and the sexual plant developed from a gamete is of the same sex. *Enteromorpha* has an isomorphic alternation of generations. The chromosome number is doubled at the time of gametic union and halved just prior to zoospore formation.[2] There are also species where no sexual generation has been found and where a quadriflagellate zoospore gives rise to a thallus producing quadriflagellate zoospores.[3]

Enteromorpha has been collected from several inland brine lakes and salt springs in the United States. Marine species of the Pacific Coast are also common in rivers flowing into the ocean and, as in the Carmel and Salinas rivers in California, are to be found 20 miles or more inland and at an elevation of 200 ft. or better. The general habits of growth of species of *Enteromorpha* vary so greatly in different environments, or with age, that shape of a thallus is of less service in making specific distinctions than are shape and arrangement of the cells. The species found in fresh and inland salt waters in this country are *E. acanthophora* Kütz., *E. clathrata* (Roth) C. A. Ag., *E. compressa* (L.) Grev., *E. crinata* (Roth) C. A. Ag., *E. marginata* J. G. Ag., *E. micrococca* Kütz., and *E. prolifera* (Fl. Dan.) J. G. Ag. For descriptions of them, see Collins (1903).

2. **Monostroma** Thuret, 1854. Young thalli of *Monostroma* are sessile and with a saccate structure quite like that of *Enteromorpha*. Sooner or later, there is a splitting at the apex of the saccate thallus, ultimately extending to the base, that divides the tube into an expanded sheet. The expanded sheet may remain sessile and be attached to the substratum by rhizoidal protuberances from the lowermost cells, or it may be free-floating. This sheet-like thallus is one cell in thickness except in the region of the holdfast. The cells may be parenchymatous and angular by mutual compression; or they may be rounded, *Gloeocystis*-like, and in groups of four or more separated from one another by intervening gelatinous material. All cells of a thallus, including those of the holdfast, are uninucleate.[4] They contain a single parietal laminate chloroplast that encircles the greater part of a protoplast.

Vegetative multiplication may take place by an accidental tearing of a thallus or by an abscission of proliferous shoots. Thalli of most species thus far investigated have been found[5] to be exclusively gametophytic and producing biflagellate gametes. Gametes are formed by repeated bipartition of a protoplast[6] and are liberated through a pore in the parent-cell

[1] Moewus, 1938*A*; Ramanathan, 1939*A*.
[2] Hartmann, 1929; Kylin, 1930; Ramanathan, 1939*A*. [3] Bliding, 1933, 1938.
[4] Carter, N. 1926.
[5] Carter, N., 1926; Moewus, 1938*A*; Schreiber, 1942; Suneson, 1947; Yamada, 1932; Yamada and Kanda, 1941; Yamada and Saito, 1938. [6] Carter, N., 1926.

wall. All species have been found to be heterothallic. Some species are isogamous and others are anisogamous. After swarming for a very short time, a zygote loses its flagella and secretes a wall. It may enlarge directly 15 to 20 times its original diameter,[1] or its protoplast may escape from the original wall and secrete a new wall before enlarging.[2] When fully enlarged, a zygote is more or less ovoid, and the chloroplast contains several pyrenoids. Sooner or later the zygote's contents divide to form many quadriflagellate zoospores which are liberated through a pore in the wall.[3] The zoospores grow into gametophytes, and it has been shown[4] that half

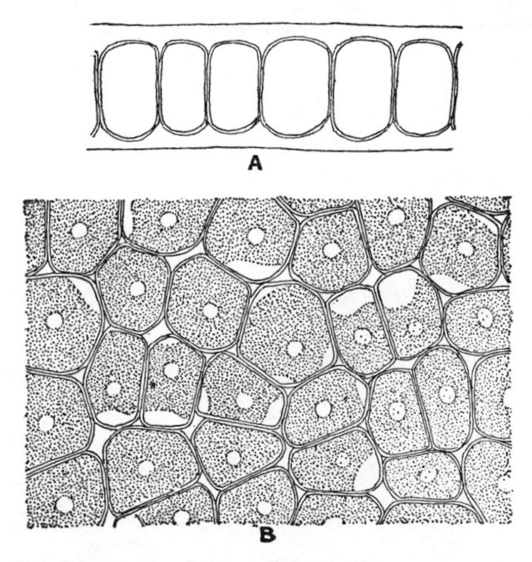

FIG. 116. *Monostroma latissimum* (Kütz.) Wittr. *A*, cross section. *B*, surface view. (× 800.)

of the zoospores from a germinating zygote develop into gametophytes of one sex and half into gametophytes of the opposite sex. Two species have thalli in which the only reproductive body is a quadriflagellate zoospore.[5] In one species, the zoospore germinates directly into a thallus producing zoospores; in the other species, a zoospore develops into a cyst-like body closely resembling a mature zygote. The contents of these cysts divide to form many quadriflagellate zoospores which develop into thalli producing quadriflagellate zoospores.

Monostroma is more often found in brackish water than in the ocean. One species is exclusively fresh-water, but some of the marine and brackish-water spe-

[1] MOEWUS, 1938*A*; SCHREIBER, 1947.
[2] CARTER, N., 1926; YAMADA and SAITO, 1938.
[3] MOEWUS, 1938*A*; YAMADA and SAITO, 1938.
[4] MOEWUS, 1938*A*.
[5] YAMADA and KANDA, 1941.

cies are occasionally found in fresh waters. The three species found in fresh waters in this country are *M. amorphum* Collins, *M. latissimum* (Kütz.) Wittr. (Fig. 116), and *M. quaternarium* (Kütz.) Desm. For descriptions of them, see Collins (1909, 1912).

FAMILY 2. SCHIZOMERIDACEAE

This family contains but one genus, *Schizomeris*. Some phycologists[1] think it is merely a developmental form of *Ulothrix* unworthy of generic rank; others[2] think it a valid genus and one belonging to the Ulotrichaceae. Still others[3] think that its affinities are more with the Ulvales than with the Ulotrichales.

1. **Schizomeris** Kützing, 1843. During early stages of its development, this alga is an unbranched uniseriate filament with an acuminate apical cell and a somewhat elongate basal cell terminating in a discoid holdfast. Later in development of a thallus, there may be vertical divisions at right

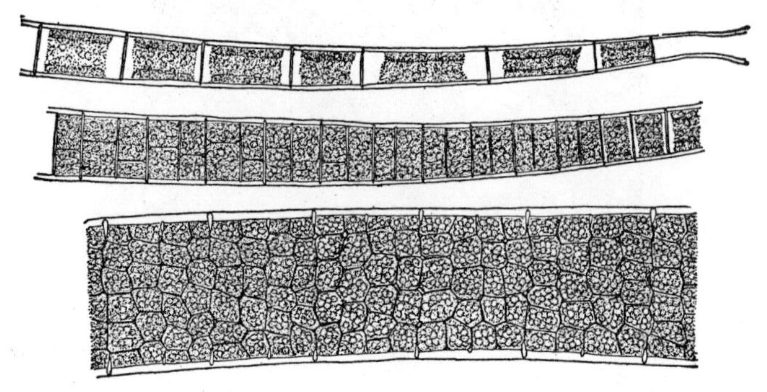

FIG. 117. *Schizomeris Leibleinii* Kütz. Portions of a thallus at three different levels. (× 325.)

angles to each other in all cells but those toward the base of a filament. Continued division in all planes results in the upper portion of a thallus becoming a solid cylinder of brick-like cells. The cylinder may have parallel sides, or it may be constricted at infrequent and irregular intervals. Cells of simple filaments of *Schizomeris* have fairly thick lateral walls and are separated from one another by ring-like transverse walls.[4] In young filaments, there are alternate rings which extend to, and which do not extend to, the surface. These rings persist after vertical cell division sets in and separate, one from another, portions of the cylindrical thallus derived from a single cell of the filamentous stage. Chloroplasts of the fila-

[1] PRINTZ, 1927; WILLE, 1890. [2] COLLINS, 1909; FRITSCH, 1935; HEERING, 1914.
[3] HAZEN, 1902; KORSHIKOV, 1927A; WEST, G. S., 1916.
[4] WATSON and TILDEN, 1930.

mentous stage are ulotrichoid and encircle about two-thirds of the protoplast. They usually contain several pyrenoids. Cells of cylindrical thalli have more massive chloroplasts which fill most of a protoplast.

Vegetative multiplication may take place by fragmentation of old thalli. The region at which a break occurs is almost always a constricted portion of a thallus and may possibly be due to disintegration of the transverse ring of wall material persisting from the filamentous stage. Asexual reproduction is by means of quadriflagellate zoospores formed in the upper part of a thallus.[1] Most[2] of those who have observed liberation of zoospores record a breaking down of the cross walls in the region of zoospore formation and an escape of zoospores through the apex of a thallus, but liberation of zoospores by a gelatinization of lateral walls has also been reported.[3]

The reported[4] germination of zoospores to form short uniserial filaments whose cells produce one or two biflagellate gametes is in need of confirmation. The zygote formed by union of these gametes is said to have a resting period of about a month and then a direct germination into a thallus producing zoospores. According to this interpretation, *Schizomeris* has a heteromorphic alternation of generations.

Although the single species, *S. Leibleinii* Kütz. (Fig. 117) has been found to be widely distributed in the United States, it is not a common alga. For a description of it, see Collins (1909).

ORDER 5. SCHIZOGONIALES

Cells of algae belonging to this order have stellate chloroplasts in which there is a single centrally located pyrenoid. The construction of the plant body is the same as in the Ulvales; the thallus may be a simple filament, an expanded sheet, or a solid cylinder. An even more important difference between the Schizogoniales and the Chlorophyceae thus far discussed is the complete lack of zoospores and of sexual reproduction among the Schizogoniales. Reproduction is by means of akinetes and aplanospores.

The first cell divisions in the development of the thallus are always transverse and result in a simple filament. The filamentous stage may at times have a false branching. This results from the death of one or two cells in the filament and the growth of the adjoining portions through the sheath investing the filament.[5] Later in the development of the simple filament, there may be longitudinal divisions. Subsequent divisions may be almost exclusively transverse and so cause the development of a ribbon-

[1] Korshikov, 1927*A*; Wood, 1872. [2] Hazen, 1902; Wolle, 1887; Wood, 1872.
[3] Korshikov, 1927*A*. [4] Kawasaki, 1937. [5] Hodgetts, 1920*A*.

shaped thallus two or a few cells broad; or longitudinal and transverse divisions may be about equal in number and so cause the development of an expanded sheet one cell in thickness. According to the interpretation of generic limits, the filamentous and laminate thalli are considered separate genera[1] or are both placed in the same genus.[2] Cell division may also take place in three planes and result in a solid cylindrical thallus. Most phycologists place Schizogoniales with a cylindrical thallus in a distinct genus, *Gayella*. All three types of thalli may have rhizoidal outgrowths. These may be formed by any cell of the plant, but most of the rhizoids develop from cells at the base of the thallus.[3]

Cells of vigorously growing thalli have a single, central, stellate chloroplast with a pyrenoid at the center. The nature of the food reserves formed by the Schizogoniales is in dispute; some[4] affirm that starch is formed, others deny[5] that it is present. The cells are uninucleate and with the nucleus excentric in position.[6]

Vegetative multiplication is of common occurrence in thalli of Schizogoniales. Simple filaments may have a fragmentation into *Stichococcus*-like segments containing one to four cells,[4] or the whole filament may dissociate into a *Protococcus*-like mass of spherical cells[7] which readily separate from one another. Vegetative multiplication of species with a laminate thallus may also take place by the abscission of small proliferous outgrowths.

Asexual reproduction is usually by means of akinetes. These may be formed directly from vegetative cells, or akinete formation may be preceded by vegetative divisions that make the thallus two cells in thickness in the region where akinete formation is to take place. Portions of the thallus two cells in thickness may have a direct development of the cells into akinetes[8] or may have the cells dividing into four daughter cells that become akinetes.[9] Akinetes are liberated by a softening of the thallus and may develop directly into new plants, or they may become aplanosporangia which contain several aplanospores.[10]

The anisogamous sexual reproduction by means of biflagellate gametes, reported for *Prasiola*,[11] is in need of confirmation before it can be accepted.

There is but one family in the order.

[1] Chodat, 1902; Collins, 1909; Gay, 1891; Setchell and Gardner, 1920.

[2] Brand, 1914; Printz, 1927; West, G. S., 1916; West and Fritsch, 1927; Wille, 1901, 1906*A*.

[3] Brand, 1914; Gay, 1891; Wille, 1901. [4] Gay, 1891; Wille, 1901.

[5] Brand, 1914. [6] Wille, 1901.

[7] Borzi, 1895; Gay, 1891.

[8] Setchell and Gardner, 1920*A*. [9] Lagerheim, 1892. [10] Wille, 1901, 1906*A*. [11] Yabe, 1932.

FAMILY 1. SCHIZOGONIACEAE

As already noted, several phycologists group all species of the order in one genus. However, pending careful cultural studies on the various species, it seems best to recognize *Schizogonium*, *Prasiola*, and *Gayella* as distinct. The two genera in the fresh-water flora of this country differ as follows:

1. Mature thalli filamentous or ribbon-like .... 1. **Schizogonium**
1. Mature thalli blade-like .... 2. **Prasiola**

1. **Schizogonium** Kützing, 1843. The thallus of this alga is usually filamentous, but at times there may be longitudinal cell divisions which result in a plant body two or a few cells broad. Narrow ribbon-like thalli

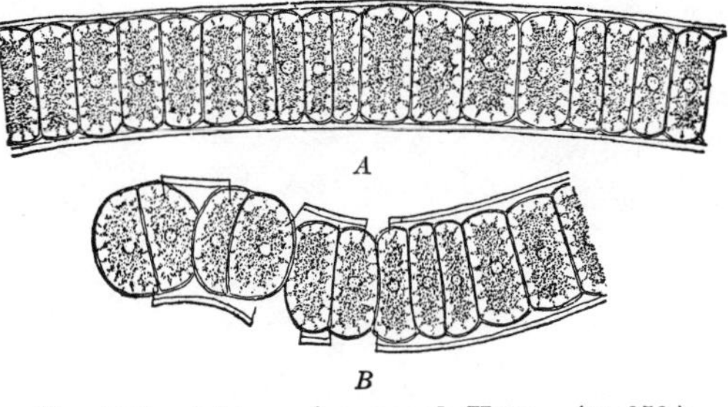

FIG. 118. *Schizogonium murale* Kütz. (× 650.)

often have rhizoidal outgrowths from certain of the cells. Cells of filamentous thalli are cylindrical and with a length greater or less than the breadth. Cells of ribbon-like thalli are approximately cubical. *Schizogonium* has uninucleate cells containing a central stellate chloroplast with one pyrenoid.

Vegetative multiplication by fragmentation is the usual method of reproduction and may be by a death of certain cells in a filament or by a dissociation into short segments containing one to four cells (Fig. 118). There may also be a formation of akinetes.

The two species found in this country are *S. crenulatum* (Kütz.) Gay. and *S. murale* Kütz. (Fig. 118). For descriptions of them, see Collins (1909).

2. **Prasiola** Meneghini, 1838. Although the thallus of *Prasiola* may be a simple filament or a narrow ribbon, it is usually an irregularly expanded sheet, one cell in thickness, that rarely attains a diameter of more than 1 cm. The thalli are sessile and are attached by rhizoidal outgrowths from the margin, or are attached by a thickened stipe. The cells have a tendency to lie in groups of four that, in turn, lie in larger groups separated

from one another by narrow or broad intervening spaces running in definite directions through a blade-like thallus. The cells are uninucleate and have a central stellate chloroplast containing one pyrenoid.

Vegetative multiplication is by an abscission of proliferous shoots which attach themselves to the substratum and grow into independent plants.[1] There may also be fragmentation at the *Schizogonium* stage of development. Asexual reproduction is by a formation of akinetes directly from vegetative cells, or by division of vegetative cells into four or more daughter cells which become akinetes.[2] Akinetes may develop directly into new thalli, or they may become sporangia which contain several aplanospores.[3]

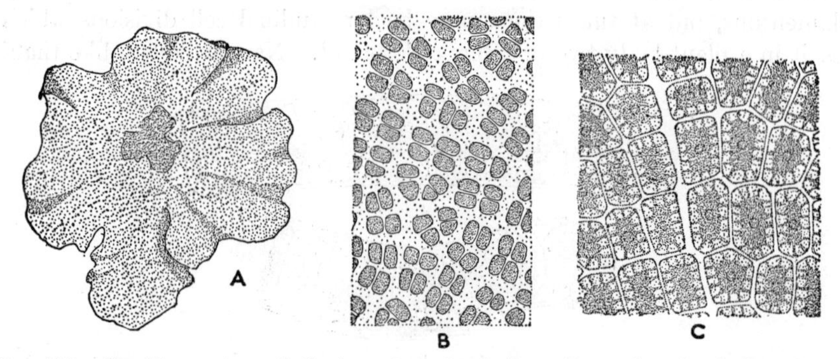

FIG. 119. *Prasiola mexicana* J. G. Ag. *A*, entire thallus. *B*, portion of a thallus. *C*, portion of a thallus of *P. meridionalis* Setchell and Gardner, a marine species. (*A*, × ½; *B*, × 485; *C*, × 975.)

Mature thalli of *Prasiola* resemble immature thalli of *Ulva* or *Monostroma*. Some of the species found in this country are marine, but there are more that are fresh-water, and either aerial or aquatic. Certain species of *Prasiola* grow only where the substratum is rich in nitrogenous compounds, as on rocks covered with the droppings of sea birds. Other species grow only in cold swiftly flowing mountain streams. Specific differences are based upon both shape of the thallus and arrangement of the cells.[4] The fresh-water species found in this country are *P. calophylla* (Spreng.) Menegh., *P. crispa* (Lightf.) Menegh., *P. fluviatilis* (Sommerf.) Aresch., *P. mexicana* J. G. Ag. (Fig. 119*A–B*), and *P. nevadense* Setchell and Gardner. For a description of *P. nevadense*, see Setchell and Gardner, 1920*A*; for descriptions of the others, see Collins (1909).

ORDER 6. OEDOGONIALES

The Oedogoniales have cylindrical uninucleate cells joined end to end in simple or branched filaments. The method of cell division is unique (see

[1] COLLINS, 1909; SETCHELL and GARDNER, 1920.
[2] LAGERHEIM, 1892; SETCHELL and GARDNER, 1920*A*.
[3] WILLE, 1901, 1906*A*.
[4] IMHAUSER 1889.

page 198). Motile reproductive cells are distinctive in having a transverse whorl of flagella at the anterior end.

Asexual reproduction is by zoospores formed singly within a cell. Sexual reproduction is always oögamous, and certain species of all genera produce their antherozoids in peculiar dwarf filaments.

There are but three genera, and of these *Oedogonium* is the only one with unbranched filaments. In the two genera with branched filaments, *Bulbochaete* has terminal and lateral one-celled hollow setae with bulbous bases, and *Oedocladium* lacks them. *Oedogonium* and *Bulbochaete* are aquatic, and *Oedocladium* is terrestrial in habit. *Oedogonium* and *Bulbochaete* are sessile when young and may remain so throughout their entire development. An apical-basal polarity is evident in many of their cells and is maintained even if filaments break away and become free-floating. *Oedocladium* is not sessile but has a similar apical-basal polarity.

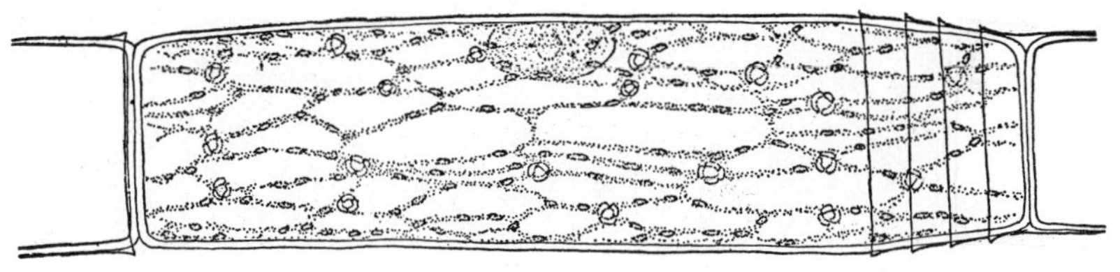

FIG. 120. Vegetative cell of *Oedogonium crassum* (Hass.) Wittr. (× 485.)

Cells of Oedogoniales have walls that seem to be homogeneous but which, except in the basal cell of sessile genera, consist of three concentric parts.[1] The portion next to the protoplast is composed of a layer of cellulose, external to this is a zone of pectose, and the outermost portion has chitin as the predominating substance. The chloroplast (Fig. 120) is a reticulate sheet extending from pole to pole and completely encircling the protoplast. According to the species, the strands of the reticulum are broad or narrow, but in either case the majority of the strands are parallel to cell's long axis. The pyrenoids, of which there are usually many in the chloroplast, lie at the intersections of the reticulum. Each pyrenoid is surrounded by a sheath of starch plates. Starch plates formed by the pyrenoids may migrate to, and accumulate in, the strands of the reticulum until the reticulate nature of the chloroplast is completely obscured by this "stroma" starch. The single nucleus usually lies midway between the ends of the cell and just within the chloroplast. It is of large size, biscuit-shaped, and with a well-defined chromatin-linin network and one or more nucleoli.[2]

The unique method of cell division, resulting in the formation of distinctive "apical caps" at the distal end of certain cells, has been repeatedly

[1] TIFFANY, 1924; WURDACK, 1923.

[2] OHASHI, 1930; STRASBURGER, 1880; TUTTLE, 1910; VAN WISSELINGH, 1908.

investigated, especially in *Oedogonium.*[1] The first indication of cell division is a migration of the nucleus toward the distal end of the cell, until it lies about two-thirds the distance from the proximal end. After elongating somewhat, the nucleus divides mitotically. During the prophases of mitosis, there is the appearance of a ring of wall material (Fig. 121*A*) that completely encircles the inner face of the lateral wall just below the distal end of the cell. The ring, which is thought to consist of hemicellulose,[2] increases in thickness until it is several times thicker than the rest of the

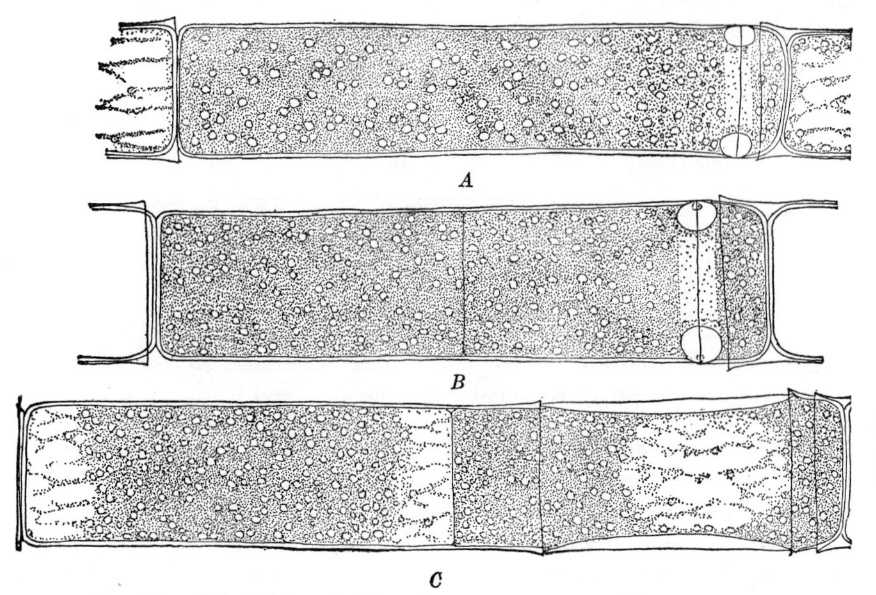

Fig. 121. Cell division of *Oedogonium crassum* (Hass.) Wittr. (× 485.)

lateral wall. There is next a formation of a small groove completely encircling the portion of the ring adjoining the lateral wall. After this groove has been developed, a transverse rent appears in the portion of the lateral wall external to the groove. Mitosis is completed by the time that the ring is fully developed, and, shortly after the two daughter nuclei are reconstructed, there is a transverse cytokinesis of the protoplast by an annular furrowing of the plasma membrane midway between the ends of the cell. There is no elongation of the cell during these stages of division (Fig. 121*B*); after the transverse division of the protoplast, each daughter protoplast elongates to about the same length as that of the mother cell (Fig. 121*C*).

[1] Hirn, 1900; Kraskovits, 1905; Ohashi, 1930; Pringsheim, N., 1858; Steinecke, 1929; Strasburger, 1880; Tuttle, 1910; van Wisselingh, 1908, 1908*A*.

[2] Steinecke, 1929.

This elongation takes but a short time and is often completed within 15 min. The lower daughter protoplast elongates until its distal end is level with, or slightly above, the former level of the hemicellulose ring. The wall lateral to this protoplast is therefore the side wall of the old parent cell. Meanwhile, the upper daughter protoplast has been elongating to about

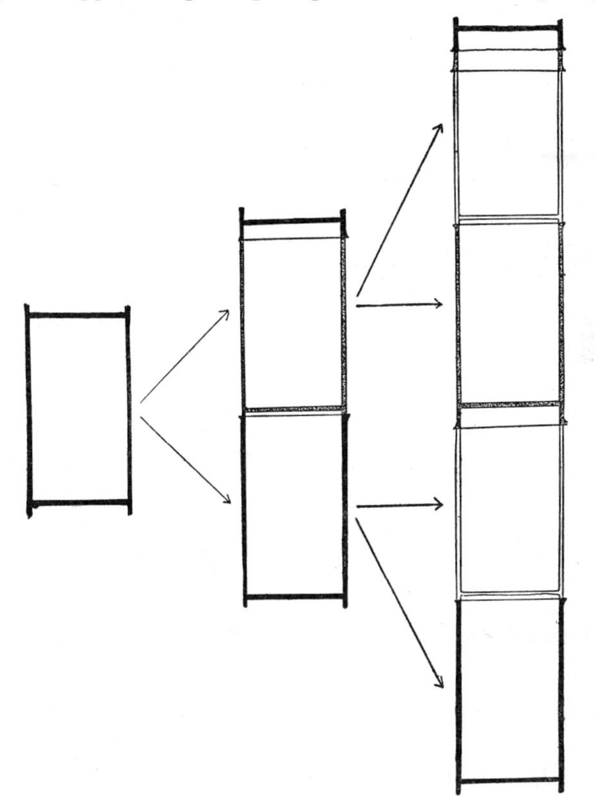

FIG. 122. Diagram showing the distribution of portions of parent-cell wall to daughter cells in *Oedogonium*. Walls of the first cell generation are shaded black, those of the second cell generation are in stipple, and those of the third cell generation are unshaded.

the same extent. The wall lateral to this elongated protoplast is formed by a stretching of the hemicellulose ring, except for the persistent portion of the mother cell wall at the distal ends, the apical cap. After the daughter cells have completed their elongation, there is the secretion of a transverse wall which separates them from each other (Fig. 122).

Cell division is intercalary in *Oedogonium*. Division of every cell in the filament and repeated division of the daughter cells would result in alternate cells with and without caps. Cells with one, two, three, and more caps would also have a definite disposition with respect to one another (Fig. 122).

This theoretical condition rarely obtains in nature, and repeated division of the distal daughter cell of previous divisions may result in filaments in which a cell with several apical caps is successively followed by several cells without caps. In *Oedocladium*, division usually takes place only in the terminal cell of the branches. As a result, the terminal cell is often the only one with caps. The succession of cell division in *Bulbochaete* will be discussed in connection with the germination of its zoospore (page 201).

Vegetative multiplication by an accidental breaking of the filament is of common occurrence in certain species of *Oedogonium* but is rarely found in either *Bulbochaete* or *Oedocladium*. Asexual reproduction by means of zoospores is characteristic of all three genera. Zoospores are formed singly

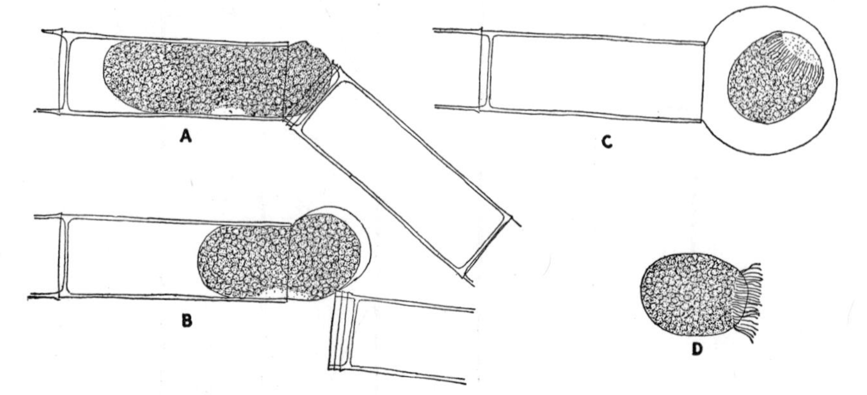

FIG. 123. Liberation of zoospores of *Oedogonium*. *A–B*, amoeboid migration from parent-cell wall. *C*, swarming of zoospore within the vesicle. *D*, free-swimming zoospore. (× 325.)

within a cell and usually only by cells with apical caps. Preparatory to zoospore formation in *Oedogonium*,[1] the nucleus retracts slightly from the chloroplast, and a hyaline region appears between the wall and nucleus. A ring of blepharoplast granules then appears around the margin of the hyaline area, and it is quite probable that each blepharoplast granule gives rise to one flagellum. After the zoospore is formed, there is a transverse splitting of the lateral wall at the apical cap, and the zoospore, enclosed by a delicate vesicle, emerges through this aperture (Fig. 123). Liberation of the zoospore and vesicle takes about 10 min. It has been thought[2] that the transverse splitting of the wall and the pushing out of the zoospore result from the pressure caused by the imbibitional swelling of gelatinous substances secreted by the protoplast. After its liberation from the wall the vesicle around the zoospore increases in size but this is soon followed by its disappearance. The zoospore within the vesicle assumes an ovoid or pyri-

[1] HIRN, 1900; KLEBS, 1896; OHASHI, 1930; PRINGSHEIM, N., 1858; STRASBURGER, 1892.

[2] STEINECKE, 1929.

form shape and moves sluggishly for a short time, but about the time the vesicle disappears it begins to move actively. The period of swarming of *Oedogonium* zoospores usually lasts but an hour or so, after which the zoospores come to rest with the hyaline end downward, retract their flagella, and develop a holdfast that attaches them to the substratum. The type of holdfast depends upon both the species concerned and the nature of the substratum. *Oedogonium* species with rhizoidal holdfasts have been shown[1] to form a simple holdfast if the substratum is smooth, and a holdfast that is more or less branched if the substratum is rough. Shortly after the zoospore becomes sessile, there is a secretion of a cell wall, but one different from that enclosing ordinary vegetative cells since it lacks the superficial layer of chitinous material.[2] Zoospores that have ceased swarming and have not become affixed to some object may develop into germlings enclosed by a wall, but most of these immediately form zoospores at the one-celled stage.[3] One-celled sessile germlings of *Oedogonium* develop into many-celled filaments, according to one of two general methods. Most species have the formation of a transverse ring at the apex of the one-celled germling and an elongation of the ring similar to that found in vegetative division.[4] After the ring has elongated vertically, there is a transverse division of the protoplast. The division of the distal cell, and the division and redivision of its daughter cells, result in a many-celled filament; the basal cell formed by the first division does not divide. A few species have no formation of a ring during division of the one-celled stage. Germlings of this type are hemispherical and with an apical cap.[5] Division of these germlings begins with the protrusion of a cylindrical outgrowth through the region of the cap, and, after the cylinder has attained a certain length there is a transverse division of the protoplast and a formation of a cross wall at the juncture of cylinder and hemisphere.

Zoospores of *Bulbochaete* produce one-celled germlings similar to those of *Oedogonium*. The first division (Fig. 124) is without the formation of a hemicellulose ring, and the distal protoplast develops into a seta with a swollen base.[6] The basal cell formed by this first division divides with a formation of a hemicellulose ring. The upper cell resulting from this second division, and from subsequent ones, rarely divides. Consequently the terminal cell of a branch is the oldest. Any cell of the filament, except the basal one, may divide obliquely and cut off a cell that becomes the initial of a side branch. This branch initial divides in the same manner as a one-celled germling, that is, without the formation of a ring. Subsequent divisions in derivatives from branch initials have a regular formation of a

[1] PEIRCE and RANDOLPH, 1905. [2] TIFFANY, 1924.
[3] FRITSCH, 1902*B*; WILLE, 1887*A*. [4] HIRN, 1900.
[5] FRITSCH, 1904; SCHERFFEL, 1901*A*.
[6] HIRN, 1900; PRINGSHEIM, N., 1858.

ring during division. Cells of branches may, in turn, produce initials of tertiary branches and so on through further branching of the thallus. Most cells in a filament also have an oblique division at their distal end, similar to that producing initials of branches, and the daughter cell cut off by the oblique division developing into a long seta.

One-celled germlings from zoospores of *Oedocladium* do not have a holdfast.[1] The rhizoidal part of a thallus may be developed by the first cell of a thallus, from one of the daughter cells of the first division, or there may be a development of a simple filament of several cells before the appearance of rhizoidal branches (Fig. 130*A*, page 210).

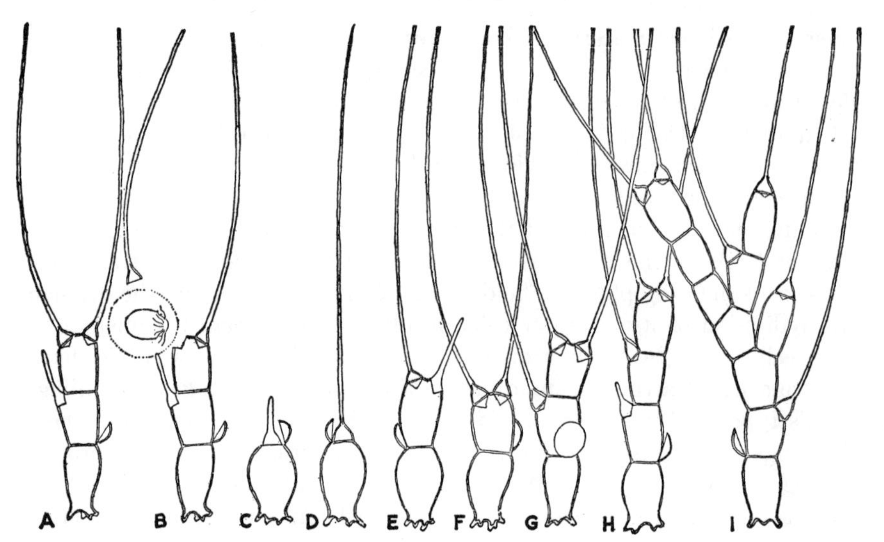

FIG. 124. *Bulbochaete*. *A–B*, a three-celled germling before and after liberation of a zoospore from the apical cell. *C–I*, successive stages in development of a young filament. (× 325.)

Oedogonium may form aplanospores singly within a cell.[2] Both *Oedogonium* and *Oedocladium* form akinetes. In *Oedogonium* they are formed in chains of 10 to 40 and superficially resemble chains of oögonia. In *Oedocladium* they are formed by cells of the rhizoidal system, and either singly or in short chains.[3] Akinetes germinate directly into new filaments.

Sexual reproduction is always oögamous but is so unlike the oögamy of other Chlorophyceae that a special terminology is necessary. If a species is one in which antheridia are produced by special dwarf male filaments, it is *nannandrous*. Species in which there are no dwarf male filaments are *macrandrous* and may be homothallic or heterothallic. The dwarf males are produced by germination of special zoospores (*androspores*) that are

[1] STAHL, 1891; TIFFANY, 1930. [2] RANDHAWA, 1937. [3] STAHL, 1891.

produced within *androsporangia.* Androsporangia are similar in appearance to antheridia of macrandrous species. If a nannandrous species is one in which the androsporangium is borne on the same filament as the oögonium, it is *gynandrosporous*; if androsporangia and oögonia are borne on separate filaments, the species is *idioandrosporous.*

Oögonia develop in much the same fashion in both nannandrous and macroandrous species of a genus and owe their development to transverse division of an *oögonial mother cell* that may be terminal or intercalary. The distal daughter cell always develops into an oögonium. The proximal daughter cell in *Oedocladium* never divides again and so always develops into a *suffultory* or supporting cell. In *Oedogonium* the proximal daughter cell may become a suffultory cell or it may function as an oögonial mother cell. This genus may, therefore, have the oögonia solitary or in series. In *Bulbochaete* the proximal daughter cell divides to form two suffultory cells that are of equal or unequal length.

Cells maturing into oögonia become more or less globose and have a diameter greater than that of vegetative cells in the filament. The oögonial wall in *Oedocladium* and *Oedogonium* has the apical-cap rings persisting from previous cell divisions localized at the upper end. In *Bulbochaete* these rings often lie in the equator of the oögonial wall. As the oögonium approaches maturity, there is the formation of a small pore or the formation of a transverse crack in the wall. Just before fertilization there is an exudation of gelatinous material from the pore or from the crack (Fig. 127*A*, page 206). The shape and position of this opening are quite typical for a species and are characters of diagnostic importance in separating species from one another. The protoplast within an oögonium develops into a single egg. In the early stages of its development, the nucleus is centrally located within the egg,[1] but shortly before fertilization it lies just internal to the opening in the oögonial wall. Eggs ready for fertilization retract slightly from the oögonial wall and have a hyaline receptive spot external to the nucleus.

Antheridia of macrandrous species are either terminal or intercalary and are formed by the division of an antheridial parent cell. This division is quite similar to that of a vegetative cell, except that the upper cell, which is the antheridium, is much shorter than the lower cell.[2] The lower cell may, in turn, divide repeatedly and so give rise to a series of 2 to 40 antheridia (Fig. 125*A*). The protoplast of the antheridium may give rise to a single antherozoid, but usually it divides vertically or transversely to form two daughter protoplasts, each of which become antherozoids. Division of the nuclei is in the transverse axis of the cell, but they may come to lie above each other before cytokinesis. The liberation of the antherozoids (Fig.

[1] Klebahn, 1892; Ohashi, 1930. [2] Gussewa, 1930; Ohashi, 1930.

125*B*) is by the same annular splitting of the wall that is found in the liberation of zoospores, and the antherozoids are, likewise, enclosed within a vesicle when first liberated. Antherozoids of most species are like zoospores, except for their smaller size and fewer flagella; but those of some species have flagella longer than the body of the antherozoid[1] (Fig. 125*C*).

Androsporangia of nannandrous Oedogoniales are similar in appearance to antheridia of macrandrous species and result from a similar unequal division of a mother cell. Only one androspore is formed within an androsporange[2] and, when it is first liberated, it is surrounded by a vesicle. After the vesicle disappears, the androspore swims freely in all directions until it comes in the vicinity of an oögonium, where it ceases swarming and becomes affixed to the suffultory cell or to the oögonium itself. More rarely the

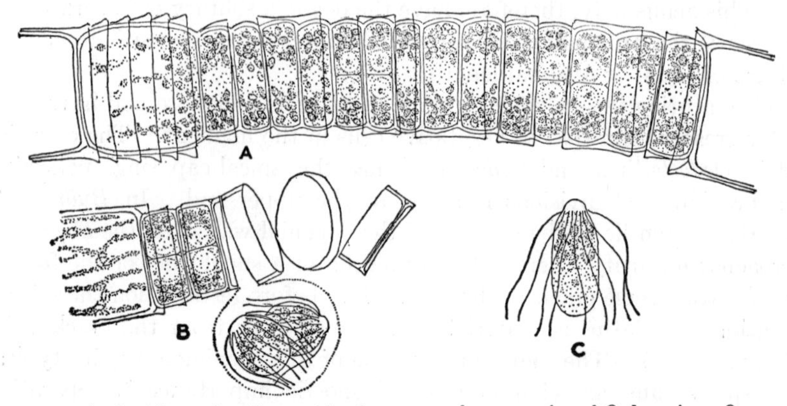

FIG. 125. Antheridia and antherozoids of a macrandrous species of *Oedogonium*, *O. crassum* (Hass.) Wittr. *A*, antheridia. *B*, liberation of antherozoids from an antheridium. *C*, free-swimming antherozoid. (× 485.)

androspore comes to rest upon a vegetative cell. One-celled germlings developed from androspores are, except for their smaller size, quite like germlings developed from zoospores and they show, from species to species, much the same range in structure of the holdfast. The one-celled germlings from androspores of most species function as antheridial parent cells and cut off one or more antheridia at their apices. The lower portion of this antheridial parent cell is never completely used up in antherium formation and persists as a stipe supporting the antheridia (Fig. 126). It is generally agreed that nannandrous Oedogoniales have been evolved from macrandrous Oedogoniales. Some phycologists think that this has been brought about by a gradual reduction in size of male filaments of heterothallic macrandrous species. The occurrence of macrandrous species with somewhat smaller male filaments[3] and the precocious formation of anther-

[1] SPESSARD, 1930. [2] HIRN, 1900. [3] HIRN, 1900; WEST, G. S., 1912*A*.

idia by young filaments of heterothallic species of *Oedogonium*[1] are held to be evidence for this. The similarity in structure and development of androsporangia and antheridia of macrandrous species indicates, however, that the former have been evolved from the latter. Androspores are, in a sense, macrandrous antherozoids which always develop parthenogenetically,[2] but which still retain enough of their gametic nature to swim to, and germinate upon, the oögonium or suffultory cell.

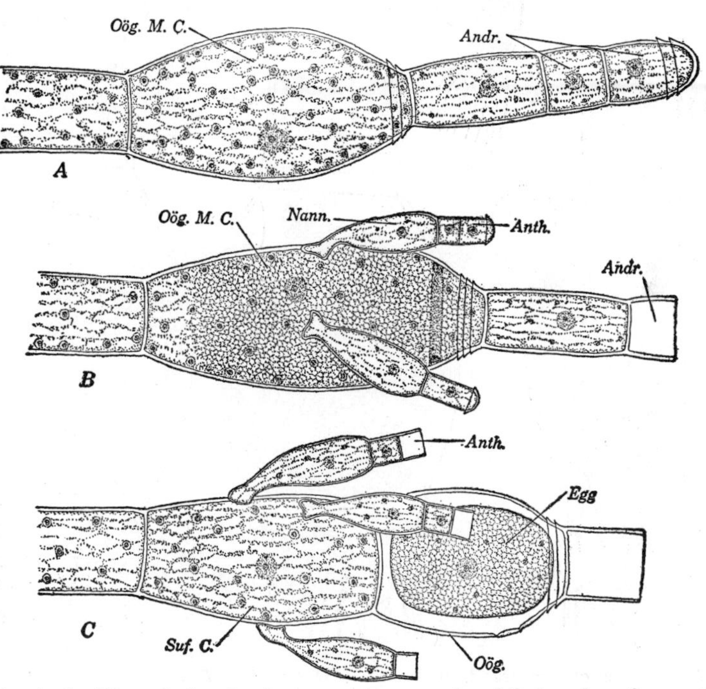

Fig. 126. Antheridia and oögonia of a nannandrous species of *Oedogonium*, *O. concatenatum* (Hass.) Wittr. (*Andr.*, androsporangium; *Anth.*, antheridium; *Nann.*, nannandrium; *Oög.*, oögonium; *Oög. M. C.*, oögonial mother cell; *Suf. C.*, suffultory cell.) (× 325.)

Fertilization (Fig. 127) in both nannandrous and macrandrous Oedogoniales is by the antherozoid swimming through the opening in the oögonial wall and entering the egg at the hyaline receptive spot.[3] Gametic fusion takes place soon after the entrance of the antherozoid. The zygote, which is somewhat retracted from the oögonial wall and often of a different shape, begins to secrete a wall as soon as it is formed. Walls of mature and female nuclei unite with each other in a resting condition,[4] and their

[1] Fritsch, 1902*B*. [2] Schaffner, 1927.

[3] Hirn, 1900; Klebahn, 1892; Ohashi, 1930; Pringsheim, N., 1858.

[4] Gussewa, 1930; Klebahn, 1892; Ohashi, 1930.

zygotes are usually composed of three layers, but some Oedogoniales have a wall with two layers only. The layer outside the innermost may be smooth, but more often it is ornamented with pits, scrobiculations, reticulations, or costae. During the later stages in the development of the zygote, the color changes from a green to a brown or a red, largely because of an accumulation of a reddish oil within the protoplast.

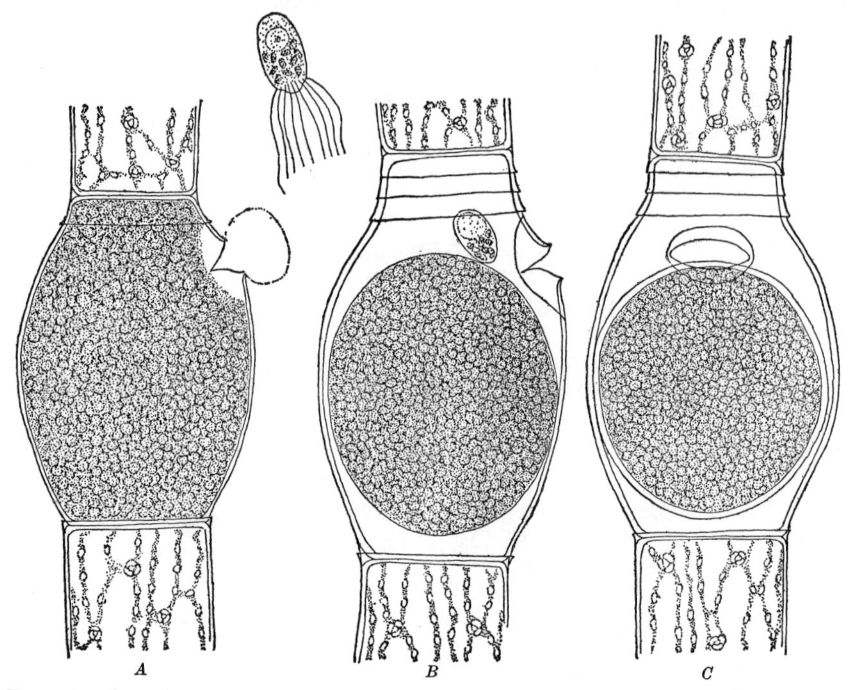

FIG. 127. Oögonia of a macrandrous species of *Oedogonium, O. crassum* (Hass.) Wittr. *A*, with the egg ready for fertilization. *B*, just after fertilization: the antherozoid within the oögonium is probably a supernumerary one. *C*, with the zygote beginning to form a wall. (× 485.)

Some species have a regular development of unfertilized eggs into parthenospores; other species have a disintegration of eggs that are not fertilized. Parthenospores have a zygote-like wall, but they may be distinguished from zygotes by the fact that they completely fill the oögonial cavity and have the same shape.[1] The statement[1] that parthenospores germinate soon after they are formed is probably erroneous.

The zygote is eventually liberated from the filament by the decay of the oögonial wall. The zygote usually undergoes a further period of rest before it germinates, and germination may be delayed for a year or more.[1] Dur-

[1] MAINX, 1931.

ing the ripening of the zygote there is a reductional division of the zygote nucleus to form four haploid nuclei.[1] Shortly before germination, the protoplast becomes green and divides to form four daughter protoplasts, each of which becomes a zoospore.[2] When the zoospores are first liberated by a bursting of the zygote wall, they are surrounded by a common vesicle, which soon disappears. The swarming and subsequent development of zoospores into filaments are identical with those of zoospores produced from vegetative cells. In the case of one heterothallic macrandrous species of *Oedogonium*, it has been shown* that two of the four zoospores develop into male filaments and two into female filaments. Under certain cultural conditions, this species had an equational division of the zygote nucleus into two daughter nuclei and a formation of two diploid zoospores both of which grew into female filaments, but filaments double the size of haploid ones.

There is but one family, the Oedogoniaceae.

FAMILY 1. OEDOGONIACEAE

All three genera of the Oedogoniaceae are found in this country. The characters separating these genera from one another are sharply defined, and the genera can be distinguished from one another even at the one-celled stage. Specific determinations in the various genera can be made only from ripe fruiting material and are based upon shape and size of the mature zygote and the organs connected with its formation. Hirn's monograph[3] is the official starting point for the nomenclature of the family.

The three genera may be distinguished as follows:

1. Filaments unbranched .... 1. **Oedogonium**
1. Filaments branched 2
   2. Branches with long setae .... 2. **Bulbochaete**
   2. Branches without setae .... 3. **Oedocladium**

1. **Oedogonium** Link, 1820. Sterile specimens of *Oedogonium* may be recognized by the unbranched filaments of cylindrical cells in which certain cells have transversely striate walls at the distal end. The basal cell of a filament is modified to form a holdfast, and the apical cell is usually broadly rounded or acuminate. Cells of *Oedogonium* are sometimes slightly enlarged at their upper ends, and they may have straight, nodulose, or undulate sides. Cell division is either terminal or else intercalary and it may take place in any cell but the basal one. The cells are uninucleate and have a single reticulate chloroplast completely encircling the protoplast.

[1] GUSSEWA, 1930; MAINX, 1931.
[2] GUSSEWA, 1930; JURÁNYI, 1873; MAINX, 1931; PRINGSHEIM, N., 1858.
* MAINX, 1931. [3] HIRN, 1900.

The chloroplast usually has many pyrenoids, one at each of the larger intersections in the reticulum.

Vegetative multiplication may result from an accidental breaking of a filament, but there is never a dissociation of a filament into fragments except at the time of zoospore formation. Asexual reproduction is by means of zoospores that are always formed singly within a cell. The zoospores are ovoid to pyriform, with a hyaline anterior end and a whorl of flagella just beneath the hyaline region. They usually lack an eyespot but sometimes have one. The one-celled germling resulting from germination of the

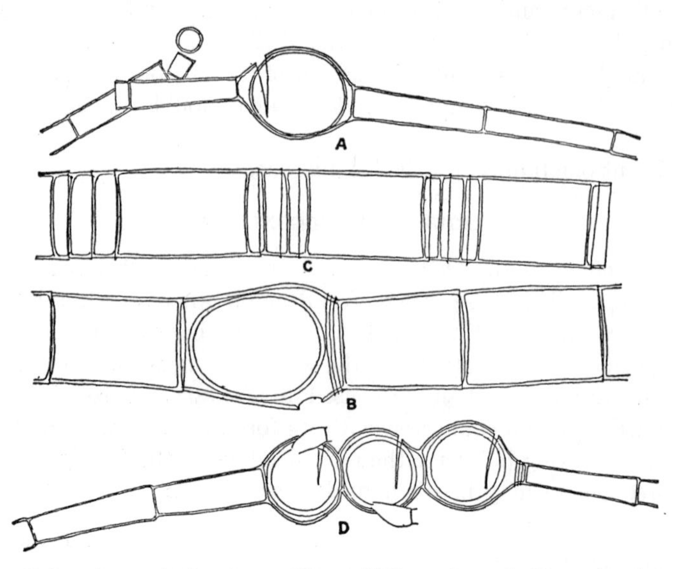

FIG. 128. *Oedogonium*. *A*, *O. crispum* (Hass.) Wittr., a homothallic macrandrous species. *B*, *O. capillare* (L.) Kütz., a heterothallic macrandrous species. *C*, *O. macrandrium* forma *aemulans* Hirn, a nannandrous species. (× 325.)

zoospore is more or less ovoid and attached to the substratum by a hapteron, or it may be hemispherical and with the flattened side next the substratum. Asexual reproduction may also take place by means of aplanospores and akinetes.

Sexual reproduction is oögamous and macrandrous or nannandrous. Macrandrous species may be homothallic (Fig. 128*A*) or heterothallic (Fig. 128*B* and *C*). Antheridia and oögonia are formed by the direct division of a parent cell. The antheridia are discoid and solitary, or in series of 2 to 40. Each produces one or two antherozoids. The oögonia are ellipsoidal to subspherical, solitary or in short series. They contain but a single egg, and the zygote resulting from fertilization has a thick wall that may be smooth or variously ornamented. Nannandrous species (Fig. 128*D*) may be idioandrosporous or gynandrosporous, but in either case the androsporangium gives rise to one androspore only. The dwarf males generally lie on the suffultory cell of the oögonium, and the dwarf males of each species have a characteristic shape. The oögonia and zygotes of nannandrous species have much the same range in shape and structure as is found in the macrandrous species. Germinating zygotes of *Oedogonium* give rise to four zoospores, each of which may develop into a new filament.

Oedogonium is always aquatic and usually found in small permanent bodies of water such as pools and ponds. It is rarely found in a fruiting condition when growing in streams, unless the flow is very sluggish.[1] The filaments may be in free-floating masses of considerable extent, or they may be epiphytic upon the leaves and stems of submerged water plants and upon Cladophoraceae or larger species of *Oedogonium.* The number of epiphytic individuals upon macrophytes is generally limited, and but very few filaments are usually present in a mount made for microscopical examination.[2] For this reason, many of the smaller species are often overlooked by the collector gathering material or by one examining collections. Many of these smaller species are often so densely incrusted with lime that even generic determination is impossible. If these encrusted filaments are placed on a slide with a few drops of lactic acid and gently heated, the entire encrustation disappears without injury to the cells. Tiffany[3] lists and describes the 223 species occurring in the United States. These species can be distinguished from one another only when fruiting, a condition most abundant during May and July in the North Central states.[1]

2. **Bulbochaete** Agardh, 1817. The filaments of *Bulbochaete* are unilaterally branched and with the majority of cells bearing a long seta that is

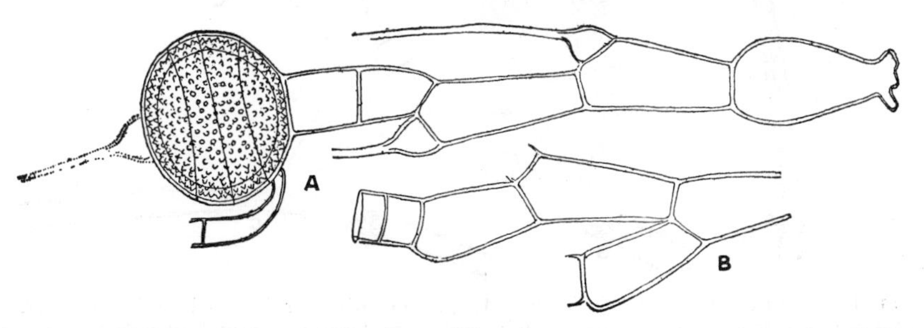

FIG. 129. *Bulbochaete gigantea* Pringsh., an idioandrosporous species. *A*, portion of filament with oögonia. *B*, portion of a filament with androsporangia. (× 325.)

swollen at the base. The cells are broader at their upper ends, and the transverse striation of the wall, characteristic of the family, is usually found only on the terminal cell of a branch. Mature thalli of *Bulbochaete* are usually sessile and with the basal cell having a holdfast at its lower end. Cell division is usually restricted to the basal cell, and each division inter-

[1] TIFFANY and TRANSEAU, 1927. [2] TIFFANY, 1924*A*. [3] TIFFANY, 1937.

calates a new cell between the basal cell and the one above (see page 202). The structure of the protoplast is similar to that of *Oedogonium.*

Vegetative multiplication resulting from the accidental breaking of the filament is of much rarer occurrence than in *Oedogonium,* and one rarely finds free-floating mats of *Bulbochaete.* Asexual reproduction by means of zoospores is quite frequent and these are formed and liberated in the same manner as in *Oedogonium.*

Sexual reproduction is oögamous (Fig. 129). A few species are macrandrous; the great majority are nannandrous and most of these have gynandrous androsporangia. The oögonia are always solitary and subtended by two suffultory cells.

Bulbochaete is found in the same type of habitats that have been noted for *Oedogonium.* Fruiting filaments may become detached and free-floating, but the great majority are always affixed. Similar to *Oedogonium,* sexual reproduction is found in greatest abundance during mid-spring and late summer in the North Central states.[1] For descriptions and distribution of the 42 species found in this country, see Tiffany (1937).

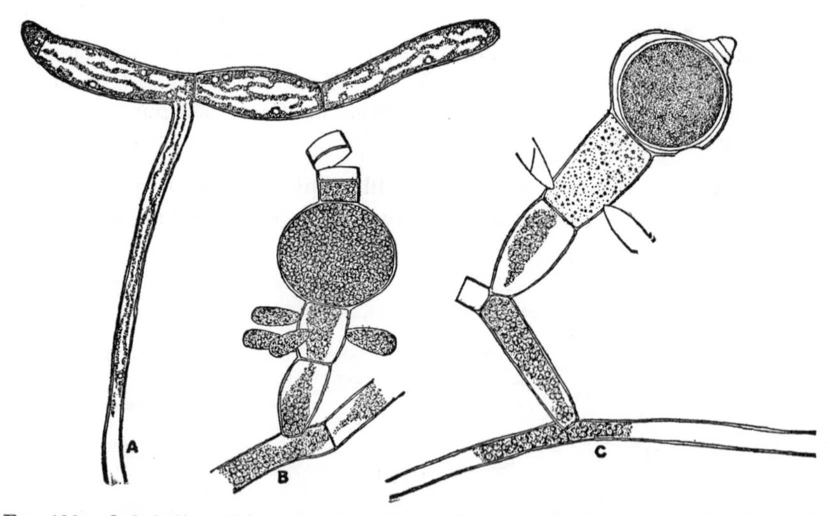

FIG. 130. *Oedocladium Hazenii* Lewis. *A,* germling with a rhizoidal branch. *B,* portion of a filament with empty androsporangia, immature dwarf males, and an unfertilized egg. *C,* portion of a filament with an oögonium containing a ripe zygote. (× 325.)

3. **Oedocladium** Stahl, 1891. Thalli of this alga are freely branched, without setae, and with rhizoidal branches composed of much narrower and longer cells. Most of the species are terrestrial. These species have colorless rhizoidal branches penetrating the soil; the portion above ground is erect or prostrate and composed of relatively short cells with chloroplasts.

[1] TIFFANY, 1928.

Aquatic species also have rhizoidal branches, but these contain chloroplasts (Fig. 130*A*). Cell division in branches is more often terminal than intercalary, hence apical caps are found chiefly upon conically pointed cells at the tips of branches. The cells are uninucleate and with the reticulate chloroplast typical of the family.

Asexual reproduction is by means of zoospores with an anterior whorl of flagella, but a zoospore does not form a holdfast when it germinates. The one-celled germling formed by a germinating zoospore may develop a rhizoidal branch, or formation of rhizoidal branches may be delayed until the germling is several cells in length. Akinetes develop on rhizoidal branches only[1] and may be solitary or in catenate series of three or four. They have fairly thin walls, a reddish protoplast, and develop directly into green filaments.

Oedocladium is a rare alga and in this country known chiefly from the southeastern portion of the United States. It is usually found growing intermingled with mosses and liverworts on banks of sandy loam. Certain species are aquatic. The following species have been found in the United States: *O. albermarlense* Lewis, *O. Hazenii* Lewis (Fig. 130), *O. Lewisii* Whitford, *O. medium* Lewis, and *O. Wettsteinii* Knapp. For a description of *O. Lewisii*, see Whitford (1938); for descriptions of the others, see Tiffany (1937).

ORDER 7. CLADOPHORALES

Genera belonging to this order have multinucleate cylindrical cells united end to end in simple or branched filaments. Each cell contains a reticulate chloroplast, with numerous pyrenoids, or the chloroplast may be broken up into numerous disk-like fragments.

Asexual reproduction is by quadriflagellate zoospores, aplanospores, and akinetes.

Sexual reproduction is isogamous and by a union of biflagellate gametes.

Genera with branching thalli usually have the branches free from one another. Genera with anastamosing branches never have them laterally united in pseudoparenchymatous thalli, such as are found among the Ulotrichales. Thalli of Cladophorales are usually sessile and either attached to the substrate by rhizoidal outgrowths from elongate basal cells (*Hormiscia, Cladophora*) or with multicellular rhizoidal systems developed from the lower part of the thallus. In some genera, there is little growth of thallus portions which become detached and free-floating; in other genera (*Pithophora, Rhizoclonium*), detached portions of the thallus may continue to grow indefinitely.

Growth of the thallus may be restricted to the terminal portions of the branches (*Cladophora, Pithophora*), or it may be intercalary (*Rhizoclonium,*

[1] Stahl, 1891.

Chaetomorpha). Cell division is independent of nuclear division, and the annular furrow which divides the protoplast into two parts is usually formed midway between the ends of the cell.

Cells of Cladophorales have thick stratified walls composed of three concentric zones. The outermost zone consists chiefly of chitin, the middle zone is rich in pectic material, and the innermost zone is composed of cellulose. The absence of a sheath of pectic material external to the chitinous wall layer makes the cells of Cladophorales a particularly favorable substratum for the growth of epiphytic algae, and one generally finds old filaments of members of the family densely clothed with diatoms and other epiphytic algae. There is a fairly thick layer of cytoplasm between the cell wall and the central vacuole. The chloroplast or chloroplasts lie toward the exterior of the cytoplasm. Usually there is a sheet-like reticulate chloroplast in which the intersections of the network are connected by strands of varying thickness. In some cases, there is a pyrenoid at every intersection of the reticulum; in others the majority of intersections are without a pyrenoid. Sometimes complete obliteration of the strands connecting the intersections of the reticulum results in a number of small discoid chloroplasts, the majority of which are without pyrenoids. Pyrenoids of the Cladophorales are usually surrounded by a conspicuous sheath of starch plates. Old cells often have so much stroma starch that but little can be made out concerning the structure of the protoplast.

The nuclei lie internal to the chloroplast. The average number of nuclei in a cell depends more upon the volume of the cell[1] than upon the genus concerned. If the cell is small, there are but few nuclei, sometimes only two or three; if the cell is large, the number is correspondingly larger and sometimes 50 or more. Increase in the number of nuclei is by mitosis, and as a rule all nuclei in a cell divide simultaneously.

Asexual reproduction by means of quadriflagellate zoospores occurs in most members of the family. Most genera have ovoid zoospores, but *Urospora* has spindle-shaped ones in which the posterior end is acutely pointed. Zoospore formation may take place in all cells of a filament but the holdfast (*Chaetomorpha*, *Urospora*), or it may be restricted to cells toward the extremities of branches (*Cladophora*). There is usually an increase in number of nuclei just before progressive cleavage of the cell contents into angular uninucleate protoplasts that metamorphose into zoospores. Coincident with cytoplasmic cleavage there is a development of a small lens-shaped area near the apex (*Cladophora*) or the equatorial region (*Chaetomorpha*) of the lateral cell wall. Gelatinization and dissolving of this area result in a small circular pore through which the zoospores escape. Akinete-like resting cells are occasionally formed by genera

[1] BRAND, 1898; CARTER, N., 1919*C*.

which regularly produce zoospores (*Cladophora*). Akinetes are of regular occurrence in *Pithophora*.

Sexual reproduction is isogamous and by means of biflagellate gametes formed and liberated in the same manner as zoospores. Some species of *Chaetomorpha*,[1] *Cladophora*,[2] and *Spongomorpha*[3] are heterothallic and have a union of gametes only when the two have been produced by different thalli. Certain heterothallic species have a disintegration of gametes which do not fuse to form a zygote; others have a parthenogenetic germination of gametes that fail to fuse with another gamete.

A zygote may germinate directly into a new plant without a period of rest. Among such Cladophoraceae, division of the zygote nucleus has been shown to be equational in *Chaetomorpha aerea* (Dillw.) Kütz.[1] and in certain species of *Cladophora*.[2] The diploid plant produced by a germinating zygote is asexual and forms zoospores only. The nuclear division immediately preceding zoospore formation is meiotic, except in the case of *Cladophora glomerata* (L.) Kütz.,[4] and the zoospores germinate to form a haploid gamete-producing generation identical in appearance with the diploid generation. In *C. glomerata* the zoospores develop into diploid thalli in which meiosis immediately precedes the formation of gametes. In *Urospora*, and possibly in *Spongomorpha*, there is no multicellular diploid generation.

There is but one family, the Cladophoraceae.

FAMILY 1. CLADOPHORACEAE

The fresh-water Cladophoraceae found in this country differ as follows:

1. Filaments freely branched 2
1. Filaments with few branches 3
    2. Without akinetes .... 1. **Cladophora**
    2. Regularly with akinetes .... 4. **Pithophora**
3. Lateral branches confined to vicinity of holdfast .... 3. **Basicladia**
3. Lateral branches short, rhizoid-like .... 2. **Rhizoclonium**

1. **Cladophora** Kützing, 1843. The cells of *Cladophora* have a length 5 to 20 times their breadth and are united end to end in freely branched filaments. The branching is usually lateral but often appears to be dichotomous because of the pushing aside (*evection*)[5] of the original axis of the branch. Branches originate as lateral outgrowths from the upper end of a cell and are usually formed only from cells near the end of a filament. The first cross wall formed in a branch is laid down close to the point of origin of the outgrowth. Thalli of *Cladophora* are usually sessile and attached to the substratum by fairly long rhizoidal branches, some of

[1] HARTMAN, 1929. [2] FÖYN, 1929, 1934. [3] SMITH, G. M., 1946.
[4] LIST, 1930. [5] BRAND, 1901*A*.

which arise adventitiously from cells near the base of the thallus. Certain of the rhizoidal branches develop short thick cells from which grow bushy short-celled branches.[1] Many species of *Cladophora* are perennial, the thallus dying back to the prostrate rhizoidal system, whose cells are filled with food reserves. In the following growing season, certain of these irregularly shaped cells give rise to new erect filaments.

The cells have thick stratified walls consisting of an inner cellulose zone, a median pectic zone, and an outer chitinous zone.[2] There is a fairly thick layer of cytoplasm next the wall and usually one large central vacuole, but sometimes there are several vacuoles in the central portion of the protoplast. The chloroplast may be a reticulate sheet completely encircling the protoplast, with pyrenoids here and there in the reticulum; or there may be numerous discoid chloroplasts most of which lie next the cell wall, but

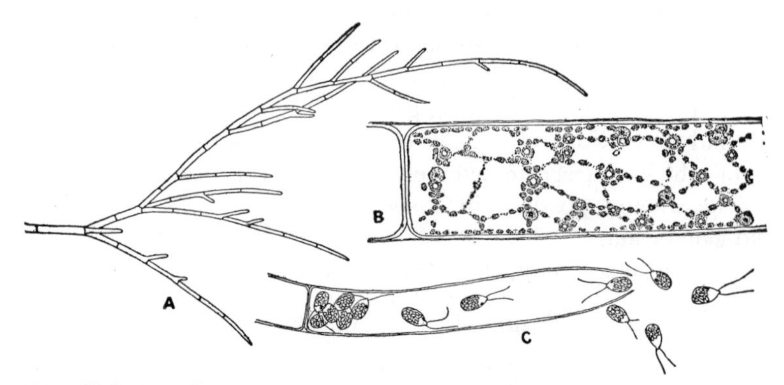

FIG. 131. *Cladophora Kuetzingianum* Grun. *A*, portion of a filament. *B*, portion of a cell. *C*, liberation of gametes. (*A*, × 30; *B*, × 325; *C*, × 220.)

a few of which may lie in cytoplasmic strands crossing the central vacuole (Fig. 131*B*). Only certain of the discoid chloroplasts contain pyrenoids. The cells are always multinucleate and with the nuclei internal to the chloroplasts. The nuclei are relatively large and with a well-defined chromatin-linin network. Chromosomes formed during mitosis may all be of the same length or of different lengths.[3] There is no distinct spindle in the division of the nuclei, and the nuclear membrane persists until division is completed. Frequently, the nucleolus is also persistent and is divided into two equal parts during mitosis.[4]

Asexual reproduction is by means of quadriflagellate zoospores which are usually formed in vigorously growing cells near the ends of filaments. Prior to zoospore formation, there is a period of active nuclear division.[5]

[1] BRAND, 1909*A*. [2] WURDACK, 1923. [3] SCHUSSNIG, 1930*C*.
[4] CARTER, N., 1919*C*; SCHUSSNIG, 1923; T'SERCLAES, 1922.
[5] CZEMPYREK, 1930.

In certain species, these divisions have been shown to be meiotic,[1] but there is no meiosis prior to zoospore formation in *C. glomerata* (L.) Kütz.[2] After the completion of the nuclear divisions there is, by a process of vacuolization,[3] a progressive cleavage into uninucleate protoplasts. The zoospores escape singly through a small circular pore in the distal end of the cell wall. The germlings developed from zoospores soon divide into two branches, one of which becomes the erect portion of the thallus and the other the rhizoidal portion.[4] Reproduction may also be effected by a disarticulation of the cells in the rhizoidal system and a development of these individual cells into new plants. Cells of the erect filaments may develop into akinete-like structures, but these often divide to form new cells without becoming detached from the filament in which they were formed.[5]

Sexual reproduction is by means of biflagellate gametes. These are formed and liberated (Fig. 131C) in much the same manner as the zoospores. Certain species are heterothallic and have a fusion of gametes only when the two come from different plants.[6] In some heterothallic species there is a disintegration of gametes that fail to unite with other gametes; other heterothallic species may have a parthenogenetic germination of the gametes to form thalli.

Most species of *Cladophora* have an alternation of a diploid asexual generation with a haploid sexual generation[7] and one in which both generations are identical in appearance. In *C. glomerata* there is an alternation of sexual and asexual generations but both are diploid, and the reduction in number of chromosomes takes place just before the gametes are formed.[2]

This is one of the largest genera of algae and is of world-wide distribution in fresh, brackish, and salt water. There are few sharply defined characters separating the species one from another; and many of the species are quite variable in form at different seasons of the year and under different conditions of growth. *Cladophora* is often found in abundance on stones in slowly flowing streams, in lakes, and attached to the sides of watering troughs. Collins[8] recognizes 13 species as occurring in the fresh-water flora of this country. The most remarkable of these species is *C. profunda* var. *Nordtsedtiana* Brand, which was found at a depth of 150 ft. below the surface of Lake Ontario.[9]

The origin and nature of "lake balls" or "*Cladophora* balls" are problems which have always aroused the interest of phycologists. *Cladophora* balls are 2 to 10 cm. in diameter, spherical to subspherical in shape, and solid or hollow (Fig. 132). They are made up of *Cladophora*-like filaments which are irregularly branched and often with certain of the cells irregularly shaped and with thick walls.[10] The older

[1] Schussnig, 1928*A*, 1930*C*. [2] List, 1930. [3] Czempyrek, 1930.
[4] Brand, 1909*A*. [5] Cholnoky, 1930. [6] Föyn, 1929, 1934.
[7] Schussnig, 1928*A*, 1930, 1930*A*, 1930*B*, 1930*C*.
[8] Collins, 1909, 1912, 1918. [9] Kindle, 1915. [10] Brand, 1902*A*, 1909*A*.

phycologists considered the algae in these balls a genus (*Aegagropila* Kützing, 1849) distinct from *Cladophora*. In recent years, certain phycologists have continued to recognize *Aegagropila* because of the lack of flagellated reproductive cells[1] but this distinction has been invalidated by the discovery of biflagellated swarmers in *Aegagropila*.[2] The species which form *Cladophora* balls generally grow unattached and as a loose felty layer on the bottom of shallow lakes. The formation of balls is not due to the mode of growth of the alga but to an undulatory movement of the water at the bottom, set up by wave action at the surface.[3] A similar formation of balls from waterlogged plant fibers, pine needles, and other filamentous bodies[4] shows that the formation of *Cladophora* balls is a passive result of the action of under-water parts of waves. *Cladophora* balls are irregular in shape when first formed and with the filaments irregularly intertwined. Growth of the filaments

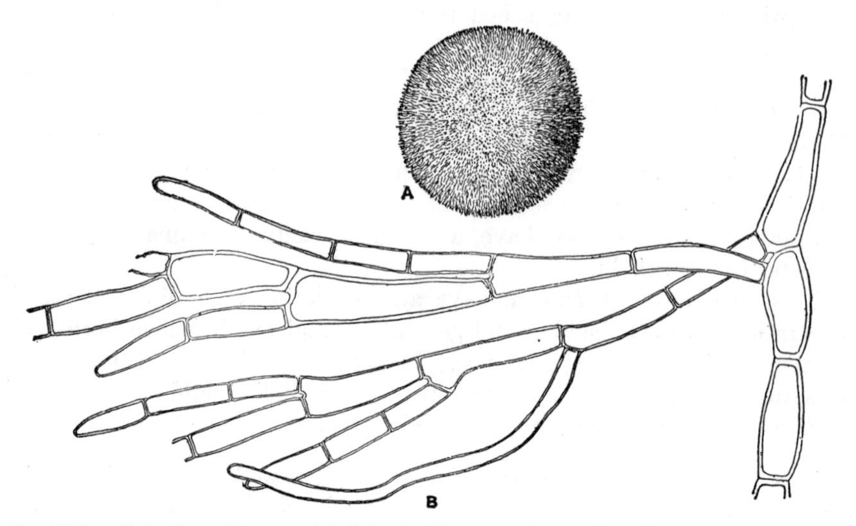

Fig. 132. *Cladophora* (*Aegagropila*) *holsatica* Kütz. *A*, *Cladophora* ball. *B*, portion of a filament. (*A*, natural size; *B*, × 160.)

that have been rolled into a ball is largely radial. The continued abrasion of the plant mass, as it swishes back and forth across the sandy bottom, results in its assuming a globular form. As a ball continues to increase in size, there is often a death and decay of filaments at the center of the ball, and gases given off during photosynthesis may accumulate in this cavity and cause the balls to rise to the surface of the lake. *Cladophora* balls are common in certain lakes of Europe, but they seem to be of rare occurrence in this country. The only published record of their occurrence is from Massachusetts,[5] but Prof. L. A. Kenoyer writes that he has found them in Minnesota.

2. **Rhizoclonium** Kützing, 1843. The cells of this alga have a length several times the breadth and are united end to end in filaments bearing

[1] Printz, 1927. [2] Nishimura and Kanno, 1927. [3] Acton, 1916*A*.
[4] Ganong, 1905, 1909; Hayren, 1928; Kindle, 1934. [5] Collins, 1909.

one-celled or few-celled rhizoidal branches. Cell division is intercalary and may take place in any cell but the basal holdfast. The cells have a stratified wall similar to that of *Cladophora*. The chloroplast is parietal, reticulate, and with pyrenoids at certain intersections of the reticulum. Species with narrow filaments have a small number of nuclei in each cell; those with broad cells have many nuclei per cell. The nuclei are relatively large and divide in a manner similar to those of *Cladophora*.[1]

Reproduction is chiefly by an accidental breaking of filaments, and severed portions of a filament may continue growth indefinitely. Biflagellate zoospores, which escape through a pore in the side wall, have been recorded but are of rare occurrence among fresh-water species of the genus. Akinetes may also be formed, but these do not differ much from irregularly shaped vegetative cells with somewhat thickened walls.[2]

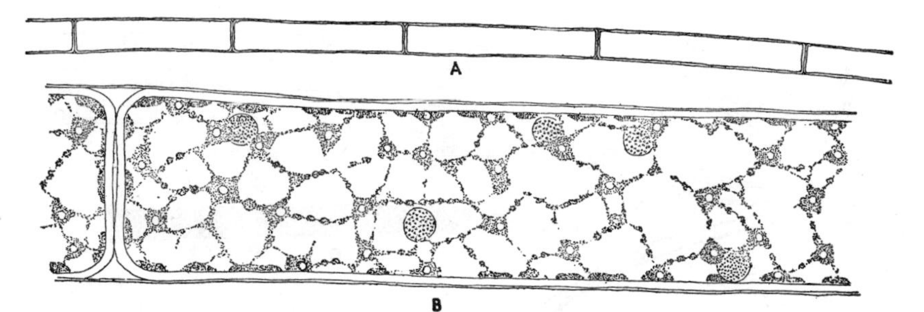

FIG. 133. *Rhizoclonium hieroglyphicum* (Ag.) Kütz. *A*, portion of a filament. *B*, portion of a cell. (*A*, × 125; *B*, × 650.)

Fresh-water species of *Rhizoclonium* are generally found in quiet waters. Frequently the cells are so densely packed with starch that the structure of the protoplast is obscured. The following five species have been found in fresh water in this country: *R. crassipellitum* W. and G. S. West, *R. crispum* Kütz., *R. fontanum* Kütz., *R. hieroglyphicum* (Ag.) Kütz. (Fig. 133), and *R. Hookeri* Kütz. For descriptions of them, see Collins (909).

3. **Basicladia** Hoffmann and Tilden, 1930. One feature distinguishing this alga from others of the family is that it has been found growing only on the backs of fresh-water turtles. The thallus is attached to the turtle's carapace by rhizoidal outgrowths at the lower end of an elongate basal cell, or by means of a coralloid system of prostrate branches composed of subrectangular cells. The erect filaments are composed of cells that become progressively shorter and broader toward the apex of a filament. Basal cells of erect filaments generally bear one or more delicate lateral branches in which the first cross wall is some distance from the insertion of the

[1] CARTER, N., 1919*C*; PETERSCHILKA, 1924.
[2] BRAND, 1898.

branch. In very rare cases, lateral branches may arise from cells some distance above the basal cell.[1] The cells are multinucleate.[2]

Biflagellate swarmers (zoospores ?) are formed by progressive cleavage and may be formed in any cell but the basal one.[2] They escape through a small pore in the lateral wall.[3] There may also be a formation of aplanospores.[4]

Two species, *B. Chelonum* (Collins) Hoffmann and Tilden (Fig. 134) and *B. crassa* Hoffmann and Tilden, have been reported from several states east of the Rocky Mountains. A third species, **B. sinensis** (Gardner) comb. nov. (*Chaetomorpha sinensis* Gardner), has been found in an aquarium in California on the back of a turtle brought from China.[5] It is very probable that this species is not native to the United States. For descriptions of *B. Chelonum* and *B. crassa*, see Hoffmann and Tilden (1930); for a description of *B. sinensis*, see Gardner (1937).

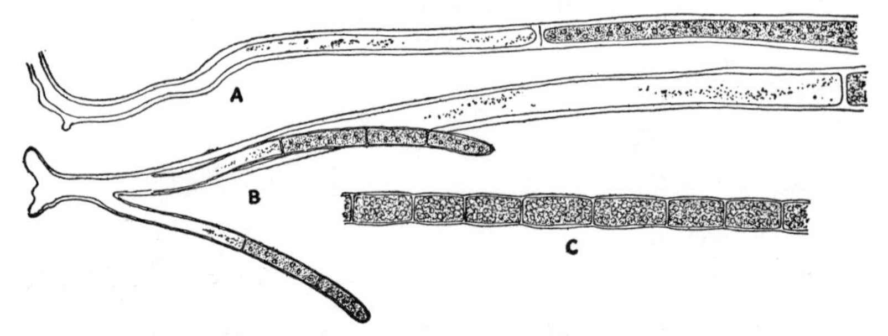

FIG. 134. *Basicladia Chelonum* (Collins) Hoffmann and Tilden. *A–B*, basal portion of filaments. *C*, upper portion of a filament. (× 160.)

4. **Pithophora** Wittrock, 1877. *Pithophora* has the general appearance of a *Cladophora* but is easily recognized by the large terminal and intercalary akinetes densely packed with reserve foods. The filaments are freely branched and with the branches originating a short distance below the distal end of the cell producing them. The branches are mostly solitary, but sometimes they are in opposite pairs. In some species, there are rhizoid-like processes at the tips of certain branches or at the base of a filament.[6] The cells are cylindrical, and those at the ends of branches are longer than others of a thallus. The walls are fairly thick but without the lamellation so often evident in *Cladophora*. The protoplast is multinucleate, and the chloroplast is a reticulate parietal sheet in which the intersections of the reticulum are large, each usually with a pyrenoid.

The only method of reproduction, aside from an accidental breaking of filaments, is by means of akinetes. In the formation of an akinete, most

[1] HOFFMANN and TILDEN, 1930. [2] LEAKE, 1939. [3] THOMPSON, 1938.
[4] LEAKE, 1946. [5] GARDNER, 1937. [6] WITTROCK, 1877.

of the cell contents moves to the upper end of a cell, and this is followed by a transverse division that divides the cell into a short akinete and a considerably longer vegetative cell. Liberation of akinetes is by a circumcissal break of the upper part of the wall of the cell beneath it.[1] When an akinete germinates, there is a transverse division into two daughter cells, one of which develops into a rhizoid and the other into a branched filament.[2]

Pithophora, although much less frequently encountered than *Cladophora* or *Rhizoclonium*, is widely distributed in the United States. The following species have been found in this country: *P. Cleveana* Wittr., *P. kewensis* Wittr., *P. Mooreana*

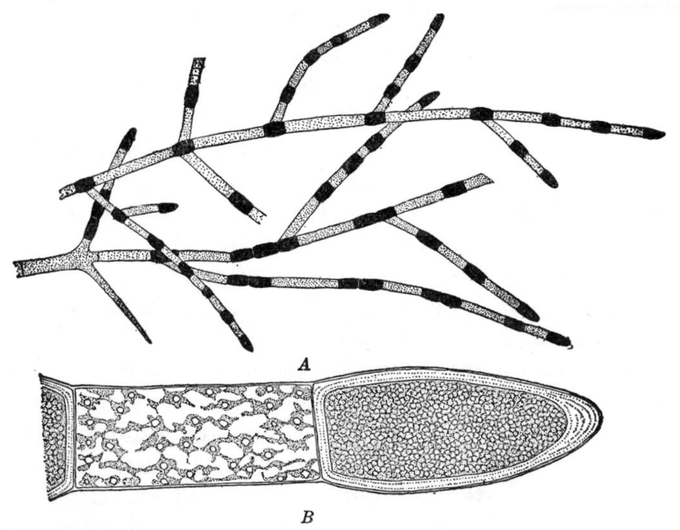

FIG. 135. *Pithophora Oedogonia* (Mont.) Wittr. *A*, upper portion of a filament. *B*, vegetative cell and an akinete. (*A*, × 20; *B*, × 215.)

Collins, *P. Oedogonia* (Mont.) Wittr. (Fig. 135), *P. Roettleri* (Roth) Wittr., and *P, varia* Wille. For a description of *P. kewensis*, see Wittrock, (1877); for descriptions of the others, see Collins (1909, 1912).

ORDER 8. CHLOROCOCCALES

The Chlorococcales have cells which may be solitary or united in nonfilamentous colonies of definite size and shape. The cells may be uninucleate or multinucleate, but in neither case do they divide vegetatively.

Asexual reproduction is by a formation of zoospores or autospores.

Sexual reproduction is usually isogamous, but it may be anisogamous or oögamous.

The third of the evolutionary tendencies from the motile unicellular condition (see page 8) has resulted in the Chlorococcales. In this

[1] RAMANATHAN, 1939. [2] ERNST, 1908; WITTROCK, 1877.

evolutionary series, there has been an obliteration of cell division except for the formation of zoospores, gametes, or other reproductive bodies. However, the plant body may be more complex than a unicell, since colonies may result either from a fusion of gelatinous envelopes around cells, or from coherence of all zoospores or autospores produced by a single cell. Disappearance of cell division in this phylogenetic line has not always been accompanied by a disappearance of nuclear divisions, and there are multinucleate members of the order with a definite or an indefinite number of nuclei.

The diversity of forms among Ulotrichales is due in large part to the fact that their thalli are multicellular. One might expect that lack of cell division in the Chlorococcales would result in an evolutionary series with the various genera more or less alike. Such is not the case, because there are cells of almost every conceivable shape among the Chlorococcales. The genera with spherical cells are probably the simplest members of the order. There are a number of genera with elongate cells, in which the shape is reniform, vermiform, or acicular. Other Chlorococcales have angular cells that are bilaterally or radially symmetrical. This multiplicity of cell types and the organization of many of these cell types into nonfilamentous colonies result in an order in which there are many genera.

Most Chlorococcales with uninucleate cells have a typical chlamydomonad chloroplast with one pyrenoid. Others have parietal chloroplasts; either a single longitudinal band next to one side of a cell (*Scenedesmus*), or two more transverse bands partially girdling the protoplast (*Oöcystis*), or numerous discoid parietal chloroplasts (*Eremosphaera*). Most genera with parietal chloroplasts have but one pyrenoid per chloroplast, but certain of them, as *Hydrodictyon*, have a chloroplast with many pyrenoids. Except when densely packed with starch or other reserve foods, the chloroplasts are usually well defined at all stages in cell development. Coenocytic Chlorococcales may have a limited number of nuclei that is always a multiple of two because of simultaneous divisions (*Characium*, *Pediastrum*); or they may have a large indefinite number of nuclei (*Hydrodictyon*). Genera with a limited number of nuclei may have a sudden increase in the number just before reproduction (*Tetraëdron*, *Pediastrum*), or the number may gradually increase as a cell grows in size (*Characium*).

Some Chlorococcales reproduce by means of zoospores. These may be formed by a repeated bipartition, by a simultaneous cleavage, or by a progressive cleavage of the protoplast. Zoospores are usually biflagellate. They may be liberated by a breaking of the parent-cell wall or through a pore in the wall. When discharged in a mass, the zoospores are frequently surrounded by a vesicle that may soon disappear or, as in *Pediastrum*, may persist throughout the entire swarming period of the zoospores. In the latter case, the zoospores become organized into an autocolony at the end of the swarming period.

Many Chlorococcales never form zoospores. These algae reproduce by means of *autospores* (aplanospores with a shape similar to that of the vegetative cell). Autospores of certain genera become separated from one another when liberated by rupture or gelatinization of the parent-cell wall (*Tetraëdron, Chodatella*). Some genera do not have an escape of autospores from the gelatinous matrix resulting from gelatinization of the parent-cell wall (*Kirchneriella, Quadrigula*) and thus have the cells united into gelatinous colonies. Other genera have an escape of autospores from the parent-cell wall but no separation of autospores one from another. The autocolonies formed in this manner always have the cells arranged in a specific manner (*Scenedesmus, Crucigenia*).

The phylogenetic connection between zoospores and autospores is shown by genera with both types of spore. The autospores of these genera may be so much like zoospores that they have contractile vacuoles and an eyespot(*Marthea*),[1] or they may have contractile vacuoles and lack an eyespot (*Desmatractum*). However, none of the genera reproducing solely by means of autospores has them with contractile vacuoles or eyespots. In a few cases, as in *Pediastrum*, the entire protoplast of a cell develops into a single aplanospore.

Some zoosporic members of the order also produce flagellated gametes. Gametic union among them is usually isogamous, but it may be anisogamous. Members of the order which do not produce zoospores may reproduce sexually, and all cases thus far discovered among them are oögamous. Germination of the zygote has been noted in only a few genera of the order. In the Hydrodictyaceae a germinating zygote produces zoospores each of which becomes an angular cell that enlarges greatly and then forms zoospores which unite to form a colony. In *Chlorochytrium* the zygote enlarges to form a large vegetative cell that produces gametes. Cytological studies show[2] that this cell has a diploid nucleus, and that meiosis takes place just before formation of gametes. It must not be inferred that this is true for all Chlorococcales in which the zygote enlarges into a vegetative cell because the available evidence indicates that *Protosiphon* has haploid nuclei.

The families described on succeeding pages show something of the relationships between the various genera of the order. The Hydrodictyaceae and Scenedesmaceae are very natural families. Others of the families,

[1] Pascher, 1918*A*.
[2] Kurssanow and Schemakhanova, 1927.

as the Oöcystaceae and Chlorococcaceae, are more or less artificial and include genera that have but little in common aside from the method of reproduction.

Family 1. Chlorococcaceae

The Chlorococcaceae are unicellular zoosporic Chlorococcales in which the cells are solitary but sometimes gregarious and apposed to form an extended stratum.

The cell shape varies from genus to genus and is spherical, subspherical, or fusiform. The cell wall may be smooth and evenly or unevenly thickened, or it may be variously ornamented. Protoplasts may be uninucleate or multinucleate, and with a cup-shaped parietal chloroplast or with a stellate axial chloroplast. Cup-shaped chloroplasts of young cells are massive and with one pyrenoid; those of old cells may contain several pyrenoids and may tend to fragment into smaller portions. Genera with a stellate chloroplast rarely have more than one pyrenoid within it.

Cells of Chlorococcaceae may grow to many times their original size, but they never divide vegetatively. All genera in the family reproduce by means of zoospores. Some genera (*Chlorococcum*, *Trebouxia*) usually have the protoplast dividing into a large number of zoospores; other genera rarely form more than four or eight zoospores (*Desmatractum*). Liberation of zoospores is generally through a pore at one side of the parent-cell wall, and they may escape singly or all be discharged simultaneously within a vesicle. Most of the genera also reproduce by means of autospores. In some cases, these autospores (aplanospores) are obviously zoospores that have failed to escape from the parent-cell wall; in other cases, they cannot be interpreted as arrested zoospores.

All cases of sexuality thus far noted in the family are an isogamous union of biflagellate gametes. Gametes failing to unite to form a zygote may or may not germinate parthenogenetically.

The Chlorococcaceae found in this country differ as follows:

1. Free-living.......... 2
1. Endozoic in eggs of salamanders.......... 5. **Oöphila**
    2. Cells spherical to subspherical.......... 3
    2. Cells spindle-shaped, longitudinally costate.......... 4. **Desmatractum**
3. Chloroplast parietal in young cells.......... 1. **Chlorococcum**
3. Chloroplast central, more or less stellate.......... 4
    4. Cell wall of uniform thickness.......... 2. **Trebouxia**
    4. Cell wall irregularly thickened.......... 3. **Myrmecia**

1. **Chlorococcum** Fries, 1820 (*Cystococcus* Nageli, 1849). The cells of this unicellular alga may be solitary, or they may be gregarious and in pulverent masses or embedded in a gelatinous matrix. One of the features

distinguishing this genus from other unicellular green algae is the striking variation in size between various cells when the alga grows in an expanded stratum. Young cells are thin-walled and spherical or somewhat compressed. Old cells have thick walls that are often irregular in outline because of local button-like thickenings. The thickened portions of a wall are often distinctly stratified. Chloroplasts of young cells are parietal massive cups, completely filling the cell except for a small hyaline region at one side. They contain one pyrenoid. As a cell increases in size, the chloroplast usually becomes diffuse and contains several pyrenoids. The cells are uninucleate until shortly before reproduction.[1]

Reproduction by means of zoospores may take place at almost any stage in enlargement of a cell.[2] Small cells usually form 8 or 16 zoospores; large

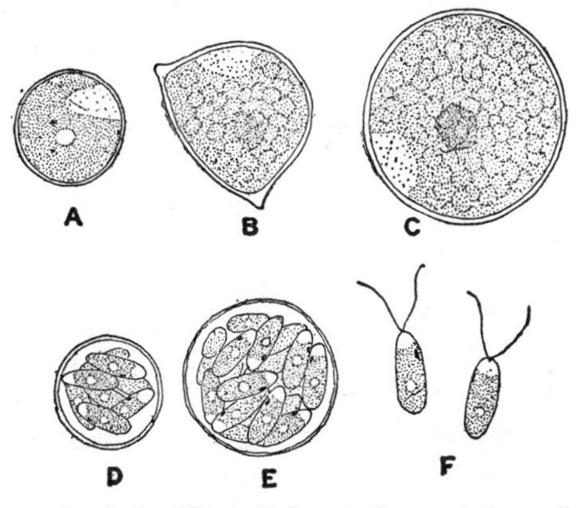

FIG. 136. *Chlorococcum humicola* (Näg.) Rab. *A–C*, vegetative cells. *D–E*, fertile cells with zoospores. *F*, zoospores. (× 800.)

cells produce many of them. The formation of zoospores has been described as by successive bipartition of the protoplast,[3] and as by a progressive cleavage.[4] The zoospores are biflagellate (Fig. 136*D*) and are liberated through an aperture in the parent-cell wall. Instead of developing into zoospores, the divided contents of old cells may develop into autospores (aplanospores), that usually remain within the old parent-cell wall until it gelatinizes to form a palmella stage. Cells of the palmella stage divide to form two or four naked daughter cells that become flagellated and fuse in pairs.[4]

Sexual reproduction may also be due to gametes formed by division of protoplasts of ordinary vegetative cells. These gametes are formed in the same manner as zoospores.

[1] BOLD, 1931; BRISTOL-ROACH, 1918. [2] ARTARI, 1892; BRISTOL-ROACH, 1918.
[3] BRISTOL-ROACH, 1918. [4] BOLD, 1931.

Chlorococcum is a subaerial alga that sometimes occurs in abundance on damp soil or on brickwork. It is also one of the algae frequently isolated from soil samples collected below the surface of the soil. Certain species are aquatic. The two species found in this country are *C. humicola* (Näg.) Rab. (Fig. 136) and *C. infusionum* (Schrank)Menegh. For descriptions of them, see Collins (1929).

2. **Trebouxia** de Puymaly, 1924. This alga is usually found only in thalli of lichens. It is not the only unicellular green alga in lichens,[1] but it is the one most often encountered. The cell shape varies from spherical to ovoid or pyriform. The cell walls are always thin and never with the irregular thickenings found in *Chlorococcum*. *Trebouxia* differs from most genera of the family in having a massive axial chloroplast that extends nearly to the cell wall. The outline of a chloroplast is somewhat irregular and lobed. There is a single pyrenoid at the center of a chloroplast, and the cell's nucleus lies at one side of the chloroplast and next to the cell wall.

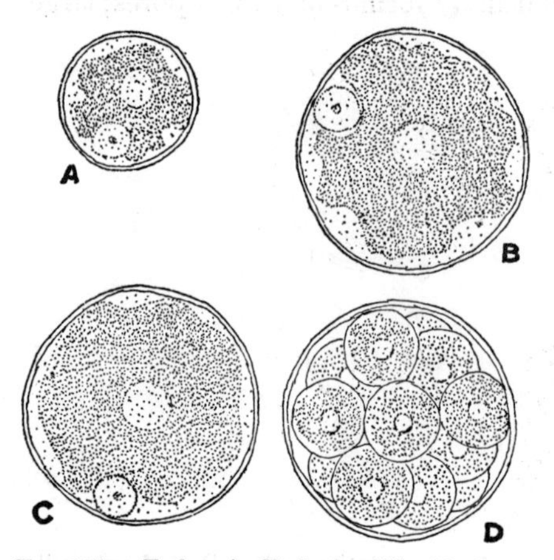

Fig. 137. *Trebouxia Cladoniae* (Chod.) G. M. Smith, from the thallus of a species of *Parmelia*. *A–C*, vegetative cells. *D*, a cell containing autospores. (× 1950.)

Reproduction by means of zoospores has been repeatedly observed when the alga is cultivated in a liquid medium,[2] and it is very probable that zoospores are formed under natural conditions during rainy periods. Zoospores are subspherical to subellipsoidal, biflagellate, and with a chloroplast at the posterior end. Their liberation is through an opening at one side of the parent-cell wall, and they escape singly or are extruded in a mass.[3] *Trebouxia* also form 8 to 200 or more autospores per cell, and they may be liberated soon after their formation or retained within the parent-cell wall until they have grown to a size equal to that of adult vegetative cells.

Sexual reproduction is by a fusion of biflagellate pyriform gametes of equal or unequal size.[3]

Trebouxia has been found in many of the commoner lichens, notably *Xanthoria*, *Cladonia*, *Parmelia*, and *Usnea*. Pure culture studies of *Trebouxia* by Chodat and his students have shown that strains isolated from different genera and species of

[1] Chodat, R., 1913.

[2] Chodat, R., 1913; Famintzin and Boranetzky, 1867, Famintzin, 1914; Jaag, 1929.

[3] Jaag, 1929.

lichens may show constant differences in size and shape. Strains that appear to be morphologically identical may be physiologically different. The species of *Trebouxia* found in a majority of lichens cannot be distinguished morphologically from *T. Cladoniae* (Chod.) G. M. Smith (Fig. 137). Several lichens collected in this country have been found to contain a *Trebouxia* agreeing morphologically with *T. Cladoniae*. For a description of it as *Cystococcus Cladoniae* Chod., see R. Chodat (1913).

3. **Myrmecia** Prinz, 1921; emend., G. M. Smith, 1933. The cells of this genus are solitary. Young cells are globose-pyriform; old cells are subpyriform and with the narrow pole broadly truncate. The cell wall is without stratification. Young cells have a mamillate thickening at one side of the wall; old cells have this thickened portion laterally expanded. The chloroplast is massive, axial, and with a more or less conspicuous indentation opposite the thickened portion of the wall. There is no pyrenoid, but there are numerous minute starch granules clustered about the centrally located nucleus.

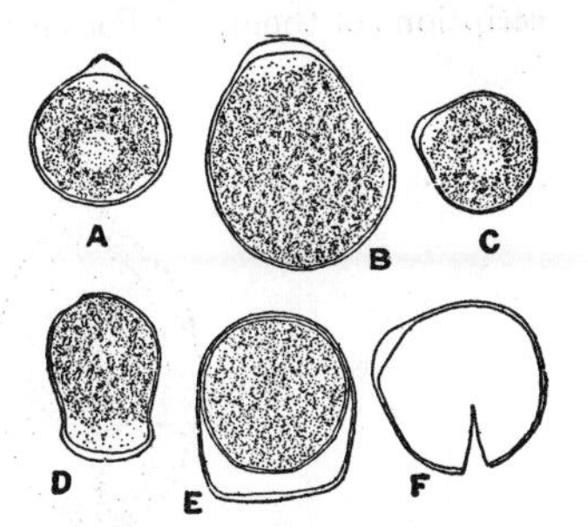

FIG. 138. *Myrmecia aquatica* G. M. Smith. *A–D*, vegetative cells. *E*, a cell containing an aplanospore. *F*, empty cell wall after the escape of the zoospores. (× 975.)

Reproduction is by successive bipartition of the protoplast into a number of zoospores that are liberated by a lateral rupture of the parent-cell wall. The entire protoplast may round up to form a single aplanospore (Fig. 138*D*).

The only species thus far found in this country is *M. aquatica* G. M. Smith, (Fig. 138), known only from a single station in California.[1] For a description of it see G. M. Smith (1933).

4. **Desmatractum** W. and G. S. West, 1902; emend., Pascher, 1930. The cells of *Desmatractum* are spherical and surrounded by a broad spindle-shaped brownish envelope. The envelope consists of two longitudinally costate halves firmly united with each other in the equatorial region of the spindle. The cell wall is hyaline and thin, and the protoplast within it has a single cup-shaped chloroplast containing one or two pyrenoids.[2]

Reproduction may be by a division of the protoplast to form two or four biflagellate zoospores with a cup-shaped chloroplast and an eyespot.[3] The zoospores are liberated by a transverse rupture of the envelope in the equatorial zone where the two halves are united with each other. Zoospores that are not liberated from a parent-cell wall may become aplanospores.

[1] SMITH, G. M., 1933.
[2] GEITLER, 1924*C*; KORSHIKOV, 1928; PASCHER, 1930*A*.
[3] PASCHER, 1930*A*.

These develop the characteristic spindle-shaped envelope before liberation by a gelatinization of wall and envelope of the parent cell.[1]

Desmatractum is usually found in acid waters and in the gelatinous coating on submerged phanerogams or on other algae. *D. bipyramidatum* (Chod.) Pascher (Fig. 139) and *D. idutum* (Geitler) Pascher have been found in this country. For descriptions of them, see Pascher (1930*A*).

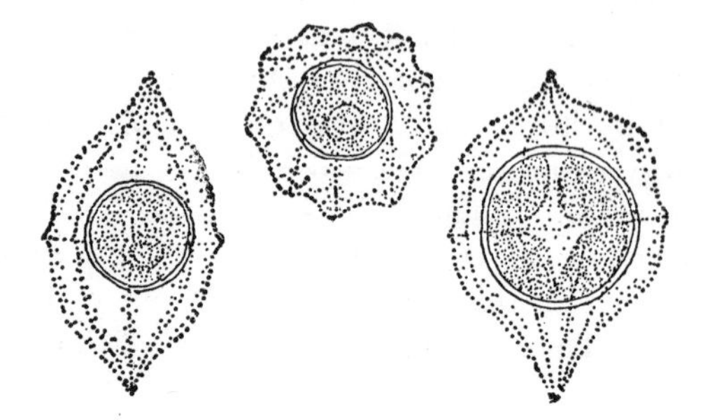

Fig. 139. *Desmatractum bipyramidatum* (Chod.) Pascher. (*After Korshikov,* 1928.) (× 1000.)

5. **Oöphila** Lambert ex Printz, 1927. The spherical to ellipsoidal cells of this alga are endozoic and grow within the envelopes surrounding eggs of the salamander (*Amblystoma*). Walls of young cells are thin and of uniform thickness; those of older cells are much thicker and with a few pits on the inner face. At first the chloroplast is irregularly lobed and axial; later it appears as if parietal and broken into four or five massive fragments. Pyrenoids are lacking, but there are numerous small starch granules in both the chloroplast and the cytoplasm.

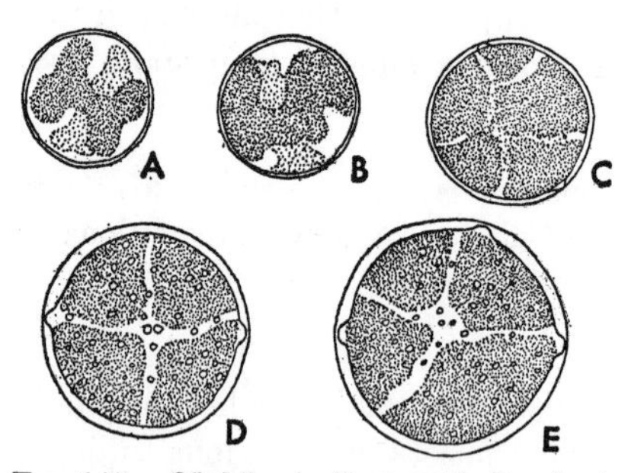

Fig. 140. *Oöphila Amblystomatis* Lambert. *A–C,* juvenile cells. *D–E,* mature cells. (× 975.)

Bi- and quadriflagellate swarmers have been found in association with this alga.[2] Presumably the former are gametes and the latter are zoozygotes.

The single species, *O. Amblystomatis* Lambert (Fig. 140), has been found in Connecticut, New York, North Carolina, Virginia, Illinois, and California. The alga is usually present in sufficient abundance to color old egg masses a grass-green. The mode of entrance of the alga into an egg is unknown, but it has been established[3]

[1] Korshikov, 1928. [2] Gilbert, 1942. [3] Gilbert, 1944.

that the relationship between the two is symbiotic. For a description of *O. Amblystomatis*, see Gilbert (1942).

Family 2. Endosphaeraceae

The Endosphaeraceae differ from the Chlorococcaceae in having larger, more irregularly shaped cells. Much of the irregularity of the cell outline is due to a development of localized thickenings on the cell wall. Most genera belonging to this family are endophytic within tissues of musci, angiosperms, or marine algae. Some of these endophytes are space parasites; others, as *Rhodochytrium* are true parasites and have rhizoidal processes that penetrate deep into the host. The Endosphaeraceae have no vegetative cell division, but the cells may become multinucleate as they increase in size. The chloroplasts of mature cells are usually axial and with numerous radial projections to the cell wall. These chloroplasts usually contain several pyrenoids. In one genus (*Rhodochytrium*), the cells are without chloroplasts, and reserve starch accumulating within a cell is obviously the result of a parasitic mode of nutrition.

Reproduction is by formation of many biflagellate or uniflagellate (?) swarmers. These may be formed by a simultaneous division, or by repeated bipartition of the protoplast. In some cases, there is always a fusion of swarmers to form a zygote[1]; in other cases, there is never a fusion[2]; in still other cases, there may or may not be a fusion to form a zygote.[3] Although our knowledge of these algae is still too fragmentary to warrant a definite statement, it is not impossible that all swarmers are gametic in nature. If this is true, the cases where swarmers germinate directly into vegetative cells must be interpreted as parthenogenesis rather than as asexual reproduction by means of zoospores. There is also the possibility that all these algae are similar to *Chlorochytrium Lemnae* Cohn and have diploid cells.

The genera found in this country differ as follows:

1. Free-living .... 2. **Kentrosphaera**
1. Endophytic in tissues of other plants 2
   2. Protoplasts green 3
   2. Protoplasts orange colored .... 3. **Rhodochytrium**
3. Cells globose .... 1. **Chlorochytrium**
3. Cells with tubular processes .... 4. **Phyllobium**

1. **Chlorochytrium** Cohn, 1872. Fresh-water numbers of this genus are endophytic and grow within tissues of mosses and phanerogams. Marine species are endophytic within thalli of various red and brown algae. Cells of *Chlorochytrium* are irregularly globose or ellipsoidal. Cell walls of

[1] Gardner, 1917; Klebs, 1881. [2] Bristol-Roach, 1917; Reichardt, 1927.
[3] Griggs, 1912; Klebs, 1881.

mature cells may be thick and stratified, or thin and homogeneous. Either type of wall may have localized lamellated thickenings. Chloroplasts of very young cells are parietal and cup-shaped,[1] but as a cell increases in size, the chloroplast becomes radially vacuolate and fills the entire lumen of a cell. Mature cells have chloroplasts with numerous radial projections that extend to the cell wall. Young cells of *C. Lemnae* Cohn are uninucleate,[2] and as the cell increases in size the nucleus increases in volume but does not divide. The division of this nucleus is meiotic.[2] There is a division of the protoplast after the first nuclear division, and further bipartitions of the cytoplasm follow each nuclear division until 256 uninucleate protoplasts are formed. These are then metamorphosed into biflagellate gametes that escape from the old parent-cell wall.[3] The gametes fuse in pairs to form quadriflagellate zygotes that swarm for a short time and then settle down upon the host and secrete a thin wall. After this the zygote grows into the large green cell that constitutes the vegetative phase

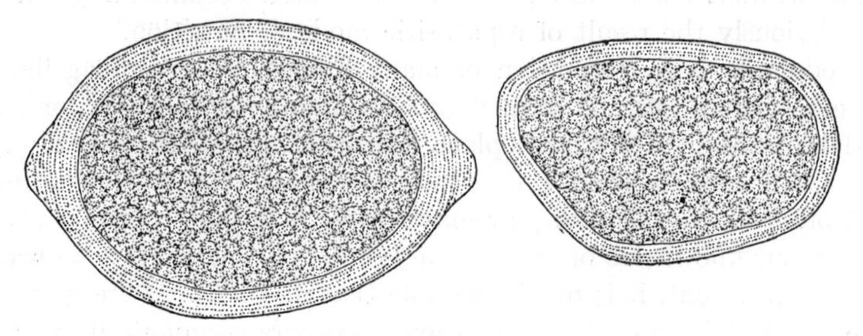

FIG. 141. *Chlorochytrium biennis* (Klebs) G. S. West. From dead leaves of *Avena*. (× 400.)

of the life cycle. Enlargement begins with formation of a tubular protrusion,[1] from the side next to the host, that grows between the host cells. The protoplast of the zygote eventually comes to lie wholly within the protrusion, and the original lumen of the zygote becomes filled with wall material.

Old cells of *Chlorochytrium* frequently enter upon an akinete-like condition in which the structure of the protoplast within the thickened wall is obscured by densely packed starch grains.

The three species found in this country are *C. Lemnae* Cohn (including *C. Archerianum* Hieron. and *C. Knyanum* Cohn and Szymanski), *C. biennis* (Klebs) G. S. West (Fig. 141), and *C. paradoxum* (Klebs) G. S. West. For descriptions of *C. Lemnae*, *C. biennis* (under *Endosphaera biennis* Klebs), and *C. paradoxum* (under *Scotinosphaera paradoxa* Klebs), see Brunnthaler (1915).

[1] COHN, 1872. [2] KURSSANOW and SCHEMAKAHNOVA, 1927.
[3] COHN, 1872; KLEBS, 1881; KURSSANOW and SCHEMAKAHNOVA, 1927.

2. **Kentrosphaera*** Borzi, 1883. Members of this genus are distinguished from *Chlorochytrium* by their free-living habit. Species of *Kentrosphaera* are usually found growing on damp soil, but they may be aquatic and intermingled with, or in the gelatinous envelopes of, other algae. The general appearance of the cells is much the same as in *Chlorochytrium*, and the walls often have irregular thickenings in certain portions. The chloroplast is axial and with numerous palisade-like processes at the periphery. The cells may be uninucleate during the period of enlargement and then become multinucleate by simultaneous mitoses just before reproduction,[1] or the number of nuclei may gradually increase during enlargement of a cell.[2]

Reproduction is by division of the protoplast into a large number of zoospores that escape through a pore in one side of the parent-cell wall.[3]

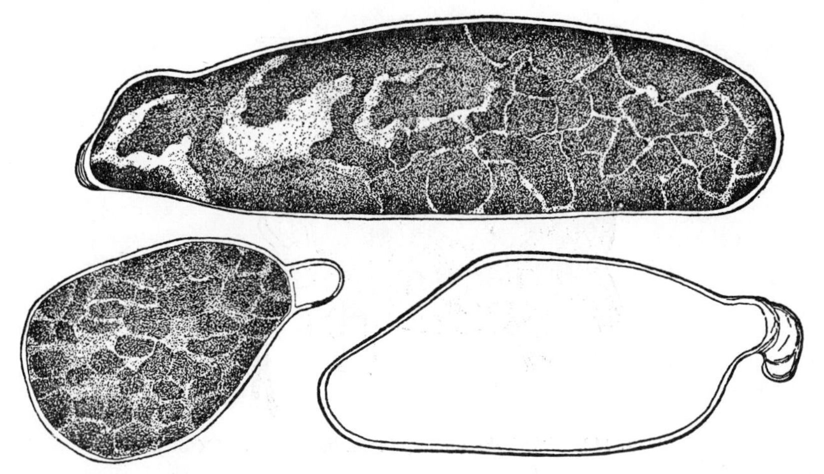

FIG. 142. *Kentrosphaera Bristolae* G. M. Smith. (*From Bristol-Roach,* 1920*A*.) (× 825.)

These come to rest after swarming, secrete a wall, and enlarge directly into vegetative cells. There may also be a division of the protoplast into numerous spherical aplanospores that develop directly into vegetative cells after they are set free by dissolution of the parent-cell wall.[1] Swarming reproductive cells of *Kentrosphaera* have never been found fusing with each other.

The free-living habit and the lack of sexual reproduction seem to be a justification for separating *Kentrosphaera* from *Chlorochytrium.* However, there are those[1] who place all species of *Kentrosphaera* in *Chlorochytrium.* *K. Bristolae* G. M. Smith

* Most British and American phycologists change Borzi's spelling to *Centrosphaera.*

[1] BRISTOL-ROACH, 1917. [2] REICHARDT, 1927.

[3] BORZI, 1883; BRISTOL-ROACH, 1917; REICHARDT, 1927.

(Fig. 142) has been found in Missouri[1] and *K. facciolae* Borzi has been found in Massachusetts.[2] For a description of *K. Bristolae*, see G. M. Smith (1933); for *K. facciolae*, see Collins (1909).

3. **Rhodochytrium** Lagerheim, 1893. This parasitic unicellular alga has been found upon a wide variety of hosts,[3] but in the United States it is most frequently found upon *Ambrosia artemisiifolia* L. The parasite attacks all aerial parts of the host but is most abundant in tissues adjoining vascular bundles. Mature cells have a flask-shaped body with a somewhat twisted neck and the basal portion bearing numerous branched rhizoids that tend to parallel vascular bundles of the host. The protoplast of *Rhodochytrium* is a bright red color throughout all stages of a cell's development and is more or less densely packed with starch. The

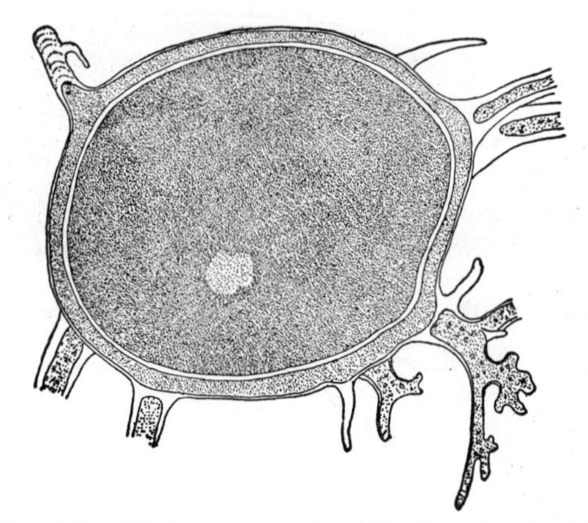

FIG. 143. *Rhodochytrium spilanthidis* Lagerh. (× 300.)

cells are uninucleate throughout the entire vegetative phase of the life cycle.[4]

Reproduction is by a formation of biflagellate gametes.[4] These result from a repeated series of nuclear divisions followed by a cleavage of the multinucleate protoplast into uninucleate protoplasts. Gametes are liberated by gelatinization of the neck-like portion of the cell facing the surface of the host.[5] Gametes may unite to form a zygote that develops directly into a vegetative cell, or they may develop parthenogenetically into vegetative cells. The first division of nucleus in a vegetative cell may be meiotic.

The seasonal life cycle of *Rhodochytrium* in the southern part of the

[1] MOORE and CARTER, 1926. [2] COLLINS, 1909. [3] PALM, 1924.
[4] ATKINSON, 1908; GRIGGS, 1912; LAGERHEIM, 1893. [5] LAGERHEIM, 1893.

United States involves several successive generations before the first of July. After this date, there is but little reproduction, and mature vegetative cells develop into akinetes that do not germinate until the following spring.[1] Development of akinetes begins with a formation of septa cutting off the rhizoids from the globular portion of a cell. Later on, the protoplast in the globular portion becomes surrounded by two additional layers of wall material (Fig. 143).

R. spilanthidis Lagerh. (Fig. 143) has been found in most of the Southeastern states and in Kansas. It is usually found on *Ambrosia artemisiifolia* L. in the Southeast but has also been found on *Solidago*. The Kansas specimens were parasitic upon *Asclepias pumila* (Gray) Vail. For a description of *R. spilanthidis*, see Griggs (1912).

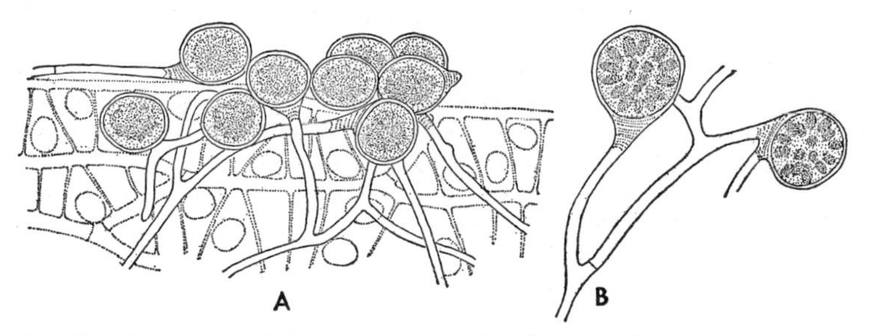

Fig. 144. *Phyllobium sphagnicola* G. S. West. *A*, edge of a *Sphagnum* leaf with empty vegetative cells and akinetes of *P. sphagnicola*. *B*, portion of a thallus isolated from a leaf of *Sphagnum*. (× 250.)

4. **Phyllobium** Klebs, 1881. *Phyllobium* is an endophyte whose simple or irregularly branched tubular cells have globosely inflated tips. Frequently one finds that all the protoplasm has migrated into the inflated cell tips and become resting cells surrounded by thick lamellated walls (Fig. 144).[2] The protoplasts contain a large number of radiately arranged spindle-shaped chloroplasts.

Reproduction is by formation of biflagellate gametes. Gametic union is anisogamous, and when two gametes unite flagella of male gametes disappear while those of the female gamete persist.[3] The biflagellate zygote swarms for a short time and then comes to rest, loses its flagella, and develops into a tubular vegetative cell.

P. sphagnicola G. S. West (Fig. 144), which grows within leaves of *Sphagnum*, has been reported from Indiana.[4] For a description of it, see G. S. West, 1908.

[1] Griggs, 1912. [2] Klebs, 1881; West, G. S., 1908. [3] Klebs, 1881.
[4] Smith, B. H., 1932.

FAMILY 3. MICRACTINIACEAE

Members of this family have spherical cells bearing one to several long setae with a length several times the diameter of the cell. The cells may be solitary or united in colonies (coenobia) with a definite number of cells. Each cell contains a cup-shaped chloroplast with a single pyrenoid.

Asexual reproduction may be by zoospores or autospores. Coenobia result from a failure of autospores formed by a single cell to separate from one another, and it is not unusual to find successive generations united in compound coenobia.

Sexual reproduction, so far as found in members of the family, is oögamous. The oögamy is unusual for Chlorophyceae in that the egg may be extruded from the parent-cell wall before fertilization.

The genera found in the United States differ as follows:

1. Cells solitary.. 2
1. Cells united in colonies.. 3
    2. Setae thickened at base.............................. 2. **Acanthosphaera**
    2. Setae not thickened at base................................ 1. **Golenkinia**
3. Cells with a single seta... 4. **Errerella**
3. Cells with more than one seta.............................. 3. **Micractinium**

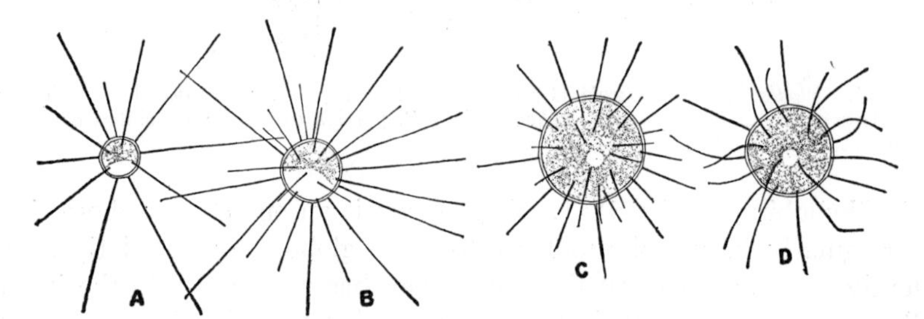

FIG. 145. *A–B, Golenkinia radiata* Chod. *C–D, G. paucispina* W. and G. S. West. (× 666.)

1. **Golenkinia** Chodat, 1894. The cells of this free-floating alga are usually solitary, though sometimes in temporary pseudocolonies resulting from an entangling of the setae. The cells are always spherical and with a thin wall bearing several long delicate setae that are not thickened in the portion next to the cell wall. Sometimes there is a hyaline gelatinous sheath outside the cell wall, but most cells lack such a sheath. The sheaths and setae are generally colorless, but they may be brownish because of an impregnation with iron compounds.[1] The chloroplast is single, parietal, cup-shaped, and filling the major portion of a protoplast. There is usually one pyrenoid within a chloroplast.[2]

[1] SUESSENGUTH, 1926. [2] CHODAT, R., 1894*B*: SMITH, G. M., 1920.

Asexual reproduction is usually by a formation of two, four, or eight autospores that are liberated by a fragmentation or by a gelatinization of the parent-cell wall. The reproduction by means of quadriflagellate zoospores described for this alga[1] has been questioned by most phycologists, but the demonstration of zoospores in the closely related *Acanthosphaera* indicates that the reported occurrence of them in *Golenkinia* is not incorrect.

Sexual reproduction is oögamous.[2] Eight or sixteen biflagellate antherozoids may be formed within a cell and, according to the species, they are naked or with a wall when liberated by fragmentation of the parent-cell wall. A female cell contains a single egg, a portion of which protrudes through a pore in the wall. An antherozoid swims to and fuses with the protruding portion of an egg. Shortly after this, the naked zygote moves out through the pore but remains attached to the old parent-cell wall. Here it secretes a thick wall with numerous verrucae or minute spines. Germination of the zygote has not been observed.

Golenkinia is a strictly planktonic genus which is of more frequent occurrence in the plankton of small shallow lakes and ponds than in that of larger lakes. The three species found in this country are *G. maxima* Tiffany and Ahlstrom, *G. paucispina* W. and G. S. West (Fig. 145*C–D*), and *G. radiata* Chodat (Fig. 145*A–B*). For a description of *G. maxima*, see Tiffany (1934); for descriptions of the others, see G. M. Smith (1920).

2. **Acanthosphaera** Lemmermann, 1899. In its spherical free-floating solitary cells with the wall bearing long setae, this genus is similar to *Golenkinia.* It differs from *Golenkinia* in having the basal portion of each seta uniformly thickened and then tapering abruptly to a very delicate hair.[3] The number of setae is usually 24, and these are regularly arranged in six tiers of four setae each.[4] The chloroplast is cup-shaped and contains one pyrenoid. A chloroplast may be massive or delicate and with smooth or irregularly lobed margins.

Reproduction is by division of the protoplast into 2, 4, or 8 zoospores that are liberated during the early morning hours.[4] The zoospores are biflagellate and have four contractile vacuoles and two (?) eyespots. They escape by a sudden rupture of the parent-cell wall and, when first liberated, are surrounded by a common vesicle. Swarming lasts for about half an hour, after which the zoospores cease moving and become spherical. Juvenile cells develop setae before they secrete a wall. The characteristic thickening at the base of setae does not appear until some time after the setae are formed. There may also be a division of the protoplast into four

[1] Chodat, 1894*B*. [2] Korshikov, 1937.
[3] Lemmermann, 1899*A*; Smith, G. M., 1920. [4] Geitler, 1924*C*.

aplanospores. These have thick walls that are without setae. The method by which they give rise to vegetative cells is unknown.

A. Zachariasi Lemm. (Fig. 146), the only species, is another strictly planktonic genus. It has been found in several of the states in the Mississippi Valley. For a description of it, see G. M. Smith (1920).

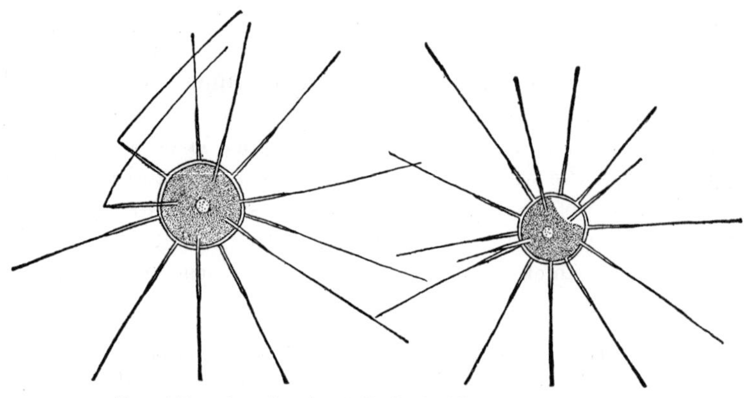

FIG. 146. *Acanthosphaera Zachariasi* Lemm. (× 1000.)

3. **Micractinium** Fresenius, 1858. The cells of this alga are spherical to broadly ellipsoidal and are usually quadrately united in four-celled coenobia. The coenobia, in turn, are almost always united with other

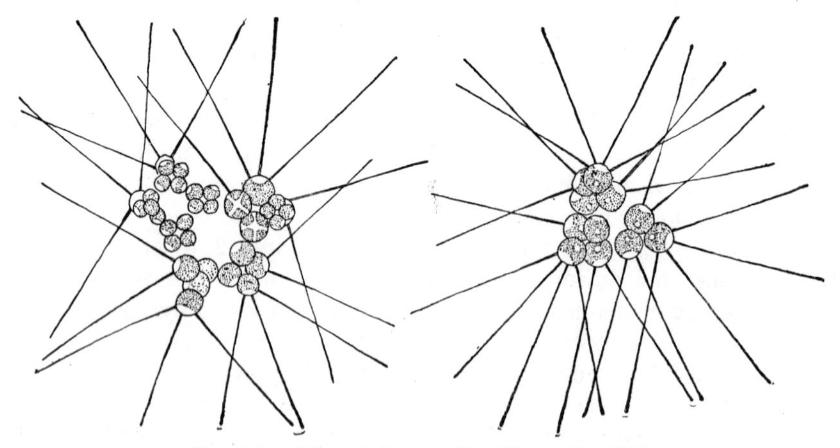

FIG. 147. *Micractinium pusillum* Fres. (× 600.)

coenobia to form a multiple coenobium that may contain 100 or more cells. The free face of each cell in a coenobium has two to seven delicate setae with a length several times the diameter of the cell. Each cell contains a single cup-shaped chloroplast with one pyrenoid.

Asexual reproduction is by division of the cell contents into four (rarely eight) autospores. These are liberated as an autocolony by a breaking of the parent-cell wall into four symmetrical parts. The newly formed coenobia, which rarely separate from the parent coenobium, do not develop the characteristic setae until some time after liberation from the parent-cell wall.[1]

Sexual reproduction is oögamous[2] and is identical with that in *Golenkinia.*

Micractinium is another strictly planktonic genus of wide distribution in this country. The three species found in this country are *M. eriense* Tiffany and Ahlstrom, *M. pusillum* Fres. (Fig. 147), and *M. quadrisetum* (Lemm.) G. M. Smith. For a description of *M. eriense*, see Tiffany (1934); for descriptions of the other two, see G. M. Smith (1920).

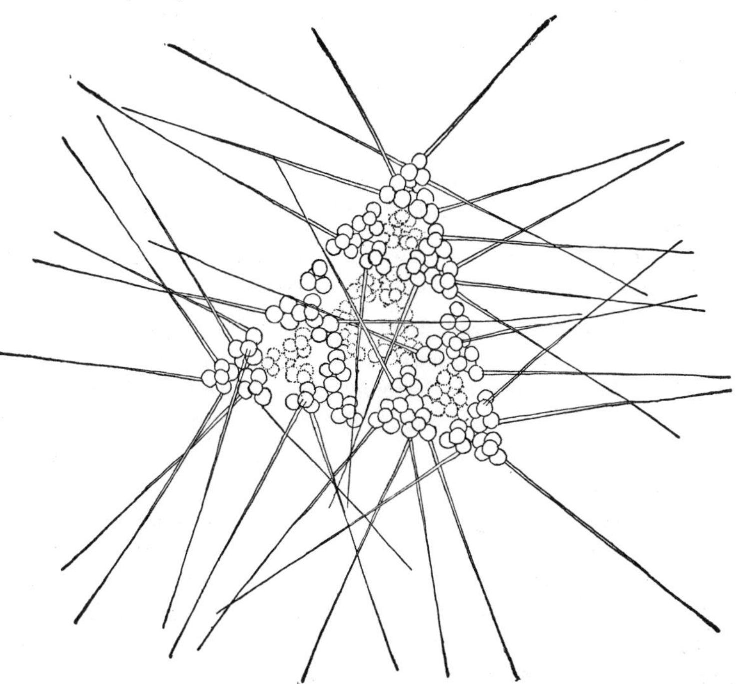

FIG. 148. *Errerella bornhemiensis* Conrad. (× 600.)

4. **Errerella** Conrad, 1913. This genus is quite like *Micractinium* in general appearance but differs from it in at least two respects. There is but a single long stout spine on the free face of a cell, and the coenobia always have the cells pyramidately arranged. Compound coenobia are of frequent occurrence and have the same pyramidate arrangement as is

[1] LEMMERMANN, 1898; SMITH, G. M., 1916C, 1920. [2] KORSHIKOV, 1937.

found in simple coenobia. The number of cells in a compound coenobium is a multiple of four and 16, 64, or 256. Each cell contains a single cup-shaped chloroplast with or without a pyrenoid.

Reproduction is by division of the cell contents into four autospores that remain pyramidately arranged after liberation from the parent-cell wall.[1]

The only known species, *E. bornhemiensis* Conrad (Fig. 148), has been found in several states in this country. For a description of it, see G. M. Smith (1926).

Family 4. Dictyosphaeriaceae

The Dictyosphaeriaceae have colonies whose cells lie at the tips of the ultimate branches of a radiating, cruciately or dichotomously, branched series of threads derived from cell walls of previous generations. A colony usually is embedded in a watery gelatinous matrix. The cells are globose to reniform and with a single chloroplast containing one pyrenoid.

Asexual reproduction is by division of the cell contents into four autospores which remain attached to the two or four connected segments into which the parent-cell wall splits. The reported reproduction by means of zoospores is open to question.

Sexual reproduction is oögamous.

The two genera found in this country differ as follows:

1. Cells of a colony all the same shape........................ 1. **Dictyosphaerium**
1. Cells of a colony of two different shapes.................... 2. **Dimorphococcus**

1. **Dictyosphaerium** Nägeli, 1849. Cells of *Dictyosphaerium* are spherical, ovoid, or reniform, and borne terminally on a cruciately or dichotomously branched system of flattened threads, the persisting parent-cell walls of previous cell generations. The colony thus formed is embedded in a rather copious, colorless, gelatinous matrix. Each cell contains a single parietal cup-shaped chloroplast with one pyrenoid.

Asexual reproduction is by division of a cell's protoplast into two or four autospores. The parent-cell wall partially splits into two or four segments, and the autospores migrate to the tips of the segments and there develop into vegetative cells. Repetition of autospore formation results in the characteristic branched system of threads at the center of a colony.[2] Reproduction by means of zoospores has been recorded,[3] but it is uncertain whether the swarmers were antherozoids or zoospores.

Sexual reproduction is oögamous.[4] Each cell of a male colony produces 16 or 32 biflagellate antherozoids which are surrounded by a gelatinous vesicle when first liberated by a transverse splitting of the parent-cell wall. The antherozoids escape from the vesicle soon after extrusion from the

[1] Conrad, 1913*A*; Smith, G. M., 1924, 1926. [2] Senn, 1899. [3] Massee, 1891.
[4] Iyengar and Ramanathan, 1940.

parent-cell wall. Each cell of a female colony produces two globose eggs which are liberated by a transverse splitting of the parent-cell wall. Antherozoids may swarm about an egg in sufficient number to cause the egg to spin in the water. Eventually one antherozoid becomes laterally attached to an egg and fuses with it to form a zygote which secretes a wall.

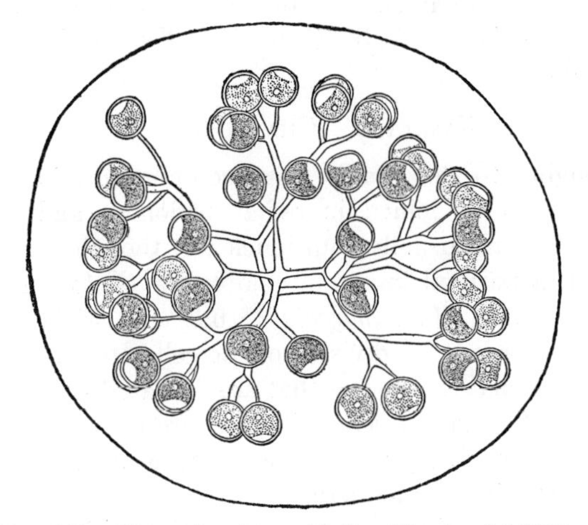

FIG. 149. *Dictyosphaerium pulchellum* Wood. (× 666.)

Dictyosphaerium is a widespread inhabitant of lakes, ponds, and semipermanent pools. The three species found in this country are *D. Ehrenbergianum* Näg., *D. planctonicum* Tiffany and Ahlstrom, and *D. pulchellum* Wood (Fig. 149). For a description of *D. planctonicum*, see Tiffany (1934); for descriptions of the other two, see G. M. Smith (1920).

2. **Dimorphococcus** A. Braun, 1855. The cells of *Dimorphococcus* are arranged in groups of four, and these tetrads are united to one another in irregularly shaped free-floating colonies by the branching remains of the old parent-cell walls. Some species have the colony surrounded by a gelatinous envelope; other species lack one. The four cells of a tetrad lie in a flat or curved plate in which the cells are cruciately arranged. Two of the cells are ellipsoidal to cylindrical and with broadly rounded ends. These cells lie end to end. The other two cells are reniform to cardioid and lie on either side of the plane of juncture between the ellipsoidal cells. Young cells have a single parietal laminate chloroplast with one pyrenoid; old cells have the chloroplast completely filling the cell and often have the pyrenoid obscured by granules of starch.

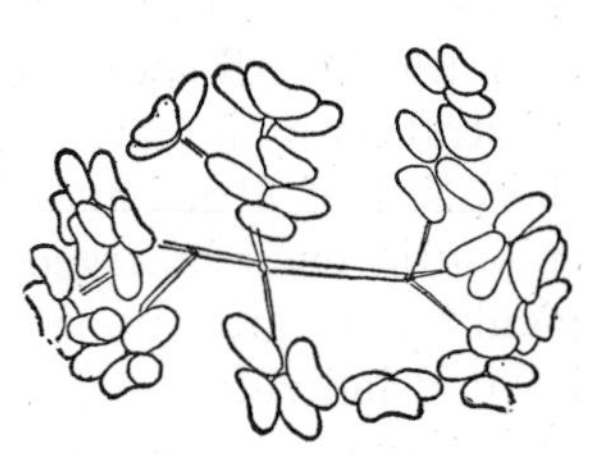

FIG. 150. *Dimorphococcus lunatus* A. Br. (× 400.)

Reproduction is by division of the protoplast into four autospores that remain attached to the remnants of the parent-cell wall.[1]

Dimorphococcus is a widely distributed plankton organism in this country and sometimes occurs in considerable quantity. It is also found sparingly intermingled with nonplanktonic algae. The two species found in this country are *D. cordatus* Wolle and *D. lunatus* A. Br. (Fig. 150). For descriptions of them, see Brunnthaler (1915).

Family 5. Characiaceae

Genera belonging to this family have zoospore-forming elongate cells that may be solitary or joined in radiate colonies, and sessile or free-floating. The cells are usually multinucleate, though sometimes uninucleate, and with a parietal laminate chloroplast containing one or more pyrenoids. One genus, *Hyalocharacium*,[2] has cells without chloroplasts.

Asexual reproduction is by division of the cell contents into 2, 4, 8, 16, 32, 64, or 128 biflagellate zoospores that are liberated through an apical or lateral opening in the parent-cell wall. In rare cases, the potential zoospores develop into aplanospores. There may also be a development of the entire protoplast into an akinete.

Sexual reproduction is by an isogamous fusion of biflagellate gametes, formed in much the same manner as zoospores.

The genera found in this country differ as follows:

1. Cells solitary 2
1. Cells radiately united in colonies .... **3. Actidesmium**
   2. Cells sessile .... **1. Characium**
   2. Cells free-floating .... **2. Schroederia**

1. **Characium** A. Braun, 1849. Cells of *Characium* (Fig. 151) may be subspherical or ovoid, but more often they are elongate and fusiform or cylindrical. All species of the genus are sessile and usually affixed to the substratum by a more or less elongate stipe, expanded into a small disk at the point of attachment. *Characium* may be epiphytic on other algae, epiphytic on submerged phanerogams, epizoic on members of the zooplankton, or grow upon submerged stones and woodwork. There may be isolated individuals upon the substratum, or the alga may be present in such abundance that it forms a continuous stratum.[3] Young cells are uninucleate and with a parietal laminate chloroplast. As a cell grows older, there may be repeated simultaneous nuclear divisions until 16, 32, 64, or 128 nuclei are present in a cell,[4] or the cell may remain uninucleate until just before reproduction.[5]

[1] Bohlin, 1897*B*; Crow, 1923*A*. [2] Pascher, 1929*C*. [3] Braun, A., 1855.
[4] Smith, G. M., 1916*A*. [5] Carter, N., 1919.

Asexual reproduction is by division of the cell contents into 8, 16, 32, 64, or 128 biflagellate zoospores. Multinucleate cells have a progressive cleavage into uninucleate protoplasts; cells that are uninucleate at maturity have repeated nuclear divisions just before reproduction and a bipartition of the cytoplasm after each mitosis.[1] Zoospores are liberated through an opening at apex, or at one side, of the parent-cell wall; and they may escape singly or may be discharged in a mass surrounded by a delicate vesicle. At the end of the swarming period, a zoospore becomes affixed to some firm object, retracts its flagella, and secretes a wall.

Sexual reproduction is by division of the protoplast into biflagellate gametes. In certain species, fusing gametes are quite different in size.[2]

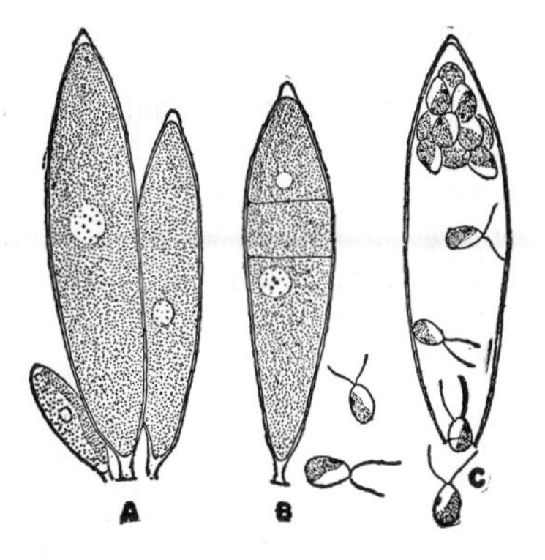

FIG. 151. *Characium angustatum* A. Br. *A*, vegetative cells. *B*, early stage in cleavage to form zoospores. *C*, liberation of zoospores. The individual drawn is probably an accidentally broken cell. (× 650.)

About 20 species have been recorded from the United States. For descriptions of most of the species of the genus, see Brunnthaler (1915).

2. **Schroederia** Lemmermann, 1899. The cells of this alga are solitary, free-floating, acicular to fusiform, and straight or curved. At both poles of a cell, the wall is continued as a long spine. Both spines may terminate

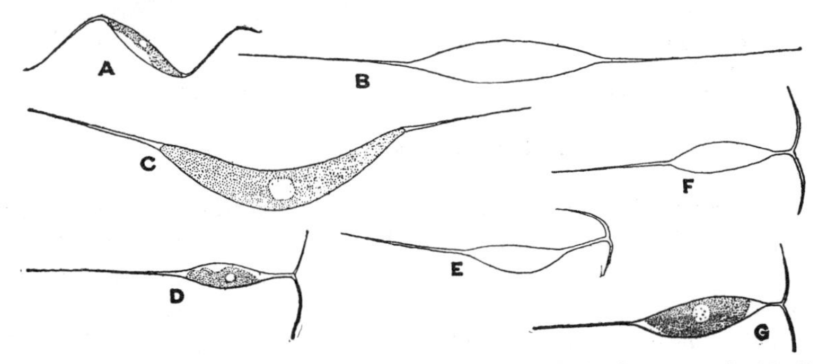

FIG. 152. *A–C*, *Schroederia setigera* (Schröder) Lemm. *D–G*, *S. ancora* G. M. Smith. (× 600.)

in an acute point; or one may terminate in a point, and the other in a small disk or in a recurved bifurcation. There is a single chloroplast, H-shaped in optical section, that extends the entire length of a cell. It usually con-

[1] CARTER, N., 1919; SMITH, G. M., 1916*A*. [2] SCHILLER, 1924.

tains a single pyrenoid, but may contain two or three of them.[1] Vegetative cells of one species have an eyespot and contractile vacuoles.[2]

Asexual reproduction is by means of zoospores.[3] Four or eight biflagellate zoospores are formed within a cell and are liberated by a transverse splitting of the parent-cell wall. The zoospores germinate to form vegetative cells without becoming affixed to some firm object.

This genus is known only from the plankton. The three species found in this country are *S. ancora* G. M. Smith (Fig. 152*D–G*), *S. Judayi* G. M. Smith, and *S. setigera* (Schröder) Lemm (Fig. 152*A–C*). For a description of *S. ancora* as *S. setigera* var. *ancora*, see G. M. Smith (1926); for descriptions of the other two, see G. M. Smith (1920).

Fig. 153. *Actidesmium Hookeri* Reinsch. (× 400.)

3. **Actidesmium** Reinsch, 1891. The spindle-shaped cells of this alga are radiately joined to one another in stellate free-floating colonies. At the center of a colony is a stellate group of stalks, and at the apex of each stalk is a stellate group of 4, 8, or 16 vegetative cells. Sometimes the stalks are in two orders of branching, the vegetative cells constituting a third order. A cell may have one or two parietal chloroplasts covering most of the lateral cell wall, or the chloroplast may be diffuse.[4] The chloroplasts are without pyrenoids. There is a conspicuous vacuole at the center of each cell.[5]

[1] Fott, 1942; Lemmermann, 1898; Smith, G. M., 1920, 1926.
[2] Korshikov, 1924*A*.
[3] Fott, 1942; Korshikov, 1924*A*; Whitford, 1943.
[4] Miller, V., 1906.
[5] Miller, V., 1906; Reinsch, 1891.

Reproduction is by successive bipartition of the cell contents into 4, 8, or 16 biflagellate zoospores that are liberated through an opening at the distal end of the parent-cell wall. The zoospores do not swarm freely but remain at the apex of the old parent-cell wall and within a minute or two begin to develop into vegetative cells that are epiphytic upon the gelatinized remains of the old parent-cell wall.[1] There may also be a discharge of the entire protoplast from the cell wall and a development of the liberated protoplast into a spherical aplanospore with granulate walls.[2] Each mature aplanospore lies at the end of a gelatinous stalk.

The type species, *A. Hookeri* Reinsch (Fig. 153), has been found in Maine[3] and California.[4] For a description of it, see Whelden (1939).

Family 6. Protosiphonaceae

Members of this family have more or less tubular multinucleate cells reproducing by means of biflagellate zoospores or by means of biflagellate gametes.

Certain phycologists[5] consider the family a member of the Siphonales rather than a member of the Chlorococcales. This certainly does not hold for *Protosiphon* since it has been shown[6] to lack the distinctive xanthophyll characteristic of Siphonales.

Protosiphon is the only member of the family in the fresh-water flora of this country.

1. **Protosiphon** Klebs, 1896. This coenocytic terrestrial alga has a green, bladder-like to tubular, aerial portion, and a colorless subterranean rhizoidal portion. The green color of the aerial portion is due to a single parietal chloroplast with several irregularly shaped perforations.[7] Mature cells have many pyrenoids, and the chief food reserve is starch.

Juvenile cells may multiply vegetatively by cutting off proliferous outgrowths by transverse septa.[8] Drying of the soil or too intense illumination may cause the protoplast to divide into a variable number of aplanospores (coenocysts).[7] Aplanospores may develop directly into vegetative cells, but usually there is a cleavage of their contents into biflagellate gametes.

Flooding of cells of any age is soon followed by a progressive cleavage of the multinucleate protoplast into biflagellate gametes.[7] Gametic union is isogamous, and a cell producing gametes may be homo- or heterothallic.[9]

[1] Miller, V., 1906. [2] Reinsch, 1891. [3] Whelden, 1939.
[4] Smith, G. M., 1933.
[5] Bold, 1933; Fritsch, 1935; Setchell and Gardner, 1920; West, G. S., 1916.
[6] Strain, (in press). [7] Bold, 1933. [8] Bold, 1933; Klebs, 1896.
[9] Moewus, 1935.

Gametes which do not fuse with other gametes develop parthenogenetically into vegetative cells. The zygote is minute, stellate, and with an orange-colored protoplast. A germinating zygote develops directly into a vegetative cell.[1] Genetic analysis of gametes from juvenile four- or eight-nucleate cells shows that division of the zygote nucleus is meiotic.[2]

Protosiphon is usually found growing on drying muddy banks of streams and ponds, or on bare damp soil. It often grows intermingled with *Botrydium*, but the two can always be distinguished from each other by testing for starch; *Protosiphon* containing it, *Botrydium* never having it. *P. botryoides* (Kütz.) Klebs (Fig. 154) is widely distributed in this country. For a description of it, see Collins (1909).

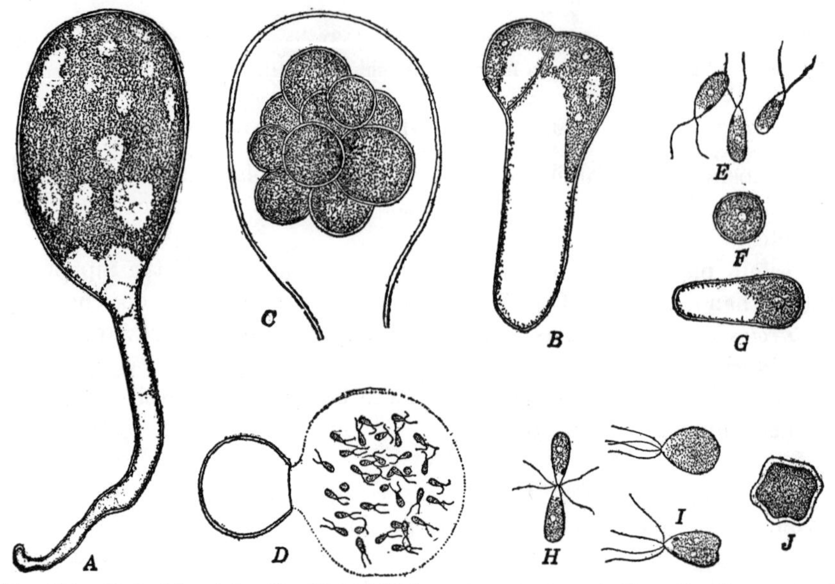

Fig. 154. *Protosiphon botryoides* (Kütz.) Klebs. *A*, mature vegetative cell. *B*, young cell cutting off a proliferation. *C*, aplanospores. *D*, germination of aplanospore to form gametes. *E*, gametes. *F–G*, parthenogenetic germination of gametes. *H*, gametic union. *I*, young zygotes. *J*, mature zygote. (*A*, *C*, *D*, × 230; *B*, *E–J*, × 650.)

Family 7. Hydrodictyaceae

Members of this family reproduce by zoospores and have all the zoospores from a cell becoming apposed to form a coenobium at the end of the swarming period. In most genera, the number of cells in a colony ranges from 2 to 256. Generic differences are based upon the shape of the cells and the manner in which they are arranged to constitute a coenobium.

[1] Bold, 1933; Klebs, 1896; Moewus, 1933. [2] Moewus, 1935.

Sexual reproduction is isogamous and by a fusion of biflagellate gametes. The zygotes give rise to zoospores when they germinate, and the zoospores develop into large angular solitary cells. These cells form zoospores which become united to form a coenobium.

The genera found in this country differ as follows:

1. Coenobium a saccate reticulum.............................. 4. **Hydrodictyon**
1. Coenobium not a saccate reticulum.............................. 2
 2. All cells in the same plane.............................. 3
 2. Cells radiating from a common center.............................. 2. **Sorastrum**
3. Coenobium always 2-celled.............................. 3. **Euastropsis**
3. Coenobium usually 4- to 64-celled.............................. 1. **Pediastrum**

1. **Pediastrum** Meyen, 1829. The colonies of *Pediastrum* (Fig. 155) are free-floating and with 2 to 128 polygonal cells arranged in a stellate plate one cell in thickness. The coenobium may be entire or perforate.

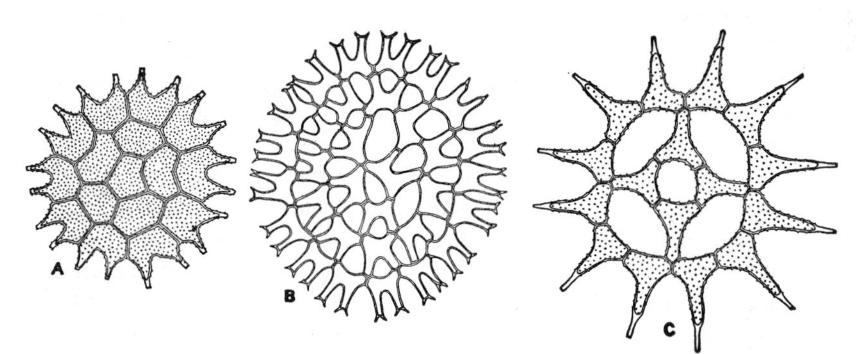

FIG. 155. *A, Pediastrum Boryanum* (Turp.) Menegh. *B, P. biradiatum* Meyen. *C, P simplex* var. *duodenarium* (Bailey) Rab. (× 333.)

If the colony has 16 or more cells, there is a tendency for them to be in concentric rings, with a definite number of cells in each ring. The occurrence or nonoccurrence of this regularity in arrangement is determined by factors affecting the extent to which the zoospores swarm at the time a colony is formed.[1] Peripheral cells of a colony often differ in shape from interior cells and may have one, two, or three processes not found on interior cells. Cell walls may be smooth, granulate, or finely reticulate. Walls of plankton species sometimes have long tufts of gelatinous bristles. Young cells have a single parietal discoid chloroplast with one pyrenoid; old cells have a diffuse chloroplast that may contain more than one pyrenoid. Mature cells may have one, two, four, or eight nuclei.[2]

Every cell in a coenobium is capable of giving rise to biflagellate zoo-

[1] HARPER, 1916, 1918, 1918*A*. [2] SMITH, G. M., 1916*B*.

spores, but there is rarely a simultaneous production of zoospores by all cells in a colony. The zoospores produced by a cell are enclosed by a vesicle as they escape from the parent-cell wall, and the vesicle persists throughout the period of swarming and for a short time after the new colony is formed.[1] During the night before reproduction, there is a two- or fourfold increase in the number of nuclei, followed by a progressive cleavage of the coenocyte into uninucleate protoplasts that are metamorphosed into zoospores.[2] The number of zoospores formed is dependent upon the physiological condition of the cells. Thus, cells in a 16-celled colony may give rise to 4- or 8-celled daughter colonies, or they may produce daughter colonies with 32 or 64 cells. Liberation of zoospores usually takes place shortly after daybreak, and it is very unusual to find swarming at any other time of day. At the beginning of the swarming period there is a sudden slit-like rupturing of the parent-cell wall and an extrusion of the spore mass surrounded by a sac-like vesicle. The vesicle is derived from the inner wall layer of the parent cell.[3] For the first 3 or 4 min. after extrusion, the zoospores swim actively and freely within the vesicle; after this, they tend to arrange themselves in a flat plate and to have their motion restricted to a writhing and twitching. Coincident with slowing down of movement, the zoospores begin to assume the shape of adult cells, and cell walls are formed within a very few minutes after swarming ceases. In very rare cases, the entire protoplast of a cell develops into a thick-walled aplanospore. These aplanospores (hypnospores) are extremely resistant to adverse conditions and have been known[4] to germinate after desiccation for 12 years.

Sexual reproduction of *Pediastrum* is isogamous and by a fusion of spindle-shaped biflagellate gametes formed in the same manner as zoospores.[5] After the spherical zygote has increased greatly in size, its contents divide to form a considerable number of biflagellate zoospores which are liberated through an opening in the zygote wall.[6] The zoospores swim freely in all directions and, upon cessation of swarming, develop into solitary *Tetraëdron*-like cells which increase greatly in size. Eventually the contents of these "polyeders" divide to form a number of zoospores which are liberated within and never escape from a vesicle. These zoospores become apposed to form a coenobium.

Isolated individuals of *Pediastrum* (Fig. 155) are often encountered in collections from permanent or semipermanent pools. The genus is almost always present in the plankton, sometimes in considerable quantity. Some 15 species have been found in this country. For descriptions of most species of the genus, see Brunnthaler (1915).

[1] Askenasy, 1888; Braun, A., 1851; Smith, G. M., 1916*B*.
[2] Smith, G. M., 1916*B*. [3] Harper, 1918. [4] Ström, 1921*A*.
[5] Askenasy, 1888; Palik, 1933. [6] Palik, 1933.

2. **Sorastrum** Kützing, 1845. This genus is distinguished from the foregoing by its spherical colonies of 8 to 128 cells. The cells are pyriform, semilunar, or reniform; and with one, two, or four stout spines on the outer face. Each cell bears a broad gelatinous stalk at its inner face. The stalks are five- to six-faceted at their base and with the bases apposed to bases of other stalks to form a gelatinous sphere at the center of the coenobium.[1] Young cells have a single parietal chloroplast, but this becomes diffuse as a cell grows older. Adult cells are multinucleate[2] probably with the number of nuclei a multiple of two.

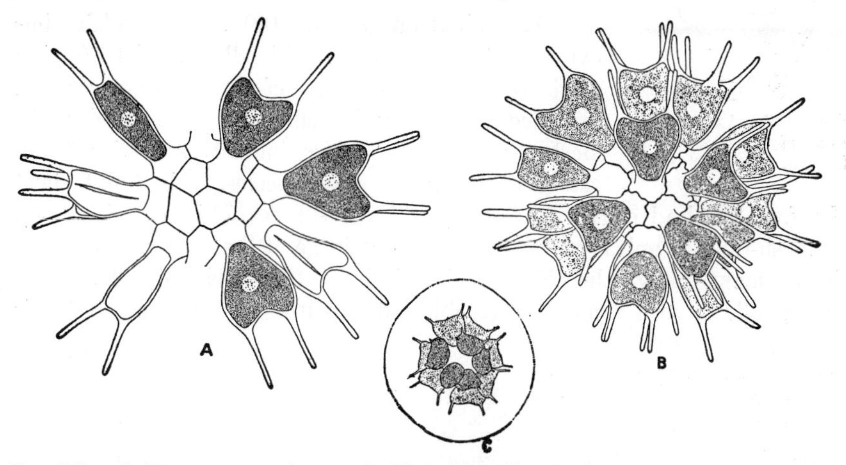

FIG. 156. *A*, *Sorastrum americanum* (Bohlin) Schmidle. *B–C*, *S. americanum* var. *undulatum* G. M. Smith. (× 666.)

Reproduction is by a formation of zoospores that swarm in the same manner as do those of *Pediastrum*.[3] The zoospores are formed by a repeated bipartition of the protoplast.

Sorastrum is found in the same habitats as *Pediastrum*, but is a much rarer alga. *S. americanum* (Bohlin) Schmidle (Fig. 156), *S. bidentatum* Reinsch, and *S. spinulosum* Näg. have been found in this country. For descriptions of them, see Brunnthaler (1915).

3. **Euastropsis** Lagerheim, 1894. *Euastropsis* has two-celled free-floating colonies, in which the apposed bases of the two cells are mutually flattened and the distal ends are deeply emarginate. Old cells have diffuse chloroplasts containing one pyrenoid.

Reproduction is by repeated bipartition of the protoplast into 4, 8, 16 or 32 zoospores that are surrounded by a common vesicle when extruded from the parent-cell wall. The vesicle persists throughout the swarming

[1] BOHLIN, 1897*B*; SMITH, G. M., 1920. [2] GEITLER, 1924*B*.
[3] GEITLER, 1924*B*; PROBST, 1916, 1926.

period and for a short time after the zoospores have become apposed in pairs to form the characteristic coenobia.[1]

It is quite possible to mistake two-celled colonies of *Pediastrum* for *Euastropsis*, but two-celled colonies of *Pediastrum* are of such rare occurrence that one is not apt to make this error. There is but one species, *E. Richteri* (Schmidle) Lagerh. (Fig. 157), and thus far in this country it has been found only in Michigan[2] and Wisconsin.[3] For a description of it, see G. M. Smith (1920).

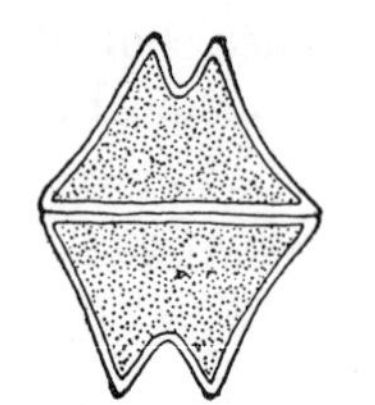

FIG. 157. *Euastropsis Richteri* (Schmidle) Lagerh. (× 2000.)

4. **Hydrodictyon** Roth, 1800. This alga has cylindrical to broadly ovoid cells united to form a meshwork in which most of the interspaces are bounded by five or six cells. In some species, the meshwork is a closed cylindrical sac with thousands of cells; in other species, it is a flat sheet of a few hundred cells. Some species have the coenobium maintaining its mesh-like organization no matter how large it becomes; other species frequently have the old nets dissociating into individual cells. Very young cells have a zonate entire chloroplast with one pyrenoid, but this

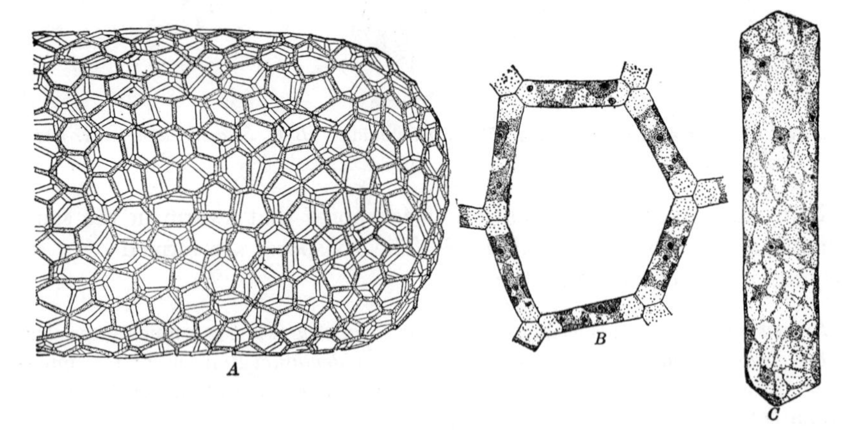

FIG. 158. *Hydrodictyon reticulatum* (L.) Lagerh. *A*, portion of a coenobium. *B*, very young cell. *C*, somewhat older cell. (*A*, × 20; *B–C*, × 430.)

soon becomes reticulate and with a pyrenoid at many of the intersections in the reticulum (Fig. 158*B–C*).[4] Still later the chloroplast becomes diffuse and completely covers the surface of the protoplast. A cell is uninucleate when first formed, but as it grows in size, it becomes multinucleate and eventually contains thousands of nuclei.[5]

[1] LAGERHEIM, 1894. [2] TAFT, 1939. [3] SMITH, G. M., 1920.
[4] LOWE and LLOYD, 1927; TIMBERLAKE, 1901. [5] TIMBERLAKE, 1902.

Some species reproduce asexually by means of zoospores, but vegetative cells of certain species do not produce zoospores.[1] In species reproducing asexually, zoospores may be formed by progressive cleavage in any cell of a colony.[2] The zoospores are biflagellate and usually swarm shortly after daybreak, but this may be delayed until later in the morning. Zoospores never escape from the parent-cell wall. For a time, they swim in all directions in the space enclosed by the wall, but toward the end of the swarming period they accumulate just within the wall and, as their motion gradually ceases, come to lie in a single layer and touching one another. Each cell secretes a wall of its own, and elongation of the cells results in the characteristic closed saccate cellular meshwork. The colony thus formed is eventually liberated by a gelatinization of the parent-cell wall.

Sexual reproduction is by means of biflagellate isogametes that are formed in the same manner as zoospores. The gametes usually escape through pores in the parent-cell wall and fuse in pairs outside the cell in which they were formed. The zygotes become spherical and form a thin wall. After a short period of rest there is a meiotic division of the zygote nucleus,[3] followed by a division of the protoplast into four zoospores. After liberation from the old zygote wall, the zoospores develop into *Tetraëdron*-shaped cells when they cease swarming. These cells (polyeders) increase greatly in size, and their contents divide to form 50 by 100 zoospores that are liberated within and never escape from a common vesicle. At the end of the swarming period, the cells become arranged in a flat plate-like meshwork. In some species, all formation of new colonies seems to be due to germination of polyeders.[1] In other species, cells of the plate-like colony give rise to saccate colonies which also reproduce asexually.

Of the two species found in this country, *H. reticulatum* (L.) Lagerh. (Fig. 148) is of widespread distribution, and *H. patenaeformis* Pocock is known from only two localities in California and one in Utah. For descriptions of these species, see Pocock (1937).

Family 8. Coelastraceae

Members of this family have the cells radially arranged in more or less globose coenobia in which the number of cells may range from 4 to 128. Reproduction is exclusively by means of autospores which become united with one another to form a coenobium before liberation from the parent-cell wall.

Coelastrum is the only member of the family found in this country.*

[1] Pocock, 1937*A*. [2] Artari, 1890; Klebs, 1891; Timberlake, 1902.

[3] Mainx, 1931*A*. * H. F. Copeland (1937) has brought forth good evidence showing that the putative alga to which Kofoid (1914) gave the generic name *Phytomorula* is in reality a compound polled grain of species of *Acacia*.

1. **Coelastrum** Nägeli, 1849. The coenobium of *Coelastrum* is a hollow sphere of 4, 8, 16, 32, 64, or 128 cells, closely apposed to one another or united by processes of varying length. The cells are spherical to polygonal, with smooth walls or one ornamented with spines or tubercules. Walls surrounding cells are composed of two parts, an inner layer of cellulose and an outer one of pectic material. The pectic portion may be of uniform thickness or may be locally thickened to form polar outgrowths or lateral processes that connect cells one to another.[1] Chloroplasts of young cells are cup-shaped and with a single pyrenoid; those of older cells are diffuse and often fill the entire lumen of a cell. Old cells of *C. proboscideum* Bohlin have been shown[2] to be multinucleate, but it is not known whether this is true of other species. The cells in a coenobium usually adhere firmly to one another throughout the life of a coenobium,

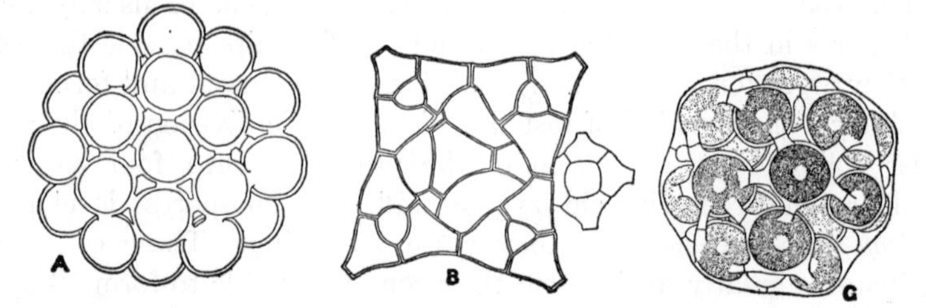

Fig. 159. *A, Coelastrum microporum* Näg. *B, C. proboscideum* Bohlin. *C, C. reticulatum* (Dang.) Senn. (× 550.)

but certain factors, as a high concentration of salts or semianaerobic conditions, may cause a dissociation into individual cells.[3]

Any cell of a coebonium may have its protoplast divide and redivide to form 4, 8, 16, 32, 64, or 128 autospores that are firmly united in an autocolony when liberated from the parent-cell wall. Liberation of daughter coenobia is due more often to a bi- or quadripartition of the parent-cell wall than to a gelatinization of the wall.[4] Species without processes between the cells usually have an immediate escape of the daughter coenobia; those with connecting processes, as *C. reticulatum* (Dang.) Senn, may not have the daughter coenobia escaping until they, in turn, have formed new coenobia. There may also be a development of the entire protoplast of a cell into an aplanospore. These are usually liberated from parent-cell walls before their contents give rise to autocolonies.[5]

Coelastrum is often found sparingly intermingled with other algae of pools and ditches, and in the plankton. The following species have been found in this

[1] Crow, 1924*A*; Rayss, 1915; Senn, 1899. [2] Geitler, 1924*B*.
[3] Rayss, 1915; Vischer, 1927. [4] Crow, 1925*A*; Senn, 1899.
[5] Chodat and Huber, 1894; Senn, 1899; Wille 1918*A*

country: *C. cambricum* Arch., *C. Chodati* Ducell., *C. cubicum* Näg., *C. microporum* Näg. (Fig. 159*A*), *C. morus* W. and G. S. West, *C. proboscideum* Bohlin (Fig. 159*B*), *C. reticulatum* (Dang.) Senn (Fig. 159*C*), *C. sphaericum* Näg., and *C. verrucosum* Reinsch. For a description of *C. Chodati*, see Ducellier (1915); for descriptions of the others, see Brunnthaler (1915).

Family 9. Oöcystaceae

The sole method of reproduction of members of the Oöcystaceae is by autospores, and these are never united in autocolonies when they are liberated from the parent-cell wall. Some genera belonging to the family have solitary cells; others have the cells united in colonies either because of an entangling of cells one with another (*Ankistrodesmus*), because of adhesions between cells (*Selenastrum*), or because of their inclusion within a common matrix resulting from gelatinization of the parent-cell wall (*Quadrigula*, *Kirchneriella*). In none of the colonial genera is there a definite orientation of the cells with respect to one another as is found in the Scenedesmaceae. The cell shape is quite variable from genus to genus and may be spherical, ovoid, reniform, acicular, lunate, triangular, quadrangular, or polygonal. Many genera have a characteristic ornamentation of the cell wall or a characteristic spinescence. Most members of the family have cells with a single chloroplast, but a few have numerous chloroplasts.

Practically all the genera are aquatic, and many of them are known only from the plankton. Wide gelatinous sheaths, long setae, and large surface in proportion to volume make Oöcystaceae with these structures especially adapted to a pelagic existence.

The relationships between certain members of the family are evident, those between other members are obscure. For this reason, no attempt has been made to follow the usual practice of splitting the family into several subfamilies.

The genera found in this country differ as follows:*

1. Cells solitary.......... 2
1. Cells in colonies.......... 28
    2. Cells angular.......... 3
    2. Cells spherical, ovoid, lunate, or elongate.......... 7
3. With distinct spines at the angles.......... 4
3. Without spines at the angles, but sometimes with angles pointed and branched. 6
    4. Each angle with several spines.......... 31. **Poleydriopsis**
    4. Each angle with one spine.......... 5
5. Spines hyaline, apex acute.......... 12. **Treubaria**

* *Thamniastrum* Reinsch has long been considered a genus of doubtful validity. Taft (1945) suggested that it is based upon the central gelatinous portion of *Gomphosphaeria lacustris* from which the cells have broken away

5. Spines brownish, apex blunt .... 13. **Pachycladon**
    6. Central portion of cell distinct from projections at angles .. 29. **Tetraëdron**
    6. Central portion of cell not distinct from projections at angles . 30. **Cerasterias**
7. Cells spherical 8
7. Cells not spherical 17
    8. With smooth walls 9
    8. With walls sculptured or with spines 13
9. Cells with one or a few chloroplasts 10
9. Cells with many chloroplasts 11
    10. With one chloroplast .... 1. **Chlorella**
    10. With more than one chloroplast .... 2. **Palmellococcus**
11. Cells with a wide gelatinous sheath .... 8. **Planktosphaeria**
11. Cells without a gelatinous sheath 12
    12. Chloroplasts discoid, tending to lie in reticulate series .... 9. **Eremosphaera**
    12. Chloroplasts inverted cones and in close contact .... 10. **Excentrosphaera**
13. Walls with long spines 14
13. Spines, if present, very short 15
    14. With several hyaline acutely pointed spines .... 11. **Echinosphaerella**
    14. With four brown truncate spines .... 13. **Pachycladon**
15. Wall alveolar .... 4. **Keriochlamys**
15. Wall not alveolar 16
    16. Growing in snow fields .... 3. **Mycanthococcus**
    16. Not growing in snow fields .... 7. **Trochiscia**
17. Cells more or less ellipsoid 18
17. Cells elongate, straight, or lunate 24
    18. Walls without setae or spines 19
    18. Walls with setae 22
19. Wall thin 20
19. Wall thick and unevenly thickened at poles or at sides ..... 10. **Excentrosphaera**
    20. Wall smooth 21
    20. Wall with spiral longitudinal ridges .... 15. **Scotiella**
21. Cells not over 10 μ long, without polar nodules .... 2. **Palmellococcus**
21. Cells over 10 μ long, often with polar nodules .... 14. **Oöcystis**
    22. Setae or spines covering entire wall 23
    22. Setae at poles or equator of wall .... 18. **Chodatella**
23. Autospores liberated by swelling of parent-cell wall .... 19. **Franceia**
23. Autospores liberated by lateral rupture of parent-cell wall ...... 20. **Bohlinia**
    24. Cells lunate 25
    24. Cells straight 27
25. Cell surrounded by a gelatinous sheath .... 26. **Kirchneriella**
25. Cell not surrounded by a gelatinous sheath 26
    26. Poles of cell with a stout spine .... 24. **Closteridium**
    26. Poles of cell without spines .... 21. **Ankistrodesmus**
27. Chloroplast with an axial row of pyrenoids .... 23. **Closteriopsis**
27. Chloroplast with one or without pyrenoids .... 21. **Ankistrodesmus**
    28. Cells lying in a gelatinous matrix 29
    28. Cells not lying in a gelatinous matrix 35
29. Cell wall with spines .... 20. **Bohlinia**
29. Cell wall without spines 30
    30. Cells spherical to broadly ovoid 31
    30. Cells elongate, length at least double breadth 32

31. Center of colony with remains of old cell walls .... 6. **Radiococcus**
31. Center of colony without remains of old cell walls .... 8. **Planktosphaeria**
  32. Cells straight 33
  32. Cells curved 34
33. Cells radiately disposed in colony .... 28. **Gloeoactinium**
33. Cells with long axes parallel .... 27. **Quadrigula**
  34. Gelatinous sheaths of colony firm .... 17. **Nephrocytium**
  34. Gelatinous sheaths of colony watery .... 26. **Kirchneriella**
35. With remains of old cell walls at center of colony .... 5. **Westella**
35. Without remains of old cell walls at center of colony 36
  36. Cells spherical to ovoid, enclosed by old parent-cell wall 37
  36. Cells longer than broad 38
37. Cells separated by a dark cruciform substance .... 16. **Gloeotaenium**
37. Cells not separated by a dark substance .... 14. **Oöcystis**
  38. Cells joined end to end .... 22. **Dactylococcus**
  38. Cells not joined end to end 39
39. Cells straight .... 21. **Ankistrodesmus**
39. Cells not straight 40
  40. Cells lunate, with convex faces apposed .... 25. **Selenastrum**
  40. Cells twisted about one another .... 21. **Ankistrodesmus**

1. **Chlorella** Beijerinck, 1890. The cells of *Chlorella* are small and spherical to broadly ellipsoidal. They have a single parietal chloroplast that is usually cup-shaped but may be a curved band. Pyrenoids are usually lacking.[1]

The only method of reproduction is by means of autospores, 2, 4, 8, or 16 of which are formed within a cell and are liberated by rupture of the parent-cell wall.

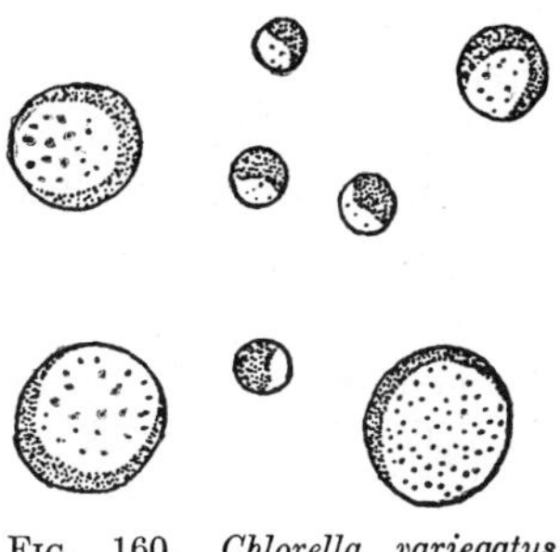
FIG. 160. *Chlorella variegatus* Beijerinck. (× 1200.)

Chlorella may be free-living or grow within the cells and tissues of invertebrates. The relationship between alga and animal varies all the way from accidental commensalism to a true parasitism.[2] Free-living members of the genus are known chiefly from material obtained by the pure culture method, and a large number of species have been described from material obtained in such manner. Many of these species are indistinguishable from one another on the basis of cellular morphology. The following species have been found in this country: *C. conductrix* (Brandt) Beijerinck, *C. ellipsoidea* Gerneck, *C. parasitica* (Brandt) Beijerinck, *C. variegatus* Beijerinck (Fig. 160), and *C. vulgaris* Beijerinck. For descriptions of *C. conductrix* and *C. parasitica*, see Beijerinck (1890); for descriptions of the others, see Brunthaler (1915).

2. **Palmellococcus** Chodat, 1894. The cells of this alga are spherical to ellipsoidal and grow singly or in an expanded stratum. Mature cells

[1] BEIJERINCK, 1890. [2] GOETSCH and SCHEURING, 1926.

contain one or more discoid chloroplasts that are generally without pyrenoids but may have them in certain cases. Sometimes the structure of the chloroplasts is obscured by an accumulation of a reddish oil within the chloroplast.

Reproduction is by division of the protoplast into 2, 4, 8, 16, or 32 autospores that are liberated by rupture of the parent-cell wall. Under certain conditions, the entire protoplast develops into a thick-walled aplanospore that lies within the old parent-cell wall.

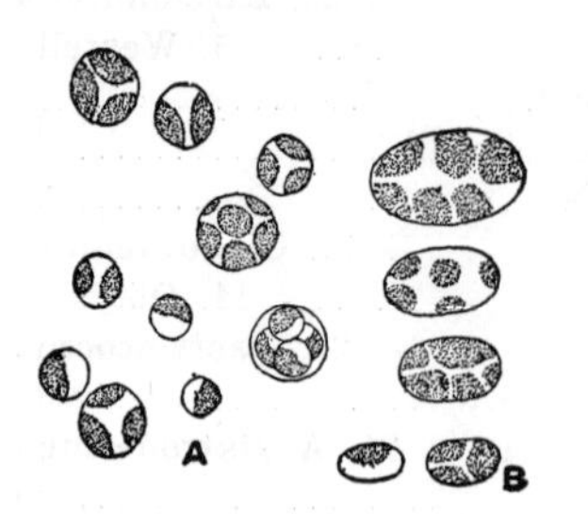

FIG. 161. *A*, *Palmellococcus miniatus* (Kütz.) Chod. *B*, *P. protothecoides* (Krüger) Chod. (*A*, × 600; *B*, × 975.)

P. miniatus (Kütz.) Chod. (Fig. 161*A*) and *P. protothecoides* (Krüger) Chod. (Fig. 161*B*) have been found in the United States. For a description of *P. miniatus*, see Collins (1909); for *P. protothecoides*, see R. Chodat (1909).

3. **Mycanthococcus** Hansgirg, 1890. The cells of this imperfectly known genus are solitary and spherical or ellipsoidal. Walls of young cells are smooth and thin; those of old cells are thick and covered with short stout spines or verrucae. The protoplast has been described as colorless,[1] but it may be a very pale green.[2]

Reproduction is by division of the cell contents into a small number of autospores.

M. antarcticus Wille (Fig. 162) has been found in considerable quantity in green snow in Yellowstone National Park.[2] For a description of it, see Kol (1941).

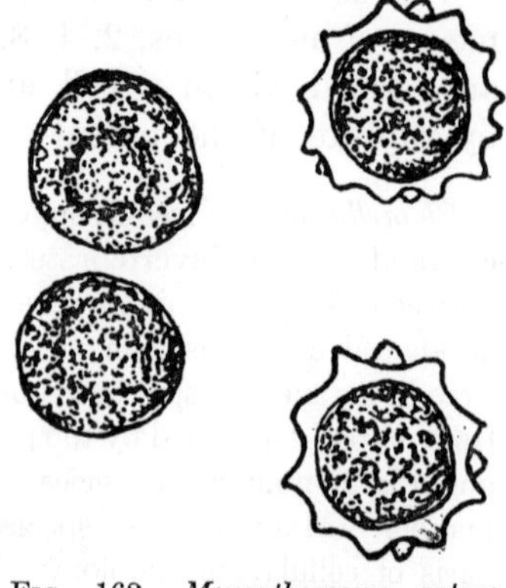

FIG. 162. *Mycanthococcus antarcticus* Wille. (*From Kol*, 1941.)

4. **Keriochlamys** Pascher, 1934. *Keriochlamys* usually has free-floating solitary cells but may have them in packets of two, four, or more because of slow development of autospores into vegetative cells while still within the parent-cell wall. Solitary cells are spherical to broadly ellipsoidal. The cell wall is the distinctive feature of the genus. It is relatively thick and looks as if it were alveolar because of the numerous, evenly spaced, minute, refractive bodies of unknown nature embedded within it. Each cell contains a single bowl-shaped chloroplast with a single pyrenoid.[3]

[1] GAIN, 1911; HANSGIRG, 1892. [2] KOL, 1941. [3] PASCHER, 1943*A*.

Reproduction is by division of the cell contents into two or four autospores with homogeneous walls. Autospores may develop into vegetative cells with characteristic walls before liberation from the parent-cell wall.

Prof. R. H. Thompson writes that he has found *K. styriaca* Pascher (Fig. 163), the only species, in Kansas. For a description of it, see Pascher (1943*A*).

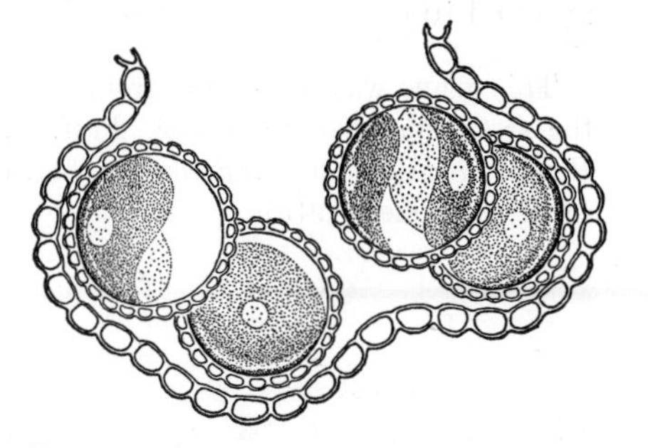

Fig. 163. *Keriochlamys styriaca* Pascher. (*Drawn by R. H. Thompson.*) (× 1650.)

5. **Westella** de Wildemann, 1897. *Westella* has spherical to subspherical cells grouped in fours (rarely eights) in irregularly shaped free-floating colonies of 30 to 100 cells. The groups of four or eight cells lie in the same plane and are quadrately disposed or are in a linear series. The groups are held together by the nongelatinized remains of old parent-cell walls. The chloroplast is parietal and cup-shaped, or entirely filling the cell. It may be with or without a pyrenoid.[1]

Reproduction is by division of the cell contents into four or eight autospores that usually remain partly surrounded by the ruptured parent-cell wall.

The two species found in this country are *W. botryoides* (W. West) de Wildm. (Fig. 164) and *W. linearis* G. M. Smith. For descriptions of them, see G. M. Smith (1920).

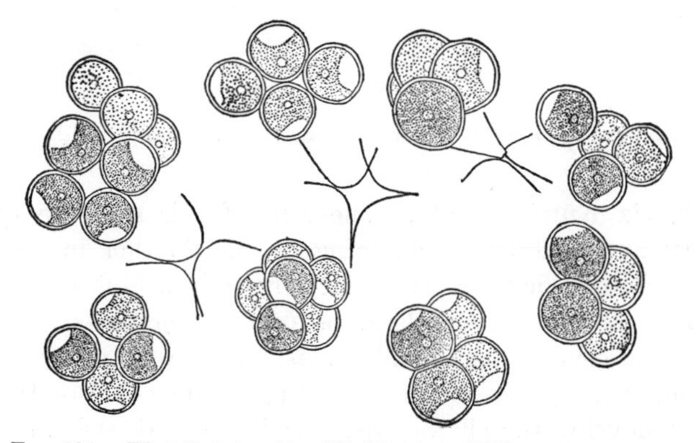

Fig. 164. *Westella botryoides* (W. West) de Wildm. (× 1000.)

6. **Radiococcus** Schmidle, 1902. This colonial free-floating alga has globose cells pyramidately grouped in fours within a broad radially fibrillar gelatinous matrix. When there is more than one group of four cells within

[1] Smith, G. M., 1920; West, W., 1892.

a colony, there are the nongelatinized remains of old parent-cell walls. Each cell has a parietal cup-shaped chloroplast with one pyrenoid.

Reproduction is by division of a cell into four autospores which remain apposed to one another as they develop into vegetative cells.

This genus was segregated from *Westella* because of the gelatinous matrix and the arrangement of the cells.[1] Prof. R. H. Thompson writes that he has found *R. nimbatus* (de Wildm.) Schmidle (Fig. 165) in Maryland. For a description of it, see Brunnthaler (1915).

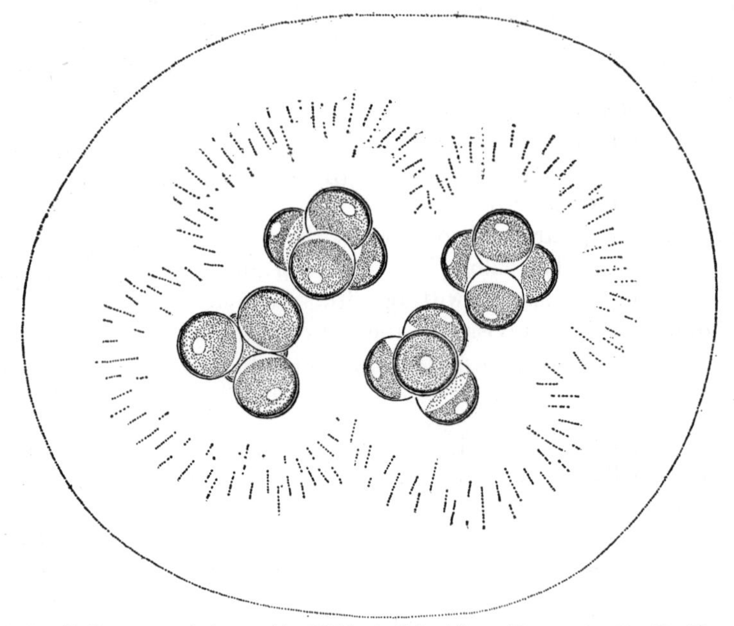

FIG. 165. *Radiococcus nimbatus* (de Wildm.) Schmidle. (*Drawn by R. H. Thompson.*) (× 1560.)

7. Trochiscia Kützing, 1845. The cells of this alga are spherical to subspherical and solitary, or adhering to one another in small clumps. The wall is fairly thick and is characteristically sculptured with areolations, spines, reticulations, ridges, or other projections. Mature cells usually contain several discoid chloroplasts, each with a pyrenoid.[2]

Reproduction is by division of the protoplast into 4, 8, or 16 autospores that do not develop the characteristic sculpturing of the wall until after liberation from the parent-cell wall.

The ornamentation of the cell wall is quite like that of walls of zygotes of certain Chlamydomonadaceae, and some species described for the genus are doubtless

[1] SCHMIDLE, 1902*A*. [2] REINSCH, 1886.

zygotes. The demonstration of reproduction by means of autospores shows that all such ornamented cells cannot be considered zygotes. *Trochiscia* is generally found sparingly intermingled with other algae in permanent and semipermanent pools. The following species have been reported from the United States: *T. aciculifera* (Lagerh.) Hansg., *T. arguta* (Reinsch) Hansg., *T. aspera* (Reinsch) Hansg., *T. erlangensis* Hansg., *T. granulata* (Reinsch) Hansg., *T. hirta* (Lagerh.) Hansg., *T. obtusa* (Reinsch) Hansg. (Fig. 166), *T. pachyderma* (Reinsch) Hansg., and *T. reticularis* (Reinsch) Hansg. For a description of *T. erlangensis*, see Prescott and Croasdale (1942); for *T. pachyderma,* see Brunnthaler (1915); for the others, see Collins (1909).

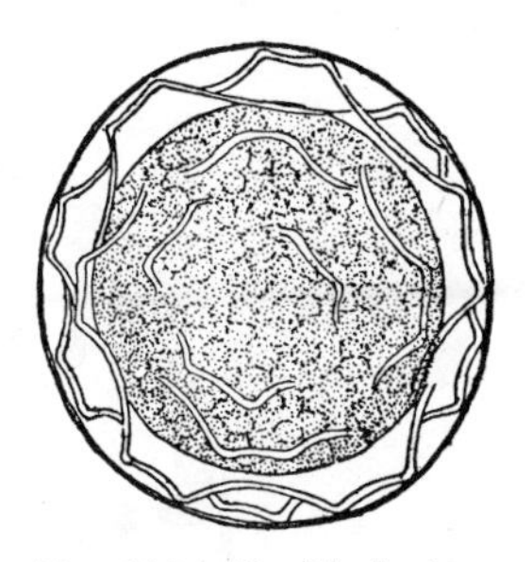

Fig. 166. *Trochiscia obtusa* (Reinsch.) Hansg. (× 800.)

8. **Planktosphaeria** G. M. Smith, 1918. The cells of this alga are spherical and embedded within a wide colorless homogeneous envelope. They may be solitary, or they may be in colonies and irregularly distributed within the gelatinous matrix. Young cells have a single cup-shaped chloroplast; mature cells have several polygonal and flattened chloroplasts, each with a pyrenoid, that are parietal in distribution.

Reproduction is by division of the cell contents into 4, 8, or 16 auto spores that immediately escape from, or may be retained within, the old parent-cell wall.[1]

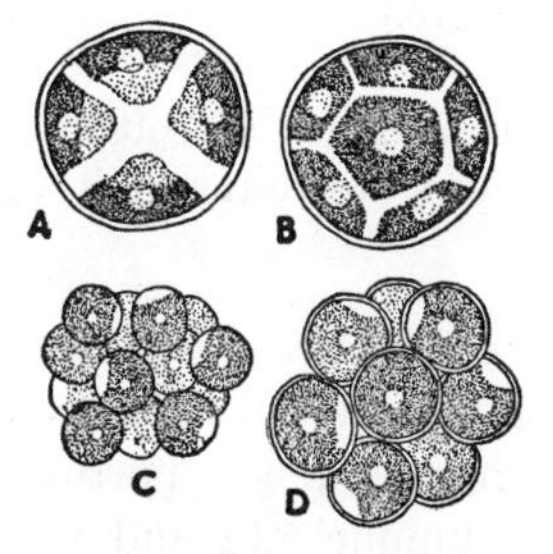

Fig. 167. *Planktosphaeria gelatinosa* G. M. Smith. *A–B*, mature cells. *C–D*, young colonies. (× 1000.)

P. gelatinosa G. M. Smith (Fig. 167), the only known species, has been found in several states of the Middle West. For a description of it, see G. M. Smith (1920).

9. **Eremosphaera** DeBary, 1858. *Eremosphaera* is sharply differentiated from other unicellular Chlorophyceae by its large thin-walled cells. The cells are uninucleate and with the nucleus held in the center of the cell by numerous strands of cytoplasm. The layer of cytoplasm immediately within the wall contains numerous discoid chloroplasts, each with a single pyrenoid. The chloroplasts are often arranged in a reticulum in which the meshes of the network have three to seven seriately disposed chloroplasts.

Reproduction is by division of the protoplast into two or four autospores that are liberated by rupture of the parent-cell wall. The first division of

[1] Smith, G. M., 1918.

the nucleus is followed by a cleavage that starts between the two daughter nuclei and then progresses outward to the plasma membrane.[1] There is a possibility that this cytokinesis may be due to a cell plate instead of a cleavage.[2] There may also be a development of the entire protoplast into a single large aplanospore with a brick-red contents.[3]

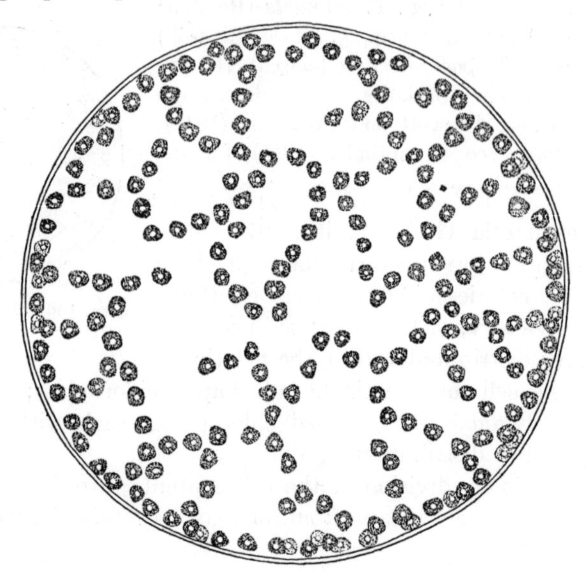

Fig. 168. *Eremosphaera viridis* DBy. (× 375.)

E. viridis DBy. (Fig. 168), the only species, is usually found in soft-water areas and in habitats favoring the growth of desmids. For a description of it, see Collins (1909).

10. **Excentrosphaera** G. T. Moore, 1901. *Excentrosphaera* differs from *Eremosphaera* in having ellipsoidal to subspherical cells in which the wall is irregularly thickened and often stratified in the thickened portions. It is like *Eremosphaera* in that the protoplast is uninucleate and with numerous angular chloroplasts just inside the cell wall. The chloroplasts may contain one or several pyrenoids.

Reproduction is by division of the cell contents into numerous small aplanospores that are liberated through an opening in the old parent-cell wall.[4]

There is but one species, *E. viridis* G. T. Moore (Fig. 169). For a description of it, see Collins, 1909.

[1] Reichardt, 1927. [2] Mainx, 1927.
[3] Moore, G. T., 1901; Playfair, 1916. [4] Moore, G. T., 1901.

11. **Echinosphaerella** G. M. Smith, 1920. The cells of this alga are always solitary and free-floating. The cell wall is spherical and with its surface densely clothed with heavy, long, hyaline, delicately tapering

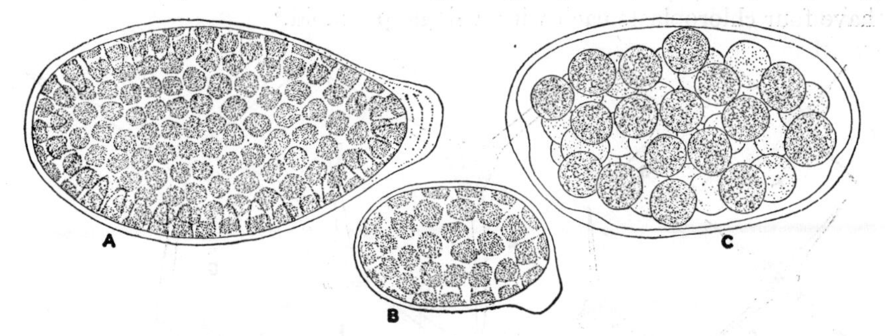

FIG. 169. *Excentrosphaera viridis* G. T. Moore. *A–B*, vegetative cells. *C*, a cell with aplanospores. (× 650.)

gelatinous spines with slightly concave sides. The chloroplast is single, parietal, cup-shaped, and with one pyrenoid.[1]

The method of reproduction is unknown.

E. limnetica G. M. Smith (Fig. 170) is a plankton alga and thus far in this country has been found only in Wisconsin.[1] For a description of it, see G. M. Smith, 1920.

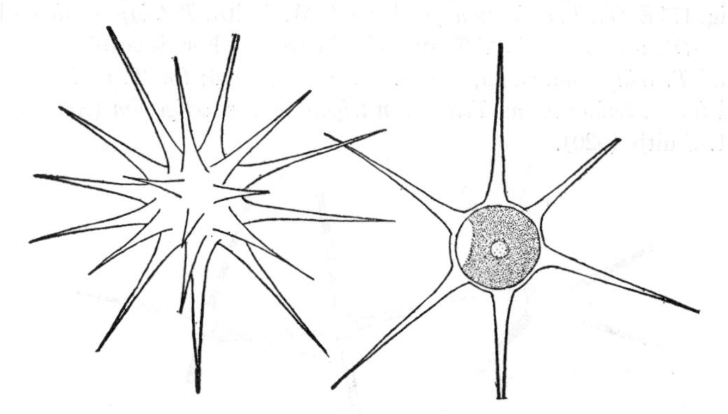

FIG. 170. *Echinosphaerella limnetica* G. M. Smith. (× 1000.)

12. **Treubaria** Bernard, 1908. The cells of *Treubaria* are usually pyramidal, with broadly rounded angles and more or less retuse sides; more rarely the cells are compressed and with the four angles lying in the same plane. Each angle bears a single stout hyaline spine, greater in length that the diameter of the cell. The spines may have subparallel sides and then taper abruptly to a sharp point, or they may gradually taper from a

[1] SMITH, G. M., 1920.

broad base to an acute apex. Young cells have a single cup-shaped chloroplast with one pyrenoid; old cells may have a massive chloroplast filling the entire cell and with a pyrenoid in each angle of the cell; or they may have four chloroplasts each with a single pyrenoid.[5]

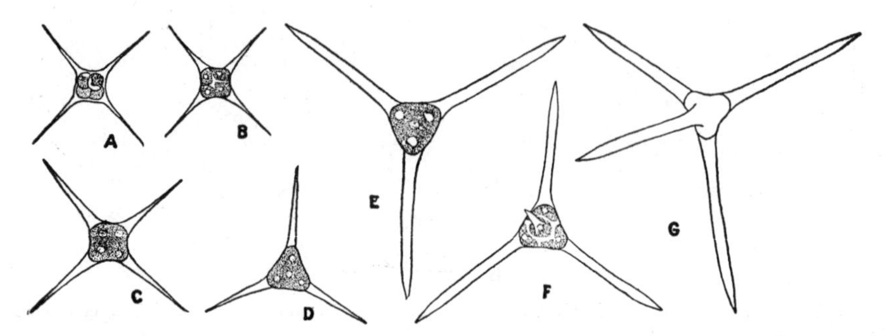

FIG. 171. *A–D, Treubaria triappendiculata* Bernard. *E–G, T. crassispina* G. M. Smith. (× 400.)

Reproduction is by division of the protoplast into four autospores that are liberated by rupture of the parent-cell wall.

Most of the records for the occurrence of *Treubaria* in this country are from the plankton. The four species found in the United States are *T. crassispina* G. M. Smith (Fig. 171*E–G*), *T. setigerum* (Archer) G. M. Smith, *T. triappendicula* Bernard (Fig. 171*A–D*), and *T. varia* Tiffany and Ahlstrom. For descriptions of *T. crassispina* and *T. triappendiculata*, see G. M. Smith, 1926; for *T. varia*, see Tiffany 1934; and for *T. setigerum* as *Tetraëdron trigonum* var. *setigerum* (Archer) Lemm., see G. M. Smith (920).

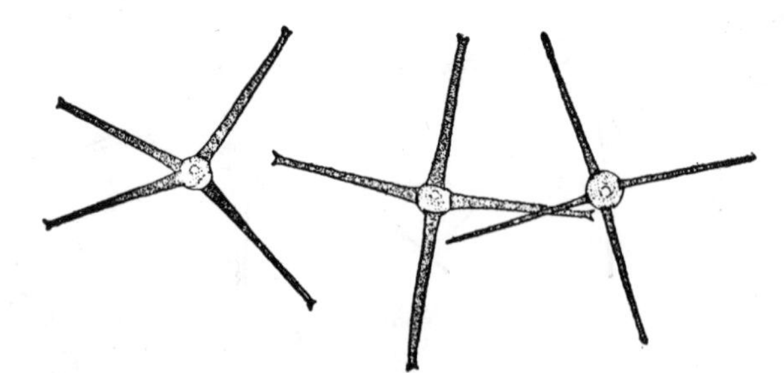

FIG. 172. *Pachycladon umbrinus* G. M. Smith. (× 400.)

13. **Pachycladon** G. M. Smith, 1924. The cells of this alga are spherical and enclosed by a thin wall that is without a gelatinous envelope. The wall bears four stout appendages that are usually quadrately arranged but sometimes are pyramidate in arrangement. The appendages are dark

[1] BERNARD, 1908; SMITH, G. M., 1926.

brown in color and taper gradually from a broad base to a blunt or bifurcate apex. The chloroplast, which fills most of the space within the wall, is cup-shaped and with one pyrenoid.[1]

This genus has considerable resemblance to *Treubaria* but differs in shape of the cell, in nature of appendages, and in the chloroplast. The single species, *P. umbrinus* G. M. Smith (Fig. 172) is known from New York,[1] North Carolina,[2] and Kentucky.[3] For a description of it, see G. M. Smith (1924).

14. **Oöcystis** Nageli, 1855. The cells of *Oöcystis* may be solitary, or 2, 4, 8, or 16 of them may be surrounded by a partially gelatinized and greatly expanded parent-cell wall. The cells are broadly to narrowly ellipsoidal, subcylindrical, or panduriform and with rounded to somewhat pointed poles. The cell wall is thin and without spines and other ornamentation except for a small nodular thickening at each pole. Many species have no nodules at their poles. A majority of the species have one to five parietal, laminate to irregularly discoid chloroplasts, with or without pyrenoids; a few species have numerous discoid chloroplasts, each with a pyrenoid.

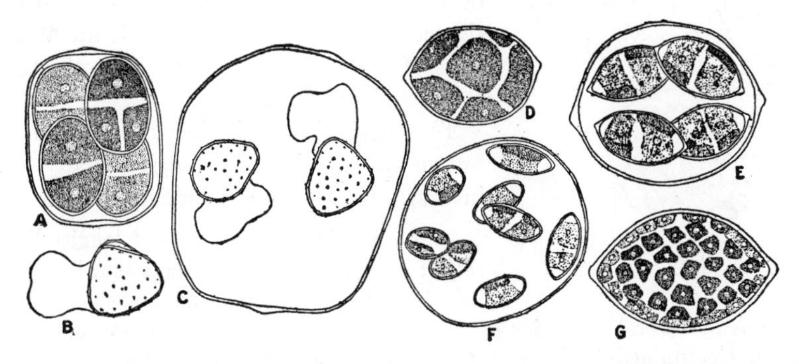

FIG. 173. *A–C, Oöcystis Borgei* Snow. *D, O. crassa* Wittr. *E, O. lacustris* Chod. *F, O. parva* W. and G. S. West. *G, O. Eremosphaeria* G. M. Smith. (× 666.)

Reproduction is by division of the cell contents into 2, 4, 8, or 16 autospores that remain for some time within the greatly expanded parent-cell wall. Sometimes the persistent parent-cell walls enclose two, three, or even four successive cell generations. There may also be an escape of the protoplast from the cell wall and its development into a *Tetraëdron*-like aplanospore with a punctate wall (Fig. 173*B–C*). When these aplanospores germinate,[4] there is a rupture of the wall and a division of the protoplast into two or four parts that develop into typical vegetative cells.

[1] SMITH, G. M., 1924. [2] WHITFORD, 1943. [3] MCINTEER, 1939.
[4] WILLE, 1908.

Oöcystis is of frequent occurrence both in the plankton of lakes and in the microflora of pools and ditches. Nineteen species have been found in the United States. For descriptions of many species of the genus, see G. M. Smith (1920), and Printz (1913).

15. **Scotiella** Fritsch, 1912. The cells of this alga are solitary in habit. They are ellipsoidal to broadly fusiform and with a wall that has several straight or spirally twisted longitudinal ridges. The ridges may lie equi-distant from one another or be in two lateral zones. In certain cases, the chloroplast is a median girdle containing a single pyrenoid, but in most cases the entire protoplast is so deeply colored with hematochrome that little can be determined concerning the structure of the cell's contents. Cells with an abundance of hematochrome contain both starch and droplets of oil.

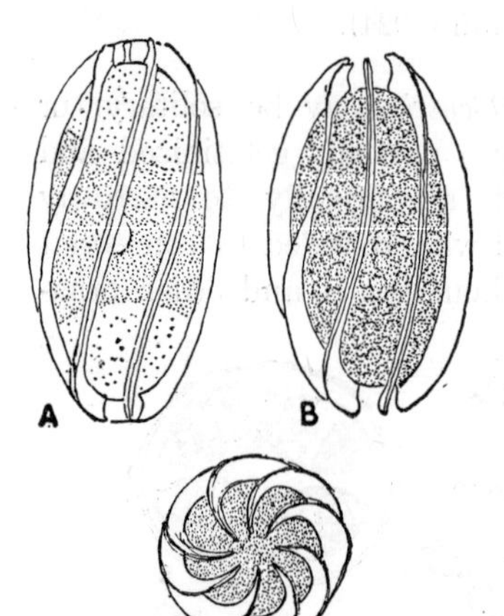

FIG. 174. *Scotiella nivalis* (Chod.) Fritsch. (× 900.)

The method of reproduction is unknown, but empty cell walls longitudinally split into two halves have been found.[1]

This genus includes a group of longitudinally ridged unicellular red-snow algae that have been found in the Antarctic and on snow-covered mountains in several countries of Europe. A specimen of red snow collected in California[2] contained a *Scotiella* tentatively identified as *S. nivalis* (Chodat) Fritsch (Fig. 174). For a description of *S. nivalis* as *Pteromonas nivalis* Chodat, see Pascher, 1927*A*.

16. **Gloeotaenium** Hansgirg, 1890. *Gloeotaenium* has spherical to ellipsoidal cells, two or four (rarely eight) of which lie within the closely fitting cell wall of the previous cell generation. The primary feature distinguishing this genus from others of the family is the presence of a dark-colored gelatinous mass separating the cells one from another within the old parent-cell wall. The dark color of the gelatinous mass is due to its impregnation with calcium carbonate.[3] In addition to the gelatinous material separating the cells one from another, there is often a small cap of gelatinous material external to each cell. Four-celled colonies are of more frequent occurrence than two- or

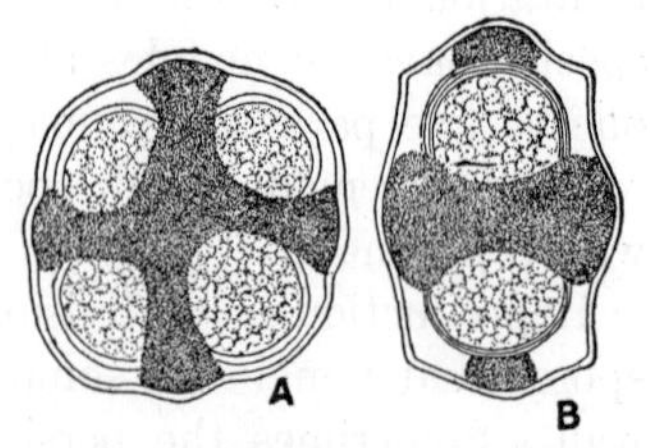

FIG. 175. *Gloeotaenium Loitlesbergerianum* Hansg. (× 400.)

[1] CHODAT, R., 1902. [2] SMITH, G. M., 1933. [3] HUBER-PESTALOZZI, 1919.

eight-celled ones. Four-celled colonies may have all cells in the same plane or may have them arranged in a pyramid. The parent-cell wall enclosing the four cells is fairly thick and often more or less folded. The walls surrounding the individual cells are thick and often distinctly stratified. Within the cell is a single massive chloroplast, with or without a pyrenoid.[1]

Reproduction is by division of the cell contents into two, four, or eight autospores. The newly formed autospores are liberated from the parent colony soon after they are formed, but the parent-cell wall surrounding them remains intact for a long time after liberation. In very rare cases,[2] there is an *Oöcystis*-like retention of more than one cell generation within the old parent-cell wall. Akinetes have also been recorded for this alga.

G. Loitlesbergerianum Hansg. (Fig. 175), the only species, is widely distributed in this country but is rarely found in abundance. For a description of it, see Collins (1909).

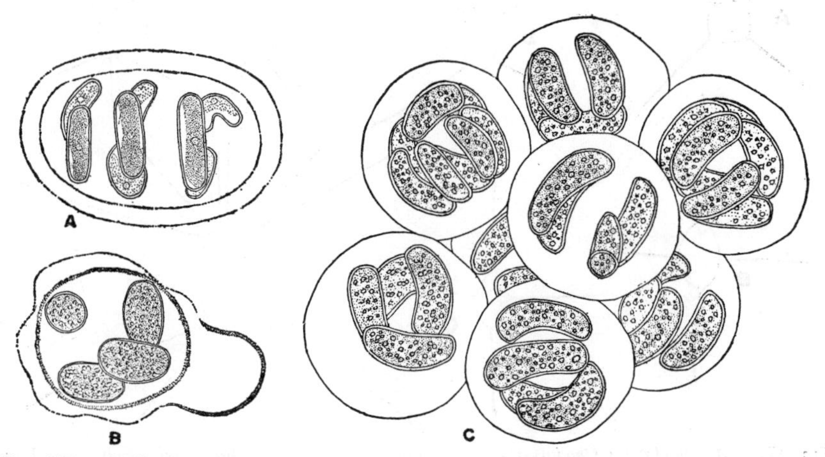

FIG. 176. *A, Nephrocytium Agardhianum* Näg. *B, N. ecdysiscepanum* W. and G. S. West. *C, N. limneticum* G. M. Smith. (*A, C*, × 1000; *B*, × 300.)

17. **Nephrocytium** Nägeli, 1849. This is another of the Oöcystaceae in which all autospores formed by a cell are retained for some time within the old parent-cell wall. In the case of *Nephrocytium*, the old parent-cell wall is partially gelatinized and somewhat expanded. Its cells are asymmetrical, and reniform or oblong-ellipsoidal. Colonies usually contain eight cells, and young colonies often have the cells spirally arranged (Fig. 176*A*), but this spiral arrangement is not evident in old colonies. Young cells have a parietal plate-like chloroplast running the length of a cell and containing one pyrenoid. The chloroplast of an old cell is diffuse and completely fills the cell.

[1] HUBER-PESTALOZZI, 1919; TRANSEAU, 1913*B*. [2] HUBER-PESTALOZZI, 1924.

Reproduction is by division of the cell contents into 2, 4, 8, or 16 autospores.

Nephrocytium is often found sparingly intermingled with other free-floating algae in small bodies of permanent water. The following species are found in this country: *N. Agardhianum* Näg. (Fig. 176*A*), *N. ecdysiscepanum* W. and G. S. West (Fig. 176*B*), *N. limneticum* G. M. Smith (Fig. 176*C*), *N. lunatum* W. West, and *N. obesum* W. West. For a description of *N. limneticum* as *Gloeocystopsis limneticum* G. M. Smith, see G. M. Smith (1920); for descriptions of the others as species of *Nephrocytium*, see Brunnthaler (1915).

18. **Chodatella** Lemmermann 1898 (*Lagerheimia* Chodat 1895, not *Lagerheimia* Saccardo, 1892). The cells of this alga are always solitary and

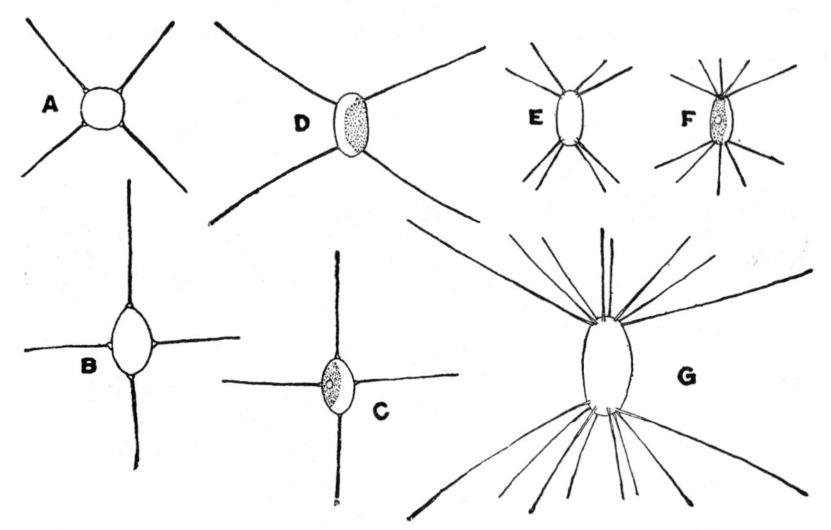

Fig. 177. *A, Chodatella Chodati* (Bernard) Ley. *B–C, C. wratislawiensis* (Schröder) Ley. *D, C. quadriseta* Lemm. *E–F, C. subsalsa* Lemm. *G, C. longiseta* Lemm. (× 800.)

free-floating. They are citriform, ellipsoidal, subcylindrical, or subspherical and with a delicate wall that is without an external gelatinous layer. The cells are ornamented with long setae that are subpolar, or both subpolar and equatorial, in insertion. The protoplasts contain one to four or more laminate to discoid chloroplasts, each usually with a single pyrenoid.

Reproduction is by formation of two, four, or eight autospores which are generally liberated at once, but which may remain within the old parent-cell wall for a time. There is no development of setae until after autospores are liberated from the parent-cell wall.

Chodatella is a strictly planktonic genus and one found more frequently in shallow than in deep lakes. The genus is better known under the name of *Lagerheimia*, but it has been shown[1] that *Lagerheimia* Chodat is a homonym of *Lagerheimia* Saccardo and that *Chodatella* is the first validly published generic name. The following species have been found in this country: *C. Chodati* (Bernard) Ley (Fig. 177*A*), *C. ciliata* (Lagerh.) Lemm., *C. cingula* (G. M. Smith) Ley, *C. citriformis* Snow, *C. longiseta* Lemm. (Fig. 177*G*), *C. quadriseta* Lemm. (Fig. 177*D*), *C. subglobosa* (Lemm.) Ley, *C. subsalsa* Lemm. (Fig. 177*E–F*), and *C. wratislawiensis* (Schröder) Ley (Fig. 177*B–C*). For descriptions of them as species of *Lagerheimia,* see G. M. Smith (1920, 1926).

19. **Franceia** Lemmermann, 1898. The cells of *Franceia* are free-floating and solitary, or in colonies of two, three, four, or more. The cells are ellipsoidal and with broadly rounded poles. The cell wall is thin and densely clothed with delicate setae that may have tubercules at their base or may lack them. There may be a single parietal laminate chloroplast within a cell, or there may be two or three parietal chloroplasts. Each chloroplast usually contains a single pyrenoid.[2]

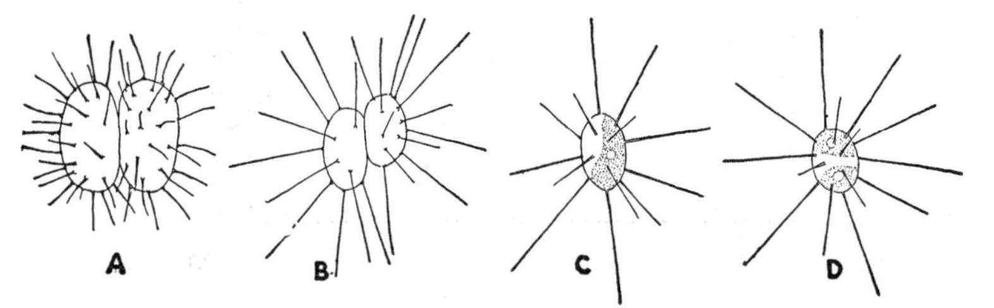

FIG. 178. *A–B. Franceia tuberculata* G. M. Smith. *C–D, F. Droescheri* (Lemm.) G. M Smith. (× 600.)

Reproduction is by a formation of two, four, or eight autospores that may separate from one another after liberation from the parent-cell wall or may remain closely apposed after liberation.

This genus is also known only from the plankton. It differs from *Chodatella* in having setae covering the entire wall, and in the tendency of the cells to remain united in colonies. The three species found in this country are *F. Droescheri* (Lemm.) G. M. Smith (Fig. 178*C–D*), *F. ovalis* Francé, and *F. tuberculata* G. M. Smith (Fig. 178*A–B*). For descriptions of them, see Tiffany, (1934).

20. **Bohlinia** Lemmermann, 1899; emend., G. M. Smith, 1933. The cells of this alga may be solitary, or they may be embedded in a common gelatinous matrix to form a flocculent mass containing thousands of cells. The individual cells are ellipsoidal and with setae scattered over the entire wall but less numerous in the equatorial region than at the poles. The cells contain one to four (rarely eight) laminate parietal chloroplasts, each with a single pyrenoid.

[1] LEY, 1948. [2] LEMMERMANN, 1898· SMITH, G. M., 1920, 1926.

Reproduction is by division of the cell contents into two or four autospores that are liberated by a slight swelling of the parent-cell wall followed by its lateral rupture.[1]

Thus far in this country only the type species, *B. echidna* (Bohlin) Lemm. (Fig. 179), has been found and only in Massachusetts[2] and California.[3] For a description of it, see Prescott and Croasdale (1942).

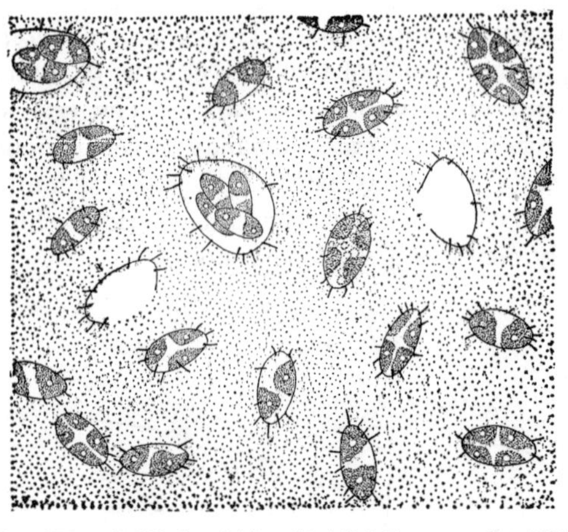

FIG. 179. *Bohlinia echidna* (Bohlin) Lemm. (× 600.)

21. **Ankistrodesmus** Corda, 1838; emend., Ralfs, 1848. The cells of *Ankistrodesmus* are acicular to spindle-shaped, gradually tapering to a point at the ends, and usually with a length several times the breadth. They may be solitary and straight, lunate, or sigmoid; they may be twisted about one another; or they may be in loose aggregates in which the cells are without definite arrangement. *Actinastrum*-like stages have also been found[4] when the alga is grown in pure culture. Cells of *Ankistrodesmus* have a thin wall without an external gelatinous layer. The protoplast may contain a single laminate chloroplast with or without a pyrenoid, or there may be more or less fragmentation of the chloroplast.

Reproduction is by division of the cell contents into two, four, or eight autospores.

Although *Ankistrodesmus* is found in a wide variety of habitats, it is rarely present in abundance. Aquariums and other receptacles containing water that have been standing in a laboratory for some time may contain a practically unialgal culture of this alga. The species found in this country are *A. Braunii* (Näg.)

[1] BOHLIN, 1897*A*; SMITH, G. M., 1933.

[2] PRESCOTT and CROASDALE, 1942. [3] SMITH, G. M., 1933. [4] VISCHER, 1919.

Collins, *A. falcatus* (Corda) Ralfs (Fig. 180), and *A. spiralis* (Turn.) Lemm. For descriptions of them, see Brunnthaler (1915).

22. **Dactylococcus** Nägeli, 1949. The cells of *Dactylococcus* are more or less spindle-shaped and with attenuated apices. They may be solitary or joined end to end in branching colonies that are without a gelatinous envelope. Young cells have a laminate parietal chloroplast at one side of the cell; old cells have the chloroplast completely filling the protoplast. A single pyrenoid is sometimes evident in the chloroplast.

Reproduction is by division of the protoplast into autospores, usually four, that may remain attached end to end after liberation from the parent-cell wall.

There is considerable dispute among phycologists as to whether or not this is a valid genus. Early in the study of algae by the method of pure culture there was a demonstration of the formation of extensive *Dactylococcus*-like stages by a strain of

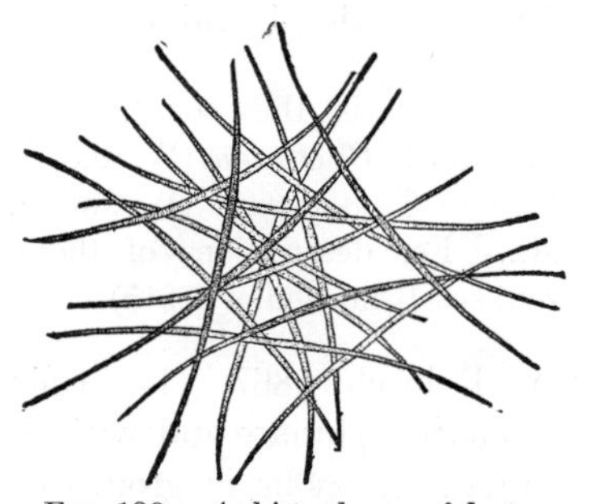

FIG. 180. *Ankistrodesmus falcatus* (Corda) Ralfs. (× 666.)

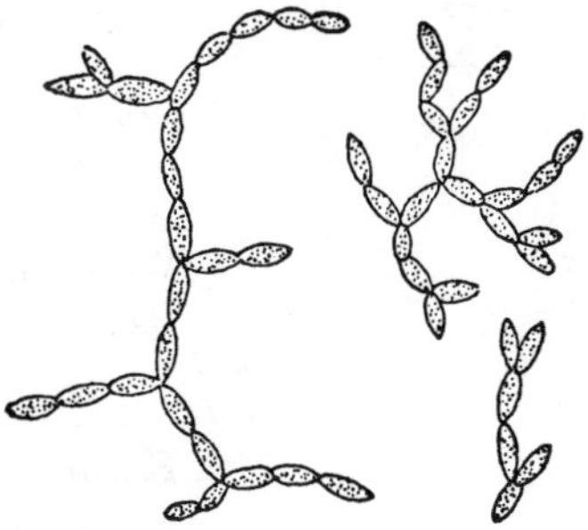

FIG. 181. *Dactylococcus infusionum* Näg. (× 210.)

Scenedesmus obliquus (Turp.) Kütz.[1] Other strains of *S. obliquus* and *S. dimorphus* (Turp.) Kütz. have been shown[2] to have growth stages suggestive of *Dactylococcus* though none has the indefinite branching characteristic of the true *Dactylococcus*. Because of this there has been a tendency[3] to hold that *Dactylococcus* is merely a growth form of certain species of *Scenedesmus*. The small number of records for the occurrence of *Dactylococcus* in this country is probably due to the fact that colonies break up very readily and that isolated cells, or two- or three-celled fragments, have been identified as *Ankistrodesmus* or *Ourococcus*. A strain of *D. infusionum* Näg. (Fig. 181) isolated in unialgal culture in material collected in Wisconsin[4] developed typical many-celled colonies and lacked *Scenedesmus*-like stages. For a description of *D. infusionum*, see Nägeli (1849).

23. **Closteriopsis** Lemmermann, 1898. This genus resembles *Ankistrodesmus* in its solitary acicular cells that are without a gelatinous sheath.

[1] GRINTZESCO, 1902. [2] CHODAT, R., 1913, 1926.
[3] PRINTZ, 1927; WEST, G. S., 1904, 1916. [4] SMITH, G. M., 1914.

It differs from *Ankistrodesmus* in its much longer cells and in the chloroplast having an axial row of a dozen or more pyrenoids. The genus also resembles certain species of *Closterium*, but their chloroplasts are always interrupted in the equatorial region of a cell.

C. longissima Lemm. (Fig. 182) has been found in a few localities in this country. For a description of it, see Tiffany (1934).

Fig. 182. *Closteriopsis longissima* var. *tropica* W. and G. S. West. (× 400.)

24. **Closteridium** Reinsch, 1888. The cells of this alga are solitary and free-floating. They are arcuate to lunate and with a short stout spine at either pole. The cell wall is relatively thick, and the single chloroplast completely fills the cell. It usually contains one pyrenoid.[1]

The method of reproduction is unknown.

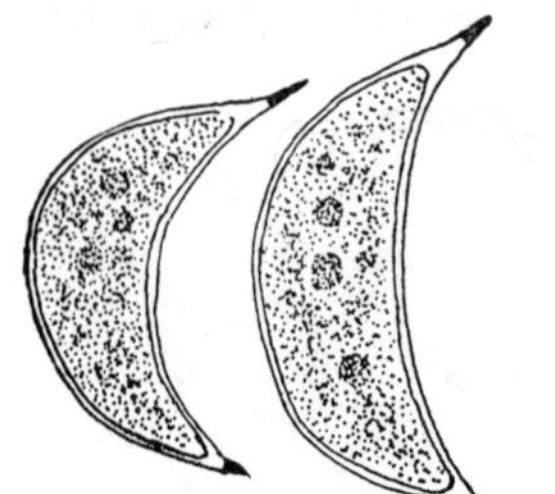

Fig. 183. *Closteridium lunula* Reinsch. (*After Reinsch, 1888.*)

The three species found in this country are *C. lunula* Reinsch (Fig. 183), *C. obesum* (W. and G. S. West) G. M. Smith, and *C. siamensis* (W. and G. S. West) G. M. Smith. For descriptions of them as species of *Tetraëdron*, see Brunnthaler (1915).

25. **Selenastrum** Reinsch, 1867. The cells of *Selenastrum* are arcuate to lunate and with their apices acutely pointed. They lie in groups of 4, 8, or 16, with their convex faces apposed, and the group is without a gelatinous envelope. Several groups may, in turn, be joined to one another to form a colony containing 100 or more cells. Young cells contain a single laminate parietal chloroplast, that lies on the convex side of a cell; old cells have the chloroplast entirely filling the cell. There is usually a single pyrenoid.[2]

Fig. 184. *A*, *Selenastrum gracile* Reinsch. *B*, *S. Bibraianum* Reinsch. (× 1000.)

[1] Reinsch, 1888. [2] Reinsch, 1867; Smith, G. M., 1920.

Reproduction is by the division of the cell contents into 4, 8, or 16 autospores that usually remain apposed to one another after liberation from the parent-cell wall.

Selenastrum is widely distributed as a plankton organism in this country. It is also found sparingly intermingled with other free-floating algae in pools and other quiet waters. The following species have been found in the United States: *S. Bibraianum* Reinsch (Fig. 184*B*), S. *gracile* Reinsch (Fig. 184*A*), *S. minutum* (Näg.) Collins, and *S. Westii* G. M. Smith. For a description of *S. minutum*, see Collins (1909); for descriptions of the others, see G. M. Smith (1920).

26. **Kirchneriella** Schmidle, 1893. The cells of this alga may be lunate to sickle-shaped and with their apices almost touching; or they may be

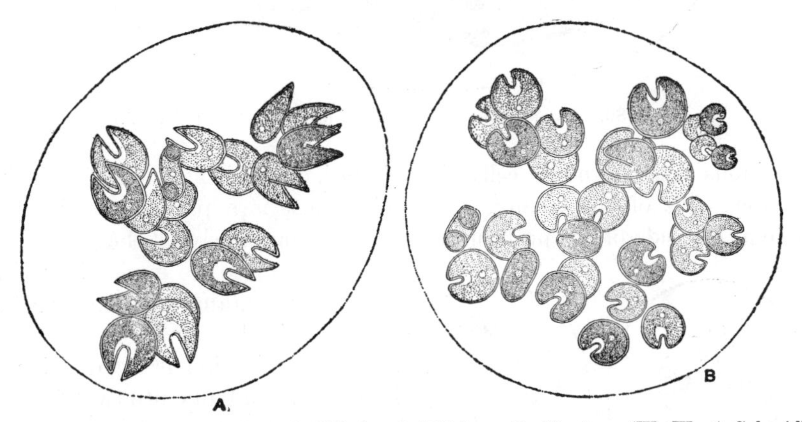

Fig. 185. *A*, *Kirchneriella lunaris* (Kirchner) Möbius. *B*, *K. obesa* (W. West) Schmidle. (× 666.)

arcuate to irregularly curved cylinders with broadly rounded ends. The cells lie in groups of four or eight within a wide homogeneous gelatinous matrix. There is a certain tendency for the cells in each group of four or eight to have their convex faces apposed, but this tendency is not so strongly marked as in *Selenastrum*. The chloroplast is single, parietal, and on the convex side of a cell. There is usually a single pyrenoid within a chloroplast.

Reproduction is by division of the cell contents into four or eight autospores that separate somewhat from one another as the parent-cell wall gelatinizes. Colony reproduction is by dissociation of the colonial matrix.

This is another of the genera known chiefly from the plankton. The following species have been found in this country: *K. contorta* (Schmidle) Bohlin, *K. elongata* G. M. Smith, *K. lunaris* (Kirchner) Möbius (Fig. 185*A*), *K. obesa* (W. West) Schmidle (Fig. 185*B*), and *K. subsolitaria* G. S. West. For a description of *K.*

subsolitaria, see G. S. West (1908); for descriptions of the others, see G. M. Smith (1920).

27. **Quadrigula** Printz, 1915. The straight or slightly curved cells of *Quadrigula* are broadly spindle-shaped to subcylindrical and with more or less pointed ends. The cells, with a length 5 to 20 times the breadth, lie parallel to one another in groups of four or eight within a fairly copious

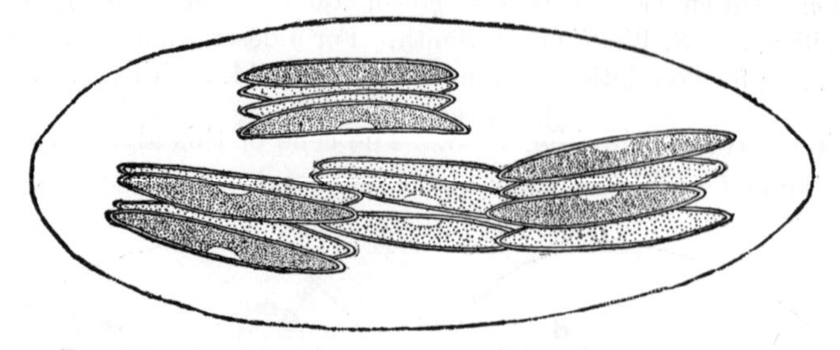

FIG. 186. *Quadrigula closterioides* (Bohlin) Printz. (× 1000.)

gelatinous matrix, and all cells of a colony have their long axes paralleling the long axis of the colony. The chloroplast is a parietal longitudinal plate at one side of the protoplast, or it may completely fill the protoplast. According to the species, it is without or contains either one or two pyrenoids.[1]

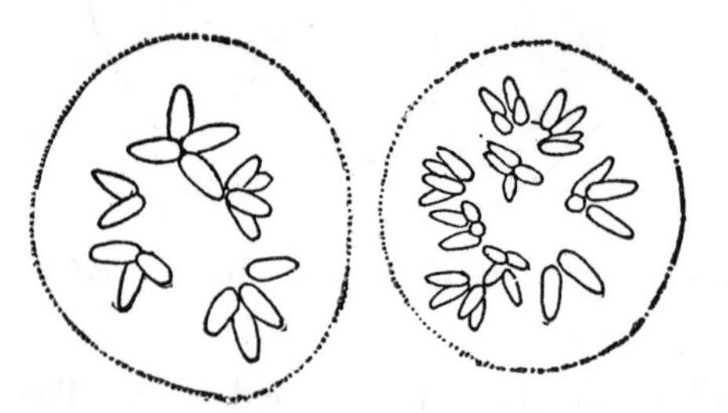

FIG. 187. *Gloeoactinium limneticum* G. M. Smith. (× 8000.)

Reproduction is by division of the cell contents into two, four, or eight autospores. The autospores separate somewhat from one another as the parent-cell wall gelatinizes and merges with the colonial matrix.

This genus stands in much the same relationship to *Ankistrodesmus* that *Kirchneriella* does to *Selenastrum*. It is a common plankton alga in this country. The species found in the United States are *Q. Chodatii* (Tanner-Fullman) G. M. Smith. *Q. closterioides* (Bohlin) Printz (Fig. 186) and *Q. lacustris* (Chod.) G. M. Smith. For descriptions of them, see G. M. Smith (1920).

28. **Gloeoactinium** G. M. Smith, 1926. The cells of this alga are narrowly ovate-cuneate and apposed at their bases in radiating groups of two or four; several of these groups lie toward the periphery of a wide homogeneous gelatinous matrix. The basal poles of cells are very broadly rounded; the apical poles are narrower and more acute. All cells in a

[1] PRINTZ, 1915, 1927; SMITH, G. M., 1920.

colony are with their long axes radiating from a common center. The chloroplast is laminate and parietal, or completely filling the cell; it lacks a pyrenoid.[1]

Reproduction is by the formation of two or four autospores that remain embedded within the colonial matrix.

G. limneticum G. M. Smith (Fig. 187), the type species, is known only as a plankton alga from Iowa. For a description of it, see G. M. Smith (1926).

29. **Tetraëdron** Kützing, 1845. The cells of *Tetraëdron* (Fig. 188) are always solitary and free-floating. They may be flattened or isodiametric,

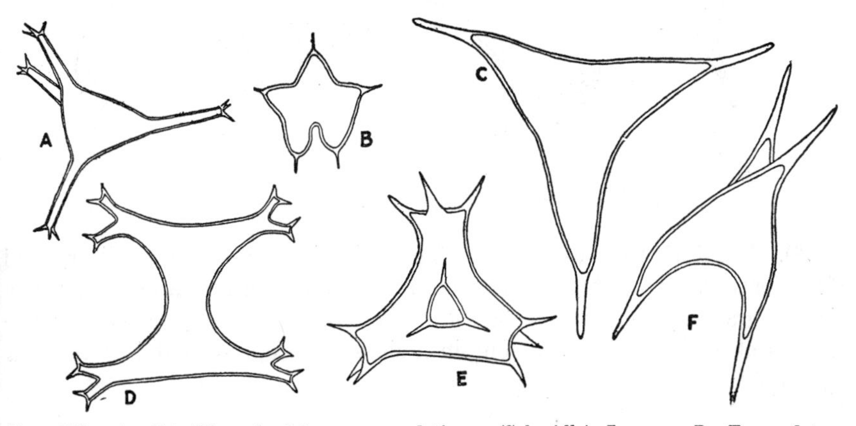

Fig. 188. *A, Tetraëdron hastatum* var. *palatinum* (Schmidle) Lemm. *B, T. caudatum* (Corda) Hansg. *C, T. trigonum* var. *gracile* (Reinsch) DeToni. *D, T. constrictum* G. M. Smith. *E, T. lobulatum* (Näg.) Hansg. *F, T. victoriae* var. *major* G. M. Smith. (× 1000.)

triangular, quadrangular, or polygonal. The angles of the cells may be simple or produced into simple or furcate processes. Species with the angles produced usually have a rather sharp transition from body of the cell to the process. The cell wall is relatively thin, and smooth or verrucose. The cells may contain one to many parietal discoid to angular chloroplasts, or there may be a single chloroplast entirely filling the cell. Chloroplasts are with or without pyrenoids. Young cells are uninucleate; mature ones may contain two, four, or eight nuclei.[2]

Reproduction is by successive bipartition of the protoplast into 2, 4, 8, 16, or 32 autospores that are immediately liberated by rupture of the parent-cell wall.[2] The reported reproduction by means of zoospores[3] is open to question.

Approximately 40 species have been found in this country. Some of them are open to suspicion since they have not been found producing autospores and so may

[1] Smith, G. M., 1926. [2] Smith, G. M., 1918*A*. [3] Probst, 1926.

be stages in the life history of other algae. Species with small cells are usually found intermingled with other unattached algae in pools and ditches. Many species with large cells are known only from the plankton. For descriptions of most species of the genus, see Brunnthaler (1915), Reinsch (1888), and G. M. Smith (1920, 1926).

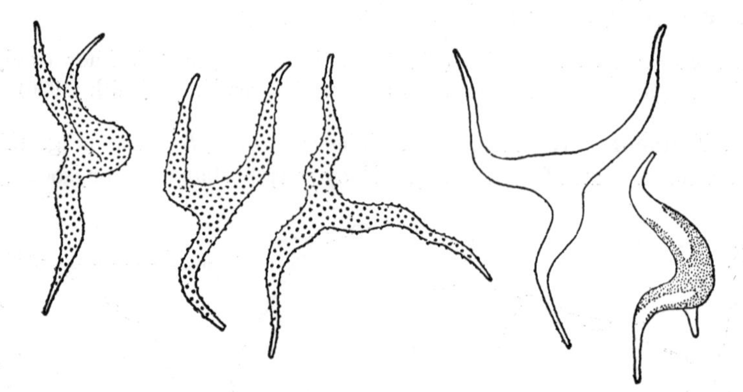

Fig. 189. *Cerasterias irregulare* G. M. Smith. (× 800.)

30. Cerasterias Reinsch, 1867. The cells of *Cerasterias* are solitary and free-floating. The chief distinction between this genus and *Tetraëdron* is the gradual transition from the central body to the processes at the angles of a cell. The cells contain a single chloroplast that is without a pyrenoid.

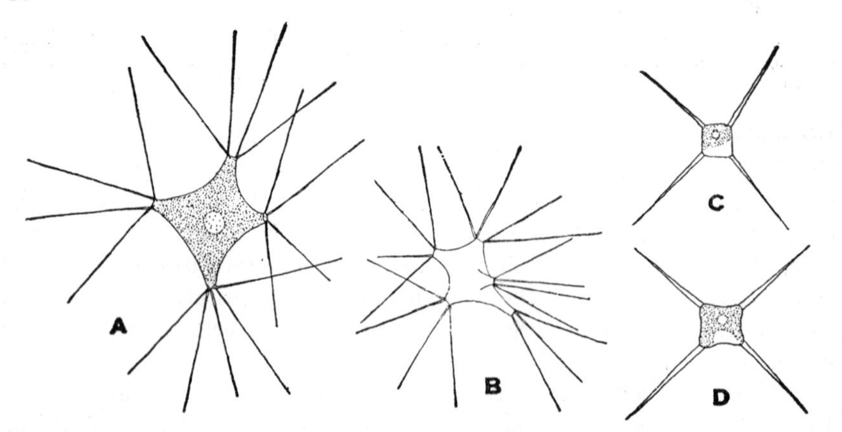

Fig. 190. *A–B, Polyedriopsis spinulosa* Schmidle. *C–D, P. quadrispina* G. M. Smith. (× 800.)

Certain "species" referred to this genus have been shown[1] to be spores of fungi instead of unicellular algae. Others are unquestionably algae. One of these, *C. irregulare* G. M. Smith (Fig. 189) has been found in the plankton in Ohio[2] and Iowa.[3] For a description of it, see G. M. Smith (1926).

[1] Karling, 1935; Lowe, 1927. [2] Lackey *et al.*, 1943. [3] Smith, G. M., 1926.

31. **Polyedriopsis** Schmidle, 1899. *Polyedriopsis* has solitary free-floating cells that are usually tetragonal and compressed but sometimes are five-angled and pyramidate. Each of the angles may bear a tuft of four to six delicate long setae, or there may be a single spine-like seta at each angle. The single chloroplast is cup-shaped, fairly massive, and with one pyrenoid.

Reproduction is by the formation of two, four, or eight autospores that do not develop setae until after they are liberated from the parent-cell wall.[1]

Two species, *P. quadrispina* G. M. Smith (Fig. 190*C–D*) and *P. spinulosa* Schmidle (Fig. 190*A–B*), have been found in this country. For descriptions of them, see G. M. Smith (1920, 1926).

Family 10. Scenedesmaceae

The Scenedesmaceae are Chlorococcales reproducing only by means of autospores and in which the autospores produced by any cell always remain attached to one another in autocolonies in which the cells have a definite orientation. The number of cells in a coenobium is always a multiple of two, and generally two, four, or eight. Sometimes the coenobia of several successive generations remain attached to one another in compound coenobia (*Crucigenia*), but more often the daughter coenobia immediately separate from one another (*Scenedesmus, Tetrastrum*).

Cells of Scenedesmaceae may be spherical, ellipsoidal, acicular, triangular, or trapezoidal and with the walls smooth or ornamented with spines or ridges. All cells of a coenobium may lie in the same plane or the cells may be pyramidately or radiately arranged. Some genera regularly form only four-celled coenobia (*Crucigenia, Tetrastrum*); in others the number of cells in a coenobium ranges from 2 to 32. In these latter genera, the number of cells in a coenobium depends in part upon physiological vigor of the parent cell, and there is a general tendency for cells growing in a favorable environment to produce daughter coenobia with a larger number of cells than do cells growing under unfavorable conditions.

The genera found in this country differ as follows:

1. Cells globose 2
1. Cells not globose 3
   2. Coenobia cubical .... 9. **Pectodictyon**
   2. Coenobia not cubical .... 8. **Coronastrum**
3. Long axes of cells parallel 4
3. Long axes of cells not parallel 6
   4. Long axes in one plane 5
   4. Long axes not in one plane .... 2. **Tetradesmus**
5. Cells quadrately arranged .... 3. **Crucigenia**

[1] Smith, G. M., 1918.

5. Cells not quadrately arranged.................................. 1. **Scenedesmus**
   6. Cells lunate.. 7
   6. Cells not lunate.. 8
7. Poles of cells touching one another.............................. 5. **Tetrallantos**
7. Poles of cells not touching one another.......................... 6. **Tomaculum**
   8. Cells quadrately arranged..................................... 9
   8. Cells radiately arranged...................................... 7. **Actinastrum**
9. Cells with spines.. 4. **Tetrastrum**
9. Cells without spines... 3. **Crucigenia**

1. **Scenedesmus** Meyen, 1829. The coenobium of *Scenedesmus* (Fig. 191) is a flat, rarely curved, plate of ellipsoidal to spindle-shaped cells arranged in a single, alternating, or double series with their long axes parallel to one another. The number of cells in a coenobium is always a multiple of two and usually four or eight, though sometimes 16 or 32.

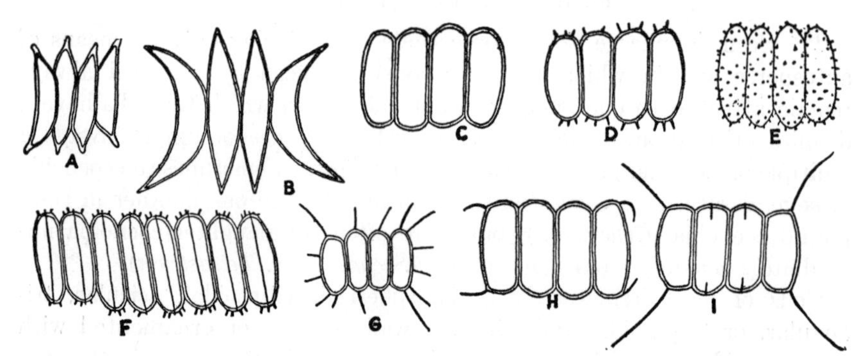

Fig. 191. *A, Scenedesmus obliquus* (Turp.) Kütz. *B, S. dimorphus* (Turp.) Kütz. *C, S. bijuga* (Turp.) Lagerh. *D, S. denticulatus* var. *linearis* Hansg. *E, S. hystrix* Lagerh. *F, S. brasiliensis* Bohlin. *G, S. abundans* (Kirchner) Chod. *H, S. quadricauda* var. *quadrispina.* (Chod.) G. M. Smith. *I, S. armatus* (Chod.) G. M. Smith. (× 1000.)

According to the species, the cell wall is smooth, corrugated, granulate, or spicate; and with or without marginal or lateral teeth or spines. Young cells have a single longitudinal laminate chloroplast containing one pyrenoid; chloroplasts of old cells usually fill the entire cell cavity. The cells are uninucleate.

Each cell in a coenobium is capable of giving rise to a daughter coenobium, but there is rarely a simultaneous formation of daughter coenobia by all cells. Daughter coenobia are formed by transverse and longitudinal divisions of a protoplast to form 2, 4, 8, 16, or 32 autospores.[1] These remain united after their liberation by a longitudinal splitting of the parent-cell wall. The number of cells in a daughter coenobium is not necessarily the same as that in the parent coenobium.

[1] Smith, G. M., 1914*A*.

Most collections from aquatic habitats contain one or more species of this genus. About 30 species have been found in this country. For descriptions of most of the species of *Scenedesmus*, see R. Chodat (1926), and G. M. Smith (1916*D*).

2. **Tetradesmus** G. M. Smith, 1913. The free-floating coenobia of this alga are always four-celled and with the cells in two planes. The cells are narrowly to broadly fusiform and with their long axes parallel. Some species have the apices of the cells divergent. When viewed from the top, the coenobia have the cells quadrately arranged. There is a single laminate chloroplast that often fills the entire cell. It contains one pyrenoid.[1]

FIG. 192. *Tetradesmus wisconsinensis* G. M. Smith. (× 1000.)

Reproduction is by transverse and longitudinal division of the cell contents to form four autospores that remain united in an autocolony after liberation by a longitudinal splitting of the parent-cell wall.

The two species found in this country are *T. Smithii* Prescott and *T. wisconsinensis* G. M. Smith (Fig. 192). For a description of *T. Smithii*, see Prescott (1944); for *T. wisconsinensis*, see G. M. Smith (1920).

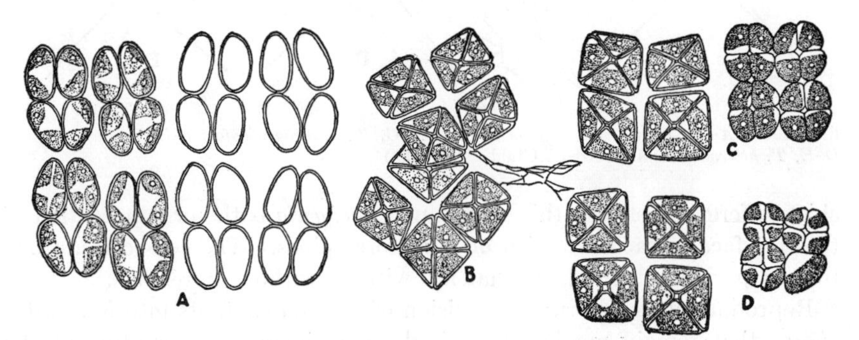

FIG. 193. *A, Crucigenia rectangularis* (Näg.) Gay. *B, C. tetrapedia* (Kirchner) W. and G. S. West. *C–D, C. quadrata* Morren. (× 666.)

3. **Crucigenia** Morren, 1830. The cells of this alga are united in free-floating four-celled coenobia that are solitary or joined to one another to form plate-like multiple coenobia of 16 or more cells. Multiple coenobia are due to the presence of a gelatinous envelope or to a persistence of portions of parent-cell walls that unite the coenobia one to another. The four cells in a coenobium are quadrately arranged, and there is usually a large or a small quadrangular space at the center of a coenobium. The cells are elliptical, triangular, trapezoidal, or semicircular in surface view, and

[1] SMITH, G. M., 1913; WEST, G. S., 1915.

the cell wall is without ornamentation, except for the polar thickenings found in certain species. The cells contain one to four parietal, discoid to laminate, chloroplasts, each of which usually contains a single pyrenoid.

Reproduction is by cruciate division of the protoplast into four autospores that remain quadrately apposed. There may be a liberation of the daughter coenobium after rupture of the parent-cell wall, or the daughter coenobium may remain partially surrounded by the old wall. There is a single record of the formation of akinetes in *Crucigenia*.[1]

The majority of the 12 species found in the United States are known only from the plankton. For descriptions of these species, see G. M. Smith (1920, 1926).

4. **Tetrastrum** R. Chodat, 1895. The coenobia of *Tetrastrum* have the same quadrate arrangement of cells that is found in *Crucigenia*, but they are very rarely joined to one another to form compound coenobia. The

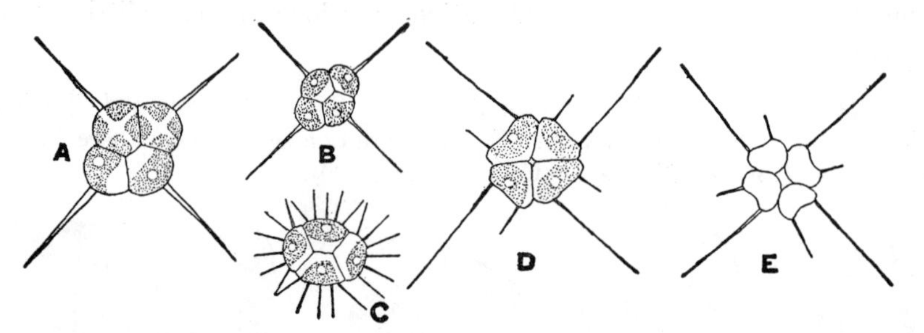

FIG. 194. *A–B, Tetrastrum elegans* Playfair. *C, T. staurogeniaeforme* (Schröder) Lemm. *D–E, T. heterocanthum* (Nordst.) Chod. (× 800.)

chief difference between this genus and *Crucigenia* is the ornamentation of the free face of each cell with one or more spines. The cells contain one to four parietal chloroplasts that are with or without pyrenoids.

Reproduction is by cruciate division of the cell contents into four autospores that remain quadrately united when liberated by rupture of the parent-cell wall.

Tetrastrum is another genus that is rarely found outside the plankton. The following species have been found in this country: *T. elegans* Playf. (Fig. 194*A–B*), *T. glabrum* (Roll) Ahlstrom and Tiffany, *T. heterocanthum* (Nordst.) Chodat (Fig. 194*D–E*), and *T. staurogeniaeforme* (Schröder) Lemm. (Fig. 194*C*). For descriptions of them, see Ahlstrom and Tiffany (1934).

5. **Tetrallantos** Teiling, 1916. The sausage-shaped cells of *Tetrallantos* are strongly curved and usually united in four-celled coenobia, in which the cells have a definite orientation with respect to one another. Two of

[1] SCHMIDLE, 1900.

the cells lie in the same plane; these face each other and touch only at the poles. The two remaining cells are vertical to the apposed pair and joined to them where they abut on each other. The colonies are often surrounded by a broad gelatinous matrix, and four or more of them may lie within a common matrix. Chloroplasts are single, parietal, with one pyrenoid.

Reproduction is by division of the cell contents into two, four, or eight autospores that are liberated by a rupture of the parent-cell wall. The autospores remain united in a coenobium that may separate from the parent coenobium, or may remain attached to it by remnants of the parent-cell wall.[1]

T. Lagerheimii Teiling (Fig. 195) has been found in the plankton in several states east of the Mississippi River. For a description of it, see Teiling (1916).

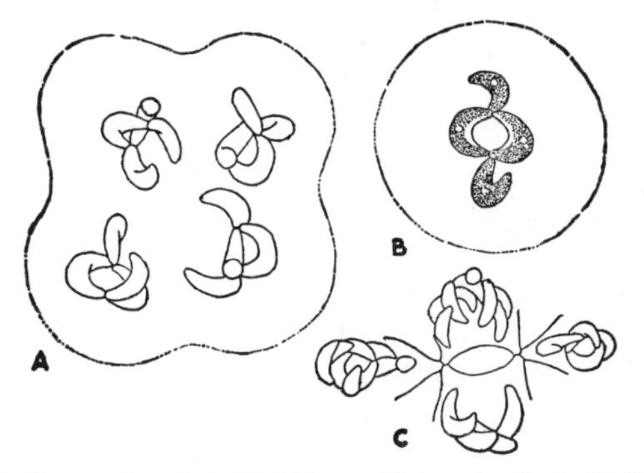

FIG. 195. *Tetrallantos Lagerheimii* Teiling. (*Drawn by J. C. McKee.*) (× 650.)

6. **Tomaculum** Whitford, 1943. This alga has curved sausage-shaped cells that may or may not have a cylindrical projection on the concave face. The poles and the lateral projection, if present, of cells are joined to those of other cells by delicate strands to form a saccase reticulate colony that is embedded in a copious gelatinous matrix. A cell contains one or two parietal laminate chloroplasts, each with a single pyrenoid.[2]

The method of reproduction is unknown.

The single species, *T. catenatum* Whitford (Fig. 196), is known from two localities in North Carolina.[2] For a description of it, see Whitford (1943).

7. **Actinastrum** Lagerheim, 1882. The coenobia of *Actinastrum* are composed of 4, 8, or 16 (generally 8) elongate cells that radiate in all directions from a common center. Sometimes the coenobia are united in multiple coenobia of irregular shape. The individual cells have a length four to eight times the breadth and are cylindrical, fusiform, or

[1] TEILING, 1916. [2] WHITFORD, 1943.

tetaniform. The chloroplast is a longitudinal strip, partially encircling the protoplast and containing one pyrenoid.

Reproduction is by transverse and longitudinal division of the protoplast to form 4, 8, or 16 autospores. These lie in two longitudinally

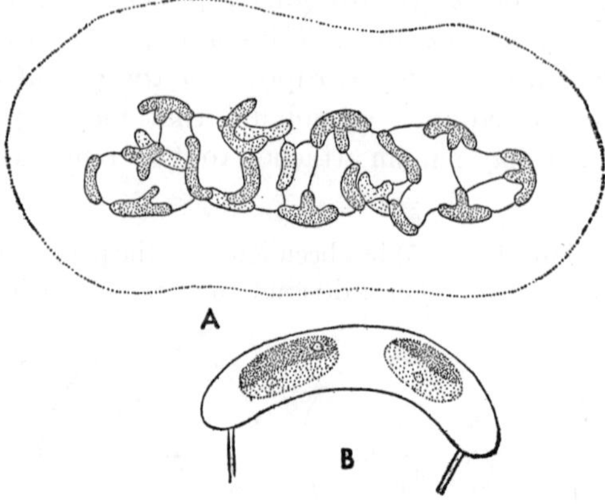

FIG. 196. *Tomaculum catenatum* Whitford. *A*, colony. *B*, cell showing chloroplasts and connecting strands. (*After Whitford*, 1943.) (*A*, × 150; *B*, × 800.)

joined fasciculate bundles at the time they are liberated by the breaking of the parent-cell wall. After a daughter coenobium is liberated, its cells become outwardly divergent from the zone where the two fascicles touch each other.[1]

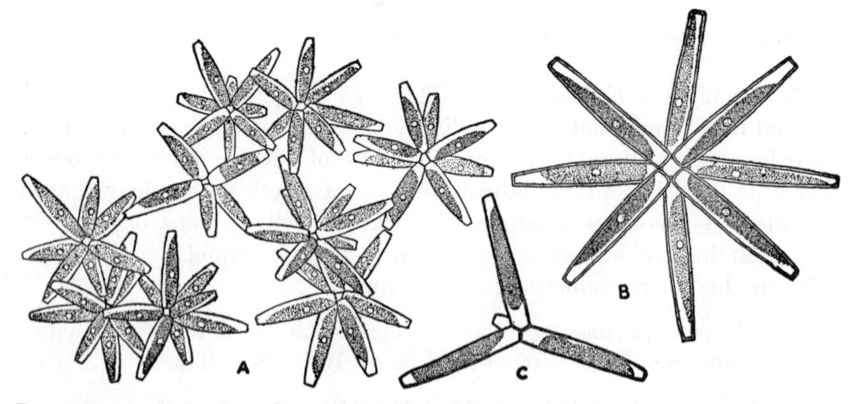

FIG. 197. *A*, *Actinastrum Hantzschii* Lagerh. *B–C*, *A. gracillimum* G. M. Smith. (*A*, × 500; *B–C*, × 1000.)

Actinastrum is of widespread occurrence in the plankton of lakes and ponds in the United States. The two species found in this country are *A. gracillimum* G. M. Smith (Fig. 197*B–C*) and *A. Hantzschii* Lagerh. (Fig. 197A). For descriptions of them, see G. M. Smith (1920).

[1] LAGERHEIM, 1882; SMITH, G. M., 1920, 1926.

8. **Coronastrum** Thompson, 1938. The coenobia of this alga are four-celled, free-floating, and solitary, or joined to one another into compound coenobia of 16 cells. The four cells of a coenobium are quadrately arranged in a flat plate and are separated from one another by strands, the persisting remains of the parent-cell wall. Each cell bears in its free face a scale-like appendage derived from the parent-cell wall. A coenobium is without surrounding gelatinous material. The cells are subglobose, and each contains a parietal chloroplast with a single pyrenoid.[1]

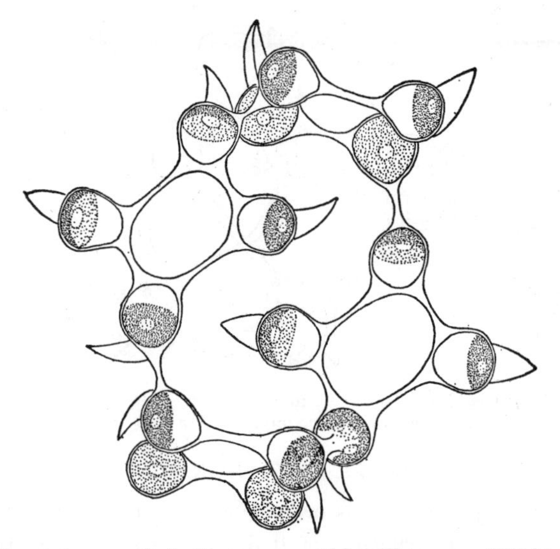

FIG. 198. *Coronastrum aestivale* Thompson. (*After Thompson,* 1938*A*.) (× 1500.)

Reproduction is by division of the cell contents into four autospores, and the cells formed by enlargement of the autospores remain permanently attached to the angles of the four-sided mesh into which the parent-cell wall splits.

There is but one species, *C. aestivale* Thompson (Fig. 198), and it is known from only a single locality in Kansas. For a description of it, see Thompson (1938*A*).

9. **Pectodictyon** Taft, 1945. The coenobia of this alga are eight-celled, free-floating, and solitary or joined to one another in compound coenobia. A coenobium has a cubical framework of gelatinous strands and has a single cell at each of the eight corners of the framework. The cells are spherical, and each contains a single massive cup-shaped chloroplast with a minute pyrenoid at its base.[2]

Reproduction is by successive bipartition of the protoplast to form eight autospores which lie in two tiers of four and are liberated as a unit. At

[1] THOMPSON, 1938*A*. [2] TAFT, 1945*A*.

first, the young coenobium is a solid gelatinous cube with a cell just within each corner (Fig. 199*B*); but later on, as the coenobium increases in size, gelatinous strands appear between adjoining corners of the gelatinous material.[1]

There is but one species, *P. cubicum* Taft (Fig. 199), and it is known from only a single locality in Ohio. For a description of it, see Taft (1945*A*).

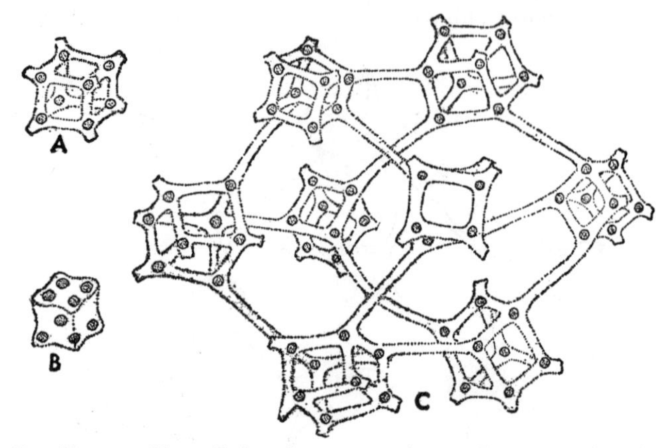

FIG. 199. *Pectodictyon cubicum* Taft. *A*, mature colony. *B*, young colony before formation of gelatinous strands. *C*, compound colony. (*After Taft*, 1945*A*.)

ORDER 9. SIPHONALES

The thallus of members of this order is a single multinucleate cell which often grows to form a structure of definite macroscopic shape. Only a few members of the order form zoospores or aplanospores. Most members of the order reproduce sexually, and gametic union is usually anisogamous but may be oögamous.

The Siphonales are among the most sharply defined of all Chlorophyceae. Although they are extremely varied as to size and external form, they are all fundamentally alike in that the entire plant body is a single multinucleate cell with numerous discoid chloroplasts.

All the fresh-water Siphonales have simple thalli consisting of elongate sparingly branched tubular cells. It does not necessarily follow that these are the most primitive of the order; in fact, *Dichotomosiphon* is the most advanced of the Siphonales as far as its method of sexual reproduction is concerned. Some marine Siphonales have sparingly branched thalli, but in the great majority of them the single cell comprising the plant body is elaborately branched. Certain marine Siphonales have the branches of the cell differentiated into erect tufts arising from a prostrate portion

[1] TAFT, 1945 *A*.

(*Derbesia*) or have the erect branches feather-like and with the main axis bearing numerous primary or secondary branchlets (*Bryopsis*). *Caulerpa* is especially remarkable since the single cell is a plant of macroscopic size that simulates in appearance the differentiation into roots, stem, and leaves found in vascular plants. Other marine Siphonales, especially the Codiaceae, have an intricate interweaving of the coenocyte's branches to form a plant body of definite external form and a decimeter or more in height.

In spite of their external complexity, the Siphonales have a relatively simple internal structure. There is a single central vacuole that may run without interruption the whole length of the coenocyte. The layer of cytoplasm between central vacuole and cell wall is relatively thin, with nuclei toward its inner face and chloroplasts toward its outer face. The chloroplasts are discoid or lenticular and each with a pyrenoid, or pyrenoids may be lacking. Genera with a simple thallus, as *Derbesia* and *Dichotomosiphon*, have a more or less uniform distribution of chloroplast. Genera with a complicated thallus often have the chloroplasts restricted to certain portions of the plant body; *Codium*, for example, having the chloroplasts localized in special palisade-like branches at the surface of the plant body. Starch is the usual reserve food accumulated by Siphonales.

The cell wall is fairly thin and with (*Codium*) or without (*Dichotomosiphon*) a gelatinous envelope. Many of the complex Siphonales, as *Codium*, have broad annular ingrowths of the wall at intervals along the coenocyte. Increase in thickness of ingrowths often completely blocks the cell cavity at intervals and results in a septation of the coenocyte. Old thalli of such Siphonales are, therefore, essentially multicellular. Portions of thalli developing into sporangia or gametangia regularly have their protoplasts blocked off from the remainder of the cell by annular ingrowths of the cell wall (*Bryopsis*, *Codium*, *Derbesia*).

Simple members of the order rarely multiply vegetatively by fragmentation; complex members frequently multiply by fragmentation. Their vegetative multiplication may be by an abscission of branchlets or by a development of proliferous shoots that become detached from the plant body.

Derbesia is the only genus reproducing by means of zoospores. These are formed within inflated sporangia resulting from enlargement of a lateral branchlet, and each sporangium contains a number of *Oedogonium*-like zoospores with a whorl of flagella at the anterior end. The striking similarity in structure of zoospores of *Derbesia* and the Odeogoniales is without phylogenetic significance. Several Siphonales form aplanospores in large numbers throughout the length of the thallus. *Dichotomosiphon* is the only member of the order that regularly forms akinetes.

Sexual reproduction of most genera in the order is anisogamous, a uniting pair may differ but slightly in size (*Bryopsis*) or differ markedly (*Codium*). These gametes may or may not be formed within gametangia of distinctive shape. *Dichotomosiphon*, the only oögamous genus, has antheridia containing many antherozoids and oögonia containing a single egg. Here fertilization takes place within an oögonium, and the zygote enters upon a period of rest before germinating directly into a new thallus.

In several of the anisogamous genera, meiosis occurs just prior to formation of gametes.[1] The zygote germinates soon after it is formed and develops directly into a coenocyte whose nuclei are diploid. The life cycle of these algae consists of an alternation of a haploid uninucleate phase with a phase in which there are many successive generations of diploid nuclei.

The Siphonales are divided into seven families, only two of which have fresh-water genera. Representatives of both these families are found in the fresh-water flora of this country.

Family 1. Phyllosiphonaceae

The Phyllosiphonaceae have sparingly to profusely and irregularly or dichotomously branched tubular thalli in which reproduction is exclusively by means of aplanospores formed throughout the length of the thallus. All members of the family are endophytic or endozoic. *Phyllosiphon* is the only genus in the fresh-water flora of this country.

1. **Phyllosiphon** Kühn, 1878. *Phyllosiphon* grows as an intercellular parasite in stems and leaves of various Araceae. The parasitism of the alga hinders development of chloroplasts by the host, hence the yellowish-green color of areas infected with the alga.[2] Later, the presence of the parasite stimulates a formation of yellowish-orange droplets of oil within cells of the host. Still later, the presence of the alga may cause a disappearance of green coloration from the entire leaf except where the *Phyllosiphon* filaments are interwoven to form a green mat. The thallus of *Phyllosiphon* is a dichotomously or irregularly branched tube in which branching is profuse and the various branches are loosely interwoven with one another. The entire coenocyte is densely packed with elliptical chloroplasts, except at the tips of growing branches. The chloroplasts are without pyrenoids and may form either starch or oil.[3]

Reproduction is by the formation of many small ellipsoidal aplanospores within all portions of the coenocyte.[4] The aplanospores grow directly into new thalli.[5] Motile reproductive stages have not been observed in this alga.

[1] Schussnig, 1930. [2] Maire, 1908. [3] Just, 1882; Tobler, 1917.
[4] Just, 1882; Maire, 1908; Tobler, 1917. [5] Tobler, 1917.

The only host within which *Phyllosiphon* has been found growing in this country is the Jack-in-the-Pulpit [*Arisaema triphyllum* (L.) Schott]. The American *Phyllosiphon* appears to be identical with the *P. Arisari* Kühn (Fig. 200) found on several species of *Arisarum* in Europe. It has been collected in Wisconsin[1] and New Hampshire.[2] For a description of it, see Collins (1909).

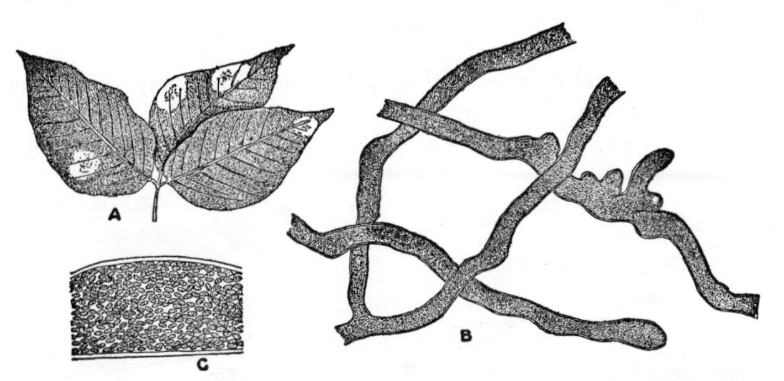

FIG. 200. *Phyllosiphon Arisari* Kühn. *A*, leaf of *Arisaema triphyllum* infected with *Phyllosiphon*. *B–C*, portions of thallus of *Phyllosiphon*. (*A*, × ⅓; *B*, × 110; *C*, × 430.)

FAMILY 2. DICHOTOMOSIPHONACEAE

The most striking difference between this and other families of the order is the oögamous sexual reproduction. *Dichotomosiphon* is the only genus in the fresh-water flora of this country.

1. **Dichotomosiphon** Ernst, 1902. The thallus of *Dichotomosiphon* is a dichotomously branched tubular coenocyte transversely constricted at each dichotomy and with constrictions between the dichotomies. The entire thallus, except for colorless rhizoidal branches at the base, contains numerous lens-shaped chloroplasts without pyrenoids and without starch. In addition there are leucoplasts containing granules of starch. The wall is said[3] to lack cellulose.

Asexual reproduction is by a formation of large tuberous akinetes that are generally borne at the ends of rhizoid-like lateral branches. The akinetes are densely packed with starch and may or may not be set off from the remainder of the coenocyte by a transverse septum. They germinate directly into new coenocytes.[4]

Sexual reproduction is oögamous. The coenocytes are homothallic and with the sex organs borne at the di-, tri-, or tetrachotomously branched end of a filament. The branches subtending each antheridium and oögonium are strongly curved. An antheridium is of the same diameter as the supporting branch and separated from it by a transverse septum. The protoplast of an antheridium divides into a large number of minute biflagellate

[1] SWINGLE, 1894. [2] COLLINS, 1909. [3] FELDMANN, 1946. [4] ERNST, 1902.

antherozoids that are liberated by an apical rupture of the antheridial wall. An oögonium is spherical and with a curved supporting branch. Just before fertilization, an oögonium develops a small beak-like opening at its apex.[1] Mature zygotes have a smooth thick wall surrounding a protoplast densely packed with starch and are retained within the oögonium for a considerable time after fertilization.

Even when in a vegetative condition, *Dichotomosiphon* may be distinguished from *Vaucheria* by the constrictions at intervals along a cell. There is but one species, *D. tuberosus* (A. Br.) Ernst (Fig. 201). European phycologists consider this

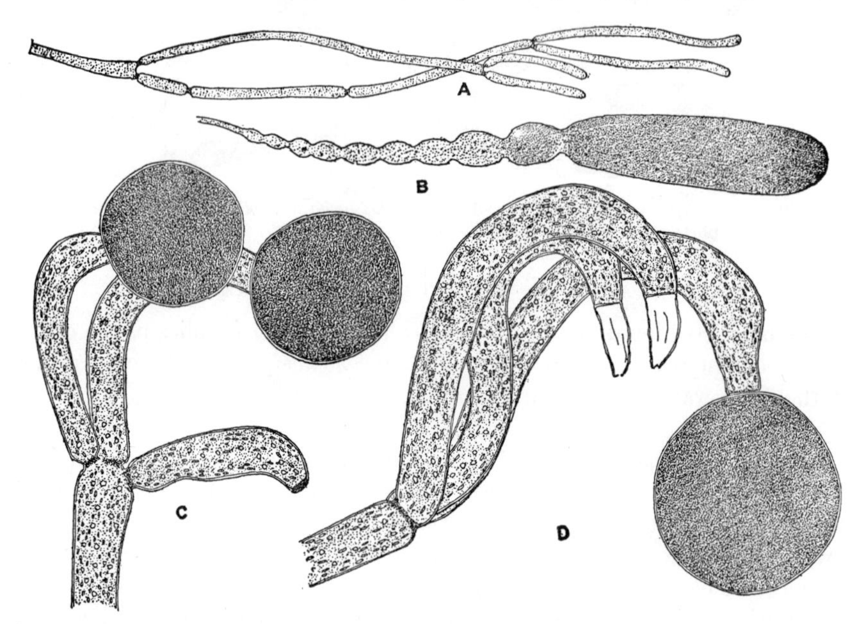

Fig. 201. *Dichotomosiphon tuberosus* (A. Br.) Ernst. *A*, portion of a thallus. *B*, akinete. *C–D*, sex organs. (*A–B*, × 6; *C–D*, × 325.)

one of the rarest of fresh-water algae, but it has been found at a number of widely separated stations in the United States. In this country it has been found both at the surface of shallow pools and 20 to 50 ft. below the surface of deep-water lakes. For a description of *D. tuberosus*, see Collins (1909).

ORDER 10. ZYGNEMATALES

The Zygnematales constitute a well-defined series, sharply delimited from other Chlorophyceae by their lack of flagellated reproductive cells and by their production of isogamous aplanogametes. The organization of the protoplast, especially the structure of the chloroplast, is also quite

[1] Ernst, 1902.

different from that of other green algae. Because of these differences, many systems of classification place the Zygnematales as a separate class coordinate with the Chlorophyceae,[1] or as a subclass of the Chlorophyceae.[2] The placing of them in a separate class is certainly too drastic a treatment since it ignores their undoubted relationships with other Chlorophyceae. Whether or not the Zygnematales are to be considered a subclass depends upon how remote they are from other Chlorophyceae. If they represent a phylogenetic series evolved directly from unicellular Volvocales, but with intermediates lost, they may possibly merit recognition as a subclass. The occasional occurrence of amoeboid instead of flagellated gametes in *Chlamydomonas*,[3] and the presence of various types of chloroplast in this genus, suggest the possibility of a derivation of Zygnematales from a one-celled motile ancestor. On the other hand, cells of Zygnematales have the same ability to divide vegetatively as is found in the tetrasporine series of Chlorophyceae. Primitive and advanced members of the tetrasporine series also have the capacity to form amoeboid instead of flagellated reproductive cells.[4] There is, therefore, the possibility that the Zygnematales are an offshoot from the tetrasporine line, that may have arisen at an early or a relatively late stage in evolution of the series. If such is the case, the Zygnematales merit no higher rank than the Oedogoniales, another offshoot of this series.

Members of the Zygnematales may have solitary cells, or the cells may be joined end to end in unbranched filaments. All genera of the order have uninucleate cells and almost always have the nucleus central in position. The chloroplasts are of three general types: peripheral spirally twisted bands extending the length of a cell; an axial plate extending the length of a cell; or two stellate chloroplasts axial to each. There are many modifications of the last-named type in the Desmidiaceae, and many members of this family have "stellate" chloroplasts from which the central mass has entirely disappeared. The wall surrounding protoplasts of Zygnematales is generally composed of two concentric layers: a cellulose layer next to the protoplast and an outer layer of pectic material. Filamentous Zygnematales are usually slippery to the touch because of the mucilaginous sheath of pectose.

Multiplication of unicellular Zygnematales is by cell division, followed by an immediate separation of the two daughter cells. In most filamentous genera, vegetative multiplication is due only to an accidental breaking; but some of the narrow-celled filamentous species have a regular dissociation into individual cells or into short fragments which, in time,

[1] Blackman and Tansley, 1902; Oltmanns, 1922; Wille, 1890. [2] Printz, 1927; West, G. S., 1916.

[3] Pascher, 1918. [4] Pascher, 1915.

may develop into new filaments. In certain genera, the protoplast may contract and become a spore-like body with the same shape and ornamentation as a zygote. These bodies have been called parthenospores, but it is better to follow those[1] who consider them aplanospores because they are not formed by gametes whose fusion has been interrupted. A few Zygnematales may also form akinetes.

Sexual reproduction is by a fusion of amoeboid gametes; formed singly within a cell, and in most genera all the protoplast is used in production of a gamete. During sexual reproduction, there may be an establishment of a tubular connection between two cells of opposite sex, or the gametes may escape from their enclosing walls at the time they fuse with each other. The zygote formed by union of two gametes develops a thick wall, usually with an ornamentation or sculpturing characteristic of the species, and enters upon a period of rest before it germinates. The fact that all Zygnematales, thus far investigated cytologically, have a meiotic division of the zygote nucleus seems to justify the assumption that vegetative cells of all members of the order are haploid. Gametes that have failed to unite with another one may develop into parthenospores that, except for their smaller volume, are identical in appearance with zygotes.

Practically all present-day phycologists divide the order into three families.

Family 1. Zygnemataceae

Members of this family have cylindrical cells, more or less permanently united in unbranched filaments. The cells have unsegmented walls without pores and contain either peripheral spirally arranged ribbon-shaped chloroplasts, or an axial laminate chloroplast, or two to four axial disk-shaped to stellate chloroplasts. Conjugating cells do not have their protoplasts escaping from the enclosing wall during gametic union.

Filaments of Zygnemataceae are usually free-floating and intermingled with one another to form a slippery mass. Species growing in quiet water may be sessile and attached to a substratum by lateral or terminal outgrowths (*haptera*) from near the end of a filament[2] or by a tendril-like coiling of a filament.[3] Species growing attached to rocks in swiftly flowing streams regularly develop haptera.[4]

Lateral walls of cells of Zygnemataceae have an inner layer of cellulose (two layers in *Spirogyra*) and an outer layer of pectose.[5] The pectose layer is usually thin, but in robust species of *Spirogyra* it may reach a thick-

[1] Taft, 1937; Transeau, 1934.

[2] Borge, 1894; Iyengar, 1923; Pickett, 1912; West, W. and G. S., 1898; Weatherwax, 1914.

[3] Iyengar, 1923. [4] Collins, 1904; de Pumaly, 1927. [5] Tiffany, 1924.

ness of 15 μ. The transverse wall separating two adjoining cells has a middle lamella of pectose and a layer of cellulose on either side of the middle lamella. Cellulose layers adjoining the middle lamella may be flattened or, as in several species of *Spirogyra*, they may have annular ingrowths—the so-called *replicate end walls*. As will be shown on a later page, replication of end walls is intimately connected with fragmentation of a filament.

Three types of chloroplast are found in Zygnemataceae. *Spirogyra* has ribbon-like spiral chloroplasts, peripheral in location and running the length of a cell. The cells may have from one to a dozen chloroplasts. Within certain limits, the number of chloroplasts and the amount of their twisting are constant for a given species. Inconstancy in number of chloroplasts is best exemplified by filaments in which some cells have a single chloroplast and others have two. This has been explained as being due to a transverse division of a chloroplast into two fragments followed by a growth of the two parts past each other.[1] Another explanation holds that it is due to the elongation of a single chloroplast until it bends back on itself, followed by a breaking at the point of bending.[2] Chloroplasts of the *Spirogyra* type contain several pyrenoids. These are seriately arranged, usually equidistant from one another, and often connected by a cytoplasmic strand. *Mougeotia* and its relatives have a single axial plate-like chloroplast, as broad as the cell and usually nearly as long. Narrow cells with this type of chloroplast may have two, three, or more pyrenoids that lie in an axial series; broad cells have several irregularly scattered pyrenoids. Chloroplasts of the *Mougeotia* type have the ability to orient themselves with respect to the direction of illumination. In diffuse light, the plane of the chloroplast is at right angles to the incident rays; in intense light, the plane parallels the rays.[3] This usually results in chloroplasts of several successive cells having the same orientation, but it is not uncommon to find cells in which the two poles of a chloroplast are at right angles to each other. *Zygnema* has two stellate chloroplasts, each with a single pyrenoid, that lie axial to each other on either side of the nucleus. There are also Zygnemataceae with this type of chloroplast, in which both chloroplasts are greatly elongate and with a rod-like pyrenoid, or in which each elongate chloroplast is transversely constricted into two chloroplasts.[4]

Protoplasts of the Zygnemataceae are always uninucleate under natural conditions, but *Spirogyra* has been induced to produce binucleate cells under controlled conditions in the laboratory. These artificially produced binucleate cells remain alive for a considerable time and have even been found conjugating.[5] Cells with a *Spirogyra* type of chloroplast

[1] Lewis, I. F., 1925. [2] Hill, 1916; Kasanowsky, 1913.
[3] Lewis, F. J., 1898. [4] Transeau, 1925. [5] Gerassimoff, 1898.

have the nucleus surrounded by a sheath of cytoplasm and suspended in the central vacuole by several radiately disposed cytoplasmic strands. The distal end of each strand usually terminates immediately beneath a pyrenoid. Cells with the *Mougeotia* type of chloroplast have the nucleus flattened against the chloroplast midway between its end. In the *Zygnema* type of cell, the nucleus lies between the two stellate chloroplasts.

Cell division follows very shortly after nuclear division. Cytokinesis has been held[1] to be due to a development of an annular furrow in the plasma membrane, midway between the poles of the cell, but cell division in *Spirogyra* may be due to a phragmoplast (cell plate) similar to that formed in cells of higher plants.[2] Division of the protoplast is followed by the formation of a thin layer of wall material, the pectose middle lamella of a cross wall separating mature cells. Cells with chloroplasts similar to those of *Spirogyra* and *Mougeotia* have their chloroplasts transversely severed by cytokinesis. Because of this, *Spirogyra* often has the spiral line of the chloroplast or chloroplasts continuous from cell to cell. Genera with stellate chloroplasts do not have them divided by cytokinesis. In *Zygnema*[3] the newly formed daughter protoplasts have one chloroplast only. Soon after cell division, the nucleus migrates to a point lateral to the chloroplast and midway between the two poles. This is followed by a division of the chloroplast into two daughter chloroplasts, each with a pyrenoid resulting from bipartition of the original pyrenoid, and the nucleus moves in between the two newly formed chloroplasts.

Cell division increases the number of cells in a filament but does not result in a direct increase in the number of filaments. Vegetative multiplication of the Zygnemataceae may be accidental and due to a severing of filaments by the action of water currents or to aquatic animals feeding upon the plant mass. The rapid increase in the number of filaments in masses of *Spirogyra* or *Zygnema* growing in quiet water shows that this is an efficient method of reproduction. Vegetative multiplication may also result from the disjunction of filaments into individual cells or into small fragments. Disjunction is generally associated with a change in the chemical nature of the middle lamella, probably a conversion of the pectose into pectin.[4] The actual disjunction of the cells has been ascribed to a shearing of the lateral walls[5] resulting from differential turgor pressure in adjoining cells rather than to a mere imbibitional swelling of the middle lamella as was once supposed.[6] This type of disjunction is found in the smaller, thin-walled species of *Mougeotia* and *Spirogyra* and is especially common in species of the latter that have replicate end walls (Fig. 202). Disjunction in larger, thick-walled species of *Spirogyra* is by an abscission

[1] Strasburger, 1785, 1880. [2] McAllister, 1931. [3] Merriam, 1906.
[4] Tiffany, 1924. [5] Lloyd, 1926. [6] Benecke, 1898.

of H-pieces near the cross walls.[1] The localized hydrolysis bringing about this abscission is induced by unfavorable conditions such as wounding, infection by fungi, or death and decay of certain cells in the filament.

Cells of Zygnemataceae may have a rounding up of the protoplast and a secretion of a thick wall around the retracted protoplast. For any given species, the structure and ornamentation of the special walls surrounding these bodies are identical with those of walls of zygotes of the species. In certain cases, these bodies obviously result from failure of a gamete to unite with another gamete and hence are appropriately called *parthenospores*. In other cases, as in *Zygnemopsis* and in certain species of *Mougeotia* and *Spirogyra*, these bodies are formed in filaments where conjugation is not taking place. Thus, in spite of morphological similarities between them and zygotes, it is very probable that they are asexual in nature and are to be regarded as in the nature of aplanospores.

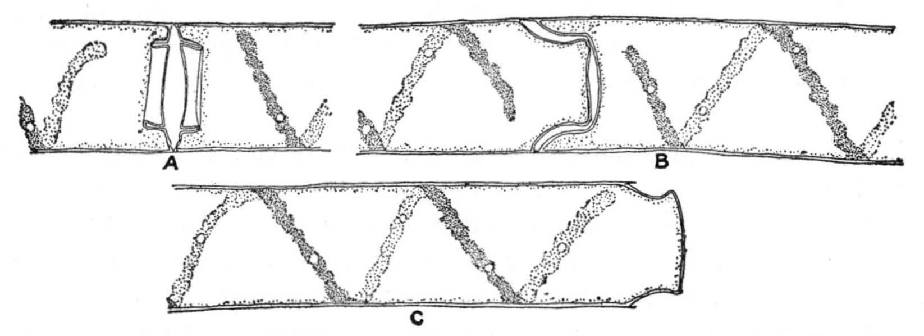

FIG. 202. Vegetative multiplication of *Spirogyra protecta* Wood, a species with replicate end walls. (× 325.)

Species of *Mougeotia* growing in densely shaded pools, or in ponds of alpine and arctic regions, may have isolated cells in a filament becoming thick-walled. Similar cells are found in filaments of *Spirogyra* growing in rapidly flowing water, and in terrestrial species of *Zygnema*. These thick-walled cells are of an akinete-like nature.

The Zygnemataceae have a marked seasonal periodicity of sexual reproduction, and each species usually fruits at a definite time of the year. Most species fruit in the spring, but there are those that fruit in the summer or in the autumn.[2] The factors inducing conjugation are not wholly connected with changes in the external environment, and fruiting cannot be induced at will by altering the conditions of illumination, temperature, and mineral content of the surrounding water. In many species of *Spirogyra*, the time of conjugation is directly correlated with the ratio between the cell's surface and volume, and cells with the largest surface in proportion to the volume fruit the earliest.[3] However, this is somewhat influenced

[1] LLOYD, 1926. [2] TRANSEAU, 1913. [3] TRANSEAU, 1916.

by external factors, especially temperature, and is accelerated or retarded as the seasonal temperature average is high or low.

All the Zygnemataceae have the vegetative cells functioning directly as gametangial cells and each cell giving rise to a single nonflagellated gamete. At the time of gametic union, the two conjugating cells are usually connected with each other by tubular outgrowths (the *conjugation tube*), but in some cases, as *Sirogonium*, there is no development of a conjugation tube between the apposed cells. The conjugation may be between cells of different filaments (*scalariform conjugation*), or between adjoining cells of the same filament (*lateral conjugation*). A majority of the genera belonging to the Zygnemataceae have all vegetative cells potentially capable of producing gametes, and conjugating filaments of such Zygnemataceae contain but few vegetative cells. *Sirogonium* and *Temnogametum* have a differentiation of the two daughter cells formed in the cell generation immediately preceding conjugation.[1] One of these daughter cells elongates considerably and is purely vegetative; the other remains short and has the ability to conjugate or to produce parthenospores.

Many species with scalariform conjugation have conjugation beginning with the lateral approximation of the filaments throughout their entire length and the secretion of a common gelatinous envelope around the two filaments.[2] While the two filaments are thus apposed, there is the development of a papilla, the primordium of the conjugation tube, from each cell. Further elongation of the papillae pushes the filaments a certain distance from each other. Zygnemataceae with a differentiation of male and female gametes often have the papillae of the male cells developing earlier than those of the female cells. Species with like gametes, and a formation of zygotes in the conjugation tube, usually have a simultaneous development of the two papillae. Sooner or later after the papillae appear there is a hydrolysis of their walls in the zone of mutual contact, the sudden bursting[3] or the gradual dissolution of this zone of contact establishing an opening through the conjugation tube connecting the two cells. Zygnemataceae with lateral conjugation have the papillate outgrowths that eventually develop into the conjugation tube of these species always arising near the transverse wall separating adjoining cells. Later stages in the growth of these papillae and the establishment of the perforation in the conjugation tube seem to be the same as in scalariform conjugation.

There are occasional irregularities in both types of conjugation. These include the fusion of papillae from two cells in one filament with the papilla from one cell in the other filament, with a resultant Y-shaped conjugation

[1] West, G. S., 1916; West, W. and G. S., 1897.

[2] Czurda, 1925; Hemleben, 1922; Saunders, 1931. [3] Lloyd, 1926*B*.

tube,[1] lateral and scalariform conjugation in the same filament, and lateral conjugation by the perforation of the transverse wall instead of by a conjugation tube.[2] Even more interesting are the cases of the establishment of conjugation tubes and gametic union between filaments of two distinct species.[3]

The entire protoplast in each pair of conjugating cells may enter into the composition of the gamete, or a small cytoplasmic residue may remain after the protoplast has developed into a gamete (*Mougeotia*). Gametes developed by a conjugating pair of cells may be strictly isogamous and without morphological or physiological differentiation. All species of a genus may show such an isogamy (*Mougeotia*, *Debarya*), or some species of a genus may be strictly isogamous and others physiologically anisogamous (*Zygnema*).

The fusion of the two conjugating protoplasts is not the same in all Zygnemataceae. In *Debarya* and *Zygnemopsis*, the movement of the gametes toward each other is purely passive and results from the accumulation of a gelatinous material (said to be a pectic and cellulose colloid[4]) within the cell cavity. In other cases (*Mougeotia* and *Zygnema*), the two gametes move toward each other in an amoeboid manner; or one of the gametes is actively amoeboid and the other passive (*Spirogyra* and *Zygnema*). In gametic union of the last-named type, differentiation into male and female gametes may be recognized at a relatively early stage by the earlier shrinkage of the protoplast developing into the male gamete. The shrinkage is due to a lowering of the osmotic pressure and is brought about by the continued development and bursting of small contractile vacuoles just beneath the surface of the plasma membrane.[5]

In both the scalariform and the lateral types of conjugation, there may be a development of cross septa in the conjugation tube that completely separates the uniting protoplasts from the gametangia (*Zygogonium*, in part *Mougeotia*); or there may be a development of septa that shut off the zygote from the ends of the gametangia (in part *Mougeotia*). More frequently (*Spirogyra*, *Zygnema*, *Debarya*), there is no formation of special walls except those surrounding the zygote. Walls of zygotes of Zygnemataceae consist of three or more layers: a thin inner of cellulose, outside this another layer of cellulose with irregular deposits of chitin, and at the outside a layer of cellulose which may also have an encircling sheath of pectose. The complete differentiation of these layers, especially the ornamentation of the median layer and the coloration of the outermost layer, is not attained until the zygote is quite old. Ripening of a zygote is

[1] Atwell, 1889*A*; Czurda, 1925; Rose, 1885. [2] Hodgetts, 1920*A*; Rose, 1885.
[3] Andrews, 1911; Bessey, 1884; Transeau, 1919. [4] Transeau, 1934.
[5] Lloyd, 1926*A*, 1926*B*.

accompanied by a disappearance of the chlorophyll and a conversion of most of the starch into a yellowish oil. Germination of a zygote is generally preceded by a redevelopment of chlorophyll.

A gamete that fails to unite with another gamete may round up and secrete a wall whose structure and ornamentation are identical with that of the zygote. Such bodies are parthenospores (azygotes). The rare cases of twin parthenospores,[1] sometimes called "twin zygotes," are all due to failure of fusion in apposed gametes and the secretion of a wall around each of them.

When two gametes unite to form a zygote, their cytoplasm fuses at once, but union of their nuclei may be delayed for a considerable time. It is rather difficult to follow the behavior of the chloroplasts in *Spirogyra*, but there is a certain amount of evidence showing that the chloroplast or

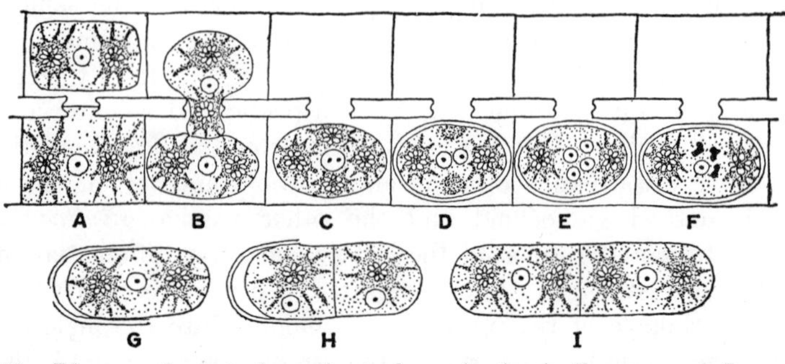

Fig. 203. Diagram of zygote formation and germination in *Zygnema*. *A–B*, conjugation. *C*, after fusion of the gamete nuclei. *D*, after division of the fusion nucleus to form two nuclei. *E*, after the formation of four nuclei. *F*, three nuclei beginning to degenerate. *G–I*, stages in germination of a zygote. (*Diagram based upon Kurssanow*, 1911.)

chloroplasts contributed by the male gamete disappear and that those from the female gamete persist.[2] The behavior of the chloroplasts is much easier to follow in *Zygnema* since the two male chloroplasts lie in the short axis of a zygote. In this case, there is an evident disintegration of chloroplasts contributed by the male gamete, and a persistence of those from the female gamete.[3] The two Zygnemataceae whose nuclear history has been carefully investigated (*Spirogyra*, *Zygnema*) have been shown to have a reduction division in the first two divisions of the zygote nucleus.[4] Three of the nuclei thus formed soon disintegrate; the fourth persists unchanged until the zygote germinates (Fig. 203).

Zygotes of the Zygnemataceae rarely germinate as soon as they appear to be mature, and it is probable that in the majority of cases there is no

[1] Transeau, 1926; West, W. and G. S., 1898; Wittrock, 1878.
[2] Chmielevsky, 1890; Lewis, I. F., 1925; Tröndle, 1911.
[3] Kurssanow, 1911. [4] Kurssanow, 1911; Tröndle, 1911.

germination until the spring following their formation. Germination begins with a rupture of the two outer wall layers of the zygote and the development of a tubular outgrowth, surrounded by a wall derived from the inner wall layer of the zygote. The outer zygote wall layers usually burst irregularly, but some species of *Mougeotia* have a regular lid-like opening of these layers.[1] The elongating tubular outgrowth may contain the nucleus and all the chloroplasts, or the chloroplasts may lie partly within the outgrowth and partly within the old zygote. After the outgrowth has attained a certain length, the nucleus divides and this is followed by a transverse cytokinesis. The plane of transverse wall formation is usually near the point where the tube emerges from the zygote (*Spirogyra*), but it may be remote from the point of emergence [*Sirogonium sticticum* (Engl. Bot.) Kütz.]. In either case, there is no division of the proximal daughter cell, and the filament produced by a germinating zygote is derived by repeated division of the distal cell and its derivatives. In a few cases, as *Zygnema pectinatum* (Vauch.) Ag., the protoplast, surrounded by the

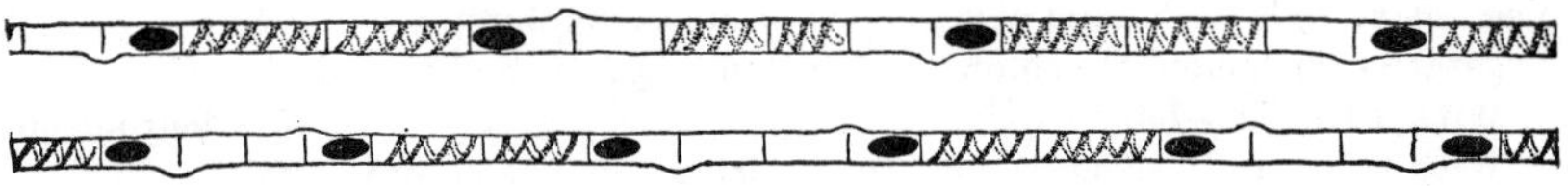

FIG. 204. A laterally conjugating species of *Spirogyra* in which the fertile cells alternate with pairs of vegetative cells. (× 80.)

innermost zygote wall layer, escapes from the outer zygote wall layers and starts development as a free-living cell.

The scanty observations on hybridism among the Zygnemataceae[2] show that the shape and ornamentation of the zygote are always derived from the female gamete. Cytological observations seem to show that there is a degeneration of the chloroplasts derived from the male gametes. It is quite possible, therefore, that the chloroplast structure and such characters as size and wall structure of cells produced by the germinating zygote are derived wholly from the female gamete.

Study of the problem of differentiation of sex in the Zygnemataceae should start with the germinating zygote but, as yet, the technique of growing Zygnemataceae in pure culture and having them conjugate[3] has not been developed to a stage where zygotes can be induced to germinate. All studies on sex have been based upon conjugation in filaments of unknown parentage. Interest in this question has centered around filaments with lateral conjugation and in those with a production of both male and female gametes by both filaments when conjugation is scalariform (*cross*

[1] DEBARY, 1858.
[2] ANDREWS, 1911; BESSEY, 1884; TRANSEAU, 1919; WEST, W. and G. S., 1898.
[3] CZURDA, 1926.

conjugation). Here it is quite clear that sex determination is not associated with meiosis and that differentiation of sexual cells from vegetative cells occurs in the haploid generation. Every cell in a filament may divide to form two cells of opposite sex, or each cell of a filament may divide to form one daughter cell whose descendants remain vegetative and one whose descendants become sexual. Differentiation of sexes in the daughter cell with fertile descendants may take place in the first or in the second cell generation derived from it (Fig. 204). Scalariform species with cross conjugation may even have sex differentiation occurring in earlier cell generations and have both filaments of a conjugating pair containing several successive male cells followed by several successive female cells.

Generic distinctions among Zygnemataceae are based both upon vegetative characters, especially structure of the chloroplast, and upon behavior of the gametangia. On this account, it is impossible to identify certain genera unless they are fruiting. The genera found in this country differ as follows:

1. Vegetative cells with axial disk-shaped or stellate chloroplasts................ 2
1. Vegetative with elongate chloroplasts extending from end to end of a cell...... 5
   2. With two disk-shaped chloroplasts........................ 7. **Pleurodiscus**
   2. With two (rarely several) stellate or cushion-shaped chloroplasts........... 3
3. Chloroplasts stellate.. 4
3. Chloroplasts cushion-shaped.................................... 6. **Zygogonium**
   4. Gametangia filled with gelatinous material after zygotes are formed.......... 5. **Zygnemopsis**
   4. Gametangia not filled with gelatinous material after zygotes are formed..... 4. **Zygnema**
5. With one axial chloroplast.. 6
5. With one to several parietal spiral chloroplasts................................. 8
   6. Chloroplast without pyrenoids........................... 3. **Mougeotiopsis**
   6. Chloroplast with pyrenoids.. 7
7. Gametangia filled with gelatinous material after zygote is formed.... 2. **Debarya**
7. Gametangia not filled with gelatinous material after zygote is formed.......... 1. **Mougeotia**
   8. Conjugation tube conspicuous................................8. **Spirogyra**
   8. Conjugation tube not conspicuous.........................9. **Sirogonium**

1. **Mougeotia** Agardh, 1824. The cylindrical cells of *Mougeotia*, which are usually at least four times longer than broad, are joined end to end in unbranched filaments. The filaments sometimes have unicellular or multicellular outgrowths (haptera) near end walls of certain cells, but these are of much rarer occurrence than in *Spirogyra*.[1] The cell walls are thin, and the layer of pectose at the outside of a filament never becomes very thick. Most species have cells with a single axial laminate chloroplast, but a few species have two axial laminate chloroplasts connected by a cytoplasmic

[1] Borge, 1893; Pascher, 1901*A*.

bridge. Chloroplasts of narrow cells have two, three, or more pyrenoids arranged in a linear series; chloroplasts of broad cells have several irregularly arranged pyrenoids. The chloroplast is attached to the cytoplasmic layer lining the wall by delicate strands of cytoplasm; the remainder of the cell cavity is filled with a cell sap which is usually colorless but which, as in *M. capucina* (Bory) Ag.[1] may be colored. Depending upon the intensity of illumination, the chloroplasts lie at right angles to, or parallel with, the incident rays of sunlight.[2] Chloroplasts of several successive cells usually have the same orientation, but sometimes the chloroplast of a single cell is

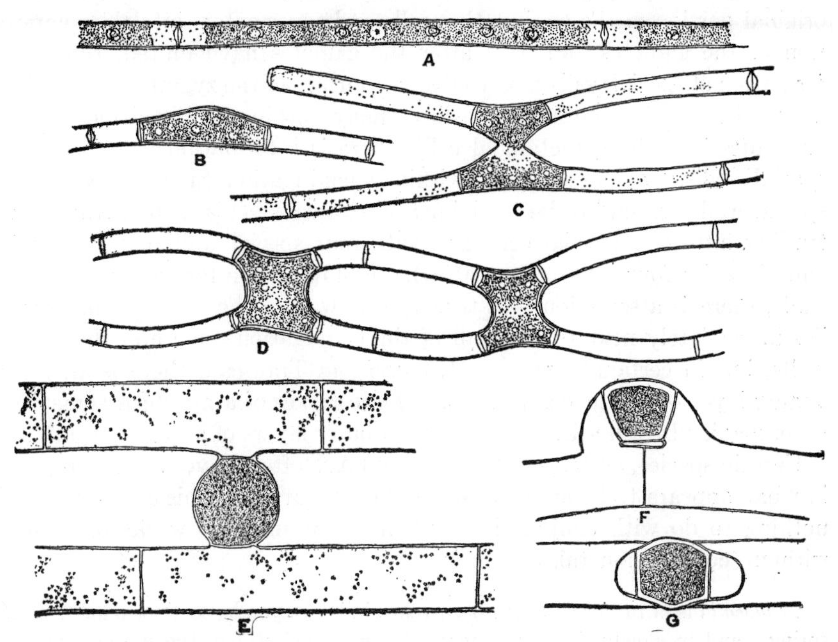

Fig. 205. *A–D, Mougeotia viridis* (Kütz.) Wittr. *A*, vegetative filament. *B*, parthenospore. *C–D*, zygotes. *E, M. scalaris* Hass. *F–G*, lateral conjugation in *M. genuflexa* (Dillw.) Ag. (× 325.)

so twisted that opposite ends are at right angles to each other. The cell contains a single nucleus midway between the poles, and it lies flattened against the chloroplast. The first wall layer formed between two recently divided cells is pectic in nature and becomes the middle lamella of the wall separating two mature cells.[3]

Asexual reproduction is by an accidental breaking of filaments, or by a dissociation into single cells or short-celled fragments. Dissociation is due to conversion of the pectose middle lamella of transverse walls into pectin.[3] Aplanospores are of more frequent occurrence in this genus than in any

[1] Lagerheim, 1895. [2] Lewis, F. J., 1898. [3] Tiffany, 1924.

other of the family. Certain species with aplanospores are not known to produce zygotes. Isolated cells in a filament may also become thick-walled akinetes.[1]

Sexual reproduction is usually scalariform, but certain species, as *M. genuflexa* (Dillw.) Ag. (Fig. 205*F–G*) and *M. scalaris* Hass., may also have lateral conjugation. In scalariform conjugation, there is a migration of the major portion of both protoplasts into the conjugation tube, and this is often accompanied by a considerable increase in diameter of the tube. A portion of the cytoplasm of both protoplasts always remains behind in the original portion of the conjugating cells and never enters into the composition of the gametes. Shortly after the gametes have united, there is a formation of special walls next to the free surface of the zygote. Depending upon whether the young zygote lies wholly within a conjugation tube or protrudes into the gametangial cells, there is a formation of two or four special walls. Zygotes wholly within a conjugation tube and with two special walls are said to be "adjoined by two cells" (Fig. 205*E*); those protruding into gametangial cells and with four special walls are said to be "adjoined by four cells" (Fig. 205*D*). Following the formation of special walls, there is a secretion of a true zygote wall. The zygote wall usually lies immediately next to the wall of the conjugation tube and the special walls; but in certain cases, as *M. americana* Transeau, there is an intervening layer of pectic compounds. Germination of a zygote is sometimes accompanied by a lid-like opening of the outer layers of a zygote wall.[2]

Certain species, as *M. genuflexa* (Dillw.) Ag., often have the cells apposed in what appears to be an early stage of conjugation. This *genuflexing* has nothing to do with conjugation and may continue for weeks or months without conjugation taking place.

Mougeotia is of rather common occurrence in lakes, ponds, semipermanent pools, springs, and in slowly flowing waters. In most collections, the number of conjugating filaments is small, and it is often necessary to search through considerable material to find fruiting specimens. Fifty-three species are known for the United States. For names and descriptions of them, see Transeau (in press).

2. **Debarya** Wittrock, 1872. The cell structure of *Debarya* is identical with that of *Mougeotia*, and it is impossible to distinguish between the two on the basis of vegetative structure. The two genera are readily differentiated from each other when conjugating because the entire protoplast of each conjugating cell of *Debarya* enters into the composition of the gamete and the young zygote is not separated from the gametangia by special walls. *Debarya* also differs from *Mougeotia* in the filling of the lumen of gametangia with a stratified pectic-cellulose colloid as the gametes move toward each other.

[1] Transeau, 1926 [2] DeBary, 1858; Wittrock, 1867.

Debarya is found in the same sort of habitats as is *Mougeotia*, but is of much rarer occurrence. Fruiting specimens are readily distinguished from *Mougeotia* by the bluish-white refractive contents of the gametangia. The three species found in this country are *D. Ackleyana* Transeau, *D. glyptosperma* (DeBary) Wittr., and *D. Smithii* Transeau (Fig. 206). For descriptions of them, see Transeau (in press).

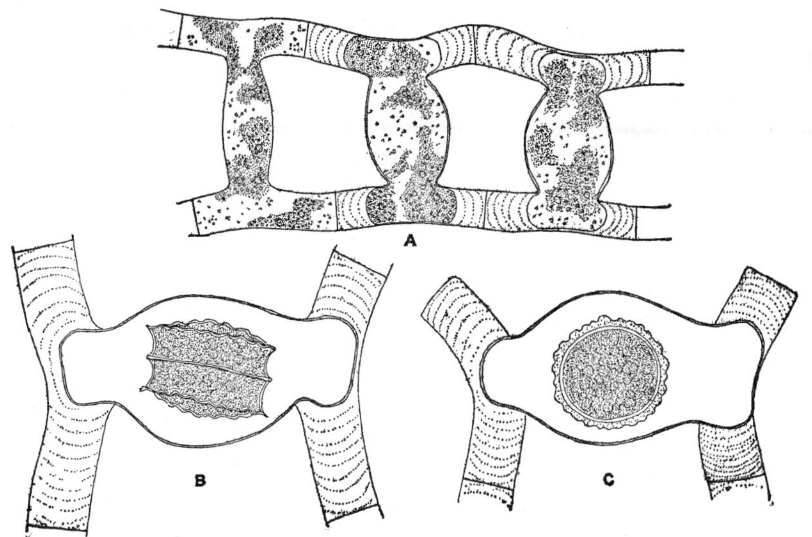

FIG. 206. *Debarya Smithii* Transeau. *A*, early stages in conjugation. *B–C*, old zygotes (*A*, × 250; *B–C*, × 325.)

3. **Mougeotiopsis** Palla, 1894. Vegetative cells of this genus have a length not more than double the breadth and an axial laminate chloroplast that is without pyrenoids.[1]

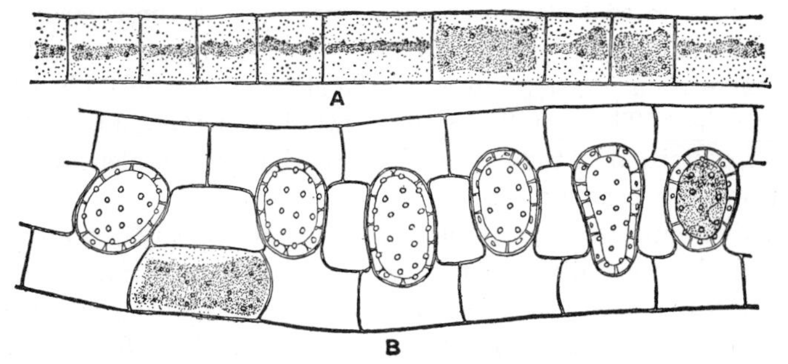

FIG. 207. *Mougeotiopsis calospora* Palla. *A*, vegetative filament. *B*, zygotes. (× 485.)

The genus resembles *Debarya* in having the entire protoplast of a conjugating cell entering into the composition of the gamete, and in the lack of special walls separating the zygote from the gametangia. It differs from

[1] PALLA, 1894.

Debarya in that there is no filling of the lumen of conjugating cells with refractive pectic-cellulose colloids.

There is but one species, *M. calospora* Palla (Fig. 207), and in this country it has been found only in Michigan and Wisconsin. For a description of it, see Transeau (in press).

4. **Zygnema** Agardh, 1824. The cylindrical cells of *Zygnema* usually have a length slightly greater than the breadth, but some species have cells two to five times as long as broad. Rhizoidal outgrowths (haptera) are very rare in this genus.[1] Lateral walls of filaments rarely have a thick pectose layer,[2] and there is never the replication of transverse walls that is sometimes found in *Spirogyra*. Protoplasts of *Zygnema* contain two stel-

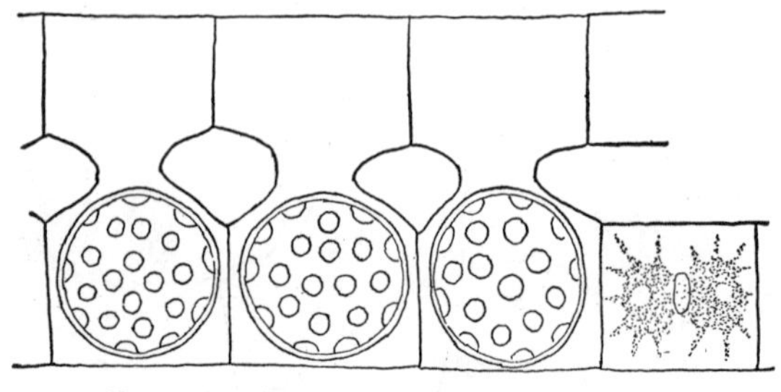

FIG. 208. *Zygnema Collinsianum* Transeau.

late chloroplasts that lie axial to each other in the longitudinal axis of a cell. The chloroplasts usually have numerous delicate strands extending to the plasma membrane. Each chloroplast has a single massive pyrenoid at its center. A nucleus lies midway between the two chloroplasts (Fig. 208).

In cell division each daughter cell receives one of the chloroplasts from the parent cell. Shortly after cytokinesis is completed, each daughter nucleus takes up a position lateral to the chloroplast and midway between the cell's poles. This is followed by a division of the chloroplast, accompanied by a division of the pyrenoid, after which the nucleus migrates to a position midway between the two.[3]

Vegetative multiplication of filaments is almost always by an accidental breaking and practically never by a dissociation into individual cells or into short fragments as in *Spirogyra* and *Mougeotia*. Aplanospores are not formed in abundance by *Zygnema*. A terrestrial form of *Z. peliosporum* Wittr. has been reported[4] as forming akinetes.

Most species of *Zygnema* have scalariform conjugation only, but certain

[1] COLLINS, 1904. [2] TIFFANY, 1924. [3] ESCOYEZ, 1907; MERRIAM, 1906.
[4] DE PUYMALY, 1922*A*.

of them may also conjugate laterally. Some species with scalariform conjugation are strictly isogamous and with both gametes migrating toward each other and uniting in the conjugation tube; other species have one of a pair actively amoeboid and the other passive. Mature zygotes have a wall composed of three layers in which the median one is often colored and sculptured. During ripening of a zygote, there is a disintegration of chloroplasts derived from the male gamete[1] and a meiotic division to form four haploid nuclei, three of which disintegrate (Fig. 203, page 290). In germination of a zygote there is a rupture of the two outer zygote wall layers and an escape of the protoplast surrounded by the innermost zygote wall layer, or there may be only a partial escape from the outer layers of the zygote wall.[2]

Zygnema occurs in much the same type of habitat as does *Spirogyra* and is often found intermingled with it. In some parts of the United States, as in Southern California, *Zygnema* is more abundant than *Spirogyra*, but in most regions it is considerably rarer. For names and descriptions of the 46 species known to occur in this country, see Transeau (in press).

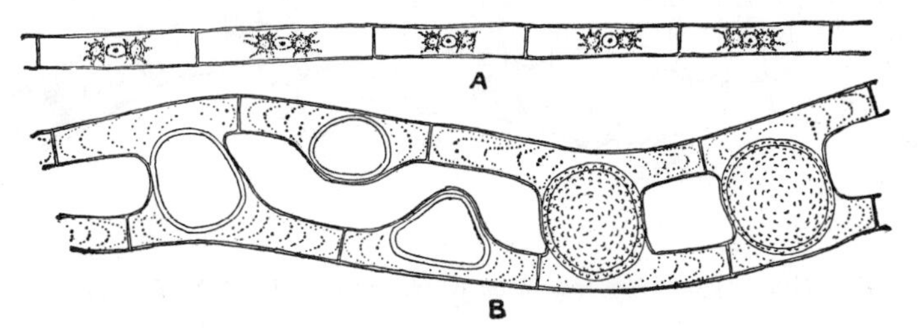

Fig. 209. *Zygnemopsis americana* Transeau. (× 325.)

5. **Zygnemopsis** Skuja, 1929; emend., Transeau, 1934. The general organization of the filaments and the structure of the cells of *Zygnemopsis* are similar to that of *Zygnema*.

When species form aplanospores, there is the same filling of empty portions of the cell lumen with pectic-cellulose colloids as there is when zygotes are formed. A formation of akinetes has also been found in *Z. decussata* Transeau.[3]

All cases of conjugation thus far observed are scalariform, with a formation of zygotes in the conjugation tube and with the gametangia becoming filled with a stratified pectic-cellulose colloid as in *Debarya*. When parthenospores are formed, there is also a filling of the cell with pectic-cellulose colloids (Fig. 209).

[1] Kurssanow, 1911. [2] DeBary, 1858; Kurssanow, 1911.
[3] Transeau, 1915, 1925.

The following species have been found in this country: *Z. americana* Transeau (Fig. 209), *Z. decussata* Transeau, *Z. desmidioides* (W. and G. S. West) Transeau, *Z. floridiana* Transeau, *Z. minuta* Randhawa, *Z. spiralis* (Fritsch) Transeau, and *Z. Tiffaniana* Transeau. For descriptions of them, see Transeau (in press).

6. **Zygogonium** Kützing, 1843. The cylindrical cells of this alga have a length from half to several times the breadth and are united in unbranched filaments. The cell walls may be thin and colorless, or thick and yellowish to brownish. In extreme cases, the thickness of walls is greater than that of the cell lumen. A cell contains two more or less cushion-shaped chloroplasts that may have short irregular processes and so resemble chloroplasts of *Zygnema*. The cell sap may be colorless, or colored purple to violet by a water-soluble pigment (phycoporphyrin).[1]

A formation of aplanospores is of frequent occurrence in *Zygogonium*, and sexual reproduction ha; not been observed in certain species regularly

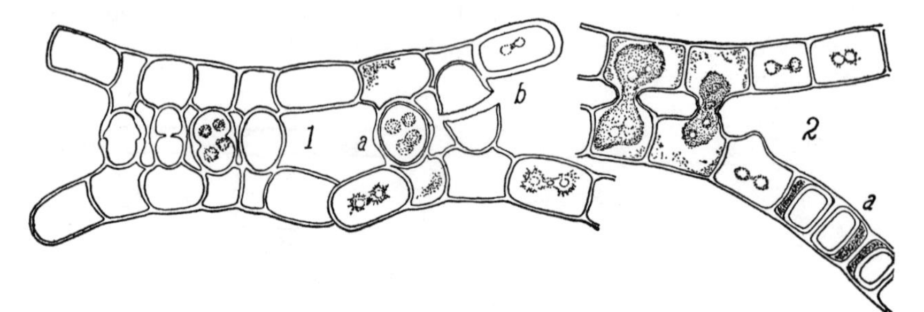

Fig. 210. *Zygogonium ericetorum* Kütz. (*Drawn by E. N. Transeau.*)

forming aplanospores. There is also frequently a formation of akinetes and certain species with thick walls have been considered[2] to be in a permanent akinete condition.

Conjugation is usually scalariform, and early in formation of the conjugation tube there is a migration of the major portion of the two protoplasts into the papillae which eventually become the conjugation tube. As is the case with *Mougeotia*, special transverse walls are formed that separate the fusing gametes from the major portion of the gametangia. Similarly, also, there is a residual portion of cytoplasm that does not enter into composition of a gamete. Shortly after the special walls are formed, each gamete becomes completely surrounded by a new wall; later on, there is a disappearance of these walls where they abut on each other and after this a fusion of the two gametes. The wall layer formed about the gametes becomes the outermost wall layer of the mature zygote. Sometimes there is a failure of the gamete walls to dissolve at the zone of contact. When

[1] Lagerheim, 1895. [2] Fritsch, 1916*A*.

this happens, there is a development of two parthenospores, in the conjugation tube, each of which secretes a thick wall similar to that surrounding a zygote.

Unlike other members of the family, *Zygogonium* is primarily aerial and grows on moist acid soils, rocks, and peat. When growing on soil, it may develop in patches several yards in diameter. The remarkable capacity of *Zygogonium* for taking up and returning large quantities of water plays no inconsiderable role in colonization of bare soil by the smaller phanerogams.[1] The following species have been found in this country: *Z. ericetorum* Kütz. (Fig. 210), *Z. pectosum* Taft *Z. punctatum* Taft, and *Z. Stephensiae* Transeau. For descriptions of them, see Transeau (in press).

7. **Pleurodiscus** Lagerheim, 1895. The cylindrical cells of this alga have a length less than double the breadth. Their walls are fairly thick. The filaments are unbranched, but rhizoidal outgrowths from them may be so well developed that a filament appears to be branched. Each cell of a

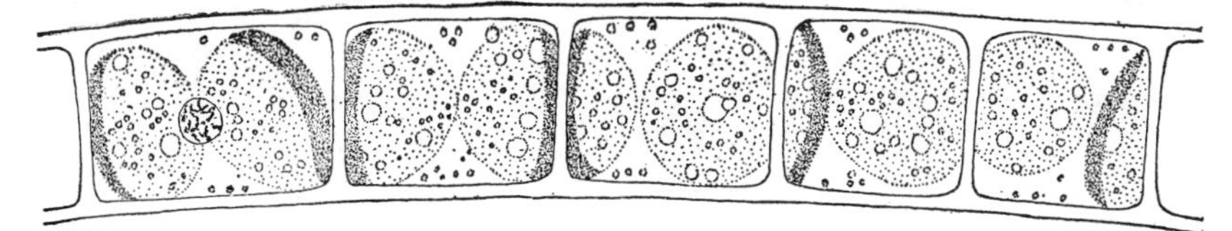

Fig. 211. *Pleurodiscus purpureus* (Wolle) Lagerh. (*Slightly modified from Lagerheim,* 1895.)

filament contains two disk-shaped chloroplasts, one at each end of the cell and usually at an angle with each other. The chloroplasts are connected by a cytoplasmic bridge in which lies the nucleus[2] (Fig. 211). The cell contents may or may not be colored a deep purple by a pigment (phycoporphyrin) dissolved in the cell sap.

Conjugation is scalariform and with a formation of the zygote in the conjugation tube.[3] Gamete formation is similar to that of *Mougeotia* and *Zygogonium* in that a residue of cytoplasm remains behind in the gametangia.

P. borinquenae Tiffany is the only completely described species.[3] Since only vegetative specimens of *Pleurodiscus* have been found in this country, it is impossible to determine whether or not they should be referred to this species.

8. **Spirogyra** Link, 1820. Filaments of *Spirogyra* may have cells nearly as broad as long, or the length may be several times the breadth. The lateral walls of cells consist of three layers: the two interior ones of cellulose,[4] and the outermost of pectose. In many species the pectose layer is

[1] West, G. S., 1904. [2] Lagerheim, 1895. [3] Tiffany, 1936.

thin, but in large-celled species it may be 10 to 15 μ in thickness. Transverse walls in a filament have a middle lamella of pectose and a layer of cellulose on either side of the lamella. Species with replicate end walls (Fig. 212*A*) have an annular ingrowth of the cellulose layer. A few species have been shown[1] to have the middle lamella expanded into an H-piece similar in appearance to H-pieces of *Tribonema* and *Microspora*. The chloroplasts are elongate bands extending from end to end of a cell and embedded in the layer of cytoplasm just internal to the cell wall. Some species, especially those with one or two chloroplasts, have them coiled in close spirals; species with several chloroplasts usually have them making less than one complete turn. Each chloroplast contains several pyrenoids that lie equidistant from one another and in a linear series. The cells are uninucleate and with the nucleus surrounded by a sheath of cytoplasm and suspended in the middle of the central vacuole by cytoplasmic strands whose distal ends usually terminate just beneath pyrenoids.

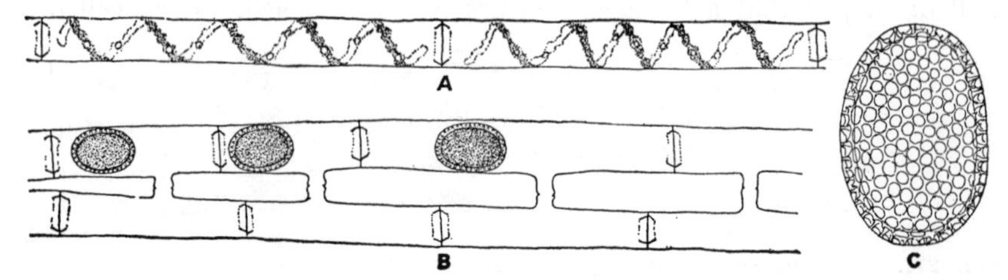

FIG. 212. *Spirogyra protecta* Wood. *A*, vegetative filament. *B*, conjugating filaments. *C*, zygote. (*A*, × 145; *B*, × 325.)

Vegetative multiplication may take place by an accidental breaking of the filaments, or the filaments may dissociate into single cells or into short-celled fragments (Fig. 202, page 287). Dissociation is most common in species with narrow cells and replicate end walls. The dissociation has been ascribed to a conversion of the pectose middle lamella into pectin,[2] followed by a differential shearing of the lateral walls as a result of turgor differences in adjoining cells.[3] Certain large-celled thick-walled species have the dissociation due to a development of H-pieces near the end walls.[4] Thick-walled resting cells are of very rare occurrence and are known chiefly from species growing in rapidly flowing water.[5]

Conjugation is always physiologically anisogamous and with a migration of the male gamete into the gametangial cell containing the female gamete. Most species have practically all cells in a filament producing gametes, but *S. punctata* Cleve, *S. Collinsii* (Lewis) Printz, and certain other species have a more or less regular alternation of sterile and fertile cells.[6] Conjugation

[1] HODGETTS, 1920*A*; LLOYD, 1926. [2] TIFFANY, 1924. [3] LLOYD, 1926.
[4] HODGETTS, 1920*A*; LLOYD, 1926. [5] DE PUYMALY, 1927. [6] LEWIS, I. F., 1925.

may be scalariform or lateral. In scalariform conjugation, all the cells in a filament may develop gametes of the same sex, or there may be a production of both male and female gametes in each of the conjugating filaments[1] (cross conjugation). Scalariform conjugation often begins with a lateral approximation of two filaments and their becoming enveloped in a common gelatinous matrix.[2] In most species, both cells contribute equally to the formation of the conjugation tube, but some species, as *S. punctata*, have the tube formed entirely or in large part by the male cell. Lateral conjugation is usually effected through a conjugation tube developed next the abutting ends of the conjugating pair, but it may also take place through a perforation in the cross wall separating the two cells.[3] Shortly after the conjugation tube is formed, there is a contraction of the male gamete, a contraction due to a lowered osmotic pressure resulting from the repeated development and bursting of small contractile vacuoles just within the plasma membrane.[4] This is followed by a similar contraction of the female gamete, as the male gamete migrates through the conjugation tube. In large-celled species, the shrinkage of the gametes is so great that the zygote resulting from their union does not cause a distention of the old female cell wall; narrow-celled species, on the other hand, often have the zygote distending the old female cell to double its original diameter. The wall secreted around the zygote is the usual three-layered structure found in the Zygnemataceae, with the ornamentation and coloration restricted to the median layer. In rare cases,[5] there is a disintegration of male chloroplasts before the gametes unite, but in most cases the chloroplasts derived from male gametes do not disintegrate until after gametic union.[6] During ripening of a zygote, there is a meiotic division of the zygote nucleus to form four haploid nuclei, three of which disintegrate.[7] When a zygote germinates, there is a rupturing of the two outermost wall layers and a development of the innermost layer into a wall surrounding the tubular outgrowth sent out by a zygote. The first transverse division is usually near the region of emergence of the tube; further development is entirely from the distal cell formed by this cell division.

Parthenospores are not infrequent because of a failure of union of gametes. There is not the regular formation of aplanospores in nonconjugating filaments so frequently found in *Mougeotia* and *Debarya*.

Spirogyra (Fig. 212) is one of the commonest of green algae in quiet waters. Most species fruit late in spring, but some fruit during summer or early autumn. Specific differences are based in large part upon structure of the zygote, and fully

[1] Hemleben, 1922; West, W. and G. S., 1898.

[2] Czurda, 1925; Hemleben, 1922; Lloyd, 1928; Saunders, 1931.

[3] Hodgetts, 1920*A*. [4] Lloyd, 1926*A*, 1926*B*, 1928. [5] Lewis, I. F., 1925.

[6] Chmielevsky, 1890; Tröndle, 1911. [7] Tröndle, 1911.

mature fruiting material is essential for critical determination of species. In his monograph describing all species of the genus, Transeau (in press) lists 143 species as occurring in the United States.

9. **Sirogonium** Kützing, 1843. Members of this genus have filaments and cells resembling those species of *Spirogyra* in which there are several chloroplasts, but chloroplasts of *Sirogonium* rarely make more than half a turn in extending from end to end of a cell. The cell wall differs from that of other Zygnemataceae in that lateral walls do not have an outer layer of pectose.

In lateral conjugation, there is a pronounced genuflexion so that fertile cells of one filament lie against fertile cells of the other.[1] A disk of gelatinous material is secreted where these cells abut on each other, but there is no formation of a definite conjugation tube prior to gametic union. Germi-

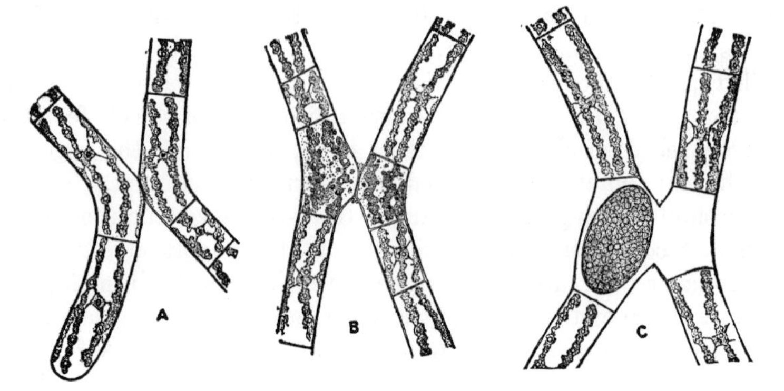

FIG. 213. *Sirogonium sticticum* (Engl. Bot.) Kütz. (× 185.)

nation of a zygote is unusual in that the first cross wall of a germling is formed at some distance from the point of emergence of the projecting tube.[2]

Sirogonium is a much rarer genus than *Spirogyra* but is widely distributed in this country. The species found in the United States are *S. floridanum* (Transeau) G. M. Smith, *S. illinoisensis* (Transeau) G. M. Smith, *S. megasporum* (Jao) Transeau, *S. pseudofloridianum* (Prescott) Transeau, *S. sticticum* (Engl. Bot.) Kütz. (Fig. 213), and *S. tenuis* (Nordst.) Transeau. For descriptions of them, see Transeau (in press).

FAMILY 2. MESOTAENIACEAE

Members of this family, often called "saccoderm desmids," differ in a number of respects from desmids of the placoderm type.[3] The cell walls

[1] DEBARY, 1858; TRANSEAU, 1914. [2] DEBARY, 1858.
[3] LÜTKEMÜLLER, 1902; WEST, W. and G. S., 1904.

of Mesotaeniaceae are without pores, and dividing cells do not have the "regeneration" of a new half-cell that is found in placoderm desmids. Conjugation is usually by means of a definite conjugation tube, whereas most placoderm desmids do not form a conjugation tube.

Cell walls of Mesotaeniaceae are unsegmented and without pores.[1] They also lack the impregnation with iron compounds so often found in Desmidiaceae.[2] *Spirotaenia*, *Mesotaenium*, and most other genera of the family have a wall composed of two concentric layers: an inner homogeneous layer of cellulose, and an external layer of gelatinous material (pectose). *Gonatozygon* and *Genicularia* have a wall composed of three layers: the innermost of pure cellulose, the median with little or no cellulose, and the outermost of pectose. The pectose gelatinous layer of the wall is of uniform thickness, except in the rare cases[3] where it functions as an organ of attachment. The gelatinous sheaths of the cells may be confluent with one another and so produce mucilaginous masses containing a large number of cells (*Cylindrocystis*, *Mesotaenium*).

The chloroplasts of Mesotaeniaceae are much simpler in form than those of Desmidiaceae,[4] and there are the same three types that are found in the Zygnemataceae. *Spirotaenia* and *Genicularia* have spiral chloroplasts of the *Spirogyra* type. Cells of *Gonatozygon* and *Mesotaenium* have a single, axial, laminate chloroplast similar to that of *Mougeotia*. The chloroplast of *Roya* may be interpreted as a modification of this type. Chloroplasts of *Cylindrocystis* are of a true *Zygnema* type and behave in an identical fashion when the cell divides;[5] chloroplasts of *Netrium* are *Zygnema*-like but have the stellate rays modified into longitudinal ribs.[6] Cells of the three foregoing types have their nuclei localized as in the corresponding types among the Zygnemataceae.

Cell division is usually followed by an immediate separation of the daughter cells. In *Gonatozygon* and *Genicularia*, the cells may remain united end to end, but such filaments readily dissociate into single cells when disturbed. The meager accounts of cell division among the Mesotaeniaceae[7] seem to show that the method of division is identical with that of Zygnemataceae, and that separation of daughter cells is due to the disintegration of a middle lamella formed shortly after cytokinesis. Increase in length seems to take place throughout the entire length of the daughter cells and not, as in Desmidiaceae, by the formation of a new "semicell."

Conjugation has been recorded for all genera of the family. It takes place more frequently in *Mesotaenium* and *Cylindrocystis* than in other

[1] LÜTKEMÜLLER, 1905. [2] HÖFLER, 1926. [3] SKUJA, 1928.
[4] CARTER, N., 1919*A*. [5] KAUFFMANN, 1914. [6] CARTER, N., 1919*A*.
[7] DEBARY, 1858; KAUFFMANN, 1914; WEST, G. S., 1915*A*.

genera belonging to this family or to the Desmidiaceae. The process is initiated by two cells becoming enveloped by a common gelatinous matrix. The pair of cells may lie parallel to, or at right angles with, each other. Conjugation is usually between fully mature cells, but in *Netrium*[1] it takes place between recently divided cells. Several species of *Spirotaenia*[2] have each of the approximated cells dividing and a production of two sets of gametes that unite to form twin zygotes. A similar formation of twin zygotes is also known for *Cylindrocystis*.[3] Early in conjugation each of the

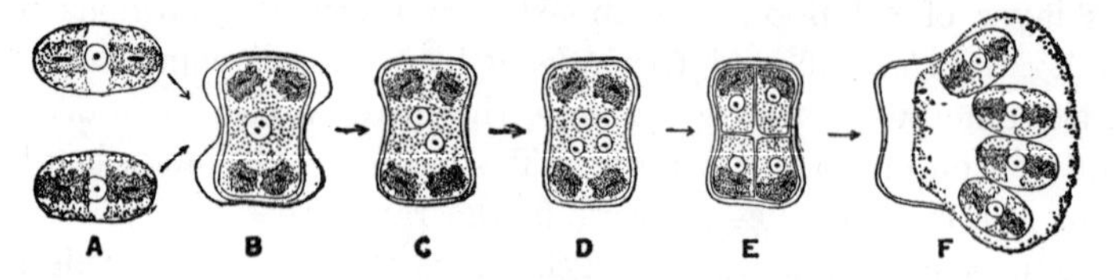

Fig. 214. Diagram of the formation and germination of zygotes of *Cylindrocystis Brebissonii* Menegh. (*Diagram based upon Kauffmann*, 1914.)

cells develops a papillate outgrowth similar to that formed in conjugating cells of Zygnemataceae. In the majority of cases[4] (*Gonatozygon*, *Genicularia*, *Mesotaenium*, *Cylindrocystis*), the papillae elongate until they touch each other, after which there is a formation of a true conjugation tube. *Roya*[5] develops no conjugation tube but has the papillate outgrowths gelatinizing to form clear-cut circular pores through which the gametes escape from the old parent-cell walls. *Spirotaenia*[6] does not form papillate outgrowths and has a complete gelatinization of the cell wall before gametic

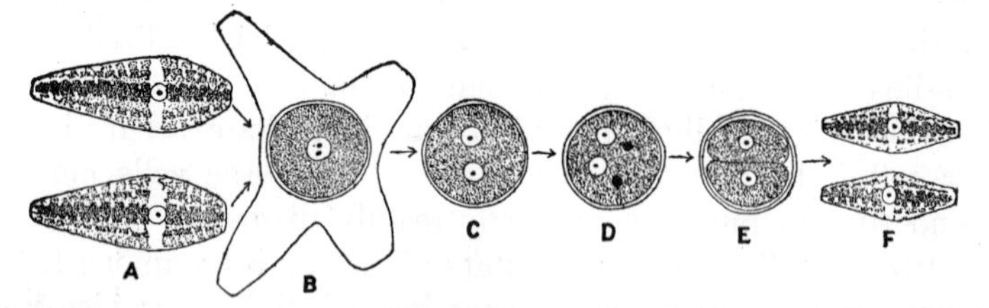

Fig. 215. Diagram of the formation and germination of zygotes of *Netrium digitus* (Ehr.) Itz. and Rothe. (*Diagram based upon Potthoff*, 1928.)

union. All genera producing a conjugation tube have the zygote formed in the tube. There may be a persistence of the tube during the ripening of the zygote, or the tube may gelatinize. Genera without a conjugation tube have naked gametes that migrate toward each other and fuse to form a zygote midway between the old parent-cell walls. Zygotes of Mesotaeniaceae have a thick wall, usually composed of three layers. However,

[1] Potthoff, 1928. [2] Archer, 1867; Potthoff, 1928.
[3] Archer, 1874; Lundell, 1871. [4] DeBary, 1858; Kauffmann, 1914.
[5] Hodgetts, 1920*C*. [6] Potthoff, 1928.

there are not the elaborate sculpturing and spinescence so often found in zygotes of Desmidiaceae.

Spirotaenia and *Cylindrocystis* have been shown[1] to have a reduction division of the fusion nucleus in the ripening zygote and a formation of four functional haploid nuclei (Fig. 214). *Netrium*[2] is the only member of the family thus far known to produce two macronuclei and two micronuclei in a germinating zygote (Fig. 215). The nuclear divisions are followed, respectively, by a division of the protoplast into four or two daughter protoplasts. This cell division may take place before rupture of the outer wall layers of a zygote, or it may take place after the protoplast has been extruded.

1. Chloroplasts parietal, spirally twisted .... 6. **Spirotaenia**
1. Chloroplasts axial, straight or spirally twisted 2
    2. Chloroplast spirally twisted .... 6. **Spirotaenia**
    2. Chloroplast not spirally twisted 3
3. Chloroplast laminate 4
3. Chloroplast not laminate 5
    4. Poles of cell flattened, length of cell many times the breadth .... 2. **Gonatozygon**
    4. Poles of cell rounded, length of cell not over five times the breadth .... 1. **Mesotaenium**
5. With one chloroplast extending entire length of cell .... 5. **Roya**
5. With two or four chloroplasts 6
    6. Chloroplasts with stellate rays .... 3. **Cylindrocystis**
    6. Chloroplasts with radiating longitudinal plates .... 4. **Netrium**

1. **Mesotaenium** Nägeli, 1849. The cells of *Mesotaenium* are cylindrical to subcylindrical and with broadly rounded poles. They may be solitary or embedded in large numbers within a common watery gelatinous matrix. The cellulose layer next to the protoplast is homogeneous and without evidence of pores.[3] A cell contains a single axial laminate chloroplast, with one to several pyrenoids, and flattened against the chloroplast is the nucleus. The cells frequently contain droplets of an oily nature. Several species are violet or purple because of phycoporphyrin dissolved in the cell sap.[4]

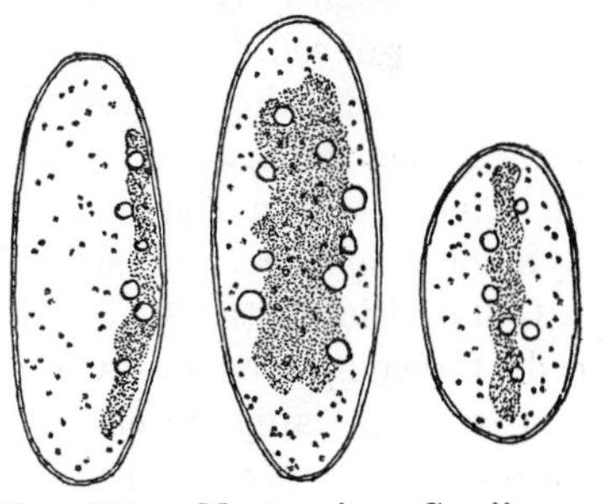

FIG. 216. *Mesotaenium Greyii* var. *breve* W. West. (× 485.)

Asexual reproduction is by cell division. There may also be a lid-like abscission of the wall at one pole of a cell and an escape of the protoplast, followed by its development into a spherical aplanospore of a reddish-brown color.[5]

[1] KAUFFMANN, 1914; POTTHOFF, 1928. [2] POTTHOFF, 1928.
[3] LÜTKEMÜLLER, 1902. [4] LAGERHEIM, 1895. [5] ARCHER, 1864.

In sexual reproduction there is an establishment of a very broad conjugation tube between two cells and a development of a thick-walled zygote that fills most of the space within the walls. A germinating zygote may produce two or four new cells.[1] *Mesotaenium* may also have the contents of a cell contracting slightly and developing into a parthenospore whose wall ornamentation is identical with that of a zygote.[2]

Mesotaenium is generally found in mucilaginous masses on damp rocks or on dripping cliffs, but sometimes the cells are solitary and free-floating. The species found in this country are *M. aplanosporum* Taft, *M. chlamydosporum* DBy., *M. clepsydra* (Wood) Wolle, *M. Endlicherianum* Näg., *M. Greyii* Turn. (Fig. 216), and *M. macrococcum* (Kütz.) Roy and Biss. For a description of *M. aplanosporum*, see Taft (1937); for *M. clepsydra*, see Wolle (1884*A*); for the others, see W. and G. S. West (1904).

2. **Gonatozygon** DeBary, 1856. The cells of this desmid are cylindrical and with a length many times the breadth. Their sides are parallel,

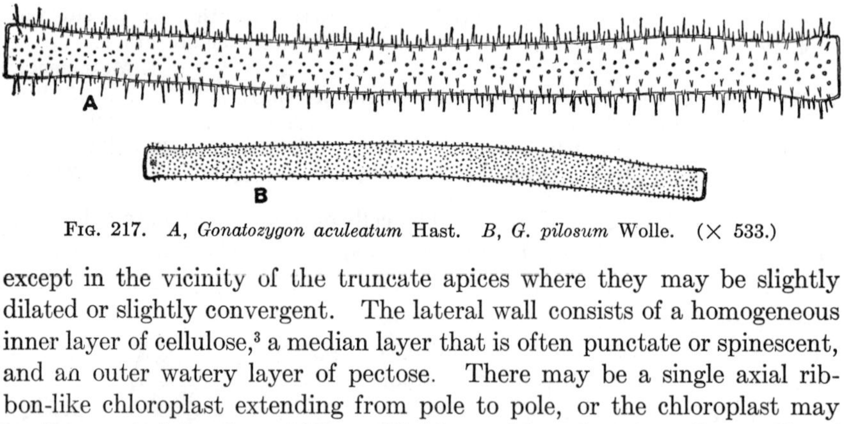

A

B

FIG. 217. *A, Gonatozygon aculeatum* Hast. *B, G. pilosum* Wolle. (× 533.)

except in the vicinity of the truncate apices where they may be slightly dilated or slightly convergent. The lateral wall consists of a homogeneous inner layer of cellulose,[3] a median layer that is often punctate or spinescent, and an outer watery layer of pectose. There may be a single axial ribbon-like chloroplast extending from pole to pole, or the chloroplast may be interrupted in the middle. Within a chloroplast is a linear file of pyrenoids spaced equidistant from one another.

Cell division is transverse and, during subsequent elongation of daughter cells, the spiny median layer of the parent-cell wall persists as a cap at one end of each daughter cell. The daughter cells often remain terminally united to one another in unbranched filaments, but such filaments readily dissociate into solitary cells when disturbed.

Sexual reproduction is by establishment of a conjugation tube, followed by formation of a spherical zygote in the tube.[4] During ripening of a zygote, there may be a gelatinization of the old conjugation tube.

[1] DEBARY, 1858; WEST, G. S., 1915*A*; TAFT, 1937. [2] TAFT, 1937.
[3] LÜTKEMÜLLER, 1902. [4] DEBARY, 1858.

Gonatozygon is generally free-floating, but sometimes it is sessile and attached to submerged water plants by a gelatinous disk at one end of a cell.[1] The species found in this country are *G. aculeatum* Hast. (Fig. 217*A*), *G. asperum* (Bréb.) Rab., *G. Kihnahani* (Arch.) Rab., *G. leiodermum* Turn., and *G. pilosum* Wolle (Fig. 217*B*). For descriptions of *G. aculeatum* and *G. pilosum*, see G. M. Smith (1924); for *G. leiodermum*, see Turner (1892); for the others, see W. and G. S. West (1904).

3. **Cylindrocystis** Meneghini, 1838. The cylindrical cells of this desmid usually have a length two to three times the breadth; broadly rounded poles; and may or may not have a slight constriction in the equatorial region. The cell wall consists of an inner homogeneous cellulose layer and an outer structureless gelatinous layer.[2] Each semicell contains a single stellate chloroplast with a spherical to rod-shaped pyrenoid. The nucleus is central in position and lies between the two chloroplasts. *C. Brebissonii* Menegh. has a yellow pigment dissolved in the cell sap.[3]

Division of the nucleus is followed by transverse division of both chloroplasts and a migration of a daughter nucleus to a point midway between each pair of newly formed chloroplasts.[4] The zone of cell division lies midway between the cell's poles.

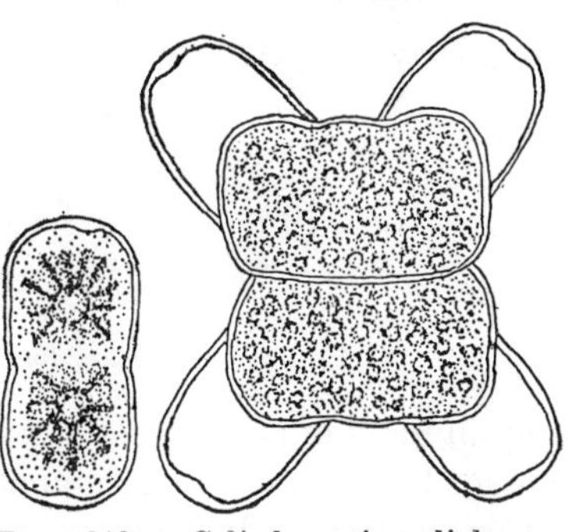

FIG. 218. *Cylindrocystis diplospora* Lund. (*After Lundell*, 1871.) (× 400.)

Sexual reproduction begins with an establishment of a broad conjugation tube between a pair of cells, and the spherical to subquadrangular zygote fills most of the cavity within the conjugating cells. Conjugation usually takes place between fully mature cells, but in certain cases[5] the cells divide just before conjugation. Since these latter cells are connected to each other, there is a formation of twin zygotes. Shortly before a zygote germinates (Fig. 214, page 304), there is a meiotic division of the zygote nucleus into four nuclei of equal size. The four chloroplasts derived from the two gametes are persistent, and meiosis is followed by a cytoplasmic quadripartition in such a manner that each of the four protoplasts contains one nucleus and one chloroplast. A wall is formed around each protoplast, the chloroplast within it divides, and the nucleus moves to a point midway between the two daughter chloroplasts. There then follows a rupture of the zygote wall and a liberation of the four cells within a common gelatinous matrix.[6]

[1] SKUJA, 1928. [2] LÜTKEMÜLLER, 1902. [3] LAGERHEIM, 1895.
[4] KAUFFMANN, 1914. [5] ARCHER, 1874; LUNDELL, 1871.
[6] DEBARY, 1858; KAUFFMANN, 1914.

Cylindrocystis is usually found growing on damp soil or on moist cliffs. The species found in this country are *C. americanum* W. and G. S. West, *C. angulatum* W. and G. S. West, *C. Brebissonii* Menegh., *C. crassa* DBy., *C. diplospora* Lund (Fig. 218), *C. minutissima* Turn., and *C. splendida* Taft. For a description of *C. splendida*, see Taft (1942); for descriptions of the others, see W. and G. S. West, (1904).

4. **Netrium** Nägeli, 1849; emend., Lütkemüller, 1902. *Netrium* has fairly large cells with a length at least three times the breadth. The cells may be fusiform, oblong-cylindrical, or cylindrical, and with rounded or truncate poles. There is no suggestion of a median constriction. The cell wall consists of a homogeneous inner layer of cellulose and an outer layer of pectose.[1] One species has four chloroplasts; all others have two. The chloroplasts are axial, and with 6 to 12 radiating longitudinal plates that often have notched margins.[2] Each chloroplast usually contains a single rod-like pyrenoid, but this may fragment into an axial row of spherical to irregularly shaped ones. Certain species have terminal vacuoles that

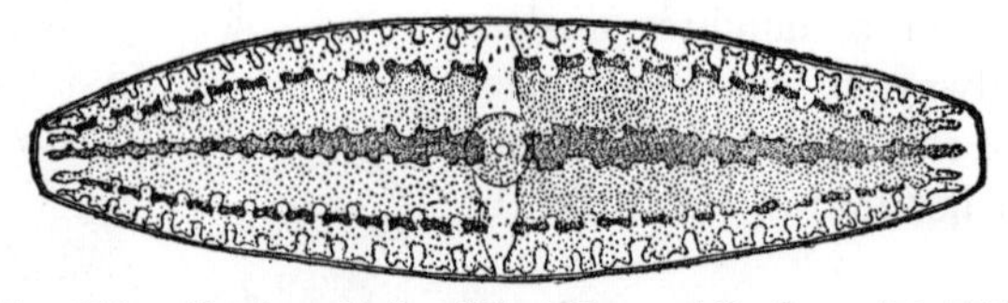

FIG. 219. *Netrium digitus* (Ehr.) Itz. and Rothe. (× 400.)

contain moving particles of gypsum,[3] and crystals of gypsum may also be present between the longitudinal plates of a chloroplast.

Conjugation is between recently divided cells that have not become wholly mature.[4] The conjugating cells form a broad conjugation tube and produce a spherical thick-walled zygote within the tube. During ripening of a zygote (Fig. 215, page 304), there is a formation of two macro- and two micronuclei, followed by a division of the protoplast into two parts. Germination is by rupture of the outer zygote wall layers and a liberation of the two daughter cells surrounded by the innermost zygote wall layer.

This genus is usually found sparingly intermingled with other desmids. The species known for this country are *N. digitus* (Ehr.) Itz. and Rothe (Fig. 219), *N. interruptum* (Bréb.) Lütkem., *N. Naegelii* (Bréb.) W. and G. S. West, and *N. oblongum* (DBy.) Lütkem. For descriptions of them, see W. and G. S. West (1904).

5. **Roya** W. and G. S. West, 1896; emend., Hodgetts, 1920. The cylindrical cells of *Roya* are slightly attenuated toward the poles and

[1] LÜTKEMÜLLER, 1902. [2] CARTER, N., 1919*A*. [3] FISCHER, A., 1884.
[4] POTTHOFF, 1928.

straight or slightly arcuate. Their walls consist of an inner homogeneous layer of cellulose and an outer sheath of pectose.[1] The chloroplast is axial and with or without longitudinal ridges; it extends the entire length of a cell but has a lateral indentation in the middle where the nucleus is lodged.[2] A chloroplast contains an axial series of four to six pyrenoids that lie equidistant from one another. Some species have the chloroplast completely interrupted in the middle. There may be a vacuole at either end of a cell, but these lack the gypsum crystals usually found in terminal vacuoles.[2]

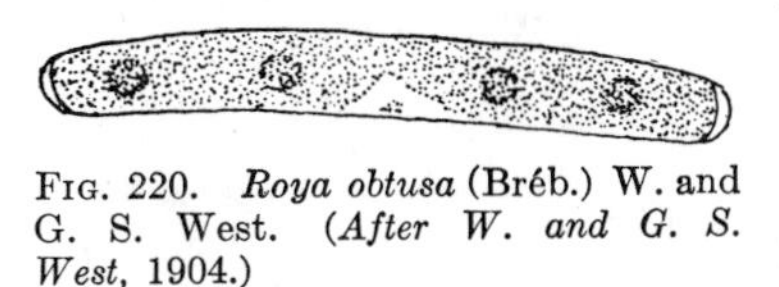

FIG. 220. *Roya obtusa* (Bréb.) W. and G. S. West. (*After W. and G. S. West*, 1904.)

During sexual reproduction, each of two conjugating cells puts out a protuberance which becomes a circular pore through which the cell contents escape. The amoeboid gametes meet midway between the empty cell walls and fuse to form a zygote.[2] Germination of zygotes has not been observed.

Roya is a genus of relatively rare occurrence and is usually found intermingled with other desmids. The two species found in this country are *R. anglica* G. S. West and *R. obtusa* (Bréb.) W. and G. S. West (Fig. 220). For a description of *R. anglica*, see Hodgetts (1920*C*); for *R. obtusa*, see W. and G. S. West (1904).

6. **Spirotaenia** de Brébisson 1848. *Spirotaenia* has straight or slightly curved cylindrical to fusiform cells with rounded poles. The cell wall has a homogeneous inner layer composed of cellulose and an outer gelatinous layer.[3] This genus is readily distinguishable from other desmids by its uninterrupted spirally twisted chloroplast running the length of a cell. In most species, the chloroplast is a parietal ribbon with several close spirals, but some species have an axial chloroplast with spirally twisted ridges. The nucleus is excentric; it lies internal to the chloroplast in cells with a parietal chloroplast, and external to the chloroplast in cells with an axial one.

Sexual reproduction is often preceded by a division of each of two cells lying within a common gelatinous matrix.[4] The four cells within the matrix have a complete gelatinization of their walls and a conjugation in pairs to form twin zygotes. During ripening of a zygote, there is a meiotic division of the zygote nucleus and a formation of four haploid nuclei.[5] When the zygote germinates, there is a division of the protoplast into four uninucleate germlings.

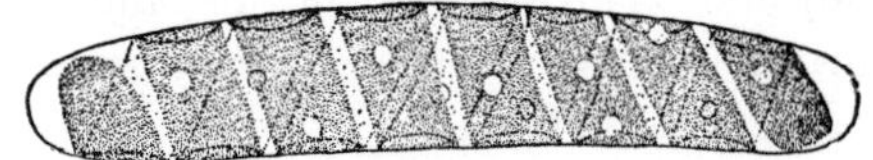

FIG. 221. *Spirotaenia condensata* Bréb. (× 400.)

[1] LÜTKEMÜLLER, 1902. [2] HODGETTS, 1920*C*. [3] LÜTKEMÜLLER, 1902.
[4] ARCHER, 1867; POTTHOFF, 1928. [5] POTTHOFF, 1928.

Spirotaenia occurs sparingly intermingled with other desmids. It is often overlooked in preserved material because the most distinctive feature, the spirally twisted chloroplast, is not conspicuous after the chlorophyll has been extracted. The species found in this country are *S. bispiralis* W. West, *S. condensata* Bréb. (Fig. 221), *S. endospira* (Kütz.) Arch., *S. minuta* Thur., *S. obscura* Ralfs, and *S. parvula* Arch. For descriptions of them, see W. and G. S. West (1904).

FAMILY 3. DESMIDIACEAE

Desmids belonging to this family have cell walls in which there are vertical pores through the innermost and median wall layers. Cell division is also distinctive. It always takes place at a definite point, each daughter cell receiving an unaltered half of the parent-cell wall and "regenerating" a new half-wall with the same characteristic shape and ornamentation. The most conspicuous diagnostic character of Desmidiaceae, and one found in all genera but *Closterium*, *Spinoclosterium*, and *Penium*, is the median constriction (*sinus*) dividing the cell into two distinct halves (*semicells*) that are joined together by a connecting zone (*isthmus*). Three genera have been shown to have a formation of two functional and two degenerating nuclei in a germinating zygote, but these data are too meager to justify the assumption that this is characteristic of the family as a whole.

Most of the Desmidiaceae are unicellular, but certain genera have the cells united in unbranched filaments of indefinite length, and two genera have the cells in amorphous colonies. Desmids are found sparingly intermingled with free-floating algae everywhere, but collections rich in species and in number of individuals are usually made only where the waters have a pH of five to six.

The beautiful symmetry and complexity of shape found in these algae have long attracted the attention of microscopists, and desmids have been favorite objects of study for those delighting in beautiful microscopic objects. The members of this family are so various in shape, and so diverse in ornamentation, that it is impossible to attempt to describe them in a paragraph or two.

All Desmidiaceae have a wall of the placoderm type:[1] one in which there are two concentric layers internal to the gelatinous sheath and which differ from each other in chemical composition. The innermost layer is structureless, except for the pores, and composed entirely of cellulose. The layer external to this has a substratum of cellulose that is impregnated with pectic compounds. The iron salts, so often found in walls of Desmidiaceae, are localized chiefly in the median wall layer. The iron may be uniformly distributed throughout the entire wall, or it may be localized in transverse bands, especially in the region of the isthmus. It has been

[1] LÜTKEMÜLLER, 1902.

held[1] that the iron salts accumulate as a result of chemical attraction rather than from a physical cause, such as adsorption. Both of the wall layers are perforated by pores, usually arranged in a definite pattern. Pores may occur on all parts of the wall but the isthmus, or they may be localized in definite parts of each semicell. The pores are filled with a pectic material that is often of tougher consistency than, and extends into, the more watery sheath of pectic material at the exterior of the wall. Sometimes the gelatinous material extending through the pores is evident in living cells, but more often it is evident only when the walls have been stained with special reagents.[2] The locomotion known to occur in many desmids, best shown by their migration toward the illuminated side of an aquarium, is intimately connected with a localized secretion of gelatinous material through pores at one end of the cell.[3] Direct observation of locomotion has shown that the movement is by a series of jerks.

In the vast majority of Desmidiaceae there is at least one chloroplast in each semicell.[4] A few species with very small cells have one chloroplast extending the entire length of the cell. Semicells containing one chloroplast have it axial in position. Those with two chloroplasts usually have them axial and lateral to each other. Species with four or more chloroplasts in each semicell always have them parietal in position. There is great diversity in the profile of the chloroplast from species to species, and in many cases the outline is further complicated by parietal outgrowths. In some species there is little variation from individual to individual; in other species there is so marked a tendency to vary that the majority of individuals do not conform to any given type. Because of the variation found in certain species, it is impossible to segregate genera or subgenera on the basis of chloroplast structure alone. There is a similar variation in number and position of the pyrenoids. Semicells with a small chloroplast regularly have but one or two pyrenoids. Those with large and massive chloroplasts usually have numerous pyrenoids, indiscriminately scattered throughout the entire chloroplast. In *Closterium*, and in several species of *Penium* and *Pleurotaenium*, there is a conspicuous vacuole at each end of the cell. The vacuoles contain one or more moving granules. These particles have been shown to be crystals of gypsum,[5] and it has been held[6] that they function as statoliths.

The nucleus always lies at the isthmus and often in contact with the chloroplasts. It has been suggested that there is a more or less intimate connection between nucleus and chloroplast,[7] because of the finding of chloroplast strands running to the nucleus and the persistence of this connection even when the nuclei have been laterally displaced by centri-

[1] LÜTKEMÜLLER, 1902. [2] HÖFLER, 1926.
[3] KLEBS, 1885; KOL, 1927; SCHRÖDER, 1902.
[4] CARTER, N., 1919*A*, 1919*B*, 1920, 1920*A*.
[5] FISCHER, A., 1884. [6] STEINECKE, 1926. [7] CARTER, N.. 1919*A*.

fuging. In the few cases where the nuclear structure has been studied (*Closterium*,[1] *Hyalotheca*,[2] *Cosmarium*[3]), the nucleus has a conspicuous nucleole and a more or less well-defined chromatin-linin network. In the only desmid studied in detail (*Closterium*), the nucleolus does not function as an endosome during mitosis.

There has been no cytological study of division of desmids with a conspicuous isthmus. It is well known that desmids of this type have a division of the nucleus, followed by an elongation of the isthmus (Fig. 222). There is then a transverse division at the elongated isthmus, after which the portion of the isthmus attached to each semicell develops into a new semicell. Division of the chloroplast or chloroplasts in each of the original semicells takes place during development of the new semicells. New pyrenoids may be formed by a division of those in the chloroplast (*Hyalotheca*),[4] or new pyrenoids may be formed *de novo*.[5] Occasionally

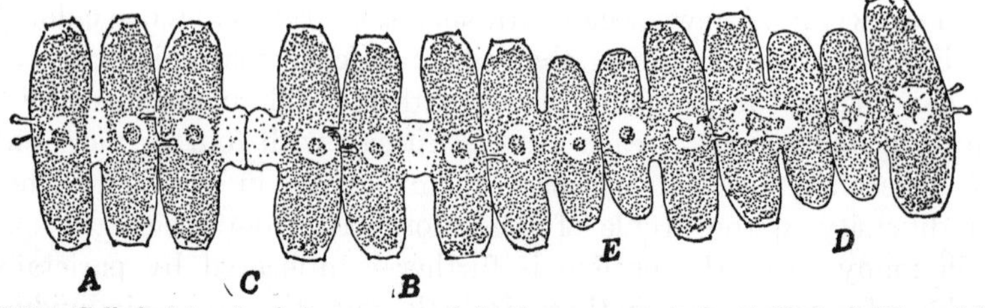

FIG. 222. Cell division in *Sphaerozosma Aubertianum* var. *Archerii* (Gutw.) W. and G. S. West. *A*, undivided cell. *B–E*, successive stages in cell division. (× 730.)

there is a complete or incomplete failure of cell division as the isthmus starts to elongate. The continued enlargement of the isthmus results in monstrous cells in which an abnormally shaped structure is intercalated between the two semicells.

In *Closterium*, a genus in which most species are unconstricted, each chloroplast begins to divide transversely before the nucleus starts to divide.[6] However, mitosis is completed before the chloroplasts are fully divided and a daughter nucleus migrates to the region where each chloroplast is dividing. Cell division is in a plane midway between the two original chloroplasts and takes place in essentially the same manner as in *Spirogyra*. The newly formed wall in this portion of the daughter cell is at first a flat plate, but it soon becomes a conical protuberance that continues to enlarge until it is identical in size and shape with the original semicell. Certain species of *Closterium* and *Penium* have the cell wall of the new semicell developing in two distinct portions (Fig. 223); a first-formed

[1] KLEBAHN, 1891; LUTMAN, 1911. [2] ACTON, 1916*B*; POTTHOFF, 1927.
[3] KLEBAHN, 1891. [4] ACTON, 1916*B*. [5] CARTER, N., 1919*A*. [6] LUTMAN, 1911.

portion covering the new semicell's apex and a later formed girdle portion adjacent to the isthmus.[1] The line of juncture between the two portions is clearly evident as a transverse striation of the cell wall. Sometimes the terminal and girdle portions of the semicell wall are different in color because of a differential impregnation with iron.

Aplanospores have been recorded for a few desmids, but most of these so-called aplanospores are parthenospores.[2] Possibly the spherical spore-like bodies formed in cells of *Hyalotheca*[3] are true aplanospores, since their shape is not the same as that of its zygotes.

Zygotes have been described for many of the Desmidiaceae, but they are of infrequent occurrence and one is rarely fortunate enough to collect material rich in fruiting specimens. Conjugation usually takes place between fully mature cells, but sometimes it is between recently divided ones. In some cases (*Micrasterias denticulata* Bréb.,[4] *M. angulosa* Hant-

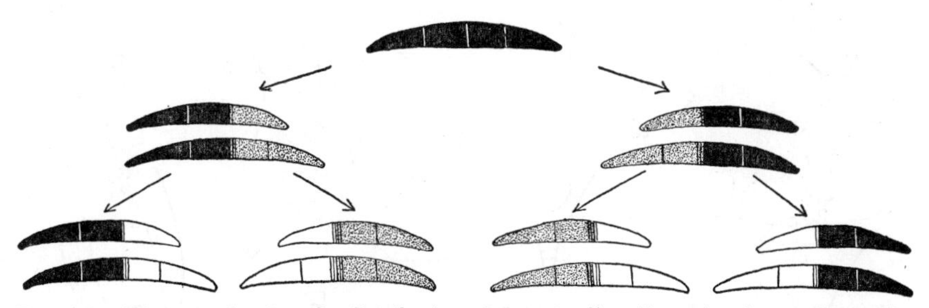

FIG. 223. Diagram showing the distribution of parent-cell walls to daughter cells in *Closterium*. Walls of the first cell generation are in black, those of the second cell generation are in stipple, and those of the third cell generation are unshaded. (*Diagram based upon Lütkemüller*, 1902.)

zsch[5] and *Closterium Ehrenbergii* Menegh.),[6] the juvenile conjugating cells are sister cells. These form a single zygote; other species with a conjugation of juvenile cells, as *Penium didymocarpum* Lund., have a pair of cells becoming apposed to another pair and a production of twin zygotes within the common gelatinous envelope. It is not clear whether the twin zygotes characteristic of *Closterium lineatum* Ehr. and *C. Ralfsii* var. *hybridum* Rab. are formed in this manner or by a division of the protoplast to form two gametes.

In the conjugation of solitary free-floating species, two cells become invested with a common gelatinous envelope. This is usually followed by a breaking of each cell at the isthmus and an amoeboid migration of the protoplasts to each other. Sometimes, as in certain *Closteria*,[7] the ap-

[1] LÜTKEMÜLLER, 1902. [2] WEST, G. S., 1916; WEST, W. and G. S., 18[illegible]
[3] WEST, W. and G. S., 1898, 1907. [4] WEST, W. and G. S., 1896.
[5] ROY and BISSETT, 1893 to 1894. [6] LUTMAN, 1911. [7] SCHERFFEL, 1928.

proximated cells put out papillate projections toward each other and develop a more or less complete conjugation tube. Filamentous genera may dissociate into single cells prior to conjugation (*Sphaerozosma*, *Onychonema*) or may have two filaments approximating throughout their entire length (*Desmidium*, *Hyalotheca*). When conjugation is between cells united in filaments, there is usually a formation of definite conjugation tubes. Species with a conjugation tube generally have the zygote formed in the tube, but cases are known,[1] regularly in *Desmidium Grevillii* (Kütz.) DBy and occasionally in *Hyalotheca dissiliens* (Smith) Bréb., where there is a formation of the zygote within the old female cell. Certain filamentous species have been described as conjugating laterally [*Spodylosium planum* (Wolle) W. and G. S. West,[2] *Sphaerozosma excavata* Ralfs.,[3] *Hyalotheca dissiliens* var. *tatrica* Racib.],[4] but none of them has a formation of a

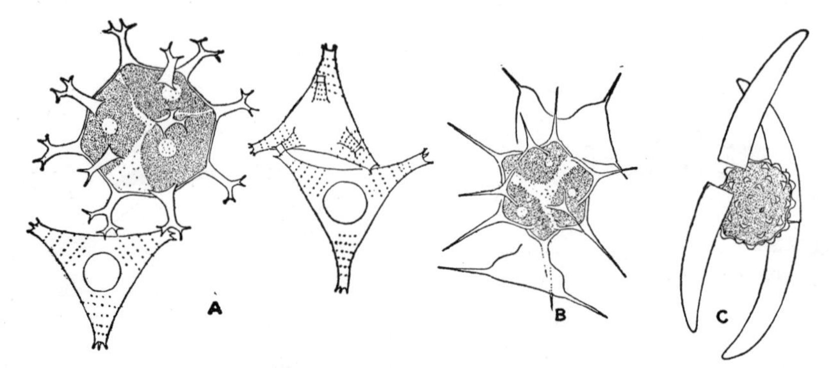

FIG. 224. Zygotes of Desmidiaceae. *A*, *Staurastrum furcigerum* Bréb. *B*, *Arthrodesmus incus* var. *extensus* Anderss. *C*, *Closterium calosporum* Wittr. (*A*, *C*, × 300; *B*, × 600.)

conjugation tube as in laterally conjugating Zygnemataceae. Zygotes of Desmidiaceae have a wall composed of three layers and in which the outermost layer often develops spines or verrucae distinctive of the species (Fig. 224).

Union of gametes is followed by a fusion of the two nuclei. Species with a single chloroplast contributed by each gamete have one chloroplast disintegrating as the zygote matures;[5] those with two chloroplasts from each gamete have two chloroplasts disintegrating.[6] *Closterium*, *Cosmarium*, and *Hyalotheca* have been shown[7] to have two successive divisions of the fusion nucleus, followed by a partial disintegration of two nuclei formed by the second division (Fig. 225). The division of the fusion nucleus has been shown to be meiotic in *Hyalotheca*, and the same is

[1] CARTER, N., 1923. [2] WEST, W. and G. S., 1898. [3] RALFS, 1848.
[4] GRÖNBLAD, 1924. [5] POTTHOFF, 1927. [6] KLEBAHN, 1891.
[7] KLEBAHN, 1891; POTTHOFF, 1927.

probably true of the two other genera just mentioned. In all three of these genera, the protoplast of the germinating zygote divides into two daughter protoplasts, each of which contains a single chloroplast, a functional nucleus, and a degenerating nucleus. *Staurastrum*[1] has been described as having the formation of four nuclei within the ripening zygote and a division of the protoplast into one, two, three, or four cells upon germination.

Parthenospores have been found in a few Desmidiaceae. Those observed in unicellular species, as *Cosmarium bioculatum* Bréb.,[2] are obviously the result of a disturbance of conjugation. Parthenospores of filamentous species, as *Spondylosium nitens* (Wallich) Arch., may be formed in filaments that show no sign of conjugation.[3] Cytological observations on ripening parthenospores of *Cosmarium*[4] show a division of the nucleus into four daughter nuclei. One of these is functional, the three others are degenerate.

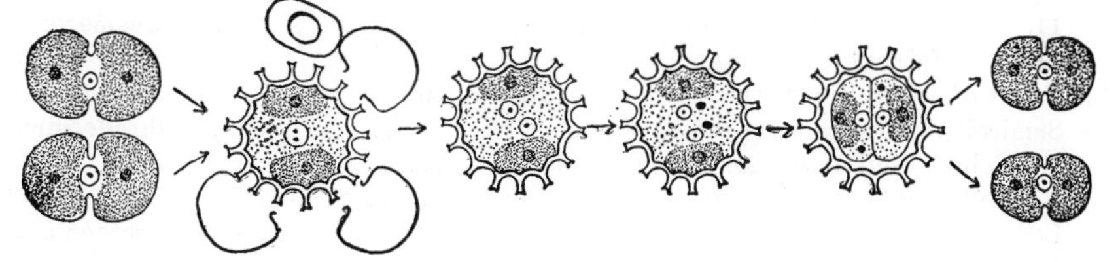

FIG. 225. Diagram of the formation and germination of zygotes of *Cosmarium*. (*Diagram based upon Klebahn*, 1891.)

When these parthenospores germinate, there is a division of the protoplast to form two daughter cells.

Generic and specific distinctions within the Desmidiaceae are based entirely upon the shape and structure of vegetative cells. Most genera may be distinguished from one another without recourse to front, vertical, and side views. Specific distinctions within certain genera, especially *Cosmarium* and *Staurastrum*, require front and vertical views before the species can be determined with accuracy. Ralfs' "British Desmidieae" (1848) is the official starting point for the nomenclature of the family.

The genera found in the United States differ as follows:

1. Cells solitary, occasionally end to end in pairs 2
1. Cells in many-celled colonies 16
 2. Cells without a median constriction 3
 2. Cells with a median constriction 5
3. Long axis straight .... **3. Penium**
3. Long axis curved 4
 4. Poles of cell with a solid spine .... **2. Spinoclosterium**
 4. Poles of cell lacking a solid spine .... **1. Closterium**

[1] TURNER, C., 1922. [2] NIEUWLAND, 1909. [3] WALLICH, 1860.
[4] KLEBAHN, 1891.

5. Length several times the breadth 6
5. Length not over three to four times the breadth 9
  6. Apices of semicells incised 7
  6. Apices of semicells not incised 8
7. Lateral walls with transverse rings of spines or verrucae .... 6. **Triploceras**
7. Lateral walls without transverse rings of spines or verrucae .... 7. **Tetmemorus**
  8. Bases of semicells longitudinally plicate .... 5. **Docidium**
  8. Bases of semicells not longitudinally plicate .... 4. **Pleurotaenium**
9. Cells not compressed 10
9. Cells compressed 11
  10. End view circular .... 3. **Penium**
  10. End view 3 to 12 radiate .... 14. **Staurastrum**
11. Apices of semicells incised 12
11. Apices of semicells not incised 13
  12. Apical and lateral incisions shallow .... 8. **Euastrum**
  12. Apical and lateral incisions deep .... 12. **Micrasterias**
13. Apices of semicells with two divergent processes .... 14. **Staurastrum**
13. Apices of semicells without processes 14
  14. Without conspicuous spines .... 9. **Cosmarium**
  14. With fairly long spines 15
15. Front of semicell wall thickened in median portion .... 13. **Xanthidium**
15. Semicell wall of uniform thickness .... 15. **Arthrodesmus**
  16. Cells united in unbranched filaments 17
  16. Cells not united in filaments 27
17. Length of cells several times breadth .... 4. **Pleurotaenium**
17. Length of cells not over three to four times breadth 18
  18. Apices of semicells incised .... 12. **Micrasterias**
  18. Apices of semicells not incised 19
19. Cells connected by means of apical processes 20
19. Cells not connected by means of apical processes 21
  20. Processes long, overlapping adjoining cells .... 16. **Onychonema**
  20. Processes short, often tuberculate .... 17. **Sphaerozosma**
21. End view circular or elliptical 22
21. End view triangular or quadrangular 25
  22. End view elliptical 23
  22. End view circular 24
23. Cells strongly compressed .... 18. **Spondylosium**
23. Cells not strongly compressed .... 21. **Desmidium**
  24. Lateral walls longitudinally striate .... 22. **Gymnozyga**
  24. Lateral walls not longitudinally striate .... 19. **Hyalotheca**
25. Apices of young semicells infolded .... 21. **Desmidium**
25. Apices of young semicells not infolded 26
  26. End view triangular .... 18. **Spondylosium**
  26. End view quadrangular .... 20. **Phymatodocis**
27. Gelatinous envelope dichotomously branched .... 11. **Oöcardium**
27. Gelatinous envelope not dichotomously branched .... 10. **Cosmocladium**

1. **Closterium** Nitzsch, 1817. The cells of this desmid are elongate, without a median constriction, and always, usually markedly, attenuated at the poles. Most species are distinctly lunate or arcuate. The cell

wall has delicate pores, demonstrable by special methods,[1] and many species have walls that are longitudinally striate. Several species have semicells with the wall region next to the isthmus, the *girdle piece*, interpolated subsequent to cell division (see page 313). The limits of the girdle piece are marked by transverse striae in the wall. Walls of *Closteria* are often brownish or yellowish brown, because of an impregnation with iron compounds.[2] There are two chloroplasts, one in each semicell. Chloroplasts may be entire or with longitudinal ridges radiating from a comparatively slender axis.[3] The pyrenoids are usually few in number and arranged in an axial series, but sometimes they are numerous and irregularly distributed. At each pole of the cell, there is a hyaline cytoplasmic region containing a conspicuous vacuole in which there are one or more particles of gypsum.[4] The nucleus lies in a bridge of cytoplasm connecting the two chloroplasts. It has a conspicuous nucleolus and a well-defined chromatin-linin network.[5]

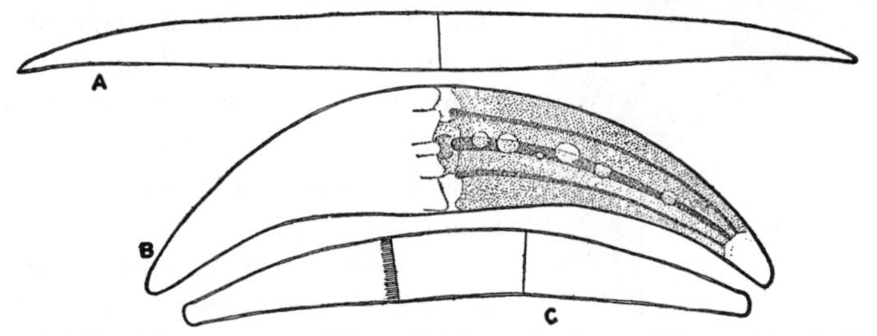

FIG. 226. *A*, *Closterium acerosum* (Schrank) Ehr. *B*, *C. moniliforme* (Bory) Ehr. *C*, *C. subtruncatum* W. and G. S. West. (× 233.)

Multiplication is by transverse division, division of nucleus and the chloroplasts preceding cell division.

Conjugation may take place between mature cells or between recently divided ones. In certain cases, as *C. Ehrenbergii* Menegh., the conjugating pair are probably sister cells. Three species regularly form twin zygotes. Some species develop a rudimentary conjugation tube (*C. parvulum* Näg.),[6] but most of them have a breaking of the cell wall at the isthmus and a migration of naked amoeboid gametes toward each other. A majority of species have spherical or ovoid zygotes with smooth walls; a few species have quadrangular zygotes or zygotes with ornamented walls (Fig. 224*C*). During ripening of a zygote, there is a division of the fusion nucleus into four daughter nuclei,[7] followed by a partial disintegration of two of them. Presumably these divisions are meiotic, but this has not

[1] LÜTKEMÜLLER, 1902. [2] HÖFLER, 1926. [3] CARTER, N., 1919*A*.
[4] FISCHER, A., 1884. [5] LUTMAN, 1910, 1911.
[6] DEBARY, 1858; SCHERFFEL, 1928. [7] KLEBAHN, 1891.

been demonstrated for *Closterium*. When a zygote germinates, the contents divide to form two protoplasts, each containing a single chloroplast, a functional and a degenerating nucleus. Shortly after this, the two protoplasts begin to assume the shape characteristic of the genus, and sooner or later each secretes a wall of its own.

Closterium (Fig. 226) is one of the most frequently encountered desmids, and several species are to be found in hard waters. Some 70 species occur in the United States, and the differences between them are based upon cell shape, amount of curvature of a cell, ornamentation of cell wall, and structure of the chloroplast. For monographic treatments of the genus, see Kriger (1935), and W. and G. S. West (1904).

2. **Spinoclosterium** Bernard, 1909 (*Closterioides* Prescott 1937). The elongate lunate cells of this genus resemble those of certain species of *Closterium*, but differ in that there is a stout solid spine at each broadly rounded pole of a cell. Within each semicell is a longitudinally ribbed chloroplast with many regularly or irregularly arranged pyrenoids. There is not a terminal vacuole at each pole of the cytoplasm as there is in *Closterium*.[1]

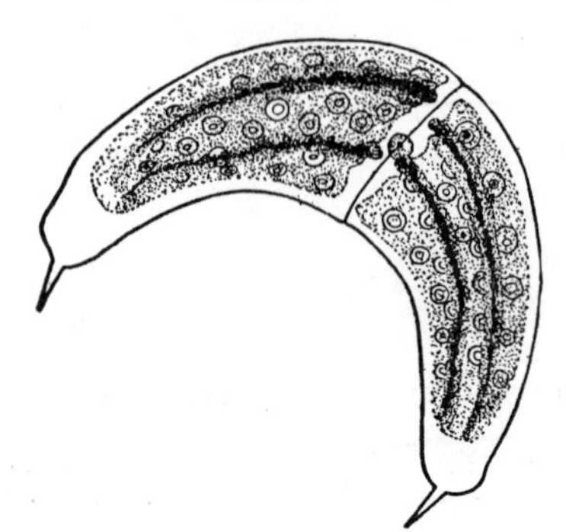

FIG. 227. *Spinoclosterium curvatum* var. *spinosum* Prescott. (*From Prescott*, 1937.)

Reproduction is by transverse division and with a transverse break in the equatorial region of the cell wall. Zygotes are unknown for this genus.[1]

S. curvatum var. *spinosum* Prescott (Fig. 227) has been found in Michigan,[1] Wisconsin,[1] and Maine.[2] For a description of it as *Closterioides spinosus* Prescott, see Prescott (1937).

3. **Penium** de Brébisson, 1844. The cells of *Penium* (Fig. 228) are cylindrical, and with parallel sides and rounded poles, or attenuated at the ends and truncate. A few members of the genus have cells that are

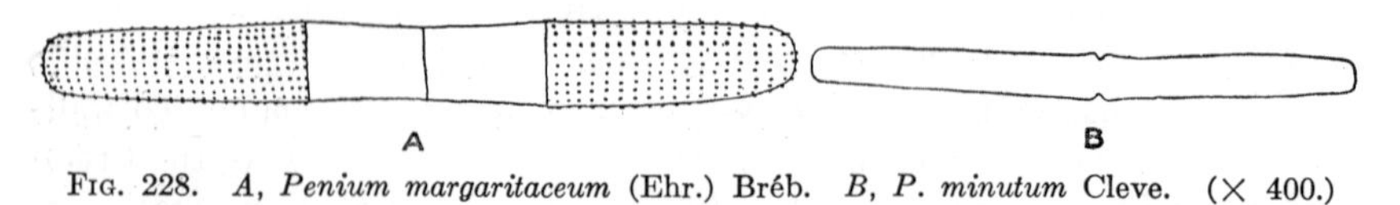

FIG. 228. *A*, *Penium margaritaceum* (Ehr.) Bréb. *B*, *P. minutum* Cleve. (× 400.)

slightly constricted in the middle, but most of them have unconstricted cells. A majority of the species have a length several times the breadth. *Penium* and *Closterium* are the only Desmidiaceae with a girdle piece in the cell

[1] PRESCOTT, 1937. [2] WHELDEN, 1943.

wall but, unlike *Closterium* (Fig. 223, page 313), *Penium* may have more than one girdle piece in each semicell. The cell wall is often longitudinally striate or punctate; it is also frequently impregnated with iron compounds and, therefore, of a brownish color. There are two chloroplasts, one in each semicell, which usually have a slender central axis and several radially disposed longitudinal plates.[1] Most species have a single spherical to rod-shaped pyrenoid within a chloroplast, but some with an elongate pyrenoid may have it fragmented in an axial series of many small, spherical to irregularly shaped, globules. A few species have a definite axial row of pyrenoids.

Zygotes are spherical or quadrangular and usually with a smooth wall. *P. didymocarpum* Lund. regularly forms twin zygotes.

Twenty species of *Penium* (Fig. 228) have been recorded for the United States. For descriptions of most species of the genus, see Krieger (1935) and W. and G. S. West (1904).

4. **Pleurotaenium** Nägeli, 1849. Cells of this alga are usually quite large, straight, and with a length several times the breadth. They have a well-defined median constriction but one that is never deep. The cylindrical semicells are never compressed and have sides that are parallel throughout or somewhat attenuated at the poles. The poles are always truncate and almost always with a whorl of mammillate to conical tubercules. The bases of the semicells are always inflated next the isthmus. Sometimes the side walls above the basal inflations are undulate. The cell wall is without transverse segmentation and is usually finely to coarsely punctate. In some species, as *P. trochiscum* W. and G. S. West, it is not of uniform thickness but has thin areas arranged in a more or less definite pattern. Such iron as accumulates in the cell wall is largely restricted to the isthmus.[2] Most species have numerous parietal, straight to undulate, band-shaped chloroplasts extending the length of the semicell. There are several pyrenoids within the chloroplast. In a few cases,[3] the chloroplast is axial, with longitudinal ribs and with an axial row of pyrenoids. Terminal vacuoles, similar to those of *Closterium*, are often present in this genus.

All species in which conjugation has been observed have a formation of spherical zygotes.

Most species of *Pleurotaenium* (Fig. 229) have solitary free-floating cells. *P. subcoronulatum* var. *detum* W. and G. S. West has the cells permanently united in filaments by an interlocking of the apical tubercules.[4] The 28 species found in this

[1] Carter, N., 1919*A*. [2] Höfler, 1926. [3] Carter, N., 1919*A*.
[4] Smith, G. M., 1924*A*.

country occur sparingly intermingled with other desmids. For descriptions of most of the species of the genus, see Krieger (1937) and W. and G. S. West (1904).

5. **Docidium** de Brébisson, 1844; emend., Lundell, 1871. The general shape of cells of *Docidium* is similar to that of *Pleurotaenium*, but they are of somewhat smaller size. *Docidium* differs from *Pleurotaenium* in having

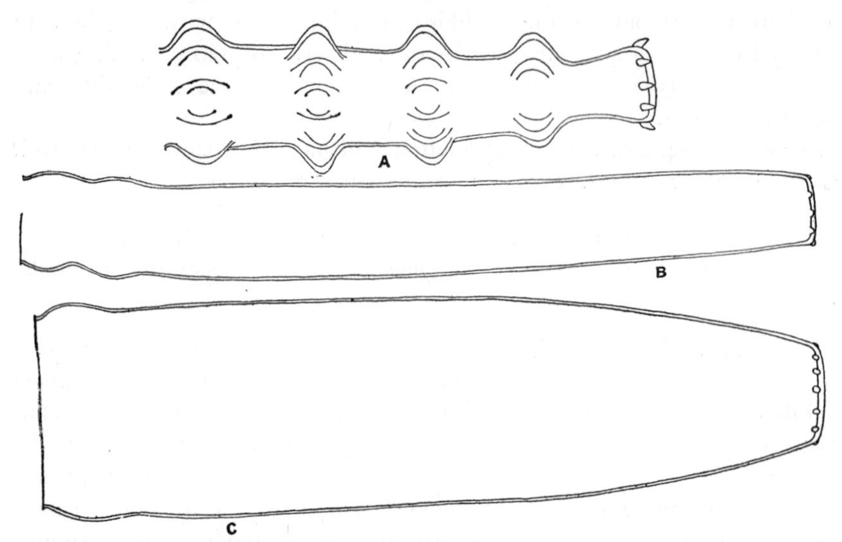

FIG. 229. *A*, *Pleurotaenium nodosum* (Bailey) Lund. *B*, *P. Ehrenbergii* (Bréb.) DBy. *C*, *P. truncatum* (Bréb.) Näg. (× 400.)

the cell wall vertically plicate at the basal inflation of each semicell, and each plication is usually subtended by a small granule. Apices of semicells also differ from those of *Pleurotaenium*, since they are without a whorl

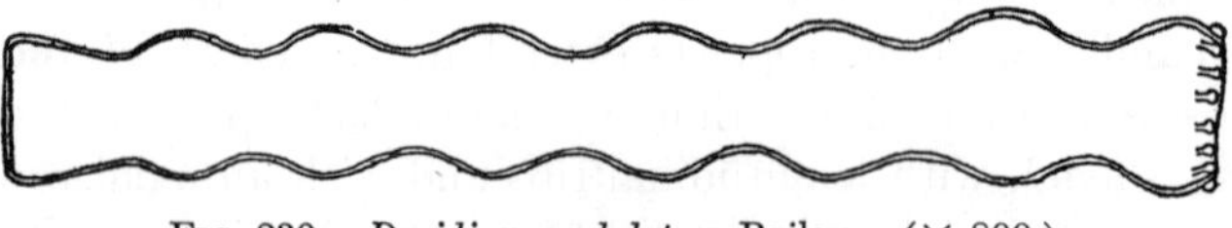

FIG. 230. *Docidium undulatum* Bailey. (× 800.)

of tubercules. Semicells of *Docidium* contain a single axial chloroplast with irregularly disposed longitudinal ridges and an axial series of six to eight pyrenoids.

Docidium and *Pleurotaenium* were not sharply delimited from each other until it was shown[1] that the two differ in ornamentation of semicell bases and in apical tuberculation. *D. baculum* Bréb. and *D. undulatum* Bailey (Fig. 230) are the only

[1] LUNDELL, 1871.

species found in this country. For descriptions of them, see W. and G. S. West, (1904).

6. **Triploceras** Bailey, 1851. *Triploceras* has the same type of elongate cell as the two preceding genera, but differs from them in a number of respects. The sides of the semicells are undulate and with transverse whorls of mamillate protuberances, in which each protuberance terminates in a single spine or in a broad emarginate verruca. The apices of the semicells are flattened and with two long upturned, obliquely disposed processes whose truncate ends bear two or three short sharp spines. Between the bases of the polar processes, there is a small mamillate protuberance bearing one or two erect spines.[1] Each semicell has an axial chloroplast, with longitudinal ribs from base to apex, that contains an axial row of pyrenoids.

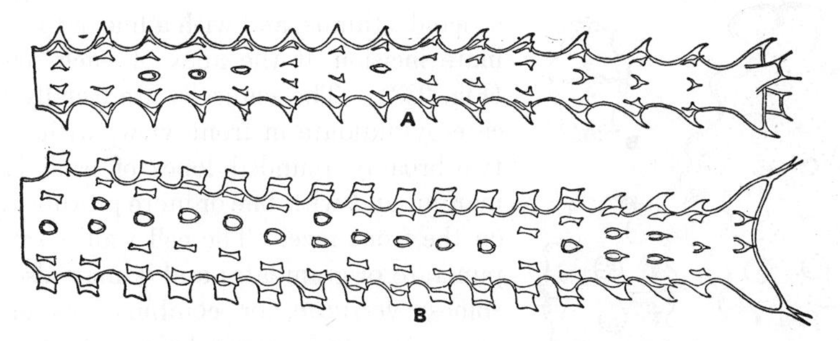

Fig. 231. *A, Triploceras gracile* Bailey. *B, T. verticillatum* Bailey. (× 400.)

Triploceras is the most striking of solitary desmids with elongate cylindrical cells. Both of the two species found in this country, *T. gracile* Bailey (Fig. 231*A*) and *T. verticillatum* Bailey (Fig. 231*B*), are widely distributed and at times occur in abundance in collections. For descriptions of them, see G. M. Smith, (1924*A*).

7. **Tetmemorus** Ralfs, 1844. Cells of this alga are relatively large, cylindrical to subfusiform, and with a length two to eight times the breadth. There is a well-defined, shallow to deep, isthmus in which the sinus is always open. The apical portion of a semicell is somewhat compressed and with a conspicuous vertical incision. The cell wall is colorless, and smooth, punctate, or minutely scrobiculate. Each semicell contains a single axial chloroplast with 8 to 10 longitudinal radiating plates that are simple or bifurcate where they abut on the cell wall.[2] The pyrenoids are fairly numerous, spherical to bacillar, and in a linear series.

Zygotes are spherical to ovoid and with or without a compressed quadrangular sheath.[3]

[1] Smith, G. M., 1924*A*; West, G. S., 1909. [2] Carter, N., 1919*A*.
[3] Ralfs, 1848.

This is another very distinctive genus. The four species found in the United States are *T. Brebissonii* (Menegh.) Ralfs (Fig. 232), *T. granulatus* (Bréb.) Ralfs, *T. laevis* (Kütz.) Ralfs, and *T. minutus* DBy. For descriptions of them, see W. and G. S. West (1904).

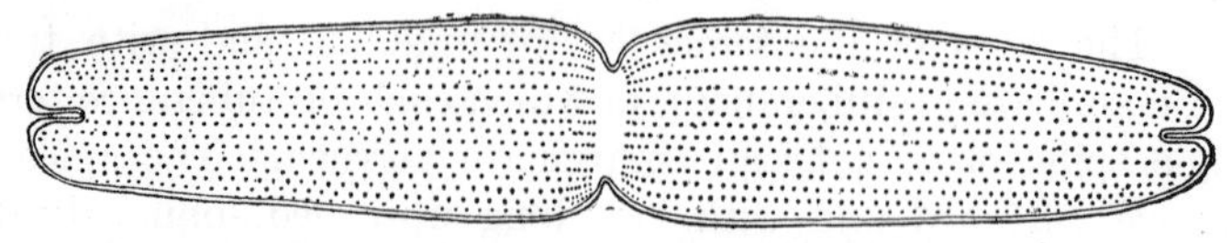

FIG. 232. *Tetmemorus Brebissonii* (Menegh.) Ralfs. (× 400.)

8. **Euastrum** Ehrenberg, 1832; emend., Ralfs, 1844. Some species belonging to this genus have relatively small cells, other species have large cells; but all have cells with a length about double the breadth. The cells are always compressed, with a deeply constricted isthmus, and with a linear to emarginate incision at the apex of each semicell (Fig. 233). The semicells are usually truncate-pyramidate in front view, with one or two broadly rounded lobes of each lateral margin, and with one or more protuberances on the front face. The cell wall is smooth, punctate or granulate; and it may also bear spines, verrucae, or combinations of the two. As seen in vertical view, the semicells are more or less elliptical, with rounded poles, and with a definite protuberance between the poles. The lateral view of semicells is narrowly truncate-pyramidate and with prominent inflations in the basal portion. Small-celled species have the semicells containing an axial chloroplast with simple lobes; semicells of large-celled species have an axial chloroplast with two radial sheets to the front and back, each sheet terminating in a broad parietal expansion.[1] *E. verrucosum* Ehr. has two chloroplasts in each semicell. According to the size of the chloroplast, pyrenoids are single, few, or many.

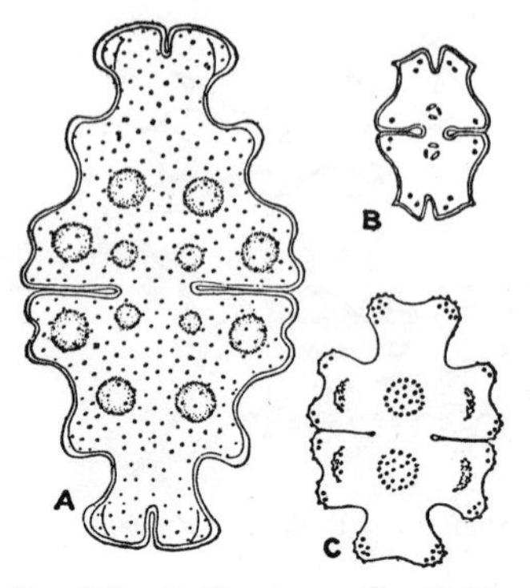

FIG. 233. *A, Euastrum affine* Ralfs. *B, E. elegans* (Bréb.) Kütz. *C, E. gemmatum* Bréb. (× 400.)

Zygotes of *Euastrum* are spherical to ellipsoidal, and with smooth, mamillate, or short-spined walls.

Euastrum (Fig. 233) has the same apical incision of semicells as is found in *Tetmemorus* but differs in the marked compression of semicells and in the different proportions between length and breadth. Certain of the small-celled species with an emarginate apical incision are rather difficult to distinguish from certain *Cosmaria*. About 70 species are known to occur in the United States. For descrip-

[1] CARTER, N., 1919*A*.

tions of most of the species of the genus, see Krieger (1937, 1937*A*), and W. and G. S. West (1905).

9. **Cosmarium** Corda, 1834. Most species of this unicellular genus have small compressed cells, with a length only slightly greater than the breadth, and a deep median constriction. A few species have cells that are circular in end view and with a relatively shallow sinus. The cell wall is smooth, or ornamented with granules of minute verrucae that are usually arranged in a definite pattern. Walls of *Cosmaria* are without spines. Many species have a localized accumulation of iron compounds in the isthmus portion of a wall or in the verrucae.[1] The front view of semicells may be semicircular, elliptic, reniform, trapezoidal, or subquadrate and with or without a

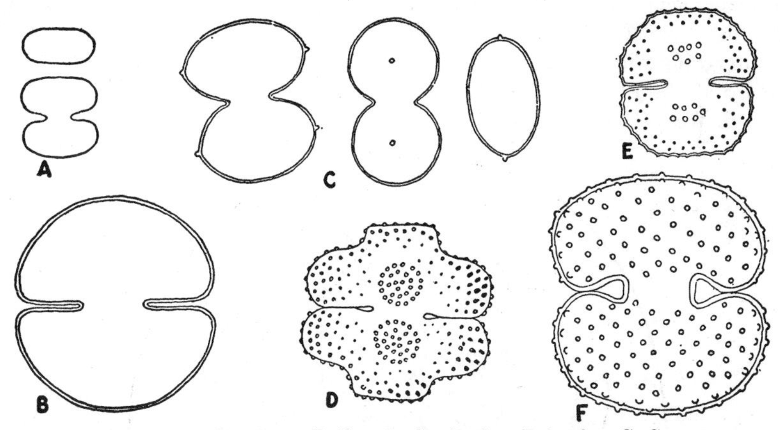

FIG. 234. *A*, *Cosmarium bioculatum* Bréb. *B*, *C. circulare* Reinsch. *C*, *C. contractum* var. *papillatum* W. and G. S. West. *D*, *C. protractum* (Näg.) DBy. *E*, *C. punctulatum* var. *subpunctulatum* (Nordst.) Börgesen. *F*, *C. reniforme* (Ralfs) Arch. (× 600.)

localized tumescence on the front face (Fig. 234). The vertical view is general elliptical in outline and often with lateral elevations midway between the poles. Most species have semicells that are circular in outline when viewed from the side. Each semicell usually contains a single axial chloroplast with four radiating plates and the pyrenoids localized in the axial portion.[2] Some species have two chloroplasts in a semicell and each with several radiating plates. A few species have four parietal chloroplasts in each semicell.

When conjugation takes place, each of the two cells surrounded by a common gelatinous sheath breaks at the isthmus and has an amoeboid escape of the protoplast. The zygote formed midway between the conjugating cells is usually globose, though sometimes angular, and with a smooth, papillate, or spiny wall. During the ripening of the zygote,[3]

[1] HÖFLER, 1926. [2] CARTER, N., 1920. [3] KLEBAHN, 1891.

there is a division of the fusion nucleus to form four daughter nuclei, two of which are functional and two degenerate (Fig. 225, page 315). The protoplast of a germinating zygote divides into two equal portions, each with a single chloroplast, one functional nucleus, and one degenerating nucleus. Each of the daughter protoplasts develops into a vegetative cell after it is liberated by a rupture of the old zygote wall. Parthenospores are sometimes formed as a result of an interruption of conjugation. Germinating parthenospores have one functional and three degenerate nuclei. Division of the functional nucleus of a parthenospore, just before germination, results in the production of two vegetative cells.

There have been several attempts to break up this large and cumbersome genus into smaller genera but without success. Certain species (Fig. 234) are difficult to distinguish from *Euastrum*. Other species, especially those with triradiate varieties, intergrade with *Staurastrum*. Approximately 280 species are known for the United States. The most extensive series of descriptions of species is that of W. and G. S. West (1905, 1908, 1912).

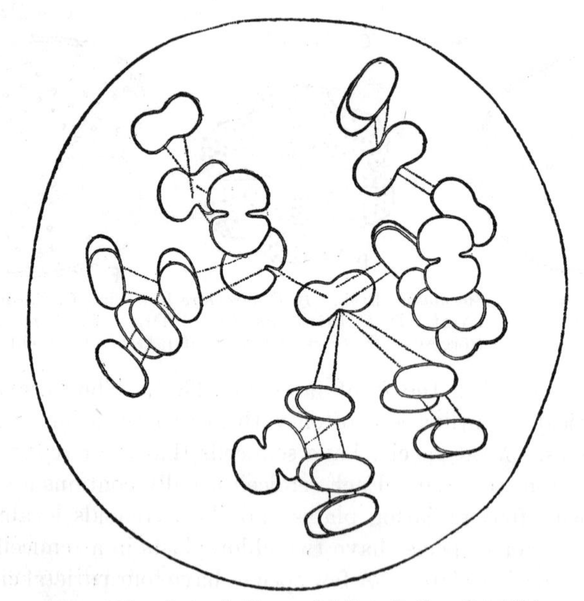

FIG. 235. *Cosmocladium pulchellum* Bréb. (× 400.)

10. **Cosmocladium** de Brébisson, 1856. The cells of *Cosmocladium* are united to one another by bands of gelatinous material, and the colonies resulting from this union have a wide, homogeneous, spherical to ellipsoidal, gelatinous envelope. Sometimes a colony develops into a *Tetraspora*-like mass over a decimeter in length.[1] The cells are small and quite

[1] BULLARD in Phycotheca Boreali-Americana No. 2120.

similar in appearance to those of *Cosmarium.* Most species have compressed cells with a deep median constriction, but there are also species with subcylindrical cells and a shallow constriction. The gelatinous strands, connecting cells one to another, are secreted through two series of minute pores on either side of the isthmus, but restricted to one face of the cell.[1] Unlike most Desmidiaceae, there is but one chloroplast in a cell. This is axial, usually with a single central pyrenoid, and with four parietal lobes.[2]

The zygotes are spherical to subangular and with several short stout spines.[3]

This genus, obviously related to *Cosmarium,* is of relatively rare occurrence. The five species found in this country are *C. constrictum* Arch., *C. Hitchcockii* (Wolle), G. M. Smith, *C. pulchellum* Bréb. (Fig. 235) (*C. saxonicum* DBr.), *C. pusillum* Hilse, and *C. tuberculatum* Prescott. For a description of *C. tuberculatum,* see Prescott and Magnotta (1935); for the others, see Heimanns (1935).

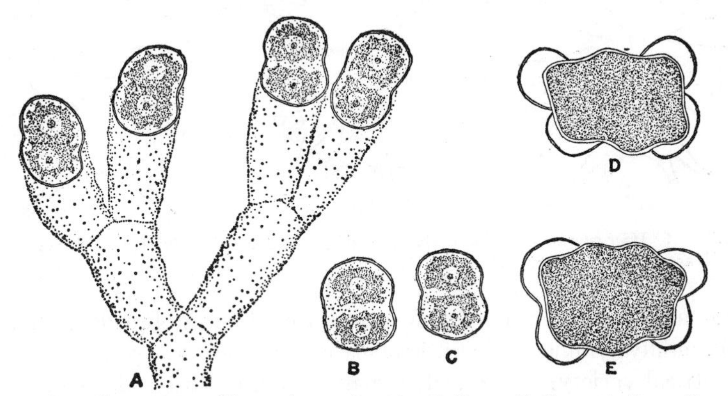

FIG. 236. *Oöcardium stratum* Näg. *A,* portion of a thallus. *B–C,* vegetative cells. *D–E,* zygotes. (× 650.)

11. **Oöcardium** Nageli, 1849. The cells of *Oöcardium* are united in sessile colonies 1 to 2 mm. in diameter, and the colonies are usually heavily impregnated with lime. The cells have much the same shape as those of moderately compressed species of *Cosmarium* with a broad isthmus and open sinus. The cell wall is smooth. Each semicell contains a single chloroplast with one pyrenoid. The colonial matrix consists of more or less parallel, dichotomously branched, broad gelatinous tubes.[4] There is a single cell at the end of each ultimate dichotomy, and the cells are usually so oriented that their long axes are at right angles to the long

[1] LÜTKEMÜLLER, 1902; SCHRÖDER, 1900. [2] CARTER, N., 1923.
[3] HOMFELD, 1929; ROY and BISSETT, 1893 to 1894.
[4] LÜTKEMÜLLER, 1902; SENN, 1899.

axis of the tubes. The lime is deposited in a cylinder external to the gelatinous tube.

Zygotes are quadrangular and with several mamillate protuberances.

This very rare genus is unusual in that it is restricted to limestone regions. It is usually found in waterfalls or attached to stones and rocks in swiftly flowing streams. There is but one species, *O. stratum* Näg. (Fig. 236). It is known from several parts of Europe, but thus far for the United States it has been found at only two localities in California.[1] For a description of *O. stratum*, see N. Carter (1923).

12. **Micrasterias** Agardh, 1827. With the exception of one filamentous species, all members of this genus are unicellular. Most species have cells

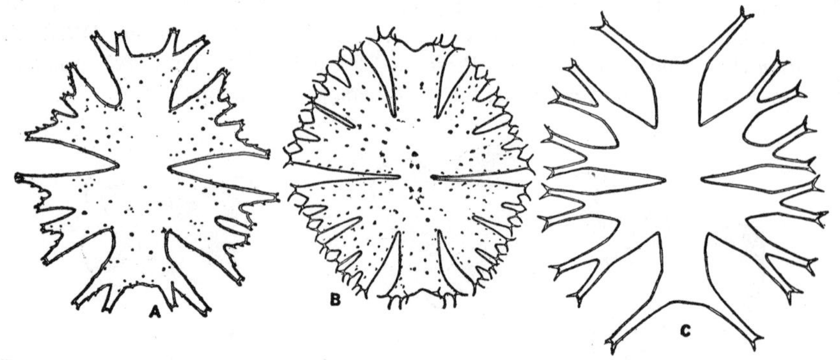

FIG. 237. *A*, *Micrasterias americana* (Ehr.) Ralfs. *B*, *M. apiculata* (Ehr.) Menegh. *C*, *M. radiata* Hass. (*A*, *C*, × 233; *B*, × 126.)

of large size, with a length somewhat greater than the breadth. The cells are bilaterally symmetrical and strongly compressed, except for one radially symmetrical variety. The median constriction is always deep and usually linear to sublinear. The outline of semicells as seen in front view is semicircular to hexagonal. The semicells are always incised and with two or four incisions. Semicells with two incisions have a central (polar) and two lateral lobes; those with four incisions have one polar and four lateral lobes. The polar lobe generally has an expanded apex that may be entire or emarginate. Some species have asymmetrically disposed processes at the apex of polar lobes (Fig. 237*A*). The lateral lobes are rarely entire; usually they are once, twice, or thrice divided into secondary lobelets with emarginate or spiniferous apices. Cells of *Micrasterias* usually have smooth walls, although some species have rows of marginal spines or spines over the entire wall. A few species have the spines or the isthmus portion of the wall impregnated with iron compounds.[2] Semicells contain a single, massive, plate-like, axial chloroplast with the same contour as the semicell, that

[1] SMITH, G M., 1933. [2] HÖFLER, 1926.

often has several small vertical ridges.[1] Pyrenoids are numerous (30 to 100) and evenly distributed throughout the chloroplast.

Conjugation is without development of a conjugation tube, usually takes place between adult cells, but may take place[2] between immature sister cells. The zygotes are spherical and ornamented with long stout radially disposed spines, terminating in simple to quadrifid apices.

There is considerable variation in shape from species to species (Fig. 237), but all have a distinct differentiation into polar and lateral lobes. Large cells are the rule and not the exception in *Micrasterias*, and many of these have a striking bilateral symmetry. Approximately 40 species are found in the United States. For descriptions of most of the species of the genus, see Krieger (1939), and W. and G. S. West, (1905).

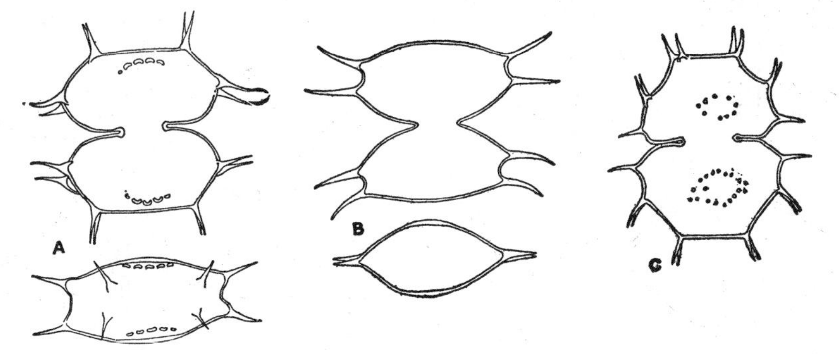

FIG. 238. *A, Xanthidium antilopaeum* var. *polymazum* Nordst. *B, X. subhastiferum* W. West. *C, X. cristatum* var. *uncinatum* Bréb. (× 400.)

13. **Xanthidium** Ehrenberg, 1837. Most species belonging to this genus have cells of medium size and a length somewhat greater than the breadth. There are always a deep median constriction and a linear to acute-angled sinus. All species have more or less compressed semicells, but certain of these have varieties that are radially symmetrical. The semicells are usually polygonal in outline as seen in front view, and with the apex flattened but not incised. Lateral margins of semicells have one or more heavy spines that are usually simple but may be furcate. The cell wall is smooth, except for a thickened protuberant area in the middle of the front side. This thickened area is often scrobiculate and of a brownish color because of a localized accumulation of iron compounds.[3] As seen in vertical view, the outline is elliptical; in lateral view, it is subcircular to polygonal. A semicell may contain two laminate axial chloroplasts,[4] or it may contain four of them more or less parietal in position. Each chloroplast usually contains but one pyrenoid.

[1] CARTER, N., 1919*B*. [2] WEST, W. and G. S., 1896.
[3] HÖFLER, 1926. [4] CARTER, N., 1919*A*.

The zygotes are spherical and with scrobiculate walls or with the wall bearing numerous simple to furcate spines.

Xanthidium (Fig. 238) may be recognized by its angular semicells, with spines at the angles, and the thickened central area on the front of a semicell. The latter character is the only one distinguishing certain species from *Arthrodesmus*. Twenty-one species are known for the United States. For descriptions of many species of the genus, see W. and G. S. West (1912).

14. **Staurastrum** Meyen, 1829. The range in size and shape among the multitude of species belonging to this genus is extremely varied. A majority of the species have cells that are radially symmetrical and usually

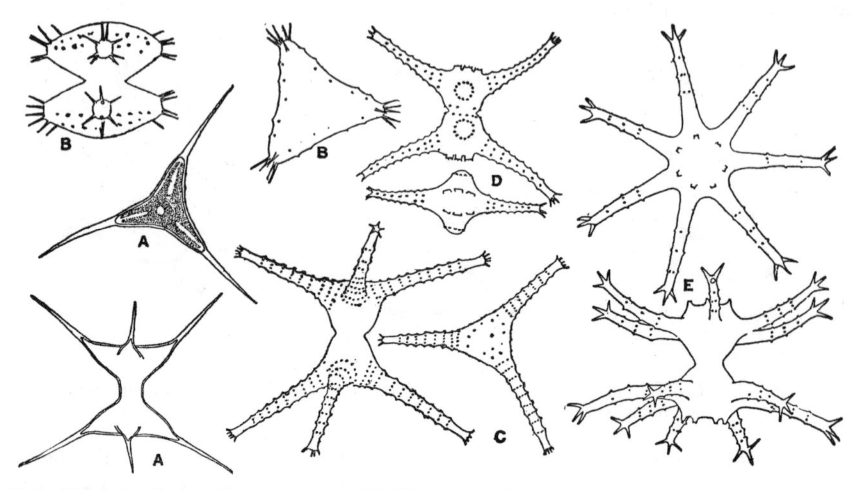

FIG. 239. *A*, *Staurastrum curvatum* W. West. *B*, *S. setigerum* var. *brevispinum* G. M. Smith. *C*, *S. paradoxum* Meyen. *D*, *S. natator* var. *crassum* W. and G. S. West. *E*, *S. limneticum* var. *cornutum* G. M. Smith. (× 400.)

triangular in end view; but there are many species with strongly compressed bilaterally symmetrical cells (Fig. 239*D*). Practically all species are deeply constricted and with an acute-angled sinus. The cell wall may be smooth, or it may be ornamented with granules, denticulations, simple to emarginate verrucae, or spines (Fig. 239). Species with ornamented walls have the ornamentation arranged in a symmetrical pattern. The front view of semicells may be elliptical, semicircular, cyathiform, triangular, quadrangular, or polygonal in outline. In many species, the superior angles of semicells are continued in processes, usually quite long, that are variously ornamented and terminate in truncate ends with short divergent spines. A few species have two transverse whorls of processes on each semicell. The semicells usually contain a single axial chloroplast with a deeply incised lobe to each angle of the semicell (Fig. 239*A*). There may be a single pyrenoid in the axial portion of a chloroplast, or several pyrenoids in each of the lobes.[1] In very rare cases, the chloroplasts are parietal.

The zygotes are more often spherical than angular and usually with conical elevations terminating in long spines with simple or furcate apices (Fig. 224*A*). In ripening of a zygote, there is a division of the fusion nucleus into four daughter nuclei and a formation of one, two, three, or four vegetative cells when the zygote germinates.[2]

There are several general types of cell shape among the numerous species of this genus (Fig. 239), but there are so many intergrades among the various types that it is impossible to divide the genus into subgenera. Species with cells of the simplest form resemble triradiate smooth-walled *Cosmaria*; in fact, it is very difficult to distinguish between certain varieties of *Cosmarium* that are triradiate in end view and certain species of *Staurastrum*. Likewise, the small compressed smooth-walled species with simple spines are extremely difficult to differentiate from *Arthrodesmus*. Practically every collection containing desmids has one or more species of *Staurastrum*. The genus is the commonest of desmids in the freshwater plankton, and several species are strictly planktonic in habit. Some 245 species are known for the United States. For descriptions of many of the species of the genus, see W. and G. S. West (1912, 1923).

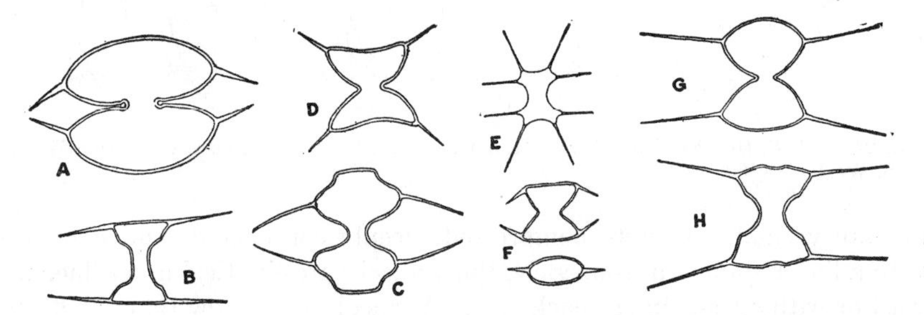

FIG. 240. *A*, *Arthrodesmus convergens* Ehr. *B*, *A. incus* var. *extensus* Anderss. *C*, *A. michiganensis* Johnson. *D*, *A. phimus* Turn. *E*, *A. octocorne* Ehr. *F*, *A. Ralfsii* W. West. *G*, *A. subulatus* Kütz. *H*, *A. triangularis* var. *subtriangularis* (Borge) W. and G. S. West. (× 400.)

15. **Arthrodesmus** Ehrenberg, 1838. Cells of *Arthrodesmus* (Fig. 240) are always small and with length and breadth about equal. Except for certain varieties, the cells are always strongly compressed, deeply constricted, and with a widely open to linear sinus. The cell wall is smooth, of a uniform thickness throughout, and with simple straight or strongly curved spines at the angles. Most species have semicells that are triangular when seen in front view; but a few have a subrectangular, trapezoidal, or elliptical outline. The cells are often embedded in a copious

[1] CARTER, N., 1920*A*. [2] TURNER, C., 1922.

gelatinous sheath with radial strands of a denser consistency. Each semicell has an axial laminate chloroplast with one or two pyrenoids.

The zygotes are spherical and with smooth walls or with the walls bearing long sharp spines (Fig. 224*B*, page 314).

On the one hand, certain species of *Arthrodesmus* intergrade with *Staurastrum*; on the other, certain species approach the smaller species of *Xanthidium* but lack the localized thickening of the cell wall characteristic of *Xanthidium*. About 20 species are known for the United States. For descriptions of many of the species of the genus, see W. and G. S. West (1912).

16. **Onychonema** Wallich, 1860. Members of this genus have the cells permanently united in unbranched filaments. Each semicell has two asymmetrically disposed, fairly long, capitate, apical processes overlapping the cell next to it and so firmly uniting the cells one to another. The cells

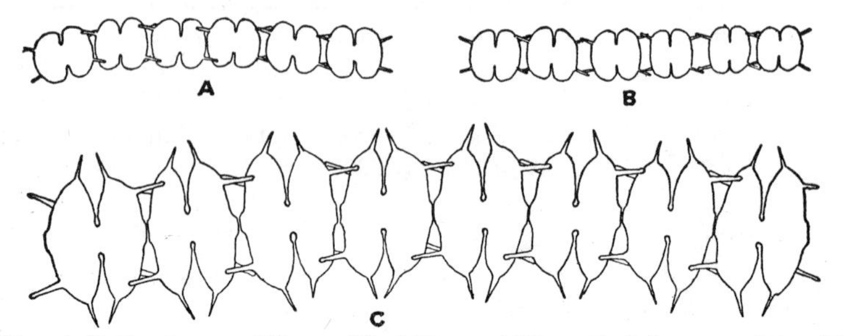

Fig. 241. *A–B*, *Onychonema filiforme* (Ehr.) Roy and Biss. *C*, *O. laeve* var. *latum* W. and G. S. West. (× 600.)

are fairly small, strongly compressed, deeply constricted, and generally with a linear sinus. In front view, the semicells are elliptical in outline, and with or without a spine at each side. Vertical views show that the apical processes evident in front view are alternately inserted. Each semicell contains a single, axial, laminate chloroplast with one pyrenoid.

Conjugation is always between solitary cells that have become disatriculated from filaments.[1] The zygotes are spherical and ornamented with short spines. Parthenospores have also been recorded[2] for *Onychonema*.

Isolated filaments of *Onychonema* are usually present in collections from localities rich in desmids. The two species found in this country are *O. filiforme* (Ehr.) Roy and Biss. (Fig. 241*A–B*) and *O. laeve* Nordst. (Fig. 241*C*). For descriptions of them, see G. M. Smith (1924*A*).

17. **Sphaerozosma** Corda, 1834. The chief distinction between this and the foregoing genus is that the apical processes of the semicells are short

[1] West, W. and G. S., 1895*A*. [2] West, W. and G. S., 1907.

and interlocked with one another, instead of overlapping adjoining cells. As in the preceding genus, the filaments may attain a considerable length. The cells are small, compressed, moderately constricted, with an open sinus, and often with the apical processes reduced to granules. Some species have smooth walls; others have granulate walls and the granules arranged in a definite pattern. The chloroplasts are axial, one in each semicell, and with a single pyrenoid.

Zygotes have been described for several species, and in all cases they are formed by a conjugation of isolated cells. The zygotes are spherical to subrectangular and with smooth or spiny walls.

Sphaerozosma is another filamentous genus frequently encountered in collections. The five species found in this country are *S. Aubertianum* W. West (Fig. 242*A*), *S. excavata* Ralfs (Fig. 242*B*), *S. exiguum* Turner, *S. granulatum* Roy and Biss.,

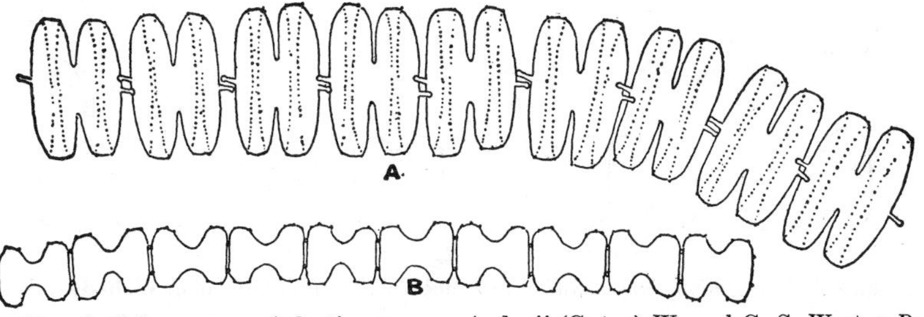

FIG. 242. *A*, *Sphaerozosma Aubertianum* var. *Archerii* (Gutw.) W. and G. S. West. *B*, *S. excavata* Ralfs. (× 600.)

and *S. vertebratum* (Bréb.) Ralfs. For a description of *S. granulatum*, see N. Carter (1923); for *S. vertebratum*, see Ralfs (1848); for the others, see G. M. Smith (1924*A*).

18. **Spondylosium** de Brébisson, 1844. The one characteristic separating this genus from the two preceding ones is the absence of processes from apices of semicells. There is more variation in size of cell from species to species in this genus, and the cells may be strongly compressed or radially symmetrical. As seen in front view, the semicells are elliptical, triangular, or rectangular in outline and with the apex flattened or elevated in the median portion. The cell wall may be smooth or punctate, but in the latter case the ornamentation is rarely in a definite pattern. The median constriction is always relatively deep, and the sinus may be widely open to linear. The chloroplasts are axial and usually with a single pyrenoid.

Conjugation is between cells that are not joined in filaments. The zygotes are usually spherical and with smooth walls. The aplanospores recorded for two species have always been found in cells united in filaments.[1]

[1] TURNER, W. B., 1892; WALLICH, 1860.

Filaments of *Spondylosium* (Fig. 243) are found sparingly intermingled with other algae in collections rich in desmids. Ten species have been found in the United States. For descriptions of many species of the genus, see N. Carter (1923).

19. **Hyalotheca** Ehrenberg, 1841. The median constriction in cells of this genus is so slight that one is apt to think that the alga does not belong to the Desmidiaceae. The cells are cylindrical to discoidal and with flattened poles that are without projections. The cell wall is without ornamentation, except for delicate transverse ridges just below the cell apices. Chloroplasts are axial and with several longitudinal lobes extending to the cell walls; pyrenoids are single and central. Mature cells have a chloroplast in each semicell, but young cells contain only a single chloroplast for some time following cell division. The two chloroplasts

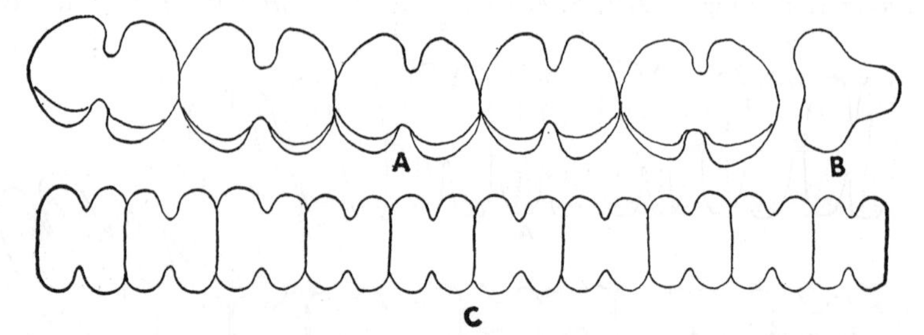

FIG. 243. *A–B, Spondylosium moniliforme* Lund. *C, S. planum* (Wolle) W. and G. S. West. (× 600.)

of an adult cell result from transverse division of the chloroplast and pyrenoid of a young cell.[1]

Unlike the preceding genera, *Hyalotheca* has no dissociation into individual cells prior to conjugation. Conjugating cells of a filament send out mamillate protuberances that meet each other and develop into a definite conjugation tube.[2] The zygote is usually formed in the conjugation tube, but it may be formed in one of the conjugating cells.[3] Conjugation may be scalariform or lateral. During ripening of a zygote, there is a meiotic division of the zygote nucleus to form two functional and two degenerate nuclei.[2] Newly formed zygotes contain four chloroplasts, those about to germinate contain two. When germination takes place, there is a division of the zygote's contents into two equal parts; each with a chloroplast, a functional nucleus, and a degenerate nucleus. There may also be a formation of aplanospores whose shape is somewhat different from that of zygotes.[4]

[1] ACTON, 1916*B*. [2] POTTHOFF, 1927. [3] CARTER, N., 1923.
[4] WEST, W. and G. S., 1898, 1907.

Hyalotheca is one of the commonest of filamentous desmids and is often present in abundance in collections. The six species found in this country are *H. dissiliens* (J. E. Smith) Bréb. (Fig. 244*B*), *H. indica* Turner, *H. laevicincta* Taylor, *H. mucosa* (Dillw.) Ehr. (Fig. 244*A*), *H. neglecta* Racib., and *H. undulata* Nordst. For descriptions of *H. dissiliens* and *H. mucosa,* see G. M. Smith (1924*A*); for *H. indica,* see W. B. Turner (1892); for *H. neglecta* and *H. laevicincta,* see Taylor (1935); for *H. undulata,* see Irénee-Maire (1939).

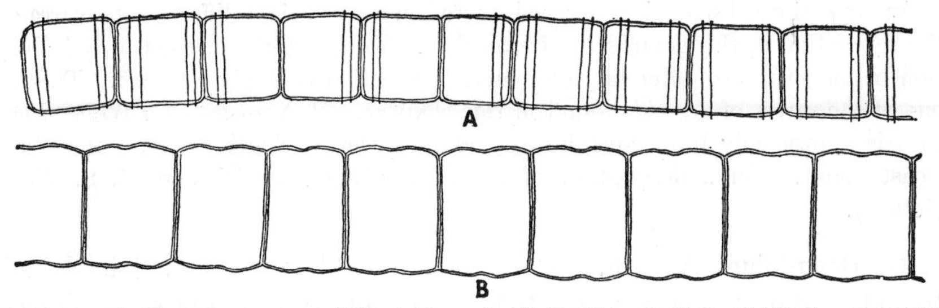

Fig. 244. *A, Hyalotheca mucosa* (Dillw.) Ehr. *B, H. dissiliens* (J. E. Smith) Bréb. (× 600.)

20. **Phymatodocis** Nordstedt, 1877. The cells of this desmid are joined end to end in unbranched filaments that may be straight or spirally twisted. The length of the cells is equal to, or less than, the breadth. The cells are quadrangular in outline as seen in front view, with a deep median con-

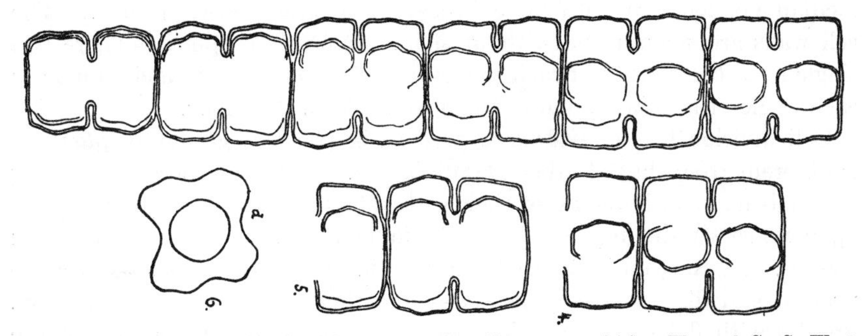

Fig. 245. *Phymatodocis Nordstedtiana* var. *minor* Börgesen. (*After W. and G. S. West,* 1895A.)

striction, and with a linear to sublinear sinus. As seen in vertical view, the semicells are irregularly to regularly quadrate and with a broadly rounded arm at each of the four corners. The four arms of one semicell may lie directly above those of the other semicell or be somewhat alternate with them. Each semicell contains a single axial chloroplast with a small central mass and two radially laminate blades to each of the four arms.[1]

When cell division takes place, there is a transverse cytokinesis at the

[1] Schmidle, 1902.

isthmus and a gradual enlargement of the new semicells, without any ring-like infolding of their walls.

Conjugation probably takes place between isolated cells. The zygote is irregularly quadrangular and with irregularly thickened walls.[1] Quadrangular parthenospores may be formed by cells that have not become separated from the filament.[2]

As far as the cell shape is concerned, there is not a sufficient difference to warrant the separation of this genus from *Desmidium*. However, the method of cell division in the two is so different that generic separation is amply justified. This is one of the rarest of desmids found in this country. *P. Nordstedtiana* Wolle (Fig. 245) is known only from New Jersey and from certain South Atlantic and Gulf Coast states. For a description of *P. Nordstedtiana*, see W. and G. S. West (1895*A*).

21. **Desmidium** Agardh, 1825. *Desmidium* has the cells joined end to end in unbranched filaments that are spirally twisted and with a broad gelatinous envelope. The cells are broader than long and never with a deep median constriction. Most species have radially symmetrical triangular or quadrangular cells, but some have compressed cells. The front view of semicells is trapezoidal in outline. Poles of adjoining cells may be flattened where they abut on each other, or they may be depressed in the median portion and with an elliptical open space between them. The cell walls are smooth and without striae, puncate, or spines. There is a single axial chloroplast in each semicell, and it has a deeply incised lobe to each angle of a semicell and a single pyrenoid in each lobe.

Cell division in *Desmidium*, as well as in *Gymnozyga* and *Steptonema*, is different from that of other Desmidiaceae. In these genera, developing semicells have an infolding of their end walls (Fig. 246*A*), quite similar in appearance to the replication found in end walls of certain *Spirogyra* species. As the young semicells increase in length, the infoldings become everted and disappear completely by the time the semicells are mature. Species with flattened semicell apices develop but one replication; those with mamillate apices develop as many replications as there are protuberances.[3]

Conjugation is between cells united in filaments and with the formation of a true conjugation tube. The zygote is usually formed in the conjugation tube, but one species regularly has a formation of the zygote in the female gametangium.[4] Zygotes of *Desmidium* are spherical to ellipsoidal and with a smooth wall. There may also be a formation of parthenospores.[5]

[1] Borge, 1918. [2] Gutwinski, 1902. [3] Carter, N., 1923.
[4] Couch and Rice, 1948; West, G. S., 1904.
[5] Couch and Rice, 1948.

The only distinction between this genus and the foregoing one is the replication of end walls in developing semicells. *Desmidium* is the most abundant of all filamentous genera of Desmidiaceae, and at times is to be found in practically pure stands. The six species found in this country are *D. Aptogonum* Bréb. (Fig. 246*D–E*), *D. asymmetricum* Gronbl., *D. Baileyi* (Ralfs) Nordst. (Fig. 246*B–C*), *D. Grevillii* (Kutz.) DBy. (Fig. 246*A*), *D. gymnozygoforme* Prescott and Scott, and *D. Swartzii* Ag. For a description of *D. asymmetricum*, see Grönblad, (1920); for *D. gymnozygoforme*, see Prescott and Scott, (1942); for the others, see G. M. Smith (1924*A*).

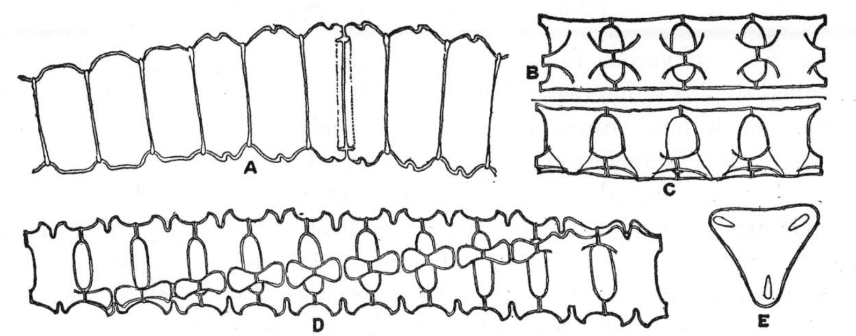

FIG. 246. *A, Desmidium Grevillii* (Kütz.) DBy. *B–C, D. Baileyi* (Ralfs) Nordst. *D–E, D. Aptogonum* Bréb. (× 400.)

22. Gymnozyga Ehrenberg, 1841. *Gymnozyga* has the cells united in long untwisted filaments that are often with a gelatinous sheath. The cells are uncompressed, barrel-shaped, and with a length about double the breadth. There is a very slight median constriction, but the basal portion

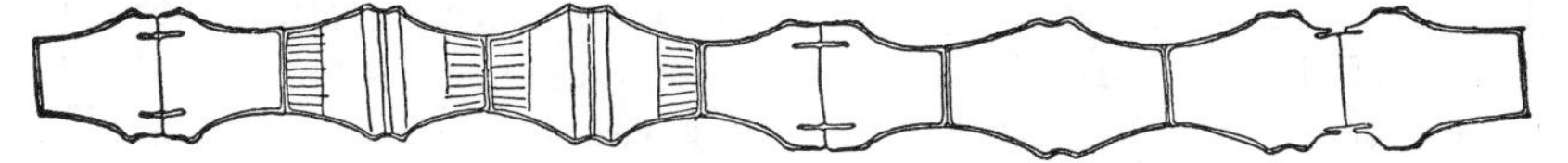

FIG. 247. *Gymnozyga moniliformis* Ehr. (× 600.)

of semicells is distinctly inflated. The apices of semicells are flattened and without protuberances. The cell wall may be smooth throughout, or with longitudinal striae near the poles and basal inflations of semicells. Each semicell contains a single axial chloroplast with six to eight delicate laminate vertical processes to the cell wall. There is a single pyrenoid at the center of a chloroplast.

Dividing cells of *Gymnozyga* have the same infolding of end walls in young semicells that is found in *Desmidium*, but eversion of the replications is greatly delayed.

Sexual reproduction begins by two filaments becoming apposed to each other and establishing conjugation tubes between cells of each. The filaments are irregularly twisted about each other and with comparatively few of the cells forming conjugation tubes. The zygote is formed in the tube and is globose.

Gymnozyga, similar to other filamentous desmids, is found intermingled with free-floating unicellular Desmidiaceae. The three species found in this country are *G. confervacea* W. and G. S. West, *G. delicatissima* (Wolle) Nordst., and *G. moniliformis* Ehr. (Fig. 247). For a description of *G. confervacea*, see W. and G. S. West (1895*A*); for *G. delicatissima*, see Wolle (1884), under *Bambusina delicatissima*; for *G. moniliformis*, see G. M. Smith (1924*A*).

CLASS 2. CHAROPHYCEAE

The Charophyceae, familarly known as stoneworts, have an erect branched thallus differentiated into a regular succession of nodes and internodes. Each node bears a whorl of branches of limited growth (the "leaves"), but branches capable of unlimited growth may arise axillary to the leaves. Sexual reproduction of Charophyceae is oögamous. The oögonia are one-celled, solitary, surrounded by a sheath of sterile cells and are always borne on the leaves. The antheridia are one-celled, united in uniseriate filaments, of which several are surrounded by a common spherical envelope composed of eight cells. Envelopes surrounding antheridia are also borne only on leaves.

The stoneworts are universally recognized as related to the Chlorophyceae, but there is great diversity of opinion concerning the degree of relationship. They have been considered an order of the Chlorophyceae, a class coordinate with the Chlorophyceae, and a division standing just below the level of the Bryophyta. The vegetative structure and the sterile sheaths surrounding sex organs are such distinctive features that the placing of them as an order of the Chlorophyceae seems too conservative a treatment. On the other hand, they cannot be considered a group standing above the algal level because their sex organs are one-celled. The best solution of the problem seems to be that of interpreting them as an offshoot from the Chlorophyceae, but a series so far removed that it should be put in a separate class.

Occurrence. Most of the Charophyceae grow submerged in fresh standing water and upon a muddy or a sandy bottom. A few species grow in brackish water. When growing in fresh-water ponds or lakes, they frequently form extensive subaquatic meadows that extend downward to a considerable depth below the surface of the water. They thrive best in clear hard waters, but aerated water is not essential. Many species, especially those of *Chara*, become encrusted with calcium carbonate, and the continued presence of the alga from year to year may result in the deposition of considerable calcareous material upon the lake bottom.

The calcareous material accumulating around a thallus may remain intact after death of the plant and decay of the organic remains, and many fossil stoneworts have been described from such calcareous casts. The structure of the fructifications, especially that of the female fructification, is so distinctive that there is no doubt concerning the nature of the group of plants producing these casts. Stoneworts are known from as far back as the Devonian.[1] Female fructifications in some fossil genera have the enveloping cells spirally twisted; in other genera, they are not arranged in this manner. All the present-day stoneworts have female fructifications with the envelope twisted in a left-hand spiral, and among fossil genera this type is known as far back as the Pennsylvanian. Spirally twisted envelopes are also known from the Mississippian and the Devonian, but here the spirals turn to the right.

Vegetative Structure. The thallus is an erect branched axis attached to the substratum by rhizoids. The rhizoids are uniseriately branched filaments, with or without a differentiation into nodes and internodes (Fig. 251*G*). The erect axis has an *Equisetum*-like differentiation into nodes and internodes (Fig. 248*A*). Each node bears a whorl of several branches, the leaves, that cease to grow after they have attained a certain length. The leaves may be simple or divided, and with or without a differentiation into nodes and internodes.

In some genera, as *Nitella*, an internode always consists of a single cell, many times longer than broad. In other genera, as *Chara*, a majority of the species have this internodal cell ensheathed (corticated) by a layer of vertically elongated cells of much smaller diameter. The ensheathing layer (cortex) of an internode is always one cell in thickness.

Terminal growth of an axis is initiated by a single dome-shaped *apical cell* which cuts off derivatives at its posterior face (Fig. 248*B*). Each derivative cut off by an apical cell develops into a node and its underlying internode. This begins with a transverse division of a derivative soon after it is cut off from the apical cell. The inferior daughter cell remains undivided, elongates to many times its original length, and matures into an internodal cell. The superior derivative, the *nodal initial*, divides and redivides to form the nodal tissue and also the corticating tissue of species in which the internode is corticated. The first division of a nodal initial is vertical, and the two daughter cells also divide vertically and in a plane intersecting the first plane of division.[2] Successive divisions are also vertical and in planes intersecting the preceding plane of division. The nodal tissue produced by these divisions consists of two central cells and an encircling ring of 6 to 20 peripheral cells. The central cells may remain undivided or may divide two or three times. All the peripheral cells divide periclinally. The inner daughter cells produced by these periclinal

[1] Peck, 1946. [2] Giesenhagen, 1896, 1897, 1898.

divisions may remain undivided or may divide periclinally (Fig. 248*C–E*). The outer daughter cells function as initials that give rise to leaves.

Among species with corticated internodes, half of the corticating tissue of an internode is derived from the node above, and the other half is derived from the node below. The basal cell of each leaf of a node produces a single ascending corticating initial and a single descending one. Each ascending and descending corticating initial is an apical cell, that lies closely applied to the internodal cell, and each apical cell produces a

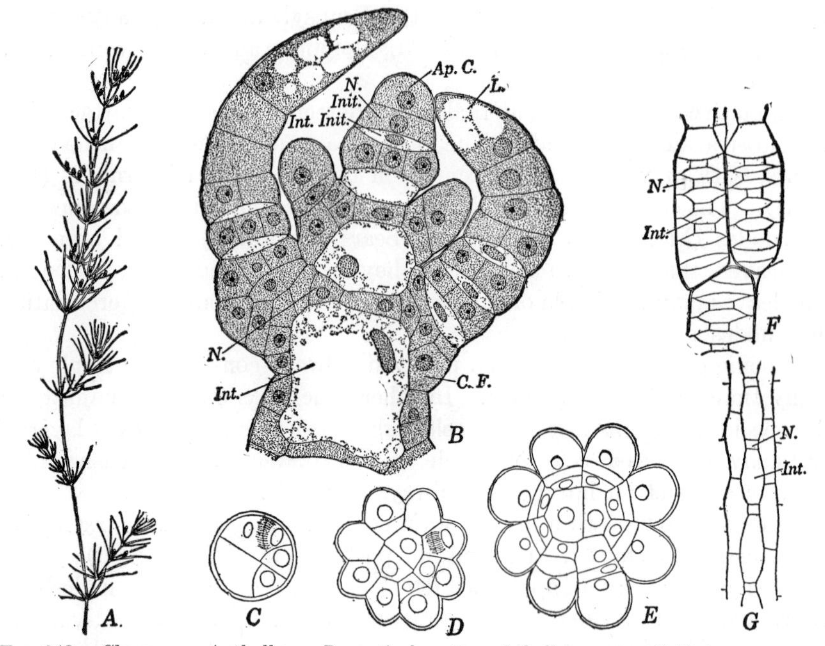

FIG. 248. *Chara* sp. *A*, thallus. *B*, vertical section of thallus apex. *C–E*, transverse sections of second, third, and fourth internodes. *F*, young corticating branches. *G*, portion of a mature corticating branch. (*Ap.C.*, apical cell; *C.F.*, corticating filament; *Int.*, internode; *Int. Init.*, internodal initial; *L*, leaf; *N*,, node; *N. Init.*, nodal initial.) (*A*, × 2/3; *B*, × 210; *C–F*, × 145; *G*, × 105.)

corticating filament differentiated into three-celled nodes and one-celled internodes. All nodal and internodal cells of a young corticating filament are at first approximately the same length, but eventually the two lateral nodal cells of a node and the internodal cell elongate to many times their original length (Fig. 248*F–G*). The median cell of a node does not elongate, and it may or may not form one-celled spines.

Leaves develop from the peripheral cells of a node. Some genera (as *Nitella* and *Tolypella*) have leaves that consist of a simple or branched uniseriate row of cells (Fig. 252). Other genera, as *Chara*, have leaves with the same structure as the axis. Here the apical cell of a leaf cuts off derivatives in the same manner as an apical cell of the main axis. The apical cell of a leaf becomes conical and ceases division after it has cut off 5 to 15 derivatives. In *Chara*, the first derivative from a leaf's apical cell becomes the basal node in all leaves; all other derivatives divide to form a nodal initial and an internodal cell. Internodal cells of a leaf mature in the same manner as those of an axis, except that they do not become so long. Development of nodal tissue from a nodal initial is much the same as in an axis. Nodes of leaves have but one central cell, and the peripheral cells never function as apical cells. Instead, all or certain of embryonic peripheral cells mature into one-celled spine-like structures. Cortication of a leaf may be similar to that of an axis, or the corticating initials at a leaf node may elongate without dividing.

Cell Structure. Cells near a branch apex are without conspicuous central vacuoles and are always uninucleate. Greatly enlarged cells of mature regions, as those of an internode, have a large central vacuole and may have a few large irregularly shaped nuclei because of nuclear division by constriction (amitosis).[1] The cytoplasm external to the central vacuole contains many small ellipsoidal chloroplasts that lie in longitudinal spirally twisted parallel series. The portion of the cytoplasm next to the central vacuole streams continuously in a longitudinal direction. There are ascending and descending longitudinal streams of cytoplasm laterally separated from each other by a motionless streak of cytoplasm without chloroplasts.

Asexual Reproduction. None of the Charophyceae produces zoospores, but many of them produce asexual reproductive bodies of a vegetative nature. Vegetative propagation may be effected by (1) star-shaped aggregates of cells developed from the lower nodes, and frequently called *amylum stars* because they are densely filled with starch; (2) bulbils developed upon rhizoids; and (3) protonema-like outgrowths from a node.

Sexual Reproduction. All genera reproduce sexually. The male and female fructifications are usually called, respectively, antheridia and oögonia, but these names are inappropriate because the structures so designated include both the sex organ (or organs), and an enveloping multicellular sheath derived from cells below the sex organ. According to the old terminology,[2] the male fructification is a *globule* and the female is a *nucule*. These names are more appropriate since they do not imply that the entire fructification is a sex organ.

In almost every case, the fructifications are borne on the leaves. The vast majority of species are homothallic, and only relatively few are heterothallic. Homothallic species usually have the two kinds of fructi-

[1] Sundaralingam, 1948. [2] Sachs, 1875.

fication borne adjacent to each other. There may be one of each and the globule above, *Nitella* (Fig. 253*A*), or below, *Chara* (Fig. 250*A*), the nucule. In other cases, a nucule is flanked on either side by a globule, as in *Lychnothamnus*; or there are several nucules flanking a single globule, as in *Tolypella* (Fig. 253*B*).

Chara may be taken as illustrative of the manner in which globules and nucules develop. Here the fructifications are borne at nodes of leaves and on the adaxial side (Fig. 253*C*). In homothallic species, a nucule always lies above a globule, and development of the two may be simultaneous or development of the globule may be somewhat in advance of that of the nucule. A superficial cell on the adaxial side of a leaf functions as an

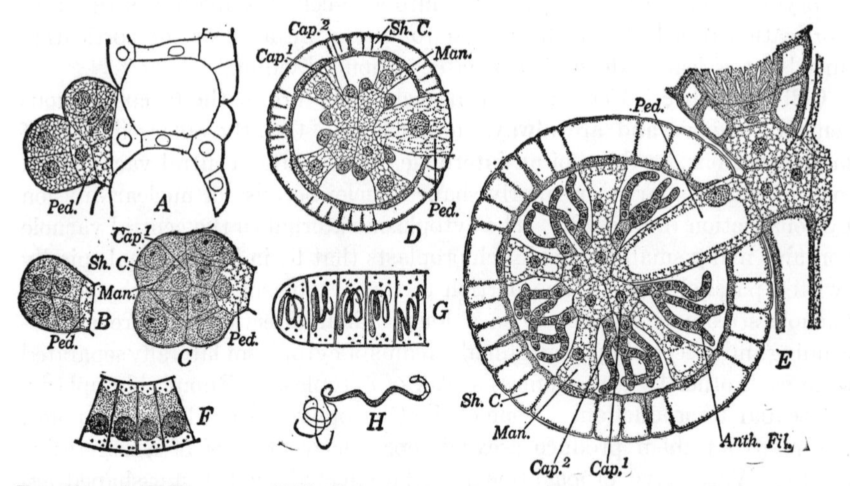

FIG. 249. *A–E*, *Chara* sp. *F–H*, *Chara foetida* A. Br. *A–E*, development of globule. *F–G*, antheridial filaments. *H*, antherozoid. (*Anth. Fil.*, antheridial filament; *Cap.*,[1] primary capitulum; *Cap.*,[2] secondary capitulum; *Man.*, manubrium; *Ped.*, pedicel; *Sh.C.*, shield cell.) (*F–H*, *After Belajeff*, 1894.) (*A–C*, × 210; *D–E*, × 145; *F–G*, × 575; *H*, × 290.)

apical cell which cuts off a derivative. The derivative divides and redivides to form nodal tissue. The apical cell then cuts off a second derivative which eventually enlarges greatly and becomes the *pedicel* cell of the globule (Fig. 249*A*). Next, by two successive vertical divisions, the apical cell divides to form four quadrately arranged cells each of which divides transversely (Fig. 249*B*).[1] Each cell of the octad divides periclinically, and the eight outer cells also divide periclinally (Fig. 249*C*). The outer of the three cells derived from each octad cell is a *shield cell*, the median is a handle cell or *manubrium*, and the inner is a *primary capitulum cell.* Maturing shield cells expand laterally; as a result, cavities appear within the

[1] CAMPBELL, 1902; SACHS, 1875.

globule. The manubria enlarge radially as the cavities enlarge, but the primary capitulum cells remain apposed to one another at the center of the globule. There is also an upward growth of the pedicel cell toward the interior of the globule (Fig. 249*D–E*). The outer periclinal wall of a maturing shield cell develops radial ingrowths which incompletely divide it into a number of compartments. Hence the outer layer of a maturing globule seems to be many cells in perimeter when viewed in cross section. Mature globules are a bright yellow or red because of a change in color of chloroplasts within the shield cells.

Each primary capitulum within a globule cuts off six *secondary capitulum cells* (Fig. 249*D*), and these may or may not cut off tertiary and quaternary capitulum cells.[1] The secondary capitulum cells usually cut off initials of antheridial filaments, but such initials may also be produced upon primary, tertiary, or quaternary capitulum cells. An antheridial initial develops into an antheridial filament that may be branched or unbranched (Fig. 249*E*). The number of cells in a filament varies greatly, and even in the same species it may range from 5 to 50.[1] Each cell of a fully developed filament is an *antheridium* whose protoplast metamorphoses into a single *antherozoid.* The nucleus of a metamorphosing protoplast moves toward the side wall, elongates, and becomes spirally coiled (Fig. 249*F–G*).[2] Meanwhile, there has been a differentiation of a spirally coiled blepharoplast, just within the plasma membrane, and it causes a formation of two long flagella a short distance back from the anterior end of the coiled antherozoid (Fig. 249*H*). When the antherozoids are mature, the shield cells of a globule separate from one another and expose the antheridial filaments and the manubrium to which they are attached. A manubrium, with its attached capitular cells and antheridial filaments, resembles a many-thonged whip. Soon after the filaments are exposed, there is an escape of antherozoids through pores in the antheridial wall. Liberation of anthrozöids generally takes place in the morning, and the swarming may continue until evening.[3]

A globule has been interpreted[4] as a metamorphosed branch in which the terminal cell divides into octants. Each octant is divided into a basal node (the shield cell), an internode (the manubrium), and an upper nodal cell (the primary capitular cell). The filamentous outgrowths (antheridial filaments) from the upper node are not differentiated into nodes and internodes.

An adaxial cell of the basal node of a globule functions as the initial of a nucule. This initial divides transversely to form a row of three cells,[5]

[1] Karling, 1927. [2] Belajeff, 1894; Mottier, 1904. [3] Sachs, 1875.
[4] Goebel, 1930.
[5] Campbell, 1902; Debski, 1898; Goetz, 1899; Sachs, 1875.

the median of which is nodal and the other two internodal in nature. The lower internodal cell remains undivided and enlarges to form the *pedicel* subtending the nucule (Fig. 250*B*). The nodal cell divides and redivides vertically to form five lateral initials encircling a single central cell. The upper internodal cell is an oögonial parent cell which elongates vertically and then divides transversely to form a short *stalk cell* and a vertically elongate *oögonium* (Fig. 250*C*). A mature oögonium contains a single *egg* with a nucleus at the base and with the cytoplasm densely packed with starch. Even before elongation of the oögonial parent cell, there is an upgrowth of the five lateral initials of the nodal tier to form a protective

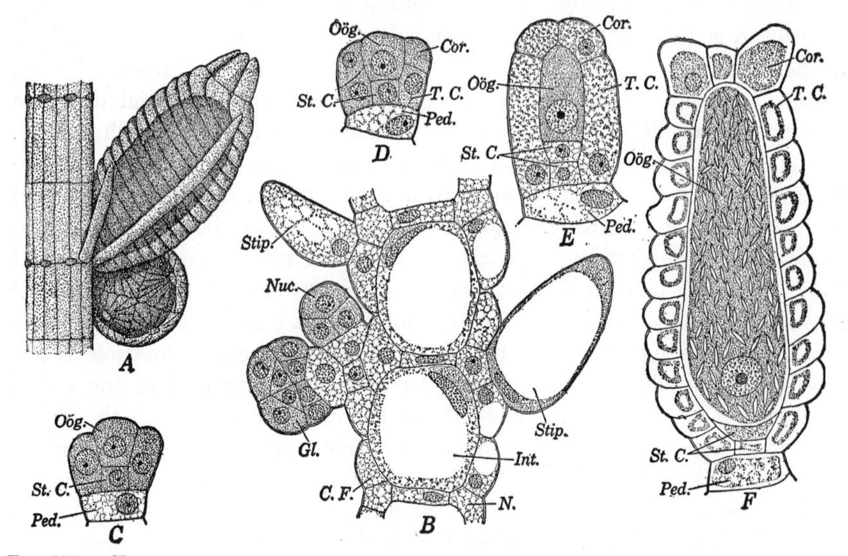

FIG. 250. *Chara* sp. *A*, portion of a leaf bearing a mature globule and nucule. *B*, vertical section of a leaf bearing a very young globule and nucule. *C–F*, stages in development of a nucule. (*Cor.*, corona; *C. F.*, corticating filament; *Gl.*, globule; *Int.*, internode; *N*, node; *Nuc.*, nucule; *Oög.*, oögonium; *Ped.*, pedicel; *St.C.*, stalk cell; *Stip.*, stipule; *T.C.*, tube cell.) (*A*, × 50; *B–E*, × 210; *F*, × 145.)

sheath enclosing the oögonial parent cell. The sheath soon becomes transversely divided into two tiers of five cells each (Fig. 250*D*). Cells of the upper tier elongate but little and mature into the five-celled *corona* capping a mature nucule; those of the lower tier, the *tube cells*, elongate to many times their original length and become spirally twisted about the oögonium (Fig. 250*E–F*). In certain genera, including *Nitella*, the corona consists of two tiers and has five cells in each tier.

The spirally twisted tube cells of a mature sheath separate from one another just below the corona to make five small angular slits.[1] Anthero-

[1] DEBARY, 1871.

zoids swim through these openings in the sheath of a nucule and down to the oögonium (Fig. 251*A*). One of them penetrates the gelatinized wall and unites with the egg. Male gamete nuclei have been observed within the egg of *Nitella*,[1] and it is thought that in *Chara* there is the same union of gamete nuclei at the base of an egg.

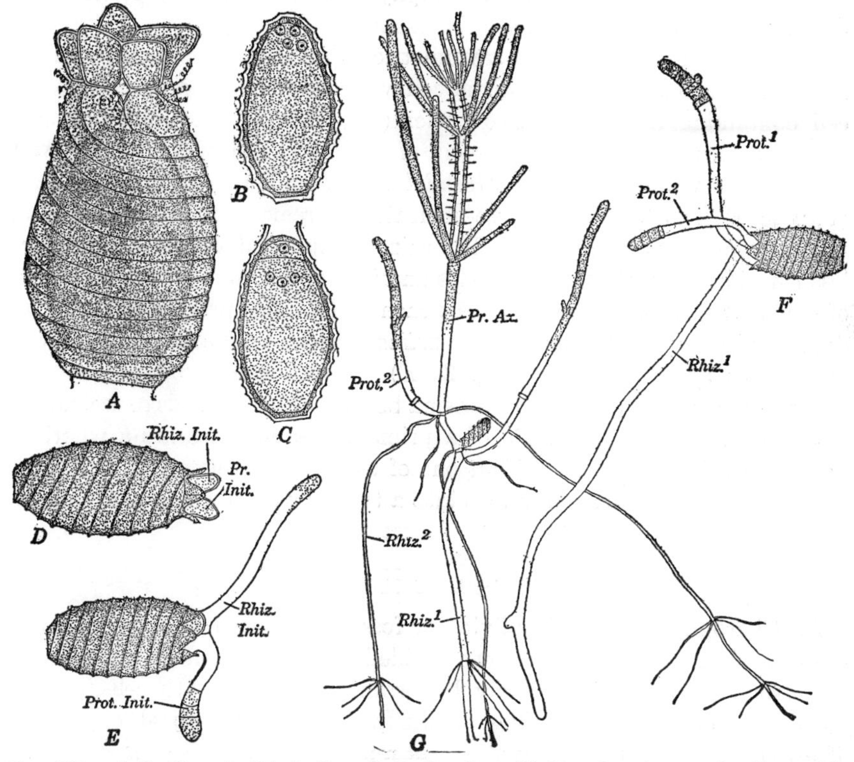

Fig. 251. *A–C*, *Chara foetida* A. Br. *D–G*, *C. crinata* Wallr. *A*, entrance of antherozoids into a nucule. *B–C*, diagram of longitudinal sections of germinating zygotes. *D–E*, surface views of germinating zygotes. *F*, germling at protonematal stage. *G*, young plant after development of first nodal branch. (*Pr. Ax.*, primary axis; *Prot.*,[1] primary protonema; *Prot.*,[2] secondary protonema; *Prot. Init.*, protonematal initial; *Rhiz.*,[1] primary rhizoid; *Rhiz.*,[2] secondary rhizoid; *Rhiz. Init.*, rhizoidal initial.) (*A*, *after DeBary*, 1871; *B–C*, *based upon DeBary*, 1875 *and Oehlkers*, 1916; *D–F*, *after DeBary*, 1875.) (*A*, × 70; *B–C*, × 45; *D–E*, × 50; *F*, × 25; *G*, × 4.)

The Zygote and Its Germination. The zygote secretes a thick wall, and the inner periclinal walls of tube cells also become thickened. Other portions of walls of a sheath decay, leaving the thickened inner tube cell walls projecting from the zygote like the threads on a screw. The zygote, with the surrounding remains of the sheath, falls to the bottom of the pool

[1] Goetz, 1899.

and there germinates after a period of a few weeks. The zygote nucleus migrates to the apical pole of a zygote and there divides[1] into four daughter nuclei (Fig. 251*B*). This division into four nuclei suggests that division is meiotic. Confirmatory evidence for this supposition is found in the absence of meiosis prior to formation of gametes.[2] According to such an interpretation, the thallus is gametophytic and the zygote is the only diploid cell in the life cycle.

Germination begins with an asymmetrical division of the quadrinucleate zygote into a small lenticular distal cell with one nucleus and a large basal cell containing the other three nuclei (Fig. 251*C*). The lenticular cell soon becomes exposed by a cracking of the distal end of the zygote wall and divides vertically into a *rhizoidal initial* and a *protonematal initial* (Fig. 251*D–E*). The large three-nucleate cell remains undivided, and its nuclei eventually disintegrate. The rhizoidal initial develops into a colorless rhizoid[3] differentiated into nodes and internodes, and one with a whorl of secondary rhizoids growing out at each node. The protonematal initial develops into a green filament (the *primary protonema*) also differentiated into nodes and internodes (Fig. 251*F*). Appendages produced by the lowermost node of a primary protonema become either rhizoids or secondary protonema. The second node of a primary protonema bears a whorl of appendages (Fig. 251*G*). All but one of them are simple green filaments; the remaining appendage develops into a typical axis in which growth is as in an adult plant.

ORDER 1. CHARALES

All Charophyceae, whether living or fossil, are placed in a single order, the Charales. The order has been divided[4] into four families distinguishable from one another primarily by arrangement of sheath cells of the nucules. Three of the families are known only in the fossil condition. The fourth family, the Characeae, has representatives in the present-day flora and has a fossil record extending as far back as the Pennsylvanian.

FAMILY 1. CHARACEAE

All genera referred to the Characeae have nucules which have five tube cells, and these are twisted in a left-hand spiral and capped by a corona with the cells in a single or in two tiers.

Generic and specific differences within the family are based upon both vegetative and reproductive structures. Taxonomists especially interested in Characeae have long used a special terminology for certain structures of a thallus or have a special meaning for certain familiar, widely used, botani-

[1] OEHLKERS, 1916. [2] LINDENBEIN, 1927; SUNDARALINGAN, 1948.
[3] DEBARY, 1875. [4] PECK, 1946.

cal terms. For this reason, the nonspecialist in Characeae often has difficulty in understanding technical descriptions of them. Wood (1947) gives definitions of the terminology used for the Characeae.

FIG. 252. Portions of thalli of Characeae. *A*, *Nitella gracilis* (J. E. Smith) Ag. *B*, *Tolypella prolifera* (Ziz.) von. Leonh.

The three genera found in this country may be distinguished as follows:

1. Cells of corona with a single tier of cells .... 3. **Chara**
1. Cells of corona with a double tier of cells 2
   2. Fertile leaves furcately branched .... 1. **Nitella**
   2. Fertile leaves not furcately branched .... 2. **Tolypella**

1. **Nitella** Agardh, 1824. The axis and branches of *Nitella* are differentiated into nodes and internodes. All species are uncorticated and without a whorl of spine-like cells at the nodes (Fig. 252*A*). Branches

arise in the axils of leaves, and two or more may arise at any node. Leaves, both sterile and fertile, are filamentous, once to repeatedly furcate, and with two, three, four, or more furcations at each point of branching.

Globules and nucules are borne only on leaves. The nucules have a coronula with two tiers of five cells each. Some species are homothallic, others are heterothallic. Male thalli of homothallic species have the globules borne singly and terminally at a point of furcation; female thalli have the nucules borne laterally just below a point of furcation, and singly or in clusters of two to several. Homothallic species have the globules borne singly and terminally at points of furcation and subtended by a single nucule or a cluster of two to several of them (Fig. 253*A*).

Nitella is the only genus found in this country in which all leaves are branched. Thirty-four species have been found in this country. For names and descriptions of them, see Wood (1948).

2. **Tolypella** Leonhardi, 1863. The axis and branches of *Tolypella* are differentiated into nodes and internodes. All species are uncorticated and without a whorl of spine-like branches at the nodes. Branches arise in the axils of leaves and two or three of them may arise at any node. Both the sterile and the fertile leaves are not differentiated into nodes and internodes. The sterile leaves are filamentous (Fig. 252*B*) and usually unbranched, but they may have a few filaments at the apex of cells in the lower portion. The fertile leaves are also filamentous and with an evident axis in which two or more lateral filaments are borne at the apex of each cell in the lower portion.

Nucules of *Tolypella* have a corona with two tiers of five cells each. Most species are homothallic and with the globules borne singly at the apex of cells comprising the axial row of a fertile branch. Nucules are borne below a globule and there are usually several of them below each globule. Both globules and nucules are subtended by a fairly long one-celled stipe (Fig. 253*B*).

Tolypella is the only genus in this country in which a stem bears both branched and unbranched leaves. At least 10 species occur in the United States. For an account of the species found in this country, see Allen (1883).

3. **Chara** Valliant, 1719. The primary axis, branches, and leaves of *Chara* are differentiated into nodes and internodes. A few species are uncorticated; the majority have corticated branches and leaves or have the cortication restricted to the branches. Branches arise in the axils of leaves, and there is usually but a single branch at a node. There is a whorl of 6 to 16 leaves at every node and, unlike *Nitella* and *Tolypella*, each leaf is subtended by a single or a pair of spine-like cells. The leaves are always

unbranched, with 5 to 15 nodes and with a whorl of spine-like leaves at each node.

Nucules have a corona with a single tier of five cells. The thalli may be heterothallic or homothallic, but in either case the fructifications are restricted to the adaxial side of leaves and are borne at the nodes. Heterothallic species have the fructifications borne singly at the nodes. Sometimes homothallic species may have the globules and nucules borne singly

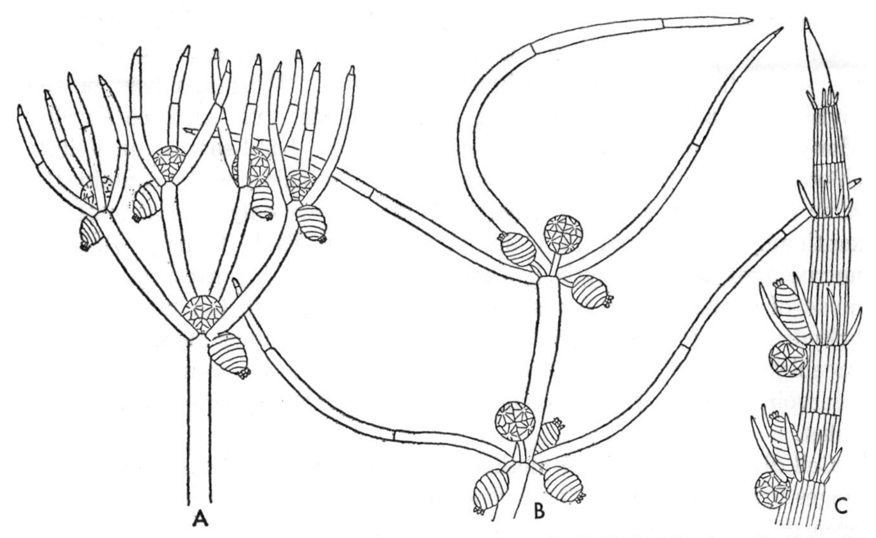

FIG. 253. Fertile leaves of Characeae. *A*, *Nitella gracilis* (J. E. Smith) Ag. *B*, *Tolypella prolifera* (Ziz.) var. Leonh. *C*, *Chara intermedia* A. Br.

at different nodes, but usually there is one of each at a node and with the nucule always borne above the globule (Fig. 253*A*).

Vegetative thalli of *Chara* may be distinguished from other Characeae of the United States by the fact that all leaves are unbranched and by the spine-like cells just below the leaves. In fruiting material, *Chara* may be differentiated from other genera in this country by the singe tier of five cells at the top of a nucule. Approximately 35 species occur within the United States. For a monograph describing a majority of the species of this country, see Robinson (1906).

CHAPTER 5

DIVISION EUGLENOPHYTA

The euglenoid algae constitute a well-defined series in which there are beginnings of an evolution toward an algal organization, but in which there are no higher types than palmelloid colonies. Most members of the division are naked free-swimming cells with one, two, or three flagella. Many of the genera have grass-green, discoid, band-shaped or stellate chloroplasts, with or without pyrenoids. The chloroplasts contain the same chlorophylls as Chlorophyceae, beta-carotene only, and at least one xanthophyll not found in Chlorophyceae (see Table I, page 3). The nutrition may be holophytic, holozoic, or saprophytic, but irrespective of the mode of nutrition, the foods reserved are paramylum (an insoluble carbohydrate related to starch) and fats. There are one or more contractile vacuoles at the anterior end of a motile cell, which are connected with a reservoir which, in turn, is connected with the cell's exterior by a narrow gullet.

Cell division is the usual method of reproduction. Thick-walled resting stages (cysts) are known for several genera. Sexual reproduction has not been demonstrated beyond all doubt for the group.

Occurrence. Euglenoids are most often found in small pools rich in organic matter. The pigmented forms, especially *Euglena*, are frequently present in sufficient abundance to color the water. The saprophytic colorless forms are rarely present in quantity, and they grow most abundantly when a considerable amount of putrefaction is taking place. Some euglenoids grow on damp mud along the banks of rivers, estuaries, and salt marshes where they may be in sufficient abundance to color the mud.[1] Sessile species grow upon algae, plant debris, and small plankton crustaceans. Members of the group found in unusual habitats include the *Euglena* that has been found[2] to be one of the organisms causing green snow, and those that grow endozoic in the intestinal tracts of amphibia.

Cell Structure. All free-swimming genera have naked cells and have the exterior portion of the cyotoplasm differentiated into a periplast. The periplast may be so rigid that the cells have a fixed shape, or it may be flexible and continously changing in shape as a cell moves through the water. Some species have a smooth periplast; others have a periplast

[1] JAHN, 1946. [2] KIENER, 1944.

with spiral or straight longitudinal striae or have one with spiral ridges. Four genera, three of which are free-swimming and one of which is sessile, have the protoplast surrounded by a lorica. The lorica is always open at the anterior end and with the flagella projecting through the opening. A lorica is composed of a firm gelatinous substance, without a trace of cellulose,[1] and is frequently brownish because of impregnation with iron compounds. The shape and ornamentation of the lorica are characteristic for any given species, and these are the chief characters differentiating species within a genus.

Practically all members of the family Euglenaceae have cells with chloroplasts. Chloroplasts are green in color and discoid, band-shaped, or stellate. The chief product of photosynthesis is paramylum, an insoluble polysaccharide not found in other algae. Paramylum is starch-like but does not respond to the usual tests for starch.[2] The shape of the paramylum bodies (spherical, discoid, bacillar, or annular) is often characteristic for a particular species. Many species with chloroplasts have one or more conspicuous pyrenoids, either within or external to the chloroplasts. Pyrenoids within chloroplasts are more or less intimately associated with paramylum bodies,[3] but those external to chloroplasts seem to have no relationship to paramylum bodies. Nutrition of Euglenaceae with chloroplasts may also be holozoic or saprophytic.[4] Ingestion of solid foods is through the cytostome, a differentiated softer portion of the periplast. The gullet at the anterior end functions as the cytostome of most genera, but some of them, as *Peranema*,[5] have a cytostome distinct from the gullet.

The gullet at the anterior end of a motile cell is usually flask-shaped and differentiated into a narrow neck, the *cylopharynx*, and an enlarged posterior portion, the *reservoir*. The reservoir is adjoined by one or more large contractile vacuoles formed by fusion of several small vacuoles.[6] The reservoir has been interpreted[7] as an accessory structure assisting in discharge of liquids from the contractile vacuoles, but its function is probably passive since, from the morphological standpoint, it is merely an invagination of the cell apex.

Most genera of the Peranemaceae have *rod organs* (*pharyngeal rods*) adjacent to the fullet. These rod-like structures lie parallel with the long axis of a gullet, and with their lower extremities level with the base of the reservoir (*Peranema*) or extending to the posterior end of a cell (*Entosiphon*). Certain genera with pharyngeal rods have them terminating beneath a cytostome entirely distinct from the gullet (*Entosiphon*, *Hetero-*

[1] Klebs, 1883. [2] Czurda, 1928. [3] Günther, 1928; Mainx, 1928.
[4] Doyle, 1943; Günther, 1928; Hall, 1939; Mainz, 1928; Tannreuther, 1923.
[5] Brown, V. E., 1930; Pitelka, 1945. [6] Jahn, 1946. [7] Wager, 1899

nema, *Peranema*). The function of the rods has been a matter of controversy. Many hold that its function is that of a trichite which serves as a supporting organ for the distended cytostome when a cell is ingesting solid foods.[1]

Flagella of motile cells are inserted in the base of the reservoir and project through the cytopharynx. The projecting portion is a "tinsel" type of flagellum; namely, it consists of an axial filament surrounded for its entire length by a sheath to which are attached diagonally inserted cilia along one side.[2] These structures of a flagellum can be demonstrated only by special methods. A majority of the genera investigated have been shown to have a neuromotor apparatus of the blepharoplast-rhizoplast-centriole type (Fig. 254). Uniflagellate genera assigned to the Euglenaceae have been shown[3] to have the flagellum bifurcating with the reservoir and to have a granular swelling at the point of forking. Euglenaceae with more than one flagellum have a granular swelling some distance above the blepharoplast but no bifurcation of the flagella.[4] Thus far, members of other families have been found to be without a granular swelling or bifurcation at the base of their flagella. All these genera have been shown to have a blepharoplast at the base of each flagellum, but in several cases there is no rhizoplast.[5]

Most of the species with chloroplasts, and certain of the colorless species have an eyespot near the anterior end of a motile cell. The eyespot consists of a simple mass of pigment granules arranged in a single layer.[6] It has also been held[7] that there is a hyaline lens exterior to the pigmented plate. Species with eyespots show phototactic responses; those without eyespots do not.

The nucleus is a prominent structure and one easily recognized without staining. All genera are uninucleate under normal conditions, but sometimes they become multinucleate if cytokinesis is inhibited.[8] The extensive literature on nuclear structure and mitosis among Euglenophyta[9] shows that practically all genera have a conspicuous karyosome, a well-defined membrane, and chromatic granules. The nuclear membrane persists throughout mitosis, and the endosome, which does not contribute material to the formation of chromosomes, divides into two portions one of which goes to each daughter nucleus.

[1] HALL, 1933; HALL and POWELL, 1928; HYMAN, 1936; RHODES, 1926; SCHAEFFER, 1918.

[2] DEFLANDRE, 1934; MAINX, 1928; PETERSEN, 1929; VLK, 1938.

[3] HALL and JAHN, 1929. [4] STEUER, 1904; WENRICH, 1924.

[5] HALL and JAHN, 1929; HALL and POWELL, 1928.

[6] FRANZÉ, 1894; WAGER, 1899. [7] MAST, 1928.

[8] KRICHENBAUER, 1937; MAINX, 1928. [9] See JAHN (1946) for numerous references.

Asexual Reproduction. Multiplication is by cell division and may take place while the cells are actively motile or after they have come to rest. Cytokinesis is longitudinal (Fig. 254) and begins at the anterior end of a cell. During later stages of mitosis in species with a single flagellum, the blepharoplast divides into two daughter blepharoplasts. The flagellum usually remains attached to one of the daughter blepharoplasts, and a new flagellum grows from the other daughter blepharoplast. In biflagellate species, both flagella may go to one daughter cell, or each of the daughter cells may receive one of the old flagella.

Species dividing while in motion have an immediate separation of the two daughter cells. If the motile species has a lorica, cell division takes place within the lorica, after which one daughter protoplast escapes and secretes a new lorica. Species with naked cells that come to rest before division often have a cell secreting a gelatinous sheath. Sometimes the

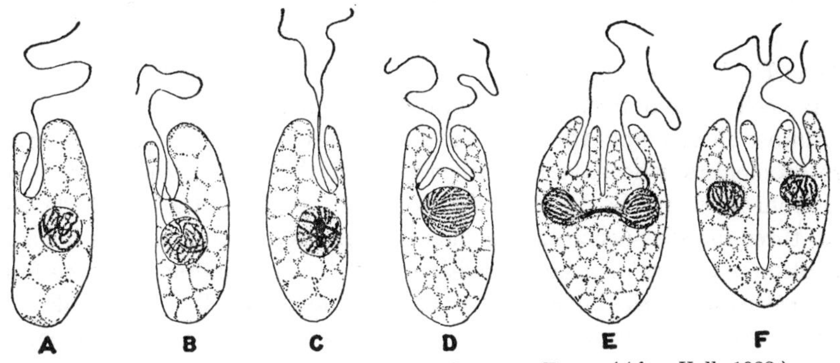

FIG. 254. Cell division of *Rhabdomonas incurvum* Fres. (*After Hall*, 1923.)

daughter protoplasts do not escape from the gelatinous matrix before they divide again. In such cases, as in *Phacus pleuronectes* (O.F.M.) Duj.[1] and *Euglena gracilis* Klebs,[2] there is a development of a temporary palmelloid colony in which the cells may return to a motile condition at any time.

Thick-walled resting stages (cysts) are common in many genera. Germinating cysts usually have the protoplast developing into a single motile cell.

Sexual Reproduction. Syngamy has been described for a few Euglenophyta,[3] but all cases noted are considered extremely dubious. Autogamy (the fusion of two nuclei both derived from the same parent cell) has been described for *Phacus*,[4] but doubt has been expressed[5] concerning this re-

[1] DANGEARD, P. A., 1901A. [2] TANNREUTHER, 1923.

[3] BERLINER, 1909; DOBELL, 1908; HAASE, 1910.

[4] KRICHENBAUER, 1937. [5] JAHN, 1946.

ported case. Evidence for autogamy is based upon the finding of binucleate cells with the two nuclei at the base of, and in the longitudinal axis of, a cell, and upon the finding of cells with large nuclei and two karyosomes. The finding of diakinesis stages indicates that the fusion nucleus divides meiotically, and the finding of quadrinucleate cells dividing to form four cells indicates vegetative cells are haploid.

CLASS 1. EUGLENOPHYCEAE

The first formal recognition of the euglenoids as a division of the plant kingdom[1] divides them into two classes, each with a single order. A more logical treatment seems to be that of placing them in a single class, the Euglenophyceae, because differences between the two orders are of the same magnitude as in orders of Chlorophyceae, Chrysophyceae, Xanthophyceae, and other classes of algae.

ORDER 1. EUGLENALES

The Euglenales include all the Euglenophyceae in which the flagellated motile cell is the dominant phase in the life cycle.

Almost all attempts to divide the Euglenales into families have been by protozoologists, who have used diverse criteria in segregating the families one from another. Some[2] stress the mode of nutrition and group Euglenales into three families according to their holophytic, saprophytic, or holozoic nutrition. Others disregard nutrition entirely and differentiate into families according to the bilateral or radial symmetry of cells,[3] or according to the presence of a rigid or a plastic periplasm.[4] Still others utilize the presence of chloroplasts or eyespots and the number of flagella in differentiating between families.[5]

Each of the foregoing systems results in an erection of more or less artificial families. The detailed structure of the flagellum has been shown[6] to be a significant character in segregating into families. Supplemental characters, such as number of flagella, rigidity or plasticity, type of locomotion, mode of nutrition, presence or absence of pharyngeal rods, are helpful but not absolute criteria in delimiting families.

FAMILY 1. EUGLENACEAE

Cells of Euglenaceae, irrespective of whether uni-, bi-, or triflagellate, have a granular swelling on the portion of each flagellum within the reservoir.[6] The basal portion of a flagellum may or may not be bifurcate. Most genera have chloroplasts, and genera without chloroplasts are to be

[1] PASCHER, 1931. [2] KLEBS, 1892; LEMMERMANN, 1913; SENN, 1900.
[3] DOFLEIN-REICHENOW, 1929. [4] ELENKIN, 1924.
[5] CALKINS, 1926; KUDO, 1946. [6] HALL and JAHN, 1929.

distinguished from other colorless genera by the presence of an eyespot. A majority of the genera have naked cells, but certain of them have the cell surrounded by a lorica. Cells of Euglenaceae are variously shaped, rigid or plastic, and with a smooth or ornamented periplast.

Cell division usually takes place while a cell is immobile and enveloped by a gelatinous sheath. Temporary palmelloid colonies are sometimes formed.

The genera found in this country may be distinguished as follows:*

1. Free-living 2
1. Inhabiting intestinal tracts of frogs .... 8. **Euglenamorpha**
 2. With one flagellum 3
 2. With two flagella .... 7. **Eutreptia**
3. Protoplast without a lorica 4
3. Protoplast surrounded by a lorica 7
 4. Cells plastic .... 1. **Euglena**
 4. Cells rigid 5
5. With two elongate chloroplasts .... 4. **Cryptoglena**
5. With numerous chloroplasts 6
 6. Cells compressed .... 3. **Phacus**
 6. Cells radially symmetrical .... 2. **Lepocinclis**
7. Cells free-swimming .... 5. **Trachelomonas**
7. Cells sessile .... 6. **Ascoglena**

1. **Euglena** Ehrenberg, 1838. This is the only green uniflagellate genus of the family in which free-swimming cells are continually changing in shape as they move through the water. The cells are fusiform to acicular and with the posterior end more or less pointed. Most species have the periplast ornamented with delicate striae or rows of punctae. A cell has a gullet at the anterior end and one or more contractile vacuoles adjoining the reservoir. The single flagellum is bifurcate at its lower end and with a granular swelling at the point of branching. Each branch terminates in a blepharoplast, one of which is connected to an extranuclear centriole by a delicate rhizoplast.[1] Most species have an eyespot at the anterior end. The chloroplasts are numerous and discoid to band-shaped. They may be with or without pyrenoids, but in either case there is a formation of paramylum bodies of characteristic shape. Some species, as *E. sanguinea* Ehr., often develop a red pigment in such quantities that the cell contents are obscured. A few species lack chloroplasts, but all of them have an eyespot and have the lower end of the flagellum bifurcate and with a

* JAHN and MCKIBBON (1937) segregate the colorless species of *Euglena* in a separate genus which they call *Khawkinea*. Since other members of the family (*Phacus* and *Trachelomonas*) have colorless species, it does not seem logical to make this generic distinction based solely upon absence of chloroplasts.

[1] HALL and JAHN, 1929.

granular swelling. Mitosis is intranuclear and with a division of the karyosome.[1]

Division may take place while the cells are motile or after they have come to rest. Division in the motile condition is longitudinal and begins at the anterior end.[2] Immobile dividing cells are usually surrounded by a gelatinous envelope; sometimes the daughter cells are retained within the envelope and redivide to form a temporary palmelloid colony.[3] Thick-walled resting cells (cysts) are common in *Euglena*. One species has been reported[4] as conjugating, but the proof of this is inconclusive.

Euglena (Fig. 255) is a common organism of waters rich in organic matter, as pools in barnyards, and frequently occurs in such abundance as to color the water a deep green. More than 50 species have been recorded for this country. For descriptions of many of the species found in the United States, see Johnson (1944).

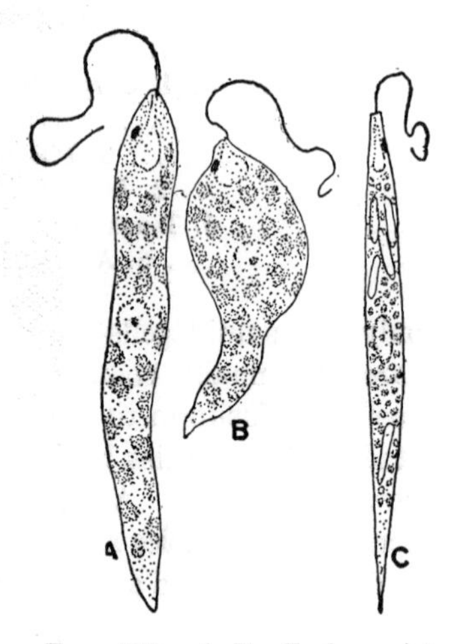

FIG. 255. *A–B, Euglena intermedia* (Klebs) Schmitz. *C, E. acus* Ehr. (× 430.)

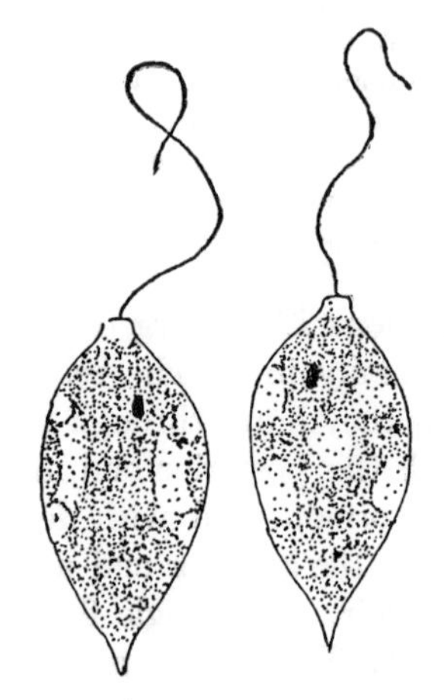

FIG. 256. *Lepocinclis Steinii* Lemm. (× 1300.)

2. **Lepocinclis** Perty, 1849. Members of this motile genus have rigid naked cells in which the periplast usually has numerous longitudinal or spiral striae. The cells are radially symmetrical, broadly ellipsoidal to ovoid, and sometimes with the posterior pole more or less pointed. The gullet and vacuolar system are similar to those of *Euglena*. The single flagellum is bifurcate at the lower end and with a granular swelling,[5] but no rhizoplast has been demonstrated. The cells usually contain numerous

[1] BAKER, 1926; RATCLIFFE, 1927. [2] TANNREUTHER, 1922.
[3] DANGEARD, P. A., 1901*A*; TANNREUTHER, 1922. [4] HAASE, 1910.
[5] HALL and JAHN, 1929.

parietal discoid chloroplasts and two large, laterally located, ring-shaped paramulum bodies.

Reproduction is by cell division and takes place after cells have come to rest.[1] Thick-walled resting stages have not been recorded for this genus.

Fifteen species of *Lepocinclis* (Fig. 256) have been reported from the United States. For the latest monograph of the genus, see Conrad (1934).

3. **Phacus** Dujardin, 1841. The naked cells of *Phacus* are solitary, free-swimming, and with a rigid shape. They are conspicuously flattened and somewhat, or markedly, twisted. The periplast is ornamented with striae, punctae, or denticulations that lie in longitudinal or in spiral series. Numerous transverse striae have been demonstrated between the longitudinal series, but these are evident only after the cells have been specially stained.[2] The gullet, vacuolar system, and bifurcation of the single flagellum are similar to those of *Euglena*.[3] The protoplast contains numerous discoid chloroplasts, and paramylum may accumulate in either one or two conspicuous ring-shaped bodies, or in several discoid ones. Some species have an eyespot; others lack one.

Division may take place either while the cell is motile or after it has become immobile and affixed by a basal gelatinous cushion.[4] Resting cells (cysts) have been described for several species, but these are of much less frequent occurrence than in *Euglena*.

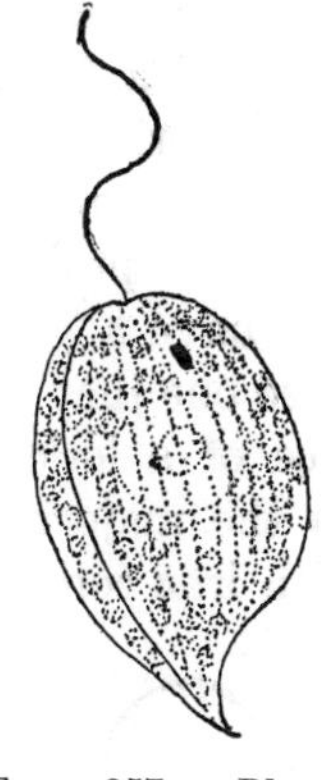

FIG. 257. *Phacus acuminatus* Stokes. (× 975.)

An autogamous fusion of sister nuclei has been described.[4] The fusion nucleus divides meiotically to form four nuclei after which there is a quadripartition of the four-nucleate cell into four uninucleate vegetative cells.

Phacus (Fig. 257) does not have the same preference for stagnant waters as does *Euglena*. It is found in a wide variety of habitats, but rarely in abundance. Some 32 species have been recorded for the United States. Allegre and Jahn (1943) give descriptions of many species from the United States and Pochmann (1942) gives descriptions of all known species.

4. **Cryptoglena** Ehrenberg, 1831. This imperfectly known genus has free-swimming, rigid, ovate naked cells that are somewhat compressed. The anterior pole of a cell is broadly rounded and slightly indented at the point of emergence of the single flagellum; the posterior pole is acute. There is a gullet at the anterior end, but the structure of this and that of

[1] DANGEARD, 1901*A*. [2] DEFLANDRE, 1931. [3] HALL and JAHN, 1929.
[4] KRICHENBAUER, 1937.

the flagellum's base are unknown. The cells have two laminate longitudinal chloroplasts, one on either side of a cell. The nucleus lies toward the posterior end of a cell, and an eyespot toward the anterior end. Reproductive stages have not been observed in *Cryptoglena.*

The single species, *C. pigra* Ehr. (Fig. 258) has been found at several widely separated localities in this country. For a description of it, see Lemmermann (1913).

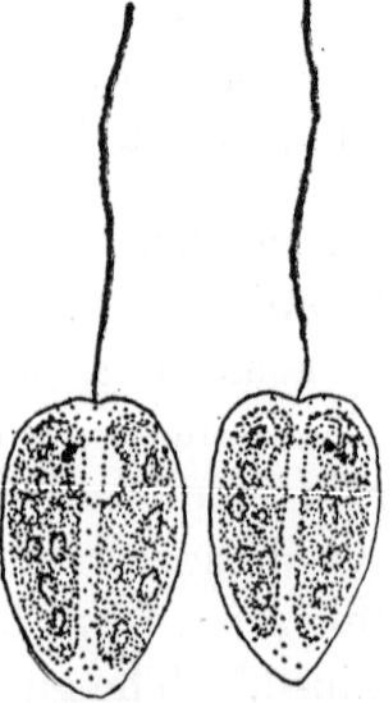

FIG. 258. *Cryptoglena pigra* Ehr. (*After Klebs,* 1892.) (× 1500.)

5. **Trachelomonas** Ehrenberg, 1833. *Trachelomonas* is the only member of the family in which the cells are free-swimming and surrounded by a lorica. The lorica (Fig. 259) always has a circular pore at the anterior end and may or may not have the portion adjacent to the pore elevated into a collar. The flagellum projects far beyond the collar. Most species have a globose or an ellipsoidal lorica; a few have one that is campanulate or fusiform. The surface of a lorica may be smooth, punctate, spiny, reticulate, or striate. In the majority of cases, a lorica is brown because of an impregnation with iron compounds, and the color may be so deep that it is impossible to see the protoplast. The protoplast, which is not attached to the lorica, has a plastic periplast. It has a gullet at the anterior end, and the single flagellum has the bifurcate base and granular swelling typical of Euglenaceae.[1] There are 2 to 15 discoid parietal chloroplasts, and these may or may not have pyrenoids. One species, *T. reticulata* Klebs, lacks chloroplasts. The nucleus is usually central in position.

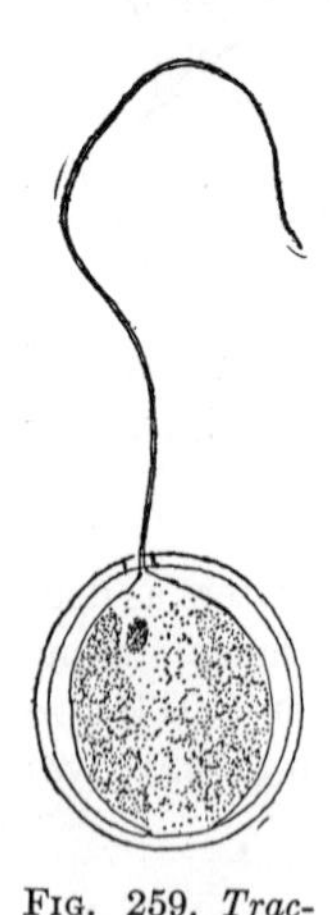

FIG. 259. *Trachelomonas volvocina* Ehr. (× 975.)

Cells of *Trachelomonas* become immobile before division of the protoplast into two daughter protoplasts. One daughter protoplast escapes and secretes a new lorica; the other remains within the old lorica.[2] Unfavorable conditions may cause the protoplast to emerge from the lorica. If such protoplasts divide, they usually develop into palmelloid colonies. Several species are known to form thick-walled resting cells.

The two most recent monographs of the genus[3] recognize approximately 150 species that are differentiated from one another on the shape and ornamentation of the lorica. Some 70 species have been recorded for the United States.

[1] HALL and JAHN, 1929. [2] DANGEARD, 1901*A*.
[3] DEFLANDRE, 1926; SKVORTZOW, 1925*A*.

6. **Ascoglena** Stein, 1878. The cells of *Ascoglena* are sessile and affixed to the substratum by a cylindrical to urn-shaped lorica that is open at the upper end. Except for a collar-like zone at the upper end, a lorica is impregnated with ferric compounds[1] and brownish in color. The protoplast, whose structure is similar to that of *Euglena*, is attached to the base of the lorica by a short cytoplasmic stalk, the only portion of the protoplast that is in contact with the lorica.

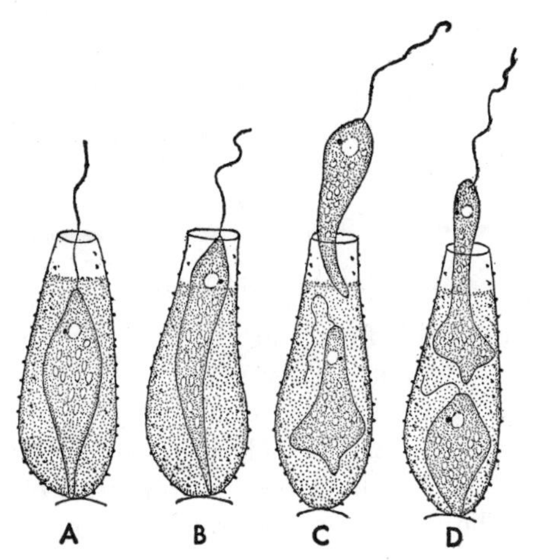

FIG. 260. *Ascoglena vaginicola* Stein. (*After Stein*, 1878.) (× 650.)

Reproduction is by division into two daughter protoplasts. One of the daughter protoplasts remains within the old lorica; the other swims away, becomes affixed to some object, and secretes a new lorica.[2]

A. vaginicola Stein (Fig. 260) has been found in Ohio. For a description of it, see Lemmermann (1913).

7. **Eutreptia** Perty, 1852. The cells of *Eutreptia* are free-swimming and with a pronounced plasticity. There are two flagella of equal length at the anterior end, and each of them has an elongate granular swelling in the basal portion.[3] The cells are fusiform, with the anterior end blunter, and have a finely striate periplast. The gullet is conspicuous and adjoined by small vacuoles,[4] and a discoid eyespot lies adjacent to the reservoir portion of the gullet.[5] A cell contains numerous parietal discoid chloroplasts, and the nucleus lies in the posterior portion of a cell.

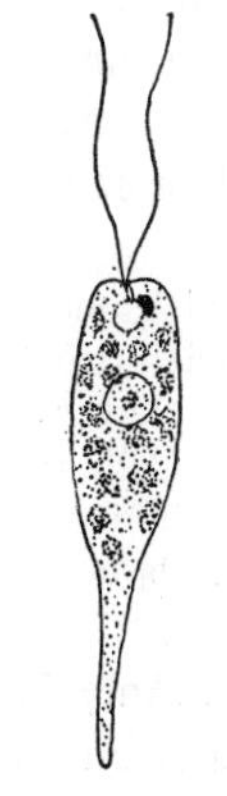
FIG. 261. *Eutreptia viridis* Perty. (*After Klebs*, 1883.) (× 300.)

Cell division takes place while a cell is actively motile, is logitudinal, and cytokinesis begins at the anterior end. One species has been found with thick-walled resting cells.[6]

The types species, *E. viridis* Perty (Fig. 261), has been recorded from several widely separated stations in this country. For a description of it, see Lemmermann (1913).

8. **Euglenamorpha** Wenrich, 1923. *Euglenamorpha* is the only member of the family that has three flagella. The number of flagella and

[1] KLEBS, 1883. [2] STEIN, 1878. [3] STEUER, 1904. [4] KLEBS, 1883.
[5] KLEBS, 1883; STEUER, 1904. [6] STEUER, 1904.

their lack of bifurcation at the point of insertion are the only morphological features separating this genus from *Euglena*.[1]

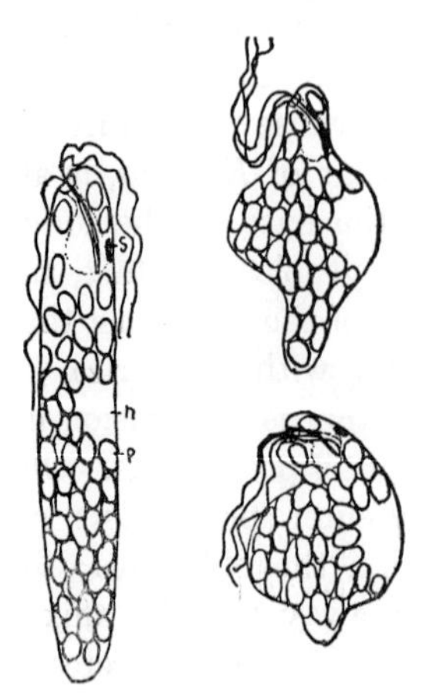

FIG. 262. *Euglenamorpha Hegneri* Wenrich. (*From Wenrich*, 1924.)

Euglenamorpha is endozoic in habit and has been found in the intestinal tract of tadpoles of *Rana*. Probably it is widespread, but thus far it has been found[1] only in New York, Pennsylvania, and Massachusetts. Both the pigmented and colorless individuals are referred to a single species, *E. Hegneri* Wenrich (Fig. 262). For a description of it, see Wenrich, (1924).

FAMILY 2. ASTASIACEAE

Members of this family are without chloroplasts or an eyespot. They may have one or two flagella, but the bases of the flagella are without granules, and in only one species has the base of the flagellum been found to be bifurcate. The cells never move with a gliding or a creeping motion. None of the genera referred to the family has been found to have a pharyngeal rod apparatus.

The Astasiaceae have been combined with the Euglenaceae to form a single family,[2] but differences in flagellar structure of the two show[3] that this is unjustified.

The genera found in this country differ as follows:

1. With one flagellum 2
1. With two flagella .... 4. **Distigma**
  2. Cells strongly plastic .... 1. **Astasia**
  2. Cells rigid or slightly plastic 3
3. Cells radially symmetrical .... 3. **Rhabdomonas**
3. Cells not radially symmetrical .... 2. **Menoidium**

1. **Astasia** Ehrenberg, 1830; emend., Stein, 1878. Cells of *Astasia* are free-swimming, uniflagellate, and with more or less change in shape as they swim through the water. There are no chloroplasts. Eyespots have been found in certain species referred to *Astasia*, but the structure of the flagellum shows[3] that they should be referred to *Euglena*. The flagellum is without a granular swelling in the basal portion,[3] and in only one species has the basal portion of the flagellum been found[4] to be bifurcate. Recently divided cells have the blepharoplast at the flagellar base connected

[1] WENRICH, 1924. [2] DOFLEIN-REICHENOW, 1929. [3] HALL and JAHN, 1929.
[4] LACKEY, 1934.

with an extranuclear centriole by a delicate rhizoplast.[1] The periplast may have delicate striae. Nutrition of many species is saprophytic, and the size and number of paramylum bodies vary greatly from species to species.

Division is longitudinal and takes place while the cells are motile. Certain species form pentagonal cysts.[2]

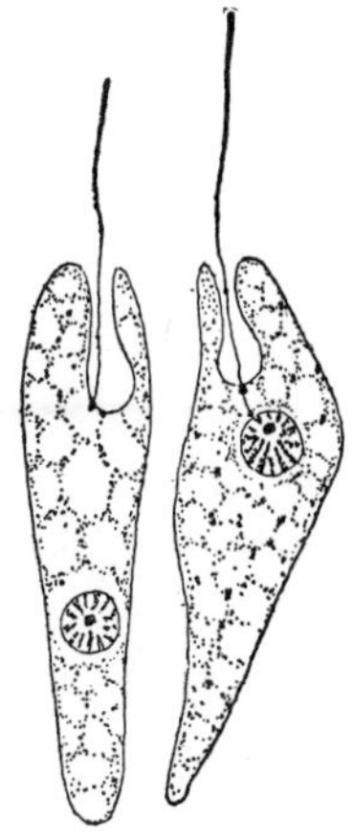

Astasia is generally found in stagnant waters rich in organic matter. *A. Dangeardii* Lemm. (Fig. 263) and *A. Klebsii* Lemm. have been found in this country. For descriptions of them, see E. G. Pringsheim (1942).

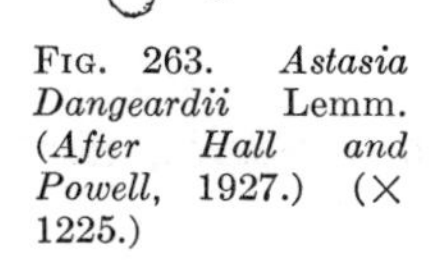

FIG. 263. *Astasia Dangeardii* Lemm. (*After Hall and Powell*, 1927.) (× 1225.)

2. **Menoidium** Perty, 1852. The rigid colorless cells of this genus have a single flagellum. In front view, the outline of a cell has one side approximately straight and the other convex.[3] In cross section, a cell is triangular with the straight side of the front view constituting the front angle of the triangle. The shape of a cell has been compared[3] to a segment of an orange. The nutrition is saprophytic. There are several large rods or elongate rings of paramylum in the anterior half of a cell, and numerous small spherical paramylum bodies in the posterior half.

M. falcatum Zach., *M. pellucidum* Perty (Fig. 264) and *M. tortuosum* Stokes have been found in this country For descriptions of them, see Lemmermann (1913).

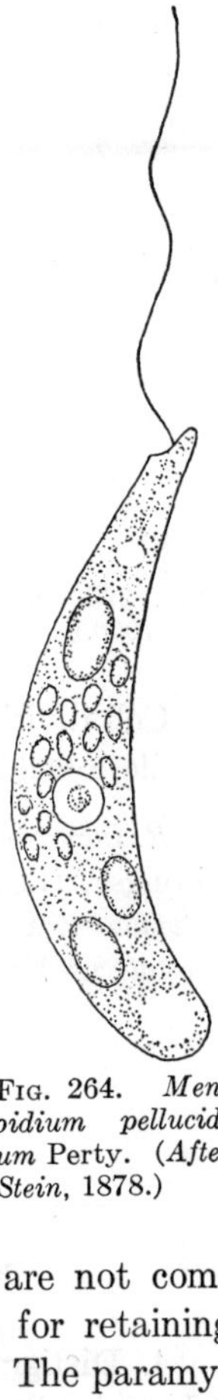

FIG. 264. *Menoidium pellucidum* Perty. (*After Stein*, 1878.)

3. **Rhabdomonas** Fresenius, 1858. In its rigid, uniflagellate, colorless, spirally ridged cells, *Rhabdomonas* closely resembles *Menoidium*. It differs in that the cells are not compressed, and this has been deemed[3] a sufficient justification for retaining it as a separate genus instead of uniting it with *Menoidium*. The paramylum bodies are similar to those of *Menoidium*. There is a well-developed gullet, but no contractile vacuoles have been demonstrated[4] adjacent to the reservoir. The single flagellum is inserted in the reservoir and is without bifurcation or a granular swelling at the base. The blepharoplast

[1] HALL and POWELL, 1927. [2] LACKEY, 1934. [3] PRINGSHEIM, E. G., 1942.
[4] HALL, 1923.

at the base of flagellum is connected with an extranuclear granule by a delicate rhizoplast.[1] The nucleus is conspicuous and generally centrally located.

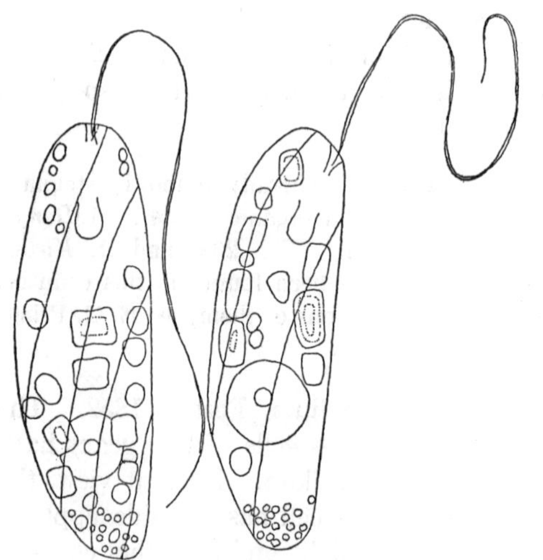

FIG. 265. *Rhabdomonas costata* (Korshikov) Pringsh. (*After E. G. Pringsheim*, 1936.) (× 1050.)

Cell division is longitudinal, begins at the anterior end and takes place while a cell is actively motile (Fig. 254, page 351).

R. incurva Fres. is widely distributed in this country, and *R. costata* (Korshikov) Pringsh. (Fig. 265) has been found in North Carolina. For a description of *R. incurva* as *Menoidium incurvum* (Fres.) Klebs, see Lemmermann (1913); for *R. costata* as *Menoidium longum* Pringsh., see E. G. Pringsheim (1936).

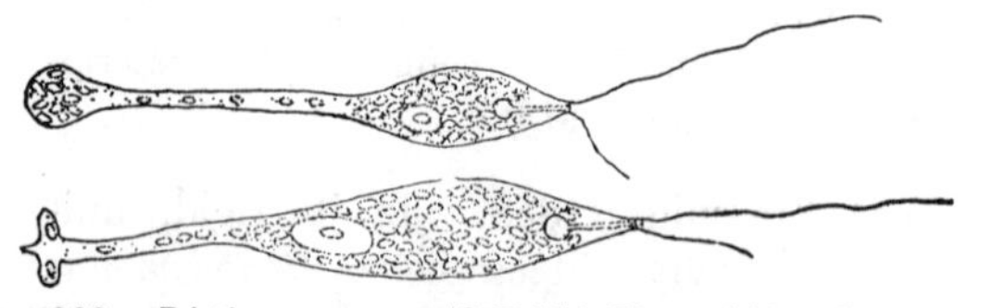

FIG. 266. *Distigma proteus* (O.F.M.) Ehr. (*After Stein*, 1878.)

4. **Distigma** Ehrenberg, 1832. *Distigma* has colorless, biflagellate, spindle-shaped cells in which the two flagella are markedly different in length. The longer flagellum is directed forward, the shorter is bent to one side,[2] and both flagella are without a granular thickening in the basal portion.[3] The protoplasm of most species is very fluid, and a cell is continually chang-

[1] HALL, 1923; HALL and JAHN, 1929.
[2] PRINGSHEIM, E. G., 1942. [3] LACKEY, 1934.

ing in shape as it swims through the water. The periplast has very delicate spiral striae. The gullet extends some distance back into a cell and is adjoined by several contractile vacuoles. According to the species, the paramylum bodies are rod-shaped or ovoid.

Division is longitudinal and takes place while a cell is motile.[1]

D. proteus Ehr. (Fig. 266) has been reported from several localities in this country. For a description of it, see E. G. Pringsheim (1942).

Family 3. Peranemaceae

The Peranemaceae have colorless cells that may be rigid or strongly plastic. Locomotion is usually a creeping or a gliding movement. Some genera have one flagellum; others have two. Biflagellate genera have one flagellum (the swimming flagellum) pointing straight ahead and the other trailing. Neither uni- nor biflagellate genera have a granular thickening in the basal portion of flagella. Most genera have the gullet adjoined by a pharyngeal rod apparatus. In several genera, the lower end of this structure is level with the base of the reservoir; in some genera, as *Entosiphon*, it extends to the posterior end of a cell. Nutrition of Peranemaceae is usually holozoic, and certain genera show a pronounced selectivity with respect to the foods ingested.

Division usually takes place while the cells are motile. Thick-walled resting cells are known for certain genera but are of rare occurrence.

There is not a single consistent character separating this family from other Euglenales. Typical members of the family have a pharyngeal rod apparatus, are holozoic, and move in a distinctive manner.

The genera found in this country differ as follows:

1. With one flagellum 2
1. With two flagella 6
   2. Body rigid .... 12. **Petalomonas**
   2. Body more or less plastic 3
3. Cells more or less spindle-shaped 4
3. Cells flask-shaped .... 11. **Urceolus**
   4. With pharyngeal rods 5
   4. Without pharyngeal rods .... 10. **Euglenopsis**
5. Feeding exclusively upon diatoms .... 2. **Jenningsia**
5. Not feeding exclusively upon diatoms .... 1. **Peranema**
   6. Body rigid 7
   6. Body more or less plastic 13
7. Trailing flagellum the shorter 8
7. Trailing flagellum the longer 10
   8. Periplast longitudinally ridged 9
   8. Periplast not longitudinally ridged .... 9. **Notosolenus**
9. Ridges straight .... 8. **Sphenomonas**

[1] Lackey, 1934.

9. Ridges curved .... 7. **Tropidoscyphus**
    10. Body markedly compressed .... 6. **Anisonema**
    10. Body not markedly compressed 11
11. Pharyngeal rod reaching posterior of cell .... 5. **Entosiphon**
11. Pharyngeal rod, if present, not reaching posterior of cell 12
    12. Trailing flagellum very inconspicuous .... 1. **Peranema**
    12. Trailing flagellum conspicuous .... 3. **Dinema**
13. Cells compressed .... 6. **Anisonema**
13. Cells not compressed .... 4. **Heteronema**

1. **Peranema** Dujardin, 1841. *Peranema* has markedly plastic cells which, when fully extended, gradually broaden from a subacute apex to a broadly rounded base. The cells are colorless and biflagellate. The swimming flagellum is conspicuous; the trailing flagellum is so closely applied to the cell[1] that its existence was overlooked for a long time.

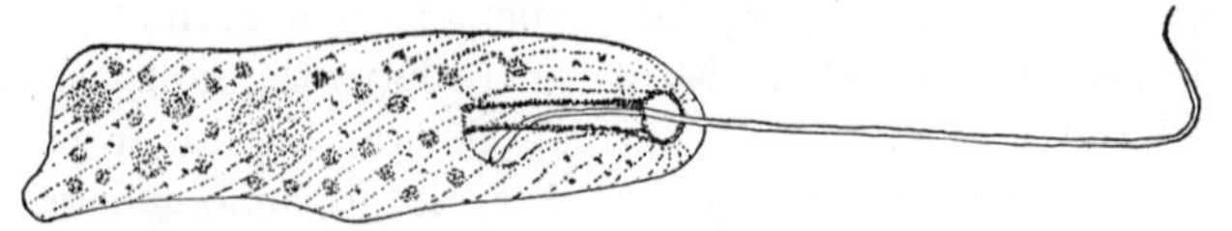

FIG. 267. Diagrammatic sketch of *Peranema trichophorum* (Ehr.) Stein. (*After Hall and Powell*, 1928.) (× 1000.)

The basal portion of both flagella is without a granular swelling.[2] The periplast is ornamented with spiral striae extending backward from the anterior end. The gullet terminates in a large reservoir which is adjoined by small contractile vacuoles, and by two pharyngeal rods whose bases lie slightly below the base of the reservoir. The periplast external to the upper end of the rods is differentiated into a cytostome distinct from the gullet. Nutrition of *Peranema* is holozoic.

Reproduction is by longitudinal division and takes place while a cell is motile.

P. granulifera Penard and *P. trichophorum* (Ehr.) Stein (Fig. 267) have been found in the United States. For descriptions of them, see Lemmermann (1913).

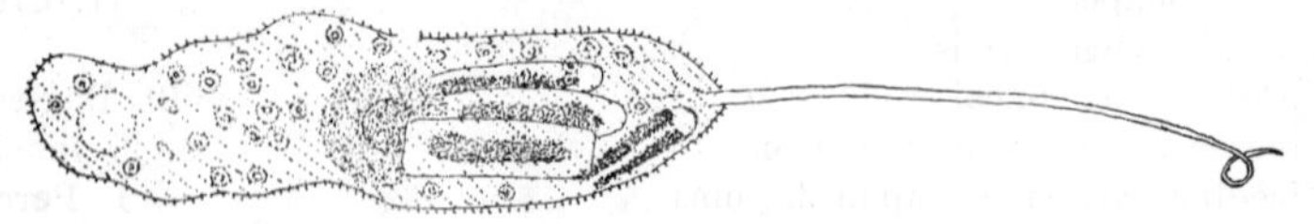

FIG. 268. *Jenningsia diatomophaga* Schaeffer. (*After Schaeffer*, 1918.)

2. **Jenningsia** Schaeffer, 1918. *Jenningsia* is usually recognized as being distinct from *Peranema*, but there are few morphological differences between the two. The distinctive features of *Jenningsia* include a pharyn-

[1] CHADEFAUD, 1938; HALL, 1934; LACKEY, 1933; PITELKA, 1945.
[2] LACKEY, 1933; PITELKA, 1945.

geal rod apparatus with several rods, ring-shaped paramylum bodies, and a marked selectivity of diatoms at the source of food.[1]

There is but one species, *J. diatomophaga* Schaeffer (Fig. 268), and it is known only from the original description[1] of material collected in Tennessee.

3. **Dinema** Perty, 1852. *Dinema* has rigid ellipsoidal cells that, when plasmolyzed, show a separation of the spirally striate periplast from the cytoplasm internal to it.[2] The cells are biflagellate and with the trailing flagellum longer and stouter than the swimming flagellum. A longitudinal

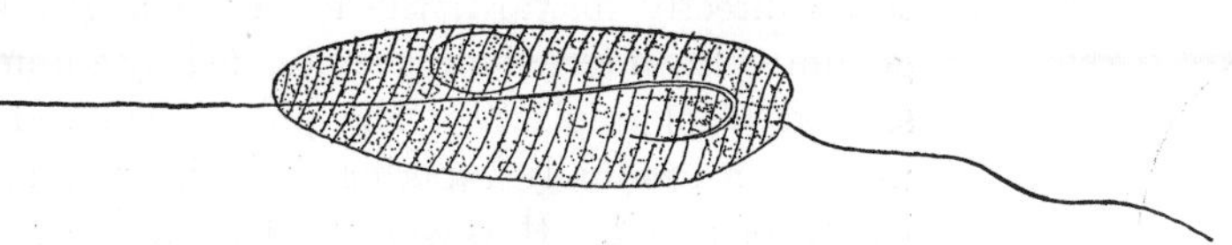

FIG. 269. *Dinema griseolum* Perty. (*After Klebs*, 1892.) (× 660.)

furrow extends a short distance backward from the base of the flagella. The gullet is sac-shaped and adjoined by a single contractile vacuole. Posterior to the gullet is a pharyngeal rod apparatus similar to that of *Peranema*. The nucleus is conspicuous and lies about two-thirds the distance from anterior to posterior ends of a cell.

D. griseolum Perty (Fig. 269) has been recorded[3] from Ohio. For a description of it, see Lemmermann (1913).

4. **Heteronema** Dujardin, 1841; emend., Stein, 1878. The cells of *Heteronema* are biflagellate and markedly plastic. When fully extended, they are cylindrical to spindle-shaped, with a smooth surface or with the surface elevated in spiral ridges. Species with a smooth surface usually have the periplast with spiral striae. The swimming flagellum is directed forward and is considerably longer than the trailing flagellum. The flagella are inserted in the base of the reservoir and terminate in blepharoplasts, but no rhizoplasts are evident in mature or dividing cells. A pharyngeal rod apparatus lies next to the reservoir but is independent from it.[4] The presence of a cytostome external to a curved member (falcate trichite) of the apparatus has been affirmed and denied.[5] The nucleus is conspicuous but is variable in position. Nutrition of *Heteronema* is generally holozoic, and certain of the species have a marked selectivity with respect to the foods ingested.[6]

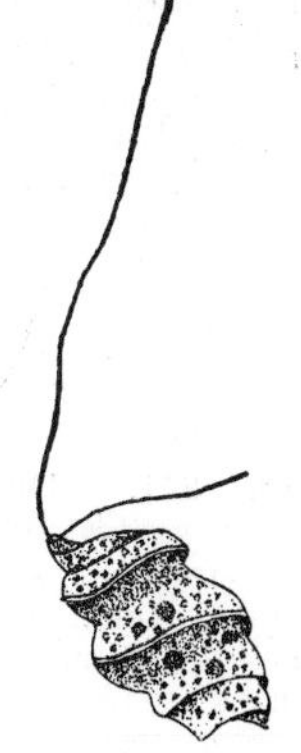

FIG. 270. *Heteronema spirale* Klebs. (*After Klebs*, 1892.) (× 475.)

[1] SCHAEFFER, 1918. [2] KLEBS, 1892. [3] LACKEY, 1938. [4] LOEFER, 1931.
[5] LOEFER, 1931; RHODES, 1926. [6] RHODES, 1926.

The following species are known for this country: *H. acus* (Ehr.) Stein, *H. globiferum* Stein, *H. mutabile* (Stokes) Lemm, and *H. spirale* Klebs (Fig. 270). For descriptions of them, see Lemmermann (1913).

5. **Entosiphon** Stein, 1878. The rigid cells of *Entosiphon* are ovoid, slightly compressed, and longitudinally furrowed with 6 to 12 grooves that are most clearly evident at the anterior end of a cell. There are two flagella inserted in the base of the reservoir and extending through the neck of the gullet. The shorter of these is a swimming flagellum and extends directly forward; the longer is a trailing flagellum. Each flagellum is subtended by a dumbbell-shaped blepharoplast.[1] The pharyngeal rod apparatus is a gradually tapering funnel, extending the length of a cell. It is separated from the reservoir and has a distinct cytostome at the anterior end.[2] The gullet has a short narrow neck and a rather large reservoir which is adjoined by several contractile vacuoles. The nucleus is centrally located and has a large irregularly shaped endosome. Locomotion is generally with a gliding motion and is induced by active vibration of the distal end of the swimming flagellum. It has been held[2] that the "siphon" does not function in the mass ingestion of foods.

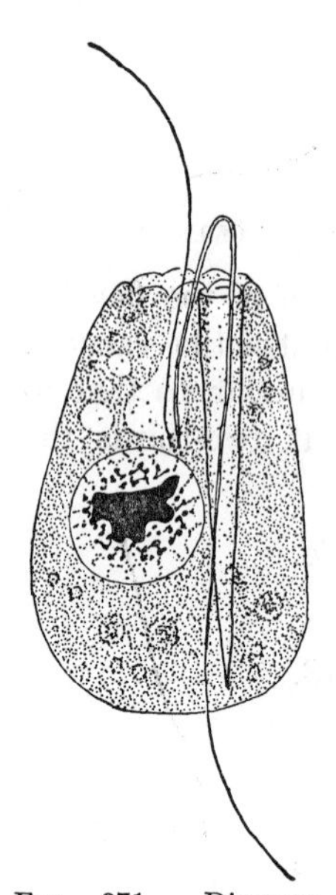

FIG. 271. Diagrammatic ventral view of *Entosiphon sulcatum* (Duj.) Stein. (*Slightly modified from Lackey,* 1929.)

Cell division is longitudinal and takes place while a cell is in motion.

E. sulcatum (Duj.) Stein (Fig. 271) has been reported from several localities in the United States. For a description of it, see Lemmermann (1913).

6. **Anisonema** Dujardin, 1841; emend., Stein, 1878. Certain species of *Anisonema* have rigid cells; others have cells that are plastic.[3] The cells are oval in outline, markedly compressed, and with a longitudinal furrow running from the insertion of flagella to the posterior end of a cell. All species are biflagellate and with the trailing flagellum longer than the swimming flagellum. There is a single contractile vacuole at the side of the reservoir, and the nucleus is excentrically placed with respect to the reservoir. *Anisonema* has the same elongate "siphon" as *Entosiphon*, but this is not so prominent a structure in unstained cells.[4]

[1] LACKEY, 1929. [2] LACKEY, 1929; PROWAZEK, 1903.
[3] KLEBS, 1892.
[4] DOFLEIN-REICHENOW, 1929.

The following five species have been found in this country: *A. acinus* Duj., *A. emarginata* Stokes, *A. ovale* Klebs (Fig. 272), *A. pusillum* Stokes, and *A. truncatum* Stein. For descriptions of them, see Lemmermann (1913).

7. **Tropidoscyphus** Stein, 1878. This genus has colorless, rigid, fusiform to ovoid, biflagellate cells with several curved longitudinal ribs. The swimming flagellum is directed forward and is much longer than the trailing flagellum. The flagella are inserted in an evident gullet, but there is no pharyngeal rod apparatus.[1] The gullet is adjoined by one or more small vacuoles. The nucleus is conspicuous and lies midway between the ends of a cell. The cytoplasm contains numerous globules of a food reserve whose chemical nature is unknown (Fig. 273).

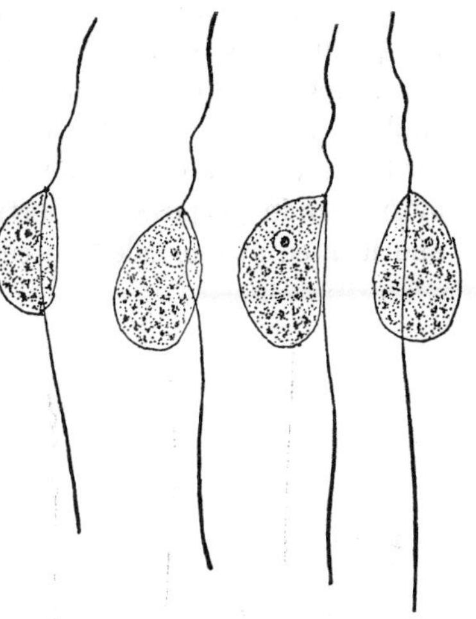

Fig. 272. *Anisonema ovale* Klebs. (× 1300.)

T. quadrangularis Stein has been recorded from Ohio.[2] For a description of it, see Lemmermann (1913).

8. **Sphenomonas** Stein, 1878. The rigid colorless biflagellate cells of *Sphenomonas* are spindle-shaped, and with one to four straight ribs running longitudinally from anterior to posterior ends of a cell. The periplast may have a smooth or a delicately striate surface.[3] The swimming flagellum is markedly longer than the trailing flagellum. There may or may not be an evident gullet adjoined by either one or two vacuoles. If there are two vacuoles, only one of them is contractile.[4] The cells do not contain paramylum bodies or droplets of oil. At the center of a cell is a large sphere of gelatinous material, probably of the nature of a food reserve.

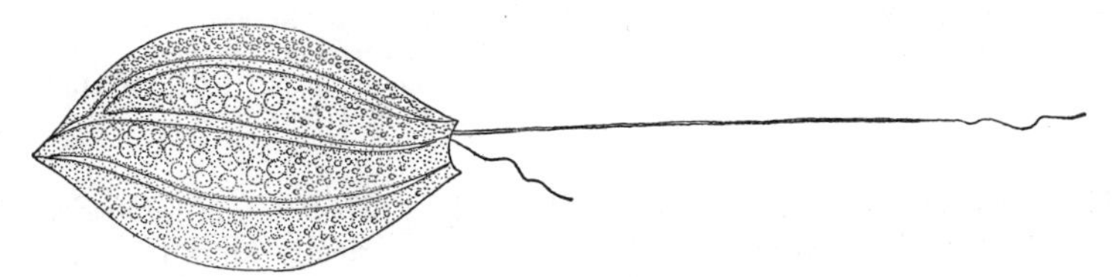

Fig. 273. *Tropidoscyphus octocostatus* Stein, a species not known for the United States. (*After Stein*, 1878.) (× 750.)

[1] Skuja, 1934. [2] Lackey, 1938. [3] Skuja, 1926. [4] Klebs, 1892.

S. quadrangularis Stein (Fig. 274) is known from Ohio.[1] For a description of it, see Lemmermann (1913).

9. **Notosolenus** Stokes, 1884. The rigid cells of this genus are ovoid to campanulate, have longitudinal costae, and are somewhat compressed in the anterior portion. There is a long swimming flagellum and a short trailing flagellum. A pharyngeal rod apparatus has not been recorded for this genus, and the structure of the gullet has not been described. The cytoplasm in the posterior portion of a cell is granular; the nucleus is obscure and located at one side of a cell.[2]

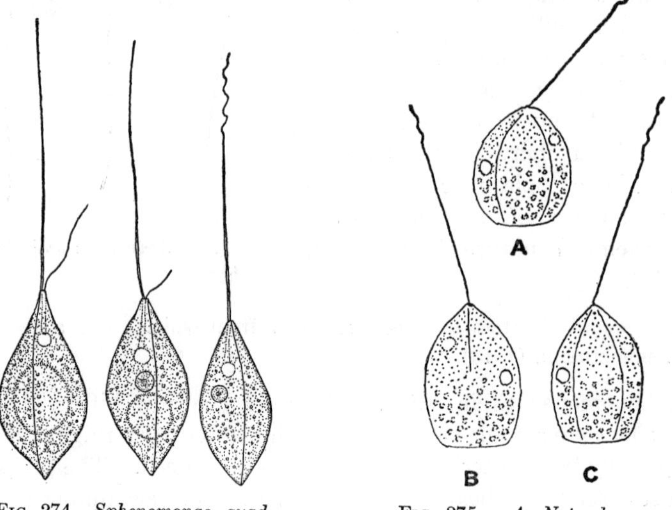

FIG. 274. *Sphenomonas quadrangularis* Stein. (*After Stein,* 1878.) (× 750.)

FIG. 275. *A, Notosolenus orbicularis* Stokes. *B–C, N. apocamptus* Stokes. (*After Stokes* 1884.) (*A,* × 1320; *B,* × 1750.)

The three species, *N. apocamptus* Stokes (Fig. 275*B–C*), *N. orbicularis* Stokes (Fig. 275*A*), and *N. sinuatus* Stokes, of this imperfectly known genus have not been found outside of the United States. For descriptions of them, see Stokes (1884).

10. **Euglenopsis** Klebs, 1892. The uniflagellate cells of *Euglenopsis* are colorless, spindle-shaped, and slightly plastic. Delicate spiral striae are sometimes evident on the periplast's surface. The single flagellum is inserted at the base of an excentric gullet, lacks, a basal bifurcation or a granular swelling, and has a blepharoplast at its base.[3] There is a small cytostome near the point of emergence of the flagellum, and this seems to be

[1] LACKEY, 1938. [2] STOKES, 1884. [3] HALL and POWELL, 1927.

independent from the gullet. The nutrition of *Euglenopsis* may be either saprophytic or holozoic and often results in a considerable accumulation of paramylum.[1] The nucleus is fairly large and lies near the center of a cell.

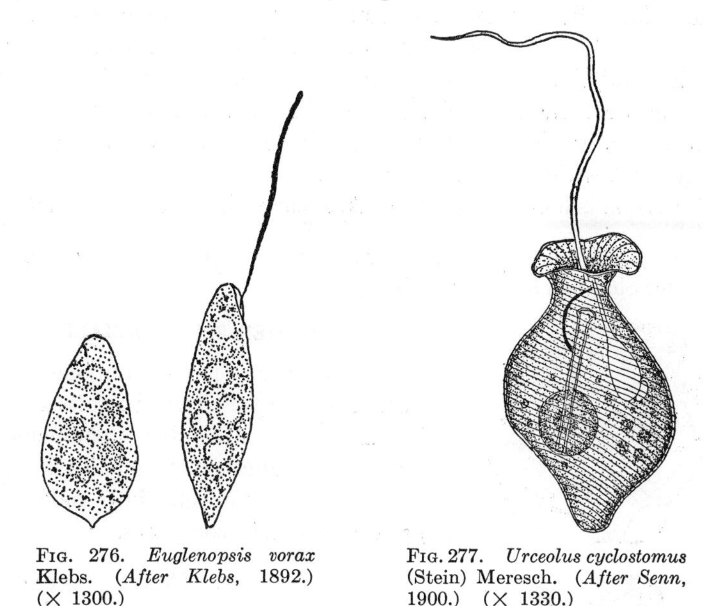

FIG. 276. *Euglenopsis vorax* Klebs. (*After Klebs,* 1892.) (× 1300.)

FIG. 277. *Urceolus cyclostomus* (Stein) Meresch. (*After Senn,* 1900.) (× 1330.)

The single species, *E. vorax* Klebs (Fig. 276) has been found in this country. For a description of it, see Lemmermann (1913).

11. **Urceolus** Mereschkowsky, 1877. Members of this genus have flask-shaped cells with a narrow or a widely flaring mouth above the constricted neck. The cells are colorless and extremely plastic. The periplast has coarse or delicate striae that extend spirally backward from the anterior end. There is a single long flagellum which is inserted in a laterally located reservoir. The reservoir is adjoined by a single contractile vacuole. A pharyngeal rod apparatus, with a flaring upper end, lies somewhat removed from the reservoir. The nucleus is excentric and is toward the posterior end of a cell.[2]

The two species known for this country are *U. cyclostomus* (Stein) Meresch. (Fig. 277) and *U. sabulosus* Stokes. For descriptions of them, see Lemmermann (1913).

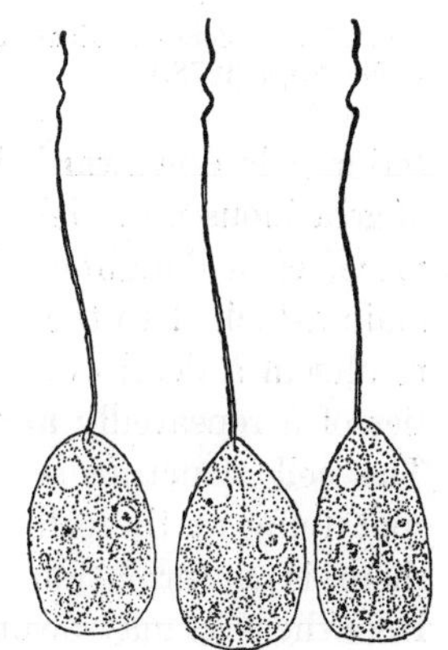

FIG. 278. *Petalomonas abscissa* (Duj.) Stein. (× 975.)

[1] KLEBS, 1892. [2] SENN, 1900.

12. **Petalomonas** Stein, 1859. Cells of *Petalomonas* (Fig. 278) are rigid, more or less compressed, asymmetrical, and of various shapes. Several species have one or more prominent longitudinal costae. There is a single flagellum, laterally inserted in a depressed gullet. A pharyngeal rod apparatus composed of very short rods has been described[1] for this genus. The reservoir is adjoined by a contractile vacuole. The nucleus usually lies midway between the poles of a cell and just beneath the periplast. Locomotion is usually with a gliding or creeping motion and with movement of the flagellum restricted to the distal end.

Cell division is longitudinal and takes place while a cell is in motion.

Twelve species are known for the United States. For names and descriptions of them, see Shawhan and Jahn (1947).

ORDER 2. COLACIALES

The Colaciales have immobile cells permanently encapsuled within walls and united in amorphous or dendroid palmelloid colonies. Reproduction of colonies is by metamorphosis of the cells into naked uniflagellate euglenoid zoospores.

FAMILY 1. COALIACIACEAE

This, the only family of the order, contains but one genus, *Colacium*.

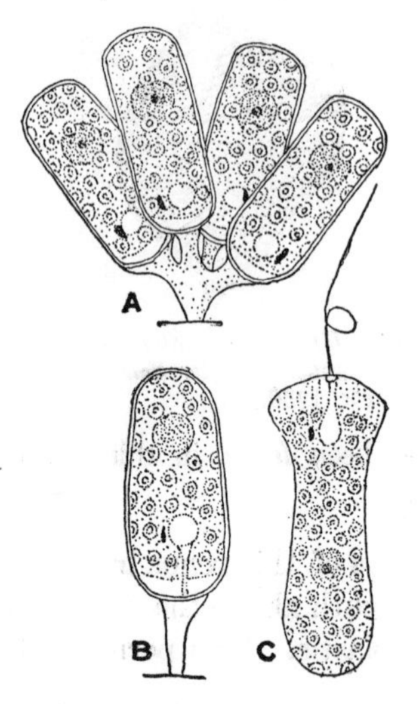

FIG. 279. *Colacium calvum* Stein. (*After Stein*, 1878.)

1. **Colacium** Ehrenberg, 1833. *Colacium* grows epizoically upon *Cyclops*, Copepods, Rotifers, and other members of the fresh-water zooplankton. When growing upon animals, the cells are surrounded by a gelatinous sheath and affixed, with the anterior pole downward, by means of stalks resulting from greater secretion of gelatinous material at the anterior end of a cell. When cell division takes place, each daughter cell secretes a stalk of its own, and these stalks remain attached to the stalk of the parent cell. Repetition of cell division results in a dendroid colony in which the cells are borne at the extremities of a repeatedly and dichotomously branched gelatinous stalk system.[2] The cell structure is much the same as in *Euglena*, the gullet and eyespot lying at the cell end adjoining the stalk. When grown in culture, there is a formation of amorphous palmelloid colonies without stalks. Here the cells may be uninucleate or with two or eight nuclei.[3] There may

[1] BROWN, V. E., 1930. [2] STEIN, 1878. [3] JOHNSON, 1934.

also be a formation of naked amoeboid stages with four to eight nuclei.[1] Multinucleate palmelloid and amoeboid cells may cut off uninucleate cells.

A cell of a dendroid or palmelloid stage may have its protoplast developing a single flagellum and escaping as a naked zooid that swims for but a few hours before becoming sessile and secreting a gelatinous envelope. A flagellum of the motile phase is not bifurcate at the base but does have a granular swelling near the base.[1]

Colacium is widely distributed in this country. Two species, *C. calvum* Stein (Fig. 279) and *C. vesiculosum* Ehr., have been found in this country. For descriptions of them, see Lemmermann (1913).

[1] JOHNSON, 1934.

CHAPTER 6

DIVISION CHRYSOPHYTA

The Chrysophyta have their pigments localized in chromatophores that are yellowish green to golden brown because of a predominance of carotenes and xanthophylls. The food reserves include both leucosin, a carbohydrate of unknown structure, and oils. The cell wall is generally composed of two overlapping halves and is frequently impregnated with silica. The cells may be flagellated or nonflagellated, and solitary or united in colonies of definite or indefinite shape.

Asexual reproduction of immobile genera may be by means of flagellated or nonflagellated spores. There is a widespread, although not universal, formation of a unique type of nonflagellated spore, the *statospore.*

Sexual reproduction is usually isogamous and by a union of flagellated or nonflagellated gametes, but it may also be anisogamous or oögamous.

Pascher (1914) was the first to suggest a relationship between Xanthophyceae, Chrysophyceae, and Bacillariophyceae and to propose that these be united in a common group which he called the Chrysophyta. Previous to this, the Xanthophyceae were usually grouped with the Chlorophyceae, the Chrysophyceae with the Flagellatae, and the Bacillariophyceae were considered a series distantly related to the Phaeophyceae. Among the reasons given by Pascher (1914, 1924) for considering Xanthophyceae, Chrysophyceae, and Bacillariophyceae as related to one another were similarities in pigmentation, similarities in nature of food reserves, cell walls with two overlapping halves in vegetative cells or in spores, and formation of a distinctive type of spore, the statospore. Except for similarity of pigments all these arguments have proved valid. In the case of the pigments, a fuller knowledge of them has shown that there are certain differences in pigmentation of the three classes (see Table I, page 3).

Phycologists are generally agreed that there is a relationship between Xanthophyceae and Chrysophyceae. They are less certain about the relationships of the Bacillariophyceae to them, and all who place the Bacillariophyceae among the Chrysophyta think that their relationship to Xanthophyceae and Chrysophyceae is not so close as that between Xanthophyceae and Chrysophyceae.

CLASS 1. XANTHOPHYCEAE

The Xanthophyceae (Heterokontae) have yellowish-green chromatophores which contain chlorophyll *a*, chlorophyll *e*, beta-carotene, and but one xanthophyll. Food reserves generally accumulate as leucosin, but there may be an accumulation of oils. There is never a formation of starch. Vegetative cells of certain genera have a wall with two overlapping parts of equal or unequal size. The plant body may be unicellular or multicellular. Motile vegetative and reproductive cells have two flagella of different length and structure at the anterior end.

Asexual reproduction may be by means of zoospores or aplanospores. Certain genera are also known to have an endogenous production of statospores.

Sexual reproduction is known for a few genera. In most cases, it is isogamous, but that of one genus is regularly oögamous.

Occurrence. With a few exceptions the Xanthophyceae are fresh-water organisms. Most fresh-water species are aquatic, and free-floating or epiphytic. Many free-floating species, especially unicellular ones, occur very sparingly intermingled with other free-floating algae of semipermanent or permanent pools of soft-water areas.

Other members of the class are aerial. Some aerial genera, as *Monocilia*, grow in tree trunks, on damp walls, or intermingled with mosses and lichens. Still other members of the class are terrestrial and grow intermingled with other algae of the soil (*Botrydiopsis*); or they grow in dense stands on drying mud, especially along the banks of streams and ponds (*Botrydium*).

Organization of the Plant Body. The Xanthophyceae have evolved from a motile unicellular ancestry, and within the class are to be found practically all the types postulated in the theory of plant-body types (see page 6). Certain of the body types are but poorly represented in the Xanthophyceae; others have numerous examples.

There are relatively few of the unicellular flagellate type, and none of these is known for this country. No examples of the volvocine type are known. There are a few typical rhizopodial forms and a somewhat larger number of palmelloid ones. The number of filamentous genera is also small, but these include both the branched and the unbranched type. A large majority of the genera are of the coccoid type, and there are two siphonaceous genera.

Cell Structure. Cell walls of Xanthophyceae are usually composed of pectic compounds, either pectose or pectic acid, and with or without impregnation with some silica. Slight traces of cellulose have been found in walls of *Tribonema*[1] and the wall of *Botrydium* has been found to consist almost wholly of cellulose.[2] Many genera have cells whose walls are composed of two overlapping halves that fit together as do the two parts of a bacteriologist's Petri dish. In unicellular genera, the two parts may be of

[1] Tiffany, 1924. [2] Miller, 1927; Pascher, 1937.

equal length or markedly unequal in length. The two-parted nature of a wall cannot be made out unless the cells have been treated with certain reagents, including concentrated KOH. Detailed study[1] of the wall of one unicellular genus (*Ophiocytium*) has shown that the longer half consists of successive cup-shaped layers fitted one inside the other; the shorter half-wall, the cover, is homogeneous in structure. In filamentous genera, as *Tribonema*, the wall of a filament is composed of a linear file of pieces that are H-shaped in optical section (Fig. 280). They alternately overlap one another so that each protoplast is enclosed by halves of two successive

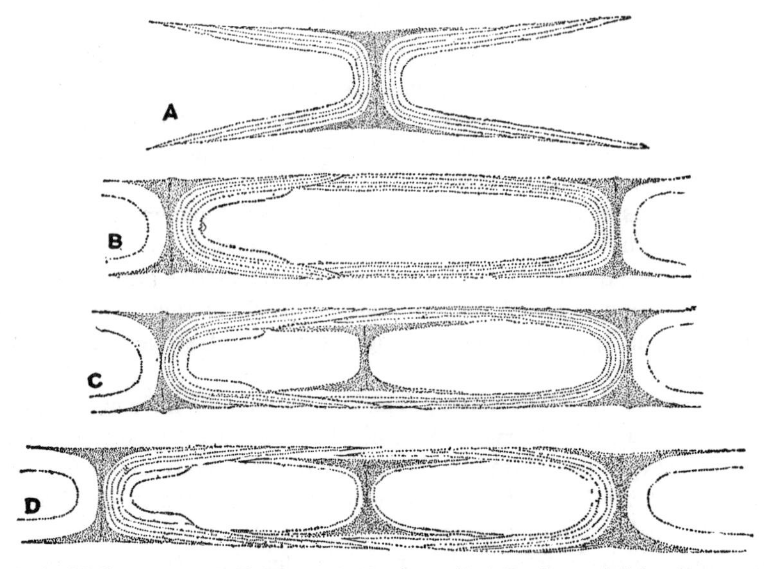

FIG. 280. Wall structure of *Tribonema bombycinum* (Ag.) Derbes and Sol., after treatment with potassium hydroxide. *A*, H-piece. *B*, two H-pieces articulated to enclose a single protoplast. *C–D*, recently divided cells showing the intercalation of a new H-piece. (× 900.)

H-pieces. Each segment (H-piece) of a wall consists of two cup-shaped open cylinders with a common base that constitutes a transverse wall of the filament. When first formed, an H-piece is homogeneous in structure (Fig. 280*C*). Later on, as a cell grows in length, successive layers of wall material are deposited along portions of the H-piece next to the protoplast and in such a manner that each successively formed layer projects beyond the free edge of the previously formed one (Fig. 280*A*).

According to the genus, the protoplast of a cell contains one, two, a few, or innumerable chromatophores. These are parietal in position and almost always disk-shaped. Chromatophores of most species are without pyrenoids; but those of certain species, as *Botrydium*, may have evident ones.

[1] BOHLIN, 1897.

Pyrenoids, when present, are of the "naked" type and are not intimately concerned with the accumulation of reserve foods.

Food reserves accumulate either as droplets of oil or as a whitish insoluble substance called leucosin. The presence of oils is readily demonstrated by customary microchemical tests with Sudan III or osmic acid. There are no specific reagents for demonstrating the presence of leucosin. Leucosin is thought to be a carbohydrate, but there has been no chemical proof of this assumption. Not all whitish refractive granules within a protoplast are to be considered leucosin, since some are probably excretory products.

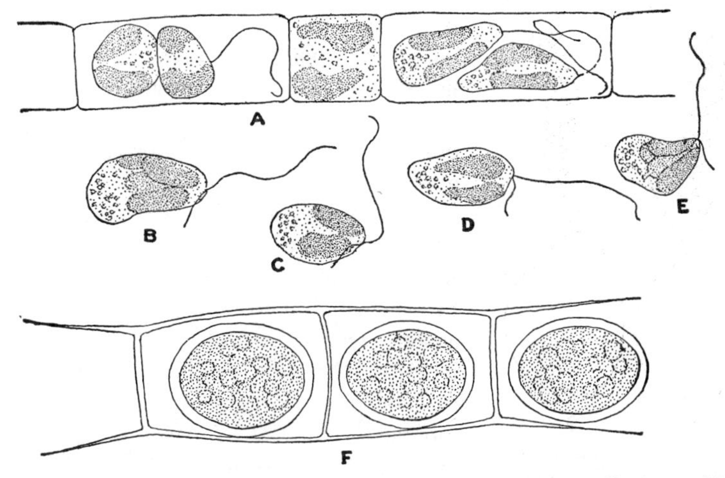

Fig. 281. *A–E*, zoospore formation in *Bumilleria sicula* Borzi. *A*, cells from which the zoospores have not been liberated; the shorter flagellum is not evident in the zoospores. *B–E*, free-swimming zoospores, showing the typical xanthophycean flagellation. *F*, aplanospores of *Tribonema bombycinum* (Ag.) Derbes and Sol. (× 900.)

Many genera have uninucleate cells, but others have multinucleate ones and with the number a multiple of two or indefinite. The nuclei are usually so small that they cannot be recognized with certainty in living cells. In the cases where nuclei have been investigated cytologically,[1] they have been found to have the structure typical of nuclei in other plants.

Asexual Reproduction. Multiplication of tetrasporine and filamentous colonies may be vegetative and due to an accidental breaking of a colony into two parts.

Most of the genera that reproduce vegetatively and all genera without vegetative multiplication produce one or more types of spore. Zoospores are formed by a majority of the genera, and they may be formed singly or in numbers within a cell. The zoospores are always biflagellate, with the

[1] Carter, 1919; Gross, 1937.

two flagella anterior in insertion and markedly different in length (Fig. 281). The longer flagellum, which is often four to six times longer than the shorter, extends straight ahead and is the propulsive organ of a zoospore. The shorter one, sometimes called the "trailing flagellum," arises from the same point as the longer one and extends backward from the point of insertion. Many of the older descriptions of xanthophycean zoospores as uniflagellate were due to a failure to observe the shorter flagellum because it lies so close to the body of the spore, and one by one genera which were thought to be uniflagellate have been shown to be biflagellate. Staining of flagella by special methods[1] has shown that the two differ in structure. The longer is of the tinsel type and beset with a double row of delicate cilia; the shorter is of the whip type and is without cilia (Fig. 282). Zoospores of Xanthophyceae are always naked and are usually pyriform. They generally have one or more contractile vacuoles, one to a few chromatophores, but rarely an eyespot.

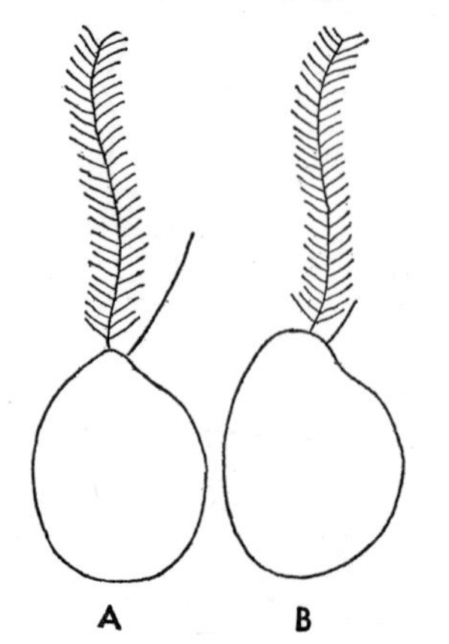

FIG. 282. Diagrams showing structure of flagella of Xanthophyceae. *A*, *Tribonema*. *B*, *Botrydium*. (*After Vlk*, 1938.)

Instead of producing zoospores, the entire protoplast may produce a single aplanospore or divide into a number of parts, each of which becomes an aplanospore (Fig. 281*F*). In some cases, environmental conditions determine whether the alga shall reproduce by means of zoospores or aplanospores. Thus, submerged thalli of *Botrydium* produce zoospores; those growing on damp soil produce aplanospores.[2] An aplanospore liberated from a parent cell may grow directly into a new plant (Fig. 314*C–E*, page 401), or it may give rise to zoospores, which, in turn, give rise to new plants.[3]

Aplanospores that have the same shape as the cell in which they are formed are frequently called autospores, and Heterococcales are the only Xanthophyceae producing autospores. More than one autospore is always formed within a parent cell, and each of them develops the characteristic shape and wall structure of the parent cell before it is liberated from the old parent-cell wall.

A few flagellated and rhizopodial Xanthophyceae are known[4] to form spores endogenously within their protoplasts. Such spores are usually called cysts, but they may also be called statospores because they seem to be homologous with the statospores of diatoms. In the formation of a

[1] VLK, 1931, 1938. [2] ROSTAFIŃSKI and WORONIN, 1877. [3] PASCHER, 1937.
[4] PASCHER, 1932, 1937.

statospore, there is an internal delimitation of a spherical protoplast that is separated from the peripheral portion of the cell's protoplast by plasma membranes only. The endogenously differentiated protoplast then secretes a wall with two overlapping halves of equal or unequal size. Germinating statospores have their contents dividing to form two or more protoplasts that may be liberated as naked amoeboid bodies or as zoospores.[1]

Vegetative cells may change directly into spore-like resting stages with much thicker walls and more abundant food reserves than vegetative cells. Such spore-like cells in which the spore wall is not distinct from the parent-cell wall are called *akinetes.* The best examples of akinetes among Xanthophyceae are generally found among filamentous genera, but their presence has also been recorded for nonfilamentous genera. In filamentous genera, only an occasional cell, several consecutive cells, or all cells in a filament may develop into akinetes.

Sexual Reproduction. A union of motile gametes has been reported for a number of genera, but subsequent investigations have shown that many of these records are erroneous. Sexual reproduction by a fusion of zoogametes is definitely established for only two genera. In one of them (*Tribonema*[2]) one gamete in a uniting pair is immobile and the other motile; in the other (*Botrydium*[3]) both of a fusing pair are motile. Gametic union in *Botrydium* may be isogamous[4] or anisogamous.[5] Sexual reproduction in the well-known genus *Vaucheria* is oögamous.

Classification. Until the beginning of the century, such Xanthophyceae as were known were placed among various orders of the Chlorophyceae. When these genera were first recognized as constituting a distinct class,[6] called the Heterokontae, the motile genera were placed in one order and all other genera in another. Shortly afterward it was pointed out[7] that evolutionary lines evident among Chlorophyceae are also evident among Xanthophyceae, and that Xanthophyceae could be segregated into orders on the same basis as the Chlorophyceae. When classified in such a manner,[7] the Heterochloridales are comparable to the Volvocales, the Heterocapsales to the Tetrasporales, the Heterotrichales to the Ulotrichales, the Heterococcales to the Chlorococcales, and the Heterosiphonales to the Siphonales. Later there was an establishment of the Rhizochloridales,[8] an order without counterpart among the Chlorophyceae. With the exception of the Heterochloridales,* all these orders are represented in the fresh-water algal flora of this country.

[1] PASCHER, 1932. [2] SCHERFFEL, 1901. [3] MOEWUS, 1940; ROSENBERG, 1930.
[4] ROSENBERG, 1930. [5] MOEWUS, 1940. [6] LUTHER, 1899.
[7] PASCHER, 1913. [8] PASCHER, 1914.

* *Chlorochromonas* Lewis, originally discovered in this country, has been placed among the Heterochloridales (G. M. Smith, 1933), but Pascher (1937) holds that it is a chrysomonad and referable to the genus *Ochromonas.*

ORDER 1. RHIZOCHLORIDALES

The Rhizochloridales include those Xanthophyceae in which the cells have a plasmodial organization. The amoeboid protoplast may be naked, or it may secrete a lorica within which it lives during the vegetative portion of the life cycle (*Stipitococcus*). Naked protoplasts may be uninucleate or multinucleate, and solid or reticulate. Large plasmodia result from enlargement of small uninucleate ones and not from a fusion of several small amoeboid ones as in Myxomycetae.

Reproduction may be due to a division of the protoplast into two or more amoeboid parts, or there may be a division into a number of zoospores. There may also be an endogenous formation of statospores.

Family 1. Stipitococcaceae

This family includes those Rhizochloridales in which the plasmodial protoplast is surrounded by a lorica. *Stipitococcus* is the only genus found in this country.

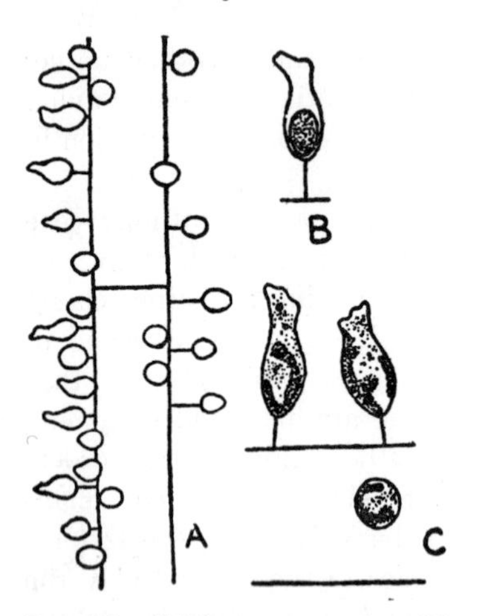

Fig. 283. *Stipitococcus urceolatus* W. and G. S. West. (*After G. S. West*, 1904.) (*A*, × 750; *B–C*, × 1040.)

1. **Stipitococcus** W. and G. S. West, 1898. The protoplasts of *Stipitococcus* are surrounded by a campanulate lorica, open at the distal end and attached to some substratum by a minute stipe that terminates in a minute disk at the point of attachment. The protoplast, which does not usually fill the lorica, contains one or more chromotophores and may have one or more delicate pseudopodia projecting through the open end of the lorica.[1]

Reproduction may be by division of the protoplast into two zoospores that escape singly; or, the entire protoplast may escape from a lorica and develop flagella during or after its escape. Zoospores have been shown to have the flagellation typical of Xanthophyceae.[2]

Stipitococcus is a gregarious epiphyte which is most frequently found upon Zygnematales. *S. apiculatus* Prescott, *S. capensis* Prescott, *S. crassistipitatus* Prescott, *S. urceolatus* W. and G. S. West (Fig. 283) and *S. vasiformis* Tiffany are the species known for the United States. For a description of *S. capensis*, see Prescott and Croasdale (1937); *S. apiculatus* and *S. crassistipitatus*, see Prescott (1944); for *S. urceolatus*, see Pascher (1925*A*); for *S. vasiformis*, see Tiffany (1934).

[1] Pascher, 1937; Poulton, 1930; West, G. S., 1904. [2] Poulton, 1930.

ORDER 2. HETEROCAPSALES

The Heterocapsales have palmelloid colonies in which the cells divide vegetatively. The cells, which may have such characteristics of motile cells as contractile vacuoles and an eyespot, have the ability to return directly to a motile condition.

Reproduction may be by fragmentation of a colony or by a formation of zoospores. Thick-walled akinetes may also be formed.

Family 1. Chlorosaccaceae

Members of this family have free-floating, or sessile, colonies with an indefinite number of cells that may be irregularly distributed through a homogeneous gelatinous matrix, or regularly arranged at the periphery of a matrix. *Gloeochloris* is the only member of the family found in this country.

1. **Gloeochloris** Pascher, 1932. The colonies of this alga, which may be 20 mm. or more in diameter, are spherical to subspherical, and grow free-floating or attached to submerged aquatics. Their color is a very pale yellowish green. Increase in size of a colony is by vegetative cell division. The cells are ellipsoidal to subspherical and lie irregularly or radially distributed throughout a colorless gelatinous matrix. At times, individual cells, or pairs of cells, are surrounded by a gelatinous sheath of denser consistency. The protoplasts contain a few disk-shaped parietal chromatophores without pyrenoids, droplets of oil, and granules of leucosin. Vegetative cells of certain species have a contractile vacuole and an eyespot.[1]

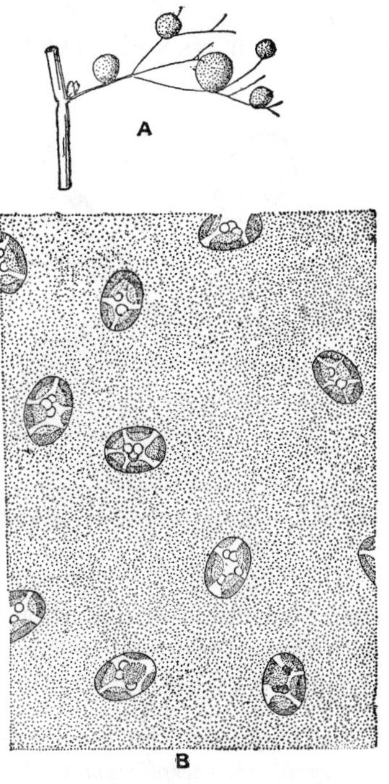

Fig. 284. *Gloeochloris Smithiana* Pascher. *A*, colonies epiphytic upon *Ranunculus aquatilis* L. *B*, portion of a thallus. (A, $\times \frac{1}{2}$; B, $\times$ 600.)

Reproduction is by direct metamorphosis of vegetative cells into zoospores. There may also be a formation of thick-walled spores with a wall composed of two overlapping halves.[1]

G. Smithiana Pascher (Fig. 284) is known from California where it was erroneously identified as *Chlorosaccus fluidus* Luther.[2] This species has also been recorded from North Carolina.[3] For a description of *G. Smithiana*, see Pascher (1937).

[1] Pascher, 1932*C*. [2] Smith, G. M., 1933. [3] Whitford, 1943.

FAMILY 2. MALLEODENDRACEAE

The Malleodendraceae have dendroid, dichotomously branched, sessile colonies, with a cell at the free end of each of the ultimate dichotomies. There is but one genus, *Malleodendron*.

1. **Malleodendron** Pascher, 1937. This sessile colonial alga has the cells terminal at the free ends of a two- to fourfold, dichotomously branched, gelatinous stalk. Branches of the gelatinous stalk system may be transversely stratified. A cell at the tip of an ultimate dichotomy is globose to pyriform and with the portion next to the stalk rounded or markedly flattened. Growth of a colony is by cell division in a vertical plane followed by secretion of gelatinous material from the inferior side of each daughter cell. The protoplast of a cell contains one to three discoid chromatophores and droplets of oil.

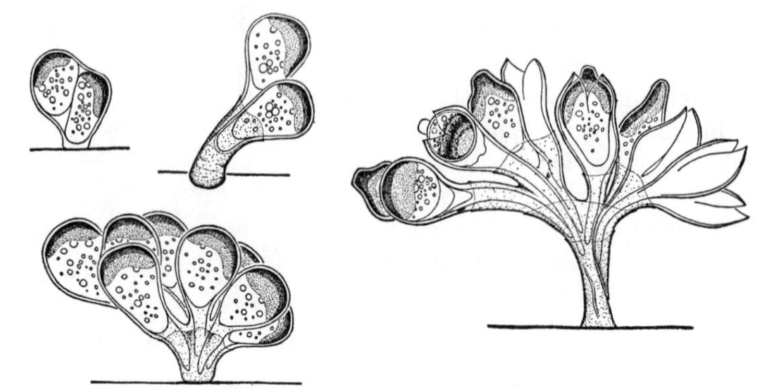

FIG. 285. *Malleodendron caespitosum* Thompson. (*Drawn by R. H. Thompson.*) (× 800.)

Reproduction is by metamorphosis of recently divided cells into zoospores which, after liberation and swimming about for a time, come to rest with the anterior end downward and begin forming gelatinous material on the side facing the substratum.[1]

The only record for the occurrence of *Malleodendron* in this country is the discovery of *M. caespitosum* R. H. Thompson sp. nov.* (Fig. 285) near Solomons, Maryland.

ORDER 3. HETEROCOCCALES

The Heterococcales comprise the nonfilamentous Xanthophyceae in which the vegetative cells do not have the capacity of returning directly to a

[1] PASCHER, 1937.

* *Malleodendron caespitosum* R. H. Thompson sp. nov. Colonia profuse et dense ramosa est. Cellulae ovatae aut ellipticae; 8–18 μ longae, 4–11 μ latae sunt. Quisque cellula chromatophora viridis badius aut oleaginus et paucae vel multa grana, refringendi vim habens, continet

motile condition and which rarely, if ever, divide vegetatively. The plant body may be strictly unicellular, or it may be multicellular and with the cells held together by a watery or a cartilaginous gelatinous matrix. Some of the aquatic genera are free-floating, others are sessile.

As in the homologous order (Chlorococcales) of the Chlorophyceae, there are certain genera that reproduce exclusively by means of zoospores, others that reproduce by means of zoospores or autospores, and still others that reproduce only by means of autospores.

The number of genera and species in this order exceed the combined number of all other orders. Pascher (1937, 1938, 1939) divides the Heterococcales into 10 families, all but two of which are represented in the algal flora of this country.

Family 1. Pleurochloridaceae

Members of this family have solitary free-floating cells. The cells may be globose, ellipsoidal, angular, or spindle-shaped; but when ellipsoidal or spindle-shaped the length is not markedly greater than the breadth. The cell wall may be smooth or ornamented and is usually homogeneous and not composed of two overlapping halves.

Reproduction is by a formation of zoospores or autospores.

This is a more or less artificial family with many genera. The genera found in this country differ as follows:

1. Cells tetrahedral or triangular 12
1. Cells neither tetrahedral nor triangular 2
    2. Cells hemispherical .... 11. **Chlorogibba**
    2. Cells not hemispherical 3
3. Cells spherical to subspherical or ovoid 4
3. Cells not spherical to subspherical or ovoid 8
    4. Cell wall smooth 5
    4. Cell wall sculptured .... 8. **Arachnochloris**
5. Mature cells ovoid .... 3. **Leuvenia**
5. Mature cells spherical to subspherical 6
    6. Cells epiphytic .... 4. **Perone**
    6. Cells not epiphytic 7
7. Cells varying greatly in size .... 2. **Botrydiopsis**
7. Cells not varying greatly in size .... 1. **Diachros**
    8. Cells cylindrical 9
    8. Cells spindle-shaped 11
9. Cell wall smooth .... 5. **Monallantus**
9. Cell wall ornamented 10
    10. Wall with small areolae .... 9. **Trachychloron**
    10. Wall with large areolae .... 10. **Chlorallanthus**
11. Poles of cells not thickened .... 7. **Chlorocloster**
11. Poles of cells thickened or with delicate spines .... 6. **Pleurogaster**
    12. Cells tetrahedral .... 12. **Tetraedriella**
    12. Cells compressed and triangular .... 13. **Goniochloris**

1. **Diachros** Pascher, 1937. The spherical free-floating cells of this alga are usually solitary but at times may lie in a small colony surrounded by the gelatinous remains of the old parent-cell wall. The cell wall is colorless to reddish brown, thin to relatively thick, and composed of two overlapping hemispheres. Protoplasts contain one to several disk-shaped chromatophores.

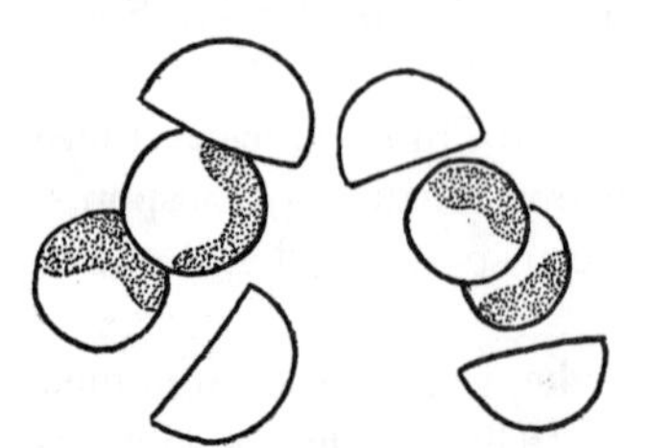

FIG. 286. *Diachros simplex* Pascher. (*Drawn by G. W. Prescott.*)

Reproduction is by formation of two, four, or eight autospores which are liberated by a spreading apart of the two halves of the parent-cell wall.[1]

Prof. G. W. Prescott writes that he has found *D. simplex* Pascher (Fig. 286) in Michigan. For a description of it, see Pascher (1937).

2. **Botrydiopsis** Borzi, 1889. The cells of *Botrydiopsis* are spherical, free-living, and not united in colonies. Their size is extremely variable, but the wall is relatively thin in proportion to the size. When young, a cell contains but one or two chromatophores but, as it grows in size, the chromatophores increase in number until there are many within a cell. The

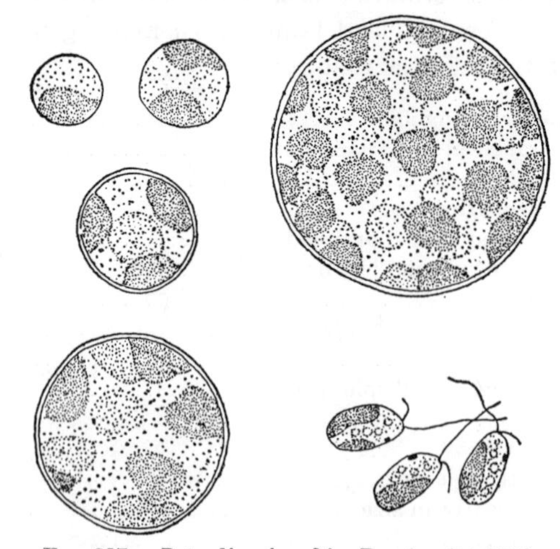

FIG. 287. *Botrydiopsis arhiza* Borzi. (× 900.)

chromatophores of adult cells are disk-shaped and evenly spaced from one another.

Zoospore formation may take place at any stage in the growth of a cell. When produced by young cells, only four or eight are formed; when pro-

[1] PASCHER, 1937.

duced by mature cells, there may be 200 or more. Zoospores have two flagella of unequal length.[1] Frequently, and possibly because of environmental conditions, a cell forms aplanospores instead of zoospores. These, similar to zoospores, grow directly into vegetative cells. Sometimes aplanospores develop very thick walls and have a greater amount of reserve foods.[2] Such hypnospores appear to enter into a rest period before their contents divide into zoospores or aplanospores.

B. arhiza Borzi (Fig. 287) is a widely distributed terrestrial alga in this country, but one frequently overlooked because it rarely occurs in quantity sufficient to form a conspicuous growth. One generally finds it intermingled with such other terrestrial algae as *Vaucheria* and *Microcoleus*. For a description of *B. arhiza*, see Pascher (1925*A*). The aquatic *B. eriensis* Snow described[3] from Lake Erie is a questionable species.

3. **Leuvenia** Gardner, 1910. Young cells of *Leuvenia* are spherical and have one or two parietal discoid chromatophores. As a cell becomes older

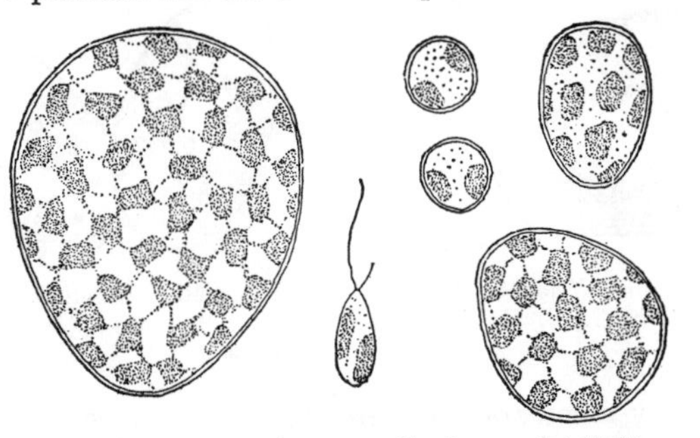

FIG. 288. *Leuvenia natans* Gardner. (× 975.)

the shape becomes ovoid or pyriform, and there is a great increase in number of chromatophores which are connected to one another by delicate strands of cytoplasm. Young cells are uninucleate; older ones are multinucleate. The cells are solitary, free-floating, and without a gelatinous sheath.

Juvenile cells produce one zoospore with two flagella of unequal length. Adult cells have a division of the protoplast into a number of zoospores, each with two chromatophores and two flagella of unequal length. At the beginning of the swarming period, the zoospores are pyriform; toward the end of the period, they are of amoeboid shape. They come to rest on the surface of the water, assume a spherical shape, and secrete a wall.[4]

This alga was originally described as a flagellate, and the immobile vegetative cells were thought to be cyst-like in nature.[4] The alga has also been found embedded in large numbers within branched gelatinous strands,

[1] LUTHER, 1899; POULTON, 1925; PASCHER, 1937.
[2] BORZI, 1895; POULTON, 1925. [3] SNOW, 1903. [4] GARDNER, 1910.

and because of this it has been placed in the Heterocapsales.[1] It is very probable that the palmelloid stage is of such rare occurrence that the alga should be considered coccoid in nature and so placed among the Heterococcales.

Leuvenia grows as a green, somewhat oily, film on the surface of semipermanent pools. When present at any station, it usually develops in abundance and covers the surface of the water. There is but one species, *L. natans* Gardner (Fig. 288). For a description of it, see Gardner (1919).

4. **Perone** Pascher, 1932. *Perone* is a unicellular epiphytic alga in which the mature cells are more or less biscuit-shaped. The cell wall is smooth and without ornamentation. The protoplast contains several vacuoles, and the cytoplasm between the lies in a meshwork which contains a number of small discoid chromatophores. Young cells are uninucleate; older ones are multinucleate.

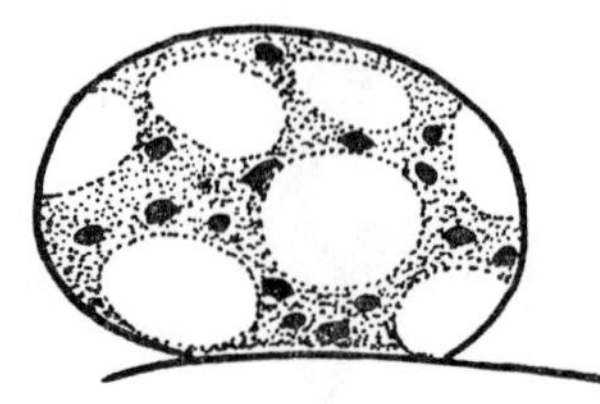

Fig. 289. *Perone dimorpha* Pascher. (*Drawn by G. W. Prescott.*)

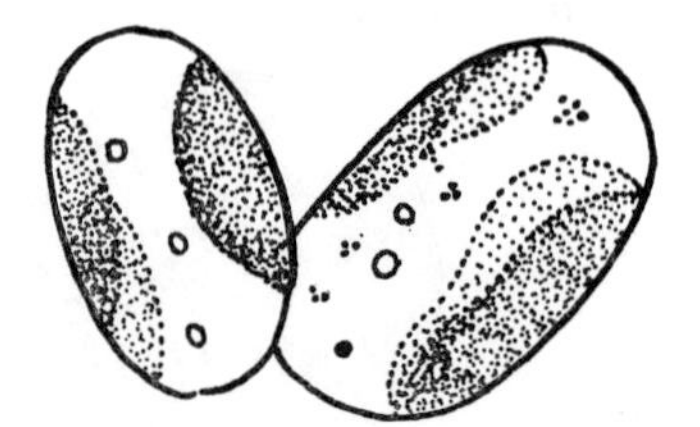

Fig. 290. *Monallantus brevicylindrus* Pascher. (*Drawn by G. W. Prescott.*)

Reproduction by means of zoospores may take place in either juvenile or adult cells, and older cells may produce as many as 200 zoospores. When liberated from a parent cell, a zoospore may develop directly into a vegetative cell after it comes to rest upon a host; or it may lose its flagella and become amoeboid. The amoeboid phase may develop into a vegetative cell or into a plasmodial mass with numerous hair-like rhizopodial processes.[2]

The type species *P. dimorpha* Pascher (Fig. 289) grows epiphytically upon the green cells of *Sphagnum*. Prof. G. W. Prescott writes that he has found *P. dimorpha* in Michigan. For a description of it, see Pascher (1932*A*).

5. **Monallantus** Pascher, 1937. This unicellular free-floating alga has straight cylindrical cells with a length never more than double the breadth. The poles of a cell are broadly rounded. The cell wall is homogeneous and without sculpturing or other ornamentation. The protoplast contains one to four discoid to laminate chromatophores, with or without pyrenoids. Reddish droplets of oil are frequently present in the cytoplasm.

[1] Pascher, 1925*A*. [2] Pascher, 1932*A*.

Reproduction is by division of the cell contents to form two, rarely four, zoospores or autospores.[1] Liberation of spores is by softening and rupture of the parent-cell wall.

This genus closely resembles *Bumilleriopsis* but differs in having straight, much shorter, cells that contain a smaller number of chromatophores. Prof. G. W. Prescott writes that he has found *M. brevicylindrus* Pascher (Fig. 290) in Michigan. For a description of it, see Pascher (1937).

6. **Pleurogaster** Pascher, 1937. This unicellular free-floating alga has plump, asymmetrical, spindle-shaped cells in which the poles may be prolonged into short stout spines or long delicate ones. One side of a cell is strongly convex; the opposite side is slightly convex to concave. As seen in transverse section, the outline of a cell is circular to broadly elliptical, never angular. The cell wall is smooth and homogeneous. The protoplast contains one to four pale discoid chromatophores and reddish droplets of oil.

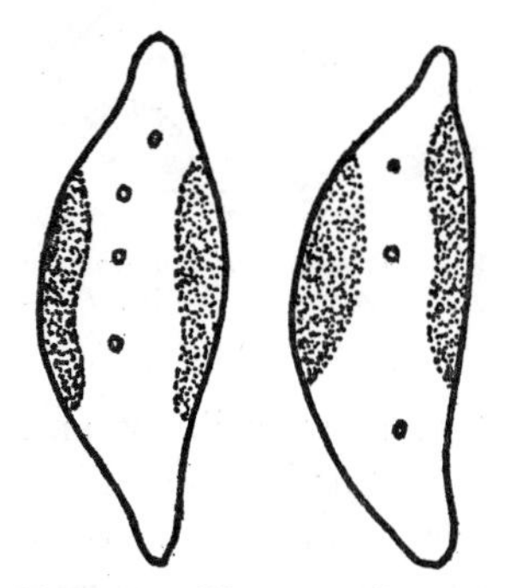

FIG. 291. *Pleurogasterl unaris* Pascher. (*Drawn by G. W. Prescott.*)

Reproduction is generally by a division of the cell contents into two or four autospores, but there may be a formation of zoospores.[1]

Prof. G. W. Prescott writes that he has found *P. lunaris* Pascher (Fig. 291) in Michigan. For a description of it, see Pascher (1937).

7. **Chlorocloster** Pascher, 1925. The cells of this alga are solitary and free-floating. They are narrowly to broadly, and symmetrically, spindle-shaped; straight, arcuate or S-shaped, and with a delicate homogeneous wall. Within a cell are a few to several parietal chromatophores with or without pyrenoids (Fig. 292).

Reproduction is by division of the cell contents into two, four, or eight autospores which, at times, may have contractile vacuoles.[1] One to four thick-walled globose aplanospores may be formed within a cell.

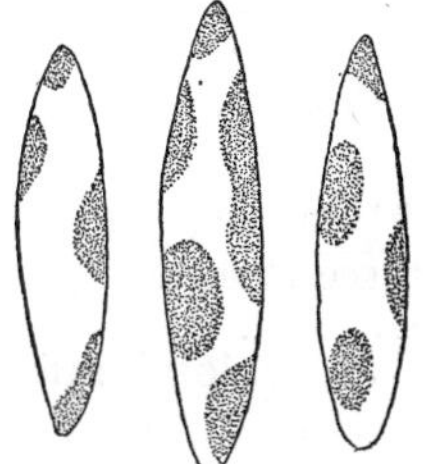

FIG. 292. *Chlorocloster terrestris* Pascher, a species not known for the United States. (*After Pascher*, 1925*A*.)

Prof. G. W. Prescott writes that he has found *C. pyrenigera* Pascher in Michigan. For a description of it, see Pascher (1937).

8. **Arachnochloris** Pascher, 1930. The cells of this unicellular free-floating alga are broadly ellipsoidal to spherical and may attain a relatively large size. The cell wall, which is often strongly silicified, is thin and regularly reticulate with a

[1] PASCHER, 1937

series of small circular thinner areas. There is a single large parietal chromatophore which lies at one side of a cell and has several band-like lobes projecting to the opposite side of the cell.

Reproduction may be by a formation of two zoospores, or by a formation of two or four autospores.[1]

Prof. G. W. Prescott writes that he has found *A. minor* Pascher (Fig. 293) in Michigan. For a description of it, see Pascher (1930*C*).

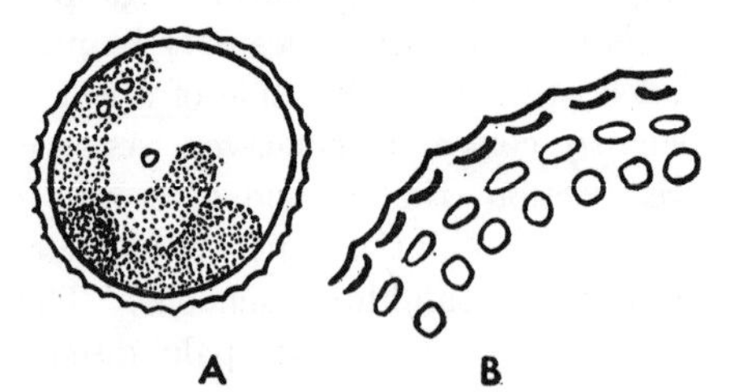

Fig. 293. *Arachnochloris minor* Pascher. *A*, optical section of a cell. *B*, portion of cell wall. (*Drawn by G. W. Prescott.*)

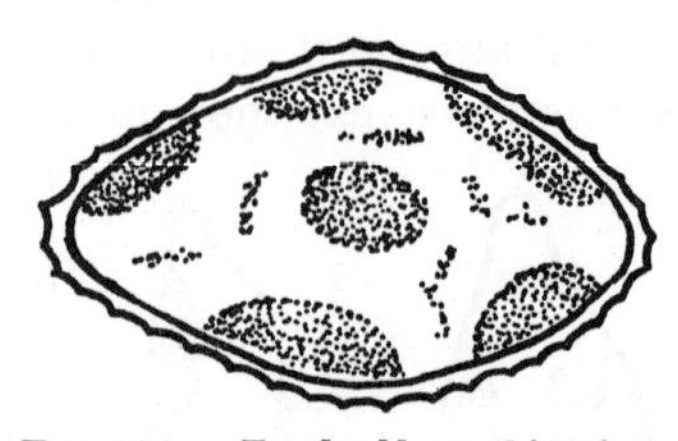

Fig. 294. *Trachychloron biconicum* Pascher. (*Drawn by G. W. Prescott.*)

9. **Trachychloron** Pascher, 1938. Cells of this genus are solitary and free-floating. Most species have ellipsoidal cells; but some have biconical ones, more or less rhomboidal in transverse section, and with broadly rounded poles. The cell wall has a finely reticulate sculpturing. There may be a single chromatophore that is H-shaped in optical section; or there may be several small disk-shaped chromatophores.

A division of cell contents to form two or four autospores is the only method of reproduction thus far observed.[2]

Prof. G. W. Prescott writes that he has found *T. biconicum* Pascher (Fig. 294) in Michigan. For a description of it, see Pascher (1938).

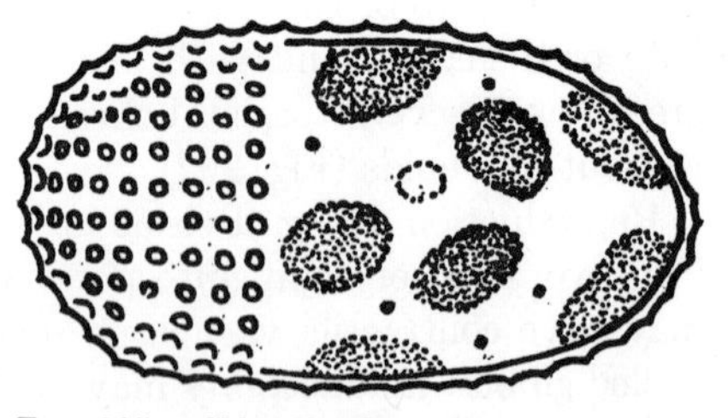

Fig. 295. *Chlorallanthus oblongus* Pascher. (*Drawn by G. W. Prescott.*)

10. **Chlorallanthus** Pascher, 1930. *Chlorallanthus*, is unicellular, free-floating, and with cylindrical to ellipsoidal cells with broadly rounded poles. The cell wall is of variable thickness from cell to cell, but always is sculptured with regularly arranged intersecting rows of small cirular pits. The protoplast contains several parietal disk-shaped chromatophores, droplets of oil, and leucosin.

[1] Pascher, 1930*C*. [2] Pascher, 1938.

Reproduction is by a formation of either zoospores or autospores. These are liberated by a separation of the parent-cell wall into two halves.[1]

Prof. G. W. Prescott writes that he has found *C. oblongus* Pascher (Fig. 295) in Michigan. For a description of it, see Pascher (1930*C*).

11. **Chlorogibba** Geitler, 1928. This alga has solitary, free-floating, more or less hemispherical cells with a crenulate outline. The cell wall is homogeneous and without ornamentation. There may be a single chromatophore or a few parietal chromatophores. At times the chromatophores are a very pale yellowish-green color.

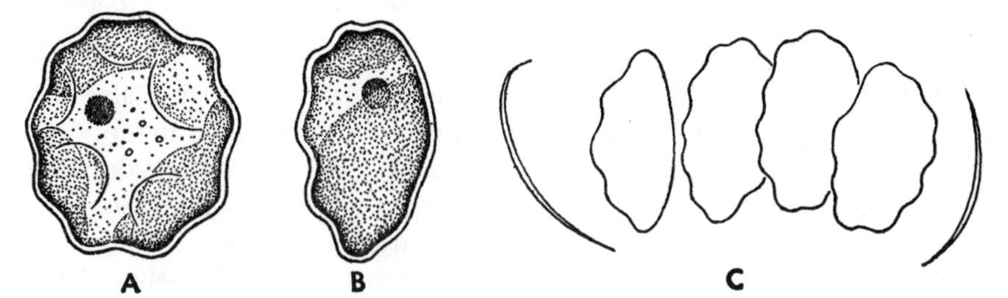

FIG. 296. *Chlorogibba trochisciaeformis* Geitler. *A*, top view. *B*, side view. *C*, liberation of autospores. (*Drawn by R. H. Thompson.*) (× 3230.)

Reproduction is by a division of the protoplast to form two or four zoospores.[2] There may also be a formation of autospores (Fig. 296). Liberation of spores is by a separation of the parent-cell wall into two halves.

Prof. R. H. Thompson writes that he has found *C. trochisciaeformis* Geitler (Fig. 296) in Maryland. For a description of it, see Geitler (1928).

12. **Tetraedriella** Pascher, 1930 (including *Tetragoniella* Pascher, 1930). This unicellular free-floating alga has tetrahedral cells in which the four faces may be concave or convex. Some species have spines at the corners of the cells; other species[3] have them along the ridges connecting the corners of the cells. The cell wall is sculptured with circular pits that are not arranged in a definite manner. The protoplast contains a small number of discoid chromatophores, droplets of oil, and leucosin.

The only method of reproduction thus far observed[1] is a division of the cell contents to form four tetrahedrally arranged autospores.

Since the only distinction between *Tetraedriella* and *Tetragoniella* is the size of cells and number of chromatophores, it seems best to unite the two. Prof. R. H Thompson writes that he has found *T. acuta* Pascher (Fig. 297) in Kansas; and

[1] PASCHER, 1930*C*. [2] PASCHER, 1938. [3] PASCHER, 1932*B*.

Prof. G. W. Prescott writes that he has found *T. gigas* (Pascher) comb. nov. (*Tetragoniella gigas* Pascher) in Michigan. For descriptions of the two, see Pascher (1930*C*).

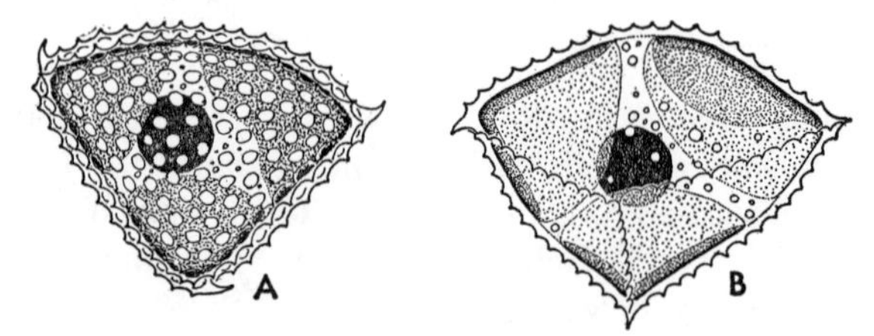

Fig. 297. *Tetraedriella acuta* Pascher. *A*, surface view. *B*, optical section. (*Drawn by F. H. Thompson.*) (× 2200.)

13. **Goniochloris** Geitler, 1928. This unicellular free-floating alga has markedly compressed cells that are triangular or quadrangular in top view, and with or without small spines at the angles. When seen in side view, a cell is bisected into two symmetrical halves by a longitudinal ridge. The cell wall is sculptured with hexagonal pits arranged in intersecting rows at an angle of 60 deg. The protoplast contains several disk-shaped parietal chromatophores, droplets of oil, leucosin, and a single excentrically located nucleus.[1]

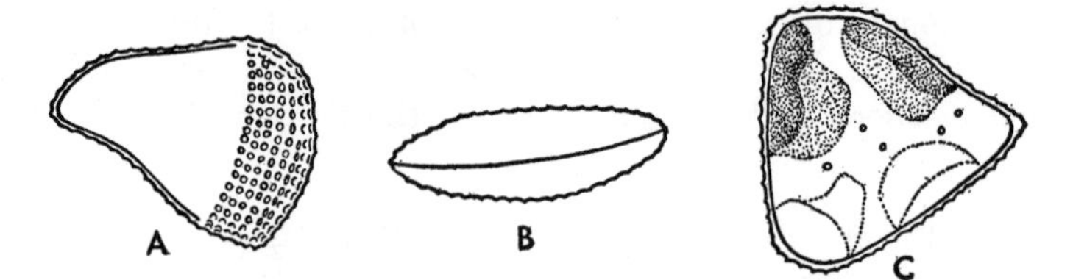

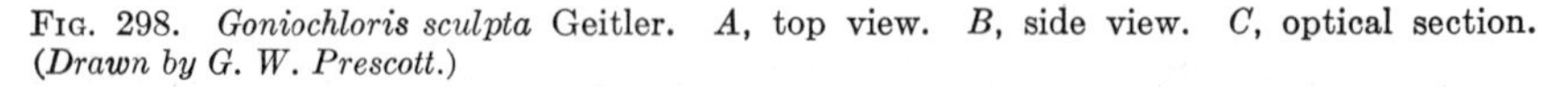

Fig. 298. *Goniochloris sculpta* Geitler. *A*, top view. *B*, side view. *C*, optical section. (*Drawn by G. W. Prescott.*)

Reproduction is by division of the cell contents into two zoospores which eventually swim through the gelatinized parent-cell wall.[1] There may also be a division of the cell contents to form four autospores.[2]

Prof. G. W. Prescott writes that he has found *G. sculpta* Geitler (Fig. 298) in Michigan. For a description of it, see Geitler (1928).

Family 2. Gloeobotrydiaceae

Members of this family are colonial and with few to many globose cells embedded within a homogeneous or concentrically stratified gelatinous matrix.

[1] Geitler, 1928. [2] Pascher, 1930*C*.

Reproduction is by division of the contents of a cell into zoospores or autospores.

The two genera found in this country differ as follows:

1. Gelatinous matrix homogeneous, with many cells .... 1. **Gloeobotrys**
1. Gelatinous matrix stratified, with few cells .... 2. **Chlorobotrys**

1. **Gloeobotrys** Pascher, 1930. *Gloeobotrys* has free-floating amorphous colonies with many globose to broadly ellipsoidal cells irregularly distributed through a colorless homogeneous gelatinous matrix. The cell walls are smooth and homogeneous. A protoplast contains three or four parietal disk-shaped chromatophores and droplets of oil.

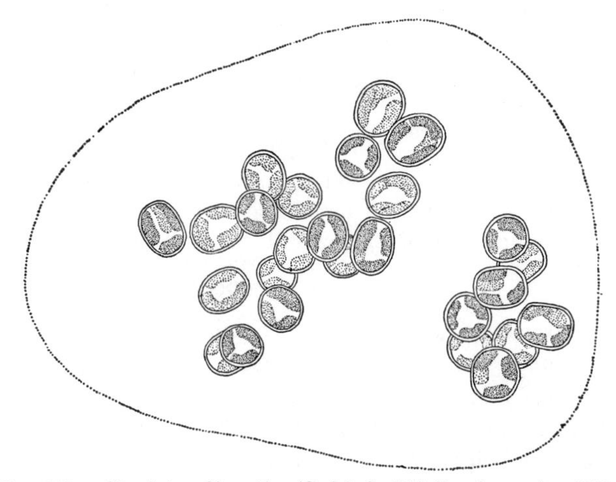

Fig. 299. *Gloeobotrys limnetica* (G. M. Smith) Pascher. (× 1000.)

Reproduction is by division of the cell contents into two autospores which remain within the colony. As the autospores develop into vegetative cells, there is a complete gelatinization of the parent-cell wall. There may also be a formation of zoospores.[1]

G. limnetica (G. M. Smith) Pascher (Fig. 299) has been found in Wisconsin, Wyoming, and Lake Erie. For a description of it as *Chlorobotrys limneticus* G. M. Smith, see G. M. Smith (1920).

2. **Chlorobotrys** Bohlin, 1901. The spherical or ovoid cells of *Chlorobotrys* are united in colonies of 2, 4, 8, 16, or more cells that are surrounded by a hyaline faintly stratified gelatinous envelope. *C. regularis* (West) Bohlin has a thin wall composed of two overlapping halves strongly impregnated with silica.[2] Proof of silicification rests upon persistence of

[1] Pascher, 1930*C*. [2] Bohlin, 1901; West, W. and G. S., 1903.

cell walls after successive treatment with concentrated hydrochloric and sulfuric acids, a procedure that caused disappearance of all other algae present in the collection but diatoms. Protoplasts of this species contain several discoid chromatophores and numerous droplets of oil. There may also be one or two conspicuous reddish spots within a cell.

Reproduction is by division of the cell contents into two or four autospores that remain equidistant from one another within a gelatinous envelope. Discoid hypnospores with silicified walls have also been noted.[1] When a hypnospore germinates, its contents divide to form two thin-walled aplanospores that are liberated by a separation on the hypnospore wall into two halves.

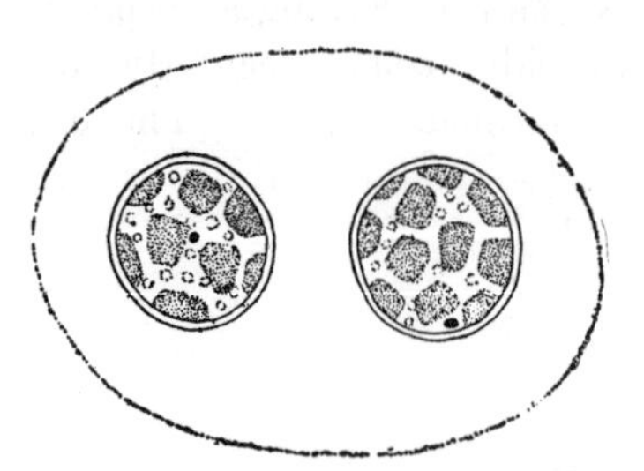

Fig. 300. *Chlorobotrys regularis* (West) Bohlin. (× 1000.)

Chlorobotrys is most frequently encountered sparingly intermingled with other algae of soft-water bogs that are generally rich in desmids. The only species known for this country is *C. regularis* (West) Bohlin (Fig. 300). For a description of it, see G. M. Smith (1920).

Family 3. Mischococcaceae

The Mischococcaceae have dendroid colonies with di- or trichotomously branched gelatinous stalks whose ultimate branches terminate in cells. Growth of a colony is by germination of aplanospores embedded in the gelatinous stalks. There is but one genus, *Mischococcus*.

1. **Mischococcus** Nägeli, 1849. Colonies of this alga are always epiphytic, generally on coarse filamentous algae, and consist of a dichotomously branched system of gelatinous tubes with spherical inflations at places where the tubes fork. The cells are borne singly or in pairs, one above the other, at the tips of the branching tubes. Each cell is surrounded by a wall, and within the protoplast are two to four yellowish-green chromatophores. Reserve foods are stored chiefly as minute droplets of oil in the cytoplasm.

Very young one-celled plants have a discoid gelatinous base by which they are attached to the substratum. Later, the cell divides to form two autospores that develop into vegetative cells without liberation from the parent plant and are pushed out of the old parent-cell wall by the development of a cylinder of gelatinous material that elongates to a length four to ten times greater than the breadth. The two daughter cells may lie vertically above each other in the same gelatinous tube, or they may be borne on separate stalks that diverge from each other. Repetition of a

[1] Bohlin, 1901.

formation of autospores and their germination eventually result in a much-branched colony whose history of development can be traced by the empty parent-cell walls persisting in the system of branched gelatinous tubes (Fig. 301). Instead of growing into a dendroid colony, there may be a development of an amorphous palmelloid colony in which the cells tend to lie in groups of two or four within an irregularly expanded gelatinous matrix.[1] Reproduction may also be by means of zoospores with the flagellation typical of the class.[2] These swim freely in all directions after liberation from the parent-cell wall and develop into new colonies after they have become affixed to a suitable substratum.

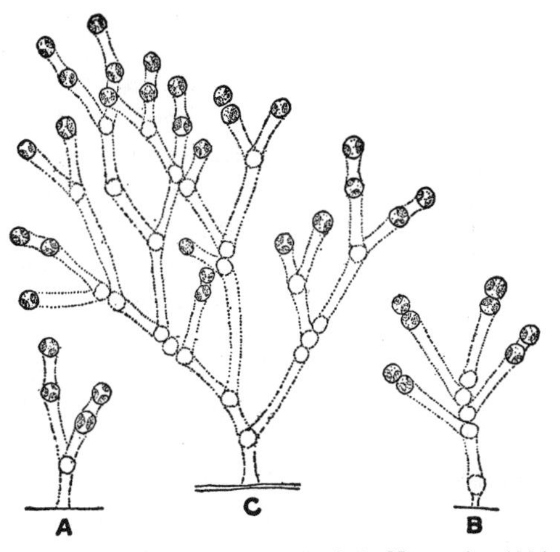

FIG. 301. *Mischococcus confervicola* Näg. (× 400.)

M. confervicola Näg. (Fig. 301), the only species known for the United States, is a comparatively rare alga but has been found in several of the states. For a description of it, see Pascher (1938).

FAMILY 4. CHARACIOPSIDACEAE

The Characiopsidaceae are unicellular and epiphytic. They may be sessile upon the host or affixed to it by a stipe-like prolongation of the cell wall. The shape of the cells varies markedly from genus to genus, but in all cases the cell wall is homogeneous and not composed of two overlapping halves.

Reproduction is by means of zoospores or autospores.

The genera found in this country differ as follows:

1. Cells shorter than stipe .... **3. Peroniella**
1. Cells sessile or longer than stipe 2

[1] BORZI, 1895. [2] PASCHER, 1938.

2. Cells angular .... 2. **Dioxys**
2. Cells not angular 3
3. Chromatophores deeply colored .... 1. **Characiopsis**
3. Chromatophores very pale or colorless .... 4. **Harpochytrium**

1. **Characiopsis** Borzi, 1895. *Characiopsis* (Fig. 302) is always sessile, generally on other algae, and the cells may be solitary or gregarious. The cell shape varies from species to species, but it is usually ovoid and with a rounded or pointed apex. At the base of a cell is a gelatinous stalk of variable length, that is always discoid in the region of attachment to the substratum and frequently with the discoid portion so impregnated with ferric compounds that it is a rusty-brown color. The cell wall is homogeneous and not composed of two overlapping halves. The protoplast usually has several chromatophores, and fats and leucosin in the cytoplasm though rarely in abundance. Young cells are uninucleate but, as a cell increases in size, the nucleus divides and redivides until 8, 16, 32, or 64 are present in an adult cell.[1]

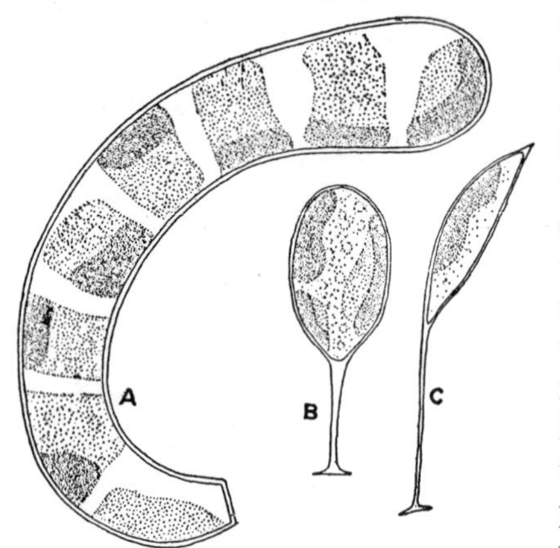

FIG. 302. *A*, *Characiopsis cylindricum* (Lambert) Lemm. *B*, *C. pyriformis* (A. Br.) Borzi. *C*, *C. longipes* (Rab.) Borzi. (× 1000.)

Reproduction is by division of the cell contents into 8, 16, 32, or 64 zoospores, and these have been shown[2] to have the flagellation typical of Xanthophyceae. There may be a formation of aplanospores instead of zoospores.

Approximately a dozen species are known for the United States. For a monographic treatment of the genus, see Pascher (1938).

2. **Dioxys** Pascher, 1932. *Dioxys* (Fig. 303) has solitary, epiphytic, compressed, triangular cells whose angles may or may not be prolonged into stout spines. The cells are affixed to the substratum by an evident stipe. The cell walls are homogeneous and not composed of two overlapping halves. The protoplast of a cell contains two to several parietal disk-shaped chromatophores, droplets of oil, and leucosin.

[1] CARTER, 1919.
[2] POULTON, 1925.

Reproduction is by division of the cell contents to form two, rarely four, zoospores with a typical xanthophycean structure.[1] There may also be a formation of two, four, or eight aplanospores.

The only record for the occurrence of *Dioxys* in this country is the discovery in Kansas of two undescribed species, *D. inermis* R. H. Thompson* sp. nov. (Fig. 303*A–C*) and *D. tricornuta* R. H. Thompson† sp. nov. (Fig. 303*D*).

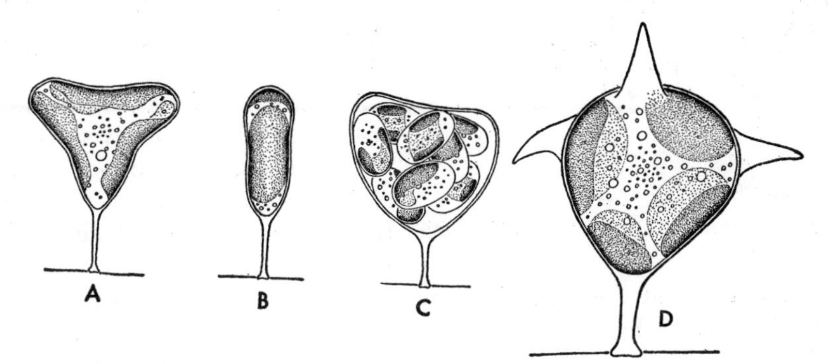

FIG. 303. *A–C*, *Dioxys inermis* Thompson sp. nov. *A*, front view. *B*, side view. *C*, a cell containing aplanospores. *D*, *D. tricornuta* Thompson sp. nov. (*Drawn by R. H. Thompson.*) (× 2080.)

3. **Peroniella** Gobi, 1887. This genus of epiphytic unicellular Xanthophyceae differs chiefly from *Characiopsis* in the smaller number of chromatophores (generally one) and in the very long and more delicate stalk that does not always terminate in a basal disk.

Reproduction is by means of zoospores.[2]

This is another rare alga known from relatively few localities in this country. The two species reported for the United State are *P. Hyalothecae* Gobi and *P. planctonica* G. M. Smith (Fig. 304). For descriptions of them, see Pascher (1925*A*).

[1] PASCHER, 1932.

* *Dioxys inermis* R. H. Thompson sp. nov. Cellulae triangulae in frontato prospectu, anguste obovatae aut cellulae formae pandurae in lato prospectu sunt, et figuntur gracilo stipite qui in levi amplificationis finitus est. Apices libere rotundi sunt. Cellulae unam aut tres gilvas virides chromatophoras iacenter in peripheria cellulae et numerosa grana refringendi vim habienta continent. Cellulae, sine stipite 11–15 μ longae, 7–12 μ latae, et 4 μ crassae sunt. Stipite 4–6 μ longus est.

† *Dioxys tricornuta* R. H. Thompson sp. nov. Cellulae tetrahedrae aut obpyriformi ferentes robustum spinam in quoque libero apice et uno apice robusto stipite, non colore qui finitur in parvo disco fuguntur. Cellula uno aut quatuor virides chromatophoras et pauca aut multa grana, refringendi vim habienta continet. Nucleus centralis est. Cellulae, sine spinis, 14–16 μ in maximo diametro sunt. Spinae 4–5 μ longae sunt et stipites tenues, et 6.5 μ longi sunt.

[2] GOBI, 1887; PASCHER, 1930*C*; SMITH, G. M., 1916*C*.

4. **Harpochytrium** Lagerheim, 1890. *Harpochytrium* has sessile, elongate, cylindrical cells which are usually bent in an arc. The lower end of a cell is gradually attenuated; the upper end may be broadly rounded or gradually attenuated. Certain species have very pale chromatophores; other species are colorless and saprophytic in nutrition. The protoplasts contain numerous droplets of oil.

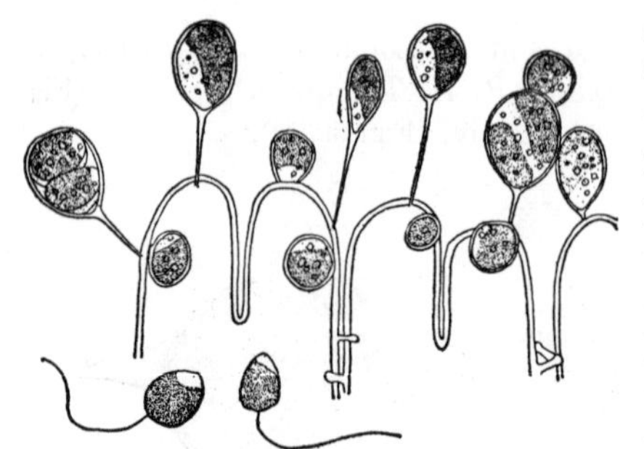

FIG. 304. *Peroniella planctonica* G. M. Smith. (× 1000.)

At the time of reproduction, the protoplast divides transversely into two portions. The upper portion divides into a number of zoospores; the lower portion remains undivided.[1] The zoospores escape through a pore at the apex of the cell wall. Only one flagellum has been recorded, but it has been suggested[2] that there is a second one.

Until the discovery of species with pale chromatophores,[3] *Harpochytrium* was placed among the fungi. Since then there has been a growing tendency to include it among the algae. *Harpochytrium* is found growing upon various algae, especially Zygnematales. The species found in the United States are *H. Atkinsonianum* Pascher (Fig. 305), *H. Hyalothecae* Lagerh., and *H. intermedium* Atkinson. For descriptions of them, see Pascher (1938).

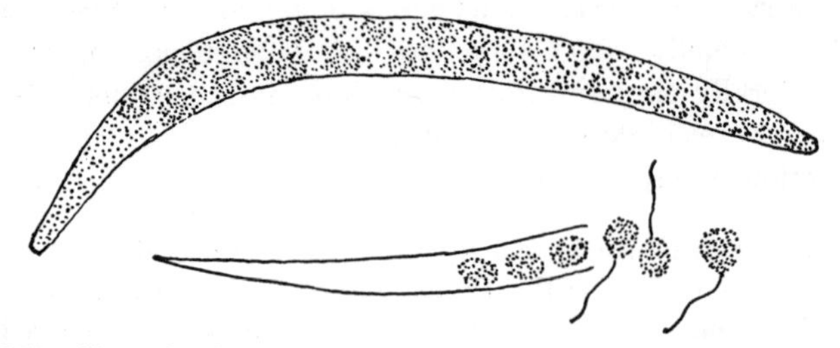

FIG. 305. *Harpochytrium Atkinsonianum* Pascher. (*After Atkinson,* 1903.)

FAMILY 5. CHLOROPEDIACEAE

The Chloropediaceae have sessile cells which may be solitary or laterally adjoined in a monostromatic layer resting directly on the host.

Reproduction is by means of zoospores which become affixed to the host and secrete a wall after they cease swarming.

Lutherella is the only representative of the family found in this country.

1. **Lutherella** Pascher, 1930. *Lutherella* is epiphytic and with solitary

[1] ATKINSON, 1903. [2] PASCHER, 1938. [3] SCHILLER, 1926.

cells, or with the cells in groups of two or four. The cells are globose and, lacking a stipe, are affixed directly to the host. Within a cell is a single large parietal chromatophore with lobed margins.

Reproduction is usually by formation of two zoospores within a cell. Occasionally there is a formation of two or four aplanospores that germinate to form vegetative cells without being liberated from the parent-cell wall. This results in groups of two or four cells.[1]

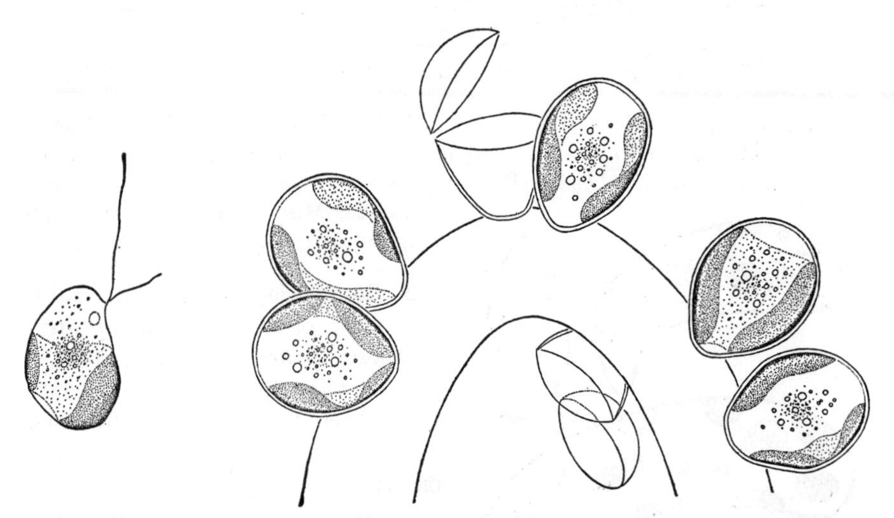

FIG. 306. *Lutherella adhaerens* Pascher. (*Drawn by R. H. Thompson.*) (× 1560.)

Prof. R. H. Thompson writes that he has found the type species, *L. adhaerens* Pascher (Fig. 306), in Maryland and Kansas. For a description see Pascher (1930*C*).

FAMILY 6. CENTRITRACTACEAE

The Centritractaceae have free-floating solitary cells with a cell wall composed of two equal or unequal overlapping halves. The cells are variously shaped and with or without spines.

Reproduction is by means of zoospores or autospores.

The two genera found in this country differ as follows:

1. Cells cylindrical, with spines at the poles .... 1. **Centritractus**
1. Cells cylindrical, without spines .... 2. **Bumilleriopsis**

1. **Centritractus** Lemmermann, 1900. Cells of this unicellular free-floating alga are cylindrical, with a length several times the breadth, and with both poles prolonged into a long straight spine. The cell wall is composed of two halves which may overlap each other or which may lie

[1] PASCHER, 1930*C*.

some distance from each other.[1] The protoplast has two more or less longitudinal laminate chromatophores and a single nucleus.

The method of reproduction is unknown.

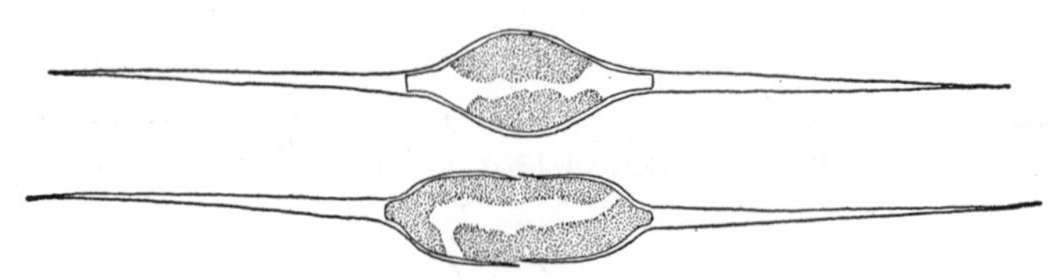

FIG. 307. *Centritractus belonophorus* (Schmidle) Lemm. (*After Schmidle*, 1900*A*.)

Centritractus resembles certain species of *Ophiocytium* but differs in structure of cell wall, in shape and number of chromatophores, and in being uninucleate. *C. belonophorus* (Schmidle) Lemm. (Fig. 307) has been found in several of the states east of the Mississippi River. For a description of it, see Pascher (1925*A*).

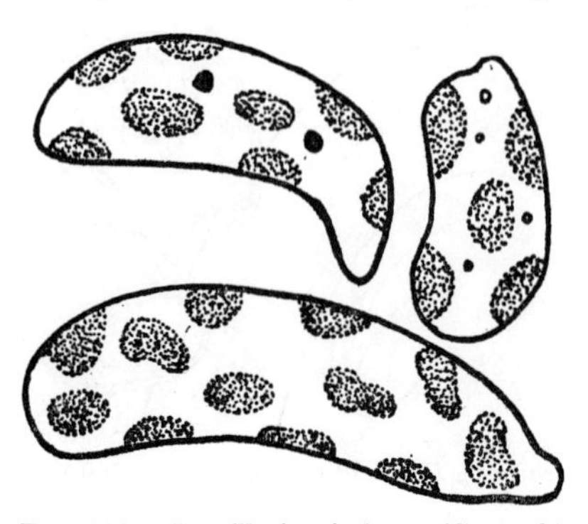

FIG. 308. *Bumilleriopsis breve* (Gerneck) Printz. (*Drawn by G. W. Prescott.*)

2. **Bumilleriopsis** Printz, 1914. The cells of this unicellular free-floating alga are cylindrical, straight to somewhat arcuate, with a length two to ten times the breadth, and without spines at the poles. At times, one of the broadly rounded poles may be narrower than the other. The cell wall is composed of two overlapping halves but without one part cover-like as in *Ophiocytium*. The protoplast is uninucleate and contains many small disk-like parietal chromatophores and droplets of oil.

Reproduction is by division of the cell contents into several zoospores which escape through an opening at one pole of a cell. There may also be a formation of globose aplanospores.[2]

Prof. G. W. Prescott writes that he has found *B. breve* (Gerneck) Printz (Fig. 308) in Michigan. For a description of it, see Printz (1914).

FAMILY 7. CHLOROTHECIACEAE

The Chlorotheciaceae have sessile or free-floating cells which are usually solitary but which may be joined in dendroid colonies. The cell wall is composed of two overlapping halves of equal or unequal size which, in certain genera, are known to be of unlike structure.

[1] SCHMIDLE, 1900*A*; SKUJA, 1934*A*. [2] PRINTZ, 1914.

Reproduction is by means of zoospores or autospores.

The two genera found in this country differ as follows:

1. Cells spherical, ellipsoidal, or pyriform .... 1. **Chlorothecium**
1. Cells cylindrical, usually with terminal spines .... 2. **Ophiocytium**

1. **Chlorothecium** Borzi, 1885. This alga has solitary, sessile, globose to pyriform cells affixed to the substratum by a disk-like expansion of the side next to the substratum. The cell wall consists of two overlapping parts, and the protoplast within it contains two to four parietal discoid chromatophores without pyrenoids.

Reproduction is by division of the cell contents to form 16, 32, 64, or more aplanospores which may germinate before liberation from the parent-cell wall.[1] A germinating aplanospore produces one to four zoospores which become affixed after swarming and develop into vegetative cells. At times, there may be a formation of globose aplanospores with thick walls and nodular thickenings at the poles.[2]

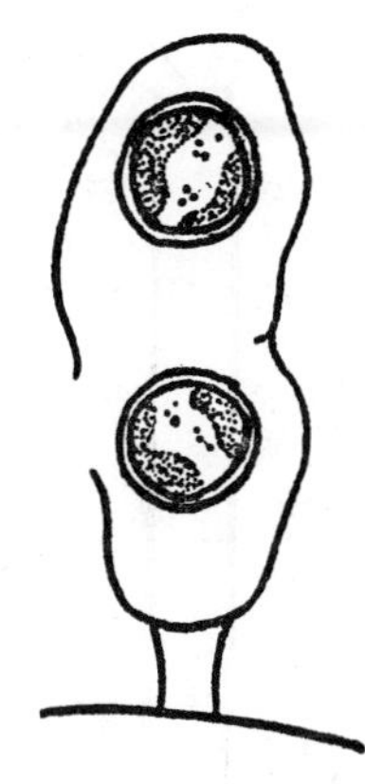

FIG. 309. *Chlorothecium Pirottae* Borzi. (*Drawn by G. W. Prescott.*)

Chlorothecium differs from *Characiopsis* in structure of the cell wall and in that there is not a direct division of the protoplast into zoospores. Prof. G. W. Prescott writes that he has found the type species, *C. Pirottae* Borzi (Fig. 309), in Michigan. For a description of it, see Pascher (1925*A*).

2. **Ophiocytium** Nägeli, 1849. Cells of this genus may be free-floating or sessile; and solitary or epiphytic upon empty cell walls of previous cell generations to form dendroid colonies. The cells are cylindrical, though sometimes with slightly dilated ends, and usually several times longer than broad. Sessile individuals are commonly straight; free-floating cells are often curved or twisted in regular spirals. Both ends of a cell are rounded and, according to the species, both ends bear a single spine, one end has a spine, or spines are lacking. Sessile species usually have a single spine, by which they are attached to the substratum, and the spine frequently terminates in a brownish disk at the point of attachment. The structure of the two parts of the wall has already been noted (see page 372). The protoplast is multinucleate[3] and has several chromatophores that lie one above the other and are either discoid or transversely elongate into bands that encircle the protoplast. Sometimes the transversely elongate chromatophores are H-shaped in optical section.[4]

[1] BORZI, 1895. [2] PRINTZ, 1914. [3] BOHLIN, 1897; BORZI. 1895.
[4] BOHLIN, 1897.

Formation of new cells is by means of zoospores or autospores. In zoospore formation, a protoplast divides into four or eight zoospores that are liberated by a shedding of the "cover" half of a cell wall.[1] Zoospores may swim away from the parent-cell wall before they germinate; or they may come to rest on the open end of the parent-cell wall. In the latter case, the result is a dendroid colony. Zoospores that do not escape from a parent-cell wall may develop into autospores.[2]

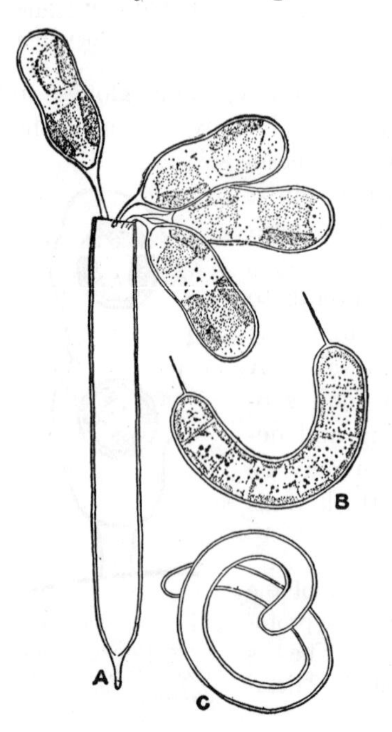

FIG. 310. *A. Ophiocytium arbusculum* (A. Br.) Rab. *B, O. capitatum* Wolle. *C, O. parvulum* (Perty) A. Br. (× 1000.)

Ophiocytium is frequently present, though never in quantity, in collections from semipermanent or permanent pools. It is also a widely distributed plankton organism in this country. The species found in the United States are *O. arbusculum* (A. Br.) Rab. (Fig. 310*A*), *O. bicuspidatum* (Borge) Lemm., *O. capitatum* Wolle (Fig. 310*B*), *O. cochleare* (Eichw.) A. Br., *O. desertum* Printz, *O. majus* Näg., *O. mucronatum* (A. Br.) Rab., and *O. parvulum* (Perty) A. Br. (Fig. 310*C*). For descriptions of them, see Pascher (1925*A*).

ORDER 4. HETEROTRICHALES

Members of this order have cylindrical cells uniseriately united end to end in branched or unbranched filaments.

Reproduction may be by means of zoospores or aplanospores, of which one, two, or more may be formed within any cell of a filament. Sexual reproduction is isogamous.

FAMILY 1. TRIBONEMATACEAE

The Tribonemataceae include all Heterotrichales with unbranched filaments.

The two genera found in this country differ as follows:

1. Cell wall thick, H-pieces clearly evident.......................... 1. **Tribonema**
1. Cell wall thin, H-pieces not clearly evident.................... 2. **Bumilleria**

1. **Tribonema** Derbes and Solier, 1856. Filaments of this alga are composed of cylindrical or barrel-shaped cells that are two to five times

[1] BOHLIN, 1897; BORZI, 1895; LEMMERMANN, 1899.
[2] BOHLIN, 1897; PRINTZ, 1914.

longer than broad. The manner in which the H-pieces of the wall are conjoined, and their structure have already been discussed (see page 372). The protoplast of a cell is uninucleate and, according to the species, contains a few or many discoid chromatophores. Pyrenoids are lacking, and the reserves from photosynthesis are stored as oils or as granules of leucosin, never as starch. Old cells often have numerous small refractive granules within the cytoplasm, the majority of which are probably waste products.

Asexual reproduction may be purely vegetative and result from accidental breaking of a filament; or it may result from disarticulation of certain H-pieces to permit escape of spores. Reproduction by means of aplanospores is of much more frequent occurrence than by means of zoospores. Aplanospores may be formed singly within a cell, or more than one may be

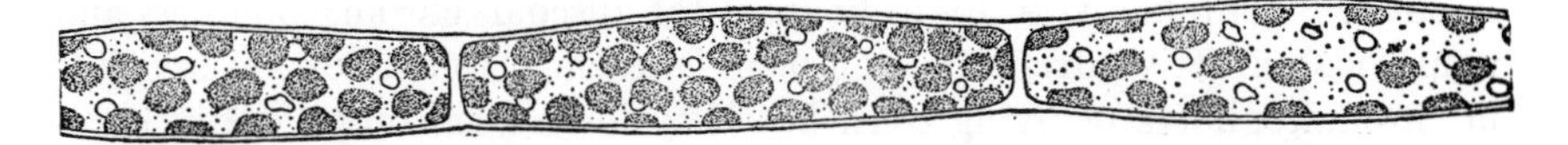

FIG. 311. *Tribonema bombycinum* (Ag.) Derbes and Sol. (× 600.)

formed.[1] An aplanospore, after being liberated by a pulling apart of the surrounding H-pieces, germinates directly into a free-floating filament. Its germination begins with a separation of the two halves of the aplanospore wall and an elongation of the protoplast into a cylindrical cell which develops a new wall.[2] Zoospores are usually formed singly within a cell. They have several chromatophores and the flagellation typical of Xanthophyceae.[3] After swarming for a time, a zoospore comes to rest on some firm object and secretes a cell wall which is attached to the substratum by a brownish discoid holdfast. One-cell germlings resemble *Characiopsis*, but this resemblance ceases after they have divided transversely and begun to grow into filaments that remain sessile until the distal portion breaks away and continues growth as a free-floating filament. Akinetes may also be formed by filaments of *Tribonema*, either singly or in short series.

Sexual reproduction is isogamous, one of a uniting pair of gametes coming to rest and withdrawing its flagella just before the other swims up to and unites with it.[4]

Tribonema and *Microspora* have much the same general appearance, since both have walls made up of H-pieces, but the two can be distinguished by the presence of starch in the latter. *Tribonema* is one of the commonest filamentous algae of standing waters and, during early spring, is frequently the only filamentous "green" alga in temporary pools. The species found in this country are *T. bombycinum*

[1] HAZEN, 1902; LAGERHEIM, 1889; POULTON, 1925. [2] LAGERHEIM, 1889.
[3] LUTHER, 1899; POULTON, 1925; PASCHER, 1925A. [4] SCHERFFEL, 1901.

(Ag.) Derbes and Sol. (Fig. 311), *T. cylindricum* Heering, *T. minus* (Wille) Hazen, *T. Raciborskii* Heering, and *T. utriculosum* (Wille) Hazen. For descriptions of *T. cylindricum* and *T. Raciborskii* , see Pascher (1925*A*); for the others, see Hazen (1902).

2. **Bumilleria** Borzi, 1895. The unbranched filaments of this alga are composed of cylindrical cells that are never barrel-shaped and rarely with a length more than twice the breadth. The cell walls are quite similar to those of *Tribonema,* but the articulated H-pieces are much more delicate and usually discernible only in cells that have formed zoospores (Fig. 312*B–C*). Unlike *Tribonema,* there are H-pieces here and there in a filament that are thicker and sometimes slightly brownish. Two such successively thicker H-pieces originally enclosed a single protoplast, but they later became separated from each other by intercalary cell divisions.[1] Each cell contains two to eight parietal discoid chromatophores and numerous refractive droplets of oil or of leucosin. Pyrenoids are invisible in chromatophores of living cells, but they have been observed[2] in chromatophores of fixed and stained cells.

Zoospore formation takes place either in recently divided or in mature cells. Short cells give rise to a single zoospore; long cells produce two or four zoospores. Zoospores usually remain within the parent-cell wall for some time after they are formed, migrating from one end of a cell to the

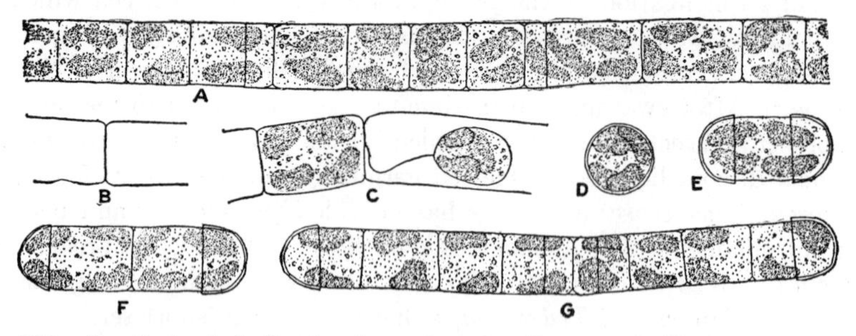

Fig. 312. *Bumilleria sicula* Borzi. *A*, portion of a filament. *B*, H-piece. *C*, zoospore. *D–G*, juvenile filaments developing from zoospores. (× 600.)

other and continously changing in shape. They are liberated by a pulling apart of the two parts of the parent-cell wall in the zone where the articulated H-pieces overlap. The zoospores are pyriform, with two or four chromatophores in the anterior end, and several refractive granules in the broadly rounded posterior end. Soon after a zoospore ceases to swim, it withdraws its flagella, becomes spherical, and secretes a wall composed of two overlapping halves (Fig. 312*D*). The protoplast of the spherical cell

[1] Borzi, 1895; Chodat, 1913; Klebs, 1896; Pascher, 1925*A*.

[2] Korshikov, 1930.

soon elongates to form a short cylinder with rounded poles and with each pole capped by a half of the original spherical wall (Fig. 312*E*). Transverse division of the cylindrical cell into two daughter cells and their repeated elongation and division result in a free-floating filament many cells in length, but one in which the terminal cells are rounded and capped by the hemispherical remains of the original spherical cell wall (Fig. 312*F–G*). One, two, or four aplanospores may also be formed within a cell, and these are liberated by a separation of the surrounding H-pieces of the old parent-cell wall.[1]

Bumilleria usually grows in pools with a clayey bottom. The two species found in this country are *B. exilis* Klebs and *B. sicula* Borzi (Fig. 312). For descriptions of them, see Pascher (1925*A*).

Family 2. Monociliaceae

This family includes the Xanthophyceae with cylindrical cells united end to end in branched filaments. Thus far but one genus is known for this country.

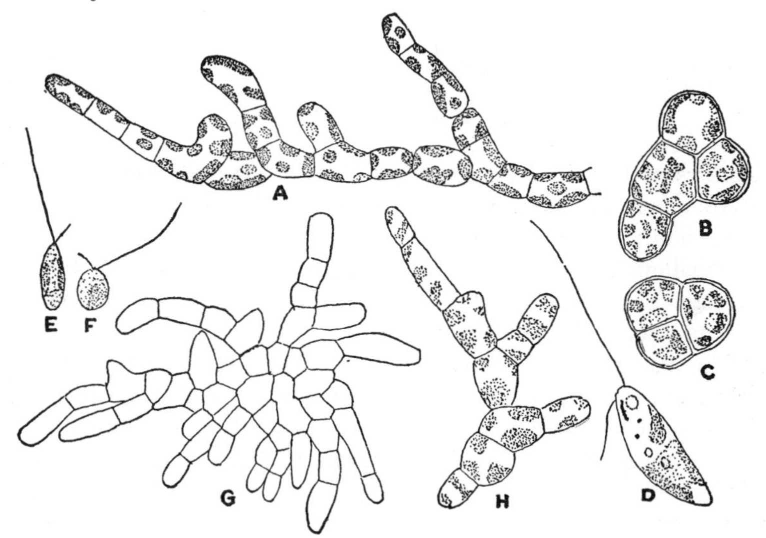

Fig. 313. *A–F, Monocilia viridis* Gerneck. *G–H, M. flavescens* Gerneck. (*A, after Gerneck*, 1907; *B–D, after Chodat*, 1909; *E–F, after Poulton*, 1925; *G–H, after Snow*, 1911.)

1. **Monocilia** Gerneck, 1907. The thallus of this alga consists of a freely branched filament, without differentiation into base and apex, that is always of microscopic size. *Monocilia* shows much the same plasticity of plant-body organization as *Stigeoclonium*, and under certain cultural

[1] Borzi, 1895; Chodat, 1913; Klebs, 1896.

conditions it may assume a palmelloid condition in which the cells are isolated or joined in small *Protococcus*-like packets.[1] When in the palmelloid state, it is scarcely distinguishable from *Botrydiopsis*, but cultural studies have shown that *Botrydiopsis* always remains unicellular, whereas *Monocilia* may be induced to return to the filamentous stage.[2] The cells of *Monocilia* are uninucleate, with several discoid chromatophores, and store their photosynthetic reserves chiefly as oil, never as starch.

Reproduction is by a formation of typical biflagellate xanthophycean zoospores.[3] The zoospores are of two different types:[4] macrozoospores with a length double the breadth and microzoospores which are approximately spherical. There is no evidence that either of the two types functions as gametes. Aplanospores are also formed in abundance, in much the same number within a cell as are zoospores.[4]

The two species found in the United States are *M. flavescens* Gerneck (Fig. 313*G–H*) and *M. viridis* Gerneck (Fig. 313*A–F*). For descriptions of them as species of *Heterococcus*, see Pascher (1925*A*).

ORDER 5. HETEROSIPHONALES

This order includes the multinucleate siphonaceous Xanthophyceae.

FAMILY 1. BOTRYDIACEAE

Members of this family have the siphonaceous multinucleate cell differentiated into an inflated aerial portion with chromatophores and a colorless tubular subterranean portion.

1. **Botrydium** Wallroth, 1815. This unicellular multinucleate terrestrial alga consists of a vesicular, globose or branched, aerial portion (containing the chromatophores) and a colorless rhizoidal portion which penetrates the soil. The shape of the vesicular portion, which may be 1 to 2 mm. in diameter, is considerably influenced by environmental conditions. It is usually elongate cylindrical when growing in shaded habitats and spherical when growing in brightly illuminated places.[5] The vesicular portion has a relatively tough wall within which is a delicate layer of cytoplasm containing many nuclei and chromatophores. The chromatophores are discoid in shape and often connected to one another by strands of dense cytoplasm. Pyrenoid-like bodies are often present in chromatophores of young cells, but there is never any starch in the protoplast, and photosynthetic reserves accumulate as oil or as leucosin. The rhizoidal portion, which may

[1] CHODAT, 1909; GERNECK, 1907; SNOW, 1911. [2] MOORE and CARTER, 1926.
[3] CHODAT, 1909; POULTON, 1925. [4] POULTON, 1925.
[5] KOLKWITZ, 1926.

be profusely or sparingly branched, is without chromatophores but contains many nuclei scattered through its dense or vacuolate cytoplasm.

Cells of *Botrydium* are incapable of vegetative division, and the only method by which new plants may be formed asexually is through a production of aplanospores or hypnospores. In the formation of aplanospores, there is a cleavage of the cell contents into uni- or multinucleate protoplasts which round up and secrete a wall.[1] These aplanospores develop directly into new plants (Fig. 314*C–E*). Hypnospores may be formed in all portions of a cell or after practically all of the protoplasm has migrated into the rhizoids and there divided to form hypnospores.[2]

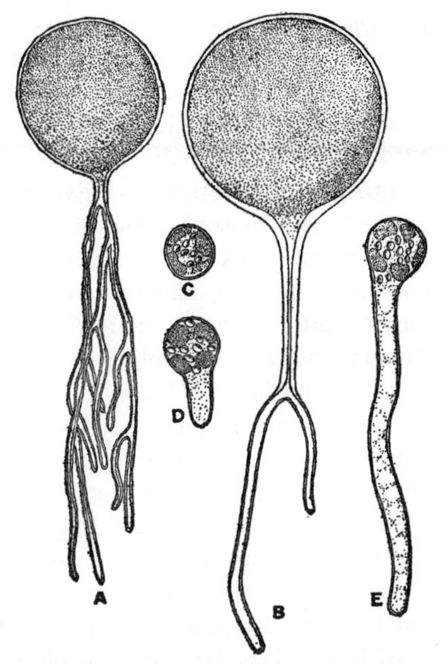

FIG. 314. *A*, *Botrydium granulatum* (L.) Grev. *B*, *B. Wallrothii* Kütz. *C–E*, germinating aplanospores of *B. granulatum*. (*A–B*, × 25; *C–E*, × 600.)

When a cell is flooded with water, this may be followed by a division of its contents into many biflagellate swarmers. These are usually called zoospores, but they have been shown[3] to be gametic in nature. Gametic union may be isogamous or anisogamous, and cells producing gametes may be homothallic or heterothallic.[3] The gametes are biflagellate, pyriform, and become apposed at their anterior ends when uniting in pairs to form a zygote. Gametes which have not fused with those of opposite sex develop parthenogenetically into thalli. A germinating zygote develops directly into a vegetative thallus.[3]

Botrydium is a terrestrial alga that grows chiefly on muddy banks of streams and ponds and on bare soil, especially soil along the margins of paths. Whenever present, the alga usually occurs in abundance and frequently in quantity sufficient to hide the underlying soil. Of the two species found in this country, *B. granulatum* (L.) Grev. (Fig. 314*A*, *C–E*) is of more frequent occurrence than *B. Wallrothii* Kütz. (Fig. 314*B*). For descriptions of these species, see Pascher (1925*A*).

[1] MILLER, V., 1927.
[2] IYENGAR, 1925; MILLER, V., 1927; ROSTAFIŃSKI and WORONIN, 1877.
[3] MOEWUS, 1940.

Family 2. Vaucheriaceae

Members of this family have tubular, sparingly branched, siphonaceous, multinucleate thalli in which sexual reproduction is oögamous. There is but one genus, *Vaucheria*.*

Vaucheria has always been placed among the Siphonales of the Chlorophyceae, but certain phycologists[1] have been uncertain about this because of the predominance of yellow pigments in the plastids, the formation of oil instead of starch, and the possibility that both flagella of a pair are not of equal length. Evidence of a more certain nature has been obtained through a careful analysis of the pigments in three species of *Vaucheria*.[2] All three of these species have the pigments typical of Xanthophyceae instead of Chlorophyceae (see Table I, page 3). Furthermore, these three species lack siphonoxanthin, a xanthophyll of universal occurrence among Siphonales, and one found only in this order. All these data indicate that *Vaucheria* should be placed among the Xanthophyceae rather than among the Chlorophyceae.

1. **Vaucheria** De Candolle, 1803. *Vaucheria* is an unseptate tubular coenocyte with a very sparse or fairly abundant branching. It may be aquatic or terrestrial in habit. Terrestrial individuals frequently have narrow colorless rhizoidal branches penetrating the soil. The cell wall is relatively thin, and just within it is a thin layer of cytoplasm with numerous elliptical chromatophores toward the exterior and many minute nuclei toward the interior. Droplets of oil, accumulating here and there throughout the cytoplasm, are the only food reserves.

Although *Vaucheria* can develop transverse septa that block off injured portions of the coenocyte, there is but little reproduction by accidental breaking of filaments. Asexual reproduction of aquatic individuals is usually by means of multiflagellate multinucleate zoospores formed singly within a terminal sporangium separated from the rest of the cell by a transverse wall.[3] The entire surface of a zoospore is usually clothed with pairs of flagella. Terrestrial individuals may have the entire contents of a sporangium developing into a single multinucleate aplanospore. Development of the sporangial contents into a zoospore or into an aplanospore is

* *Vaucheria arrhyncha* Heidinger, a species found in this country, has been made the type of a second genus (*Vaucheriopsis*) by Heering (1921). The major character for this segregation was the presumed absence of oil, but it has now been shown by Whitford (1943) that *V. arrhyncha* does contain oil. Thus there seems to be no valid reason for recognizing Heering's genus.

[1] Bohlin, 1897; Feldmann, 1946; Fritsch, 1935; Pascher in Heering (1921); Printz, 1927.

[2] Strain, 1949.

[3] Birckner, 1912; Götz, 1897; Klebs, 1896; Strasburger, 1880.

in part dependent upon the external environment. A transfer of aquatic individuals from light to darkness, or from running to quiet water, induces a formation of aplanospores.[1] Flooding of terrestrial individuals frequently induces a formation of zoospores. Terrestrial individuals may have a transverse segmentation of the entire cell into short segments, each of which becomes a hypnospore[2] (Fig. 316*A*). Aquatic individuals may have the entire contents of a cell dividing to form a large number of micro-aplanospores.[3]

Sexual reproduction is oögamous, and all the fresh-water species are homothallic. Development of sex organs is of frequent occurrence in thalli growing on damp soil or in quiet water but is infrequent in thalli growing in swiftly flowing water. The antheridia and oögonia are borne adjacent to one another and either on a common lateral branch or on ad-

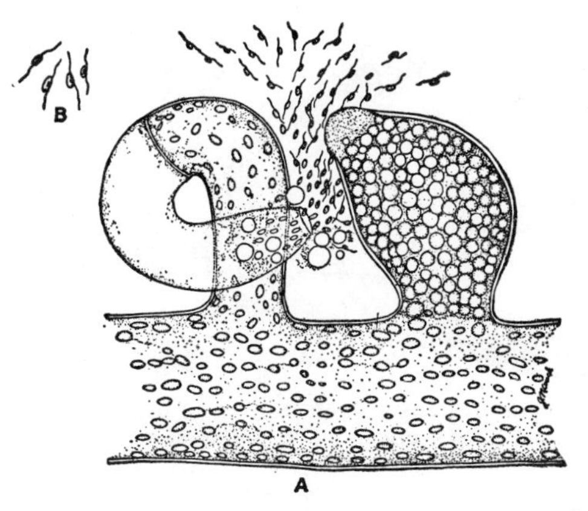

FIG. 315. *Vaucheria sessilis* (Vauch.) DC. *A*, discharge of antherozoids. *B*, antherozoids. (*Drawn by J. N. Couch.*) (*A*, × 320; *B*, × 635.)

joining lateral branches. An oögonium is separated from the subtending branch by a transverse wall and contains a single uninucleate egg.[4] An antheridium is separated from the subtending branch by a transverse wall and contains a number of small biflagellate antherozoids with what appears to be a lateral insertion of the flagella (Fig. 315). Liberation of anthero-zoids takes place before daybreak,[5] and according to the species, they are liberated through a single terminal pore, or two, three, or more lateral pores in the antheridial wall. Fertilization is effected by antherozoids swimming through a relatively large pore in the oögonial wall The

[1] KLEBS, 1896. [2] DE PUYMALY, 1922; STAHL, 1879. [3] SMITH, G. M., 1944.

[4] COUCH, 1932; DAVIS, B. M., 1904; MUNDIE, 1929; OLTMANNS, 1895; WILLIAMS, 1926.

[5] COUCH, 1932; MUNDIE, 1929; OLTMANNS, 1895.

zygote secretes a thick wall and remains within the oögonium until liberated from it by a decay of the oögonial wall. It generally enters upon a dormant period for several months before germinating directly into a new filament.[1] The somewhat inconclusive data[2] indicate that division of the zygote nucleus is meiotic.

Specific differences in *Vaucheria* are based entirely upon structure of mature sex organs and, according to the structure of the antheridium, the genus is divided[3] into seven sections, representatives of four of them being known for the fresh-water flora of this country.[4] For a list of the species known for the United States, see Prescott, 1938; for descriptions of most of them, see Heering (1921).

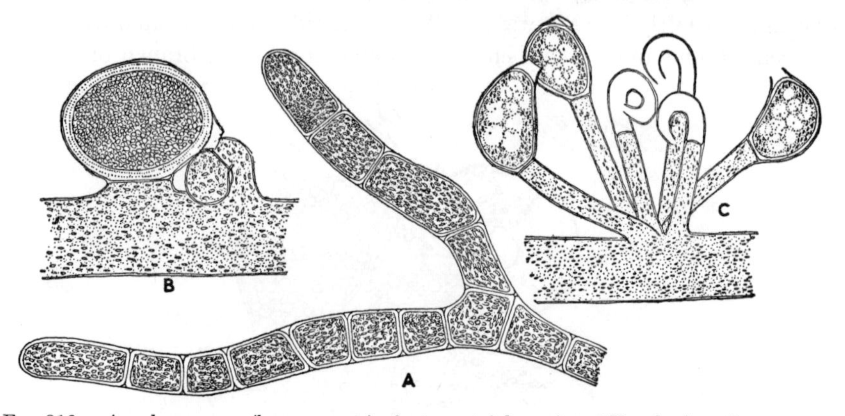

FIG. 316. *A*, aplanospores (hypnospores) of a terrestrial species of *Vaucheria*. *B*, sex organ of *V. pachyderma* Walz. *C*, sex organs of *V. Gardneri* var. *tenuis* Collins. (*A*, × 155 *B–C*, × 185.)

DOUBTFUL XANTHOPHYCEAE

1. **Botryococcus** Kützing, 1849. *Botryococcus* is a valid genus, but one in which it is uncertain whether it is a member of the Xanthophyceae or of the Chlorophyceae. The color of the plastids and the reported structure of the cell wall indicate that it is a member of the Xanthophyceae; the reported presence of starch points toward its being one of the Chlorophyceae. The thallus of *Botryococcus* is a free-floating colony of indefinite shape and with a cartilaginous, hyaline, or orange-colored envelope whose surface is wrinkled and folded. The cells lie close to one another in groups connected by broad or delicate strands of the colonial envelope. The individual cells are ovoid or cuneate and lie radiately arranged in a single layer toward the periphery of the colonial envelope. The cell wall is

[1] MUNDIE, 1929; PRINGSHEIM, N., 1855; WALZ, 1866.
[2] GROSS, 1937; HANTASCHEK, 1932; WILLIAMS, 1926.
[3] HEERING, 1921. [4] HOPPAUGH, 1930.

transversely divided into two unequal overlapping halves,[1] of which the lower half is much the longer. When a cell divides longitudinally to form autospores, the lower half of the parent-cell wall becomes a firm cup-like gelatinous mass with a stalk-like projection toward the center of the colony. The protoplasts within a cell contain a single cup-shaped chromatophore with a naked pyrenoid-like body. Minute granules of starch have been observed[2] within the chromatophore, but not in association with the pyrenoid. Cells of *Botryococcus* produce oil in quantity and often in such abundance that the cell contents are completely obscured.

There is regularly a vegetative multiplication of colonies by a breaking or dissolution of the strands connecting cell aggregates with one another.

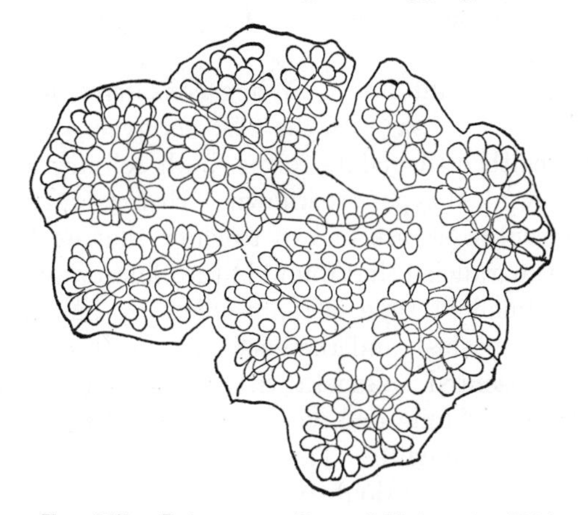

FIG. 317. *Botryococcus Braunii* Kütz. (× 500.)

Autospores are the only type of reproductive body. They generally remain within the colonial envelope, but they may be extruded singly because of pressure of the colonial envelope.[3]

Botryococcus is a widely distributed plankton alga in this country. It is also occasionally found in permanent or semipermanent pools. The three species found in this country are *B. Braunii* Kütz. (Fig. 317), *B. protuberans* W. and G. S. West, and *B. sudeticus* Lemm. For descriptions of them, see G. M. Smith (1920).

CLASS 2. CHRYSOPHYCEAE

Cells of members of this class have chromatophores of a distinctive golden-brown color because of a predominance of beta carotene and certain xanthophylls (see Table I, page 3). The chief photosynthetic reserves are leucosin, an insoluble compound of unknown composition but thought

[1] BLACKBURN, 1936; GEITLER, 1925*C*.
[2] BLACKBURN, 1936; CARLSON, 1906; CHODAT, 1896.
[3] WEST, W. and G. S., 1903.

to be a carbohydrate, and oils. According to the genus, motile vegetative and reproductive cells have one flagellum, or two flagella of equal or unequal length, or, in very rare cases, three flagella. Uniflagellate cells have a flagellum of the tinsel type; biflagellate cells have one tinsel type and one whip type of flagellum, An endogenous formation of statospores is of widespread occurrence. Sexual reproduction in members of the class has not been demonstrated beyond all doubt.

Occurrence. Many fresh-water Chrysophyceae are restricted to cold brooks, especially mountain streams, and to springs. Others are found in the plankton of lakes and in greatest abundance during spring and autumn when the water is cool, and most thrive best in water relatively free from impurities. There are only a very few Chrysophyceae that are aerial or terrestrial in habit.

Organization of the Plant Body. When considered in the light of the theory of plant bodies (see page 6), the Chrysophyceae are a class in which there is a great wealth of flagellate forms, both solitary and colonial, but one in which but few algal types are known. However, a sufficient number of immobile representatives are known to show that evolution within the Chrysophyceae has paralleled that found in the Chlorophyceae and Xanthophyceae.[1] The palmelloid type is represented by half a dozen or more genera (the Chrysocapsales), and the coccoid type (the Chrysosphaerales) is about as well represented. Genera with a true filamentous organization (the Chrysotrichales) include one with unbranched filaments and two or three with branched filaments. Thus far, no siphonaceous forms have been found, and it is rather doubtful whether such forms exist, because multinucleate palmelloid or coccoid forms, the potential ancestors of a truly siphonaceous plant, have not been discovered.

The Chrysophyceae are richer than any other class in forms that have started evolving toward an animal-like rather than a plant-like organization. Thus, there are several genera in which the amoeboid type of cell is the dominant phase in the life history, and where autotrophic nutrition is supplemented by a mass ingestion of solid foods.[2] Some of these amoeboid forms have developed to a point where they have a true plasmodial organization,[3] other rhizopodal forms have completely lost their autotrophic nutrition.[4]

Cell Structure. The majority of genera in the Chrysophyceae have protoplasts that contain but one or two golden-brown chromatophores, although protoplasts of certain genera, as *Phaeothamnion*, may have several chromatophores. Pyrenoid-like bodies are found in the chromatophores of certain genera, as *Chromulina* and *Mallomonas*, but their function is

[1] PASCHER, 1914, 1925. [2] PASCHER, 1915*B*, 1917. [3] PASCHER, 1916, 1916*A*.
[4] PASCHER, 1912.

not known. The autotrophic nutrition of the Chrysophyceae results in the accumulation of leucosin, fats, and volutin[1] but never starch. Glycogen has also been found in the cysts of *Mallomonas*.[2] Leucosin, whose chemical composition is unknown,[3] is laid down in the form of white refractive granules which generally accumulate in the cytoplasm and toward the posterior end of the cell. Heterotrophic nutrition is both by the osmotic intake of dissolved organic substances[3] from the surrounding medium, and, when the cells are amoeboid, by the mass ingestion of solid substances including bacteria and other minute organisms.

Many of the motile forms have contractile vacuoles at the base of the flagella, and these vacuoles may persist even after the flagellated phase has been metamorphosed into an amoeboid phase. Examples of this are to be seen in *Dinobryon*[4] and *Chrysamoeba*.[5]

Although the Chrysomonadales are the only members of the class in which the nuclear phenomena have been studied intensively, it is rather probable that all of the class have uninucleate cells. In the genera investigated, the nuclei have a distinct nucleolus, a nuclear membrane, and considerable chromatic material. Nuclear division is mitotic and by means of a spindle that is intranuclear in origin.[6] Centrosomes have been observed in association with the spindles, but the relationship between centrosome and neuromotor apparatus is obscure.[7]

All biflagellate genera in which the flagella have been studied in detail[8] have one flagellum of the tinsel type (with cilia along both sides) and one flagellum of the whip type. Motile members of the class have naked cells, but the surface of the cytoplasm may be covered with plates of silica (*Mallomonas*). Some motile genera have the protoplast surrounded by an open lorica to which only the basal portion of the protoplast is attached. Coccoid and filamentous genera have the cells with definite walls.

Asexual Reproduction. Cell division in solitary Chrysomonadales is always longitudinal and with an immediate separation of the two-daughter cell. Colonial Chrysomonadales, as *Dinobryon*, may form new colonies by one of the protoplasts swimming away and developing into a new colony. Multiplication of colonies may be by fragmentation both in flagellated and nonflagellated colonies. Chrysococcales and Chrysotrichales reproduce by means of uniflagellate zoospores or biflagellate zoospores with the flagella of equal length. The zoospores are naked and with one or two chromatophores.

The striking type of spore is the statospore. These are usually spherical, though sometimes ellipsoidal, and with an enclosing wall with a small

[1] Doflein, 1923. [2] Conrad, 1927. [3] Klebs, 1892; Molisch, 1913.
[4] Pascher, 1912*A*. [5] Klebs, 1892. [6] Conrad, 1927; Doflein, 1916, 1918.
[7] Conrad, 1927; Doflein, 1923. [8] Pascher, 1924; Scherffel, 1924.

circular pore that is closed by a conspicuous plug. The wall is silicified; the plug may or may not be silicified.[1] Statospores of many species have a smooth wall and a simple plug;[2] those of other species have a smooth wall but have the margin of the pore elevated into a flange-like collar (Fig. 318*A*). In still other cases, and where the pore margin is simple or elevated, the wall may be ornamented with small punctae (Fig. 318*B*), simple or forked spines (Fig. 318*C*), or with flange-like plates (Fig.

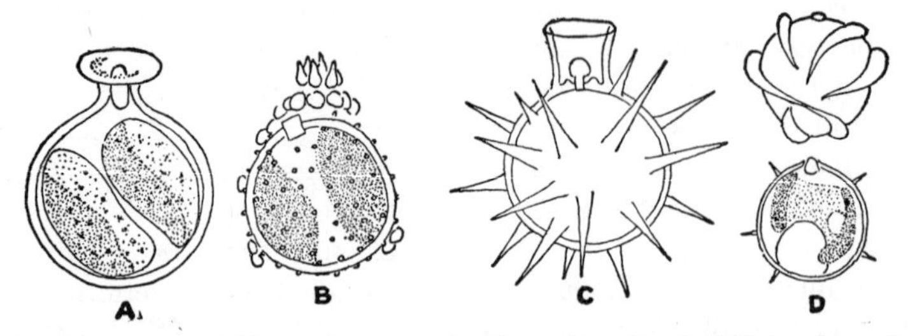

FIG. 318. Statospores of Chrysophyceae. *A*, *Chromulina Pascheri* Hofeneder. *B*, *Mallomonas coronata* Boloch. *C*, *Ochromonas stellaris* Dofl. *D*, *Celloniella palensis* Pascher. (*A*, *after Conrad*, 1926; *B*, *after Conrad*, 1927; *after Doeflein*, 1922*A*; *D*, *after Pascher*, 1929*A*.)

318*D*). As is the case with desmids, no one type of sculpturing is characteristic for a particular genus; as in *Chromulina*, there are, from species to species, marked differences in ornamentation of statospores.

The endogenous method by which statospores of Chrysophyceae are formed was first described in *Chromulina*[3] and has since been found to take place in much the same manner in several other genera.[4] There is every reason for supposing that statospores of the remaining genera for which they are known also form statospores in a similar manner. Statospores of genera with motile vegetative cells are formed after a cell has come to rest, retracted its flagella, and assumed an approximately spherical shape. The first step in the process (Fig. 319*A–E*) is the internal differentiation of a spherical protoplast that is separated only by plasma membranes from the peripheral portion of the original protoplast, which contains contractile vacuoles and droplets of fat. After the plasma membranes are dilimited, there is a secretion of a wall between them, except for a small circular area, the future pore. In formation of one type of statospore, the *Chromulina* type,[5] the cytoplasm external to the statospore wall migrates inward through the pore and fuses with the cytoplasm inside the wall, after which there is a formation of a plug that closes the pore opening. In the *Ochromonas* type of statosphore, the cytoplasm external to the wall gradually disintegrates as a statospore matures.

[1] PASCHER, 1924; SCHERFFEL, 1924. [2] CONRAD, 1922.
[3] CIENKOWSKI, 1870; SCHERFFEL, 1911.
[4] CONRAD, 1927, 1928; DOFLEIN, 1921*A*.
[5] DOFLEIN, 1923.

Statospores germinate by a dissolution of the plug, or by a separation of the plug from the statospore wall, and an amoeboid migration of the protoplast from the enclosing wall (Fig. 319*F–H*). During or after migration of

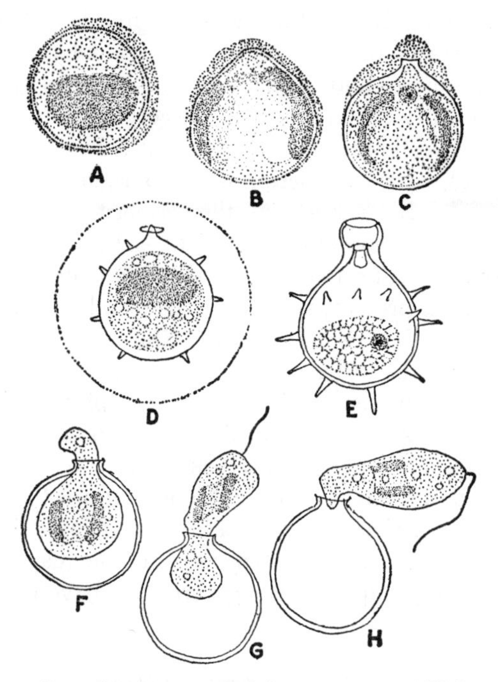

FIG. 319. *A–E*, formation of statospores of *Ochromonas crenata* Kleba. *F–H*, germination of statospores of *Chromulina freiburgensis* Dofl. (*After Doflein*, 1923.)

a protoplast from the enclosing wall there is a formation of flagella.[1] Sometimes, as in *Chromulina Pascheria* Hofeneder,[2] the protoplast within a statospore wall divides into two or four flagellated cells before the contents are liberated.

Classification. Pascher's (1912*A*, 1914, 1925) separation of the chrysophycean algae into groups homologous with the groups into which the grass-green algae are segregated has met with general acceptance ever since it was proposed. However, his giving each group of chrysophycean algae the rank of a class[3] seems unwarranted since the magnitude of differences between them does not appear to be greater than that of an order. It is a matter of opinion whether one should follow those who place the volvocine, plasmodial, and tetrasporine Chrysophyceae in separate orders[4] or place them as suborders of a single order.[5]

[1] DOFLEIN, 1923. [2] CONRAD, 1926; HOFENEDER, 1913. [3] PASCHER, 1931.
[4] SMITH, G. M., 1933. [5] FRITSCH, 1935.

ORDER 1. CHRYSOMONADALES

This order includes those genera which are motile during vegetative phases of their life cycle, and in which amoeboid or rhizopodial stages are only temporary. According to the genus, the motile cells may be solitary or united in colonies of definite shape. Individual cells have the characteristic chromatophores and other internal structures typical of Chrysophyceae. Unlike many of the genera in the corresponding order of Chlorophyceae, the motile cells are not enclosed by a wall, but the peripheral portion of the cytoplasm (the periplast) is so rigid that the cells have a characteristic shape. In some genera, the periplast contains small siliceous scales or calcareous inclusion of characteristic shape. Some genera have a characteristically shaped firm envelope (the lorica) external to the protoplast. A lorica, when present, is open at the distal end and is usually not in lateral contact with the protoplast. Statospores are known for a majority of genera in the order.

According to the number of flagella, many phycologists[1] divide the genera into three series with a rank above that of a family. Since it is uncertain whether such a segregation based solely on number of flagella is natural or artificial, it seems best for the present not to attempt a grouping into categories of higher rank than a family. It must be admitted that certain of the families recognized on pages to follow may also be artificial, but these families do group together genera that have certain features in common.

FAMILY 1. CHROMULINACEAE

Members of this family have uniflagellate free-swimming cells with an undifferentiated protoplast, in which the cells may be naked or lying within a lorica. The genera found in this country differ as follows:

1. Cells without a lorica.......... 2
1. Cells with a lorica.......... 4
    2. With one or with two chromatophores.......... 3
    2. With four chromatophores.......... 2. **Amphichrysis**
3. Chromatophore reticulate.......... 3. **Chrysapsis**
3. Chromatophore not reticulate.......... 1. **Chromulina**
    4. Lorica globose.......... 4. **Chrysococcus**
    4. Lorica vase-shaped.......... 5. **Kephyrion**

1. **Chromulina** Cienkowski, 1870. The spherical, ellipsoidal, oval, or fusiform cells of *Chromulina* have uniflagellate naked cells whose firm periplast is either smooth or finely granular. There may be one, two, or

[1] FRITSCH and WEST, 1929; FRITSCH, 1935; SMITH, G. M., 1933; PASCHER, 1931.

even more vacuoles at the base of the single flagellum.[1] A protoplast may contain one or two plate-like chromatophores; if two are present, they lie on opposite sides of a cell. In very rare cases, as *C. dubia* Dofl., there is a

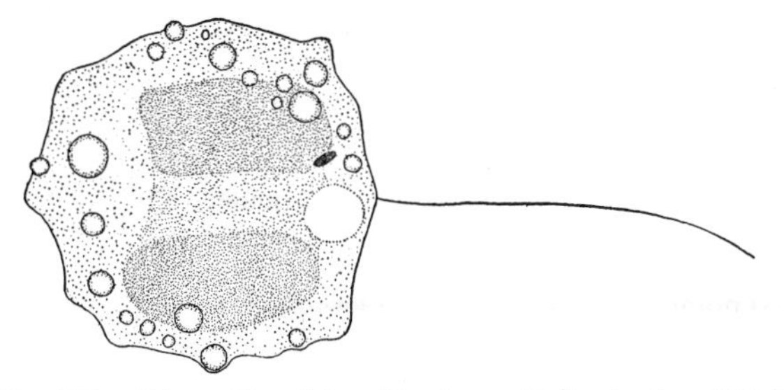

FIG. 320. *Chromulina globosa* Pascher. (*After Lackey,* 1939.)

single pyrenoid within each chloroplast. The position of the nucleus is extremely variable from species to species, and it may be in the anterior, median, or posterior portion of a cell. Some species have an eyespot near the point of insertion of the flagellum. Leucosin is usually formed in much greater abundance than fats, and it ordinarily accumulates in a single large granule at the posterior end of a cell.

Reproduction is by longitudinal division and may take place without a cell's coming to rest and retracting its flagellum.[2] Rhizopodial stages are only occasionally seen in this organism,[3] and palmella stages have been reported for one species.[4] Statospores are known for several species. They are smooth-walled, and some or all of the cytoplasm external to the statospore wall migrates inside the wall before the plug is formed.

C. globosa Pascher (Fig. 320) and *C. ovalis* Klebs have been found repeatedly in Ohio,[5] and *C. ovalis* has been found in Kansas.[6] The genus is undoubtedly of wider distribution in this country than the published record indicates. For descriptions of the two species, see Pascher (1913*A*).

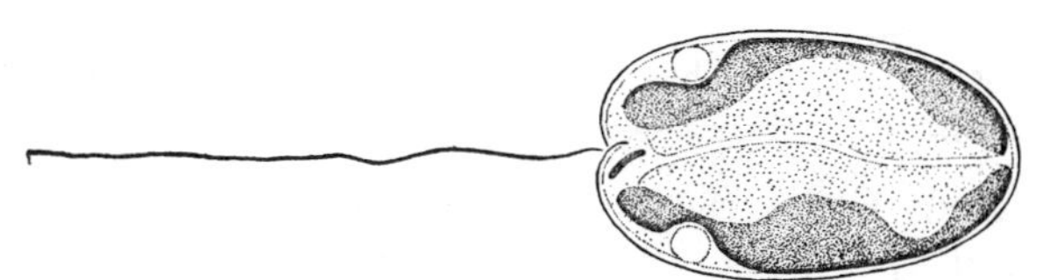

FIG. 321. *Amphichrysis compressa* Korshikov. (*Drawn by R. H. Thompson.*) (× 1080.)

2. **Amphichrysis** Korshikov, 1929. This unicellular, uniflagellate free-swimming form has ellipsoid to ovoid cells that may be slightly compressed in the longitudinal axis. The length of the flagellum is approximately

[1] DOFLEIN, 1923. [2] DOFLEIN, 1923; SCHERFFEL, 1911. [3] SCHERFFEL, 1911.
[4] PASCHER, 1910. [5] LACKEY, 1938, 1939. [6] THOMPSON, 1938.

the same as that of the cell. At the anterior end of a cell are two contractile vacuoles and a conspicuous eyespot. The protoplast contains four band-like chromatophores with lobed margins, which extend longitudinally for pole to pole of a cell and are flanked laterally by numerous granules of leucosin.

At the time of reproduction, a cell becomes immobile, but does not lose its flagellum, and divides longitudinally. Cells may also round up, become invested with a broad hyaline gelatinous sheath, and form typical endogenous chrysophycean cysts.[1]

Prof. R. H. Thompson writes that he has found *A. compressa* Korshikov (Fig. 321), the only known species, in Maryland. For a description of it, see Korshikov (1929).

3. **Chrysapsis** Pascher, 1909. *Chrysapsis* has naked, ovoid to irregularly shaped cells that may be rigid or plastic as they swim through the water.[2] There are one or two contractile vacuoles at the base of the single flagellum, and an eyespot may or may not be present. The chromatophore

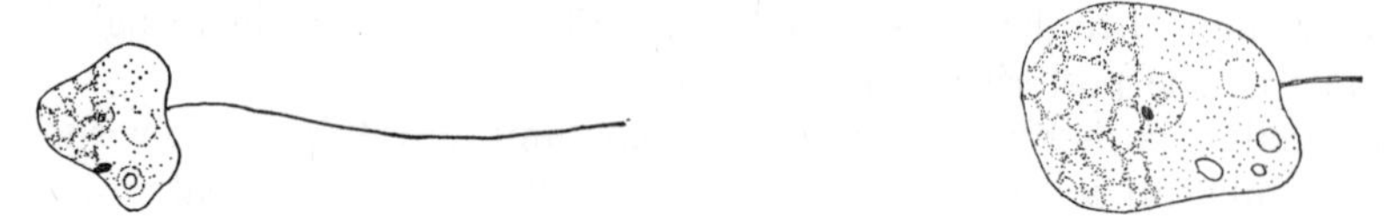

FIG. 322. *Chrysapsis sagene* Pascher. (*After Pascher*, 1909*A*.)

is the most distinctive feature of this genus. It is reticulate and may lie either at the base or at the equator of a cell. The cells usually contain several small leucosin granules.

Cell division is longitudinal and may take place while a cell is actively motile. Often the cells become rounded, lose their flagella, and develop into palmelloid colonies.[3] Typical chrysophycean cysts have been recorded for two of the species.[4]

C. sagene Pascher (Fig. 322) has been found in Ohio.[5] For a description of it, see Conrad (1926).

4. **Chrysococcus** Klebs, 1892. *Chrysococcus* has free-swimming solitary cells surrounded by a lorica. The lorica is more or less globose and with a narrow to broad circular opening at the anterior end. The protoplast is not attached to the lorica, and the single flagellum projects through the opening of the lorica. The protoplast is globose and contains one or two laminate golden-brown chromatophores, an eyespot, one or more vacuoles, and granules of leucosin.

[1] KORSHIKOV, 1929. [2] CONRAD, 1920, 1926; PASCHER 1909*A*.
[3] CONRAD, 1926; PASCHER, 1909*A*. [4] CONRAD, 1920, 1926. [5] LACKEY, 1939.

At the time of reproduction, the protoplast divides to form two daughter protoplasts, one of which escapes and secretes a new lorica, the other remains within the old lorica.[1] Chrysophycean cysts have not been recorded for this genus.

Chrysococcus is widely distributed in the drainage basin of the Ohio River and has been found in a few other localities. The following species have been found in this country: *C. amphora* Lackey, *C. asper* Lackey, *C. cylindrica* Lackey, *C. hemisphaerica* Lackey, *C. major* Lackey, *C. ovalis* Lackey, *C. rufescens* Klebs (Fig. 323), and *C. spiralis* Lackey. For descriptions of them, see Lackey (1938*A*).

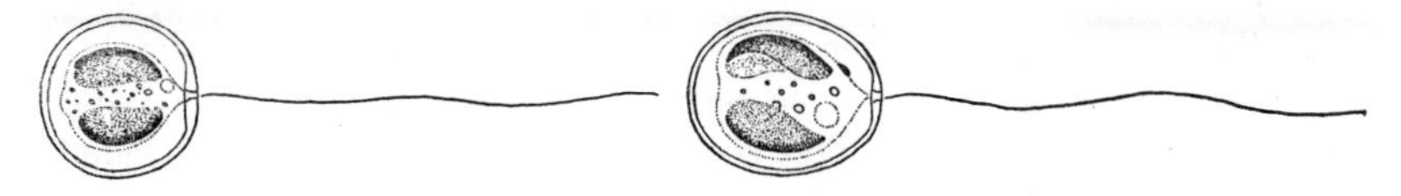

FIG. 323. *Chrysococcus rufescens* Klebs. (*Drawn by R. H. Thompson.*) (× 2160.)

5. **Kephyrion** Pascher, 1911. This genus has minute, solitary, free-swimming cells surrounded by a lorica. The lorica is spindle-shaped to ovoid and has a broad opening at the anterior end. The protoplast completely fills the lower half of the lorica, and the single flagellum projects through the opening at the anterior end.[2] The protoplast contains a single parietal chromatophore, a conspicuous nucleus, and is with or without an eyespot. Droplets of oil have been noted, but no leucosin granules.[3]

The method of reproduction is unknown.

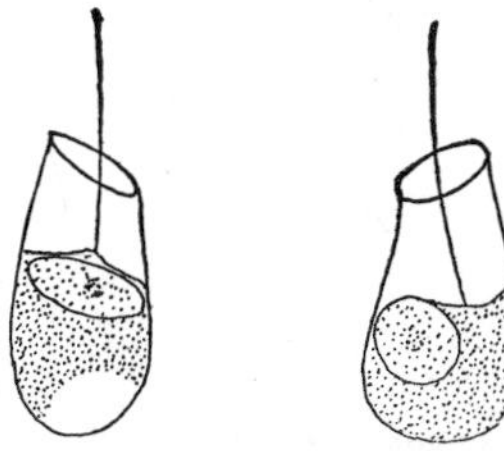

FIG. 324. *Kephyrion ovum* Pascher. (*After Pascher,* 1913*A*.) (× 3000.)

K. ovum Pascher (Fig. 324) has been reported[4] from this country but without mention of specific localities. Dr. Lackey writes that he has found it in Ohio and Alabama. For a description of *K. ovum,* see Pascher (1913).

FAMILY 2. MALLOMONADACEAE

The Mallomonadaceae have uniflagellate cells with a firm ornamented periplast and an anterior vacuolar system in which there are both contractile and noncontractile vacuoles. The two genera found in this country differ as follows:

1. Cells solitary .... 1. **Mallomonas**
1. Cells united in globose colonies .... 2. **Chrysosphaerella**

1. **Mallomonas** Perty, 1852. The solitary free-swimming cells of this chrysomonad are readily distinguishable from other uniflagellate Chryso-

[1] KLEBS, 1892. [2] PASCHER, 1911*B*. [3] LUND, 1942. [4] LACKEY, 1938*A*.

monadales by their larger size and by the siliceous plates, many of which bear a single long siliceous spine, upon the surface of the protoplast. The cells are usually longer than broad and either cylindrical, elliptical, ovoid, or fusiform in shape. The periplast is firm, though slightly elastic, and is covered with siliceous scales. According to the species, the small overlapping scales are circular, ellipsoidal, oval, or polygonal in shape and transversely, diagonally, or somewhat irregularly arranged. In some species, all the scales, in other species only the terminal scales, have a single long siliceous spine excentrically inserted on the outer surface of the scale.[1] Some species have denticulate siliceous spines at the anterior and posterior ends. The majority of species have two laterally situated golden-brown chromatophores, but a few have only one chromatophore. Two species have no chromatophores.[2] The chief photosynthetic reserve is leucosin,

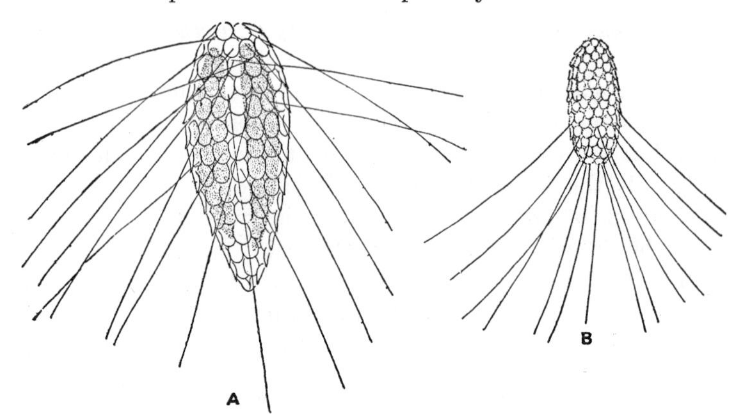

FIG. 325. *A, Mallomonas caudata* var. *macrolepis* Conrad. *B, M. producta* Iwanoff. (× 500.)

and it generally accumulates in the basal portion of the cell. The vacuolar system consists of a large noncontractile vacuole, which lies in the anterior part of the cell, and three to seven contractile vacuoles, which may be apical, median, or basal in position. The single nucleus is quite large and with a conspicuous nucleole. There is a large extranuclear granule (centrosome?) that connects with the base of the flagellum by means of a long rhizoplast.[2]

Reproduction is by longitudinal division and while the cell is motile. Naked *Chromulina*-like zoospores may also be formed. These zoospores swarm for a time and then become immobile, spherical in shape, and surrounded by a gelatinous envelope. Similar palmelloid stages have been recorded as developing from naked amoeboid protoplasts that have escaped from the sheath of silceous scales.[3]

[1] CONRAD, 1927; IWANOFF, 1900. [2] CONRAD, 1927. [3] CONRAD, 1922.

Endogenously formed statospores, of the *Chromulina* type, have been noted in several species. These are usually spherical and without spines.[1]

Mallomonas (Fig. 325) is essentially a planktonic organism and usually is found only in clear-water lakes. In many cases, the organism is to be collected in greater abundance a few meters below than at the surface. About a dozen species are known for the United States. For a monograph of the genus, see Conrad (1933).

2. **Chrysosphaerella** Lauterborn, 1896. The ellipsoidal to pyriform cells of this alga are radially united in spherical colonies of microscopic, that are partly embedded in a spherical hyaline gelatinous envelope containing

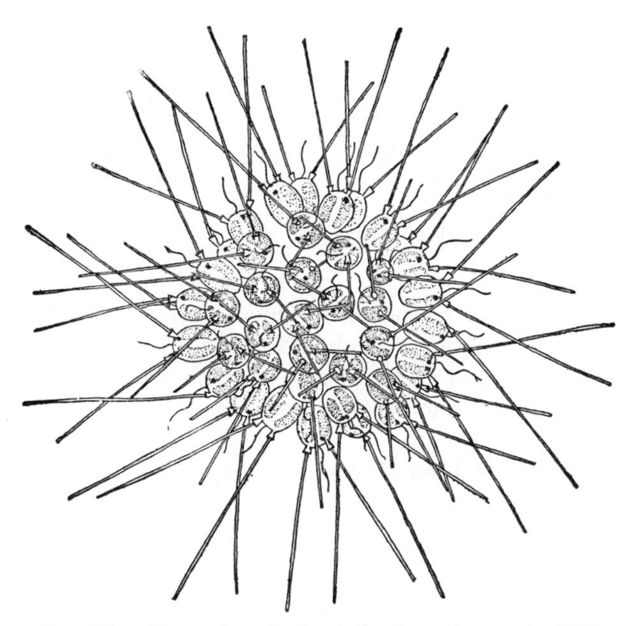

FIG. 326. *Chrysosphaerella longispina* Lauterborn. (× 540.)

numerous, tangentially placed, minute plates of silica. Each cell in a colony has a firm periplast and bears at its anterior end two short vase-shaped projections. A long straight cylindrical siliceous rod, whose length is approximately equal to the diameter of the colony, is inserted in each of the projections. At the anterior of a cell is a single long flagellum. Within a cell are two laterally placed laminate chromatophores, a single eye-spot, a central nucleus, and several vacuoles.[2]

C. longispina Lauterborn (Fig. 326) is a widely distributed plankton organism in this country. For a description of it, see G. M. Smith (1920).

[1] CONRAD, 1927. [2] LAUTERBORN, 1899.

FAMILY 3. SYNCRYPTACEAE

The Syncryptaceae (Isochrysidaceae) have biflagellate cells with flagella of equal length. The cells have an undifferentiated periplast but may be naked or with a lorica, and solitary or united in colonies.

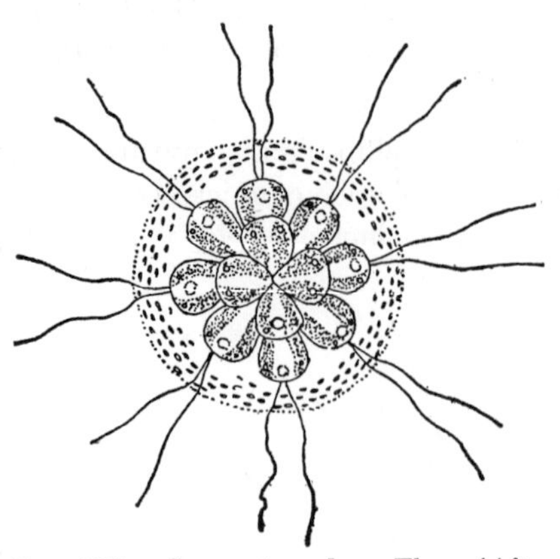

FIG. 327. *Syncrypta volvox* Ehr. (*After Stein*, 1878.)

The two genera found in this country differ as follows:

1. Cells united in colonies — 1. **Syncrypta**
1. Cells solitary and sessile — 2. **Derepyxis**

1. **Syncrypta** Ehrenberg, 1833. The individual cells of *Syncrypta* are ovoid to pyriform and radially united to one another in a microscopical, spherical, free-swimming colony. Colonies, in turn, have the mass of closely appressed cells surrounded by a hyaline gelatinous envelope in which there are numerous granular particles. The cells have two flagella of equal length on their broadly rounded anterior ends.[1] Within a cell are two laminate chromatophores on opposite sides of the cell and two contractile vacuoles. An eyespot is lacking.[2]

Statospores have not been observed in this genus.

S. volvox Ehr. (Fig. 327), the only species, is a rare organism and thus far for this country has been found only in Illinois[3] and Michigan.[4] For a description of it, see Pascher (1913).

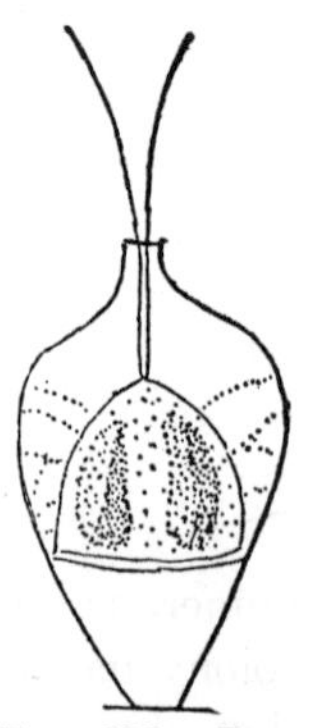

FIG. 328. *Derepyxis dispar* Stokes Lemm. (*After Pascher*, 1909*A*.)

2. **Derepyxis** Stokes, 1885. The solitary cells of *Derepyxis* have a vase-shaped lorica that is attached to the substratum by a stout gelatinous pedicel. The protoplast is spherical to broadly ellipsoidal and attached to the lorica by numerous delicate cytoplasmic strands. There may be two lateral chromatophores or a single one which lies toward the base of the protoplast. At the anterior end of the protoplast are two flagella of equal length that extend through, and for some distance beyond, the mouth of the lorica. There may be one or two vacuoles, and these may be apical or basal in position.

[1] SCHERFFEL, 1904. [2] SENN, 1900. [3] KOFOID, 1910. [4] GUSTAFSON, 1942.

Reproduction is by longitudinal division of the protoplast, after which one of the daughter protoplasts escapes through the open distal end of the lorica. The liberated biflagellate naked cell swims about for a short time and then affixes itself on some firm substratum and secretes a lorica. The statospores of *Derepyxis* are somewhat atypical in that they are larger than the cells in which they are formed.[1]

This is another genus of rare occurrence. The species known for this country are *D. amphora* Stokes, *D. dispar* (Stokes) Lemm. (Fig. 328), *D. ollula* Stokes, and *D. urecolata* (Stokes) Lemm. For descriptions of them, see Pascher (1913*A*).

Family 4. Synuraceae

The Synuraceae include those motile Chrysomonadales with two flagella of equal length in which the naked cells have a differentiated periplast.

The two genera found in this country differ as follows:

1. Cells with two parietal chromatophores .... 1. **Synura**
1. Cells with two axial chromatophores .... 2. **Skadovskiella**

1. **Synura** Ehrenberg, 1838. The cells of this alga are radially united in spherical to oblong-ovoid colonies that are not enclosed by a gelatinous sheath. The individual cells are pyriform in shape, with the anterior end broadly rounded and the posterior end prolonged into a hyaline stalk.[2] The firm periplast of the cell is covered with siliceous scales that are spirally arranged.[3] Within the cell are two laminate curved chromatophores, so placed that their concave faces are opposite. There is a single centrally situated nucleus and two or three contractile vacuoles at the base of the cell. These do not seem to be united into a single collective vacuole, as is the case with *Mallomonas*. Leucosin is the chief food reserve, and it collects in a single large granule toward the base of the cell. There is no eyespot. The two flagella at the anterior end of the cell are of the same length, but there are certain morphological and physiological differences between them.[4]

Cell division is always longitudinal. If the dividing cell is one that has recently escaped from the parent colony, the colony resulting from cell division is but two or four-celled.[2] Ordinarily, however, cell division merely adds to the number of cells in the colony. Reproduction of many-celled colonies takes place by the cells grouping themselves radially about two centers and the two parts separating from each other. Failure of the daughter colonies to separate is probably the reason for the elongate colonies of *Synura* that have been occasionally observed.[5] Reproduction

[1] Conrad, 1926; Pascher, 1909*A*. [2] Conrad, 1926. [3] Petersen, 1918.
[4] Conrad, 1926; Petersen, 1918. [5] Conrad, 1922.

may also take place by the protoplast becoming amoeboid and slipping out from its scale-covered periplast. After its escape, the protoplast may assume a rhizopodal form; or it may develop into a naked biflagellate zoospore, which, in turn, may grow into a palmella stage or into a new motile colony.[1] *Synura* also forms cysts endogenously.[2]

Synura is of widespread occurrence both in the plankton of lakes and in pools and ditches. The species known for the United States are *S. Adamsii* G. M. Smith (Fig. 329*B*), *S. caroliniana* Whitford, and *S. uvella* Ehr. (Fig. 329*A*). For a description of *S. caroliniana*, see Whitford (1942); for the other two, see Conrad (1926).

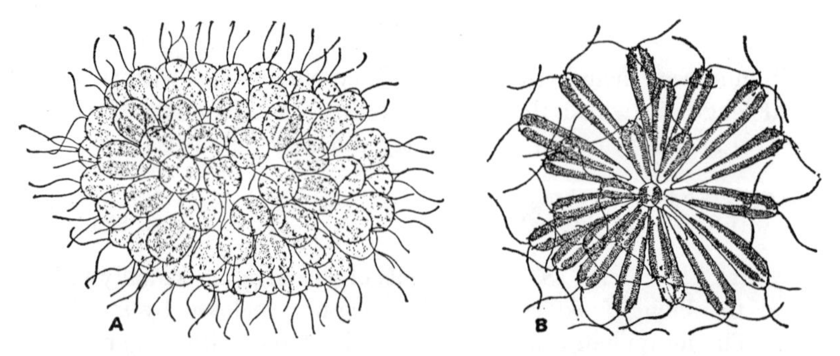

FIG. 329. *A*, *Synura uvella* Ehr. *B*, *S. Adamsii* G. M. Smith. (× 400.)

2. **Skadovskiella** Korshikov, 1927. The cells of this alga are radially united in free-swimming spherical colonies without a gelatinous envelope. The cells are ovoid and with two flagella of equal length. Unlike *Synura*, the periplast of a cell is clothed with numerous elliptical siliceous rings, each with a rod-like prolongation at one pole of the ring. The internal cell structure also differs from that of *Synura* in that the two laminate chromatophores are axial (not parietal), and leucosin granules lie between the two chromatophores.

The method of colony formation is unknown, but it is known[3] that there is a formation of statospores.

Prof. R. H. Thompson writes that he has found *S. sphagnicola* Korshikov (Fig. 330) in Maryland. For a description of it, see Korshikov (1927*B*).

FAMILY 5. COCCOLITHOPHORIDACEAE

Cells of members of this family are solitary, free-swimming, and surrounded by an envelope in which are embedded numerous small calcareous disks or rings (coccoliths).

[1] PASCHER, 1912*A*. [2] CONRAD, 1926. [3] KORSHIKOV, 1927*B*.

This distinctive family has a number of genera of widespread distribution in the ocean. A few genera are found in fresh or brackish waters,[1] and of these *Hymenomonas* has been found in fresh water in this country.

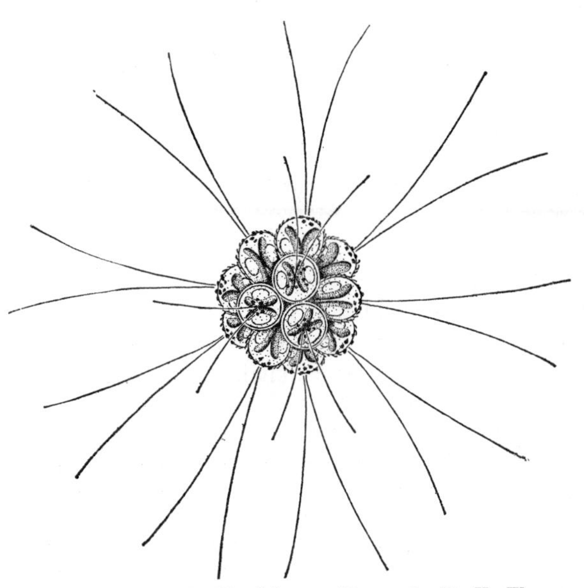

Fig. 330. *Skadovskiella sphagnicola* Korshikov. (*Drawn by R. H. Thompson.*) (× 800.)

Hymenomonas Stein, 1878. The solitary free-swimming cells of this genus are readily distinguishable from other chrysomonads by the numerous annular bodies (coccoliths) embedded in the cellular envelope. The envelope surrounding a cell consists[2] of two concentric layers, the

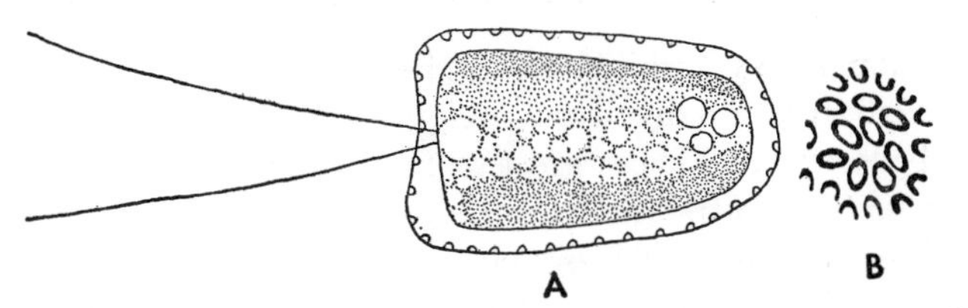

Fig. 331. *Hymenomonas roseola* Stein. *A*, optical section of a cell. *B*, surface view of coccoliths. (*After Lackey*, 1938.)

inner thin and firm, the outer broad and gelatinous. The coccoliths, which are composed of calcium carbonate,[2] are embedded at the periphery of the gelatinous layer. The cells are globose to pyriform, with two flagella of equal length, and contain two parietal golden-brown chromatophores

[1] Conrad, 1928*A*. [2] Conrad, 1914, 1926.

At the base of the flagella are about a half-dozen small contractile vacuoles that discharge their contents into a single large noncontractile vacuole. The nucleus is small and lies at the center of a cell. Leucosin accumulates at the base of a cell, and either in a single large granule or a few small granules.

Reproduction is by cell division, presumably longitudinal,[1] and takes place after a cell has become immobile and the flagella have disappeared. Statospores have also been recorded.[2]

H. roseola Stein (Fig. 331) has been found repeatedly in Ohio.[3] For a description of it, see Pascher (1913*A*).

FAMILY 6. OCHROMONADACEAE

The Ochromonadaceae include all the biflagellate chrysomonads in which there are two flagella of unequal length. The cells have an undifferentiated periplast and may be solitary or colonial, and naked or surrounded by a lorica.

1. Cells without a lorica 2
1. Cells with a lorica 5
  2. Cells solitary .... **1. Ochromonas**
  2. Cells united in colonies 3
3. Colony a flat disk .... **4. Cyclonexis**
3. Colony globose 4
  4. Center of colony with dichotomously branched gelatinous strands .... **3. Uroglena**
  4. Center of colony without gelatinous strands .... **2. Uroglenopsis**
5. Surface of lorica smooth 6
5. Surface of lorica denticulate in optical section .... **7. Hyalobryon**
  6. Cells solitary and sessile .... **6. Epipyxis**
  6. Cells in free-swimming dendroid colonies .... **5. Dinobryon**

1. **Ochromonas** Wystozki, 1887. The cells of *Ochromonas* are free-swimming, solitary, usually ellipsoidal or pyriform, and rigid or with a pronounced plasticity. There are two flagella of unequal length. There may be one or two contractile vacuoles at the base of the flagella, or there may be numerous contractile vacuoles irregularly distributed beneath the cell's surface.[4] An eyespot may or may not be present. According to the species, there are either one or two laminate golden-brown chromatophores, and either a large or a small granule of leucosin.

Cell division may take place while a cell is actively motile, or a cell may become immobile and invested with a gelatinous envelope prior to division. When surrounded by a gelatinous envelope, the cells may divide and

[1] KLEBS, 1892. [2] PASCHER, 1913*A*. [3] LACKEY, 1938, 1938*A*.
[4] CONRAD, 1926.

redivide and thus produce a palmelloid colony with a few or with many cells.[1] Many of the species are known to produce statospores.[2]

O. mutabilis Klebs (Fig. 332) has been recorded from Lake Erie[3] and *O. ludibunda* Pascher from Ohio.[4] For descriptions of them, see Pascher (1913*A*).

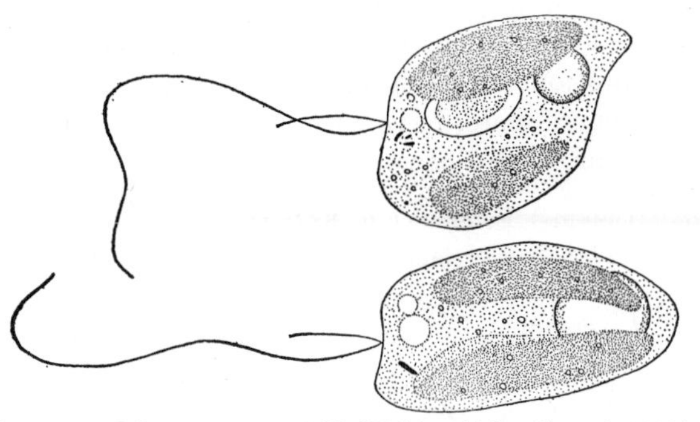

Fig. 332. *Ochromonas mutabilis* Klebs. (*After Conrad*, 1926.)

2. **Uroglenopsis** Lemmermann, 1899. The free-swimming colonies of this alga have cells that are radially embedded in a single layer at the periphery of a copious, hyaline, spherical to broadly ellipsoidal, gelatinous

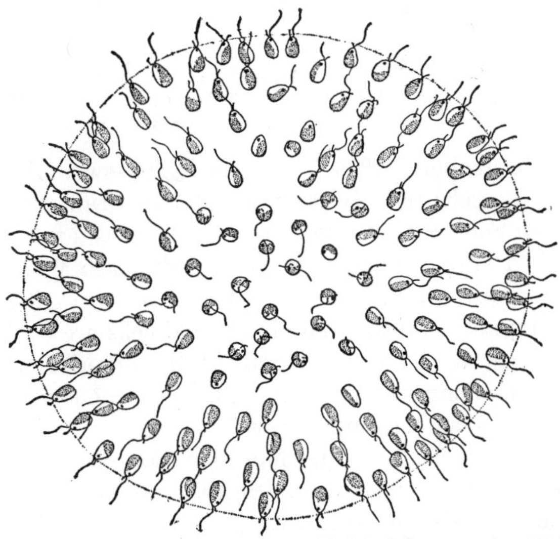

Fig. 333. *Uroglenopsis americana* (Calkins) Lemm. (× 400.)

matrix. The interior of the gelatinous matrix is farily homogeneous and lacks conspicuous dichotomoulsly branched radiating threads leading to

[1] Pascher, 1910. [2] Doflein, 1922*A*, 1923. [3] Tiffany, 1934.
[4] Lackey, 1938.

cells at the periphery.[1] The cells are spherical to narrowly ellipsoidal and have two flagella of unequal length at the anterior end. Within a cell are one or two laminate to discoid, golden-brown chromatophores, a conspicuous eyespot, and one or two vacuoles at the anterior end. Food reserves accumulate largely as minute droplets of oil.[2]

Cell division is longitudinal, and colony reproduction appears to take place by fragmentation. Resting cells, which produce four or eight zoospores when they germinate, have been observed,[3] but it is not definitely known that these are true statospores.

U. americana (Calkins) Lemm. (Fig. 333) is of widespread occurrence in reservoirs and in the plankton of lakes. For a description of it, see G. M. Smith (1920).

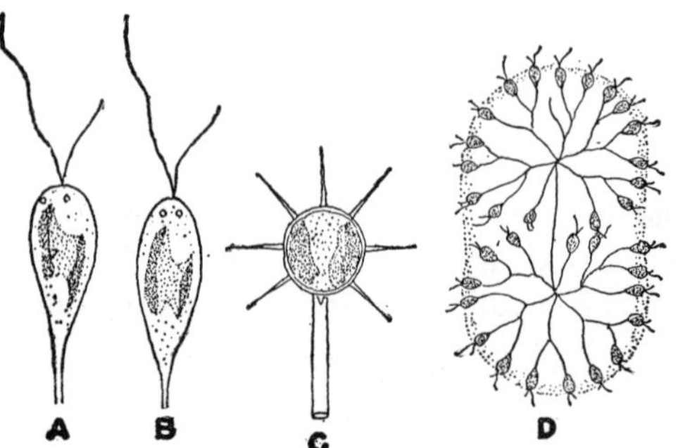

FIG. 334. *Uroglena volvox* Ehr. *A–B*, individual cells. *C*, statospore. *D*, colony. (*A–C*, *after Iwanoff*, 1900; *D*, *after Zacharias in Senn*, 1900.) (*A–C*, × 666; *D*, × 130.)

3. **Uroglena** Ehrenberg, 1833. The small free-swimming colonies of *Uroglena* have the cells radially arranged at the periphery of a spherical to ellipsoid gelatinous matrix. Within the matrix is a radiating dichotomously branched system of gelatinous threads, each free end of which terminates just below a cell. The thread system is best demonstrated by the use of appropriate stains.[4] The individual cells are ovoid to ellipsoid and with two flagella of unequal length at the anterior end. Within a cell are a single laminate chromatophore, a conspicuous eyespot, and two contractile vacuoles at the anterior end.

Cell division is longitudinal, and multiplication of a colony is by constriction into two daughter colonies. Statospores with conspicuous plugs and long spines have been observed.[5]

In this country *Uroglena* is a much rarer organism than *Uroglenopsis* but has been collected in several of the states. There is but one species, *U. volvox* Ehr. (Fig. 334). For a description of it, see Pascher (1913*A*).

[1] TROITZKAJA, 1924. [2] CALKINS, 1892. [3] MOORE, 1897. [4] ZACHARIAS, 1894. [5] IWANOFF, 1900.

4. **Cyclonexis** Stokes, 1886. The free-swimming minute colonies of this alga have the cells laterally joined to one another to form a flat discoid colony with a small open space at the center. Each colony contains 10 to 30 obovate cells, about twice as long as broad, whose anterior ends are broadly rounded. At the anterior end of each cell are two flagella of unequal length; the longer straight, the shorter usually spirally twisted. Within a cell are one or two laterally placed laminate chromatophores and two contractile vacuoles toward the anterior end.[1]

Reproduction may be due to an escape of solitary cells or to a colony dissociating into individual cells.[2] There may also be a fragmentation of a colony into two equal or unequal parts.

C. annularis Stokes (Fig. 335), the only species, has been found in Massachusetts,[2] New Jersey,[3] and Lake Erie.[4] For a description of it, see Pascher (1913*A*).

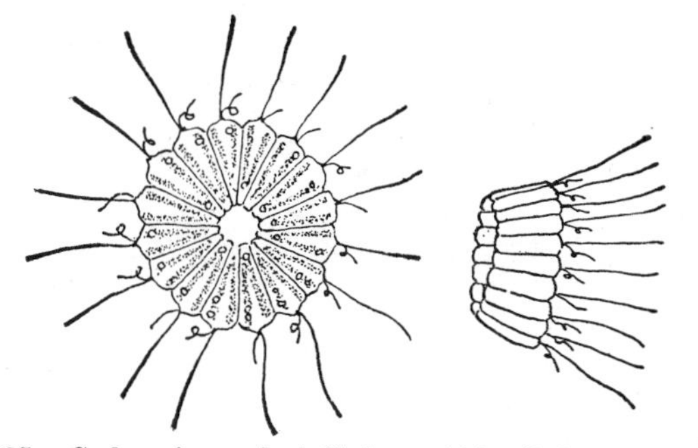

FIG. 335. *Cyclonexis annularis* Stokes. (*After Stokes*, 1886*A*.)

5. **Dinobryon** Ehrenberg, 1835. The cells of *Dinobryon* are united in arborescent free-swimming colonies. Each cell is attached to the bottom of a conical, campanulate, or cylindrical lorica that has a closed pointed base and an open cylindrical or somewhat flaring apex. The lorica may be hyaline or yellowish brown, and its surface may be smooth or spirally sculptured. It is said to contain cellulose and to be somewhat impregnated with silica.[5] The protoplast within the lorica is spindle-shaped, conical, or ovoid and is attached to the base of the lorica by a short cytoplasmic stalk. Except for the basal stalk, the protoplast is rarely in contact with the sides of the lorica. The peripheral portion of the cytoplasm is quite firm and usually smooth, though sometimes finely granulate. At the anterior end of the protoplast are two flagella of unequal length, the longer extending for some distance beyond the open mouth of the lorica, the shorter usually extending not much beyond it. Within the protoplast

[1] STOKES, 1886*A*; WHELDEN, 1939. [2] WHELDEN, 1939. [3] STOKES, 1886*A*.
[4] TIFFANY, 1934. [5] LEMMERMANN, 1900*A*.

are one or more contractile vacuoles (apical, median, or basal in position) and one or two parietal laminate chromatophores. There is usually a conspicuous eyespot near the anterior end of the cell. The photosynthetic reserves accumulate chiefly as leucosin and in a single large granule toward the posterior end of the cell. Food reserves accumulating within the cell may also be partially due to heterotrophic nutrition, since *Dinobryon* has the ability to ingest solid foods.

Reproduction is by longitudinal division into two daughter cells. In the colonial forms, one or both of the daughter cells become attached to the

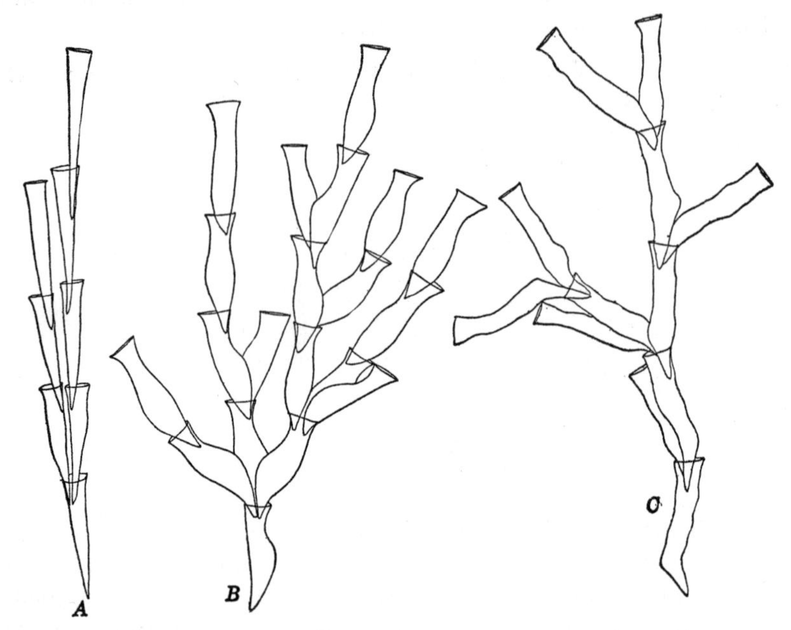

FIG. 336. *A, Dinobryon stipitatum* Stein. *B, D. sertularia* Ehr. *C, D. divergens* Imhof. (× 400.)

mouth of the old lorica and secrete a new lorica. Rhizopodial stages have been observed both within and free from the lorica.[1] There may also be a development of palmelloid stages.[2] Statospores are formed endogenously.[3] Statospores may have two nuclei, and the suggestion has been made[4] that these represent a stage in autogamy.

Dinobryon is widespread in the plankton of lakes in this country and is also frequently encountered intermingled with algae of pools and ditches. The following species occur in the United States: *D. bavaricum* Imhof, *D. campanulostipitatum* Ahlstrom, *D. cylindricum* Imhof, *D. divergens* Imhof (Fig. 336*C*), *D. pediforme*

[1] PASCHER, 1917, 1943. [2] PASCHER, 1943. [3] HAYE, 1930. [4] GEITLER, 1935*A*.

(Lemm.) Steinecke, *D. sertularia* Ehr. (Fig. 336*A*), *D. sociale* Ehr. and *D. Vanhoeffendii* Bachm. For descriptions of them, see Ahlstrom (1937).

6. **Epipyxis** Ehrenberg, 1838. The cell structure and structure of the lorica are identical with that of *Dinobryon*. *Epipyxis* differs from *Dinobryon* in that the alga is sessile and its cells are never united in colonies.

Epipyxis has frequently been placed as a section of *Dinobryon*, but in recent years there has been a growing tendency[1] to consider it generically distinct from *Dinobryon*. The five species found in this country are *E. calyciforme* (Bachm.) comb. nov. (Fig. 337*B*), *E. Stokesii* (Lemm.) comb. nov., *E. Tabellariae* (Lemm.) comb. nov. (Fig. 337*A*), *E. utriculus* Ehr., and *E. eurystoma* Stokes. For descriptions of them as species of *Dinobryon*, see Pascher (1913*A*).

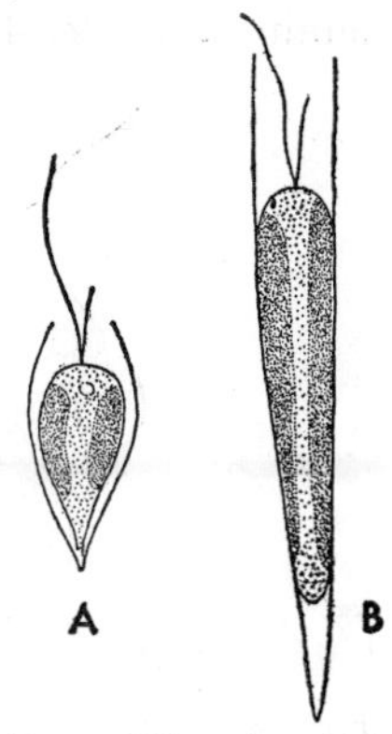

FIG. 337. *A*, *Epipyxis Tabellariae* (Lemm.) comb. nov. *B*, *E. calyciforme* (Bachm.) comb. nov. (× 1000.)

7. **Hyalobryon** Lauterborn, 1896. The cells of *Hyalobryon* may be solitary or united in branching colonies, but in either case they are epiphytic on other algae. Each protoplast is enclosed by a lorica, quite similar in shape to that of *Dinobryon*, but one made up of successive growth rings nesting one above the other.[2] As seen in optical section, the upper part of each growth ring appears as a minute denticulation of the outer surface of the lorica. The protoplast within a lorica has the same structure as that of *Dinobryon*.

There may be a formation of statospores in certain species.[3]

H. mucicola (Lemm.) Pascher (Fig. 338) is the only species recorded for this country, and it has been found only in Wisconsin.[4] For a description of it, see G. M. Smith (1920).

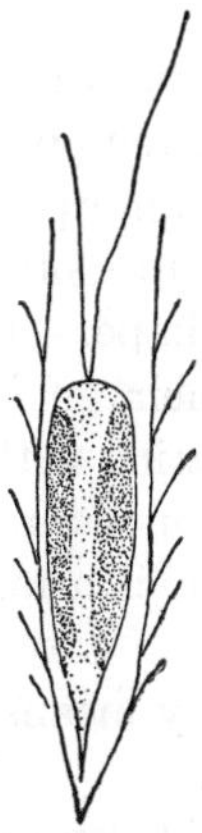

FIG. 338. *Hyalobryon mucicola* (Lemm.) Pascher. (× 1300.)

FAMILY 7. PRYMNESIACEAE

Members of this family are triflagellate and with one flagellum differing in length and orientation from the other two.

Chrysochromulina is the only genus known for this country.

1. **Chrysochromulina** Lackey, 1939. The naked solitary free-swimming cells of this chrysomonad are markedly compressed and circular in front view. There are three flagella, one longer than the other two. The longer flagellum extends directly forward from,

[1] AHLSTROM, 1937; KRIGER, 1930. [2] LAUTERBORN, 1899; PENARD, 1921.
[3] PASCHER, 1921. [4] SMITH, G. M., 1920.

and the other two project backward from, the point of their insertion. A single vacuole lies posterior to the flagella. Within a cell are two laminate golden-brown chromatophores extending longitudinally through

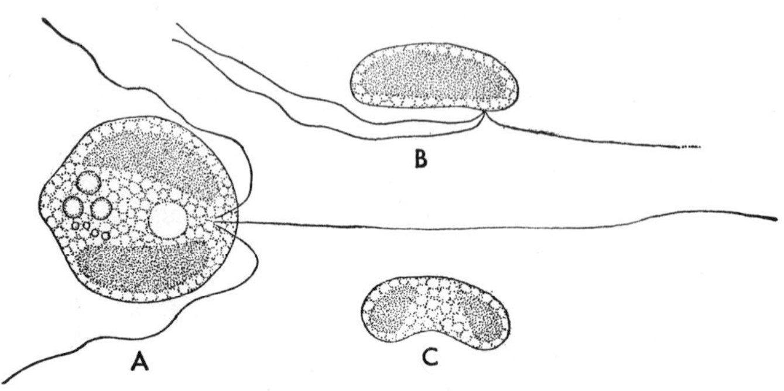

FIG. 339. *Chrysochromulina parva* Lackey. *A*, front view. *B*, side view. *C*, polar view. (*After Lackey*, 1939.)

the cell; and in the posterior end of a cell are several small refringent bodies, presumably leucosin.[1]

The method of reproduction is unknown.

There is but one species, *C. parva* Lackey (Fig. 339), known only from Ohio.[1] For a description of it, see Lackey (1939).

ORDER 2. RHIZOCHRYSIDALES

The Rhizochrysidales include those Chrysophyceae which are known only in the rhizopodial form, or in which the flagellated phase is a temporary phase in the life cycle. The cells may be solitary or united in amorphous colonies. Some genera have cells without any enveloping structure; others have the protoplast surrounded by a lorica of characteristic shape. The chrysophycean nature of these organisms is evidenced by their golden-brown chromatophores, by the types of food reserves accumulating within them, and, in a few cases, by their statospores. Their nutrition is partly autotrophic and partly heterotrophic. In the latter case, there is a mass ingestion of solid foods, especially minute organisms.

Reproduction is by cell division, colony fragmentation, and by means of zoospores. Statospores have been observed in but few genera.

The Rhizochrysidales are Chrysophyceae in which a phase of temporary occurrence in most motile forms is the dominant one. It represents an evolutionary tendency toward an animal-like rather than a plant-like organization, which may, as in *Heterolagynion*, have progressed to an animal-like condition where the chromatophores have disappeared.

For the present it is best to place all genera in the same family.

[1] LACKEY, 1939.

Family 1. Rhizochrysidaceae

The genera found in this country differ as follows:

1. Cells solitary or in temporary colonies 2
1. Cells permanently united in colonies 7
   2. Protoplast naked 3
   2. Protoplast surrounded by a lorica 4
3. With short pseudopodia .... 1. **Chrysamoeba**
3. With long pseudopodia .... 2. **Rhizochrysis**
   4. Cells free-floating .... 8. **Diceras**
   4. Cells sessile 5
5. Base of lorica with two prongs encircling host alga .... 5. **Chrysopyxis**
5. Base of lorica without prongs 6
   6. Base of lorica flattened .... 6. **Lagynion**
   6. Base of lorica rounded .... 7. **Kybotion**
7. All cells naked .... 3. **Chrysidiastrum**
7. Each cell surrounded by a perforated lorica .... 4. **Heliapsis**

1. **Chrysamoeba** Klebs, 1893. The cells of *Chrysamoeba* are generally solitary, though occasionally united in temporary colonies. For the greater part of the life cycle, a cell is in an amoeboid state and with numerous, acutely pointed, short pseudopodia radiating in all directions. The protoplast may contain either one or two golden-brown chromatophores that have or lack pyrenoids.[1] Each cell contains a single nucleus and, at times, a large granule of leucosin. Nutrition is in part photosynthetic and in part by a mass ingestion of solid foods. The change from an amoeboid to a flagellated state is accomplished by a retraction of the pseudopodia, a change to an ovoid shape, and the protusion of a single flagellum whose length is somewhat greater than that of the cell. During the motile phase, there is a contractile vacuole in the anterior end of a cell. The motile phase lasts but a short time,[2] after which the cell becomes approximately spherical and develops pseudopodia, and sooner or later the flagellum disappears.

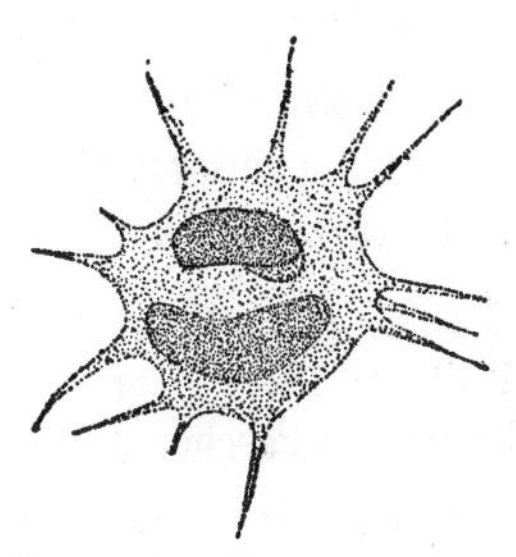

Fig. 340. *Chrysamoeba radians* Klebs. (× 1000.)

Reproduction takes place by cell division when in an amoeboid state.[3] Statospores are also formed when the cell is in an amoeboid state.

C. radians Klebs (Fig. 340) has been reported from various localities in this country. For a description of it, see Pascher (1913*A*).

[1] Doflein, 1921; Klebs, 1892; Penard, 1921.
[2] Penard, 1921. [3] Doflein, 1921.

2. **Rhizochrysis** Pascher, 1913. The primary distinction between this organism and the foregoing is that the amoeboid stage is the only one known, and that repeated observation has never revealed[1] flagellated stages. The cells are free-floating and solitary, or united in small temporary colonies. The peripheral portion of the cytoplasm is usually prolonged into numerous delicate acicular pseudopodia, but in one species the pseudopodia are short and stout. One or two golden-brown chromatophores and a single nucleus occupy the central portion of a cell. There may be a single large contractile vacuole, numerous noncontractile vacuoles, or no vacuoles. Reserve foods accumulate as leucosin and as minute droplets of oil. For the most part, nutrition is by ingestion of solid foods and bacteria; small algae and protozoa have been found within nutritional vacuoles in the cytoplasm.[2]

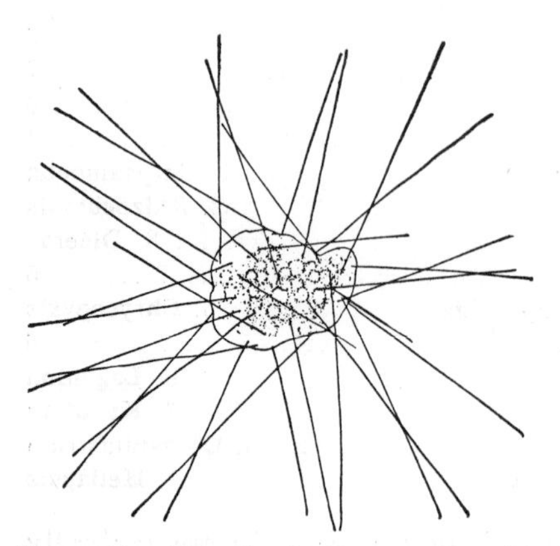

FIG. 341. *Rhizochrysis limnetica* G. M. Smith. (× 400.)

Reproduction is by cell division. Thin-walled spherical resting stages have been reported[3] for one species, but it is questionable whether these are true statospores.

R. limnetica G. M. Smith (Fig. 341) and *R. Scherffelii* Pascher are known for the United States. For a description of the former, see G. M. Smith (1920); for the latter, see Pascher (1913*A*).

3. **Chrysidiastrum** Lauterborn, 1913. This genus differs from the two foregoing in regularly having the amoeboid cells joined one to another to form free-floating linear colonies of 2 to 16 cells. The cells are spherical and with several long delicate acicular pseudopodia. The single chromatophore within a cell is discoid to laminate and centrally located.

C. catenatum Lauterborn (Fig. 342), the only species, has been found in the plankton of Wisconsin lakes.[4] For a description of it, see G. M. Smith (1920).

4. **Heliapsis** Pascher, 1940. *Heliapsis* belongs to the group of rhizopodial Chrysophyceae in which the cells are united in reticulate colonies by cytoplasmic processes from cell to cell. It differs from others of this type in that each cell is enclosed by a lens-shaped lorica. The lorica has three to seven pores in the equatorial region, and the protoplast within it is con-

[1] DOFLEIN, 1916; PASCHER, 1913*A*. [2] DOFLEIN, 1916. [3] SCHERFFEL, 1901.
[4] SMITH, G. M., 1920.

nected with other cells by delicate threads of cytoplasm extending out from each of the pores. The amoeboid protoplast within a lorica contains a single chromatophore, a contractile vacuole, and granules of reserve food.

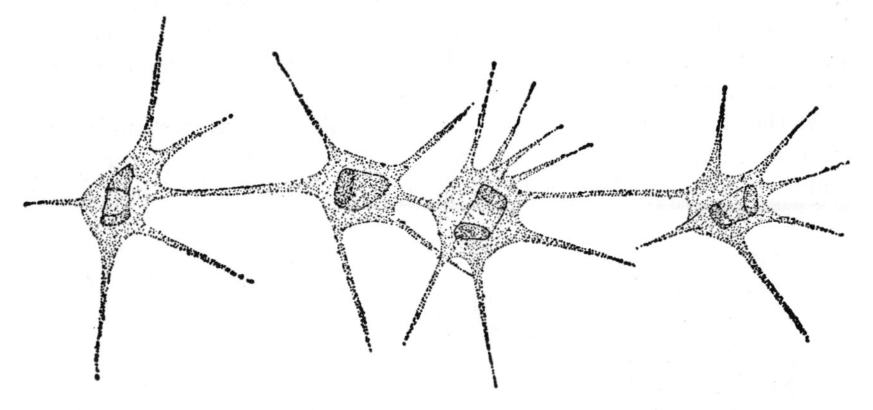

Fig. 342. *Chrysidiastrum catenatum* Lauterborn. (× 750.)

Reproduction is by division of a protoplast into two daughter protoplasts, one of which escapes by migrating through a pore in the lorica.[1] In rare cases, there may be a divison into three or into four daughter protoplasts. Zoospores and statospores have not been observed.

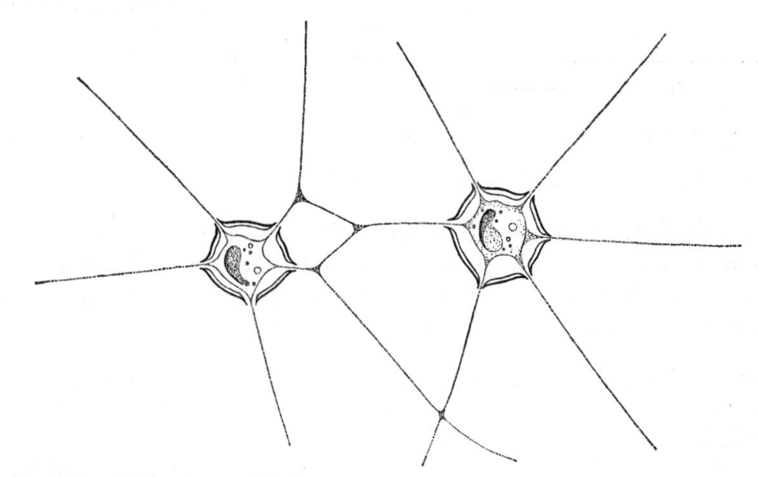

Fig. 343. *Heliapsis mutabilis* Pascher. (*Drawn by R. H. Thompson.*) (× 1650.)

Prof. R. H. Thompson writes that he has found *H. mutabilis* Pascher (Fig. 343), the only species, in Kansas. For a description of it, see Pascher (1940*A*).

[1] Pascher, 1940*A*.

5. **Chrysopyxis** Stein, 1878. The cells of *Chrysopyxis* are solitary, sessile, and surrounded by a vase-shaped lorica that is attached to the host alga by two prong-like outgrowths from the base. The protoplast, which stands free from the lorica, bears a single delicate filamentous pseudopodium that is branched at the distal end.[1] The protoplast contains one or two contractile vacuoles, a laminate golden-brown chromatophore, a nucleus, and granules of leucosin.

At the time of reproduction, a protoplast divides into two uniflagellate swarmers, one of which escapes from the lorica, swims about for a time, and then becomes affixed with its anterior end downward. The flagellum

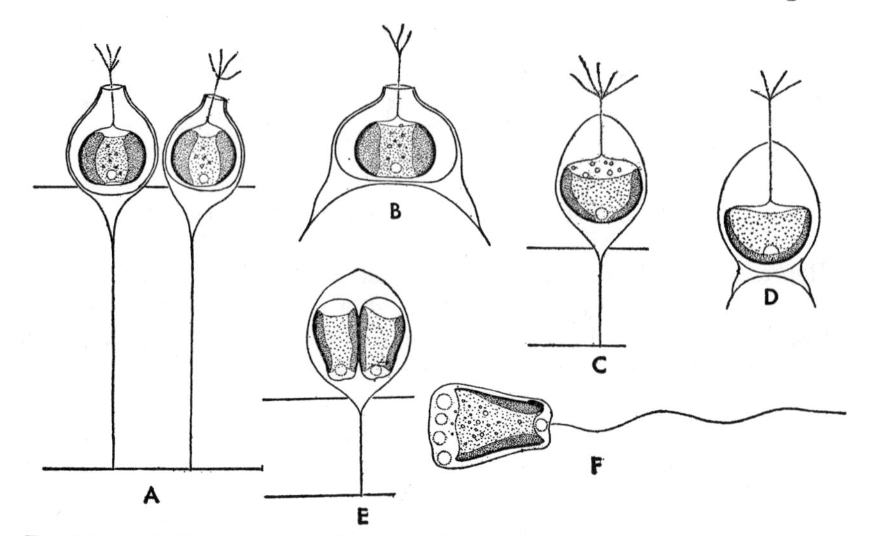

Fig. 344. *A–B*, *Chrysopyxis bipes* Stein. *A*, front view. *B*, side view. *C–F*, *C. stenostoma* Lauterb. *C*, front view. *D*, side view. *E*, dividing to form zoospores. *F*, free-swimming zoospore. (*Drawn by R. H. Thompson.*) (× 1600.)

may persist for some time, and the posterior pole of the swarmer may send forth several long pseudopodia. Eventually there is a formation of a lorica and a development of the characteristic solitary pseudopodium. There may also be a production of spherical statospores.[2]

Chrysopyxis grows epiphytic upon various filamentous Chlorophyceae. The only published record for its occurrence in this country is that for *C. bipes* Stein (Fig. 344*A–B*) from Ohio.[3] Prof. R. H. Thompson writes that he has found both it and *C. stenostoma* Lauterb. (Fig. 344*C–F*) in Maryland. For descriptions of them, see Pascher (1913*A*).

6. **Lagynion** Pascher, 1912. The solitary or gregarious cells of this epiphytic alga are each surrounded by a lorica of definite shape and one

[1] Lauterborn, 1911. [2] Iwanoff, 1900. [3] Lackey, 1939.

which is thought to be composed of cellulose.[1] The lorica, which may be hyaline or brownish, has a flattened circular base that is not attached to the substratum by prong-like projections. The lorica may be flask or bell-shaped, and at the apex has a long cylindrical neck that is open at the top. The protoplast occupies practically all of the lorical cavity and has a single long thread-like pseudopodium extending for some distance beyond the open end of the neck. Protoplasts may contain one or more chromatophores.[2]

Reproduction is by longitudinal division of a protoplast, but the further history of the daughter protoplasts is unknown.[1] Motile stages and statospores have not been observed.

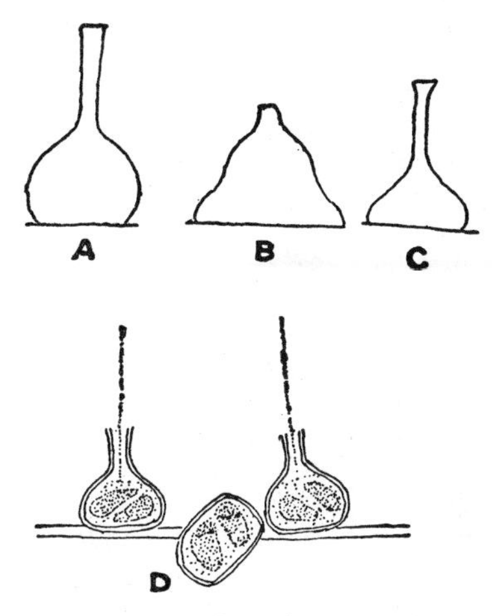

Fig. 345. *A*, *Lagynion ampullaceum* (Stokes) Pascher. *B*, *L. triangulare* (Stokes) Pascher. *C*, *L. macrotrachelum* (Stokes) Pascher. *D*, *L. Scherffelii* Pascher, a species not found in the United States. (*A–C*, *after Stokes*, 1886*B*; *D*, *after Scherffel*, 1911.)

Lagynion grows epiphytically upon other algae, especially filamentous Chlorophyceae. The following species have been found in the United States: *L. ampullaceum* (Stokes) Pascher (Fig. 345*A*), *L. macrotrachelum* (Stokes) Pascher (Fig. 345*C*), *L. reductum* Prescott, *L. reflexum* Prescott, *L. subovatum* Prescott, and *L. triangulare* (Stokes) Pascher (Fig. 345*B*). For a description of *L. reductum*, see Prescott (1944); for *L. reflexum* and *L. subovatum*, see Prescott and Croasdale (1937); for the remainder, see Pascher (1913*A*).

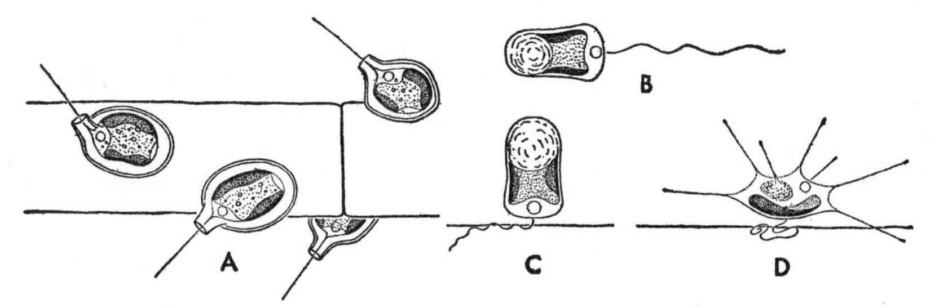

Fig. 346. *Kybotion ellipsoideum* R. H. Thompson, sp. nov. *A*, vegetative cells. *B*, free-swimming zoospore. *C*, zoospore immediately after cessation of motility. *D*, zoospore metamorphosing into rhizopodial stage. (*Drawn by R. H. Thompson.*) (× 1600.)

7. **Kybotion** Pascher, 1940. The cells of this alga are solitary, sessile, and surrounded by an ovoid lorica which may or may not be prolonged into a neck at the distal end. A lorica may or may not be encrusted with

[1] Scherffel, 1911. [2] Pascher, 1912.

a deposit of iron compounds. The protoplast lies free from the surrounding lorica and has a single, very delicate, filamentous pseudopodium projecting through the opening at the distal end of the lorica. The protoplast is more or less ellipsoidal and contains a single large parietal golden-brown chromatophore, small granules of leucosin, and either one or two contractile vacuoles.

Reproduction is by formation of zoospores with one flagellum. A liberated zoospore comes to rest with the anterior end downward and may develop several delicate pseudopodia before rounding up and secreting a lorica. A formation of typical chrysophycean cysts has also been observed.[1]

The only record for the occurrence of this genus in the United States is the finding of *K. ellipsoideum* R. H. Thompson sp. nov.* (Fig. 346) near Solomons, Maryland.

8. **Diceras** Reverdin, 1917. *Diceras* has solitary free-floating cells surrounded by a lorica. The lorica is globose to reniform and at one side has

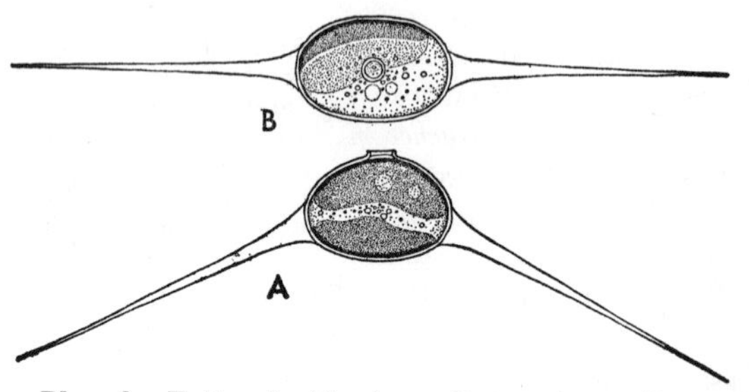

FIG. 347. *Diceras Phaseolus* Fott. *A*, side view. *B*, top view. (*Drawn by R. H. Thompson.*) (× 1600.)

a small circular pore surrounded by a low collar. At opposite poles of a lorica are two (occasionally three) long tapering spines with a length several times the diameter of the lorica. The protoplast, which completely fills the lorica, contains one or two golden-brown chromatophores, two contractile vacuoles, and several small leucosin granules.[2] At times, the protoplast may have two delicate cytoplasmic filaments projecting through the pore of the lorica.

[1] PASCHER, 1940.

* *Kybotion ellipsoideum* R. H. Thompson, sp. nov. Lorica levis, fulva, aut sine color; ovata aut elliptica desuper, hemisphaerica in latere aut in termino conspectu, est et in termino cum una tubulata cervicula, qua recta aut flexu ad utrum latus praedita. Lorica 6–8 μ longa, cervicula non continens, 5–6 μ lata et 4–5 μ alta est. Zoospores 7–8 μ longa et 4.3–5 μ lata est.

[2] FOTT, 1937.

Reproduction is by division of the protoplast into two daughter protoplasts, one of which escapes from the old lorica and secretes a new lorica.[1] In rare cases, both daughter protoplasts may escape from the lorica.

Prof. R. H. Thompson writes that he has found *D. Phaseolus* Fott (Fig. 347) in Maryland. For a description of it, see Fott (1937).

ORDER 3. CHRYSOCAPSALES

The Chrysocapsales include those Chrysophyceae in which the cells are without flagella and are united with one another in palmelloid colonies. All genera of the order are colonial, but the gelatinous colonies may have cell division localized in the apical portion or may have it taking place throughout the entire colony. As in the Tetrasporales, the homologous order of the Chlorophyceae, the vegetative cells are capable of returning directly to a motile condition. Such a formation of swarmers results in formation of a new colony, but reproduction may also take place by fragmentation of a palmelloid colony.

FAMILY 1. CHRYSOCAPSACEAE

Members of this family have colonies in which cell division may take place anywhere in a colony. The colonies may have or may lack a conspicuous gelatinous envelope, but the colonies never have long gelatinous setae. There may be a direct metamorphosis of immobile cells into motile cells which, according to the genus, are either uniflagellate, or biflagellate and with the flagella of equal or unequal length.

The genera found in the United States differ as follows:

1. Colony with a conspicuous gelatinous envelope 2
1. Colony without a conspicuous gelatinous envelope .... 3. **Phaeoplaca**
    2. Colony globose .... 1. **Chrysocapsa**
    2. Colony not globose .... 2. **Phaeosphaera**

1. **Chrysocapsa** Pascher, 1912. Colonies of this alga are free-floating, spherical to ellipsoidal, and with the cells regularly or irregularly distributed through a copious, hyaline, homogeneous, gelatinous matrix.[2] The individual cells may be spherical or ellipsoidal and with either one or two golden-brown discoid to laminate chromatophores. Sometimes the vegetative cells contain an eyespot and a contractile vacuole.[3] Certain of the species have been recorded as forming biflagellate motile cells with the two flagella of equal length,[4] and one species is said[5] to form uniflagellate motile cells.

[1] FOTT, 1937. [2] PASCHER, 1912*A*.
[3] PASCHER, 1925; WEST, W. and G. S., 1903. [4] WEST, W. and G. S., 1903.
[5] PASCHER, 1925.

Both of the two species reported for this country, *C. planctonica* (W. and G. S. West) Pascher (Fig. 348)and *C. paludosa* (W. and G. S. West) Pascher, are known only from the plankton. For descriptions of them, see Pascher (1913*A*).

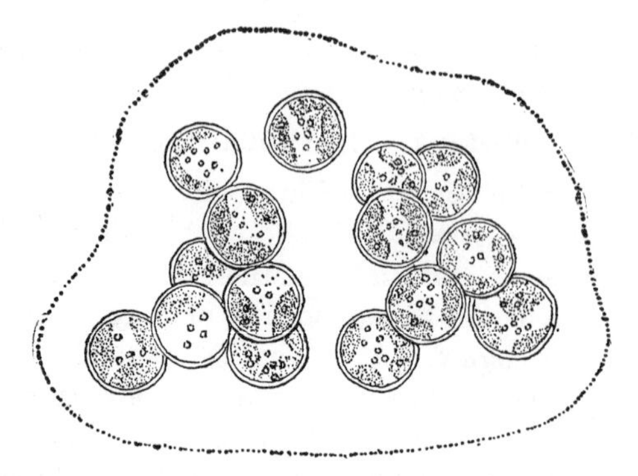

FIG. 348. *Chrysocapsa planctonica* (W. and G. S. West) Pascher. (× 1320.)

2. **Phaeosphaera** W. and G. S. West, 1903. Colonies of *Phaeosphaera* are of macroscopic or microscopic size, and simple or branched, solid or saccate, tubular cylinders. The colonial matrix is homogeneous, and within lie a large number of globose or ovoid cells. A cell contains one or

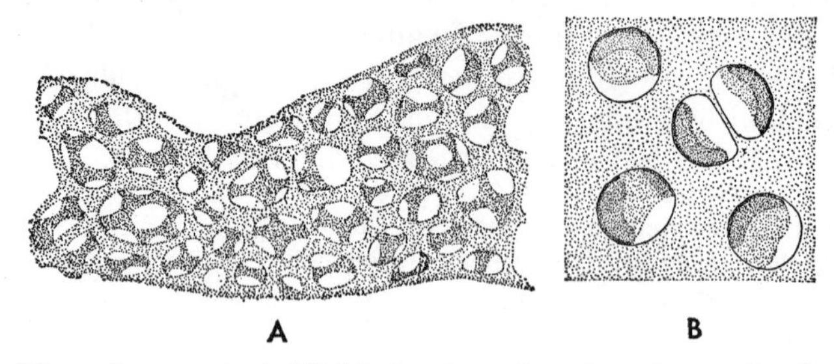

FIG. 349. *Phaeosphaera perforata* Whitford. *A*, portion of a colony. *B*, cells. (*A*, × 6; *B*, × 750.)

two parietal laminate golden-brown chromatophores, and several minute granules of food reserves, presumably oil. At times, there are one or two contractile vacuoles.[1]

The method of reproduction is unknown.

P. perforata Whitford (Fig. 349), known only from the original description,[2] has been found repeatedly at three different stations in North Carolina.

3. **Phaeoplaca** Chodat, 1925 (*Placochrysis* Geitler, 1926). *Phaeoplaca* has discoid sessile colonies with the cells closely abutting in a monostro-

[1] WEST, W. and G. S., 1903; WHITFORD, 1943. [2] WHITFORD, 1943.

matic layer. The cells, of which there are rarely more than 25 in a colony, are angular because of mutual lateral compression. Increase in size of a colony is by cell division. Each cell is surrounded by a distinct wall, and there is no gelatinous material between the cells. Within a cell are one or two yellowish-brown parietal chromatophores, droplets of oil, and granules of leucosin.[1]

Reproduction is by division of the cell contents to form four to eight zoospores.

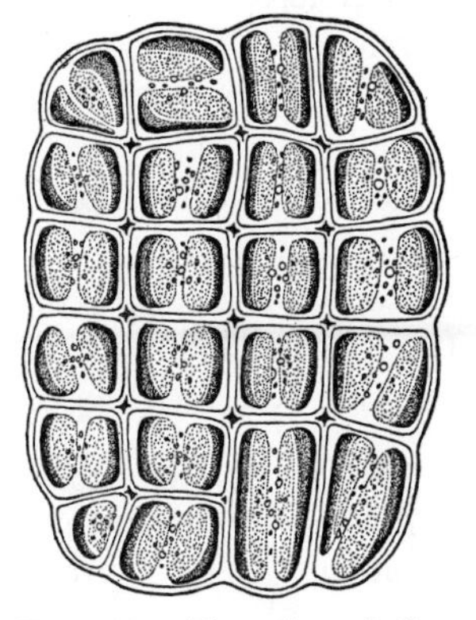

FIG. 350. *Phaeoplaca thallosa* Chod. (*Drawn by R. H. Thompson.*) (× 1600.)

Prof. R. H. Thompson writes that he has found *P. thallosa* Chodat (Fig. 350) in Maryland. For a description of it, see R. Chodat (1925).

FAMILY 2. NAEGELIELLACEAE

Members of this family have the cells united in colonies with a broad or narrow gelatinous envelope bearing conspicuous, simple or branched, long, gelatinous setae.

The two genera found in this country differ as follows:

1. Colonies sessile.. 1. **Naegeliella**
1. Colonies free floating.............................. 2. **Chrysostephanosphaera**

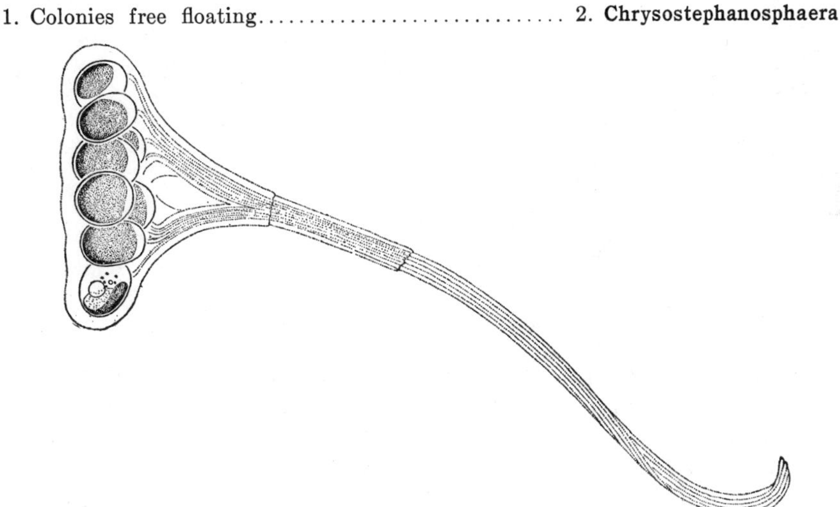

FIG. 351. *Naegeliella flagellifera* Correns. (*Drawn by R. H. Thompson.*) (× 810.)

1. **Naegeliella** Correns, 1892. This alga has epiphytic circular colonies of up to 150 cells which may be entirely monostromatic or distromatic in the central portion. The cells are more or less ovoid and surrounded by

[1] CHODAT, R., 1925; GEITLER, 1926*A*.

gelatinous envelopes which may be confluent with one another, or the individual envelopes may be more or less distinct. On the upper side of a young colony is a single cluster of very long, delicate, repeatedly branched, gelatinous setae whose basal portion is ensheathed by a common gelatinous envelope. Older colonies may have two or more clusters of setae. According to the species, the gelatinous setae are with[1] or without[2] an axial cytoplasmic filament. The cells are without a wall,[1] and each contains either one or two parietal laminate golden-brown chromatophores and leucosin granules.

Reproduction is by bipartition of the protoplasm to form two zoospores.[3] One species has been described[2] as producing biflagellate zoospores, and another described[3] as producing uniflagellate zoospores.

Prof. R. H. Thompson writes that he has found *Naegeliella* in Kansas and has tentatively identified the material as *N. flagellifera* Correns (Fig. 351). For a description of it, see Correns (1892).

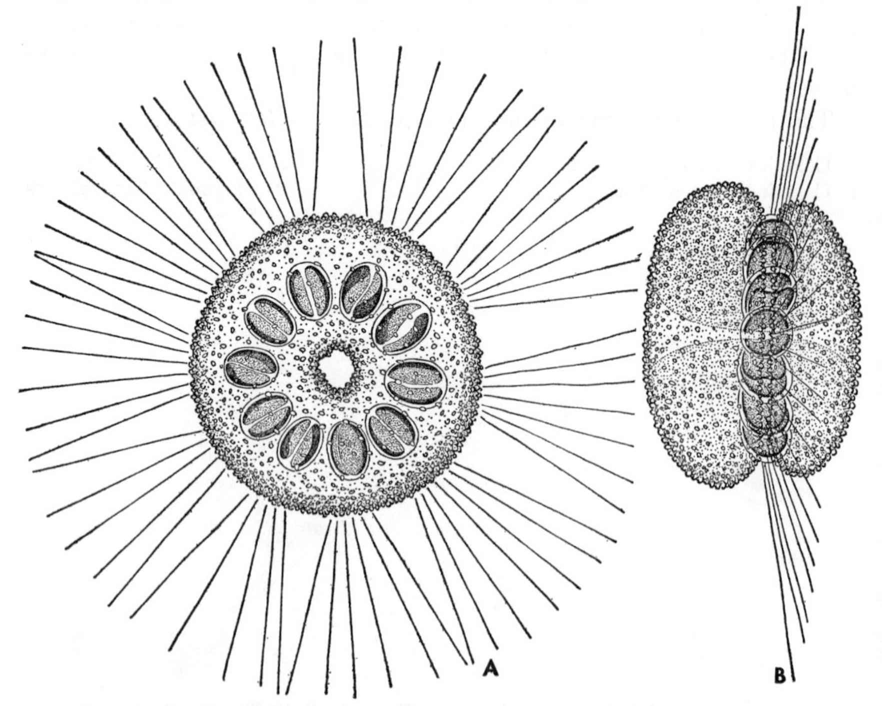

Fig. 352. *Chrysostephanosphaera globulifera* Scherffel. *A*, top view of colony. *B*, side view of colony. (*Drawn by R. H. Thompson.*) (× 300.)

2. **Chrysostephanosphaera** Scherffel, 1911. The cells of this alga are united in a colony with a broad gelatinous envelope compressed in the

[1] Godward, 1933. [2] Correns, 1892. [3] Scherffel, 1927.

longitudinal axis. As seen in polar view, a colony superficially resembles a doughnut. The surface of the envelope is covered with small granules of a denser gelatinous substance. The cells are arranged in a ring in the equatorial region of the gelatinous envelope, and with their long axes radial to the polar axis of the envelope. Each cell bears a cluster of long setae which project horizontally for some distance beyond the equator of a colonial envelope. The cells are ellipsoidal and with laminate parietal golden-brown chromatophores on opposite sides and extending from pole to pole.

At a meeting of the Phycological Society of America held in 1947, Prof. R. H. Thompson reported finding *C. globulifera* Scherffel (Fig. 352), the only species, in Maryland and suggested that it may be a palmella stage of *Cyclonexis*. For a description of *C. globulifera*, see Scherffel (1911).

Family 3. Hydruraceae

Genera referred to this family are colonial and have the cells embedded in a profusely branched gelatinous envelope. The family differs from others of the order in that cell division and growth are restricted largely to apices of branches of a colony. *Hydrurus* is the only genus found in this country.

1. **Hydrurus** Agardh, 1824. The profusely branched, penicillate, brown, gelatinous thalli of this alga are always found attached to rocks and stones in swiftly flowing cold-water streams. The gelatinous envelope in which the cells are embedded is exceedingly tough. At the apex of the envelope and in very young plants, the cells are uniseriately arranged,[1] farther back in older thalli, the branches are more than one cell in diameter. The majority of cells in a thallus are ovoid, although some may be angular because of mutual compression. Each cell contains a single golden-brown chromatophore that generally lies on the side of the cell toward the thallus apex. A chromatophore contains a conspicuous pyrenoid.[2] The colorless portion of the cytoplasm contains a single nucleus, granules of reserve food, and five to six vacuoles. Cell division may continue until a colony contains hundreds of thousands of cells and has grown to a length of 30 cm. or more. It is very doubtful whether portions of a colony accidentally broken off continue to live as independent plants for any length of time.

Reproduction by means of zoospores is of frequent occurrence and usually takes place during the early morning hours. Zoospores are formed by metamorphosis of recently divided cells at the ends of branches. Their shape is tetrahedral, with a single flagellum from the middle of one face and a single chromatophore in the corner opposite this face (Fig. 353*C*).[2] After the period of swarming, a zoospore comes to rest with its anterior

[1] Klebs, 1892: Rostafiński, 1882. [2] Klebs, 1892; Lagerheim, 1888.

face downward, retracts its flagellum, becomes ellipsoidal, and secretes a tubular gelatinous envelope. The first divisions in this germling are transverse (Fig. 353*D*). Silicified statospores (Fig. 353*E–G*) are developed from cells borne in special gelatinous stalks protruding from the branches. The stimulus causing a formation of statospores has been thought[1] to be a rise in temperature of the water.

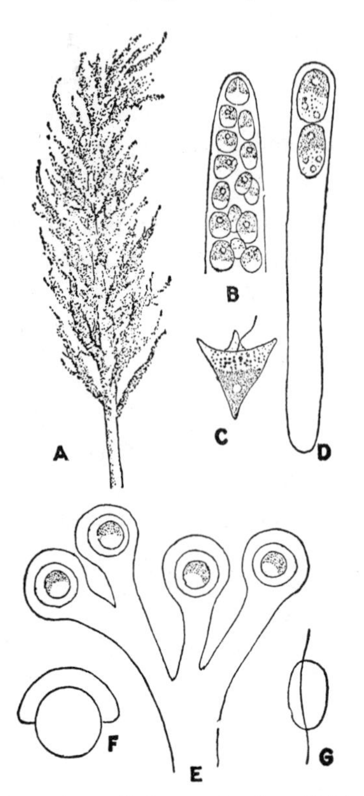

FIG. 353. *Hydrurus foetidus* (Vill.) Trev. *A*, portion of a colony. *B*, apex of a colony. *C*, zoospore. *D*, germling. *E*, statospores before liberation from a colony. *F–G*, top and side views of a statospore. (*B–G*, *after Klebs*, 1893.)

Hydrurus is essentially a cold-water organism and is found in mountain streams in both the Eastern and Western United States. When conditions are favorable, the alga often covers the entire bottom of rocky torrents. The alga may be recognized in the field by its extremely tough consistency and its acrid disagreeable odor. Although numerous forms and species have been described, it is customary to refer them all to a single species *H. foetidus* (Vill.) Trev. (Fig. 353) and not to attempt to distinguish varieties. For a description of it, see Pascher (1913*A*).

ORDER 4. CHRYSOTRICHALES

The Chrysotrichales are to be distinguished from other Chrysophyceae by their uniseriate filamentous organization. Cell structure, structure of zoospores, and especially the chrysophycean statospores show that these brown filamentous algae belong to the Chrysophyceae.

FAMILY 1. PHAEOTHAMNIACEAE

This family includes those filamentous Chrysophyceae that have a branched thallus in which the cells are never organized into flat expanded plates. *Phaeothamnion* is the only genus known to occur in the United States.

1. **Phaeothamnion** Lagerheim, 1884. The thallus of *Phaeothamnion* is composed of cylindrical to subovoid cells joined end to end in branched filaments with a conspicuous central axis and suberect lateral branches. At the base of a thallus is a prominent hemispherical cell that attaches the plant to the substratum. The basal cell is usually without chromatophores; but all other cells have one, two, or several golden-brown chromato-

[1] KLEBS, 1892.

phores and store reserve foods mainly in the form of leucosin granules.[1] Palmelloid stages are of frequent occurrence. Usually the palmelloid stage consists of branched gelatinous tubes in which the cells are spherical and uniseriately arranged (Fig. 355),[2] but the cells may be irregular in both shape and arrangement.[3]

Reproduction is by a formation of one, two, four, or eight zoospores and their liberation through a pore in the side of the parent-cell wall. Zoospores have been described as having two flagella of equal length,[4] two flagella of unequal length,[3] and as having but one flagellum. Statospores have also been recorded for this alga.[3]

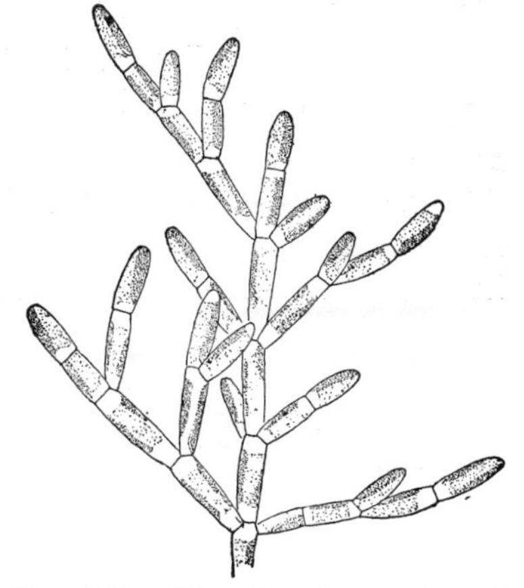

FIG. 354. *Phaeothamnion confervicola* Lagerh. (× 460.)

Phaeothamnion usually grows epiphytic upon filaments of Cladophoraceae. The only published record of its occurrence in this country is for *P. confervicola* Lagerh. (Fig. 354) which was found in California.[5] Prof. G. W. Prescott writes that he has found this species in Wisconsin. For a description of it, see Pascher (1925).

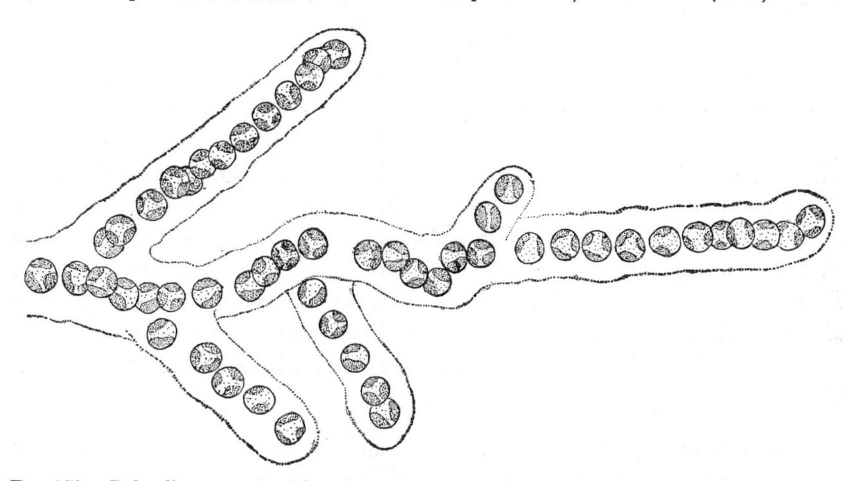

FIG. 355. Palmella stage of a *Phaeothamnion* species [*P. Borzianum* Pascher (?)]. (× 650.)

ORDER 5. CHRYSOSPHAERALES

This order, homologous with the Chlorococcales of the Chlorophyceae, has cells which do not metamorphose directly into motile stages.

[1] LAGERHEIM, 1884; PASCHER, 1925. [2] BORZI, 1892; PASCHER, 1925.
[3] PASCHER, 1925. [4] LAGERHEIM, 1884. [5] SMITH, G. M., 1933.

The cells may be solitary or aggregated in nonfilamentous colonies. Reproduction is usually by a formation of zoospores. All genera are placed in a single family.

Family 1. Chrysosphaeraceae

This family, whose characters are the same as those of the order, has several genera, only one of which has been found in this country.

1. **Epichrysis** Pascher, 1925. This unicellular epiphytic alga may have cells remote from one another, or it may have them densely aggregated to form what seems to be a crustose multicellular colony one cell in thickness.[1]

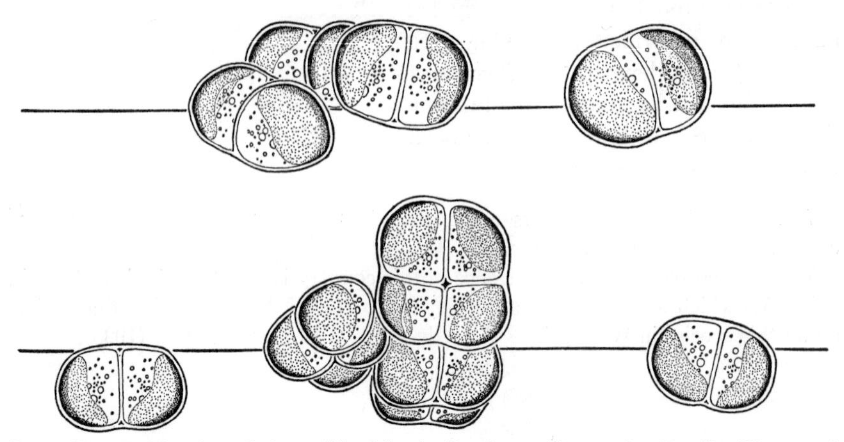

Fig. 356. *Epichrysis paludosa* (Korshikov) Pascher. (*Drawn by R. H. Thompson.*) (× 1650.)

The cells are globose, without a stipe, and are affixed directly to the host. There is a fairly thick cell wall. Within the protoplast, and on the side away from the base, is a single large disk-shaped yellowish-brown chromatophore. The cytoplasm contains droplets of oil and leucosin granules. There may also be a development of small free-floating palmelloid colonies of a few cells and with considerable gelatinous material between the cells.[2]

Reproduction is by division of the cell contents into two zoospores, each with a single flagellum.

Prof. R. H. Thompson writes that he has found *E. paludosa* (Korshikov) Pascher (Fig. 356) in Maryland. For a description of it, see Pascher (1925).

CLASS 3. BACILLARIOPHYCEAE

The Bacillariophyceae, or diatoms, include a large number of unicellular and colonial genera that differ sharply from other algae in the shape of

[1] Korshikov, 1924*B*. [2] Pascher, 1925.

their cells. The primary feature distinguishing diatoms from other algae is that the cell wall is highly silicified and composed of two overlapping halves that fit together as do the two parts of a Petri dish. The siliceous nature of the cell wall cannot be determined by microscopical examination, but diatoms may be readily recognized by the bilateral or radial markings on the wall when a cell is viewed from above. Within a cell are one to many, variously shaped, yellowish to brownish chromatophores, which contain chlorophyll *a*, chlorophyll *c*, beta-carotene, a unique carotene, and xanthophylls, certain of which are found only in Bacillariophyceae (see Table I, page 3).

Reproduction is usually by a cell dividing into two daughter cells of slightly different size. Now and then there is a formation of special rejuvenescent cells (*auxospores*) larger in size than the cells producing them. Auxospores are zygotic in nature and are formed by either autogamy or syngamy.

Occurrence. Diatoms are widely distributed in both fresh and salt waters. Certain genera are found only in fresh water; others only in the ocean. Such genera as occur in both fresh and salt water usually have species that are strictly marine or fresh-water. Marine species may be found in the brackish waters of estuaries and maritime marshes; fresh-water species, on the other hand, are seldom found in saline waters.

Marine diatoms are essentially cold-water organisms, and the same is true to a more limited extent for fresh-water forms. They are, therefore, especially common during the spring and autumn. Fresh-water diatoms are found in a wide variety of habitats, although the greater number of them are strictly aquatic. A few of these aquatic species are found only in the plankton of ponds and lakes, where during the spring and autumn they may be present in sufficient quantity to give the water a distinctly fishy odor. The algal flora of roadside ditches, semipermanent or permanent pools, and streams of rivulets always includes some diatoms. When growing in standing water, diatoms are often to be found either as a brownish sludge on the bottom of a pool, or as a coating on stems and leaves of water plants. They also grow intermingled with or epiphytic upon other algae, and filaments of *Cladophora*, *Rhizoclonium* and *Vaucheria* are often thickly covered with certain of the sessile species. When diatoms grow in rapidly flowing water, they occur mostly in a gelatinous matrix, coating rocks and stones in the bed of the stream.

Although the diatoms are not usually considered an important part of the aerial algal flora, the number of species recorded from aerial habitats is quite large.[1] Thus, an extensive survey of the aerial algae of Iceland[2] has shown that the species of diatoms far outnumber the species of

[1] Beger, 1927. [2] Petersen, 1928.

other algae growing on brickwork, rocky walls, dry cliffs, among mosses and liverworts and on the bark of trees. Diatoms from such habitats are more strictly aerial than are those growing on rocky cliffs continually moistened by the spray from waterfalls, or by the seepage of water. Many diatoms are also to be found in strictly terrestrial habitats and, when growing on bare soil, they may be on the closely packed soil of paths or on loose cultivated soil. Most of these forms, like those from aerial habitats, are of the pennate type and small-celled. Many of these terrestrial diatoms are able to withstand desiccation for weeks,[1] and one case is recorded of the resumption of growth in dry soil after 48 years.[2] Although not so important an element in the subterranean algal flora as are the Myxophyceae or Chlorophyceae, diatoms have been found growing in soil taken from a depth of over 4 ft. below the surface.[3]

Some Bacillarieae have become sufficiently acclimated to the waters of hot springs to warrant inclusion among the thermal algae. These do not grow at such high temperatures as do the Myxophyceae, but they have been recorded[4] from hot springs in Iceland (temperature, 50 to 60° C.) and from California springs.

Many species of marine diatoms, especially those of the plankton, have a limited geographical distribution, and it is possible to follow the paths of ocean currents by determining the species of diatoms in the water. Fresh-water diatoms do not have so marked a geographical distribution, although they are found in greater abundance and variety in arctic and temperate regions than in the tropics.

Diatomaceous Earth. The siliceous wall surrounding the diatom cell remains unaltered after the death and decay of the cell, and great numbers of empty walls accumulate at the bottom of any body of water in which diatoms live. Where conditions are exceptionally favorable and long-continued, such accumulations may reach a considerable thickness. If the accumulating material is in an arm of the ocean, there may be a geological change that lifts the deposit far above and inland from the ocean. Such deposits of fossil diatoms, known as *diatomaceous earth*, are found in various parts of the world. Probably the best known and most extensive deposits in the United States are those at Lompoc, California, where the beds are miles in extent and in some places over 700 ft. in thickness. The thickest deposits of diatomaceous earth thus far discovered are in the Santa Maria oil fields of California. Oil wells drilled in this region show, after correction for dip, that there is a subterranean deposit about 3,000 ft. thick. The Lompoc deposit, like most others of marine origin, is composed almost exclusively of littoral species and contains but little foreign matter. Beds of

[1] Bristol-Roach, 1920. [2] Bristol-Roach, 1919.
[3] Moore and Carter, 1926. [4] West, G. S., 1902.

fossil fresh-water diatoms have been found in California, Nevada, Maryland and several other states, but none of them is over a few feet in thickness Some of the fresh-water deposits are made up largely of plankton species and evidently were deposited in the beds of former lakes. Other fresh water deposits have nonplanktonic species predominating.

Diatomaceous earth is assuming an increasing importance as a commercial product. More than three-quarters of the world's production is from the United States and in the years 1942–1944 the production in this country was 524,872 tons.[1] The enormous quantity obtained is more readily visualized when one realizes that a single ton has a volume of 50 to 260 cu. ft. Diatomaceous earth is obtained by underground mining from one or two deposits in this country, but most of the deposits are worked as open quarries since this bulky substance can be secured much more economically by quarrying.[2] In quarrying, the overburden of soil is removed, and the diatomaceous earth is then quarried by means of hand picks which split it into slabs that break apart in parallel cleavage planes.[3] Power saws that cut the material into slabs *in situ* are also used. Diatomaceous earth is also obtained from lake beds by dredging with a suction pump and carrying the material through sluiceways to settling tanks. Material from some deposits can be utilized directly; that from other deposits must be incinerated to remove the organic substances present. Producers of diatomaceous earth market their product as powdered earth, granules, manufactured brick, and sawn brick.

The industrial uses of diatomaceous earth are varied. One of the first was as an absorbent for liquid nitroglycerin, to make an explosive, dynamite, that could be transported with comparative safety. The inert material used in the manufacture of present-day dynamite is wood meal. About half of the diatomaceous earth produced in this country is used in the filtration of liquids; about a fifth is used as a filler in the manufacture of various products; about an eighth is used in insulation of boilers, blast furnaces, and other places where a high temperature is maintained. Probably the oldest commercial application is that of a very mild abrasive in metal polishes and tooth pastes. This use is so well known that many people think that it is the major use of diatomaceous earth, but the amount used is infinitesimal compared with that used for other purposes.

Organization of the Plant Body. Diatoms are essentially unicellular organisms, and colonial diatoms never have so complex bodies as the other algae. The strictly unicellular forms may be free-floating or sessile, but in either case the outermost layer of the wall is gelatinous in nature. Sometimes, as in *Gomphonema*, the gelatinous sheath assumes the form of a

[1] METCALF and HOLLEMAN, 1948. [2] EARDLY-WILMOT, 1928.
[3] GOODWIN, 1923.

long gelatinous stalk that affixes the cell to the substratum. In other cases, the cells are united in colonies by means of their gelatinous sheaths. Frequently in littoral marine species, rarely in fresh-water species, the colonies consist of cells embedded in a gelatinous matrix, either amorphous masses in which the cells lie in all directions, or attenuated branching tubes in which all cells lie with their long axes parallel to one another. Most fresh-water colonial diatoms have their cells seriately joined face to face. The mucilaginous matter connecting cell to cell may cover the whole face (*Fragilaria*, *Melosira*), or it may be restricted to a small globule at one end of the cell. In the latter case, the cells lie in a zigzag series (*Diatoma*) or in radiating colonies (*Asterionella*).

The Cell Wall. Diatomologists have centered their attention on the structure of the wall, and their taxonomic treatment of the group is based chiefly on wall structure and ornamentation. This intense specialization is largely responsible for the special terminology currently used to designate the various parts of the wall. As already stated, the protoplast of a diatom cell is enclosed by a wall consisting of two overlapping halves. Both the diatom cell and the empty wall are called a *frustule*. In the latter usage of the term, the outer of the two half-walls is an *epitheca;* the inner a *hypotheca*.[1] The silicified portion of each half-wall, in turn, consists of a more or less flattened *valve* whose flange-like margins are attached to a *cingulum* or *connecting band*. Some diatoms have additional connecting bands (*intercalary bands*) interpolated between epitheca and hypotheca and there may be one, two, or more intercalary bands. The connecting band is usually firmly united to the valve, but in "cleaned" diatoms and in diatomaceous earth, one frequently sees connecting bands that have become separated from the valve. A cingulum is not a closed loop but an open hoop, and one with a gap between the approximated ends.[2]

When a frustule lies so that the valve side is uppermost, it is said to be in *valve view;* when the cingulum is uppermost, it is in *girdle view*. According to the genus, a water mount will show practically all the individuals in valve view (*Cyclotella*), practically all in girdle view (*Melosira*, *Tabellaria*), or indiscriminately in valve and girdle view (*Pinnularia*, *Gomphonema*).

Frustules may be perfectly symmetrical, completely asymmetrical, or with intergrades between the two. The ornamentation of the valves (page 446), taken in conjunction with the external form, is the basis for division of the Bacillariophyceae into the Centrales and Pennales. Centrales are usually radially symmetrical in valve view, and Pennales are usually bilaterally symmetrical or bilaterally asymmetrical. The pennate diatoms

[1] MÜLLER, O., 1895. [2] PALMER and KEELEY, 1900.

may have a bilateral symmetry with respect to all three planes either when the frustule is with (*Amphirora alata* Kütz.) or without [*Pinnularia viridis* (Nitzsch) Ehr.] appendages. They may also be bilaterally symmetrical with respect to one axis and not the other. The symmetry may be in the valve view and the asymmetry in the girdle view [*Achnanthes inflata* (Kütz.) Grun.], or the reverse condition may obtain (*Amphora ovalis* Kütz.).[1]

Each of the two overlapping halves of the diatom wall consists of an organic matrix that is composed in large part of pectin.[2] The wall gives no reaction for cellulose and callose. The watery gelatinous envelope surrounding many planktonic forms[3] is probably pectic acid. The valve portions and cingulum portions of the wall are silicified, and some[4] think that this silicification is not a simple impregnation with silica but a chemical combination of silicon with the organic material in the wall. Others[5] hold that there is no organic material in the silicified portion of the wall. Silicification may be demonstrated by destroying the organic matter, either by incineration, or by treatment with concentrated acids, or by treatment with oxidizing agents such as potassium chlorate. Cell walls with all organic material removed are called *cleaned* diatoms. The amount of silicification is quite variable. It is often so scanty in plankton species that the frustules cannot be cleaned by the usual methods. Nonplanktonic species usually have highly silicified wallls. The extent to which a wall is silicified is dependent in part upon the amount of siliceous materials available, and it has been shown[6] that aluminium silicate is the compound most used in silicification. An abundance of silicates favors multiplication of diatoms,[7] and a direct correlation has been found[8] between increase in number of diatoms and decrease in amount of silica in the water. Cultural studies have shown[9] that certain diatoms can multiply in an absence of silicon and have naked cells without siliceous walls.

The siliceous material deposited in the wall of a valve is not laid down as a smooth sheet. Instead, the sheet is areolate or striate and with the areolae or striae in patterns that are characteristic for a genus or species. Some species, especially those of marine Centrales, have very coarse markings; certain of the Pennales have areolae and striae that are so fine that they are revealed only by the best microscopes. The coarse markings of many marine centric diatoms are due to thinner places in the siliceous deposit, which, in turn, are bounded by ridges that lie on the inner or outer face of the wall (Fig. 357*A–B*). Areolae may have very minute vertical canals (pores) running through them or thin places (poroids) which do not entirely

[1] Müller, 1895. [2] Liebisch, 1928; Mangin, 1908. [3] Schröder, 1902.
[4] Mangin, 1908. [5] Liebisch, 1928. [6] Coupin, 1922. [7] Pearsall, 1923.
[8] Meloche *et al.*, 1938. [9] Bachrach and Lefevre, 1929; Wiedling, 1941.

perforate the wall. Pennate diatoms have areolae so small that they appear as minute dots. These lie in transverse series in two vertical rows and have a symmetrical or asymmetrical disposition with respect to the longitudinal axis of a valve. In many cases, the areolae are so minute and so close together that there seem to be transverse striae. The areolae (punctae) of pennate diatoms are thin places, not perforations, in the wall. A few Pennales (Fig. 357*C* and *D*) have true perforations that lie[1] either in the polar region (*Diatoma*) or in the median region (*Tabellaria*).

Some pennate diatoms have only a longitudinal clear space (*pseudoraphe*) laterally separating the two vertical series of transverse rows from each

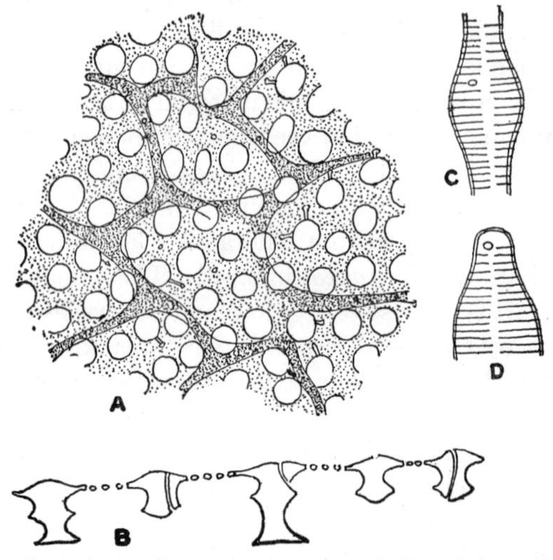

FIG. 357. *A–B*, surface view and cross section of a portion of the wall of ***Isthmia nervosa*** Kütz. *C*, median mucilage pore of *Fragilaria virescens* Ralfs. *D*, terminal mucilage pore of *Tabellaria fenestrata* Kütz. (*A–B, after Müller*, 1898; *C–D, after Müller*, 1899.)

other. The pseudoraphe is generally median in position but, as in *Achnanthes*, it may be excentric. Other pennate diatoms have the series of transverse rows longitudinally separated from each other by a vertical cleft in the valve, the *raphe*. The raphe may be straight, sigmoid, or undulate, and dividing the valve symmetrically. It is usually interrupted midway between its tow poles by a *central nodule*, and there are often similar swelling (*polar nodules*) at either end. If the central nodule is conspicuously expanded in the lateral direction, it is known as a *stauros*. The raphe is intimately connected with the movement of diatoms, and its detailed structure will be considered in the discussion of their movement (page 448). With the exception of a few genera, the epitheca and hypoth-

[1] GEMEINHARDT, 1926*A*; MÜLLER, O., 1899.

eca are similar in ornamentation. The most striking dissimilarities are found in such genera as *Achnanthes* and *Cocconeis*, where one valve has a true raphe and the other only a pseudoraphe.

Structure of the Protoplast. Immediately within the cell wall is a fairly thick layer of colorless cytoplasm in which the chromatophore or chromatophores are embedded. Internal to the cytoplasmic layer is a conspicuous central vacuole. Pennate diatoms often have the central portion of the vacuole transversely interrupted by a broad band of cytoplasm in which lies a spherical or ellipsoidal nucleus.

The chromatophores vary greatly in number and shape from species to species. Their structure is quite constant for some genera but is variable in others. For this reason, systems of classification based largely upon

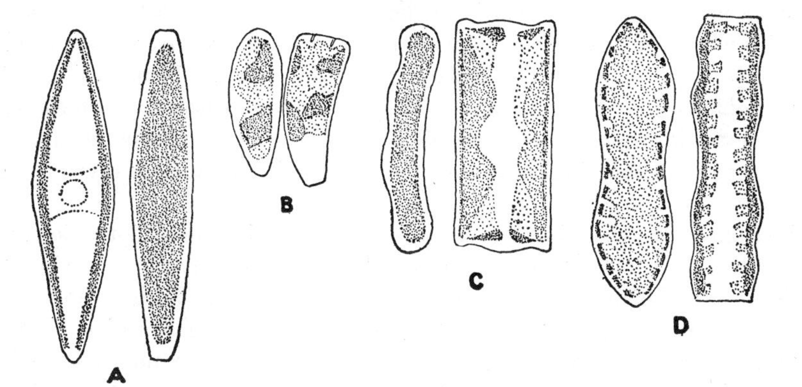

FIG. 358. Valve and girdle views of the chromatophores of various diatoms. *A*, *Navicula radiosa* Kütz. *B*, *Rhoicosphenia curvata* (Kütz.) Grun. *C*, *Eunotia diodon* Ehr. *D*, *Cymatopleura Solea* f. *interrupta*. (*After Ott*, 1900.)

chromatophores are unsatisfactory. Frustules of Centrales usually contain many discoid or irregularly shaped chromatophores; those of Pennales usually contain a single irregularly lobed or perforate chromatophore (*Gomphonema*, *Amphipleura*, *Cymbella*) or two chromatophores (Fig. 358). When there are two chromatophores, these are usually laminate and extend longitudinally along opposite sides of a frustule.[1] Typically the chromatophores are of a rich golden-brown color, but a few species have chromatophores that are a vivid green or even a bright blue.[2] The golden-brown color is due to the presence of several xanthophylls, among which fucoxanthin is present in greatest abundance (see Table I, page 3).

Chromatophores may contain one to several pyrenoids or may lack them entirely. Pyrenoids are usually ellipsoidal, biconvex, or biconcave. In most cases, they are embedded in the chromatophore, but sometimes they

[1] HEINZERLING, 1908; OTT, 1900. [2] MOLISCH, 1903*A*.

lie in a bulge on the inner face of a chromatophore or entirely separated from it.[1] Pyrenoids of diatoms are of the naked type (*i.e.*, devoid of a starch sheath), and their exact role in the metabolism of a cell is unknown. Possibly they may function as elaioplasts and be concerned in the formation of oils.

Fats are the chief food reserves formed by the photosynthetic activity of the chromatophores. These accumulate as droplets, often of considerable size, and either in the cytoplasm or in the chromatophores. The method by which fats are formed is unknown, but the fact that they serve as food reserves is demonstrated by their gradual disappearance when cells are kept continuously in a dark room. There are other food reserves in addition to fats. These food granules have been called *volutin*[2] and have been thought to be rich in nucleic acids.[3] The microchemical methods by which the nucleic acids were demonstrated were extremely crude, and one may just as well identify the volutin with the leucosin found in Xanthophyceae and Chrysophyceae. The paired plates sometimes seen in the cytoplasm immediately outside the nucleus[4] are probably structures concerned with the development of the spindle during nuclear division and not food reserves.

Diatom cells are uninucleate, and the nucleus is spherical to biconvex in shape. In centric species, it lies embedded in the cytoplasm next the wall; in most pennate species it lies in a cytoplasmic bridge across the middle of the protoplast. Numerous cytological investigations have shown that the nucleus has a definite membrane, one or more nucleoli, and a chromatin-linin network in the intervening space between the two. Some species have a centrosome lying at one side of, or in a peripheral depression of, the nucleus; other species lack a centrosome.[4] Nuclear division is always mitotic, generally with the formation of a considerable number of chromosomes.

Locomotion of Diatoms. Many of the free-living, and some of the colonial, pennate diatoms have the ability to move spontaneously. None of the centric diatoms moves independently. Movement is generally by a series of jerks and always in the direction of the major axis. After the cell has moved forward a short distance, it pauses for a short time and then, with the same jerky motion, moves backward along nearly the same path. Sometimes the motion is smooth instead of jerky, but there is always a backward and forward progression. Numerous theories have been advanced to account for the motility of diatoms; but a description of these[5] is of historical interest only, since Müller's theory of cytoplasmic

[1] Mereschowsky, 1903. [2] Meyer, A., 1904. [3] Guilliermond, 1910.
[4] Lauterborn, 1896. [5] West, 1916.

streaming[1] is now almost universally accepted as the cause of locomotion in Bacillariophyceae.

The intimate connection between movement and the presence of a raphe was brought out very clearly when it was shown that motility is restricted to those pennate forms that have a true raphe.[2] Movement is held to be due to cytoplasm streaming along the raphe from the anterior polar nodule to the central nodule; and from the central nodule to the posterior nodule. The raphe is not a simple cleft in the wall. Instead, it is an extremely complicated structure, and that of *Pinnularia* may be cited as fairly typical (Fig. 359). The raphe of *Pinnularia*, as seen in

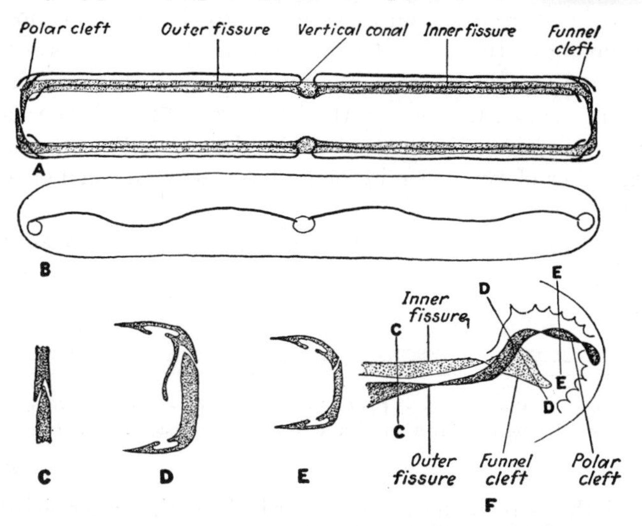

Fig. 359. Structure of the raphe of *Pinnularia*. *A*, vertical longitudinal section of a frustule. *B*, surface view of a valve. *C*, vertical section of valve wall cut in the plane *CC* of Fig. *F*. *D*, a similar section cut at *DD*. *E*, a similar section cut at *EE*. *F*, terminal portion of valve showing the inner and outer fissures in surface view. (*A–B, modified from Müller*, 1889; *C–F, after Müller*, 1896*A*.)

surface view, is a sigmoid cleft that runs from polar nodule to central nodule and thence to the other polar nodule. This cleft, as seen in vertical cross section midway between polar and central nodules (Fig. 359*C*) is not a vertical crevice, but >-shaped.[3] The upper portion of the > is called the *outer fissure*, and the lower portion is called the *inner fissure*. Near the vicinity of the polar nodule, the outer fissure bends in a semicircle and terminates in a linear expansion called the *polar cleft* (Fig. 359*F*). In the same region, the inner fissure bends in the opposite direction to the

[1] Müller, O., 1889, 1893, 1894, 1896, 1896*A*, 1897, 1908, 1909.
[2] Müller, O., 1889. [3] Müller, O., 1889, 1896*A*.

outer fissure and eventually terminates in a *funnel cleft* that opens on the inner face of the cell wall. The central nodule has a cylindrical projection toward the interior of the cell, and in this region the outer and inner fissures in each half of the valve wall are connected with each other by *vertical canals.* There is also a channel along the inner face of the central nodule that connects the inner fissures in posterior and anterior parts of the valve with each other. The rapheal system of other pennate diatoms is essentially the same as that of *Pinnularia,* but these differ from one another in minor details.[1]

According to the theory of locomotion by cytoplasmic cyclosis, there is a flow of cytoplasm from the anterior nodule to the posterior nodule (Fig. 360). Beginning at the polar cleft of the outer fissure, the stream moves backward along the outer face of the raphe, and, on reaching the vicinity of the central nodule, it moves vertically inward through the valve wall, through the anterior vertical canal. Coincident with this streaming there is an upward flow of cytoplasm in the posterior vertical canal of the central

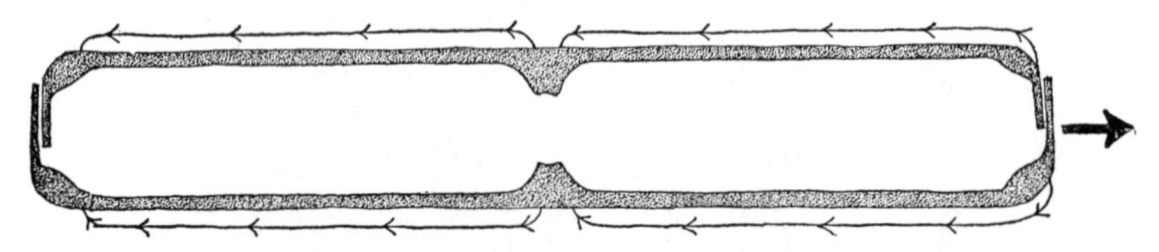

FIG. 360. Diagram of the streaming of cytoplasm on the outer face of a *Pinnularia* frustule. The heavy arrow indicates the direction of movement of the frustule. (*Modified from Müller,* 1893.)

nodule, and this stream moves backward along the outer fissure to the polar cleft in the posterior polar nodule. In both the anterior and posterior inner fissures there is a compensatory movement of cytoplasm that travels in the opposite direction to that in the outer fissures. The propulsion of the frustule is in the opposite direction to that of the streaming of cytoplasm in the outer fissures. Movement is held to be due to water currents set up by the flowing cytoplasm and to cyclonic currents established in the region of the polar nodules. The demonstration of the cytoplasmic circulation rests more upon observation of motion set up in the suspended particles, when cells are mounted in dilute india ink, than in direct observation of cytoplasmic flow. Girdle views of cells mounted in such suspensions show that there is a linear flow of particles from anterior polar nodule to central nodule, a whirlpool motion in the region of the central nodule, and a flow of particles from central nodule to posterior nodule.

[1] MÜLLER, O., 1889. 1896, 1909; HUSTEDT, 1926, 1926*A*, 1928, 1928*A*, 1929, 1929*A*.

Cell Division. Multiplication of diatoms takes place largely by division into two daughter cells of slightly different size. Most cells divide during the midnight hours, but some, as *Achnanthidium*,[1] divide between 7 and 8:30 A.M. The first indication of division is an increase in cell size at right angles to the girdle. This is followed by a mitotic division of the nucleus, the plane of division being in the short axis of the cell and perpendicular to the valves (Fig. 361). Spindles of some diatoms have a distinct centrosome at each pole;[2] other genera do not have centrosomes. Accompanying the nuclear division (generally in the time interval between prophase and anaphase), there is a division of the chromatophores. If the cell is one with a single chromatophore, its division is always longitudinal; if it has two of them, their division may be longitudinal or transverse.[3] Species with numerous chromatophores do not have a bipartition of them until after the daughter cells have been formed. Pyrenoids, at least in species with conspicuous ones, increase in number by division and not by formation

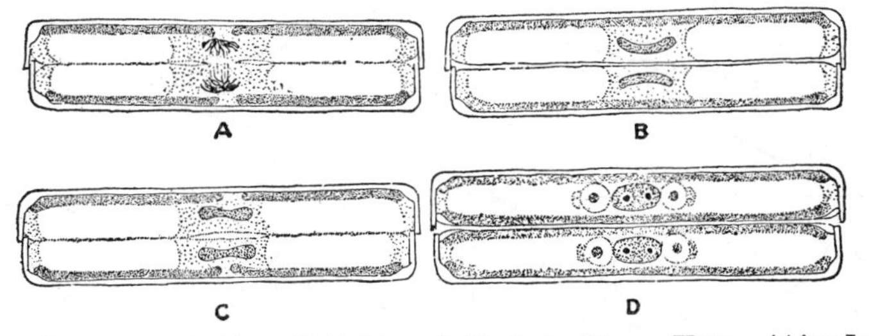

FIG. 361. Four stages in the cell division of *Navicula oblonga* Kütz. (*After Lauterborn*, 1896.)

de novo.[4] After duplication of the cell organs, there is a cytoplasmic cleavage in a plane parallel to the valves, and this results in two protoplasts, one of which lies within the epitheca and the other within the hypotheca of the parent-cell wall. There then follows a secretion of a new half-wall next to the girdle and uncovered faces of each daughter protoplast. The newly formed half-wall is always the hypotheca of the daughter cells, and the old half-wall, whether epitheca or hypotheca of the parent cell, is always its epitheca. It follows, therefore, that in a population descended from a single cell, half of the cells will have an epitheca that was secreted in the previous cell generation, a quarter of them an epitheca secreted two generations back, an eighth of them an epitheca secreted three generations back, and so on until there are two cells each with an epitheca derived from the original cell (Fig. 362).

Utilization of the two old half-walls as epithecae for the two daughter

[1] GEMEINHARDT, 1925. [2] LAUTERBORN, 1896. [3] OTT, 1900.
[4] HEINZERLING, 1908.

cells results in one cell being the same size as the parent and the other being slightly smaller (Fig. 362). Theoretically, according to Pfitzer's law,[1] continuation of this through a succession of cell generations would result in a population with certain cells appreciably smaller than the original parent cell and other cells only slightly smaller than the parent. There would also be one cell exactly the same size as the original parent cell. A corollary of Pfitzer's law is that progressive diminution in size does not continue indefinitely because cells of a certain reduced size form rejuvenescent cells (auxospores, see page 453) which give rise to vegetative cells of maximum size for the species.

The validity of Pfitzer's law has been tested by observation of material growing both under natural conditions and in clones developed from a single cell. Certain species have been shown to have the progressive reduction in size of certain cells postulated by Pfitzer's law.[2] In other

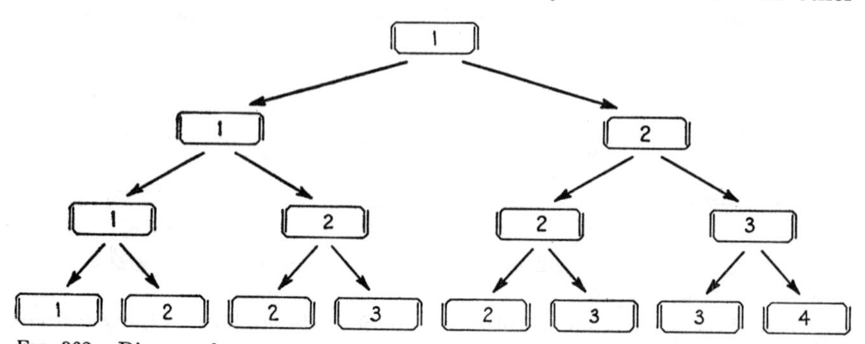

FIG. 362. Diagram showing the progressive diminution in size of certain frustules through successive cell generations of a diatom.

species, there is no appreciable diminution in size, even when cultures are carried through many generations of cells.[3] For certain of these species in which there is no reduction in size, the girdle of a half-cell has a certain amount of elasticity, and the new hypotheca is of the same size or greater than the other half-cell. Daughter cells may even become longer than the parent cell because of a "secondary growth."[4] The length of time involved in a progression from maximal to minimal size depends primarily upon the rate of cell division. Undoubtedly this varies greatly from species to species, but even under the most favorable condition it is probable that this takes a relatively long time. Some indication of the time involved has been obtained from a study of successive deposits of diatom shells laid down in lakes of Switzerland. Here it has been estimated[5] that, for certain plankton species, the time interval is from 2 to 5 years.

[1] PFITZER, 1871, 1882. [2] GEITLER, 1932A; MEINHOLD, 1911.
[3] ALLEN and NELSON, 1910, RICHTER, 1919. [4] GEMEINHARDT, 1927.
[5] NIPKOW, 1928

Craticular Stages. Pennate diatoms occasionally have the protoplast forming successive sets of new half-walls without escaping from the original wall surrounding a cell. Successively formed half-walls, which are progressively smaller, nest one within the other, and the later formed half-walls often have an imperfectly developed raphe. Such *craticular stages* are immobile and result from unfavorable environmental conditions, especially an increase in salt content of the water.[1] Return of favorable conditions induces active cell division in a craticular stage and, within a cell generation or two, the daughter cells are normal in structure and migrate from the nested half-walls of the craticular stage.

Auxospores. Sooner or later progressive diminution in size of cells of diatoms is compensated for by a production of auxospores. Auxospores are formed by cells whose size approached the minimum for the species, but not by those of minimal size.[2] Among most Pennales, auxospores are formed as a result of syngamy or autogamy, but among some they are formed parthenogenetically. Certain Centrales are also known to have an autogamous fusion of nuclei in connection with production of auxospores. Formation of an auxospore involves a liberation of a protoplast from the half-walls, a considerable enlargement of the naked liberated protoplast, and then a secretion of a silicified wall around the protoplast. This wall may be smooth or sculptured. In the latter case, the sculpturing is not identical with that of vegetative cells of the species. The enlarged auxospore divides to form vegetative cells whose size is near the maximum for the species. Auxospores are zygotic in nature, and since division of the auxospore nucleus is mitotic, the first generation of vegetative cells and all subsequent generations of vegetative cells are diploid.

Auxospores of Pennales. Auxospores of pennate diatoms may be formed in one of the following five ways: (1) by two cells conjugating to form a single auxospore, (2) by two cells conjugating to form two auxospores, (3) by two cells becoming enveloped in a common gelatinous envelope but each giving rise to an auxospore without conjugation, (4) by a single cell giving rise to one auxospore, (5) by a single cell giving rise to two auxospores. The first two of the foregoing methods are obviously sexual; superficially the last three methods appear to be asexual, but cytological study of them has shown that they are probably sexual in nature.

Cells producing auxospores according to the first method (by conjugating in pairs to form a single auxospore) may be sister cells or two that are not derived from a common parent cell. In either case they are enclosed by a common gelatinous envelope. They generally lie side by side within the envelope, but some, as *Surirella saxonica* Auersw,[3] lie end to end. Several species have been investigated cytologically and found to have a meiotic division of nuclei of conjugating cells. *Surirella saxonica*, the first

[1] GEITLER, 1927*D*; LIEBISCH, 1929. [2] GEITLER, 1932*A*. [3] KARSTEN, 1900.

diatom in which meiosis was demonstrated[1] has an enlargement of one and a degeneration of three of the nuclei resulting from meiosis. After this, the protoplasts unite to form a zygote in which the two haploid gamete nuclei fuse to form a single diploid nucleus. The zygote then elongates to form an auxospore whose long axis lies parallel to the long axes of the empty frustules of the conjugating cells. Conjugating cells of *Cocconeis pediculus* Ehr. and *C. placentula*, var. *klinoraphis* Geitler have an immedi-

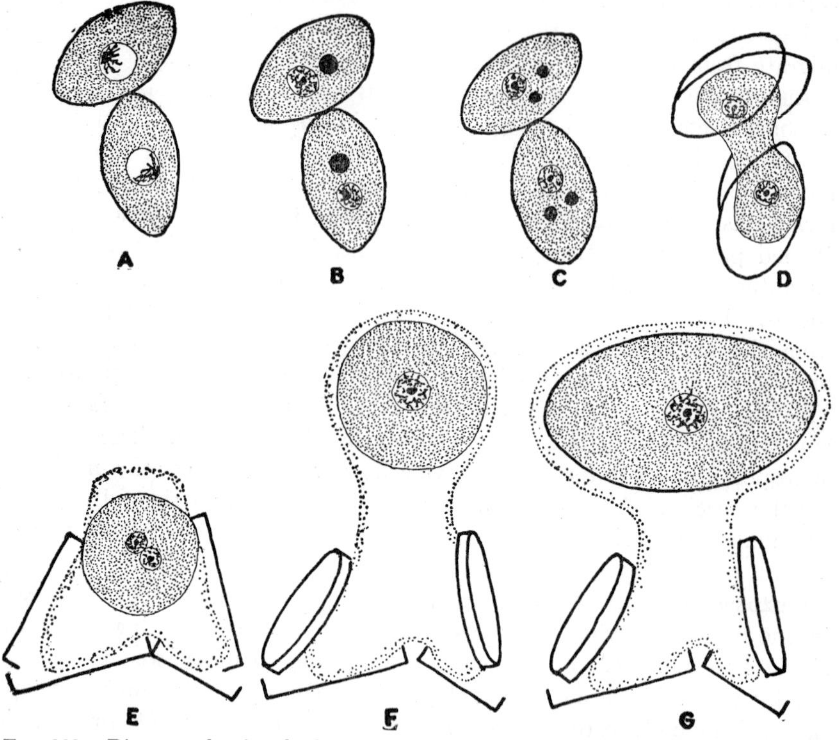

FIG. 363. Diagrams showing the formation of one auxospore by the conjugation of two cells in the pennate diatom, *Cocconeis placentula* var. *klinoraphis* Geitler. *A–D*, in top view; *E–G*, in side view. *A–C*, meiosis and degeneration of all but one daughter nucleus in each frustule. *D*, gametic union. *E*, young zygote. *F–G*, enlargement of the zygote to form an auxospore. (*Diagrams based upon Geitler*, 1927*A*.)

ate degeneration of one daughter nucleus after the first meiotic division (Fig. 363). The persisting nucleus then divides, and one of its daughter nuclei degenerates. The two protoplasts, each containing a haploid nucleus, then unite to form a binucleate zygote which later becomes uninucleate by a fusion of the two nuclei.[2] The zygote then enlarges and becomes an auxospore.

[1] KARSTEN, 1912. [2] GEITLER, 1927.

Cells producing auxospores according to the second method (by conjugation to form two auxospores) have both protoplasts of the conjugating cell dividing to form two gametes. Certain species have been shown[1] to have a meiotic division and a degeneration of two of the four haploid nuclei prior to the cytokinesis into two uninucleate gametes. Cells of other species have been shown to form four nuclei, two of which degenerate after cytokinesis. Presumably these nuclear divisions are also meiotic, but this has not been definitely established. The plane of cytokinesis in

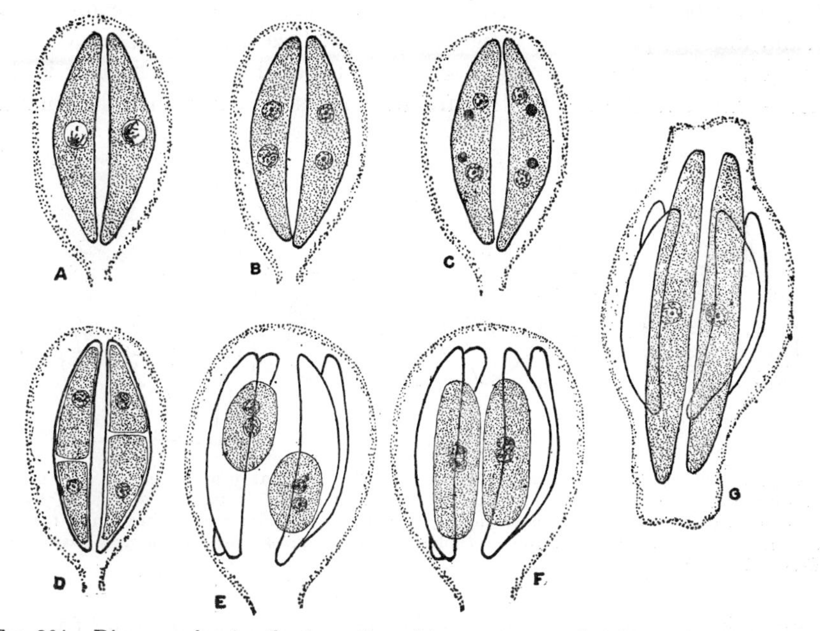

FIG. 364. Diagrams showing the formation of two auxospores by the conjugation of two cells in the pennate diatom, *Cymbella lanceolata* (Ehr.) Brun. *A–C*, meiosis and degeneration of two daughter nuclei in each frustule. *D*, after division of each protoplast into two gametes of unequal size. *E*, young zygotes. *F–G*, elongation of the zygotes to form auxospores. (*Diagrams based upon Geitler*, 1927*B*.)

gamete formation may be at right angles to the long axis of the paired cells,[2] or longitudinal.[3] The two gametes formed by a parent cell may be of equal[4] or unequal[5] size. Gametes of unequal size may result from division of the protoplast of a parent cell,[6] or from enlargement of one of the daughter protoplasts after division.[7] Even when the two gametes formed

[1] CHOLNOKY, 1928*A*; GEITLER, 1927*A*, 1928*A*; MEYER, K., 1929; SUBRAHMANYAN, 1948.

[2] KARSTEN, 1896, 1897; KLEBAHN, 1896. [3] GEITLER, 1927*A*, 1928*B*.

[4] KARSTEN, 1897*A*; KLEBAHN, 1896. [5] CHOLNOKY, 1928*A*; KARSTEN 1896.

[6] CHOLNOKY, 1928*A*. [7] KARSTEN, 1896.

by a cell are morphologically alike, there may be such physiological differences as one being motile and the other passive.[1] The two gametes formed within a cell usually unite with those of the other conjugating cell instead of with each other. In most cases, both gametes of a uniting pair are amoeboid, and their union takes place midway between the parent frustules. Less frequently, one gamete of a uniting pair is motile and the other immobile. The two gametes produced by one parent cell may be motile and the two produced by the other cell immobile,[2] or each parent cell may form one motile and one immobile gamete. The latter condition is found among species whose parent cells produce gametes of equal or of unequal size.[3] If the two gametes are of unequal size, the smaller is the active one, and the conjugating frustules are so oriented that the smaller

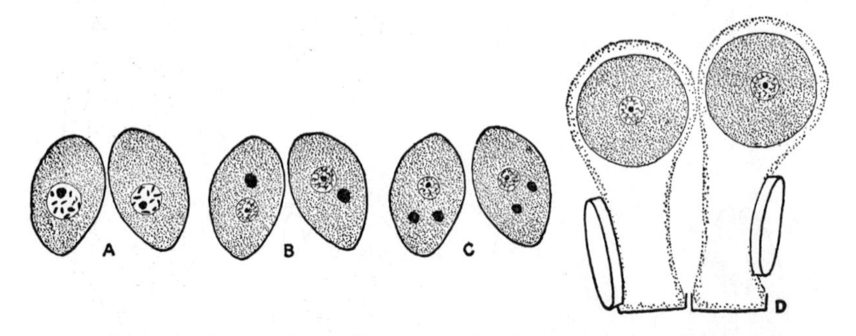

FIG. 365. Diagrams showing the parthenogenetic development of auxospores in each of two "conjugated" cells of the pennate diatom, *Cocconeis placentula* var. *lineata* (Ehr.) Cleve. *A–C*, nuclear divisions and degeneration of all but one daughter nucleus in each of the two frustules. *D*, enlargement of the protoplasts to form auxospores. (*Diagrams based upon Geitler*, 1927*A*.)

gamete in one frustule is opposite the larger gamete in the other frustule, and vice versa (Fig. 364). The pair of zygotes resulting from fusion of two sets of gametes both elongate into auxospores, considerably longer than the cells producing the gametes. Elongation of zygotes into auxospores may be in a direction parallel to the long axis of parent frustules (Fig. 364) or at right angles to their long axes.[4]

Cells producing auxospores according to the third method (a pair of cells forming two auxospores without conjugation) are of rare occurrence. Preliminary stages resemble those where conjugation takes place, and there is the same double division of the nucleus and a degeneration of all but one of the resultant nuclei (Fig. 365). Both cells of a pair have the protoplast with the persisting nucleus escaping from the frustule, enlarging, and

[1] GEITLER, 1928*A*. [2] SUBRAHMANYAN, 1948. [3] GEITLER, 1927*A*.
[4] KLEBAHN, 1896.

developing into an auxospore. In at least one case of this type, it has been shown[1] that there is no halving of the chromosome number when the nucleus divides and that the nucleus persisting after division is diploid.

Production of auxospores according to the fourth method (a single cell producing a single auxospore) may take place in various ways. One diatom has been found[2] to have a meiotic division of nuclei in solitary cells, followed by a partial degeneration of two of the four resultant nuclei. The protoplast then divides to form two gametes, each with a normal and a degenerate nucleus. These sister gametes unite to form a zygote which enlarges to become an auxospore. In another diatom where a single cell produces a single auxospore, it has been found[3] that meiotic division is not followed by cytokinesis and that there is a fusion of sister haploid nuclei to form a diploid nucleus in the enlarging auxospore. The evidence supporting such an autogamic fusion of nuclei is none too convincing.

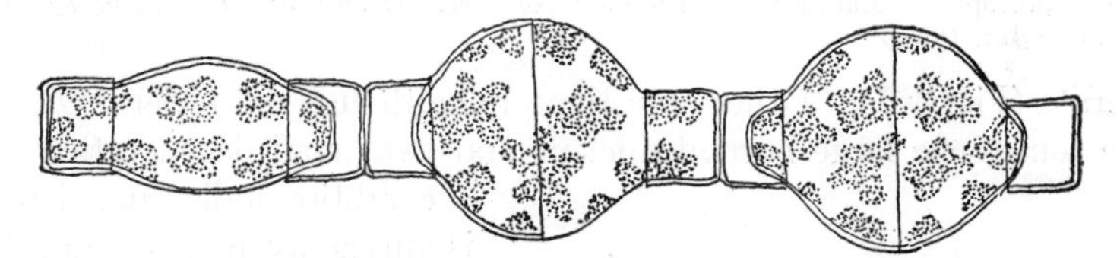

FIG. 366. Auxospores of the centric diatom, *Melosira varians* Ag. (*After Pfitzer*, 1882.)

Production of auxospores according to the fifth method (a single cell producing two auxospores) is of rare occurrence, and in the cases investigated[4] the nuclear behavior has not been worked out in detail.

Auxospores of Centrales. Auxospores are known for a wide variety of Centrales, and in all cases there is a production of a single auxospore by an old or by a recently divided cell. Among fresh-water Centrales, auxospore formation is most frequently encountered in *Melosira.* Here the two halves of a frustule pull apart, and the exposed portion of the protoplast increases in diameter. After it has swollen to two or three times its former diameter and the poles have become rounded (Fig. 366), there is a secretion of two new silicified half-walls with markings somewhat like those of vegetative cells. This auxospore usually remains attached for some time to the filament in which it was formed. It germinates by dividing transversely into two daughter cells and these, in turn, divide transversely. Daughter cells resulting from division of an auxospore are of the same diameter as the auxospore; *i.e.*, two to three times the diameter of the original vegetative cell.

For a long time, auxospores of Centrales were thought to be of an asexual nature, but discovery of nuclear divisions within developing auxospores

[1] GEITLER, 1927. [2] GEITLER, 1939. [3] GEITLER, 1928. [4] KARSTEN, 1897*A*.

has cast doubt upon this interpretation. Certain investigators[1] have found a meiotic division of the nucleus in developing auxospores. Two of the nuclei formed as a result of meiosis unite, and the other two nuclei de-

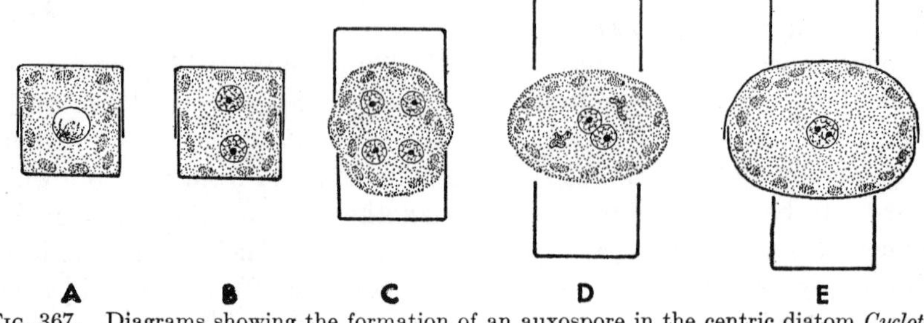

FIG. 367. Diagrams showing the formation of an auxospore in the centric diatom *Cyclotella Meneghiniana* Kütz. *A–C*, meiotic division to form four haploid nuclei. *D*, degeneration of two nuclei and beginning of fusion of the other two nuclei. *E*, nuclear fusion completed and the protoplast enlarging to form an auxospore. (*Diagrams based upon Iyengar and Subrahmanyan*, 1943.)

generate (Fig. 367). Other investigators,[2] although not observing meiosis, have found one large normal nucleus and two small degenerating nuclei, a condition indicating that there is autogamy in these developing auxospores. These relatively few cases suggest the possibility that autogamy may be widespread among Centrales, and that in this order, as in Pennales, the vegetative cells are diploid.

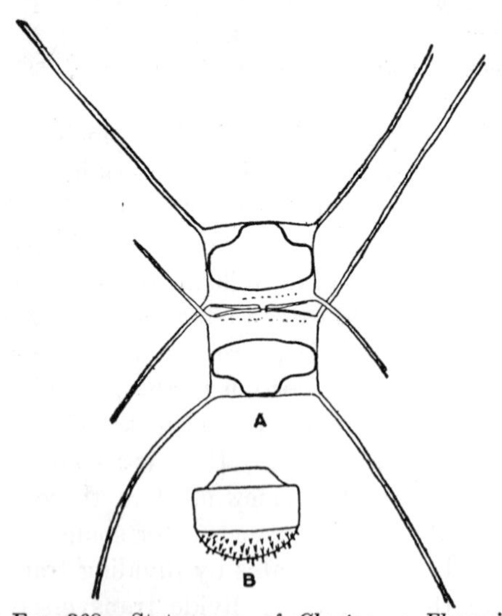

FIG. 368. Statospores of *Chaetoceros Elmorei* Boyer. *A*, two frustules, each containing an immature statospore. *B*, a mature statospore. (*After Boyer*, 1914.)

Statospores. The thick-walled spores which are formed singly within a frustule and are smaller than the parent cell are asexual in nature. These statospores are found only in centric diatoms. They are best known in marine plankton species, but they have also been found in fresh-water species of *Rhizosolenia* and *Chaetoceros* (Fig. 368). In their development, the protoplast retracts from the enclosing wall and then secretes a new wall that is composed of two overlapping halves. A statospore wall differs from that enclosing a vegetative cell, with respect

[1] PERSIDSKY, 1935; IYENGAR and SUBRAHMANYAN, 1942.
[2] CHOLNOKY, 1933; GROSS, 1937; RIETH, 1940.

both to ornamentation and to the spines it bears.[1] The cytological details of statospore formation have not been followed in full.

The pair of spores sometimes found within each cell of a filament of *Melosira italica* Kütz.[2] and the numerous small thick-walled spores found within cells of *Surierella*[3] are probably different from statospores. These resting spores, whose germination has not been observed seem to be formed at the end of a vegetative season[4] and may possibly tide the diatom over unfavorable seasons. However, resting spores are not necessary to carry diatoms through unfavorable periods, since vegetative cells of many species successfully withstand desiccation for months or years.[5]

Microspores. Many of the centric diatoms are known to produce a large number of *microspores* within a cell. A motility of microspores has been demonstrated for a sufficient number of species to warrant the assumption that microspores of all species have flagella. Microspores of certain genera are described[6] as having one flagellum; those of other species

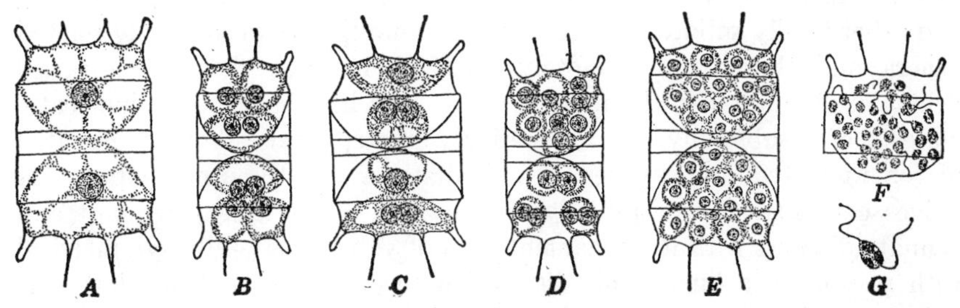

FIG. 369. Microspore formation in *Biddulphia mobiliensis* Bailey. *A*, recently divided vegetative cell. *B–E*, stages in repeated bipartition of protoplasts. *F*, microspores before liberation. *G*, free-swimming microspore. (*After Bergon*, 1907.) (*A–F*, × 200; *G*, × 1000.)

are said[7] to be biflagellate. The number of microspores within a cell is a multiple of two, and there may be 8, 16, 32, 64, or 128 of them. They may be formed by repeated simultaneous nuclear division and a cytoplasmic cleavage following the last series of nuclear divisions,[8] or a cytoplasmic cleavage may follow each nuclear division until 8, 16, 32, or more uninucleate protoplasts have been formed[9] (Fig. 369). All nuclear division during microspore formation have been described as mitotic.[10] Meiosis has also been described in connection with microspore formation, and either at the beginning[11] or at the end[12] of the series of nuclear divisions.

Some phycologists think that microspores are zoospores; others think that they are zoogametes. The evidence that they are isogametes is very

[1] CHOLNOKY, 1933; GROSS, 1937; RIETH, 1940. [2] HUSTEDT, 1927.
[3] WEST, G. S., 1912. [4] WEST, G. S., 1916. [5] BRISTOL-ROACH, 1920.
[6] BERGON, 1907; SCHILLER, 1909. [7] PAVILLARD, 1914; SCHMIDT, P., 1923.
[8] KARSTEN, 1904; SCHMIDT, P., 1929. [9] BERGON, 1907; HOFKER, 1928.
[10] KARSTEN, 1904, 1924. [11] SCHMIDT, P., 1927. [12] HOFKER, 1928.

scanty and has been found[1] in but one species. Even if the fact that microspores are not isogametes can be established beyond all doubt, there is still the possibility that they are gametic in nature. A species of *Chaetoceros* found in a plankton haul from the middle of the Atlantic Ocean and studied in a living condition was found[2] to be liberating motile microspores in abundance. These swarmed about cells with undivided contents, and this was thought to be the beginning of a gametic union. If this should prove to be the case, the microspores are to be considered male gametes (antherozoids) which escape from a parent cell and swim to a female cell containing an undivided protoplast (the egg).

Classification. Practically all treatises on diatoms written within the past half century follow the classification of Schütt (1896) which is based entirely upon shape, structure, and ornamentation of the frustules. This system has the great advantage of being equally applicable to living diatoms and to those known only in a fossil condition. Schütt places all diatoms in a single family and establishes subfamilies, tribes, subtribes, and other family subdivisions to show the affinities between closely related genera and those more remotely related. Later workers have raised Schütt's subfamilies to the rank of orders and have given his minor categories in each subfamily a correspondingly greater rank. Schütt segregated diatoms into two major series: centric diatoms with the ornamentation of valves concentrically or radially symmetrical about a central point and pennate diatoms with the ornamentation of valves bilaterally symmetrical with respect to a line. This distinction seems to be artificial, but it is quite natural and correlates with many other characters. Centric diatoms, the Centrales, usually have many chromatophores, never move spontaneously, produce statospores, form motile microspores, and never conjugate in pairs to form an auxospore. Pennate diatoms, the Pennales, usually have but one or two chromatophores, often have cells capable of spontaneous movement, lack flagellated microspores, and frequently conjugate in pairs to form auxospores.

ORDER 1. CENTRALES

The Centrales have valves that are circular, polygonal, or irregular in outline, and have an ornamentation that is radial or concentric about a central point. The valves never have a raphe or pseudoraphe. Living species generally have protoplasts with many chromatophores. There may be a production of statospores or of microspores. Auxospores are never formed by a conjugation of two cells.

As far as the number of genera is concerned, the Centrales and Pennales are about equal. However, the Centrales contain a considerably smaller number of species, most of which are found in the ocean, especially its

[1] Schmidt, P., 1923. [2] Went in Geitler, 1935.

plankton. Frustules of centric diatoms are discoid, cylindrical, or irregular in shape, and often have marginal spines, horns, or other projections. Most members of the order are free-floating and occur single or serially united with the valves apposed. Others are sessile and in filamentous or dendroid colonies in which the cells are united edge to edge in a zigzag fashion.

When seen in valve view, centric diatoms may always be recognized by the radial or concentric ornamentation with respect to a point, and by the absence of a raphe or a pseudoraphe. The centric nature is not so obvious in those species that regularly appear in girdle view when examined in water mounts. Fresh-water diatoms of this type include *Melosira*, in which the cells are united valve to valve, and *Rhizosolenia*, in which there are numerous intercalary bands.

Three of the four suborders into which the order is divided are represented in the fresh-water flora of this country.

SUBORDER 1. COSCINODISCINEAE

Genera belonging to this suborder have discoid or cylindrical cells that are usually broader than they are tall. The valve face may be flat or strongly convex, but nowhere on a valve are there local regions conspicuously elevated into horn-like processes. Margins of valves may be smooth or with numerous small spines or bristles. The valves are ornamented with radiate striations or punctations of varying degrees of coarseness, and the pattern of a valve may or may not be divided into definite sectors. The cells may be solitary or united in filaments. If in filaments, the cells may lie with their valves abutting on one another, or may lie some distance from one another and either within a common gelatinous tube or connected by small gelatinous strands. The chromatophores are usually numerous and small.

Auxospores are known for most of the genera, and a few species have been found with microspores.

Two of the families of the suborder have representatives in the fresh water algal flora of the United States.

FAMILY 1. COSCINODISCACEAE

Members of this family usually have discoid cells, but some species have cells that are taller than they are broad. The pattern on the valve face is distinctly radial but never set off into distinct sectors; the ornamentation on the valve face may continue down over the girdle, or the girdles may be smooth.

Genera found in this country may be distinguished as follows:

1. Cells united in long filaments, girdle sculptured .... 1. **Melosira**
1. Cells solitary, girdle unsculptured 2

2. Ornamentation of valve in two unlike concentric patterns.... 2. **Cyclotella**
2. Ornamentation of valve not in two unlike concentric patterns.............. 3
3. Valve surface with radiate hyaline areas..................... 3. **Stephanodiscus**
3. Valve surface without radiate hyaline areas.................. 4. **Coscinodiscus**

1. **Melosira,** Agardh. The cells of *Melosira* (Fig. 370) are cylindrical and with a length greater than the breadth. The valves may be flat or convex; if convex, they usually have a marginal ring of denticulations that help unite the cells in filaments and supplement in function the gelatinous cushions borne on the central area of valve faces. The valves are always circular in vertical view and with the ornamentation in two parts that are concentric to each other. Girdles of the half-cells often have a shallow annular constriction (*sulcus*) a short distance from where two girdles overlap. Species with a sulcus have the portion above it conspicuously ornamented and the portion below it smooth. Species without a sulcus have the entire girdle ornamented. The protoplasts contain numerous discoid chromatophores that may be so densely crowded that they obscure the markings on the wall.

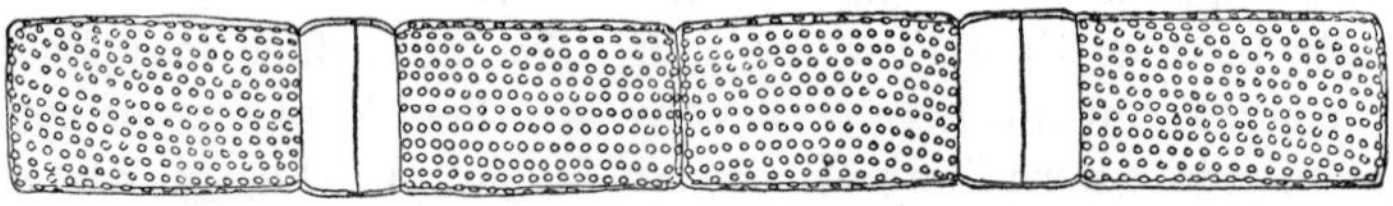

FIG. 370. *Melosira granulata* (Ehr.) Ralfs. (× 1040.)

Auxospore formation is of more frequent occurrence in this genus than in any other of the fresh-water Centrales. Microspores of a gametic nature have been recorded for one species.[1]

This genus is one in which there are numerous fresh-water and marine species and is the most frequently encountered of all fresh-water Centrales. It is often found in almost pure stands in pools, ditches, and slowly flowing streams. Several species are also found in the plankton of lakes, and these may be the dominant algae during early spring and late autumn months. The filamentous habit affords an easy method of distinguishing it from other centric forms; but the inexperienced phycologist sometimes does not recognize the diatomaceous nature of *Melosira*, because it always appears in girdle view under the microscope. About 10 species occur in this country. For descriptions of a majority of them, see Boyer (1927).

2. **Cyclotella** Kützing, 1834. The drum-shaped discoid cells of *Cyclotella* are generally circular in valve view, although a few species have valves that are elliptical in outline. Most species are solitary and free-floating, but some have cells united in straight or spirally twisted filaments, by means of gelatinous threads from valve to valve, or by means of a common gelatinous envelope. The ornamentation of a valve consists of two concentric re-

[1] SCHMIDT, P., 1923.

gions: an inner smooth region, or one that is irregularly and finely punctate, and an outer peripheral zone with radial striae or punctae. Valves, as seen in girdle view, may lie parallel to each other or be excentrically undulate. Girdles of frustules are smooth, and there are no intercalary bands between them. The protoplast of a cell contains numerous discoid chromatophores.

Auxospores are produced singly within a cell and are formed autogamously.[1]

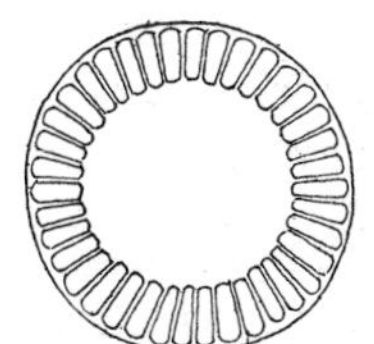
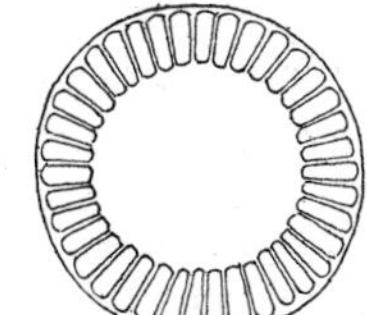

FIG. 371. *Cyclotella Meneghiniana* Kütz. (× 1300.)

Although species of *Cyclotella* (Fig. 371) are found in fresh, brackish, and salt water, the genus is primarily marine. Most of the fresh-water species in this country are planktonic and are widely distributed. Eleven species are known for the United States. For names and descriptions of them, see Boyer (1927).

3. **Stephanodiscus** Ehrenberg, 1845. Frustules of this diatom are usually discoid, though sometimes cylindrical, and occur singly and free-floating. The valves are circular in outline and radially punctate. Toward the periphery of a valve, the punctae are in multiseriate rows and alternating with linear smooth spaces; inward from each of the multiseriate rows is a single row of punctae. The transition from uniseriate to multiseriate punctation may be so abrupt that the ornamentation is divided into two concentric regions, or the transition may be so gradual that two con-

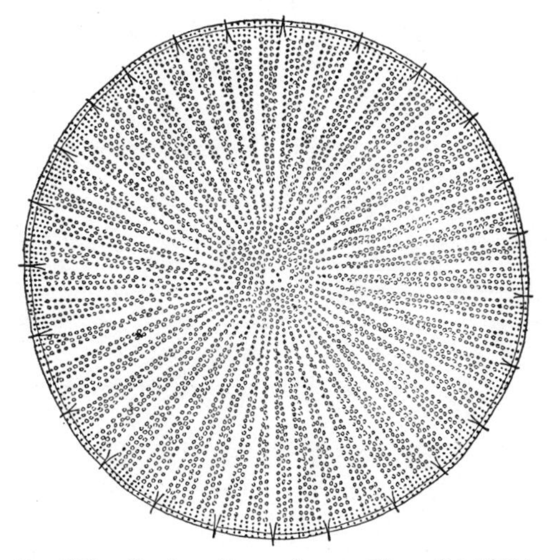

FIG. 372. *Stephanodiscus niagarae* Ehr. (× 1300.)

[1] IYENGAR and SUBRAHMANYAN, 1944.

centric regions are not distinguishable. External to each linear smooth space on a valve is a short, rather stout, spine. The girdles are smooth and without intercalary bands. There may be several small discoid chromatophores within a cell or one or two large irregularly shaped ones.

Auxospores are formed singly within a frustule and are either spherical or ellipsoidal.

Stephanodiscus is widely distributed in the plankton of lakes in the United States. The species found in this country are *S. astraea* (Ehr.) Grun., *S. carconensis* Grun., *S. Hantzschii* Grun., and *S. niagarae* Ehr. (Fig. 372). For descriptions of *S. astraea*, see Boyer (1927); for the other two, see Skvortzow (1937).

4. **Coscinodiscus** Ehrenberg, 1838. The frustules of *Coscinodiscus* are usually discoid and are always shorter in the vertical than in the transverse axis. Most species have cells that are circular in valve view, but a few are elliptical or irregular in outline. The ornamentation varies all the way from minute punctae to coarse areolae, and these markings may be irregularly distributed, in decussating series across a valve or in radial series toward

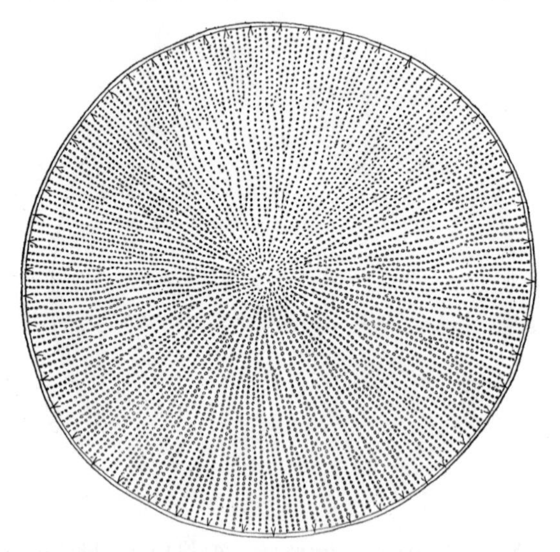

Fig. 373. *Coscinodiscus lacustris* Grun., a species known to occur in fresh water but not known for the United States. (× 1300.)

the margins of a valve (Fig. 373). A few species have small denticulations on the margin of a valve, but most species lack them. There may be numerous small discoid chromatophores within a cell, several irregularly lobed chromatophores, or two laminate chromatophores. Pyrenoids are not generally found within the chromatophores.

Auxospores, known only for a few species, are formed singly within frustules.

Coscinodiscus (Fig. 373) is a large genus with 450 or more species, practically all of which are marine and found chiefly in the plankton of oceans. *C. subtilis* Ehr. is the only species known from fresh waters of the United States. For a description of it, see Boyer (1927).

FAMILY 2. EUPODISCACEAE

Genera assigned to this family differ from others of the suborder in having mamillate protuberances near the margin of the valves. In some species, the protuberances are reduced to small rounded unsculptured areas (*ocelli*). The frustules are generally discoid and circular in valve view, though they may be elliptical or polygonal. The ornamentation is usually conspicuously radiate in arrangement and sometimes differentiated into two concentric portions. It is never bilaterally disposed with respect to a line. Protoplasts of cells usually contain numerous small chromatophores.

Auxospores have been recorded for the family.

The family is distinctly marine, with both littoral and plankton species. Only one genus, *Actinocyclus*, has representatives in fresh water.

1. **Actinocyclus** Ehrenberg, 1838. Generic characters distinguishing this diatom from other genera of the family are the presence of a single ocellus near the margin of the valve and a distinctly radiate areolation or punctation. The frustules are discoid and generally circular in valve view, though sometimes elliptical or elliptico-rhomboidal. The ornamentation usually consists of rather coarse areolae, which are often larger toward the center of the valve. Toward the periphery of the valve, the areolations lie in evenly spaced, or fasciculate, radiate series. At the extreme periphery, the valve is smooth or finely and radiately striate. Several species have minute denticulations toward the margin of the valves. The single ocellus, which is a local lenticular thickening in the valve wall, is generally located near the outer border of the radiate areolations or punctations. Within the cell are numerous discoid chromatophores.

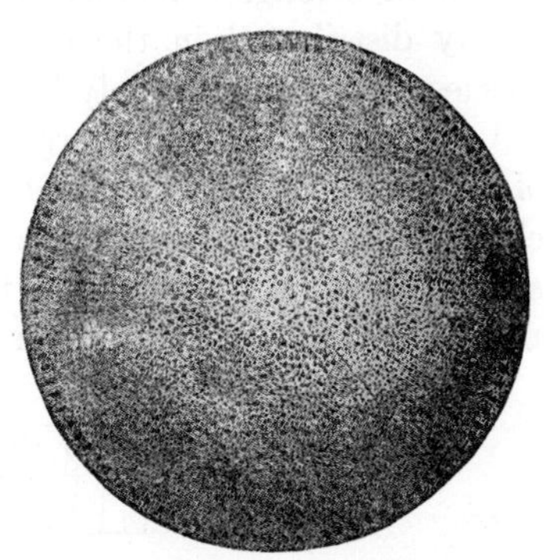

FIG. 374. *Actinocyclus niagarae* H. L. Smith. (*From H. D. Smith*, 1878.)

Actinocyclus is a genus with several marine species. The only species occurring in fresh water, *A. niagarae* H. L. Smith (Fig. 374), was found in Lake Erie and is known only from the original description by H. L. Smith (1878).

SUBORDER 2. RHIZOSOLENINEAE (SOLENOIDINEAE)

The feature distinguishing this suborder from others is the presence of a large number of intercalary bands between the girdles. As a result, the frustules are always, often many times, longer than broad. All members of the group are free-floating and occur singly or in temporary catenate series. The valves are circular or elliptical in outline and frequently bear symmetrically or asymmetrically disposed long horns or spines. Between the girdles are many intercalary bands, whose free ends interlock with one another in a vertical zigzag file or in a spiral series. Most members of the group have cell walls that are only slightly silicified, but some have a fairly heavy silicification and the development of an ornamental pattern of punctae on both the ends and sides of the cell wall. Chromatophores are usually small and discoid.

Auxospores have been recorded for most of the genera, but they are of much less frequent occurrence than are statospores. Some of the statospores are so different in shape from the vegetative cells that they have, in certain cases, been described as distinct genera.

Family 1. Rhizosoleniaceae

Genera belonging to this family are typical plankton diatoms and are widely distributed in the ocean. One genus, *Rhizosolenia*, has certain species that are found only in the plankton of fresh-water lakes.

1. **Rhizosolenia** Ehrenberg, 1843; emend., Brightwell, 1858. Frustules of this diatom are elongate cylinders, with many intercalary bands, and elliptical or circular in cross section. The cells are free-floating, and solitary or united in straight to spirally twisted chains. The valves are more frequently calyptrate or naviculate than conical. The valve apex

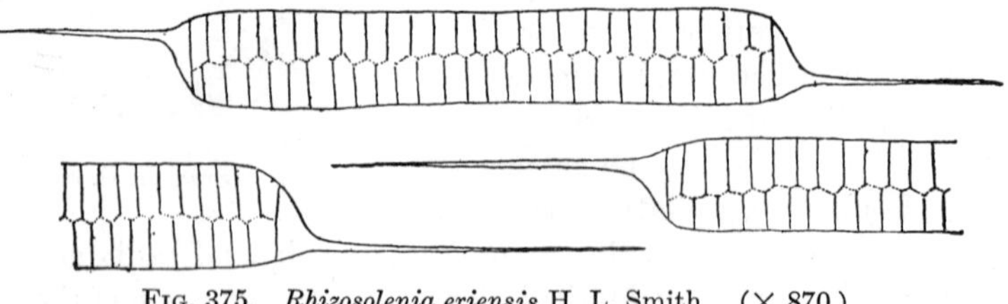

Fig. 375. *Rhizosolenia eriensis* H. L. Smith. (× 870.)

terminates in a single centric or excentric spine that is usually very long. The ends of the intercalary bands between the two girdles generally lie in a vertical imbricated series. Cell walls of *Rhizosolenia* are usually without markings, on either valve or girdle. The chromatophores within the cell are small, discoid, and some distance from one another.

There may be a formation of auxospores, statospores, or microspores.[1]

Frustules of freshwater species are so feebly silicified that they cannot be "cleaned," and the wall structure, especially that of connecting bands, is best brought out when cells have been allowed to dry on a slide without previous treatment. The two fresh-water species found in this country are *R. eriensis* H. L. Smith (Fig. 375) and *R. gracilis* H. L. Smith. For descriptions of them, see Boyer (1927).

SUBORDER 3. BIDDULPHINEAE

Cells of genera belonging to this suborder may be radially symmetrical in valve view, but more commonly they are zygomorphic. Irrespective of the shape, the valve ornamentation is always radially disposed with respect to a central point and never bilaterally symmetrical with respect to a line. Near its margin, the surface of a valve is elevated into two or more conspicuous humps or horn-like processes or has them replaced by stout spines. Frustules are usually longer than broad, because of intercalary bands between the girdles, but in most genera the development of intercalary bands is not so extensive as in the preceding suborder. According to the genus, there are numerous small discoid chromatophores, a few large irregularly shaped chromatophores, or a single large laminate chromatophore.

Auxospores, statospores, and microspores are known for the suborder, but not all genera are known to form all three.

Three of the four families into which the suborder is divided are represented in the fresh-water flora of the United States.

FAMILY 1. CHAETOCERACEAE

Frustules of genera belonging to this family are circular or elliptical in valve view. At the margin of each valve are two to several outwardly curving spine-like horns, often much longer than the valve. The horns may be solid or hollow, and smooth, punctate, or spiniferous. The two half-walls may have their girdles overlapping or separated by several intervening intercalary bands. The cells are always free-floating and usually joined in filaments by an interlocking of the horns. There may be small discoid chromatophores within a cell, several large irregularly shaped chromatophores, or a single large laminate chromatophore.

Auxospores formed by members of this family usually have their long axis vertical to the transverse axis of the parent cell. The statospores have an ornamentation very different from that of vegetative cells. Microspores of two different sizes have been recorded for one genus.

This is another family most of whose species are wholly restricted to

[1] HUSTEDT, 1929*B*.

the plankton of the ocean. The two genera with fresh-water species in this country differ as follows:

1. Frustules without intercalary bands........................ 1. **Chaetoceros**
1. Frustules with several intercalary bands......................... 2. **Attheya**

1. **Chaetoceros** Ehrenberg, 1844. Frustules of *Chaetoceros* are elliptical in valve view, and each valve has an excentrically inserted horn near each pole, the two horns being symmetrically disposed with respect to the median axis of a valve. As seen in girdle view, the two horns are widely divergent and much longer than the cell. The horns may be solid or hollow, and smooth or ornamented with punctae or spines. The girdles are overlapping and without intercalary bands. Cells of *Chaetoceros* are

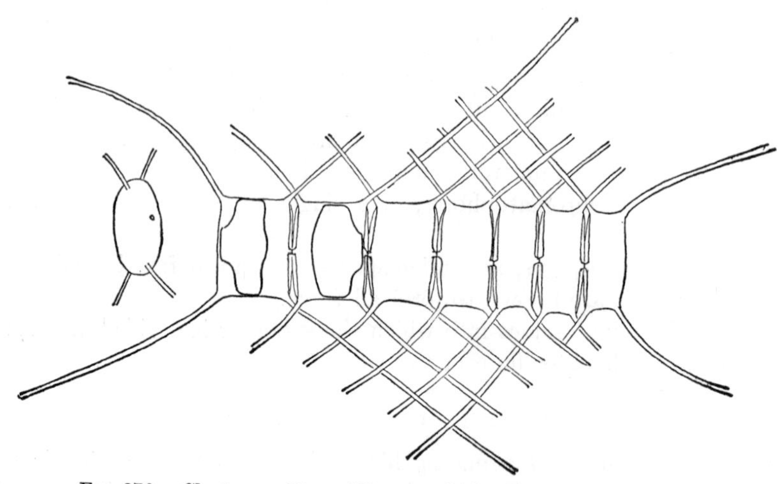

Fig. 376. *Chaetoceros Elmorei* Boyer. (*After Boyer*, 1914.) (× 600.)

usually joined in straight or spirally twisted chains because of interlocking of the horns. Cell walls of this genus are usually but little silicified and, consequently, without ornamentation. The chromatophores are quite different from species to species. There may be numerous small discoid chromatophores, both within the body of the frustule and the horns; several large chromatophores; two laminate chromatophores, one in each half-cell; or a single laminate chromatophore next to the girdle face.

When auxospores are formed, the two half-cells separate and the protoplast is extruded within a vesicle. The protoplast then develops into a single auxospore whose girdle axis is at right angles to the girdle axis of the parent cell.[1] Statospores have been found in several species. These are

[1] Ikari, 1921.

formed singly within a frustule and are enclosed by a wall whose two halves are unlike in shape and ornamentation (Fig. 368, page 458). Microspores of two different sizes have been found in one species.[1]

Chaetoceros, a genus with many species, was thought to be exclusively marine until discovery of a species, *C. Elmorei* Boyer (Fig. 376), in the plankton of Devils Lake, North Dakota.[2] Although this lake, which is located near the center of North America, has slightly brackish water, its algal flora is distinctly fresh water in nature.[3] *C. Elmorei*, therefore, must be considered a fresh-water diatom. For a description of it, see Boyer (1927).

2. **Attheya** T. West, 1860. Frustules of this diatom are markedly compressed and with many intercalary bands. As seen in girdle view, the outline is rectangular and with each angle continued in a long spine-like horn. In certain species, there is an additional horn midway between the two borne at the poles of a valve. Walls of frustules are only slightly silicified and without evident markings on valves or intercalary bands.

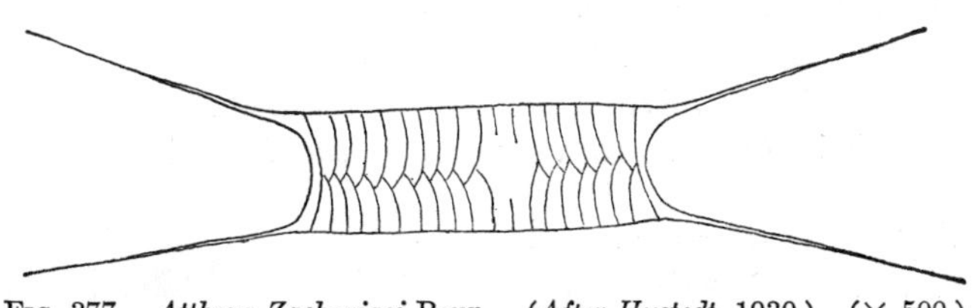

FIG. 377. *Attheya Zachariasi* Brun. (*After Hustedt*, 1930.) (× 500.)

The protoplast contains a few disk-shaped, parietally disposed, chromatophores midway between the two ends.

Statospores are known for one species.[4]

Fresh-water species of *Attheya* and *Rhizosolenia* are quite similar in general appearance, but the two have been placed[5] in separate suborders because of differences in number and arrangement of the spines. *A. Zachariasi* Brun (Fig. 377) is widely distributed in the fresh-water plankton of Europe, but in the United States has only been found in the plankton of rivers in Ohio[6] and Tennessee.[7] For a description of it, see Hustedt (1929*B*).

FAMILY 2. BIDDULPHIACEAE

Frustules of genera assigned to this family are less frequently rounded in front view than they are two-, three-, four-, or many-angled. Ornamentation of the valve is, however, centric about a point and never bilaterally symmetrical with respect to a line. Toward their margins, the valves

[1] SCHILLER, 1909. [2] BOYER, 1914; ELMORE, 1922. [3] MOORE and CARTER, 1923.
[4] HUSTEDT, 1929*B*. [5] BOYER, 1927; HUSTEDT, 1929*B*; MEISTER, 1912.
[6] LACKEY *et al.*, 1943. [7] LACKEY, 1942.

have one or more localized areas elevated into short stout horns that rarely terminate in claw-like points. In lateral view, the girdles may overlap each other, or there may be intercalary bands between them. Most of the genera belonging to the family have coarsely areolate walls and an ornamentation of valves, girdles, and intercalary bands. Some species have smooth girdles, or smooth intercalary bands, or both. Diatoms belonging to this family may be free-floating or sessile, and solitary or colonial. The formation of colonies is generally due to a development of gelatinous cushions (at the angles) that unite the cells to each other in linear or dendroid zigzag series. The cells may contain numerous small discoid chromatophores or a few large laminate ones. Auxospores, formed singly within the cells, are known for several genera. Statospores are not formed by members of this family (?). Uniflagellate microspores have been recorded.

Biddulphia is the only genus with fresh-water representatives in this country.

1. **Biddulphia** Gray, 1832. The frustules of *Biddulphia* vary so much in shape from species to species that it is difficult to describe them in general terms. However, all species have cells that are transversely compressed and have valves that bear cylindrical, conical, or globular processes at their angles. In valve view, the frustules are elliptical or

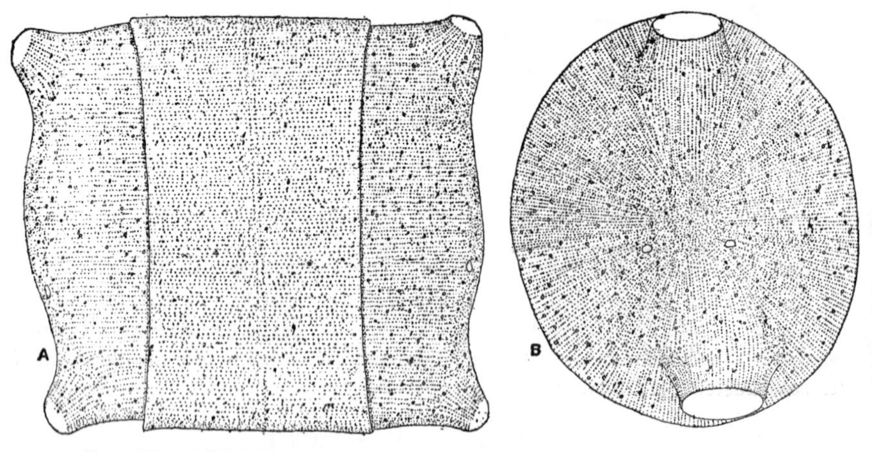

FIG. 378. *Biddulphia laevis* Ehr. *A*, girdle view. *B*, valve view. (× 650.)

angular in outline and with an ornamentation that is centric in arrangement. The walls are highly silicified, coarsely areolate but with the areolae of smaller diameter in the vicinity of the processes. There is usually an intercalary band between the girdles. The girdles and intercalary band are conspicuously ornamented, though often in different patterns. Within the cell wall, there may be internal septa, perpendicular to the valve face, but these never project into the portion of the cell encircled by the intercalary band. The chromatophores may be large or small and with a smooth or a deeply incised margin. The cells are sometimes solitary, but more often they are united to one another in linear or zigzag chains by means of gelatinous cushions that connect the processes of one frustule with those of the next frustule.

Auxospores are formed singly within a cell and are of much larger diameter than the cell producing them. Reproduction may also take place by a division of the protoplast into 32 or 64 microspores[1] (Fig. 369, page 459).

Biddulphia is a genus of littoral marine diatoms, containing 125 or more species, that grow chiefly on clayey banks or on seaweeds. *B. laevis* Ehr. (Fig. 378) has been found in fresh water in Nebraska.[2] For a description of it, see Boyer (1927).

Family 3. Anaulaceae

Features distinguishing this family from others of the order include a reduction or absence of polar processes on valves, and an internal septation of the frustules. The cells are usually compressed and either bilaterally symmetrical in valve view or asymmetrical. As far as the general outline is concerned, the frustules appear to be more of the pennate than of the centric type, but the radial or irregular disposition of the ornamentation around a central point, and the complete lack of a raphe or pseudoraphe, shows that the family belongs to the Centrales. The frustules have an intercalary band between the girdles, and there is an ornamentation of both the girdles and the intercalary bands. The internal septa lie at right angles to the long axis of a valve and extend vertically inward from the valve face to the region encircled by intercalary bands. Frustules may be solitary or united corner to corner in zigzag chains.

Auxospores, formed singly within vegetative cells, are known for a few genera. Statospores and microspores are unknown.

Terpsinoë is the only member of the family found in the fresh-water flora of the United States.

1. **Terpsinoë** Ehrenberg, 1841. Frustules of this diatom are laterally compressed and quadrangular in girdle view. The valves are elliptical or triangular in outline and with markedly undulate sides. The transverse septa within the frustule extend across the short axis of a valve. The valves are irregularly punctate or with the punctae radially disposed in the central region. The two half-walls of a frustule are separated by an intercalary band which is usually as conspicuously ornamented as the girdles.

[1] Bergon, 1907. [2] Elmore, 1922.

When seen in girdle view, the septa are perpendicular to the valve face and extend inward to the level of the intercalary bands.[1] In some cases, the internal margin of each septum is thickened and bent at an angle to the other portion. The general appearance of these septa so strongly suggests a series of musical notes that the first species to be described was called *T. musica*. Cells of *Terpsinoë* grow singly or in zigzag chains.

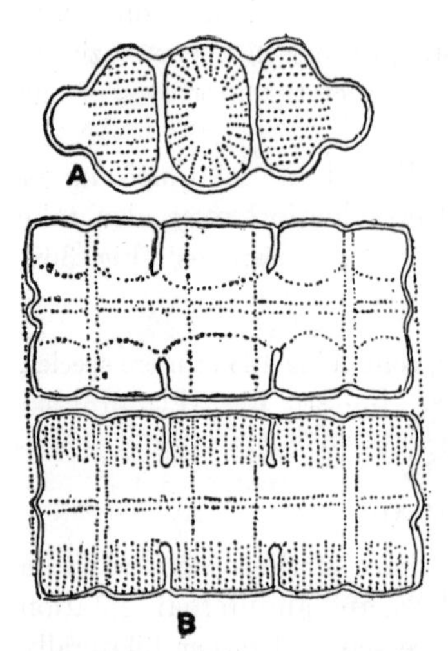

FIG. 379. *Terpsinoë americana* (Bailey) Ralfs. *A*, valve view. *B*, girdle view. (*After A. Schmidt's Atlas.*)

Auxospores are formed singly within a cell and are much larger than the parent cell.[2]

One species, *T. americana* (Bailey) Ralfs (Fig. 379), of this predominantly marine genus has been found in fresh water in this country. For a description of it, see Boyer (1927).

ORDER 2. PENNALES

Valves of Pennales have an ornamentation that is bilaterally disposed to a saggital line and never radially arranged with reference to a central point. The saggital line is medial or lateral in position and may be a narrow unornamented strip (pseudoraphe) or an unornamented strip in which there is a linear slot (raphe). A few species have valves which are nearly circular in outline; the great majority of species have elongate valves that are bilaterally symmetrical to a median axis or that are asymmetrical. The frustules are usually rectangular in girdle view, and bilaterally symmetrical or asymmetrical. Intercalary bands are usually lacking between the girdles. Internal septa are found in certain genera; these may lie parallel or vertical to the valves. The valves are without spines or processes and lack external appendages other than wings along the margin. Chromatophores may be constant in type throughout a genus, or variable. There may be numerous small discoid chromatophores or one to several laminate chromatophores with a smooth or irregular outline. Laminate chromatophores may lie parallel or perpendicular to the valve face. Perhaps the most frequently encountered condition is that of two chromatophores on opposite sides of a cell and in the long axis of the girdle. Many of the species with laminate chromatophores have pyrenoids; those with small chromatophores usually lack them. The frustules may be free-floating or sessile; and solitary, united in filaments, or united in dendroid colonies.

Statospores and microspores are unknown for the Pennales. Auxo-

[1] MÜLLER, O., 1881. [2] MÜLLER, O., 1889A.

spores are known for many genera. Where there is a production of a single auxospore by a single cell, it is due to parthenogenesis or to a union of two gametes or gamete nuclei formed by the cell. Generally, auxospores are formed by an approximation of two frustules and the direct functioning of their protoplasts as aplanogametes, or the two protoplasts may divide to form two aplanogametes which fuse in pairs to form two auxospores. All members of the order with a fusion of aplanogametes have been shown to have a reduction division immediately prior to formation of gametes.

The order is one more abundantly represented in the fresh-water flora of this country than are Centrales, and Pennales are found in all habitats where diatoms occur. The most primitive Pennales are those with a pseudoraphe and numerous chromatophores. The Surirellaceae have cells that attained the greatest complexity. Although classification of the order is based entirely upon structure of cell walls, the system is to be considered a natural and not an artificial one.

The four suborders into which the Pennales are divided are all well represented in the fresh-water flora of this country.

SUBORDER 1. FRAGILARINEAE

Genera belonging to this suborder have frustules in which both valves have a pseudoraphe or a primitive type of true raphe. The frustules are generally elongate, straight or arcuate, and with parallel sides or cuneate. They are bilaterally symmetrical in girdle view, and with or without intercalary bands. Internal septa are present in many of the genera and may completely or incompletely divide the frustule's interior. Ornamentation of valves is usually delicate and either striate or punctate. In many cases, there is no ornamentation of the girdles. Most genera of the suborder have protoplasts containing many small discoid chromatophores, but some have a few large chromatophores. The frustules may be free-floating or sessile, and solitary or united in band-like, zigzag, or stellate colonies.

A production of auxospores is rather uncommon among members of this suborder. Most genera known to form auxospores have a direct development of either one or two auxospores from a single cell. The somewhat incomplete data seem to show that these auxospores are sexual in nature.

The suborder is divided into four families.

FAMILY 1. TABELLARIACEAE

The Tabellariaceae have frustules with two to many intercalary bands and, consequently, cells that are tabular in girdle view. Parallel to the intercalary bands and between them, or between girdle and intercalary band, are incomplete or perforate longitudinal septa. As seen in valve view, the cells are narrowly linear, bilaterally symmetrical, and laterally inflated midway between the poles. The pseudoraphe is axial, and the finely punctate or striate ornamentation is symmetrically disposed with reference to it. The ornamentation may continue from valve to girdle and even to the intercalary bands, or girdles and intercalary bands may be smooth. Numerous small discoid chromatophores are characteristic of the family. Genera belonging to this family usually have the frustules conjoined in straight or zigzag chains by means of gelatinous cushions at their corners.

The three genera found in this country differ as follows:

1. Frustules with two perforate septa .... 3. **Diatomella**
1. Frustules with more than two perforate septa 2
    2. Septa straight .... 1. **Tabellaria**
    2. Septa curved .... 2. **Tetracyclus**

1. **Tabellaria** Ehrenberg, 1840. The tabular cells of this diatom are generally united in free-floating zigzag chains by gelatinous cushions at their corners. The frustules usually have numerous intercalary bands between the girdles. Internal to the wall, and between girdles and inter-

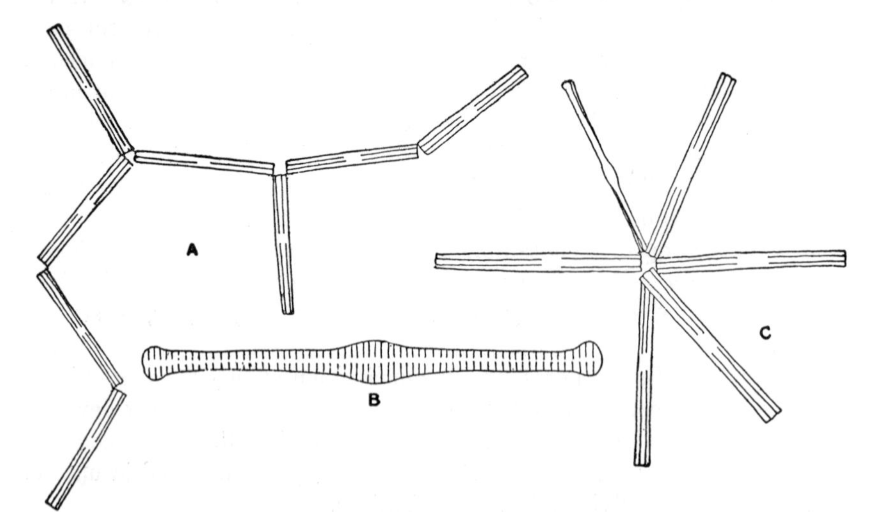

Fig. 380. *A–B*, *Tabellaria fenestrata* (Lyngb.) Kütz. *C*, *T. fenestrata* var. *asterionelloides* Grun. (*A*, *C*, × 400; *B*, × 1000.)

calary bands, are longitudinal septa, extending almost to the center of a cell. These may be opposite to, or alternate with, one another. The valves are elongate, with an evident lateral inflation in the median portion, and slightly inflated at the poles. There is a rather faint narrow pseudoraphe that is slightly broader in the middle and at the ends. Lateral to the pseudoraphe are transverse finely punctate striae. There are one or two small circular pores in the middle part of a valve.[1] Within a cell are numerous minute discoid chromatophores.

Auxospores are formed by solitary cells, but the method by which they are produced is unknown. They may be formed singly[2] or in pairs[3] within a cell.

Tabellaria is a common diatom of streams, ponds, and pools; it is also often found in the plankton of lakes. The cells are usually united in zigzag chains, but those in the plankton are sometimes stellately united with one another in much the same fashion as in *Asterionella*. The species found in this country are *T. binalis* (Ehr.) Grun. *T. fenestrata* (Lyngb.) Kütz. (Fig. 380), and *T. floccosa* (Roth) Kütz. For descriptions of them, see Boyer (1927).

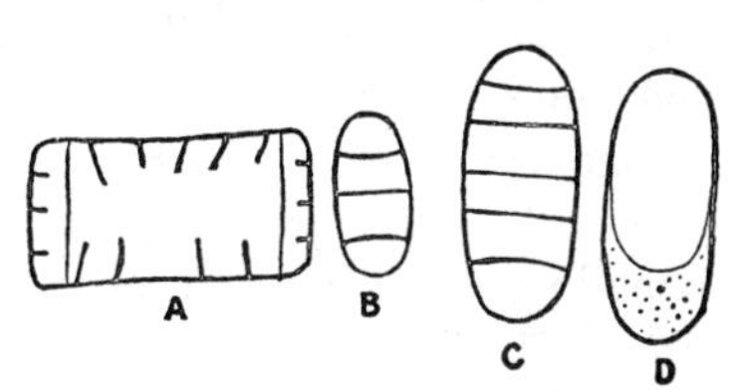

FIG. 381. *Tetracyclus rupestris* (A. Br.) Grun. *A*, girdle view. *B–C*, valve views. *D*, valve view at level of one of the septa. (× 1300.)

2. **Tetracyclus** Ralfs, 1843. In its compressed frustules joined corner to corner in zigzag chains, this genus is quite like *Tabellaria*. It differs in that the longitudinal septa extending inward from poles of the intercalary bands rarely extend more than a third of the distance to the middle of a cell and are more or less curved at their apices. Another differences is the series of conspicuous internal costae (false septa) on the valves. These are usually transverse but may be oblique or interrupted in the middle. The protoplasts have numerous small chromatophores.

Auxospores have not been found in this genus.

The two species found in fresh water in this country are *T. lacustris* Ralfs and *T. rupestris* (A.Br.) Grun. (Fig. 381). For descriptions of them, see Boyer (1927).

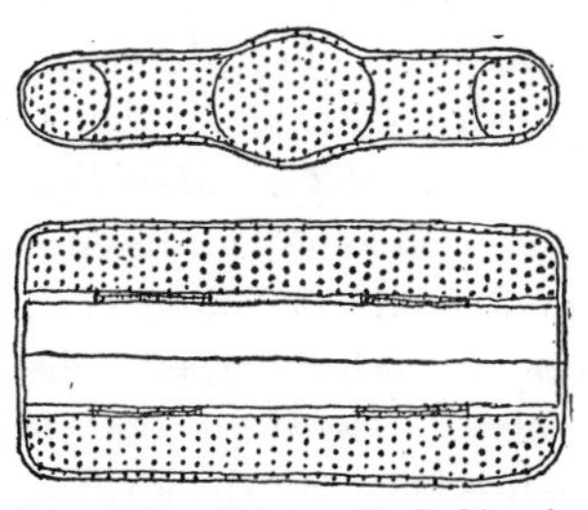

FIG. 382. *Diatomella Balfouriana* Grev. (*After W. Smith*, 1856.)

3. **Diatomella** Greville, 1855. The flattened rectangular frustules of this diatom may occur singly or adhere corner to corner in zigzag chains. There are two intercalary bands between the girdles. Frustules have two longitudinal septa that extend the whole length of a cell. As seen in valve view, each septum has a circular perforation at each pole and one of larger diameter midway between the two polar ones. The valves are ellipticolanceolate

[1] GEMEINHARDT, 1926*A*. [2] GEITLER, 1927*B*. [3] SCHÜTT, 1896.

in outline and somewhat tumid in the middle. Across their faces are delicate transversely punctate striae. There may be a smooth central area, but there is no pseudoraphe.[1]

This genus is of rare occurrence throughout the world. The only known species, *D. Balfouriana* Grev. (Fig. 382), has been found in Yellowstone National Park.[2] For a description of it, see Boyer (1927).

Family 2. Meridionaceae

Frustules of the Meridionaceae are elongate-cuneate in both valve and girdle views. The cells may be united valve to valve in fan-shaped colonies, borne on stout stipes of various length; or they may be united to form flat free-floating filaments. If the filaments are long enough, they make one or two flat spiral turns. Half-cells of the frustules are separated by one or two intercalary bands and not by several bands as is usually the case with Tabellariaceae. The frustules may have internal septa that extend part way across a cell and may be longitudinal or transverse. The ornamentation of valves consists of coarsely or finely punctate transverse striae that lie lateral to an unornamented saggital strip, the pseudoraphe. The pseudoraphe is never inflated in the median portion and only rarely so at its apices. Within a cell are numerous small discoid chromatophores. Auxospores are formed singly within solitary cells.

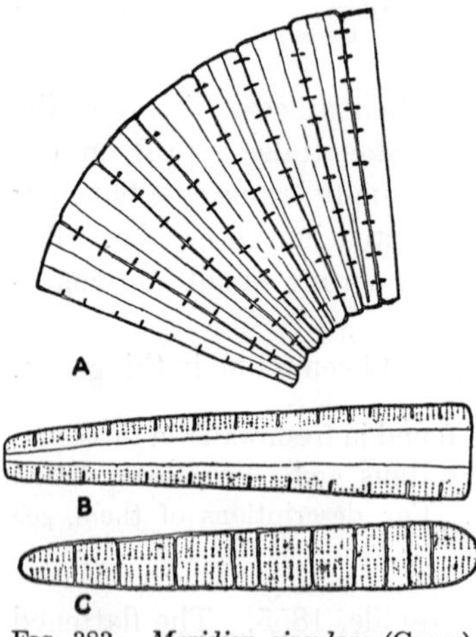

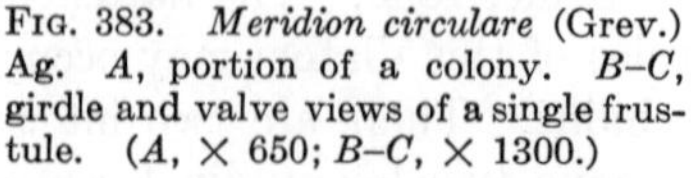

Fig. 383. *Meridion circulare* (Grev.) Ag. *A*, portion of a colony. *B–C*, girdle and valve views of a single frustule. (*A*, × 650; *B–C*, × 1300.)

Three of the four genera of the family are marine; the remaining one, *Meridion*, is found only in fresh water.

1. **Meridion** Agardh, 1824. *Meridion* differs from other diatoms in having free-floating, fan-shaped or flat-spiral, colonies made up of wedge-shaped cells joined valve to valve. There are one or two intercalary bands between the girdles. Internal to the frustule walls are transverse septa which are so rudimentary that they appear to be costae borne on the valve and girdle faces. The valves are cuneate, clavate, or obovate and with very delicate transverse striae in addition to the costae. There is a rather indistinct pseudoraphe in the sagittal axis of a valve. Within a cell are many discoid chromatophores that lie partly next to the girdle and partly next to the valve sides. Many of the chromatophores contain a single pyrenoid.

[1] Boyer, 1927; Schönfeldt. 1907. 1913. [2] Boyer, 1927.

Auxospores of rather irregular shape are formed singly within a cell.[1]

Meridion is a diatom which often develops in abundance in ditches and semi-permanent pools. The species known for the United States are *M. circulare* (Grev.) Ag. (Fig. 383), *M. constrictum* Ralfs, and *M. intermedium* H. L. Smith. For descriptions of them, see Boyer (1927).

Family 3. Diatomaceae

This family is similar to the foregoing in having transverse septa or costae across the face of a valve, but differs in having valves that are bilaterally symmetrical in both their longitudinal and transverse axes. As seen in girdle view, a frustule may be symmetrical in both axes, or symmetrical in the longitudinal axis and asymmetrical in the transverse axis. Some genera have cells that are frequently united in zigzag or flat filaments; other genera have cells that are always solitary. The valves have transverse striae in addition to costae, and there may be a conspicuous pseudoraphe down the middle of a valve, or only an indistinct one. Girdles of the frustule are usually separated from each other by intercalary bands. Most species have numerous small discoid chromatophores within a cell.

Auxospores, so far as known, are formed singly within a cell.

The two genera with fresh-water representatives differ as follows:

1. Girdle view rectangular .... 1. **Diatoma**
1. Girdle view cuneate .... 2. **Opephora**

1. **Diatoma** DeCandolle, 1805. The frustules of *Diatoma* are generally united to one another in linear to zigzag, free-floating or sessile, chains by gelatinous cushions at the corners of each frustule. Individual frustules are lanceolate to linear in valve view and sometimes slightly dilated at their apices. The two girdles may be separated from each other by one or more intercalary bands. Inward from a frustule wall are several transverse septa which appear as costae that run transversely across the valves and then down the girdle to the intercalary bands. Between the costae, on valve and girdle, are delicate punctate striations. In the middle of a valve is a narrow pseudoraphe without a lateral expansion between its two ends. Near the end of a valve is a small elliptical pore, which is probably con-

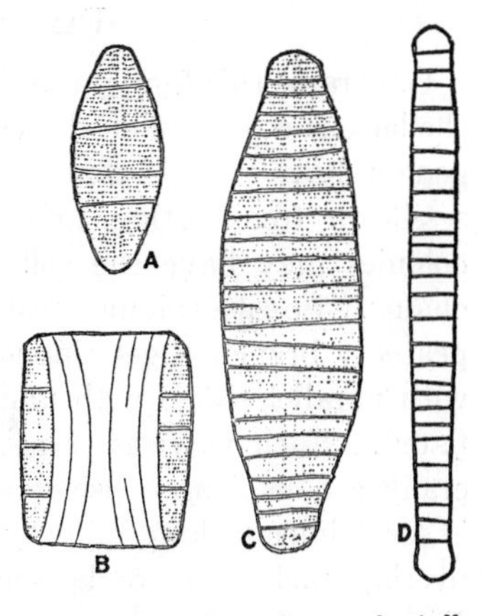

Fig. 384. *A–B*, valve and girdle view of *Diatoma hiemale* var. *mesodon* (Ehr.) Grun. *C*, valve view of *D. vulgare* Bory. *D*, valve view of *D. elongatum* (Lyngb.) Ag. (× 1300.)

[1] Geitler, 1927*B*.

cerned with the secretion of the gelatinous cushions that bind cells one to another.[1] The cells contain numerous elliptical chromatophores that lie next to both valve and girdle sides of the wall.

Auxospores are formed singly within a cell.[2]

The transversely costate frustules of *Diatoma* superficially resemble those of *Meridion* and of *Denticula*. *Diatoma* may be distinguished from *Meridion* by the transverse symmetry of the valves, and from *Denticula* by the capitate ends of the septa as seen in girdle view. The species known for the United States are *D. anceps* (Ehr.) Kirchn., *D. elongatum* (Lyngb.) Ag. (Fig. 384*D*), *D. hiemale* (Lyngb.) Herib. (Fig. 384*A*–*B*), and *D. vulgare* Bory (Fig. 384*C*). For descriptions of them, see Boyer (1927).

2. **Opephora** Petit, 1888. The frustules of this diatom are usually symmetrical in both axes when seen in valve view but are transversely asymmetrical in girdle view. *Opephora* has conspicuous transverse costae on its valves, and through the sagittal axis is a broad pseudoraphe that bisects each costa. Costate parts of the frustule wall are often finely punctate, but the remaining portions of the wall are smooth. Girdle portions of a frustule wall have costae down to the place where the girdles overlap each other or are separated by an intercalary band.

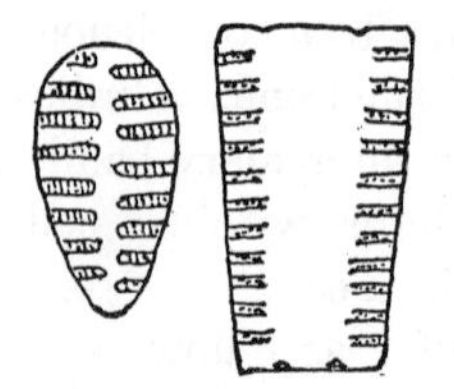

FIG. 385. *Opephora Martyi* Herib. (*After Meister*, 1912.)

This genus has a few marine and one fresh-water species. The fresh-water species, *O. Martyi* Herib. (Fig. 385), has been found in Michigan.[3] For a description of it, see Boyer (1927).

FAMILY 4. FRAGILARIACEAE

Genera of this family are distinguishable from others in the suborder by the lack of costation on the valves and by the lack of internal septa. Most genera have valves that are symmetrical in both axes. The cells are sessile or free-floating, and solitary or united in colonies. Free-floating colonies may have the cells stellately united, or in band-like to zigzag chains. Sessile colonies usually have the cells borne at the ends of repeatedly branched gelatinous stalks. The valves are elongate, and either with parallel sides, with undulate sides, or with the median portion inflated. Each valve is transversely striate or punctate, and there is generally a conspicuous pseudoraphe down the middle. The ornamentation is sometimes lacking midway between the poles of a valve. Girdles may overlap each other or be separated by one to several intercalary bands. Different species of the same genus often show marked diversity in number and shape of the chromatophores.[4]

[1] GEMEINHARDT, 1926*A*, 1928. [2] GEITLER, 1927*B*. [3] BOYER, 1927.
[4] HEINZERLING, 1908.

Auxospores are formed either singly[1] or in pairs[2] within solitary cells. The three fresh-water genera in this country differ as follows:

1. Ends of valves the same size 2
1. Ends of valves dissimilar in size .... 3. **Asterionella**
   2. Cells in filamentous, or in flat stellate, colonies .... 1. **Fragilaria**
   2. Cells solitary, or in colonies with cells radiating in all directions .... 2. **Synedra**

1. **Fragilaria** Lyngbye, 1819; emend., Rabenhorst, 1864. The frustules of *Fragilaria* are linear to fusiform in valve view, bilaterally symmetrical, and often with the poles attenuated and the sides with one or more inflations. They are rectangular in girdle view and usually with one or more intercalary bands between the girdles. The cells are united in free-floating or sessile colonies. The colonies may be band-like filaments with the cells joined valve to valve; or in zigzag filaments with the cells joined to one another by gelatinous cushions at their corners; or, in very

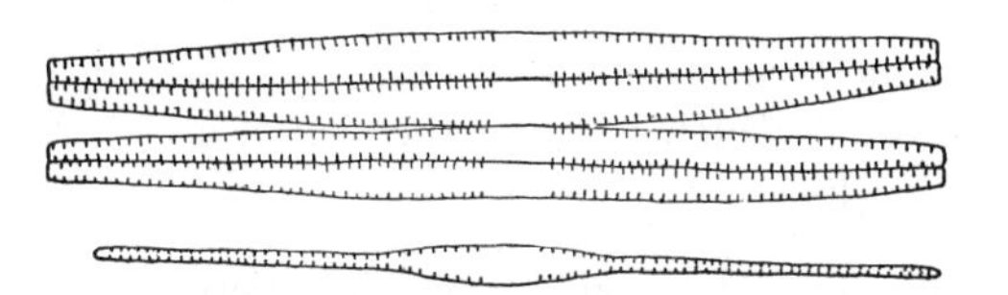

FIG. 386. *Fragilaria crotonensis* var. *praelonga* Grun. (× 1000.)

rare cases, in flat stellate colonies with the cells united to one another at their corners. The valves are ornamented with delicate transverse striae or fairly coarse transverse rows of punctae. The pseudoraphe through the sagittal axis of a valve may be delicate and indistinct, or broad and conspicuous. According to the species, there are small discoid chromatophores or one to four large laminate chromatophores with pyrenoids.[3]

Auxospores are formed singly within a cell.[4]

Fragilaria (Fig. 386) is a diatom of frequent occurrence in pools, ditches, slowly flowing streams, and in the plankton of lakes. About a dozen species are known for the United States. For names and descriptions of these species, see Boyer (1927).

2. **Synedra** Ehrenberg, 1830. The frustules of *Synedra* (Fig. 387) are usually narrow and many times longer than broad. They may be solitary and free-floating, in radiate free-floating colonies, or epiphytic and in radiating or fan-shaped colonies that are either sessile or attached to the host by a gelatinous stalk. The valves are linear to linear lanceolate and usually straight though sometimes curved. Their ends may be attenuated, capitate, or of the same diameter as the median portion. The ornamenta-

[1] KLEBAHN, 1896. [2] KARSTEN, 1897*A*. [3] HEINZERLING, 1908.
[4] KLEBAHN, 1896.

tion, which is often lacking midway between the poles, consists of transverse striae or punctae, through which runs a conspicuous pseudoraphe. Most species have one or two pores in the polar region of a valve.[1] The pores are concerned in extrusion of the gelatinous material that unites the cells in colonies. As seen in girdle view, the frustules are elongate and with truncate apices. The ornamentation of the girdle is as conspicuous as that of the valve. Fresh-water species have but two chromatophores and these lie next to the valve faces.[2] Each chromatophore usually contains three or more pyrenoids.

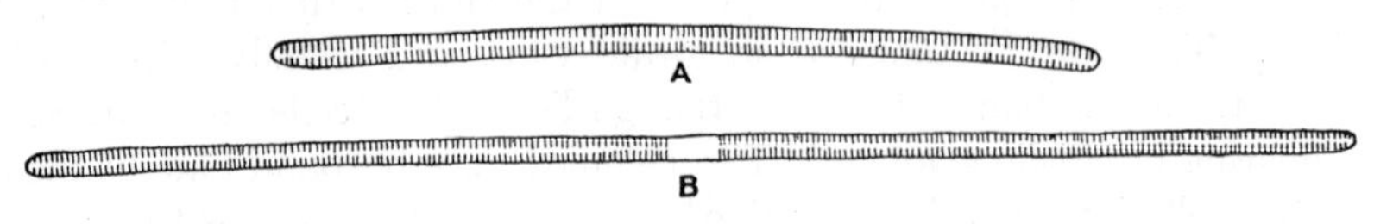

FIG. 387. *A, Synedra splendens* Kütz. *B, S. subaequalis* (Grun.) V.H. (× 400.)

A single cell may produce either one or two auxospores. Production of one auxospore is due to a fusion of two gametes formed within a frustule.[3] "Regeneration forms," which increase the size of a cell without auxospore formation, are also known.[4]

Synedra (Fig. 387) is a diatom found in a wide variety of habitats. The smaller fresh-water species are mostly sessile and grow in the form of a brownish-green coating on stones and woodwork in running water. The larger species are generally free-floating, or epiphytic upon submerged vegetation in ditches, pools, and lakes. *Synedra* is also found in the plankton of lakes. Twenty-five species have been found in the United States. For names and descriptions of them, see Boyer (1927).

3. **Asterionella** Hassall, 1850. The frustules of this diatom are linear in valve view and with inflated ends. They are joined to one another by gelatinous cushions at their edges to form flat stellate colonies in which all the cells lie in approximately the same plane. Usually the inflated ends in contact with other cells are broader than the free ends. The valves are very delicately and transversely striated and with an indistinct pseudoraphe through the sagittal axis. It has been held[5] that there are two minute pores near the pole of a valve and that the gelatinous cushions which hold the frustules in colonies are extruded through these pores. Within a cell are two chromatophores which lie axially to each other.[6]

Auxospores have not been noted in this genus.

Asterionella is a genus whose fresh-water representatives are usually found only in the plankton of lakes. Here they may be so abundant as to give the water a

[1] GEMEINHARDT, 1926*A*. [2] GEMEINHARDT, 1926. [3] GEITLER, 1939.
[4] GEMEINHARDT, 1926*A*. [5] GEMEINHARDT, 1928. [6] SCHÖNFELDT, 1907.

fishy taste.[1] The species found in the United States are *A. formosa* Hass. (Fig. 388), *A. gracillima* Heib., *A. inflata* Heib., and *A. Ralfsii* W. Smith. For a de scription of *A. gracillima*, see Meister (1912); for the others, see Boyer (1927).

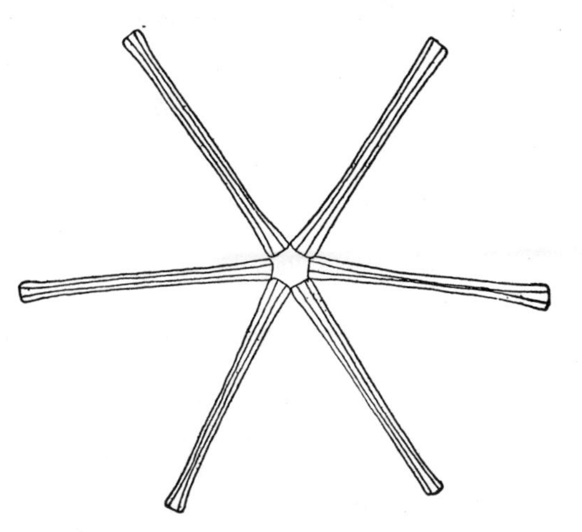

FIG. 388. *Asterionella formosa* Hass. (× 530.)

FAMILY 5. EUNOTIACEAE

Members of this family have arcuate valves with an asymmetrically disposed pseudoraphe or a primitive true raphe along the concave side. The valves are transversely striate or punctate and sometimes with the striation uninterrupted across the valve face. Costae and internal septa are lacking. The girdle view is rectangular or cuneate. The girdles are as strongly ornamented as the valves, and they may be separated from each other by smooth or more delicately sculptured intercalary bands. Frustules may be free-floating or epiphytic; and solitary, or united valve to valve in filaments or in fasciculate clusters. The cells usually contain two laminate chromatophores.

Auxospore formation is of extremely rare occurrence, but when it does take place it is by the conjugation of two cells to form a single auxospore.

The genera found in this country differ as follows:

1. Valves alike at both ends 2
1. Valves inflated at one end .... 4. **Actinella**
   2. Sides of valves dentate .... 2. **Amphicampa**
   2. Sides of valves not dentate 3
3. Concave side of valve tumid in center .... 3. **Ceratoneis**
3. Concave side of valve not tumid in center .... 1. **Eunotia**

[1] WHIPPLE, 1927.

1. **Eunotia** Ehrenberg, 1837. Frustules of *Eunotia* are free-floating or epiphytic, and solitary or united valve to valve in filamentous colonies. They are more or less arcuate in valve view and with the two poles of the same size (Fig. 389). The concave margin is usually a smooth curve.

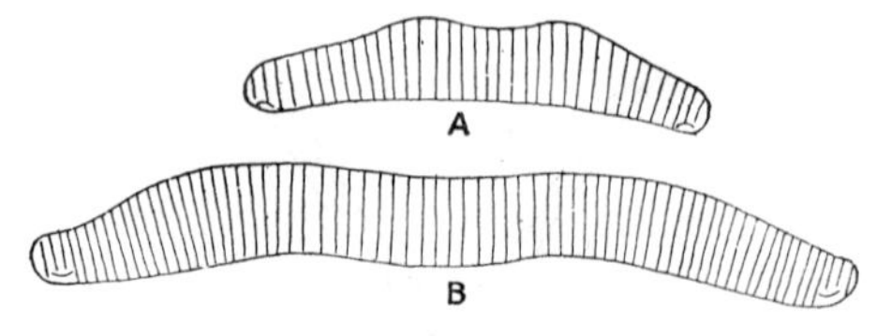

Fig. 389. *A, Eunotia pectinalis* (Kütz.) Rab. *B, E. pectinalis* var. *undulata* Ralfs. (× 1000.)

The convex margins vary all the way from smooth curves, except for the polar inflations, to a curved outline that is strongly undulate. Near each pole is a fairly conspicuous polar nodule, and diagonally from the polar nodule to the concave margin is a short raphe whose length is but a small fraction of the distance from the pole to the middle of a cell. Central nodules are lacking. Between the polar nodule and the end of a valve is a small pore through which gelatinous material is secreted.[1] There is no costation or septation of the valves. The frustules are rectangular in girdle view and with the girdles as strongly ornamented as the valves. There are usually intercalary bands between the girdles, and these may be smooth or more delicately ornamented than the girdles. Within the protoplast are two laminate chromatophores (one next to each valve) that sometimes have their margins extending down the girdle side. There are no pyrenoids.

Auxospores are formed by the approximation of two cells and conjugation of their protoplasts to form a single auxospore.[2]

Eunotia (Fig. 389) is more abundant in soft-water than in hard-water regions and is of frequent occurrence in pools and ditches but not usually in quantity. About 30 species are known from fresh waters of this country. For names and descriptions of them, see Boyer (1927).

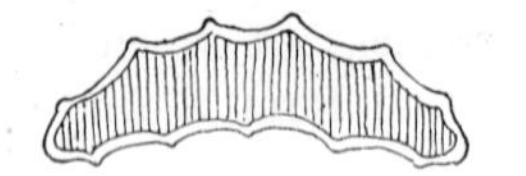

Fig. 390. *Amphicampa eruca* Ehr. (*After Ehrenberg*, 1870.)

2. **Amphicampa** Ehrenberg, 1870. This genus has arcuate cells of much the same shape as *Eunotia*. The valves differ from those of *Eunotia* in having both the concave and convex margins dentate-undulate, and in the acute instead of rounded apices of the lateral undulations. The valves have transverse striae which extend without interruption across the face of a valve. The girdle view is rectangular.

[1] Hustedt, 1926. [2] Klebahn, 1896.

A. eruca Ehr. (Fig. 390) has been recorded from California.[1] For a description of it, see Boyer (1927).

3. **Ceratoneis** Ehrenberg, 1840. The frustules of this diatom are arcuate in valve view, with rostrate-capitate apices, and with a more or less prominent tumescence (pseudonodule) in the middle of the concave side. The valves are transversely striated and have a conspicuous pseudoraphe that lies somewhat toward the concave margin. The girdle view is linear, with parallel sides and truncate ends. The frustules are usually solitary but sometimes are united valve to valve in short filaments.

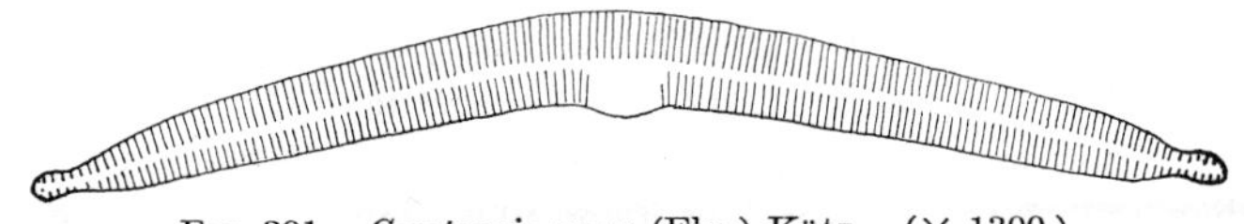

FIG. 391. *Ceratoneis arcus* (Ehr.) Kütz. (× 1300.)

Ceratoneis is a genus which, according to European diatomists, is usually found in running water, notably that of mountain streams. *C. arcus* (Ehr.) Kütz. (Fig. 391) is known for the United States. For a description of it, see Boyer (1927).

4. **Actinella** Lewis, 1863. The frustules of *Actinella* are linear in valve view and with dissimilar extremities. One extremity of a valve is broadly rounded, the other is inflated and with a retuse to apiculate apex. The sides of the valves may be smooth curves except for the polar inflation, or they may be undulate. Across the face of a valve are delicate transverse rows of punctae, and the rows are vertically interrupted by a pseudoraphe that lies toward the concave margin of a valve. The valves are sometimes ornamented with small intramarginal spines in addition to the striae. Near each extremity of a valve is a simple polar nodule, and diagonally from this and thence along the concave margin is a raphe. At the extreme apex of a frustule is a large pore.[2] As seen in girdle view, the frustules are elongate-cuneate and transversely striate next to the lateral margins only. The cells may be solitary or united valve to valve in small clusters.

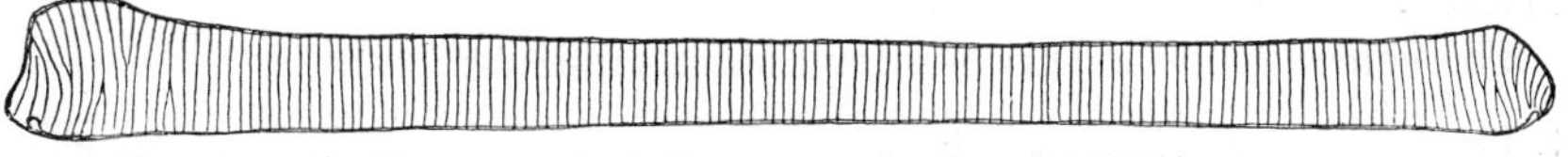

FIG. 392. *Actinella punctata* Lewis. (× 1300.)

Actinella is a rare fresh-water diatom, thus far found only in the Western Hemisphere. *A. punctata* Lewis (Fig. 392) is the only species known to occur in this country. For a description of it, see Boyer (1927).

[1] BOYER, 1927. [2] HUSTEDT, 1926.

SUBORDER 2. ACHNANTHINEAE

Genera belonging to this suborder differ from other pennate diatoms in having frustules with a pseudoraphe on one valve and a true raphe on the other. The cells are symmetrical in both axes when seen in valve view, but they are longitudinally or transversely asymmetrical when viewed from the girdle side. Members of the suborder are rarely free-floating; usually they are sessile and borne either on gelatinous stalks or with one valve apposed directly to the substratum. The ornamentation consists of transverse striae or rows of punctae, symmetrically disposed with reference to the median axis. One genus has incomplete longitudinal internal septa; other genera lack them. Some species have numerous small discoid chromatophores, but the great majority have a single laminate one, often with irregularly lobed margins, that lies next to the valve or girdle side.

There may be an approximation of two cells and a fusion of their protoplasts to form a single auxospore, or the two cells may have their protoplasts dividing into two gametes and fusing in pairs to form two auxospores. Nuclear divisions prior to auxospore formation are meiotic. Two approximated cells may also form auxospores without conjugation; in one case, this has been shown to be due to parthenogenesis.

Family 1. Achnanthaceae

All genera of the suborder are placed in a single family, the Achnanthaceae.

The genera found in this country differ as follows:

1. Frustules cuneate in girdle view.............................. 2. **Rhoiocosphenia**
1. Frustules not cuneate in girdle view.. 2
   2. Longitudinal axis of cells bent or curved........................ 1. **Achnanthes**
   2. Transverse axis of cells bent or curved.......................... 3. **Cocconeis**

1. **Achnanthes** Bory, 1822. The cells of this diatom are linear-elliptical to navicular in valve view and longitudinally bent or curved in girdle view. Sometimes the cells are free-floating, but in the great majority of cases they are attached to some firm object by means of a gelatinous stalk. Sessile forms may have the cells united valve to valve in small libriform packets or, in very rare cases, in long filaments. The two valves of a cell are dissimilar. The epitheca is always with a pseudoraphe and convex; the hypotheca is usually concave and with a raphe, rather inconspicuous polar nodules, a distinct central nodule, and sometimes a stauros. Valves may be alike in their transverse striation or punctation, or one may have transverse striae and the other somewhat radiate striae. The girdles are longitudinally bowed or bent, usually strongly ornamented, and sometimes

separated from each other by intercalary bands. There may be but a single chromatophore that lies next to the epitheca, two chromatophores next to the epitheca, or numerous small discoid chromatophores.

Auxospores are formed by the approximation of two frustules, the division of each protoplast into two gametes, and the fusion of the gametes in pairs to form two auxospores. Elongation of auxospores may be parallel with, or at right angles to, the long axes of the parent frustules.[1]

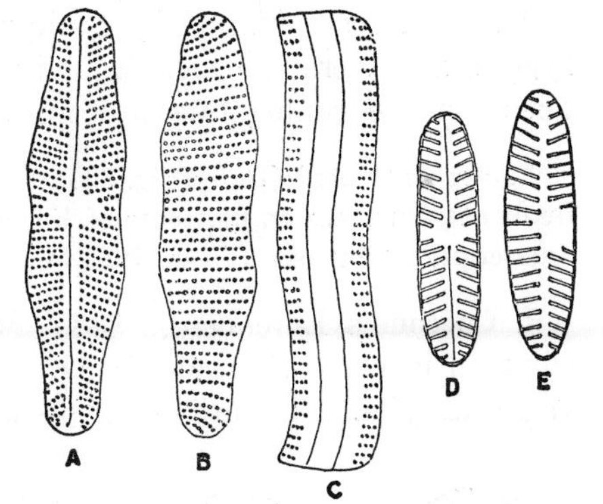

FIG. 393. *A–C, Achnanthes coarctata* (Bréb.) Grun. *A*, valve view of hypotheca. *B*, valve view of epitheca. *C*, girdle view. *D–E, A. lanceolata* (Bréb.) Grun. *D*, valve view of hypotheca. *E*, valve view of epitheca. (× 1300.)

Species of *Achnanthes* (Fig. 393) are found in both fresh and salt waters. Fresh-water species usually grow epiphytically upon filamentous Chlorophyceae or upon submerged phanerogams. About a dozen species occur in the United States. For names and descriptions of these species, see Boyer (1927*A*).

2. **Rhoicosphenia** Grunow, 1860. The wedge-shaped frustules of *Rhoicosphenia* are sessile and attached at their narrower ends to a more or less branching system of gelatinous stalks affixed to submerged phanerogams or to coarse filamentous green algae. Sometimes the stalk system is reduced to a gelatinous cushion. As seen in valve view, the frustules are oblanceolate in outline. The epitheca has a median pseudoraphe, and lateral to it are transverse rows of rather delicate striae. The hypotheca has a median raphe, with central and polar nodules, and parallel striae that are sometimes somewhat radially disposed with reference to the central nodule. When seen in girdle view, the frustules are distinctly cuneate and distinctly curved in the longitudinal axis. There are unornamented intercalary bands between the striately ornamented girdles. Within the frustules are two longitudinal septa, parallel to the valve face. Each septum has a single large oval perforation, as wide as the valve and somewhat shorter. There is a single laminate chromatophore next to one side

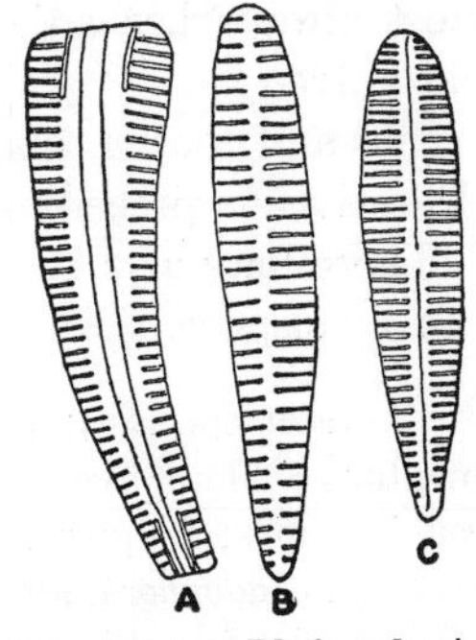

FIG. 394. *Rhoicosphenia curvata* (Kütz.) Grun. *A*, girdle view. *B*, valve view of epitheca. *C*, valve view of hypotheca. (× 1300.)

[1] KARSTEN, 1897.

of the girdle, which is often so large that it extends across the valve to the opposite side of the girdle.

Auxospores are formed by conjugation of sister cells to form a single zygote, and there is a meiotic division followed by degeneration of three nuclei in the conjugating protoplasts.[1]

R. curvata (Kütz.) Grun. (Fig. 394), the only species found in this country, is of frequent occurrence on filaments of *Vaucheria*, *Cladophora*, and *Oedogonium*. For a description of it, see Boyer (1927*A*).

3. **Cocconeis** Ehrenberg, 1838. Frustules of this diatom are broadly elliptical in valve view and transversely curved in girdle view. The two valves are similar in outline but dissimilar in structure. The epitheca has an axial pseudoraphe and lateral to it transverse striae or punctae. The hypotheca has a median, straight or sigmoid, raphe with central and polar nodule. Ornamentation of a hypotheca may be like or unlike that of the epitheca. Internal to the valves are incomplete transverse septa. There is usually a single laminate chromatophore, with one or two pyrenoids, and with a large lateral foramen in which the nucleus lies. The chromatophore almost always lies adjacent to the epithecal valve.

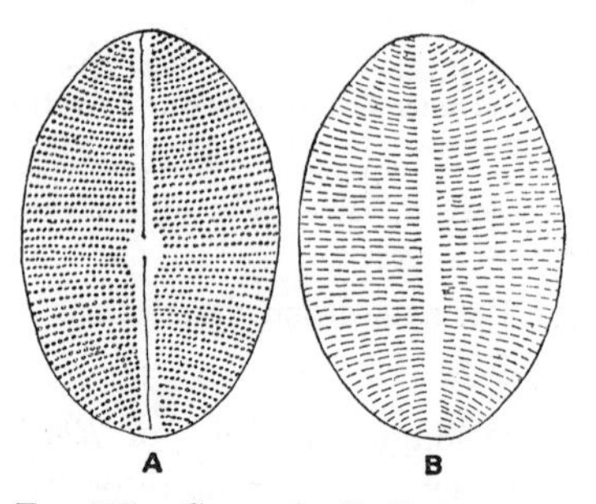

FIG. 395. *Cocconeis Pediculus* Ehr. *A*, hypotheca. *B*, epitheca. (× 1300.)

Auxospores are formed by an approximation of two cells and conjugation of their protoplasts to form a single auxospore. Nuclear division preceding gametic union halves the number of chromosomes.[2] There may also be a parthenogenetic development of each gamete into an auxospore.

A majority of the species of *Cocconeis* are marine. Fresh-water species grow epiphytic upon submerged phanerogams and upon filamentous Chlorophyceae, and with the hypotheca flattened against the host. *Cocconeis* is most frequently encountered upon old, slowly growing, filaments of *Vaucheria* and *Cladophora*, and often in such profusion as to cover the host completely. The species found in the Unites States are *C. flexulla* (Kütz.) Cleve, *C. minuta* Cleve, *C. Pediculus* Ehr. (Fig. 395) and *C. placentula* Ehr. For descriptions of them, see Boyer (1927*A*).

SUBORDER 3. NAVICULINEAE

A majority of the fresh-water Pennales belong to the suborder Naviculineae, one in which both valves have a true raphe and in which the raphe is axial on a valve and not in a marginal keel. Members of the sub-

[1] CHOLNOKY, 1927; GEITLER, 1928*B*. [2] GEITLER, 1927.

order are mostly solitary and free-floating, but some are sessile and borne at the ends of branched gelatinous stalks or they lie side by side within profusely branched tubular gelatinous envelopes. Most of the genera have frustules that are longitudinally and transversely symmetrical in valve view and with both poles of the same shape. A few genera have valves that are asymmetrical in the longitudinal or in the transverse axis. Girdle views of frustules are usually symmetrical in both axes, but they may be asymmetrical in one axis. Most genera have girdles that overlap each other and are without intercalary bands. There are usually two laminate chromatophores that lie opposite to each other next to the long sides of the girdle.

Auxospores result from the approximation of two frustules and the union of their protoplasts to form a single auxospore, or the protoplasts of each frustule may divide into two gametes which unite in pairs to form two auxospores. These two gametes may be equal or unequal in size; if of equal size, one may be amoeboid and the other immobile. Several genera are known to have a reduction division prior to gametic union. There are also cases where the two approximated frustules form auxospores without conjugation, and some of these are known to be due to parthenogenesis.

The suborder is divided into three families.

Family 1. Naviculaceae

This, the largest family of the Bacillariophyceae, has frustules whose valves are symmetrical in both axes. Both valves are alike, and they may be elliptical, lanceolate, or boat-shaped in outline. The sagittal axis is usually linear, though it may be sigmoid, and the valve face usually lies in one plane, though it may be convex or twisted. Each valve has a raphe with distinct central and polar nodules that vary considerably in shape from genus to genus. A majority of the genera have transversely punctate or striate valves, though a few have punctae that lie in decussating series. In one or two cases (as *Amphipleura*), the ornamentation is so delicate that it can be resolved only under the most favorable conditions. The girdle view is usually symmetrical in both axes. Most genera have two laminate chromatophores, symmetrically or asymmetrically disposed with respect to the girdle.

Auxospores are formed by the different methods noted for the suborder.

The genera found in this country differ as follows:

1. Frustules internally septate.................................. 15. **Mastogloia**
1. Frustules not internally septate.. 2
    2. Valves without a keel.. 3
    2. Valves with a sigmoid sagittal keel.......................... 14. **Amphiprora**
3. Ornamentation of valve interrupted by longitudinal blank spaces.............. 4
3. Ornamentation of valve not interrupted by longitudinal blank spaces......... 7

4. Longitudinal blank spaces zigzag .... 5. **Anomoeoneis**
4. Longitudinal blank spaces not zigzag 5
5. Longitudinal blank spaces near axial field .... 6. **Diploneis**
5. Longitudinal blank spaces near sides of valve 6
6. Transverse ornamentation of valve punctate .... 4. **Neidium**
6. Transverse ornamentation of valve not evidently punctate ...... 3. **Caloneis**
7. Axial field and raphe sigmoid 8
7. Axial field and raphe straight 10
8. Punctae on valve in transverse rows .... 13. **Scoliopleura**
8. Punctae on valve in intersecting rows 9
9. With transverse and oblique rows .... 12. **Pleurosigma**
9. With transverse and longitudinal rows .... 11. **Gyrosigma**
10. Raphe between siliceous longitudinal ribs 11
10. Raphe not between siliceous longitudinal ribs 13
11. Length of central nodule at least half that of valve .... 8. **Amphipleura**
11. Length of central nodule less than half that of valve 12
12. Transverse ornamentation of valve costate .... 10. **Brebissonia**
12. Transverse ornamentation of valve, when evident, finely punctate .... 9. **Frustulia**
13. Stauros extending to sides of valve .... 7. **Stauroneis**
13. Stauros, if present, not extending to sides of valve 14
14. Valve with smooth transverse costae .... 2. **Pinnularia**
14. Valve with transverse striae or transverse rows of punctae.... 1. **Navicula**

1. **Navicula** Bory, 1822. Frustules of *Navicula* (Fig. 396) are symmetrical in all three planes. The valves are elongate, usually attenuated toward the poles and with capitate, rounded, or rostrate apices. The raphe is distinct, axial, straight, and with well-defined but small central and polar

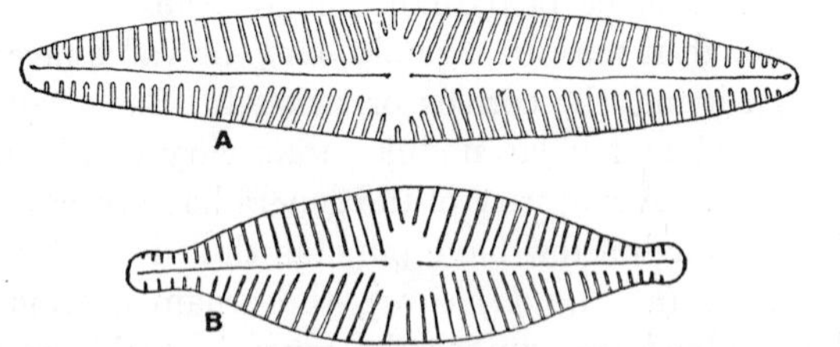

FIG. 396. *A, Navicula gracilis* Ehr. *B, N. rhyncocephala* Kütz. (× 1300.)

nodules. The axial field in which the raphe lies is fairly narrow and either without lateral expansions or expanded in the region of the polar nodules or in the region of the central nodule. The expansion in the region of the central nodule is never broad enough to be considered a stauros. Ornamentation lateral to the axial field consists of parallel striae or rows of punctae that are either strictly transverse or somewhat radiate in the region lateral to the central nodule. The frustules are rectangular in girdle view, with smooth girdles and without intercalary bands. Most species have two laminate chromatophores that lie on opposite girdle sides and sometimes overlap a portion of the valve face. More rarely there are four or eight chromatophores. Frustules of *Navicula* are generally solitary and free-floating. In certain marine species, they lie with their long axes parallel within repeatedly branched tubes that become progressively smaller with each branching.

Auxospores are formed by approximation of two cells and division of each protoplast into two gametes which fuse in pairs to form two auxospores. It is definitely known for one species[1] that meiosis precedes gamete formation.

This is a large genus for which more than 40 fresh-water species occur in this country. For descriptions of most of the species found in this country, see Boyer (1927*A*), and Cleve (1895).

2. **Pinnularia** Ehrenberg, 1840. The symmetrical frustules of *Pinnularia* (Fig. 397) have valves that are usually with rounded poles and straight parallel sides. Some species have valves that are inflated in the middle of their sides or are symmetrically undulate. The axial field in which the raphe lies is broad, sometimes over a third the diameter of the valve, and often expanded next to the central and polar nodules. The raphe is a complicated structure (see page 449) with a straight or somewhat sigmoid outer fissure. Lateral to the axial field are smooth parallel transverse costae which may be somewhat radiate near the central nodule and convergent near the polar nodules. The costae are tubular channels in the valve wall, and each of them is connected with the cell's lumen by an elongate-elliptical opening (Fig. 397). The two longitudinal lines evident

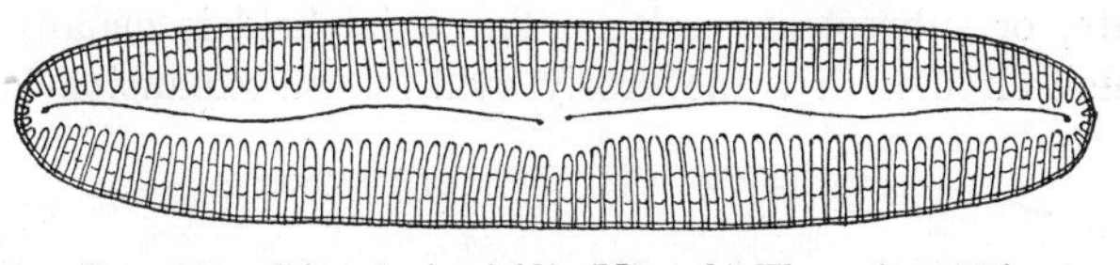

FIG. 397. *Pinnularia viridis* (Nitzsch) Ehr. (× 975.)

on either side of the axial field are successive poles of the openings. As seen in girdle view, the frustules are rectangular, with smooth girdles, and without intercalary bands. Within a cell are two laminate chromatophores, generally with pyrenoids, that lie on opposite sides of the girdle. Frustules of *Pinnularia* are usually solitary and free-floating. In very rare cases,[2] the cells lie girdle to girdle in short band-like filaments.

A majority of the species are fresh water in habit, and sometimes *Pinnularia* is present in abundance in semipermanent or permanent pools of soft-water localities Boyer (1927*A*) lists and describes 42 species for the United States.

[1] SUBRAHMANYAN, 1948. [2] HUSTEDT, 1926*A*; PALMER, T. C., 1910.

3. **Caloneis** Cleve, 1894. The frustules of *Caloneis* (Fig. 398) are quite variable in valve view and may be linear, linear-lanceolate, elliptical, or panduriform in outline. They often are laterally inflated midway between the poles. The raphe down the middle of the axial field is always straight and with rounded central and polar nodules. There are always transverse striae lateral to the axial field, and these lie parallel to one another throughout the length of a valve or are slightly radiate in the median portion of a

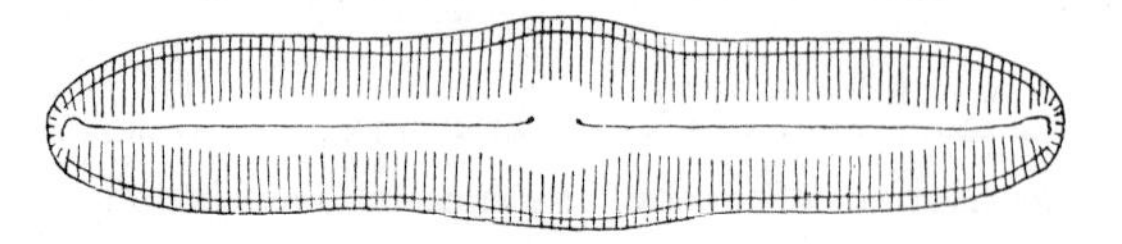

FIG. 398. *Caloneis silicula* (Ehr.) Cleve. (× 1300.)

valve. Within the lateral margins of a valve are one or more longitudinal lines or smooth areas which cross the striae at right angles. As seen in girdle view, the frustules are rectangular. Empty frustules are often yellowish brown. There are usually two chromatophores within a cell, and these lie asymmetrically disposed against opposite sides of the girdle. Each chromatophore may contain two pyrenoids.[1]

This genus has numerous species, both in fresh and salt waters. Eleven fresh-water species are known for the United States. For names and descriptions of them see Boyer (1927*A*) and Cleve (1894).

4. **Neidium** Pfitzer, 1871. Valves of *Neidium* (Fig. 399) are linear, linear-lanceolate, elliptical, or gibbous in outline and with acute, obtuse, subcapitate, or subrostrate poles. The axial field is usually narrow and with a small circular or transversely oval lateral expansion in the middle.

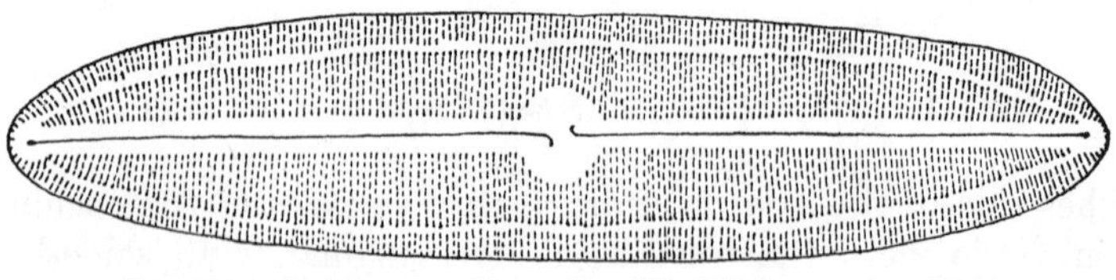

FIG. 399. *Neidium amphigomphus* (Ehr.) Pfitzer. (× 1300.)

The raphe is straight and with the ends next to the central nodule facing in opposite directions. There is often a bifurcation of the raphe in the portion next to each polar nodule. The valves are ornamented with transverse rows of punctae, and next to the valve margin are one or more blank spaces that bisect the transverse ornamentation at right angles. The frustules are rectangular in girdle view and without intercalary bands.

[1] HEINZERLING, 1908.

Empty frustules are often yellowish or brownish. A cell contains two longitudinally incised chromatophores, each with a single pyrenoid. The incision is sometimes so deep that there appear to be four chromatophores.[1]

This exclusively fresh-water genus resembles *Caloneis* in the longitudinally interrupted transverse ornamentation, but differs in having punctae instead of striae and in the bending of the raphe next to the central nodule. Boyer (1927*A*) lists and describes 10 species for the United States.

5. **Anomoeoneis** Pfitzer, 1871. Frustules of this diatom are linear, linear-lanceolate, rhombic, or elliptical in valve view; with smooth or gibbous sides and with acute, obtuse, or subcapitate apices. The valves have a narrow axial field that often has a small circular expansion midway between the two ends. The raphe is straight and with straight ends next to the central nodule. On the valve face are delicate transverse striae, each of which is interrupted by several hyaline spaces. Since the hyaline spaces in successive striae do not coincide, the result is a longitudinal pattern of zigzag clear spaces on the valve face. As seen in girdle view, the frustules

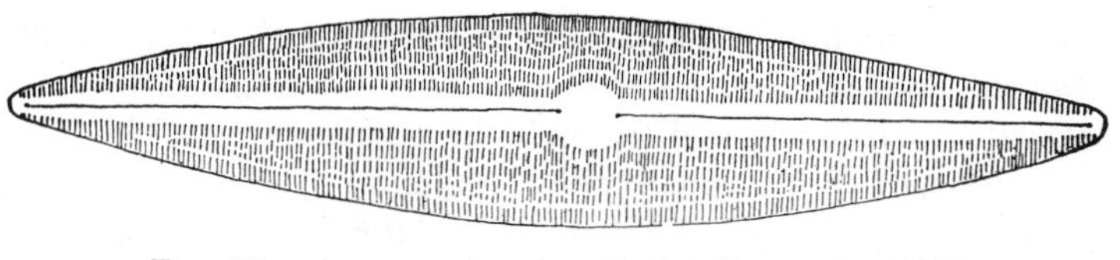

FIG. 400. *Anomoeoneis serians* (Bréb.) Cleve. (× 1300.)

are rectangular and without intercalary bands. There is a single laminate chromatophore, with deep longitudinal incisions, that lies next to one side of the girdle and partly underlies each valve.

Auxospores are formed by two sister cells that lie within a common gelatinous envelope. There is a meiotic division of the nuclei followed by a formation of two unequal-sized gametes in each cell. The gametes unite in pairs to form two auxospores.[2]

Anomoeoneis is another genus of naviculoid diatoms found only in fresh waters. It may be distinguished from other members of the family by the zigzag series of transverse dashes on the valves. The following species have been found in this country: *A. exilis* (Kütz.) Cleve, *A. follis* (Ehr.) Cleve, *A. polygramma* (Ehr.) Cleve, *A. sculpta* (Ehr.) Cleve, *A. serians* (Ehr.) Cleve (Fig. 400), *A. sphaerophora* (Kütz.) Cleve, and *A. Zellensis* (Grun.) Cleve. For descriptions of them, see Boyer (1927*A*).

6. **Diploneis** Ehrenberg, 1844. Frustules of *Diploneis* are more frequently elliptical in valve view than they are linear or with a constriction

[1] HEINZERLING, 1908. [2] CHOLNOKY, 1928*A*.

in the middle. The central nodule is more or less quadrate and with its lateral margins anteriorly and posteriorly prolonged into horns which lie on either side of the raphe. On each side of the horns and central nodule is a broad or narrow furrow, and lateral to the furrows are transverse costae or rows of punctae, that may extend to or across the furrows. The frustules are rectangular in girdle view. Within a cell are two chromatophores, with or without deep longitudinal incisions, that lie next to either the valve or girdle faces.[1]

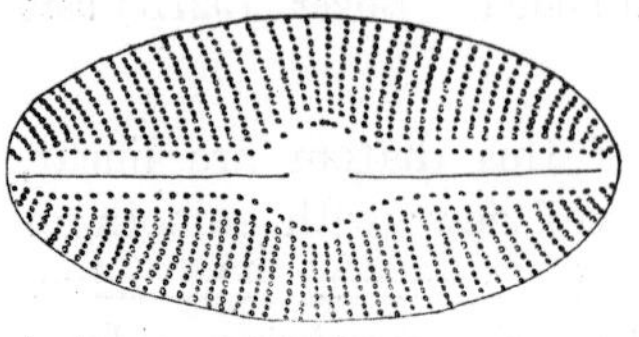

FIG. 401. *Diploneis elliptica* (Kütz.) Cleve. (× 1300.)

This genus differs from others in the horn-like processes from the central nodule. This feature is quite prominent in most marine species; but is less evident in fresh-water ones, all of which have a more delicate ornamentation of the valves. The fresh-water species of this country are *D. elliptica* (Kütz.) Cleve (Fig. 401), *D. ocula* (Bréb.) Cleve, *D. ovalis* Hilse, and *D. puella* (Schum.) Cleve. For a description of *D. ovalis*, see Cleve (1894); for the others, see Boyer (1927*A*).

7. **Stauroneis** Ehrenberg, 1843. The frustules of *Stauroneis* (Fig. 402) have much the same shape as those of *Navicula*. The axial field of a valve is narrow but conspicuous, and through the middle of the field there is a straight raphe with fairly small polar nodules. The central nodule is thickened and transversely extended to the lateral margins of the valve. There is no ornamentation in this thickened nodule, the stauros. The ornamentation of a valve consists of slightly radiate parallel striae or rows of punctae. The stauros and axial field divide the ornamentation into four parts. Within a frustule are two chromatophores that lie on opposite sides of the girdle and extend to the valve sides. Each chromatophore contains two to four pyrenoids.[2]

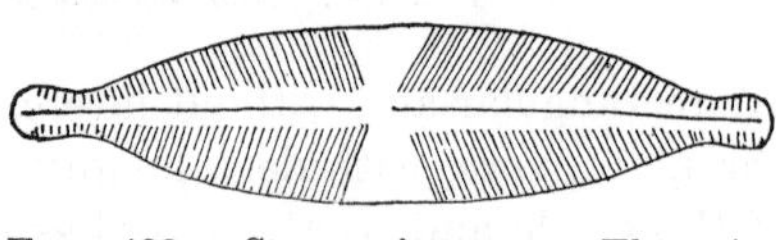

FIG. 402. *Stauroneis anceps* Ehr. (× 1300.)

Auxospores are formed in pairs between two approximated cells.[3]

Stauroneis is easily recognized because of the stauros. There are numerous species in both salt and fresh waters. Of the latter, some 15 species occur in the United States. For names and descriptions of them, see Boyer (1927*A*).

8. **Amphipleura** Kützing, 1844. The frustules of *Amphipleura* are linear-lanceolate in valve view and with the lateral margins attenuated to the rounded apices. The central nodule is greatly elongate, extending for half the length of the valve or more and terminating at each end in two parallel

[1] HEINZERLING, 1908. [2] HEINZERLING, 1908. [3] KLEBAHN, 1896.

prolongations united at their extremities with the polar nodules. Within the prongs lies a short straight raphe. The valve face appears to be smooth, but photography under special conditions of illumination shows[1] that it has transverse rows of extremely minute punctae. There are two chromatophores next to the girdle side of a cell, and these may be with or without pyrenoids.[2] *Amphipleura* is a free-floating solitary diatom.

Auxospores are formed in pairs between two cells.[3]

FIG. 403. *Amphipleura pellucida* Kütz. (× 975.)

Amphipleura has long been a favorite test object for demonstrating the resolving power of microscopes. The two fresh-water species found in this country are *A. Lindheimeri* Grun. and *A. pellucida* Kütz. (Fig. 403). For descriptions of them, see Boyer (1927*A*).

9. **Frustulia** Agardh, 1824; emend., Grunow, 1865. As seen in valve view, the frustules are linear-elliptic to rhombo-lanceolate in outline. In the middle of the sagittal axis is a rather short, vertically elongated, central nodule, and projecting axially from it are two siliceous ribs whose apices are united with the polar nodules. The raphe lies between the parallel siliceous ribs. The whole sagittal structure may be compared to two long turnbuckles connected by a very short rod. In *Amphipleura* there are two short turnbuckles connected by a long rod. The valve face is ornamented with delicate punctae that usually lie in transverse rows but sometimes in slightly radial rows in the median portion of a valve. In girdle view, the frustules are rectangular in outline and without intercalary bands.

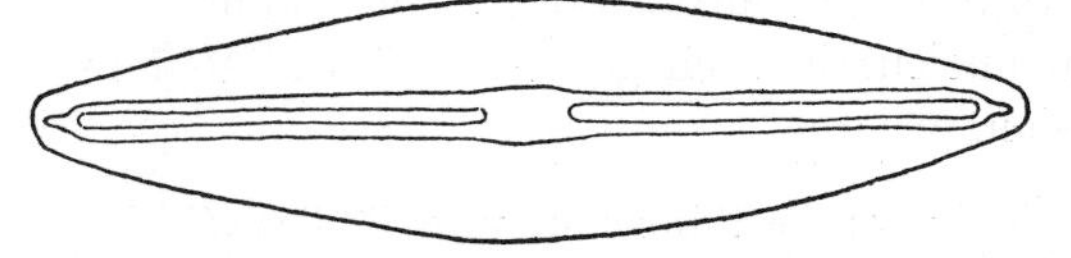

FIG. 404. *Frustulia rhomboides* (Ehr.) De Toni. (× 1300.)

There are two chromatophores, sometimes with longitudinal incisions, that lie on opposite sides of the girdle and are connected by a cytoplasmic bridge. Frustules are more often solitary and free-floating than sessile and enclosed within a gelatinous matrix. Sometimes the matrix is tubular and with the cells lying parallel to one another.

Auxospores are formed in pairs between two approximated frustules and are probably the result of conjugation of gametes.[4]

[1] GIFFORD, 1892; VAN HEURCK, 1890. [2] HEINZERLING, 1908.
[3] SCHÖNFELDT, 1907. [4] KLEBAHN, 1896.

Species of *Frustulia* are found in both fresh and brackish waters. The three fresh-water species found in this country are *F. rhomboides* (Ehr.) De Toni (Fig. 404), *F. viridula* (Bréb.) De Toni, and *F. vulgaris* (Thw.) De Toni. For descriptions of them, see Boyer (1927*A*).

10. **Brebissonia** Grunow, 1860. Valves of *Brebissonia* are rhomboidal-lanceolate and with acutely rounded or subrostrate apices. The raphe is straight, and both halves lie between inconspicuous, parallel, longitudinal ridges. On either side of the axial field are somewhat diagonal punctate-costate striae with the diagonal arrangment most pronounced at the poles of a valve. As seen in girdle view, the frustules have rectangular ends and somewhat convex or concave sides. There are usually several smooth intercalary bands between the highly ornamented girdles. A cell contains a single chromatophore that lies next to one side of the girdle and extends laterally across beneath both valves. The chromatophore has two deep longitudinal incisions in the plane of the girdle and contains a large irregularly shaped pyrenoid.[1] The frustules are always sessile and borne singly at the tips of long dichotomously branched gelatinous stalks.

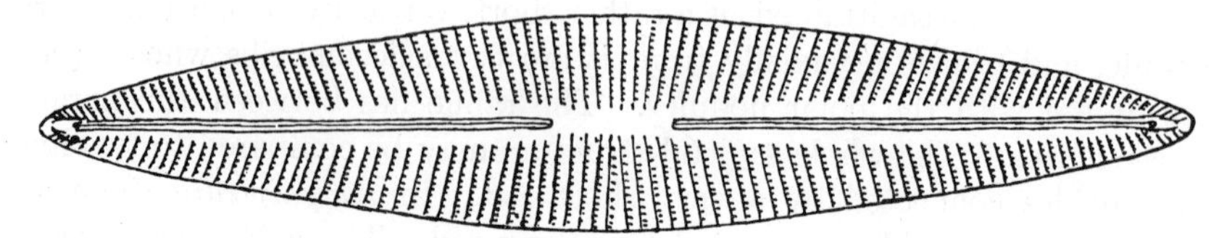

FIG. 405. *Brebissonia Boeckii* (Ehr.) Grun. (*From Boyer*, 1916.)

Auxospores are formed by the apposition of two frustules, after which there is a transverse division of each protoplast to form two gametes that fuse in pairs to form two auxospores. The finding[2] of a functional and a nonfunctional nucleus in each gamete indicates that meiosis takes place prior to division.

The two species known for this country are *B. Boeckii* (Ehr.) Grun. (Fig. 405) and *B. Palmeri* Boyer. For descriptions of them, see Boyer (1927*A*).

11. **Gyrosigma** Hassall, 1845; emend., Cleve, 1894. Valves of *Gyrosigma* (Fig. 406) are sigmoid in outline, gradually attenuated toward the acute or broadly rounded poles, and convex. In girdle view, the frustules are elliptico-lanceolate in outline and with the overlapping portion of girdles lying in a straight line between the two poles. The axial field is a narrow strip down the sigmoid sagittal axis and is usually slightly dilated near the central nodule. The raphe has the same sigmoid curvature as the axial field and has small central and polar nodules. The valve face is

[1] HEINZERLING, 1908. [2] KARSTEN, 1897.

ornamented with two systems of parallel lines that cross one another at right angles. One system of lines is longitudinal and parallel to the sigmoid axial field; the other system has lines parallel with the transverse axis of the valve. There are two chromatophores that lie on opposite sides of the girdle and partly overlap the valve face. Chromatophores may have a smooth or an irregular outline and generally contain several pyrenoids.

Twelve species of *Gyrosigma* are known from fresh waters of the United States. For names and descriptions of them, see Boyer (1927*A*).

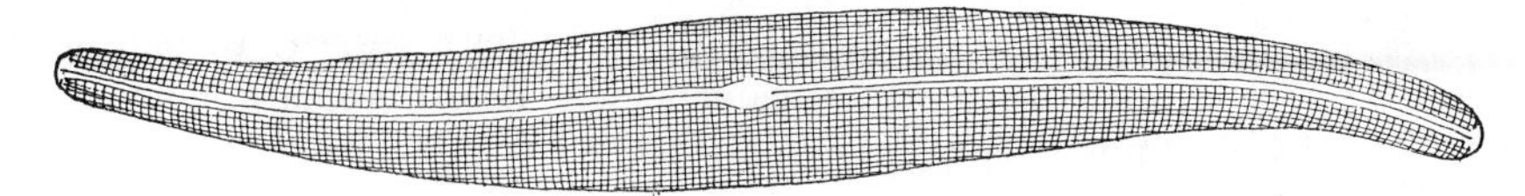

Fig. 406. *Gyrosigma acuminatum* (Kütz.) Cleve. (× 975.)

12. **Pleurosigma** W. Smith, 1852; emend., Cleve, 1894. The frustules of *Pleurosigma* (Fig. 407) have much the same general outline and the same sigmoid raphe as in *Gyrosigma*, but the parallel lines on the valve face are arranged in a different fashion. In *Pleurosigma* the striae lie in three series; one series being parallel to the transverse axis of the valve and the two others oblique to the axial field. There is more variation in chromatophores from species to species than in *Gyrosigma* since there may be two, four, or many within a cell.

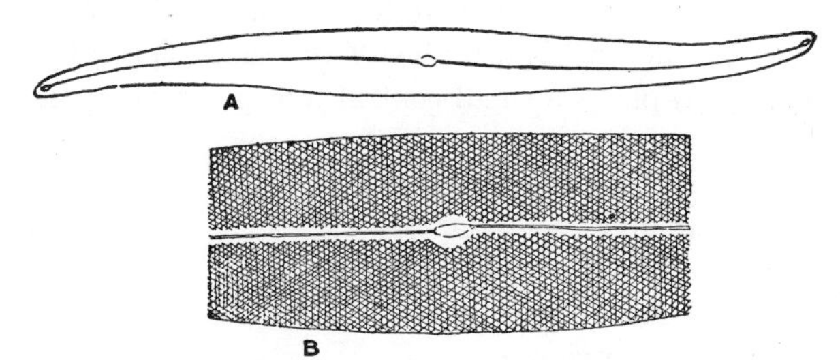

Fig. 407. *Pleurosigma delicatulum* W. Smith. *A*, valve view. *B*, ornamentation of central portion of valve. (*A*, × 495; *B*, × 1300.)

Auxospores are formed by the approximation of two cells to form two auxospores.[1]

Most species of *Pleurosigma* are found in salt or brackish water. The two species found in fresh water in this country are *P. Boyeri* Keeley and *P. delicatulum* W. Smith (Fig. 407). For descriptions of them, see Boyer (1927*A*).

13. **Scoliopleura** Grunow, 1860. Valves of this diatom are straight and naviculoid to elliptical in outline. Unlike other straight pennate diatoms,

[1] Karsten, 1899.

the axial field is distinctly sigmoid. The raphe is also sigmoid and with rather small central and polar nodules. Ornamentation lateral to the axial field consists of transverse or slightly radiate striae or rows of punctae. In certain species, the ornamentation also has a longitudinal line bordering both sides of the axial field. The frustules are elliptico-lanceolate in girdle view and with the zone where the girdles overlap somewhat sigmoid. There are four chromatophores in a frustule, two next to each long side of the girdle, and the four lie symmetrically disposed with respect to one another when seen in valve view. Each chromatophore contains a single pyrenoid.

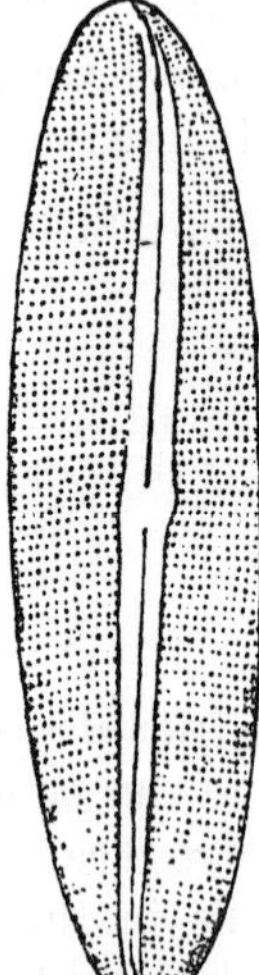

FIG. 408. *Scoliopleura peisonis* Grun. (*From Cleve*, 1894.)

Most species of the genus are marine. The type species *S. peisonis* Grun. (Fig. 408) has been found in salt marshes bordering Great Salt Lake. It is interesting to note that the type locality for this species is Neuseidlersee, a salt lake in Hungary. For a description of *S. peisonis*, see Boyer (1927*A*).

14. **Amphiprora** Ehrenberg, 1843; emend., Cleve, 1894. As seen in valve view, the frustules of *Amphiprora* are naviculoid in outline and with rather sharp poles. Longitudinally along the valve is a keel that is slightly to conspicuously sigmoid and vertical to the valve face. In the outer margin of the keel is a raphe, of the same sigmoid shape as the keel, and along the raphe are small central and polar nodules. Both the valve face and the keel are ornamented with parallel striae or rows of punctae, that are more or less parallel to the transverse axis of a valve. The frustules are broader in girdle view than in valve view and with an outline quite similar to that of an hourglass (Fig. 409). The resemblance of the outline to an hourglass is due to the sigmoid nature of the keel. Between the two girdles are several intercalary bands which may be straight or sigmoid. Both girdle and intercalary bands are usually ornamented with parallel striae or rows of punctae. In most cases, there is a single chromatophore, with irregularly incised edges, next to the girdle side of a frustule; sometimes there are two chromatophores.[1] This diatom is either free-floating

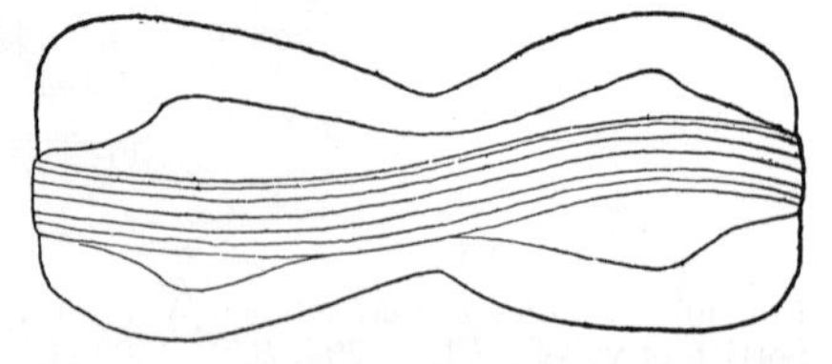

FIG. 409. *Amphiprora paludosa* W. Smith, a species known to occur in fresh water, but one not known for the United States. (× 975.)

[1] HEINZERLING, 1908.

and solitary, or sessile and embedded in a gelatinous matrix adhering to the surface of stones or woodwork.

Most species of *Amphiprora* (Fig. 409) are restricted to marine or brackish waters; a few occur in fresh water. *A. ornata* Bailey is the only fresh-water species known for the United States. For a description of it, see Boyer (1927*A*).

15. **Mastogloia** Thwaites, 1856. Frustules of *Mastogloia* are lanceolate, elliptical, or rhombic in valve view and with broadly rounded, acute, or rostrate poles. The genus differs from others of the family in having two longitudinal internal septa. Each septum has a large, centrally located, oval perforation and several small linear perforations parallel to one another and vertical to the lateral margins (Fig. 410*C*). When empty frustules are seen in valve view, the linear perforations of the septa appear as lateral canaliculi that seem to constitute a part of the valve ornamentation (Fig. 410*A*). Valves which have been separated from the rest of the frustule show that this is not the case, since they have no sign of the lateral canaliculation observable when they are attached to the frustule. Detached valves (Fig. 410*B*) show that the valves have a narrow axial field, containing a straight raphe with small central and polar nodules, and transverse striae or rows of punctae lateral to the axial field. As seen in girdle view, the frustules are rectangular in outline, with smooth girdles, and with the internal septation appearing as a file of small rectangles between the valve and the girdle. Within a cell are two chromatophores that lie next to the girdle side and have lateral projections almost completely covering the inner face of the valves. Frustules of this diatom usually occur within a copious irregularly expanded gelatinous matrix.

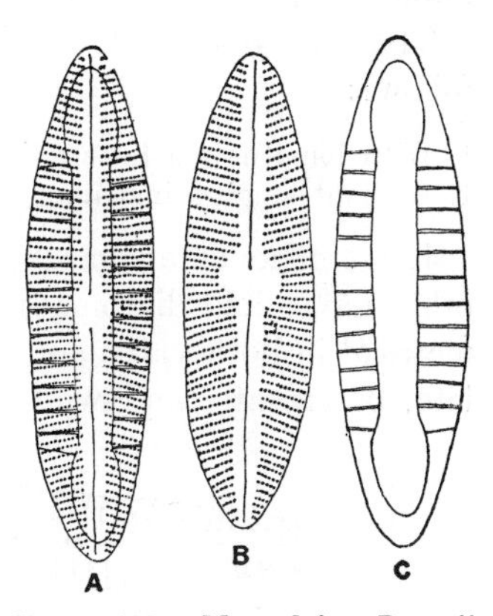

FIG. 410. *Mastogloia Danseii* Thw. *A*, valve view showing valve and internal septum. *B*, valve. *C*, internal septum. (× 1300.)

Auxospores are formed by an approximation of two frustules and a formation of a pair of auxospores.[1]

Mastogloia has many marine species, the majority of which are found in tropical or subtropical waters and are epiphytic on various seaweeds. *M. Danseii* Thw. (Fig. 410) and *M. Grevillei* W. Smith have been found in inland fresh or salt waters in this country. For descriptions of them, see Boyer (1927*A*).

[1] MEISTER, 1912.

Family 2. Gomphonemataceae

Members of this family have frustules that are longitudinally symmetrical in valve view but transversely asymmetrical. The same condition obtains in girdle view, since the frustules are broader at one end. The dendroid colonial habit is sometimes given as a second character of the family, but this is not unique for Gomphonemataceae since it is also found in certain genera of Achnanthaceae and Naviculaceae. As in other families of the suborder, both valves have a true raphe. There is but one chromatophore within a cell.

Auxospores are formed by an approximation of two cells and their forming two gametes that conjugate to form two auxospores. Certain species are known to have a meiotic division of nuclei prior to auxospore formation.

The two genera of the family, both found in this country, differ as follows:

1. With longitudinal lines adjoining valve margin................ 2. **Gomphoneis**
1. Without longitudinal lines adjoining valve margin............ 1. **Gomphonema**

1. **Gomphonema** Agardh, 1824. This genus is to be distinguished from other naviculoid diatoms by having frustules that are transversely asymmetrical in both valve and girdle views. The valves are straight, lanceolate, or clavate, and with one pole capitate or broader than the other.

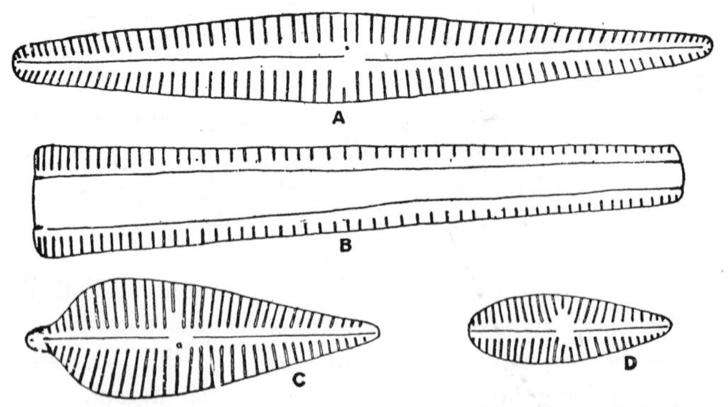

Fig. 411. *A–B*, valve and girdle views of *Gomphonema Vibrio* Ehr. *C*, *G. Augur* Ehr. *D*, *G. olivaceum* (Lyngb.) Kütz. (× 1300.)

There is a straight, rather narrow, axial field, through the center of which is a raphe with conspicuous central and polar nodules. Lateral to the axial fields are transverse or somewhat radiate rows of delicate or coarse punctae. In several species, the axial field is somewhat inflated midway between its poles, and there are one or more isolated and asymmetrically disposed punctae. As seen in girdle view, the frustules are usually cuneate in outline and with smooth girdles. There is a single chromatophore that lies next to one girdle face and has several lobes extending to the valves faces and often reaching the opposite girdle face. A chromatophore usually has a single ellipsoidal pyrenoid. Frustules of *Gomphonema* are usually epiphytic and borne at the tips of a dichotomously branched system of gelatinous stalks. Sometimes the frustules are sessile. Free-floating individuals are often encountered in collections, but these are probably individuals that have accidentally broken away from their stalks.

Auxospores are formed by an approximation of two cells and a division of their protoplasts into two gametes that fuse to form two auxospores. There is a meiotic nuclear division prior to auxospore formation.[1]

Gomphonema (Fig. 411) is a genus with many more fresh-water than marine species. About 20 species have been found in fresh waters in this country. For names and descriptions of them, see Boyer (1927*A*).

2. **Gomphoneis** Cleve, 1894. The frustules of *Gomphoneis* have the same shape as those of *Gomphonema*, and their valves have the same transverse or slightly radiate punctation. Their valves differ from those of *Gomphonema* in having a longitudinal line next to both lateral margins.

The three fresh-water species found in this country are *G. elegans* (Grun.) Cleve, *G. herculeana* (Ehr.) Cleve (Fig. 412), and *G. mamilla* (Ehr.) Cleve. For descriptions of them, see Boyer (1927*A*).

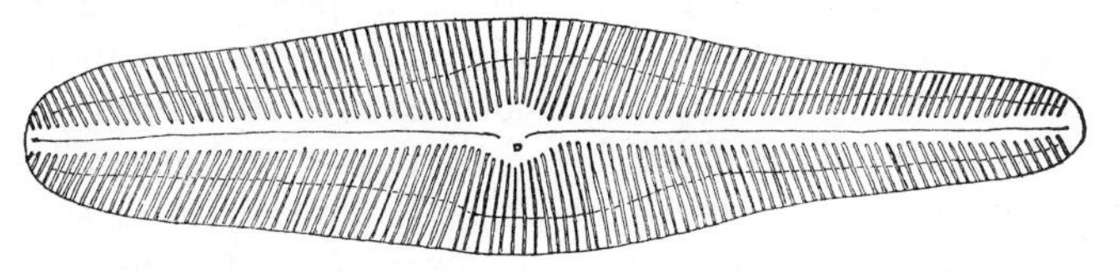

FIG. 412. *Gomphoneis herculeana* (Ehr.) Cleve. (× 975.)

FAMILY 3. CYMBELLACEAE

Any diatom with both valves bearing a raphe and with valves that are longitudinally asymmetrical is placed in the Cymbellaceae. The longitudinal asymmetry is due to the fact that one side of a valve is convex and the other side is less convex, straight, or concave. The frustules are symmetrical in all other planes. The valvular asymmetry also extends to the ornamentation, and the axial field with its included raphe is never in the sagittal axis of a valve. Ornamentation of a valve face may be like that of Naviculaceae and consists only of striae or transverse rows of punctae, or there may be transverse costae in addition. As seen in girdle view, the frustules are symmetrical in both axes, though quite variable in outline,

[1] CHOLNOKY, 1929; MEYER, K., 1929.

and they may have sides that are parallel, convex, or constricted in the middle. Some genera have intercalary bands between the girdles; others lack them. There is usually a single large chromatophore within a cell, next to the girdle side, but some members of the family have two symmetrically disposed chromatophores.

Auxospores are known for all genera, and these are formed in pairs between two approximated frustules. The two gametes formed by each cell may be equal or unequal in size. A meiotic division of nuclei during gamete formation has been shown for some genera and is to be inferred in the case of others. The two auxospores formed by conjugation of gametes in pairs may have their long axes either parallel or at right angles to long axes of the parent frustules.

The four genera of the family, all of which are represented in the fresh-water flora of this country, differ as follows:

1. Valves without transverse costae.. 2
1. Valves with transverse costae.. 3
   2. Valves flat.. 1. **Cymbella**
   2. Valves strongly convex.. 2. **Amphora**
3. Central portion of raphe acute-angled.. 3. **Epithemia**
3. Central portion of raphe not acute-angled.. 4. **Rhopalodia**

1. **Cymbella** Agardh, 1830. Frustules of *Cymbella* are more or less longitudinally asymmetrical in valve view and with a lunate, subnaviculate, or subrhombic outline. The great majority of species have lunate valves, which are gradually attenuated from the middle to the broadly rounded or acute poles, with the concave side a smooth curve or somewhat tumid in the middle. The axial field is either broad or narrow and generally laterally expanded adjacent to the central nodule. It always lies excentric to the sagittal axis and usually some distance inward from the concave margin.

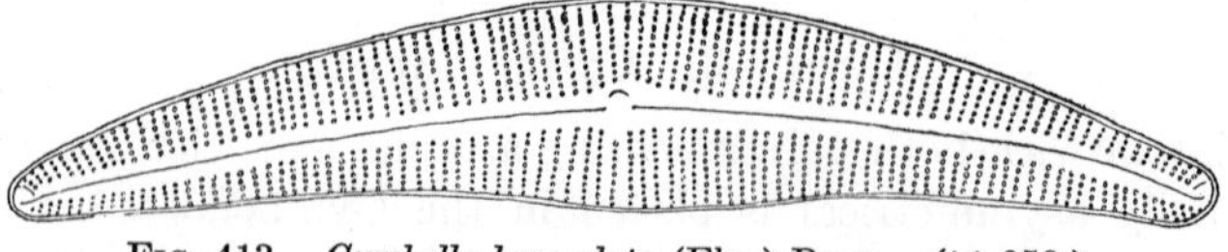

FIG. 413. *Cymbella lanceolata* (Ehr.) Brun. (× 650.)

The raphe has the same curvature as the axial field and has well-defined central and polar nodules. A raphe may extend the whole length of a valve, or its polar nodules may lie some distance in from the poles of a valve. Ornamentation of a valve is always somewhat radiate and consists of either striae or rows of punctae. Some species have one or more asymmetrically disposed punctae in the median expansion of the axial field. As seen in girdle view, frustules have parallel sides, smooth girdles, and are without intercalary bands. A cell contains a single chromatophore, that lies next to the convex girdle side and often overlaps both girdle sides. Some species are solitary and free-floating, others grow affixed. Sessile species may have the cells borne at the tips of stout gelatinous stalks, or the cells may lie seriately within sparingly branched gelatinous tubes.

Auxospores are formed in pairs between two approximated cells. In certain stalked species it has been shown[1] that the two are sister cells and that each protoplast divides into two gametes of unequal size. When conjugation takes place, there is a union of the larger gamete in one cell with the smaller in the other. Nuclear divisions prior to gamete formation are meiotic.

Cymbella (Fig. 413) is a distinctly fresh-water genus, although a few species are found in brackish waters. Thirty-three species, over half of which are widely distributed, are known for this country. For names and descriptions of them, see Boyer (1927*A*).

2. **Amphora** Ehrenberg, 1840. As seen in valve view, the frustules of *Amphora* resemble those of *Cymbella* but the axial field is more strongly excentric and always lies toward the concave side of a valve. The raphe is gibbous instead of a smooth curve and often has its central nodule very close to the concave margin of the valve. Generic differences between *Amphora* and *Cymbella* are more pronounced when the frustules are seen in girdle view. In *Amphora*, the cells are broadly elliptical in outline and with truncate ends. When the concave side lies uppermost, both raphes are visible and lie very close to the girdles. The girdles are usually separated from each other by several intercalary bands which are ornamented with punctae or striae. The width of girdles and intercalary bands is not the same throughout their perimeter, and the portion encircling the convex girdle face is usually considerably broader than that on the concave face. The greater breadth of the convex girdle side results in most frustules lying with the concave side uppermost when examined in water mounts (Fig. 414*B*). Even under such conditions, one has little difficulty in making out the valve structure and noting the features that distinguish the species one from another. Some

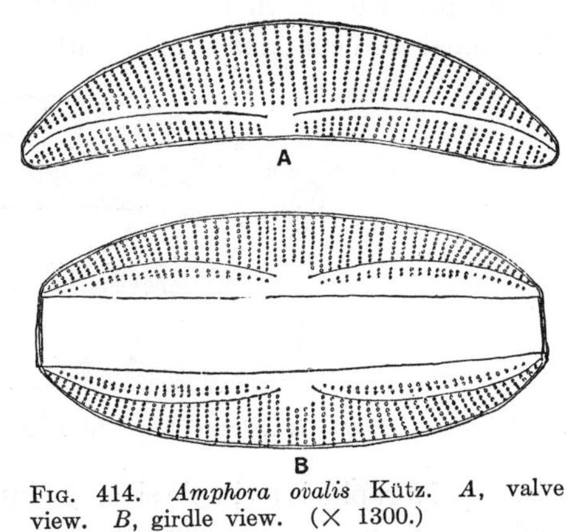

FIG. 414. *Amphora ovalis* Kütz. *A*, valve view. *B*, girdle view. (× 1300.)

[1] CHOLNOKY, 1929; GEITLER, 1927*A*.

species have a single chromatophore which lies next to the concave girdle face and project across to the opposite girdle face; others have two or four chromatophores.[1] Most species are sessile and with their concave face attached to the substratum.

Auxospores are formed in pairs between two approximated frustules.[2] A single cell may also give rise to a single auxospore, and it is thought[3] that this is due to autogamy.

Amphora is a large genus, with over 200 species, but only with a few fresh-water species. Most of the marine species are restricted to tropical waters. The species found in this country are *A. coffaeiformis* (Ag.) Kütz., *A. delphiniana* Bailey, *A. ovalis* Kütz (Fig. 414), and *A. venata* Kütz. For a description of *A. venata*, see Cleve (1895); for the others, see Boyer (1927*A*).

3. **Epithemia** de Brébisson, 1838. Frustules of this diatom are slightly arcuate in valve view and with broadly rounded or subcapitate poles. One side of a valve is strongly and convexly curved, the opposite side is slightly or strongly concave. For the greater part of its length, the axial field lies next to the concave side of a valve, but its central portion bends sharply inward from the valve margin. This is the prominent, inwardly pointed, V-shaped structure that lies midway between the poles on the concave side. The axial field contains a raphe with central and polar nodules. The outer

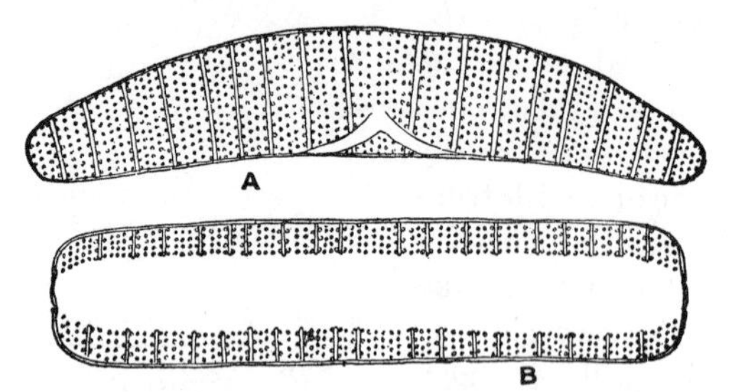

FIG. 415. *Epithemia Zebra* (Ehr.) Kützt. *A*, valve view. *B*, girdle view. (× 1300.)

fissure of a raphe is a simple slot; the inner fissure contains a number of circular pores that open toward the cell's interior.[4] Across the valve face are what appear to be transverse costae, but these are transverse septa. Between the septa are two or more transverse rows of punctae on the valve face, and these may be so large that they form a distinctly reticulate pattern. Girdle views of frustules are always rectangular in outline and with smooth girdles, or smooth girdles and intercalary bands, between the valves. Views of frustules from this side show that the transverse septa evident in valve view extend only as far inward as the juncture of valve and girdle.

[1] HEINZERLING, 1908. [2] KLEBAHN, 1896. [3] GEITLER, 1929.
[4] HUSTEDT. 1928.

Some species have a longitudinal septum at the juncture of valve and girdle and one with rounded perforations between the transverse septa. The capitate ends of transverse septa, as seen in girdle view, are the longitudinal septa. There is usually a single chromatophore next to the concave girdle side, that has irregular projections extending along both valve faces. *Epithemia* is a solitary diatom that usually is epiphytic upon submerged plants and with its concave girdle side next to the substratum.

Auxospores are formed in pairs between two frustules and, unlike most other diatoms, elongate at right angles to long axes of the empty parent frustules. Protoplasts of each frustule divide into two gametes of equal size which fuse in pairs,[1] and nuclear divisions preceding gamete formation are meiotic.[2]

Epithemia is found only in fresh and brackish waters. The fresh-water species of this country are *E. Argus* (Ehr.) Kütz., *E. gibberula* (Ehr.) Kütz., *E. Hyndmanni* W. Smith, *E. Muelleri* Fricke, *E. ocellata* (Ehr.) Kütz, *E. sorex* Kütz., *E. turgida* (Ehr.) Kütz., and *E. Zebra* (Ehr.) Kütz. (Fig. 415). For a description of *E. Muelleri*, see Meister (1912); for the others, see Boyer (1927*A*).

4. **Rhopalodia** O. Müller, 1895. This genus has frustules with broader girdle faces than valve faces and, consequently, lies girdle side up when viewed in water mounts. As seen in girdle view, the frustules are linear, linear elliptic, or clavate, inflated in the median portion, and with broadly

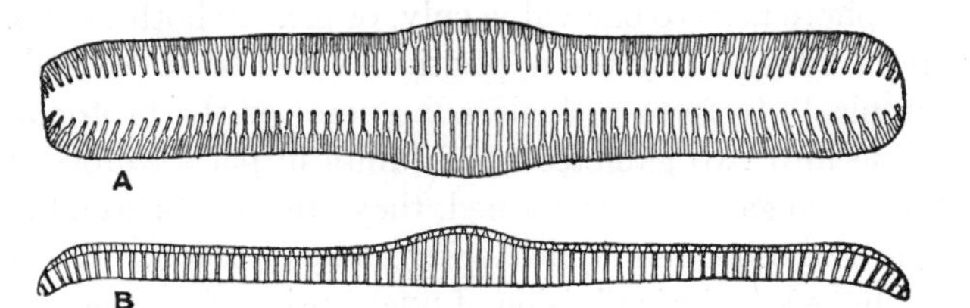

FIG. 416. *Rhopalodia gibba* (Ehr.) O. Müller. *A*, girdle view. *B*, valve view. (× 650.)

rounded poles. In valve view, the cells are lunate to reniform, often with the convex margin medianly inflated, and usually with acute apices. The axial field, which is visible throughout the whole length of a valve, lies next to the concave margin of a valve. The portion of the valve face bearing the axial field is elevated in a keel-like fashion and through the center of the field is a raphe with central and polar nodules. Outside of the axial field, the valve face is ornamented with transverse costae. Between two successive costae are one or more delicate striae. The girdle view has an unornamented girdle zone that is with or without intercalary bands between the girdles. Each cell contains, next to the girdle, a single laminate

[1] KARSTEN, 1896. [2] CHOLNOKY, 1929.

chromatophore with irregular margins. Frustules of this diatom are usually solitary and free-floating.

Auxospores are formed between two apposed cells which are generally of different size and with their concave faces opposite each other. The protoplast in each cell divides into two equal-sized gametes, which fuse in pairs. Nuclear divisions accompanying gamete formation are probably meiotic. As in *Epithemia*, the auxospores elongate at right angles to long axes of the parent frustules.[1]

All species of *Rhopalodia* are fresh-water in habit. The two species found in this country are *R. gibba* (Ehr.) O. Müller (Fig. 416) and *O. ventricosa* (Kütz.) O. Müller. For descriptions of them, see Boyer (1927*A*).

SUBORDER 4. SURIRELLINEAE

Genera belonging to this suborder have both valves alike and the raphe in each valve more or less concealed in a keel at one or both sides of the valve. The frustules may be transversely symmetrical and longitudinally symmetrical in valve view, or symmetrical in both axes. Most genera have flattened valves, but a few have curved or undulate ones. Girdle views of frustules are usually symmetrical in both axes, but they may be asymmetrical in one axis. The frustules may be rectangular or rhombic in transverse section. Chromatophores usually lie next to the valve face, and there may be a chromatophore next to one valve only, or next to both valves.

Auxospores are formed by the apposition of two frustules and the union of their protoplasts to form a single auxospore; or the protoplast in each cell may divide into two gametes which unite in pairs to form two auxospores. When two gametes are formed, they are alike in size, but one may be mobile and the other immobile during gametic union. Certain species are known to have a meiotic division of nuclei prior to conjugation, and the formation of nonfunctional nuclei in other species indicates a similar reduction in number of chromosomes.

The suborder is divided into two families.

FAMILY 1. NITZSCHIACEAE

Genera belonging to this family have a single excentric keel next to one lateral margin of a valve, and a raphe adjacent to or concealed by the keel. There is also a series of large or small dots in the region of the keel. The frustules are elongate and straight or sigmoid in valve view. The valves are symmetrical with respect to the transverse axis and asymmetrical with respect to the longitudinal axis. Frustules may be rectangular or rhombic

[1] KLEBAHN, 1896.

in transverse section and with or without an internal septation. There are usually two laminate chromatophores, one next to each side of the girdle.

Auxospores are formed in pairs between two frustules.

The three genera found in this country differ as follows:

1. Frustules without transverse septa 2
1. Frustules with transverse septa . . . . **3. Denticula**
 2. Valves with raphes diagonally opposite . . . . **1. Nitzschia**
 2. Valves with raphes opposite each other . . . . **2. Hantzschia**

1. **Nitzschia** Hassall, 1845. Frustules of *Nitzschia* (Fig. 417) are usually elongate and of extremely varied outline as seen in valve view. The valves may be straight or sigmoid; linear to elliptical in outline; with or without the lateral margins constricted in the middle; and with acute, subrostrate, or attenuated apices. Next one margin of the valve is a keel in which the raphe lies (Fig. 417). The keeled margin of one valve faces the unkeeled margin of the other valve. The raphe has small central and polar nodules,

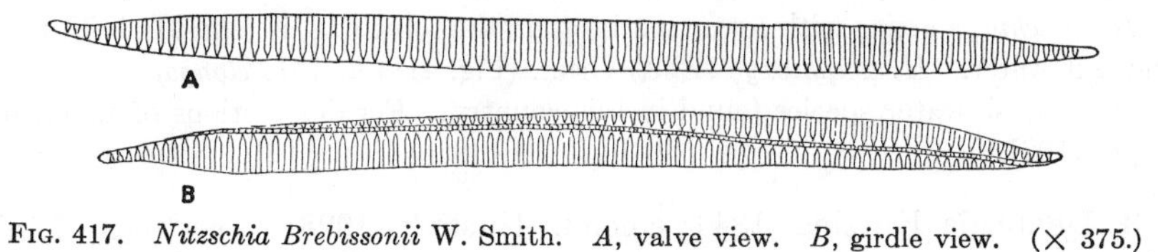

FIG. 417. *Nitzschia Brebissonii* W. Smith. *A*, valve view. *B*, girdle view. (× 375.)

and the rapheal fissure has a uniseriate row of circular pores that open toward the cell's interior.[1] These "carinal dots" are quite conspicuous and are the chief character by which one recognizes the genus. Across the face of the valve are transverse striae or rows of punctae. Unlike most other pennate diatoms, the girdle and valve sides are not at right angles to one another. Transverse sections of frustules are therefore rhombic instead of rectangular in outline. The girdle view of the frustule is elongate, straight or sigmoid, and often with the ends somewhat attenuated. Within the cell are two chromatophores that lie axial to each other. Both chromatophores are on the same girdle side and usually with convolute margins overlapping the valve face.[2] The cells may be solitary and free-floating or in dense fascicles within simple or branched gelatinous tubes.

Sexual auxospores are formed by a more or less X-like apposition of two cells which become connected to each other by a gelatinous conjugation tube. The protoplast in each cell divides longitudinally into two equal-sized gametes, and one gamete from each cell migrates through the conjugation tube and fuses with a gamete in the other cell. The two auxospores thus formed eventually become much longer than the parent cells. Nuclear divisions preceding gamete formation are reductional.[3]

[1] HUSTEDT, 1929. [2] HEINZERLING, 1908. [3] GEITLER, 1928A.

There are numerous species of *Nitzschia* (Fig. 417) in fresh, brackish, and salt waters. For names and descriptions of the 33 fresh-water species found in this country, see Boyer (1927*A*).

2. **Hantzschia** Grunow, 1880. Frustules of this diatom have much the same shape as those of *Nitzschia* and have the same marginal keel at one side of the valve. The raphe has the same structure and the same system of carinal dots as is found in *Nitzschia*.[1] The genus differs from *Nitzschia* in two respects: the frustules are rectangular instead of rhombic in transverse section, and the keeled margins of a pair of valves lie opposite instead of diagonal to each other.

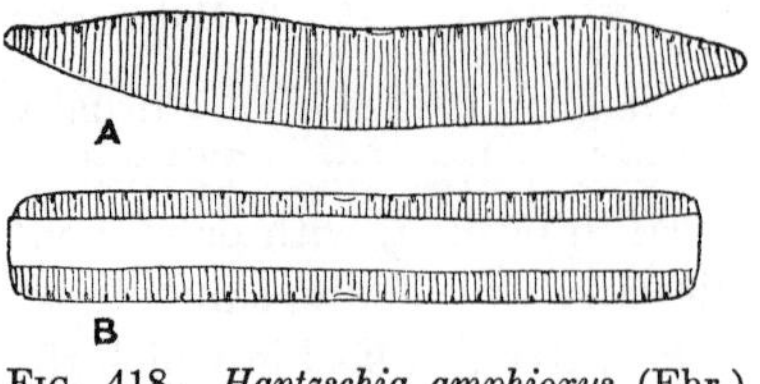

FIG. 418. *Hantzschia amphioxys* (Ehr.) Grun. *A*, valve view. *B*, girdle view. (× 1300.)

Auxospores are formed in pairs between two cells.[2]

Hantzschia, a genus with fewer species than the foregoing, is found in both fresh and salt water. *H. amphioxys* (Ehr.) Grun. (Fig. 418) and *H. elongata* Grun. are the only fresh-water species found in this country. For descriptions of them, see Boyer (1927*A*).

3. **Denticula** Kützing, 1844; emend., Hustedt, 1928. Frustules of this diatom are usually elongate and have a symmetrical outline in all axes. The valves are linear, lanceolate, or elliptical in outline and bear an almost wholly concealed keel next to one margin. Within the marginal keel is a straight raphe with small central and polar nodules. The internal face of the rapheal fissure has the same series of circular openings as in the two preceding genera, but these pores are much less evident than in *Nitzschia* and *Hantzschia*.[3] Internal to the valve face is a series of transverse parallel septa that appear as transverse costae when a frustule is seen in valve view. On the valve face between two successive "costae" are several transverse striae or rows of punctae which extend without interruption across the valve face. The frustules also contain two longitudinal septa, each with a single, large, transversely oval perforation between two successive transverse septa. As seen in girdle view, the frustules have truncate poles and somewhat convex sides. Between the girdles are several intercalary bands. The transverse septa extend to the juncture of valve and girdle and have capitate ends, which represent the only portion of the longitudinal septa visible

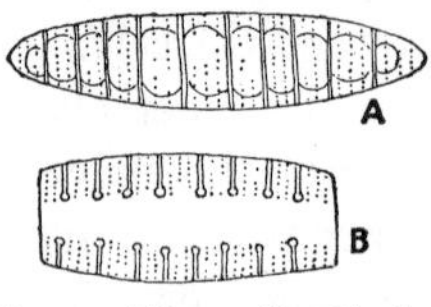

FIG. 419. *Denticula thermalis* Kütz. *A*, valve view. *B*, girdle view. (× 1300.)

[1] HUSTEDT, 1928*A*. [2] KLEBAHN, 1896. [3] HUSTEDT, 1928.

when frustules are viewed from this side. The frustules are either solitary and free-floating, or united valve to valve in short band-like filaments.

Denticula was thought to be closely related to *Tetracyclus* and *Diatomella* and placed in the Fragilariaceae until it was shown[1] to have a true raphe. Although there are several fresh-water species, only one of these, *D. thermalis* Kütz. (Fig. 419), has been found in this country. For a description of it, see Boyer (1927*A*).

Family 2. Surirellaceae

Genera belonging to this family are to be distinguished by the marginal position of the raphe and by the presence of a raphe on both margins of a valve. Even when one cannot make out the raphe with certainty, one can recognize members of the Surirellaceae by the marginal keel in which the raphe lies. The Surirellaceae may also be recognized by their distinctive costation in which the costae are much more prominent near the margins of a valve than at the center. Frustules of Surirellaceae are usually quite large and with the valves symmetrical in both axes or only longitudinally symmetrical. The valve face may be flat, or with transverse undulations, or curved in a saddle-like fashion. Girdle views are correspondingly rectangular, sinuate, or quite irregular. There is a single chromatophore.

Auxospores are formed by apposition of two cells and the union of their protoplasts to form a single auxospore. A unique feature of this family is the apposition of cells end to end instead of side by side. Meiosis has been shown to take place prior to conjugation. Auxospores may also be formed by a parthenogenetic development of an auxospore in each of two apposed cells.

The three genera found in this country differ as follows:

1. Face of valve transversely undulate .... 1. **Cymatopleura**
1. Face of valve not transversely undulate 2
   2. Face of valve flat or spirally twisted .... 2. **Surirella**
   2. Face of valve markedly bent and frustule saddle-shaped .... 3. **Campylodiscus**

1. **Cymatopleura** W. Smith, 1851. This diatom has frustules that are elliptical, naviculoid, or linear in outline when seen in valve view. The valve face is transversely undulate, a feature which shows to best advantage when cells are viewed from the girdle side. Along both sides of a valve is a marginal keel containing a raphe. The valves have broad transverse costae next to their lateral margins, but these are sometimes so short that they seem to be a marginal beading on the valve. In addition to the costation, there is a delicate transverse striation across the valve face. At times, the transverse striation is interrupted by a narrow smooth space (pseudoraphe)

[1] Hustedt. 1928.

through the sagittal axis of a valve. Many species have the pseudoraphe and striation so obscure that they cannot be made out with certainty. As seen in girdle view, the frustules are linear and with the sides markedly undulate. There seems to be a chromatophore next to each valve, but this has been interpreted[1] as a single chromatophore since the two are joined by

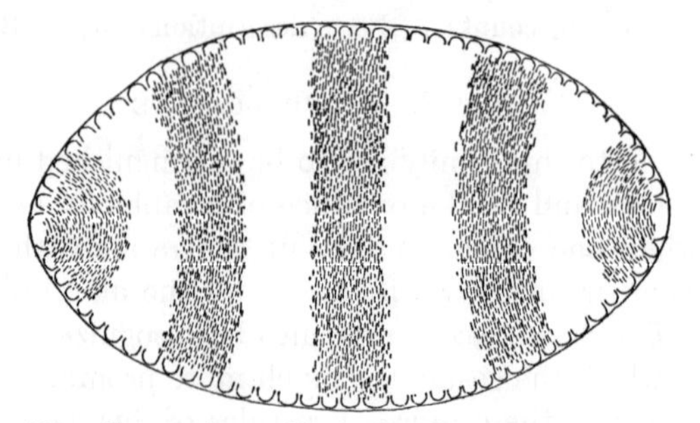

FIG. 420. *Cymatopleura elliptica* (Bréb.) W. Smith. (× 975.)

a bridge of pigmented cytoplasm. The frustules occur singly and free-floating.

Auxospores are formed by a juncture of two cells end to end and a union of their protoplasts to form a single auxospore.[2] It has also been shown[3] that two cells joined end to end may each form auxospores without conjugation. The presence of a functional and a nonfunctional nucleus in each of the two auxospores indicates that there is the same parthenogenesis, without a reduction in chromosome number, that has been definitely established for *Cocconeis*.

Cymatopleura is a genus with relatively few species, but one found in both fresh and brackish waters. The three species found in this country are *C. elliptica* (Bréb.) W. Smith (Fig. 420), *C. hibernica* W. Smith, and *C. solea* (Bréb.) W. Smith. For descriptions of them, see Boyer (1927*A*).

2. **Surirella** Turpin, 1828. The frustules of *Surirella* (Fig. 421) are linear, elliptical, or ovate in valve view and with broadly rounded to subacute poles. The entire valve face may lie in one plane, or it may be spirally twisted. Along both sides of a valve is a marginal keel, containing a raphe with small central and polar nodules. Along the inner face of the rapheal fissure is a series of circular pores opening to the cell's interior.[4] On the valve face and inward from its margins are parallel, evenly spaced, long or short, transverse costae. In addition to the costae, there are very delicate striae across the valve face, and interrupted through the sagittal axis

[1] HEINZERLING, 1908. [2] KLEBAHN, 1896. [3] KARSTEN, 1900.
[4] HUSTEDT, 1929*A*.

of a valve by a linear to lanceolate smooth space, the pseudoraphe. The girdle view is more often rectangular than it is naviculoid, cuneate, or sigmoid in outline, and it always has smooth girdles between the strongly costate valves. As in *Cymatopleura*, there is but one chromatophore. The frustules are usually solitary and free-floating.

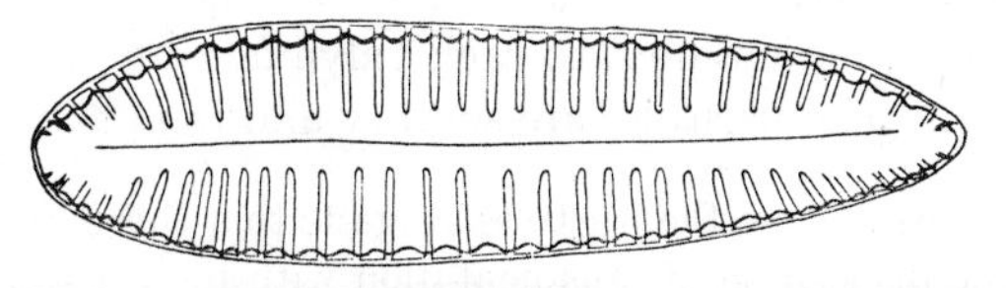

FIG. 421. *Surirella splendida* (Ehr.) Kütz. (× 400.)

Auxospores are formed by a union of two cells end to end and a fusion of their protoplasts to form a single auxospore.[1] Nuclear division prior to fusion has been shown[2] to be meiotic.

Surirella (Fig. 421) is a genus with nearly 200 species that are found in fresh brackish, and salt waters. There are about 35 fresh-water species in this country. For descriptions of most of them, see Boyer (1927*A*).

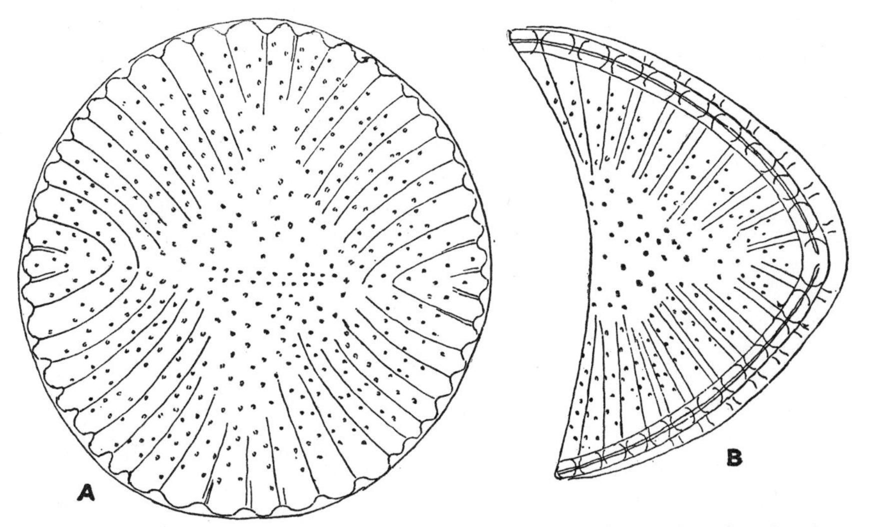

FIG. 422. *Campylodiscus hibernicus* Ehr. *A*, valve view from above. *B*, valve view from the side of a bent frustule. (× 650.)

3. **Campylodiscus** Ehrenberg, 1841. This diatom has valves which are circular to subcircular in outline and so bent that the whole frustule is distinctly saddle-shaped. Around the periphery of a valve are costae that

[1] KARSTEN, 1900. [2] KARSTEN, 1912.

converge toward the punctate or striate central portion of the valve face. There is a marginal raphe as in the two previous genera. There is, also, a more or less distinct pseudoraphe through the center of a valve, and the one in the hypotheca lies at right angles to that in the epitheca. Because of the bent frustules, the girdle view is variously shaped according to the side from which it is seen. Within a cell is a single chromatophore with a broadly expanded lamina next to each valve face and a connecting band between the two. The frustules are solitary and free-floating.

The general appearance of the frustules suggests that this diatom belongs to the Centrales, but the disposition of ornamentation with respect to a line shows that it is a member of the Pennales. There are over 100 species, only a very few of which are found in fresh water. The fresh-water species of the United States are *C. ellipticus* (Bréb.) W. Smith, *C. hibernicus* Ehr. (Fig. 422), and *C. noricus* (Bréb.) W. Smith. For descriptions of them, see Boyer (1927*A*).

CHAPTER 7

DIVISION PHAEOPHYTA

The Phaeophyta, or brown algae, have many-celled thalli that are usually of macroscopic size and distinctive shape. The brown algae differ from other algae in structure of reproductive organs, in structure of motile reproductive cells, in chemical nature of food reserves, and in pigments found in their chromatophores.

The chromatophores are a yellowish brown because xanthophylls are present in greater amount than are chlorophylls and carotenes. There are six xanthophylls, three of which are found only in brown algae (see Table I, page 3). The two chlorophylls are chlorophyll *a* and chlorophyll *c*.

Carbohydrate food reserves are stored in a dissolved state, but it is uncertain whether these accumulate in the vacuoles, in the cytoplasm, or throughout the protoplast. The two principal reserves are *laminarin*, a polysaccharide found only in brown algae, and mannitol, a hexanhydric alcohol also found in fungi and a variety of other plants.

Throughout the entire division, there is a remarkable uniformity in structure of motile reproductive cells, whether zoospores or gametes. These are pyriform and with two laterally inserted flagella of unequal length.

Reproductive organs are of two kinds, neither of which is found in any other group of algae. The one-celled or *unilocular* reproductive organ is always a sporangium and borne on a diploid thallus. At first it contains a single nucleus, and division of this nucleus is always meiotic. Following meiosis there are simultaneous mitotic divisions until there are 8, 16, 32, 64, 128, or 258 nuclei. There then follows a cleavage of the sporangial contents into uninucleate protoplasts which are metamorphosed into zoospores or, in very rare cases, nonflagellated spores. The other kind of reproductive organ is many-celled and with each cell containing a single gamete or a single zoospore. If these *plurilocular* reproductive organs are borne on haploid thalli, they are always gametangial in nature and produce gametes whose nuclei have the haploid number of chromosomes. If the plurilocular organs are borne on diploid thalli, they are sporangial in nature and produce zoospores whose nuclei have the diploid number of chromosomes.

Except for the Fucales, exemplified by *Fucus* and *Sargassum*, all other orders of Phaeophyta have a life cycle in which there is an alternation of two independent multicellular generations, one haploid, the other diploid. In some orders, the two generations are identical in size and structure; in others, the two are dissimilar in both size and structure. If the two are dissimilar, the haploid generation, the gametophyte, is always the smaller and simpler of the two. If a brown alga is one in which the sporophytes produce unilocular sporangia exclusively, the life cycle is a regular succession of haploid gametophytes and diploid sporophytes. If the brown alga is one in which the diploid generation, the sporophyte, also produced plurilocular sporangia, the spores from them are diploid and so a sporophytic generation may be succeeded by another diploid sporophytic generation.

The Phaeophyta include approximately 200 genera, all but four of which are exclusively marine. Of the genera known to occur in fresh water only one, *Heribaudiella*, is known for the United States.

1. **Heribaudiella** Gomont, 1896; emend., Svedelius, 1930. The thallus of this alga is an irregularly expanded crust growing on rocks. At first, the crust is one cell in thickness; later it becomes a dozen or more cells in thickness and consists of erect, simple or forked, vertical rows of cells laterally adjoined to one another without intercellular spaces. The cells are uninucleate and with several disk-shaped chromatophores. Some thalli have the uppermost cells of the crust enlarging greatly and developing into unilocular sporangia that lie close to one another but scarcely constitute a sorus (Fig. 423*A*). Other thalli produce plurilocular reproductive organs, one cell broad and several cells in height[1] (Fig. 423*B*). Presumably these two kinds of thalli are, respectively, sporophyte and gametophyte, but this is not definitely established.

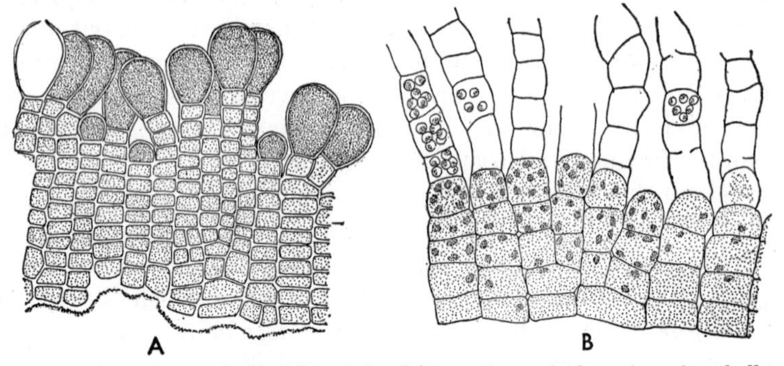

Fig. 423. *Heribaudiella fluviatilis* (Gom.) Svedelius. *A*, vertical section of a thallus with unilocular sporangia. *B*, vertical section of a thallus with plurilocular reproductive organs (gametangia ?). (*A*, *after Flahault*, 1883; *B*, *after Svedelius*, 1930.) (*A*, × 465; *B*, × 690.)

H. fluviatilis (Gom.) Svedelius (Fig. 423) has been found on stones in a brook in Connecticut.[2] Since the brook empties into ocean and the alga was found a short

[1] Svedelius, 1930.
[2] Holden in Phycotheca Boreali-Americana, No. 536.

distance above the high tide level, there is a possibility that the specimens identified as *H. fluviatilis* are in reality the marine alga *Lithoderma fatiscens* Aresch. As found in Europe,[1] *H. fluviatilis* grows in swiftly flowing streams where the streams are shaded by overhanging ledges or by overhanging vegetation. For a description of *H. fluviatilis*, see Svedelius (1930).

[1] FRITSCH, 1929*A*; SVEDELIUS, 1930.

CHAPTER 8

DIVISION PYRROPHYTA

Members of this division have their pigments localized in chromatophores which are usually greenish tan to golden brown. The pigments are chlorophyll *a*, chlorophyll *c*, beta-carotene, and four xanthophylls, three of which are found only in members of the division (see Table I, page 3). Photosynthetic reserves generally accumulate as starch or starch-like compounds, but they may also accumulate as oils. The nucleus is distinctive in that the chromatin lies in numerous bead-like threads. Cell walls, when present, generally contain cellulose.

Certain characters in common to dinoflagellates, desmokonts, and cryptomonads were first pointed out by Pascher.[1] Chief among these are two flagella unlike in movement and shape, and the formation of starch by chromatophores that are brownish. Later Pascher[2] named the three groups of organisms the Dinophyceae, the Desmokontae, and the Cryptophyceae and called the combined groups the Pyrrophyta. The relationship of the Cryptophyceae to Dinophyceae has been questioned.[3] Chief among the arguments for excluding the Cryptophyceae from the Pyrrophyta are the markedly different structure of their nuclei and the presence of a gullet.[4] If, as will be done on subsequent pages (see page 626), the Cryptophyceae are considered a class of uncertain systematic position, the Pyrrophyta contain but two classes: Desmokontae and Dinophyceae.

CLASS 1. DESMOKONTAE

Motile cells of Desmokontae lack a transverse furrow and have two apically inserted, somewhat flattened flagella that differ from each other in orientation and type of movement. Motile cells with a cell wall have the wall vertically divided into two halves (valves) that are without subdivision into definitely arranged plates. The protoplast contains brownish chromatophores.

This class contains only a few genera, all rare organisms and found almost exclusively in the ocean. There are two general types of vegetative cell: the motile flagellated cell and the immobile cell without flagella. These two types have been used[5] as a basis for dividing the class into two

[1] PASCHER, 1911. [2] PASCHER, 1914. [3] FRITSCH, 1935; GRAHAM (in press).
[4] GRAHAM (in press). [5] PASCHER, 1927.

orders, only one of which is represented in the fresh-water flora of this country.

ORDER 1. DESMOMONADALES

The Desmomonadales include all genera with flagellated motile vegetative cells. The segregation into families is based upon the presence or absence of a wall.

FAMILY 1. PROROCENTRACEAE

Members of this family have cells with a definite wall and one longitudinally divided into two halves (valves). One genus of the family is known from fresh waters in the United States.

1. **Exuviaella** Cienkowski, 1882. The cells of this flagellated unicellular alga are ellipsoidal and somewhat compressed. The protoplast is surrounded by a cellulose wall consisting of two longitudinally apposed valves which are evident when a cell is viewed from the side. There are two flagella at the anterior end, and in certain species they project through an evident pore in the wall. One flagellum projects vertically forward, and its lashing propels the cell through the water. The other flagellum stands at right angles to the propulsive flagellum: its movement is undulatory and causes a rotation of a cell as it moves through the water. The protoplast contains two brownish chromatophores, with or without pyrenoids. Reserve foods include granules, probably of a starch-like nature, and small droplets of oil. There is a conspicuous nucleus at the base of a cell.

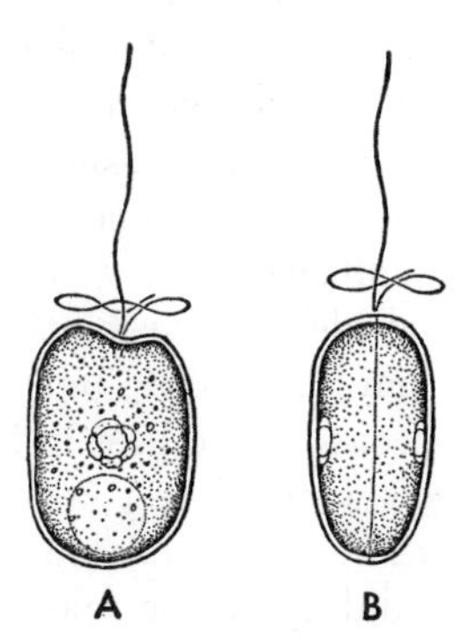

FIG. 424. *Exuviaella compressa* Ostenf. *A*, front view. *B*, side view. (*Drawn by R. H. Thompson.*) (× 800.)

Reproduction is by longitudinal bipartition. Each daughter cell receives one valve from the parent cell and secretes an entirely new one.

Prof. R. H. Thompson writes that he has found *E. compressa* Ostenf. (Fig. 424) in a swamp near Solomons Island, Maryland. *E. compressa* is a marine organism, but the salinity of the swamp water in which it was found in Maryland is so low that the water cannot even be considered brackish. For a description of *E. compressa*, see Schiller (1933).

CLASS 2. DINOPHYCEAE

The most distinctive feature of the class is the structure of motile vegetative cells and of zoospores of immobile genera. These are always completely or incompletely encircled by a transverse or by a spiral groove.

Motile cells are always biflagellate and with the two flagella inserted in the groove. One flagellum lies in the groove and encircles the cell; the other extends backward from the groove. Most members of the class have brownish chromatophores, but some are colorless and with a saprophytic or holozoic mode of nutrition. A few genera have naked protoplasts, but the great majority have cellulose walls that may be homogeneous or may consist of a definite number of articulated plates. Food reserves are stored as starch or as oil.

Reproduction of motile genera is usually by vegetative cell division, either while a cell is in motion or after it has come to rest. Motile genera may also produce aplanospores (cysts). Reproduction of immobile genera may be by means of zoospores or autospores (aplanospores).

Sexual reproduction is very infrequent.

Occurrence. The great majority of Dinophyceae are motile unicellular flagellates (dinoflagellates). Most dinoflagellates grow in the plankton of the ocean, especially that of warmer portions. Surface-dwelling marine dinoflagellates usually have chromatophores, but several of them and all deep-dwelling ones are without photosynthetic pigments. Fresh-water dinoflagellates are most abundant in pools, ditches, and small lakes with considerable vegetation. They are not uncommon in plankton catches from large lakes, but rarely occur in abundance. Some of the fresh-water species thrive best in hard waters; others are found in greatest numbers in soft waters.[1]

The nonflagellated genera (phytodinads) are rare organisms and usually found epiphytic upon the coarser filamentous Chlorophyceae. The parasitic forms, of which there are several genera, are found within and upon various animals.[2]

Organization of the Plant Body. The Dinophyceae resemble the Chrysophyceae in their richness in species with flagellated vegetative cells and in their poverty of types with a true algal organization. However, a sufficient number of Dinophyceae have been discovered to show that evolution within the class has been in accordance with the theory of plant-body types (see page 6). There are no members of the class showing evolution along the volvocine line. The first step in the tetrasporine line of evolution, the development of palmelloid colonies, is found in one species only, *Gloeodinium montanum* Klebs.[3] Two of the genera belonging to the tetrasporine series (*Dinothrix* Pascher[4] and *Dinoclonium* Pascher[5]) have a truly filamentous organization (Fig. 425). Reproduction in these filamentous genera is by each cell forming one or two *Gymnodinium*-like zoospores. The third evolutionary tendency, the chlorococcine type, has several representatives

[1] Höll, 1928. [2] Chatton, 1920. [3] Klebs, 1912; Killian, 1924.
[4] Pascher, 1914, 1927. [5] Pascher, 1927.

among the Dinophyceae. Some of these unicellular forms are free-floating (Fig. 426), others are sessile.[1] Reproduction of the chlorococcine Dinophyceae may be by means of zoospores, aplanospores, or autospores. One member of the series produces zoospores of two different sorts.[2] It is not known that any of these chlorococcine genera are multinucleate, and no genera have as yet been discovered with a truly siphonaceous organization. The fourth possibility in evolution from a unicellular motile ancestor, rhizopodial vegetative cells, has also been found in the Dinophyceae.[3]

Cell Wall. Certain of the dinoflagellates have naked protoplasts in which the cytoplasmic surface may be smooth or longitudinally ridged. Most of these are marine, but species of certain genera, as *Gymnodinium* and *Gyrodinium*, have been found in fresh waters. Genera with the protoplast en-

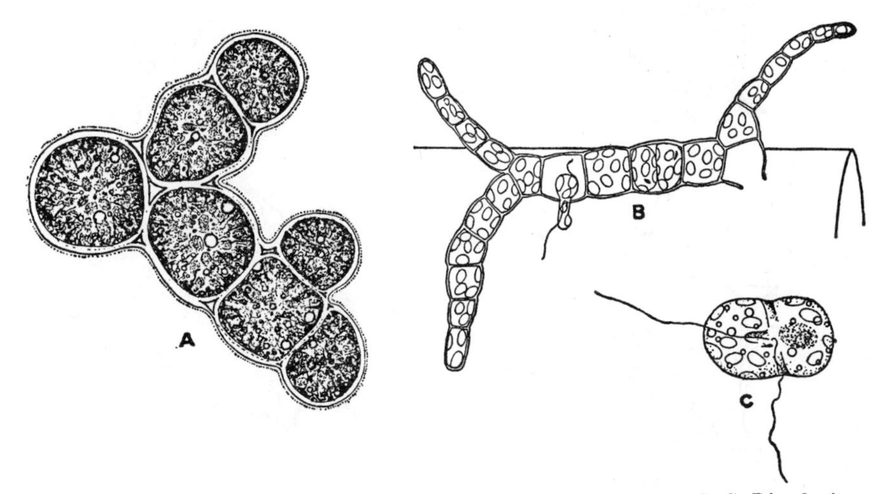

FIG. 425. Filamentous Dinophyceae. *A*, *Dinothrix paradoxa* Pascher. *B–C*, *Dinoclonium Conradi* Pascher. (*From Pascher*, 1927.)

closed by a wall may have one that is delicate or one that is relatively heavy. Walls of many species give a definite cellulose reaction, but certain species do not appear to have cellulose in their walls.[4] The cell wall may consist of a single layer, or it may be differentiated into two layers, the outer of cellulose, and the inner of unknown chemical composition.[5] Dinoflagellates rarely have an outer sheath of pectic material, but phytodinads often have the cells surrounded by a wide pectic sheath.

Walls of Dinophyceae may be homogeneous in structure or composed of interlocking plates. All the phytodinads have homogeneous walls. A few

[1] GEITLER, 1928; KLEBS, 1912; PASCHER, 1927. [2] PASCHER, 1928.
[3] PASCHER, 1915*C*. [4] SCHILLING, 1891. [5] MANGIN, 1907, 1911.

dinoflagellates have homogeneous walls, but the great majority have one composed of a specific number of plates. The number and arrangement of the plates are important in classifying the "armored" dinoflagellates, and several nomenclatorial systems have been proposed to describe their arrangement. The system of Kofoid[1] is based upon the fact that the transverse girdle divides the wall into two parts, *epitheca* and *hypotheca*, each with the plates in definite transverse bands or series (Fig. 427). Plates in the uppermost series of the epitheca are called *apical plates*; those in the

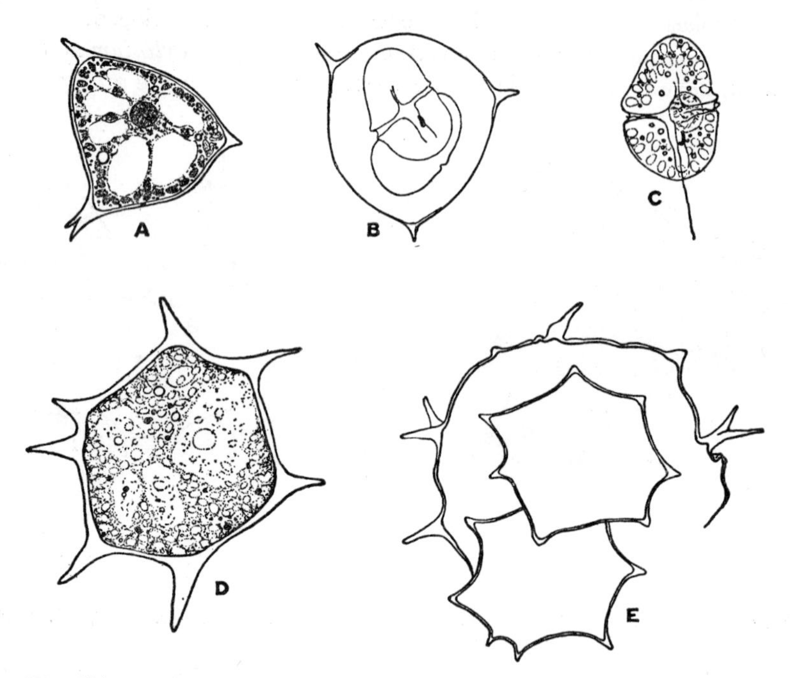

FIG. 426. Chlorococcine Dinophyceae. *A–C, Tetradinium minus* Pascher. *D–E, Dinastridium sexangulare* Pascher. (*From Pascher*, 1927.)

series adjoining the girdle are *precingular plates*. Some genera have an incomplete band of *anterior intercalary plates* between the apical and precingular series. In the hypotheca there is a series of *postcingular plates* next to the girdle and one or two *antapical plates* at the lowermost part of the cell. Occasionally there is a single *posterior intercalary plate* between these two series. Many genera have a thin membranaceous *ventral plate*, scarcely comparable to the other plates, intercalated in the girdle region and extending through the precingular and postcingular series. The maximum num-

[1] KOFOID, 1907, 1909.

ber of plates known for any species is four apicals, three anterior intercalary, seven precingular, five postcingular, one posterior intercalary, and two antapicals.

The plates are usually covered with minute spines or with a fine reticulum of small ridges. The plates of many species also have minute pores. These are not arranged in a definite pattern but are usually more numerous near the margin than at the center.[1] The lines of juncture between the plates, the *sutures*, are sometimes inconspicuous (*Hemidinium*), but usually they are strongly evident and with a longitudinal or transverse striation (*Peridinium*, *Ceratium*). As seen in cross section, the abutting margins of the plates may overlap each other, or they may be slightly infolded along the line of mutual contact.[2]

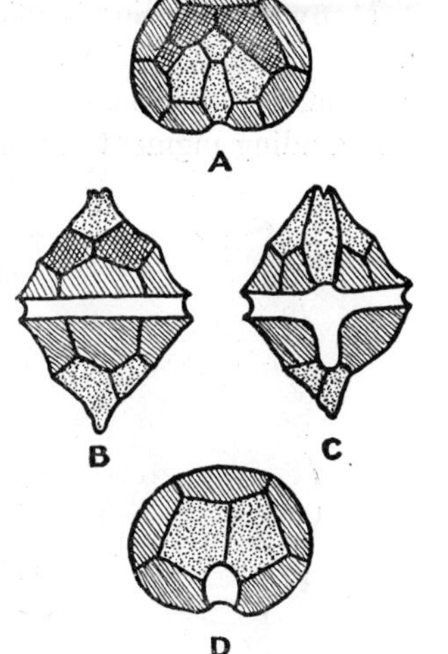

FIG. 427. Arrangement of the plates of *Peridinium wisconsinensis* Eddy. *A*, vertical view from above. *B–C*, lateral view from the front and rear. *D*, vertical view from below. Apical plates are in stipple, intercalary plates are cross-hatched, precingular and postcingular plates are shaded with diagonal lines.

Structure of the Protoplast. The chromatophores of Dinophyceae are quite variable in color and in shape. A majority of the species have rod-shaped, discoid, or irregularly band-shaped chromatophores at the periphery of the protoplasts. Some species have a stellate axial chromatophore with numerous radiating processes.[3] Many of the species have pyrenoids; these may be within the chromatophores or external to them.[4]

Cells of Dinophyceae store their photosynthetic reserves as starch or as oils. As a general rule,[5] the chief food reserve of fresh-water species is starch and that of marine species is an oil. In some cases starch formation is associated with pyrenoids; in other cases there are no pyrenoids and the starch is deposited either in the chromatophore or in the cytoplasm.[6] Mass ingestion of foods is found in both the "unarmored" and the "armored" dinoflagellates with chromatophores; this holozoic method of nutrition may be fully as important as the holophytic. Algae and protozoa are among the most easily recognized of the ingested foods, and in some of the cases reported for fresh-water dinoflagellates[7] the ingested organism has a size about half that of the dinoflagellate. The method by which

[1] KOFOID, 1909. [2] WERNER, 1910. [3] GEITLER, 1925*E*. [4] CONRAD, 1926.
[5] KLEBS, 1912; KILLIAN, 1924. [6] BÜTSCHLI, 1885; KLEBS, 1912.
[7] HOFENEDER, 1930; WOLOSZYSKŃA, 1917.

the armored species ingest food is not fully known, but it probably takes place by means of pseudopodia extruded from the girdle region.[1] Nutrition of species without chromatophores is wholly saprophytic or holozoic. Several species have short, radially disposed rodlets (*rhabdosomes*) in the peripheral portion of the cytoplasm. The precise nature of the rhabdosomes is unknown, but it has been thought that they are formed as a result of a saprophytic mode of nutrition.[2] Under certain conditions, for instance, when the cells are placed in a strong salt solution, the rhabdosomes are discharged from the cell.[3]

Many of the dinoflagellates, and most zoospores of Phytodinae, have a large eyespot of simple structure. In one family of marine forms,[2] there is a definite *ocellus* composed of two parts, a refractive hyaline lens, and a surrounding pigment mass.

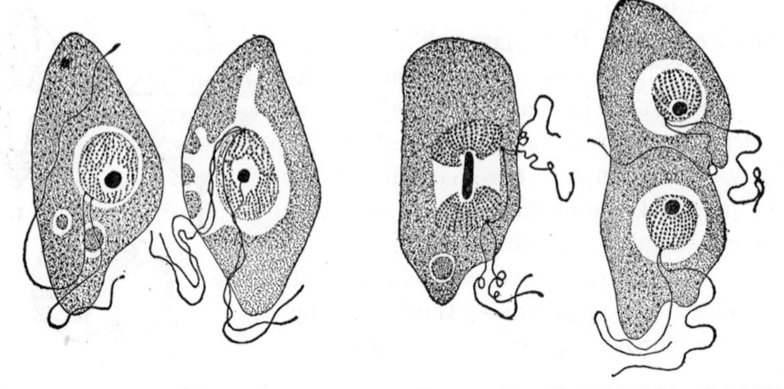

Fig. 428 Cell division of *Oxyrrhis marina* Duj. (*After Hall*, 1925*A*.) (× 1170.)

All of the dinoflagellates have a single centrally located nucleus of large size. The nucleus is surrounded by a distinct membrane and contains a conspicuous nucleole. Early karyological studies on dinoflagellates describe a reticulate arrangement of the chromatin and linin,[4] but more recent investigations[5] show that the resting nucleus has moniliform chromatin threads with a parallel or spiral arrangement. This arrangement persists even when the cells enter upon a resting aplanosporic condition. Nuclear division is mitotic and with or without a persistence of the nuclear membrane.

The two genera in which the neuromotor apparatus has been demonstrated, *Ceratium*[6] and *Oxyrrhis*,[7] have one of the blepharoplast-rhizoplast-centriole type. The centriole is extranuclear and connected with two

[1] Hofeneder, 1930. [2] Kofoid and Swezy, 1921. [3] Woloszyńska, 1927.
[4] Borgert, 1910; Jollos, 1910; Lauterborn, 1895.
[5] Entz, 1921; Hall, 1925. 1925*A*; Kofoid and Swezy, 1921.
[6] Hall. 1925. [7] Hall, 1925*A*.

widely diverging rhizoplasts, each extending to the plasma membrane and terminating in a blepharoplast (Fig. 428). One blepharoplast subtends the flagellum that encircles the cell; the other subtends the flagellum that extends longitudinally backward from the point of insertion. These two flagella always differ morphologically and in their method of movement.[1] The longitudinal flagellum is thread-like and waves in broad curves, or with an active vibration at the distal end. The transverse flagellum is usually, if not always, ribbon-like and moves in a spiral or undulatory manner. Nuclear division is accompanied by a division of the centriole into two daughter centrioles that remain connected with each other by a fine fibril, the paradesmose.[2] The two rhizoplasts, still attached to the original blepharoplasts, are distributed one to each daughter centriole, but each of the daughter centriles soon develops a second rhizoplast and blepharoplast.

The *pulsules* of dinoflagellates bear a superficial resemblance to contractive vacuoles but have a distinct membrane and are noncontractile. There are usually two pulsules, but sometimes there are more or less evanescent accessory pulsules.[1] Each pulsule apparatus consists of a sac-like vacuole that is connected with the cell's exterior by a slender canal opening into a flagellar pore of the cell wall. Pulsules are concerned with the intake of fluids into the protoplast and not, as might be supposed, with the discharge of liquids.[3]

Vegetative Multiplication. The usual method of multiplication in the dinoflagellates is by means of cell division. This may take place while the cells are actively motile, or they may come to rest before they divide. The plane of division is always more or less oblique. In the division of armored dinoflagellates, each of the daughter cells may receive a portion of the parent-cell wall, or the daughter cells may develop entirely new walls. *Ceratium* (Fig. 429) exemplifies the genera in which there is a retention of the parent-cell wall. Dividing cells of this dinoflagellate have the wall breaking along a predetermined zigzag line, in such a manner as to distribute certain precingular and postcingular plates to each daughter cell.[4]

Species with the daughter cells forming entirely new walls may have the protoplast dividing while still within the parent-cell wall, or may have it escaping from the parent-cell wall before it divides. Species in which the protoplast divides while still within the parent-cell wall may have the daughter protoplasts secreting walls before liberation, or may have them developing into *Gymnodinium*-like swarmers which do not form a wall until after liberation from the parent-cell wall.[5] Species where the proto-

[1] Kofoid and Swezy, 1921. [2] Hall, 1925*A*. [3] Kofoid, 1909.
[4] Kofoid, 1907; Lauterborn, 1895. [5] Diwald, 1938.

plast is liberated before division usually have it surrounded by a gelatinous envelope.

Asexual Reproduction. Asexual reproduction of phytodinads is by means of zoospores, aplanospores, or autospores. Both of the filamentous genera and certain of the coccoid genera have the protoplast of a cell dividing to form two, four, or eight naked gymnodinoid zoospores that are liberated through a pore in the parent-cell wall (Fig. 425) or are liberated by a gelatinization of the parent-cell wall.[1]

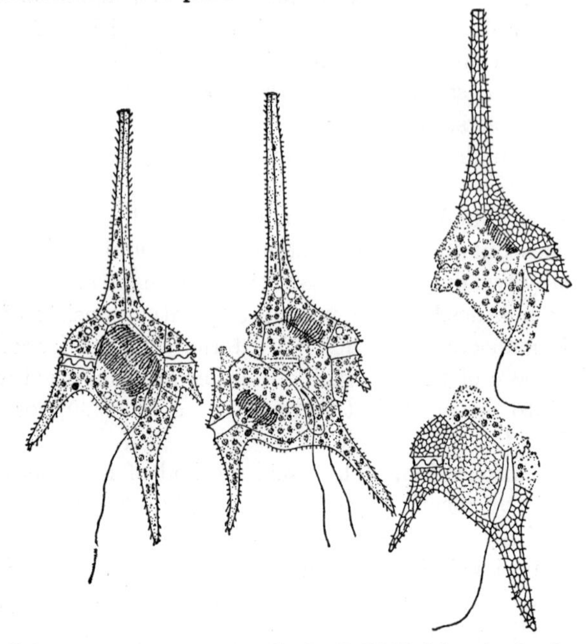

Fig. 429. Cell division of *Ceratium hirundinella* (O.F.M.) Schrank, showing the method of distribution of plates to daughter cells. (*After Lauterborn*, 1895.)

As is the case with Chlorophyceae and Xanthophyceae, division of the protoplast of a cell may be followed by a formation of aplanospores instead of zoospores. Aplanospores are usually globose, but in some cases they have a complicated outline and one quite unlike that of the vegetative. In many genera belonging to the Dinococcales, the aplanospores are of the same shape as a vegetative cell (Fig. 426). Such aplanospores are autospores.

The so-called cysts of dinoflagellates are comparable to aplanospores of phytodinads and other algae. Fresh-water dinoflagellates may form aplanospores at any time, but they usually form them in greatest abundance at

[1] Pascher, 1927.

the close of an active vegetative period.[1] A periodic formation of aplanospores has also been observed[2] in a dinoflagellate growing in culture. Aplanospore formation in marine species has been thought[3] to be correlated with holozoic nutrition and to be due to ingestion of large food bodies. Most dinoflagellates have the entire protoplast rounding up to form a single aplanospore but cases have also been reported[4] of a formation of two aplanospores. Formation of aplanospores may take place within the parent-cell wall, or the protoplast may escape from the wall before it rounds up and secretes a new thick wall. A majority of the dinoflagellates produce globose aplanospores, but certain of them form angular or lunate ones that may or may not have stout spines. The protoplast of a germinating aplanospore may develop directly into a gymnodinoid zoospore (Fig. 430*A–C*), or it may divide to form two such zoospores.

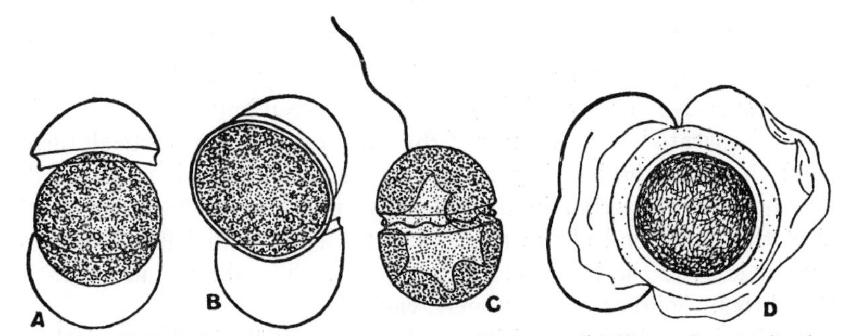

FIG. 430. *A–C*, Aplanospores of *Glenodinium uliginosum* Schilling. *D*, germination of aplanospore of *Hemidinium nasutum* Stein. (*From Woloszyńska*, 1925.)

Zoospores may be liberated by a splitting[5] or by a gelatinization[6] of the aplanospore wall (Fig. 430).

Sexual Reproduction. Relatively few students of Dinophyceae have recorded sexual reproduction. The older accounts of a fusion of aplanogametes were generally accepted with strong reservations because it was thought that they might have been based upon a misinterpretation of cell division. The description of a true conjugation of aplanogametes in *Ceratium*[7] seems fairly well substantiated. Here two cells become apposed to each other and establish a conjugation tube in which the two protoplasts unite to form a zygote.

A fusion of free-swimming gymnodinoid zoogametes has been described for *Glenodinium*.[8] This dinoflagellate was grown in culture and found to be heterothallic and to have a union of gametes only when the two came from clones of opposite sex. The zygote formed by a union of these

[1] WEST, G. S., 1909*A*. [2] DIWALD, 1938. [3] KOFOID and SWEZY, 1921.
[4] KLEBS, 1912. [5] DIWALD, 1938; WEST, G. S., 1909*A*. [6] WOLOSZYŃSKA, 1925.
[7] ENTZ, 1924; ZEDERBAUER, 1904. [8] DIWALD, 1938.

gametes is spherical and with a smooth wall. When it germinates, its protoplast divides to form four gynodinoid zoospores, and zoospore formation is preceded by a meiotic division of the zygote nucleus.

Classification. Evolution of Dinophyceae from a motile unicellular ancestor has paralleled that in Chlorophyceae, Xanthophyceae, and Chrysophyceae, and certain of the types in these classes have their counterpart among the Dinophyceae. Thus the Dinophyceae can be classified in a similar manner. The motile genera (dinoflagellates) have been segregated[1] into four orders; and the nonmotile genera (phytodinads) have also been segregated[2] into four orders. Two of the orders with motile vegetative cells and two of the orders with immobile vegetative cells are represented in the fresh-water flora of this country.

ORDER 1. GYMNODINIALES

Members of this order are without cell walls, but the naked cells may have a very firm periplast and one that is longitudinally striated. All species have a transverse furrow (girdle) that is a descending left-wound spiral with the separated ends connected to each other by a vertical furrow (sulcus) that may project beyond the upper or lower ends of the girdle. Both flagella are inserted in the sulcus, the transverse flagellum at the level of the upper end of the girdle, and the longitudinal flagellum at or below the level of the lower end of the girdle. The shape and coloration of the chromatophores are extremely variable.

Reproduction is by cell division, and this is usually in the vertical axis of a cell. Members of the order may also form aplanospores (cysts) that are spherical and surrounded by a definite wall.

All genera in the fresh-water flora of this country belong to the family Gymnodiniaceae.

FAMILY 1. GYMNODINIACEAE

This family includes the genera with cells which are not compound and which lack such specialized structures as ocelli, nematocysts, and tentacles.

The genera found in this country differ as follows:

1. Transverse furrow approximately transverse 2
1. Transverse furrow descending diagonally .... **2. Gyrodinium**
  2. Epicone and hypocone approximately equal .... **1. Gymnodinium**
  2. Epicone and hypocone unequal in size 3
3. Epicone longer and broader than hypocone .... **3. Massartia**
3. Hypocone longer and broader than epicone .... **4. Amphidinium**

[1] LINDEMANN, 1928. [2] PASCHER, 1931.

1. **Gymnodinium** Stein, 1883; emend., Kofoid and Swezy, 1921. The cells of *Gymnodinium* are approximately ovoid and with the portions above and below the transverse furrow (girdle) dissimilar in shape. The girdle is approximately equatorial in position and usually in a slight descending left spiral. The longitudinal furrow (sulcus) may extend to the cell apices, or only a short distance from the girdle. It often extends farther along the lower cell-half (hypocone) than the upper cell half (epicone). The cells are naked and with the surface of the cytoplasm smooth or longitudinally striated or ridged. There are two flagella: one transverse and encircling the cell in the girdle region, the other straight and directed backward from the hypocone portion of the sulcus. The nucleus is central or in the posterior portion of a cell. Most species have numerous, discoid, or narrowly elliptical, variously colored chromatophores, but some species are without them.

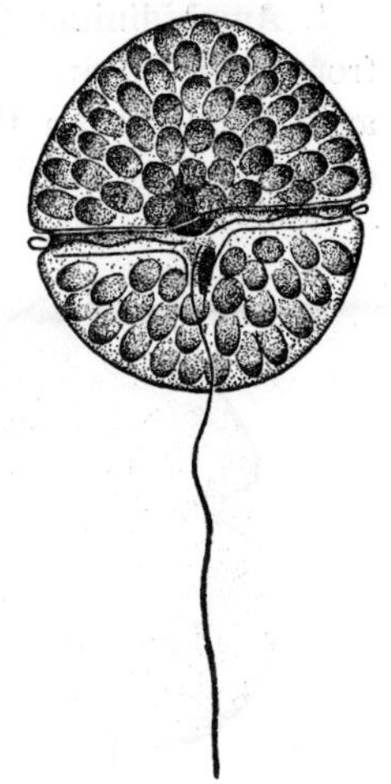

FIG. 431. *Gymnodinium neglectum* (Schilling) Lindem. (*From Thompson*, 1947.) (× 1060.)

Vegetative multiplication is by cell division and takes place while the cells are motile. The aplanospores are usually thin-walled and globose. In most cases, they are formed singly, but sometimes[1] a cell forms two of them.

A majority of the many species of *Gymnodinium* (Fig. 431) are marine, but 13 of them have been found in fresh waters in this country. For descriptions of several of them, see Thompson (1947); for a monographic treatment of the genus, see Kofoid and Swezy (1921).

2. **Gyrodinium** Kofoid and Swezy, 1921. The chief difference between this genus and other genera of naked gymnodinoid dinoflagellates is in the girdle. Here the girdle is a conspicuous descending left-hand spiral and one where the distance between upper and lower ends is at least one-fifth the length of the cell. In other respects, the cells resemble those of *Gymnodinium.*

G. pusillum (Schilling) Kofoid and Swezy (Fig. 432) has been found in Maryland.[2] For a description of it, see Thompson (1947).

3. **Massartia** Conrad, 1926. This genus differs from *Gymnodinium* and closely related genera in greater length and breadth of the upper cell-half (epicone) in comparison with the lower cell-half (hypocone).[3] In other respects, the cells are quite similar to those of *Gymnodinium.*

[1] KLEBS, 1912. [2] THOMPSON, 1947. [3] CONRAD, 1926*A*.

M. Musei (Danysz.) Schilling (Fig. 433) is known from Maryland,[1] and *M. vorticella* (Stein) Schilling has been found in Iowa.[2] For a description of the former, see Thompson (1947); for the latter, see Schilling (1913), under *Gymnodinium vorticella* Stein.

4. **Amphidinium** Claparède and Lachmann, 1858. This genus differs from other Gymnodiniaceae in that the lower cell-half (hypocone) is longer and broader than the upper cell-half (epicone). In other respects, the cells are quite similar to *Gymnodinium*.

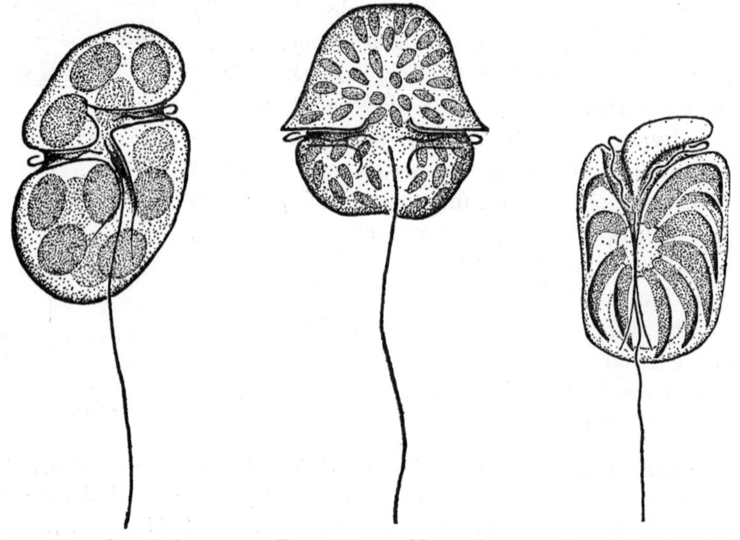

Fig. 432. *Gyrodinium pusillum* (Schilling) Kofoid and Swezy. (*From Thompson*, 1947.) (× 1615.)

Fig. 433. *Massartia Musei* (Danysz.) Schiller. (*From Thompson*, 1947.) (× 1615.)

Fig. 434. *Amphidinium Klebsii* Kofoid and Swezy. (*Drawn by R. H. Thompson.*) (× 1615.)

Dr. R. H. Thompson writes that he has found *A. Klebsii* Kofoid and Swezy (Fig. 434), a marine species, in a fresh-water pond close to the ocean in Maryland. For a description of it, see Kofoid and Swezy (1921).

ORDER 2. PERIDINIALES

The Peridiniales have biflagellate, solitary, motile, vegetative cells surrounded by a wall composed of a definite number of plates arranged in a specific manner. Most members of the order are marine, but genera belonging to four of the numerous families recognized in the system of Lindemann[3] are found in the fresh-water flora of this country.

[1] Thompson, 1947 [2] Prescott, 1927. [3] Lindemann, 1928.

Family 1. Glenodiniaceae

This family differs from others of the order in the much thinner walls, which may be so thin that the cells seem to be naked. In spite of its thinness, the wall consists of a definite number of polygonal plates of unequal size arranged in a definite manner.

The two genera found in fresh waters in this country differ as follows:

1. Transverse furrow completely encircling cell .... 1. **Glenodinium**
1. Transverse furrow incompletely encircling cell .... 2. **Hemidinium**

1. **Glenodinium** Stein, 1883. Cells of *Glenodinium* (Fig. 435) are asymmetrically globose and sometimes somewhat dorsoventrally flattened They are surrounded by a thin wall with a definite number of plates.[1] The number of precingular, anterior intercalary, and apical plates is variable from species to species. The lower half-wall of a cell (hypotheca) has

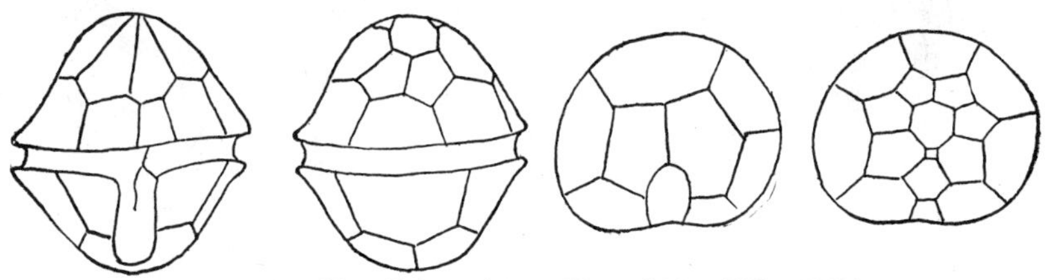

Fig. 435. *Glenodinium cinctum* Ehr. (*After Eddy*, 1930.)

five or six postcingular and two antapical plates. All plates are generally smooth and with delicate sutures between them. The girdle is median, a very slight spiral, and completely encircles the cell. The sulcus is usually restricted to the posterior half of a cell. Most species have numerous chromatophores, and some have a distinct eyespot.

Aplanospores may be globose or angular, and they usually have a heavy wall.

Sexual reproduction is isogamous and by a fusion of gymnodinoid gametes.[2] The zygote is globose, smooth-walled, and germinates to form four gymnodinoid zoospores after a meiotic division of the nucleus.

Ecologically this genus is unusual in that a majority of the 20 or more species are found in fresh waters. Thirteen species have been found in fresh water in the United States. For a monographic treatment of the genus, see Schiller (1933).

2. **Hemidinium** Stein, 1833. The cells of *Hemidinium* are ellipsoidal, somewhat compressed, and with a girdle that incompletely encircles the cell in a descending left spiral. The sulcus is restricted to the lower half-

[1] Lindemann, 1928; Woloszyńska, 1917*A*. [2] Diwald, 1938.

cell wall (hypotheca). For a long time, this dinoflagellate was thought to have a naked protoplast but it was eventually shown[1] to have a delicate wall and one composed of a definite number of plates. The upper half-wall (epitheca) has six precingular and six apical plates; the hypotheca has five postcingular, one posterior intercalary, and one antapical plates. The protoplast contains numerous yellowish to brownish chromatophores. An eyespot may or may not be present. The nucleus is ellipsoidal and lies in the lower half of a cell.

Cell division takes place while cells are motile and in an approximately transverse plane. The aplanospores are spherical and thin-walled. A germinating aplanospore[1] gives rise to a spherical nonflagellated cell surrounded by a broad, concentrically stratified, gelatinous sheath. Division of such cells may result in a *Gloeodinium*-like colony.

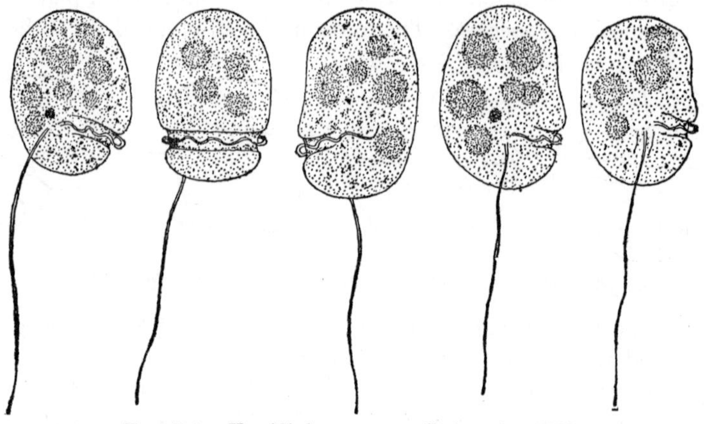

FIG. 436. *Hemidinium nasutum* Stein. (× 975.)

This genus is readily differentiated from other fresh-water dinoflagellates of this country by its incomplete girdle. The two species known for the United States are *H. nasutum* Stein (Fig. 436) and *H. ochraceum* Levander. For descriptions of them, see Thompson (1947).

FAMILY 2. GONYAULACACEAE

This is one of the families with thick walls with clearly evident plates. It differs from other families in which the wall is thick in that the hypotheca has but one antapical plate and has this adjoined by an intercalary plate.

Gonyaulax is the only genus with fresh-water species in this country.

1. **Gonyaulax** Diesing, 1866; emend., Kofoid, 1911. This genus was first segregated from *Peridinium* because of the pronounced helicoid girdle.

[1] WOLOSZYŃSKA, 1925.

The upward extension of the sulcus to the cell apex was formerly given as another distinctive character, but this is not true of all species.[1] The number and arrangement of the plates differ from that in *Peridinium*. *Gonyaulax* has a hypotheca with six postcingular, one posterior intercalary, and one antapical plate, in contrast with the five postcingular and

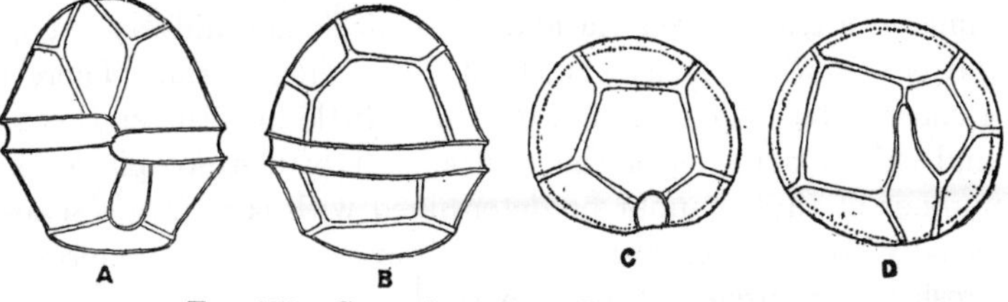

FIG. 437. *Gonyaulax palustre* Lemm. (× 650.)

two antapical plates of *Peridinium*. Epithecae of *Gonyaulax* have six precingular, one to three anterior intercalary, and one to six apical plates. The cells usually contain several discoid yellowish-brown chromatophores.

G. palustre Lemm. (Fig. 437) is the only fresh-water species thus far found in this country. For a description of it, see Eddy (1930).

FAMILY 3. PERIDINIACEAE

This is another of the families with thick walls with clearly evident plates. It differs from other families in which the wall is thick in that the hypotheca has two antapical plates.

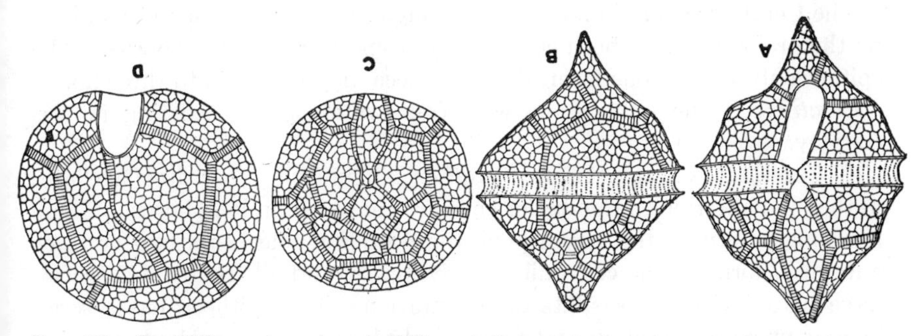

FIG. 438. *Peridinium wisconsinensis* Eddy. *A–B*, lateral views from the front and rear. *C*, vertical view from above. *D*, vertical view from below. (× 650.)

Peridinium is the only genus with fresh-water species in this country.

1. **Peridinium** Ehrenberg, 1830; emend., Stein, 1833. Cells of *Peridinium* (Fig. 438) are free-swimming and have a circular, oval, or angular outline as seen in front view. The cell wall is thick and with the sutures

[1] KOFOID, 1911.

between the plates strongly evident. Some species have small spines at the posterior end of a cell. The wall has six to seven precingular, none to eight anterior intercalary, and three to five apical plates in the epitheca; the hypotheca has five postcingular and two antapical plates. Plates of most species are conspicuously ornamented with spines or with a reticulum of small ridges. Sutures between the plates are broad and often with a longitudinal or a transverse striation. The girdle lies midway between the poles and is transverse or a very slight descending spiral. There is generally a conspicuous sulcus that extends but little into the epitheca. The girdle and sulcus portions of the wall consist of distinct plates, but the exact number of these cannot be determined with certainty. The protoplasts usually contain numerous yellowish-brown chromatophores. Some of the fresh-water species have an eyespot.

Cells of *Peridinium* become immobile before they divide. Sometimes there is a contraction and division of the protoplast while still within the wall, but more often the protoplast escaped from the wall before dividing.[1] Aplanospores of *Peridinium* are usually thick-walled and globose. Certain species produce thin-walled aplanospores that soon germinate into a *Gloeodinium*-like stage.[2]

Approximately 20 fresh-water species of *Peridinium* have been recorded for the United States. For a monograph of the fresh-water species of the genus, see Lefèvre (1932).

Family 4. Ceratiaceae

The Ceratiaceae are immediately distinguishable from other Peridiniales by the prolongation of both epitheca and hypotheca into long horns. The epitheca always has one horn; the hypotheca may have one to three horns.

Ceratium is the only genus with fresh-water species existing in this country.

1. **Ceratium** Schrank, 1793. *Ceratium* differs from other dinoflagellates in that the anterior and posterior ends are continued in long horns. There is a single horn at the apical pole; the posterior pole bears two or three antapical horns. The cell wall is fairly heavy, and all fresh-water species have plates with a reticulate ornamentation. The epitheca has a series of four precingular plates and four apical plates, the vertical extension of the latter resulting in the apical horn.[3] The hypotheca has five postcingular and two antapical plates. The girdle is transverse and interrupted by a large membranaceous *ventral plate* articulated with the precingular and postcingular plates. The protoplast has numerous discoid chromatophores and a large centrally located nucleus.

[1] Lindemann, 1919. [2] West, G. S., 1909*A*. [3] Kofoid, 1907, 1908.

Cell division takes place while the cells are motile and with an intact wall. Following the division of nucleus and neuromotor apparatus,[1] there is a diagonal division of the protoplast. One of the daughter cells (Fig. 429, page 522) is clothed with the four apical, two precingular, and three postcingular plates of the parent-cell wall; the other daughter cell receives two precingular, two postcingular, and the two antapical plates of the parent-cell wall.[2] This is followed by a reciprocal development of the missing plates and horns by each daughter cell.

Most studies of aplanospore formation have been upon a fresh-water species, *C. hirundinella* (O.F.M.) Schrank. Its aplanospores are angular, thick-walled, and formed by a retraction of the protoplast from the wall of the vegetative cell. The aplanospores are uninucleate (rarely binucleate)[3] and densely packed with reserve foods, chiefly glycogen.[4] They are strongly resistant to desiccation and remain viable for several years. When an aplanospore germinates,[5] the protoplast develops into a naked

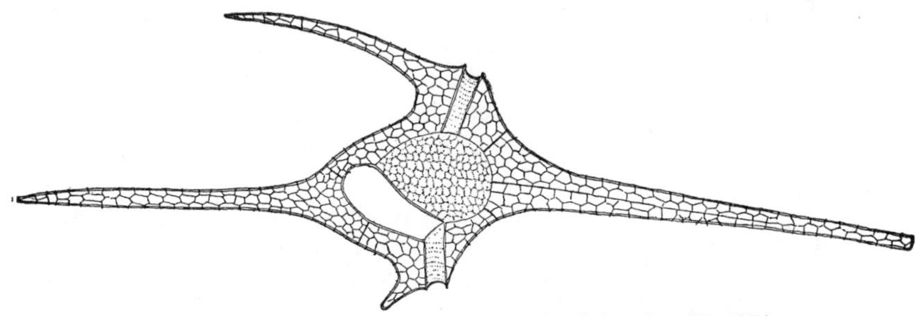

FIG. 439. *Ceratium hirundinella* (O.F.M.) Schrank. (× 485.)

zoospore that is *Gymnodinium*-like when first liberated, but which assumes a *Ceratium*-like shape within a few hours and begins to secrete a wall.

Sexual reproduction is by an apposition of two cells and an establishment of a conjugation tube in which the two protoplasts unite to form a zygote.[6]

Only two of the many species are found in fresh water and both of them, *C. caroliniana* (Bail.) Whitford and *C. hirundinella* (O.F.M.) Schrank (Fig. 439), have been found in the United States. *C. hirundinella* is widely distributed in the plankton of ponds and lakes. Seasonal changes in the environment, especially changes in temperature, have a pronounced effect upon the shape of cells of this species. For descriptions of *C. caroliniana* (as *C. curvirostre* Huitf.-Kaas) and of *C. hirundinella*, see Schilling (1913).

[1] ENTZ, 1921; HALL, 1925. [2] KOFOID, 1908; LAUTERBORN, 1895.
[3] HALL, 1925. [4] ENTZ, 1925. [5] HUBER and NIPKOW, 1922.
[6] ENTZ, 1924; ZEDERBAUER, 1904.

ORDER 3. DINOCAPSALES

The Dinocapsales are palmelloid Dinophyceae with a temporary motile gymnodinoid stage. The order corresponds to the Tetrasporales of Chlorophyceae, Heterocapsales of Xanthophyceae, and Chrysocapsales of Chrysophyceae.[1] There is but one family.

FAMILY 1. GLOEODINIACEAE

The two genera found in this country differ as follows:

1. Cellular envelope with many layers .... 2. **Urococcus**
1. Cellular envelope homogeneous or with few layers .... 1. **Gloeodinium**

1. **Gloeodinium** Klebs, 1912. *Gloeodinium* has large subspherical nonflagellated cells united in small packet-like colonies by a common, homogeneous or concentrically stratified, gelatinous envelope. The cells are uninucleate and contain many small brownish chromatophores that are sometimes radially arranged.[2] The protoplasts contain considerable starch, varying amounts of a colorless oily substance, and droplets of oil. A cell divides into two daughter cells that are retained within the parent-cell envelope for a considerable time, and often until one or both daughter cells have divided. Such colonies rarely develop beyond the four- or eight-celled stage because of gelatinization and disintegration of the original envelope.

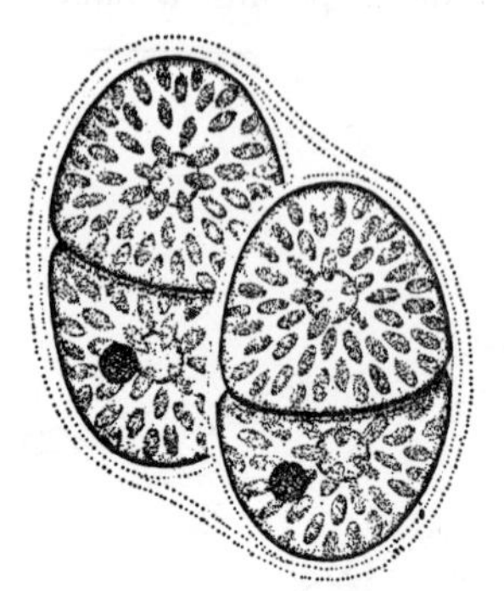

FIG. 440. *Gloeodinium montanum* Klebs. (*From Thompson*, 1949.) (× 725.)

Reproduction may also be by means of naked gymnodinoid zoospores, and this occurs more frequently at certain times of the year than at others.[3]

The only known species, *G. montanum* Klebs (Fig. 440), has been found in Maryland, Wisconsin, and Ohio. For a description of it, see Klebs (1912).

2. **Urococcus** Kützing, 1849. The cells of *Urococcus* are spherical, and each is surrounded by a firm gelatinous envelope with many concentric strata. At first, the sheath is of uniform thickness; later it becomes much thicker on one side and with concentric layers that do not completely encircle the earlier formed ones. The cells are usually solitary, but at times two to four, each with a conspicuous sheath, lie within a common

[1] PASCHER, 1927. [2] KILLIAN, 1924; KLEBS, 1912; THOMPSON, 1949.
[3] KILLIAN, 1924; THOMPSON, 1949.

sheath derived from a parent cell. The entire protoplast is brownish and with the central portion darker than the peripheral portion. It is impossible to determine whether the coloration is due to a single chromatophore or to many chromatophores so densely crowded against one another that there appears to be but one. The protoplast contains numerous granules (starch ?) and a few large to small reddish globules of oil.

Reproduction is usually by cell division, but there may also be a formation of naked gymnodinoid zoospores.

It is an almost universal practice among phycologists to refer *Urococcus* to the Chlorophyceae. In a recent letter Prof. R. H. Thompson suggests that *U. insignis* (Hass.) Kütz. (Fig. 441) should be placed among the Dinophyceae since he has discovered that it produces gymnodinoid zoospores. Other species assigned to the

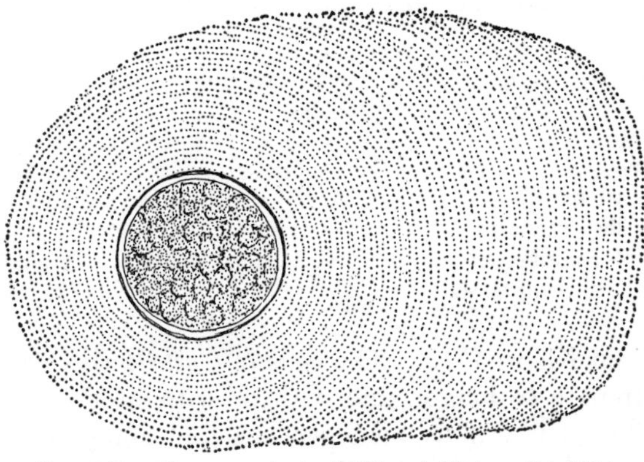

FIG. 441. *Urococcus insignis* (Hass.) Kütz. (× 650.)

same genus as *U. insignis* cannot be placed in this genus with certainty until their cells have been critically studied in a living condition. There are several records of the occurrence of *U. insignis* in this country. For a description of it, see Collins, (1909).

ORDER 4. DINOCOCCALES

The Dinococcales are unicellular, have homogeneous cellulose walls, and have globose, lunate, pyramidate, or stellate vegetative cells that are without transverse grooves or flagella. The cells contain numerous brown chromatophores and store their photosynthetic reserves as starch or as oils. The cells do not divide vegetatively, and new cells are formed by a division of the cell contents into zoospores or autospores. Zoospores bear a very close resemblance to species of *Gymnodinium* and have the same naked protoplasts, transverse furrow, eyespot, and flagellation as is found in that genus.

Immobile cells of certain genera, as *Cystodinium* and *Hypnodinium*, have been considered cyst-like in nature, and these genera have been placed[1] in the Gymnodiniales because of the structure of their motile stages. Since the motile phase in these genera is very transitory, it seems more logical to follow those[2] who interpret the motile stages as zoospores and consider these genera homologous with zoospore-producing unicellular Chlorococcales.

Family 1. Phytodiniaceae

Members of this family have variously shaped, solitary, free-living cells that may be sessile or free-floating. The protoplast of a cell is completely surrounded by a homogeneous cellulose wall.

Reproduction is by means of gymnodinoid zoospores or of autospores.

The genera found in this country differ as follows:

1. Cells free-floating 2
1. Cells sessile 3
   2. Cells lunate .... 1. **Cystodinium**
   2. Cells globose .... 2. **Hypnodinium**
3. Cells globose to ovoid .... 3. **Stylodinium**
3. Cells more or less angular 4
   4. Body of cell compressed .... 5. **Raciborskia**
   4. Body of cell not compressed .... 4. **Tetradinium**

1. **Cystodinium** Klebs, 1912. The cells of *Cystodinium* are solitary and free-floating, arcuate to lunate, and with or without the poles prolonged into spines. The cell wall is homogeneous and composed of cellulose.[3] The protoplast, which completely fills the wall contains numerous golden-brown to chocolate-brown chromatophores, a conspicuous nucleus, and a relatively large red globule of oil.

The cells do not divide vegetatively. Instead, the protoplast divides to form two or four daughter protoplasts, each with a girdle, sulcus, and an eyespot.[4] These may become zoospores that are liberated from the parent-cell wall. Liberated zoospores remain motile for a very short time (usually less than half an hour) and begin developing into typical vegetative cells immediately after they become immobile.[3] In other cases, the gymnodinoid daughter cells never develop evident flagella, assume the shape characteristic of vegetative cells while still within the parent-cell wall, and form walls before they are liberated by a gelatinization of the parent-cell wall.[5]

[1] Klebs, 1912; Kofoid and Swezy, 1921; Lindmann, 1928.
[2] Fritsch, 1935; Pascher, 1927. [3] Klebs, 1912.
[4] Klebs, 1912; Thompson, 1949. [5] Pascher, 1927; Thompson, 1949.

C. bataviense Klebs has been found in Maryland and Kansas, and *C. iners* Geitler (Fig. 442) has been found in Maryland.[1] For a description of *C. bataviense*, see Klebs (1912); for *C. iners*, see Geitler (1928*E*).

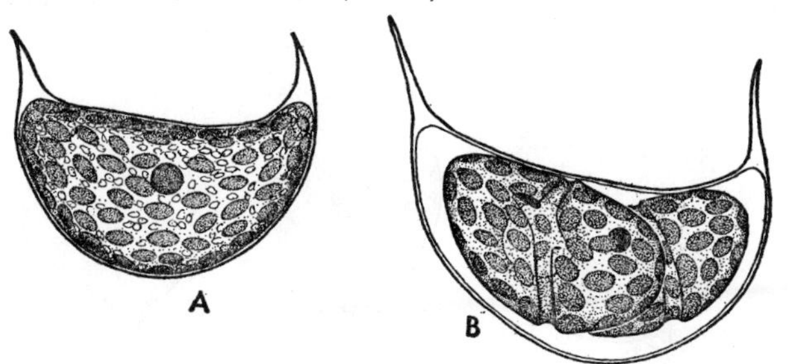

FIG. 442. *Cystodinium iners* Geitler. *A*, vegetative cell. *B*, formation of zoospores. (*From Thompson*, 1949.) (× 725.)

2. **Hypnodinium** Klebs, 1912. *Hypnodinium* has free-floating, solitary, spherical cells with a homogeneous cellulose wall. The protoplast contains numerous yellowish-brown chromatophores grouped in rosettes and the rosettes forming a reticulum.[2] The protoplast has a permanent gymnodinoid organization into girdle and sulcus, but without flagella, and usually contains a conspicuous globule of reddish oil.

At the time of reproduction, the protoplast contracts and divides to form two autospores which form a wall before liberation by a bursting of the parent-cell wall.[2]

A formation, liberation, and fusion in pairs of many small motile gymnodinoid cells have also been observed.[3]

H. sphaericum Klebs (Fig. 443) has been found in Maryland.[4] For a description of it, see Klebs (1912).

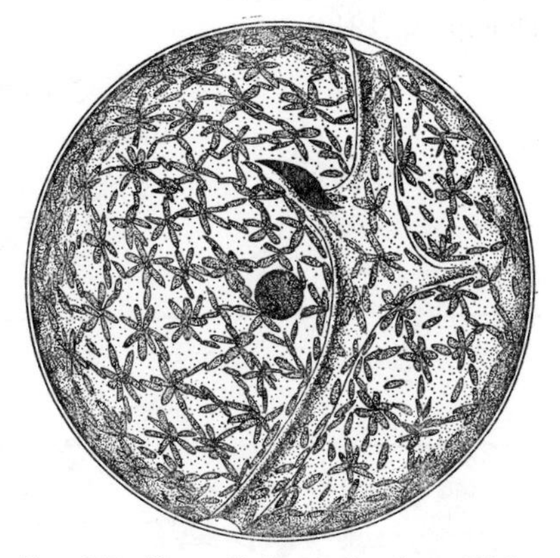

FIG. 443. *Hypnodinium sphaericum* Klebs. (*From Thompson*, 1949.) (× 1725.)

3. **Stylodinium** Klebs, 1912. *Stylodinium* has solitary, stipitate, globose to ovoid, or somewhat quadrate cells. The stipe is composed of gelatinous material, and its length varies from less than half to more than double the diameter of a cell. The protoplast contains numerous parietal, discoid to elliptical, golden-brown chromatophores and a conspicuous globule of reddish oil.

[1] THOMPSON 1949. [2] KLEBS, 1912; THOMPSON, 1949. [3] PASCHER, 1914.
[4] THOMPSON, 1949.

Reproduction may be by a bipartition into two gymnodinoid zoospores that are liberated by a rupture of the parent-cell wall.[1] There may also be a rounding up of the protoplast into a single globose aplanospore that is also liberated by rupture of the parent-cell wall.[2]

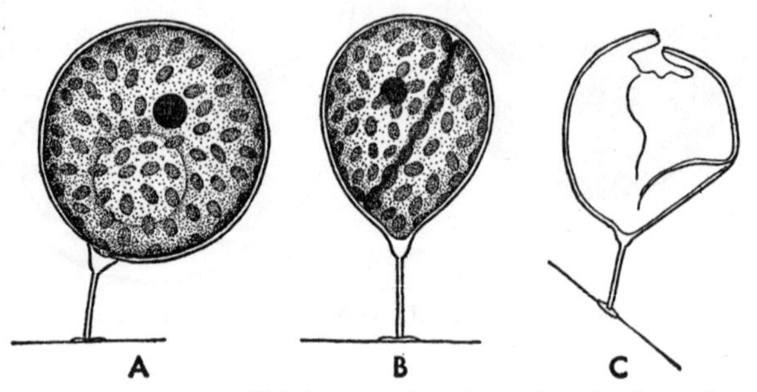

Fig. 444. *Stylodinium globosum* Klebs. *A*, vegetative cell. *B*, formation of zoospores. *C*, cell after liberation of zoospores. (*From Thompson*, 1949.) (× 725.)

S. globosum Klebs (Fig. 444) and *S. longipes* Thompson have been found in Maryland.[3] For a description of *S. globosum*, see Klebs (1912); for *S. longipes*, see Thompson (1949).

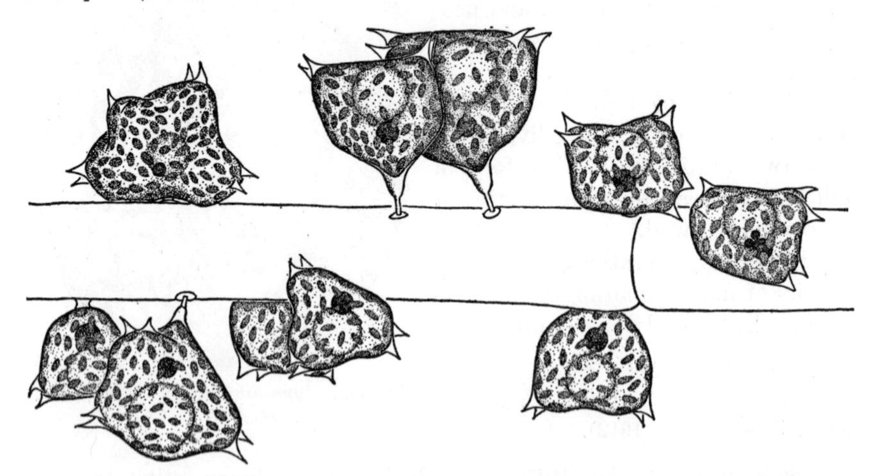

Fig. 445. *Tetradinium javanicum* Klebs. (*From Thompson*, 1949.) (× 725.)

4. **Tetradinium** Klebs, 1912. *Tetradinium* has solitary epiphytic pyramidate cells whose angles are prolonged into simple or bifurcate spines. The cell is attached to the host by a short stipe which may be so short that the cell appears to be sessile.[3] The cell wall is homogeneous and composed of cellulose. The protoplast contains numerous ellipsoidal golden-

[1] Pascher, 1927; Thompson, 1949. [2] Klebs, 1912. [3] Thompson, 1949.

brown to chocolate-brown chromatophores most of which are parietal, a large nucleus, granules of starch, and a single conspicuous globule of reddish oil.

Reproduction is by division of the protoplast into two gymnodinoid zoospores which are liberated through a rent in the parent-cell wall.[1]

T. javanicum Klebs (Fig. 445) has been found epiphytic upon filamentous algae and upon submerged phanerogams, including *Myriophyllum* and *Potamogeton*, in Maryland.[2] For a description of it, see Klebs (1912).

5. **Raciborskia** Woloszyńska, 1919. *Raciborskia* has solitary, epiphytic, compressed, more or less bean-shaped cells with a short stout spine at both poles of the major axis. The convex face of a cell is attached to the host by a short stipe midway between the poles of a cell. The cell wall is homogeneous. The protoplast contains numerous, irregularly band-shaped, parietal, brown chromatophores; or it has the chromatophores

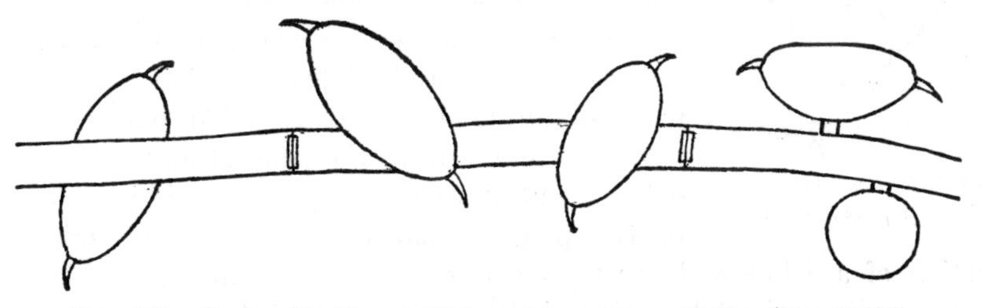

FIG. 446. *Raciborskia bicornis* Woloszyńska. (*From Woloszyńska*, 1919.)

aggregated in a parietal network.[3] Some species have a single large pyrenoid. There are also frequently numerous starch grains within a cell.

Reproduction is by division of a protoplast into two gymnodinoid zoospores which, after liberation and swimming about for a time, come to rest on a substratum and metamorphose into vegetative cells.[3]

Prof. G. W. Prescott writes that he has found the type species, *R. bicornis* Wolosz. (Fig. 446) in Wisconsin. For a description of it, see Woloszyńska (1919).

FAMILY 2. BLASTODINIACEAE

The Blastodiniaceae are always parasitic upon animals, either as ectoparasites or as endoparasites. Vegetative cells are immobile, without flagella, and amoeboid or surrounded by a wall through which a haustorium projects.

Reproduction is generally by a formation of gymnodinoid zoospores.

[1] THOMPSON, 1949; WOLOSZYŃSKA, 1919. [2] THOMPSON, 1949.
[3] PASCHER, 1932*D*; WOLOSZYŃSKA, 1919.

This family is a highly artificial one in which the various genera have relatively little in common beyond their gymnodinoid zoospores and their parasitism upon animals. Because of their naked motile swarmers, the family has been placed among the Gymnodiniales.[1] Such a treatment ignores the fact that the motile phase of the life cycle is transitory and that the immobile phase is the dominant one and vegetative in nature. *Oödinium,* the only genus known for this country, has much in common with genera obviously belonging to the Dinococcales.

1. **Oödinium** Chatton, 1912. *Oödinium* is a unicellular ectoparasite growing on various animals. The cell consists of a globose to pyriform portion, surrounded by a cellulose wall, and a naked branched rhizoidal portion whose attenuated branches penetrate tissues of the host. The portion of the protoplast within the wall contains numerous chromatophores, granules of starch, and a conspicuous nucleus. A cell increases in size to several times its original diameter, but remains uninucleate during the entire period of enlargement.[2]

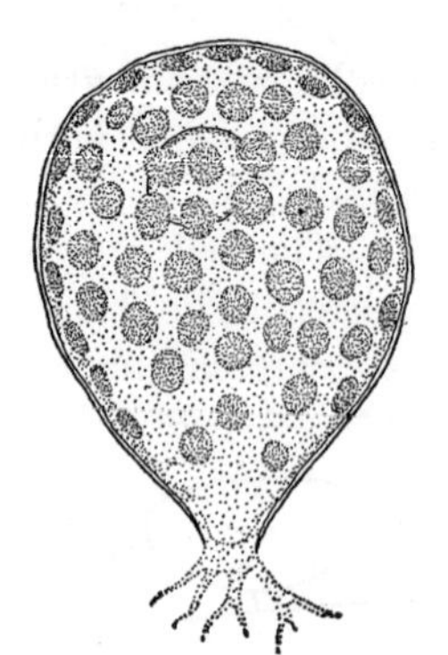

FIG. 447. *Oödinium limneticum* Jacobs. (× 750.)

Reproduction may take place at any stage during enlargement of a cell. First, there is a retraction of the pseudopodial rhizoidal portion and a secretion of a cellulose layer that closes the opening through which the pseudopodia projected.[3] There is then a division of the nucleus, a bipartition of the protoplast, and a secretion of a wall about each of the two daughter protoplasts. The two daughter cells within the old parent-cell wall divide simultaneously, and simultaneous divisions may continue until there are 128 daughter cells. The contents of each cell of the last generation then divide to form two gymnodinoid zoospores. The zoospores escape from the surrounding walls, swim freely through the water, and develop into vegetative cells after becoming attached to the host

The only known fresh-water species is *O. limneticum* Jacobs (Fig. 447), which was discovered parasitic on the epidermal covering of any part of the body of nine species of fish in Minnesota.[4] For a description of it, see Jacobs (1946).

[1] FRITSCH, 1935; GRAHAM (in press); LINDEMANN, 1928.
[2] BROWN, 1934; JACOBS, 1946; NIGRELLI, 1936.
[3] JACOBS, 1946; NIGRELLI, 1936. [4] JACOBS, 1946.

CHAPTER 9

DIVISION CYANOPHYTA

The Cyanophyta, or blue-green algae, are a distinctive group sharply delimited from other algae in a number of respects. They are the only algae in which the pigments are not localized in definite chromatophores. The pigments are localized in the peripheral portion of the protoplast and include chlorophyll *a*, carotenes, and distinctive xanthophylls (see Table I, page 3). In addition there is a blue pigment (c-phycocyanin) and a red pigment (c-phycoerythrin). Another unique feature of Cyanophyta is a primitive type of nucleus, the *central body*, which lacks a nucleolus and a nuclear membrane. Equally important, although negative in character, are the lack of flagellated reproductive cells and the total lack of gametic union in all members of the division.

All the Cyanophyta are placed in a single class, the Myxophyceae (Cyanophyceae).

Occurrence. Myxophyceae are found in a wide variety of habitats, and one can rarely examine a collection of fresh-water algae without encountering at least a few individuals of the class. Samples taken from permanent or semipermanent pools usually contain representatives of both the filamentous and the nonfilamentous genera, but neither type is ordinarily the dominant alga of the station. At times, one may encounter a pool which is practically a pure culture of some blue-green alga, usually an *Anabaena* or an *Oscillatoria*. On the other hand, temporary pools and puddles that have been standing but a week or two usually have an algal flora composed entirely of blue-green algae, and one in which a single species frequently predominates. Species of *Oscillatoria* and *Phormidium* are most frequently encountered in such pools.

Blue-green algae are always present in plankton catches from fresh-water lakes and ponds. These are usually species of *Chroococcus*, *Coelosphaerium*, *Polycystis*, or *Anabaena*. The proportion of Myxophyceae with respect to other algae is dependent on both the time of the year and the chemical composition of the water. The blue-greens usually occur in abundance only during the warm months of a year, but certain species, as *Oscillatoria prolifica* (Grev.) Gom., may also develop in quantity during the winter.[1] Soft-water lakes ordinarily have more Chroococcales than

[1] Smith, G. M., 1920.

other Myxophyceae, but these never predominate at any season of the year.[1] Hard-water lakes, on the other hand, exhibit a pronounced seasonal variation in the volume and percentage of Myxophyceae, which are usually the dominant organisms during the late summer and early fall. Not infrequently, in such lakes, there is a development of one or two species to such an extent that the water is discolored by them. This water bloom, which may also be caused by algae of other classes, may be of sporadic occurrence or may occur annually.[2] In the case of some small ponds in California, the water is in "bloom" throughout the year. Instances have been recorded where the blooming of a lake has caused the death of fish[3] or where the water is injurious to livestock drinking it,[4] but such effects are very uncommon. The disagreeable tastes and odors caused by the death and decay of the algae causing the bloom are of far greater economic importance, especially in lakes and reservoirs used for domestic water supplies, and sanitary engineers have studied at length the problem of eradicating algal growths.[5]

Blue-green algae constitute an important part of the algal flora of many subaerial habitats, and representatives of certain genera are usually found in such stations. Cliffs moistened by the spray from waterfalls and dripping rocky ledges are especially rich in Myxophyceae. Frequently there is a very definite correlation between the chemical composition of the substratum and the algae growing upon it. *Stigonema minutum* (Ag.) Hass., for example is reported[6] as occurring only on siliceous rocks, while in the same region *Schizothrix Lenormandiana* Gom. and *Symploca dubia* Gom. are restricted to moist calcareous rocks. Even more widespread is the growth of blue-greens on the surface of the soil or at some distance below its surface. These terrestrial algae are not usually conspicuous, but in certain parts of the country, especially those regions with a pronounced rainy season, the soil algae may develop to such an extent that they form patches of several square yards extent. In rare cases, the development of the soil algae is so extensive that it affects the color of the landscape.[7] The Myxophyceae are an important element in the subterranean algal flora. Subterranean Myxophyceae are generally restricted to the upper 50 cm. of the soil, although they have been found as deep as 2 m. below the surface.[8]

Blue-green algae growing within, and in the outflow from, hot springs

[1] Smith, G. M., 1924; Weisenberg-Lund, 1905; West, W. and G. S., 1909.

[2] See G. M. Smith, 1924, for literature.

[3] Baldwin and Whipple, 1906; Haine, 1918.

[4] Francis, 1878; Gillam, 1925; Nelson, 1903.

[5] See Whipple, 1927, pp. 49–70, for a discussion of this subject.

[6] Frémy, 1925. [7] West and Fritsch, 1927.

[8] Moore and Carter, 1926; Moore and Karrer, 1919.

have always excited the botanist's interest and have been studied in practically every country in which there are hot springs. Among the thousands of such springs in the western part of the United States, the best known and most thoroughly investigated are those of Yellowstone National Park. Here blue-green algae have been found growing in water with a temperature of 85° C.[1] Practically all the thermal algae are species which have become adapted to hot waters and which are not found elsewhere.

In hot springs whose waters are highly charged with soluble calcium and magnesium compounds, especially bicarbonates, the algae cause a precipitation of the calcium and magnesium salts in the form of an insoluble carbonate. The amount of carbonates thus precipitated is so considerable that the material deposited (travertine) may attain a thickness of 2 to 4 mm. during the course of a week. The terraces of travertine thus formed are usually brilliantly colored by the overlying layer of algal material. A deposition of carbonates through the agency of blue-green algae may take place elsewhere than in hot springs. Marl deposits formed at the bottom of shallow lakes are thought to be due almost wholly to the action of blue-green algae.

The action of perforating blue-green algae which grow within the shells of molluscs is just the reverse of that described in the preceding paragraph. Here, the algae cause a change of insoluble calcium and magnesium compounds into soluble ones. A similar change into soluble compounds has been ascribed to a blue-green alga growing on calcite veins in lithothamnic limestones.[2]

Many Myxophyceae regularly grow in association with other organisms. In some cases, the association is one of simple epiphytism, as *Chamaesiphon* growing on filaments of Cladophoraceae. In other cases, the alga is endophytic or endozoic and, according to the species, grows within or between cells of the organism within which it grows. The association may be one of parasitism, helotism, or symbiosis. The best examples of a true parasitism are the Oscillatoriaceae which live within the digestive tracts of man and other animals.[3] The blue-green algae growing within the protoplasts of cells of rhizopods, cryptomonads, diatoms, Tetrasporales, Chlorococclaes, and Phycomycetes are clear-cut cases of helotism. With the exception of diatoms, all these hosts are colorless, even though they belong to genera that normally contain chromatophores. The enslaved blue-green alga furnishes not only sufficient carbohydrates to meet the metabolic needs of the host, but also carbohydrates in sufficient abundance to permit accumulation of food reserves by the host. The storing of food reserves is not due to a saprophytic mode of nutrition because it has been

[1] Copeland, J. J., 1936. [2] Müller, H., 1923.
[3] Langeron, 1924; Petit, A., 1926.

shown[1] that host individuals without endocellular blue-green algae lack reserve foods.

The symbiotic relationship between a blue-green alga and another plant may be a case of "space parasitism," or the symbiosis may also involve nutrition relationships. A nutritional relationship is clearly the case with those lichens where the algal component is regularly a blue-green alga and one belonging to such genera as *Chroococcus*, *Gloeocapsa*, *Nostoc*, *Scytonema*, or *Stigonema*. The relationship is not so clear with the Nostocaceae growing in intercellular cavities within thalli of certain liverworts, or growing between cortical cells in roots of cycads. For the liverworts which regularly contain Nostocaceae within their thalli, it has been affirmed[2] and denied[3] that the presence of the alga is beneficial to the liverwort. If the relationship is one of symbiosis, it is probable that the liverwort utilizes atmospheric nitrogen fixed by the blue-green alga.[4] Possibly this is also the case with the blue-green algae growing in roots of cycads.

Organization of the Thallus. Most algae, other than the Myxophyceae, can be considered from the standpoint of the theory of plant-body types (see page 6); a theory which holds that only a limited number of basic types of body construction can be evolved from a motile, unicellular, ancestral form. Since primitive motile forms are not known for the blue-green algae (and probably have never been present in their ancestry), these cannot be used as a starting point for the evolution of body types among the Myxophyceae. The lack of a definite nucleus also precludes evolution along the siphonaceous line of development. The only types of body construction possible in the Myxophyceae are those in which immobile cells are solitary or united in filamentous or nonfilamentous colonies.

A few Myxophyceae, as certain species of *Chroococcus*, have an immediate separation of the daughter cells after cell division and are, therefore, truly unicellular. In the great majority of species, the daughter cells remain united following division, and the colonies resulting from this adhesion have either a filamentous or a nonfilamentous organization.

Organization into nonfilamentous colonies, a feature characteristic of all genera of the Chroococcales, results from a confluence of the gelatinous envelope surrounding the individual cells. The confluence may be so complete that all traces of the individual sheaths disappear (*Aphanocapsa*, *Aphanothece*, *Polycystis*) or the sheaths surrounding the individual cells may still be discernible (*Chroococcus*, *Gloeocapsa*, *Gloeothece*). Broadly speaking, the genera with distinct individual sheaths around the cells show a strong tendency toward colonial dissociation and do not produce such large colonies as do those genera in which the cells are embedded in a homo-

[1] Pascher, 1929. [2] Molisch, 1925. [3] Peirce, 1906; Takesige, 1937.
[4] Molisch, 1935.

geneous gelatinous matrix. The shape of nonfilamentous colonies is dependent upon the planes in which the cells divide. If the divisions are in two planes, the result is a layer one cell in thickness, and either a flat plate (*Merismopedia*) or a hollow sphere (*Coelosphaerium*). When the divisions are in three planes, their sequence may be so regular that there is a formation of a cubical colony (*Eucapsis*), but usually they are so irregular that there is no definite orientation of the cells within the massive colony (*Polycystis*, *Aphanothece*).

Organization into filamentous colonies results from successive divisions in the same plane. Cells organized in this fashion may be held together by walls that abut on one another (*Oscillatoria*), but commonly there is also a cylindrical sheath of gelatinous material surrounding the colony (*Lyngbya*). In such filamentous colonies, the simple row of cells is called a *trichome* and the trichome with its enclosing sheath is called a *filament*. Filaments may contain but a single trichome; or, as in *Schizothrix*, there may be several trichomes within a common sheath. Most genera have unbranched trichomes that may be straight, twisted in regular spirals, or irregularly contorted. Genera with branched trichomes, as *Nostochopsis* and *Hapalosiphon*, do not usually have the branches twisted. When there is more than one trichome within a sheath, the individual trichomes are often arranged in such a fashion as to appear to be branched. When, as in *Tolypothrix*, this arrangement exists it is called *false branching*.

The Cell Wall. The wall surrounding protoplasts of nonfilamentous Myxophyceae is composed of two concentric portions: an inner thin firm layer immediately outside of the plasma membrane and an outer more gelatinous portion (the sheath) that is often of considerable thickness. Filamentous blue-greens have the gelatinous sheath restricted to free faces of the cells. The sheath consists of pectic compounds; the inner firm portion contains a certain amount of cellulose.[1]

The sheath surrounding individual cells, or that surrounding individual trichomes, may be distinctly stratified. In other cases, sheaths surrounding individual cells or around individual trichomes show no indication of stratification. In many plankton species of *Chroococcus* and *Anabaena*, the homogeneous sheath is of so watery a consistency that it is not evident unless india ink is added to a water mount of the alga. Instead of being colorless, a sheath may be brown or red or violet. Yellow and brown coloration of a sheath is due to a mixture of the pigments fuscorhodin and fuscochlorin.[1] Red and violet coloration is due to a pigment called *gloeocapsin*.

Structure of the Protoplast. From the earliest attempts to determine structure of the myxophycean cell,[2] there has been a recognition of the fact

[1] KYLIN, 1943. [2] SCHMITZ, 1879, 1880.

that it is differentiated into a pigmented outer portion, sometimes called the *chromoplasm*, and an inner colorless portion, the *central body*. All investigations by means of cytological techniques find that all or a portion of the central body is differentially stainable in much the same fashion as is the chromatic material of true nuclei. The demonstration[1] of a positive Feulgen reaction in the central body shows that it is to be considered nuclear in nature. However, the central body is a nucleus without a definite nuclear membrane or nucleoli. Opinions vary concerning organization of the central body. There are those[2] who think that the nuclear material is

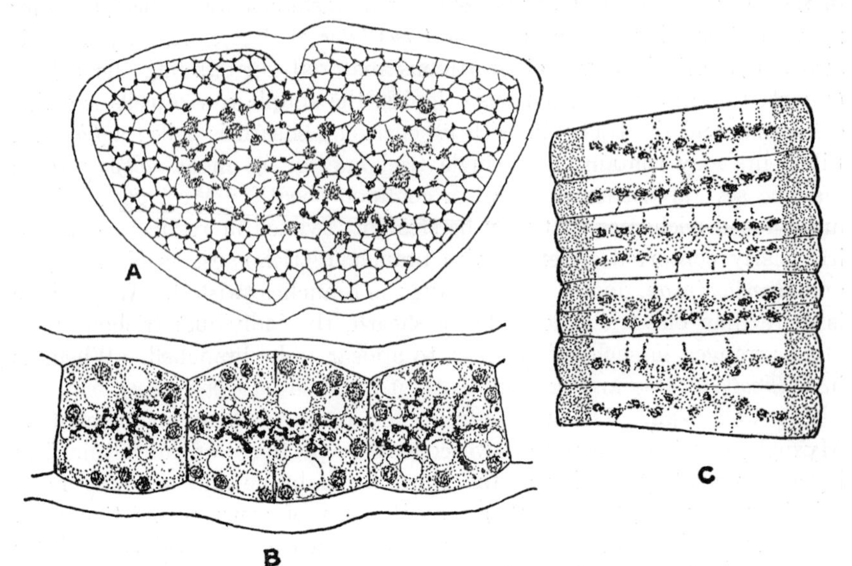

FIG. 448. Cell structure of various Myxophyceae. *A*, *Chroococcus turgidus* Näg. *B*, structure and division of cells of *Anabaena circinalis* (Kütz.) Rab. *C*, structure and division of cells of *Oscillatoria princeps* Vauch. (*A*, *after Acton*, 1914; *B*, *after Haupt*, 1923; *C*, *after Olive*, 1904.)

regularly or irregularly localized at certain juncture points of the fundamental protoplasmic reticulum. According to the species, the nuclear material may be very irregularly distributed throughout the central portion of a cell or it may be organized into a definite reticulum (Fig. 448*A*). Others hold that the central body is entirely nuclear in nature. Advocates of this interpretation of the central body are not in agreement as to the method by which it divides. Some[3] think that in division of the

[1] POLJANSKY and PETRUSCHEWSKY, 1929.

[2] ACTON, 1914; GUILLIERMOND, 1906, 1925.

[3] BAUMGÄRTEL, 1920; BROWN, W. H., 1911; LEE, 1927; OLIVE, 1904; PHILLIPS, 1904; POLJANSKY and PETRUSCHEWSKY, 1929.

central body there is a spindle apparatus resembling the spindle apparatus that effects a qualitative and a quantitative division of nuclear material. Others[1] hold that division of the central body is amitotic and results only in a quantitative division of the nuclear material.

The chromoplasm, the pigmented portion of the protoplasm external to the central body, usually has a finely alveolar structure. Embedded in it are a number of small spherical or irregularly shaped granules. In some species, these bodies are irregularly distributed; in other species, they are so regularly distributed as to be a character of taxonomic importance. It is clear that all these bodies are not of the same chemical composition and that some of them, probably the majority, are reserve food materials. Evidence for this is seen both in their greater abundance in reproductive cells and in their gradual disappearance during periods of active growth, or when plants are kept for some time in a dark room. Many phycologists think that the granules of carbohydrate food reserves

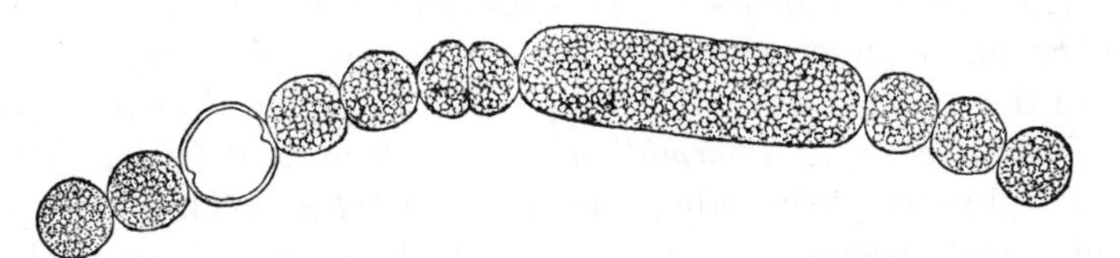

FIG. 449. *Anabaena circinalis* var. *macrospora* (Wittr.) De Toni, with pseudovacuoles in the vegetative cells and the akinete. (× 825.)

are identical with glycogen, but there is evidence that they are more closely related to starch. Because of this, it has been suggested[2] that the carbohydrate be called *cyanophycean starch.* Other bodies, the cyanophycin granules, are of a proteinaceous nature.

Pseudovacuoles. As is the case with so many other problems connected with Cyanophyta, there is diversity of opinion concerning the so-called pseudovacuoles or gas vacuoles. These are frequently present in certain plankton species of *Coelosphaerium*, *Polycystis*, and *Anabaena* (Fig. 449). Some blue-greens of the plankton, as *Coelosphaerium Kuetzingianum* Näg. and *Chroococcus limneticus* Lemm., never have them. At times, all individuals of a given species in a plankton catch will contain pseudovacuoles; at other times, certain individuals only will have them. In the latter case, the pseudovacuoles frequently appear iu the cells a few hours after collection and storage in tightly stoppered bottles. When viewed in a mass, the algae with pseudovacuoles are of a pale yellowish green. Under low powers of the microscope, the vacuoles appear as black bodies, larger than other inclusions, which are frequently present in such numbers that it is

[1] ACTON, 1914; GARDNER, 1906; GUILLIERMOND, 1906; HEGLER, 1911.
[2] KYLIN, 1943.

impossible to distinguish between central body and chromoplasm. Their reddish color when examined under high magnification is probably a refraction phenomenon. Investigations seem to prove that a partial vacuum, or pressure, will cause the vacuoles to disappear and gas bubbles to collect at the surface of the cells.[1] This has been interpreted as showing that the pseudovacuoles are gas-filled cavities, but it has also been held[2] that in reality a pseudovacuole is a cavity filled with a viscous substance. The formation of pseudovacuoles has been ascribed[3] to anaerobic respiration induced by an oxygen deficiency in water at the bottom of a lake. When this has continued for some time, the gas-filled cavities make the alga so buoyant that it rises to the surface of the water.

Pigments and Chromatic Adaptation. The chromoplast of a cell contain but one chlorophyll (chlorophyll *a*), beta-carotene and a second carotene found only in blue-green algae (flavicin), and two xanthophylls also found only in blue-green algae. In addition, the chromoplasm contains two proteinaceous pigments of the type known as *phycobilins*. One of these pigments, *c*-phycocyanin, is blue, the other, *c*-phycoerythrin, is red.

Not all these pigments are always present, or present in equal amounts. Some Myxophyceae, as *Phormidium corium* Gom. and *Oscillatoria tenuis* Ag., lack *c*-phycoerythrin; others, as *Microchaete tenera* Thur., lack or contain only small traces of *c*-phycocyanin.[4] From the theoretical standpoint, variations in the proportions of red, blue, and yellow pigments would make possible any color in the chromoplasm of Myxophyceae. The occurrence of grass-green, blue-green, olive, yellow, orange, pink, red, violet, purple, brown, and blackish Myxophyceae shows how closely this theoretical condition is found in nature. However, it should be noted that all such shades and colors are not due entirely to pigments in the chromoplasm but may be due partially to colors in the gelatinous envelope of a colony.

The causes for the development of the various pigments in different proportions are not known with certainty. The theory of complementary chromatic adaptation was first applied to Myxophyceae to explain experiments with *Oscillatoria sancta* Kütz. and *O. caldiorum* Hauck, which showed that individuals cultivated in light of different colors assumed different colors.[5] This theory, originally proposed[6] to explain variation in color of Rhodophyceae, holds that the color of light-absorbing portions of a protoplast is directly complementary to the quality (color) of the light in which the plant is growing. Such complementary color changes among Cyanophyta, sometimes called the *Gaidukov phenomenon*, have been found to

[1] KLEBAHN, 1922, 1925. [2] VAN GOOR, 1925. [3] CANABAEUS, 1929.
[4] BORESCH, 1921. [5] GAIDUKOV, 1902, 1903.
[6] ENGELMANN, 1883, 1884.

hold for certain species but not for others.[1] Thus, in one series of experiments[2] only 4 out of 18 species gave a definite Gaidukov response.

The observed chromatic changes have been thought[3] to be due to variations in the amount of pigments formed by a cell and to be limited to species capable of producing both c-phycocyanin and c-phycoerythrin in considerable amounts. Most experiments with light and changes in color of a blue-green alga have been with respect to the quality of light. That intensity of illumination must also be taken into consideration is shown by the bluish color of cultures when daylight illumination is intense and the reddish color when the intensity is reduced.[4] The best natural example of the effect of diminished intensity of light is the universal red color of Myxophyceae that grow at any considerable depth below the surface of a lake or other deep body of water. A change in color may also be due to other factors than light, chief among which is depletion of nitrogen in the substratum. Opponents of the theory of complementary chromatic adaptation[5] hold that nutritional effects are the sole cause for the color changes found in cultures of blue-green algae.

Movements of Myxophyceae. Many filamentous blue-green algae, especially members of the Oscillatoriaceae, have the ability to move spontaneously. The movement may be a forward and backward gliding of a trichome, a spiral progression and retrogression, or a slow waving of the terminal portion of a trichome. Movements of Nostocaceae are most noticeable in hormogonia or in germlings of a few cells. These movements are chiefly a forward and backward locomotion in the plane of the long axis. A waving of the terminal portion of a trichome, although reported for *Anabaena*,[6] is very rare for Nostocaceae. Oscillatoriaceae generally show both a waving and an axial movement. Axial progressions and retrogressions are frequently accompanied by a rotation in a straight or a spiral line.

Movements of Myxophyceae are markedly affected by external conditions, chief among which are light and temperature. Increased illumination is accompanied by greater activity; increases in temperature from 0 to 30° C. show a doubling of the speed for each 10 deg. of increase in temperature.[7]

Locomotion by means of a secretion of gelatinous materials from a cell has been repeatedly observed in desmids.[8] As applied to the Myxophy-

[1] Boresch, 1919, 1921*A*; Gaidukov, 1923; Harder, 1922; Sargent, 1934; Susski, 1926.

[2] Boresch, 1921*A*. [3] Boresch, 1921*A*; Kylin, 1912*A*, 1937*A*.

[4] Kylin, 1937*A*.

[5] Magnus and Schindler, 1912; Pringsheim, E. G., 1914; Schindler, 1913.

[6] Castle, 1926.

[7] Burkholder, 1934; Nienburg, 1916; Pieper, 1913; Schmid, 1918.

[8] Schröder, 1902.

ceae, it is held that locomotion along the longitudinal axis is due to a rapid swelling of gelatinous material secreted through the wall.[1] Some advocates of this view think that the gelatinous material is secreted through minute pores in the cell wall, and that rotating movements are due to the arrangement of the pores in two crossing spiral series. The terminal cells of the trichomes have been held to be of major importance in such movements, but it has been shown[2] that trichomes whose apical cells have been killed by means of sulfuric acid do not lose their capacity for locomotion. Not all recent workers have subscribed to the theory of locomotion through slime secretion, and movements of myxophycean trichomes have been ascribed to rhythmic waves of alternate expansions and contractions passing along the length of the trichome.[3]

Vegetative Reproduction. In all species of Chroococcales, the only regular method of reproduction is that of cell division. Ordinarily the

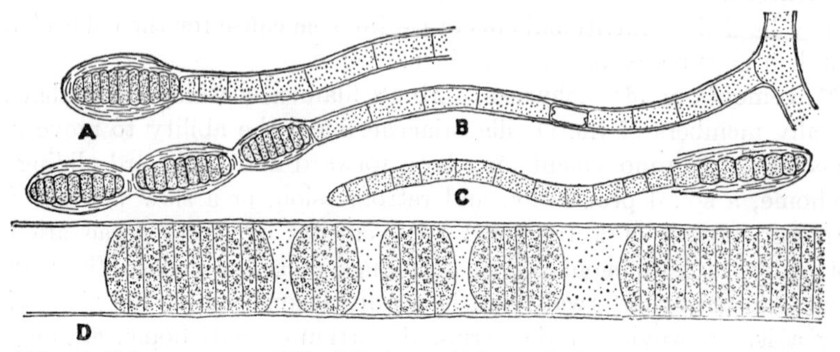

Fig. 450. *A–C*, hormospores of *Westiella lanosa* Frémy. *D*, hormogonia of *Lyngbya Birgei* G. M. Smith. (*A–C, after Frémy*, 1930.) (*A–C*, × 330; *D*, × 650.)

two daughter cells remain united to each other within a common gelatinous envelope, and the indefinite repetition of cell division may result in a colony containing many cells. Colony reproduction is a matter of chance and depends upon accidental breaking of the colonial envelope. If the envelope is soft and tends to dissolve, as *Chroococcus*, *Gloeocapsa*, or *Dactylococcopsis*, the colony never grows to a large size before it becomes separated into two or more daughter colonies. In genera with a tough envelope, as in *Aphanothece* or *Coelosphaerium*, the colony usually becomes many-celled before it breaks into smaller portions.

In filamentous species, the trichomes are, from the theoretical standpoint, capable of indefinite growth in length; but under ordinary conditions the filament sooner or later becomes broken. Breaking may result from animals feeding on the filament, from the death of certain cells in the row,

[1] Fechner, 1915; Harder, 1918; Prell, 1921; Schmid, 1918, 1921, 1923.
[2] Schmid, 1923. [3] Ullrich, 1926, 1929.

or from a weaker adhesion between certain cells than between others. Instances of the last sort are confined largely to those genera which produce heterocysts, and the zone of weak adhesion is where a heterocyst and a vegetative cell abut on each other. Short sections of the trichome specifically delimited (*hormogonia*) constitute an important method of propagation among the filamentous Myxophyceae (Fig. 450*D*). Hormogonia are delimited by a development of double concave disks of gelatinous material (*separation disks*) between two adjoining vegetative cells. The formation of separation disks is most noticeable in the larger species of *Oscillatoria* and *Lyngbya*. In these genera, the hormogonium may be but two or three cells in length or it may be several cells long. Hormogonia have an even greater capacity for locomotion than vegetative trichomes, and sooner or later after they are formed they slip away from the filament in which they are developed and grow into new filaments. Hormogonia usually develop directly into a typical plant, but occasionally, as in *Nostoc*,[1] the juvenile structure coming from the hormogonium has but little resemblance to the adult plant.

In the Scytonemataceae and the Stigonemataceae, hormogonia developed at the ends of branches may have the walls of their cells becoming much thicker (Figs. 450*A* and *B*). These multicellular spore-like bodies are called *hormospores*.[2] The hormospores germinate directly into new filaments (Fig. 450*C*).

Spore Formation. Swarm spores and flagellated gametes have never been observed in the Myxophyceae. Nonmotile resting spores are commonly formed in all families of the Hormogonales except the Oscillatoriaceae. The development of these spores begins with the enlargement of certain cells of the trichome and an accumulation of food reserves within them. During the later stages of their development, there is an appreciable thickening of the cell wall, a thickening which often results in the differentiation of distinct exospore and endospore wall layers. This type of resting spore, which contains the entire protoplast of the parent cell and in which the original wall of the vegetative cell comprises a part of the spore wall, is called an *akinete*. Akinetes of Myxophyceae are usually formed singly or in pairs, but in rare cases, as in *Nodularia*, they are formed in chains of half a dozen or more. An akinete may be formed at a specific place in the trichome, or its location may not be predetermined. When akinetes are developed in predetermined places, they always lie next a heterocyst, either at the end of the trichome (as in *Cylindrospermum* or *Gloeotrichia*), or in the middle of the trichome (as in *Anabaena Lemmermanni* P. Richt.). Akinetes that develop from cells which do not abut on heterocysts may be close to, or remote from, the heterocyst.

[1] GEITLER, 1921. [2] BORZI, 1914; GEITLER, 1930*A*.

Resting spores of the akinete type are structures for tiding the alga over unfavorable periods, and they usually germinate into vegetative filaments as soon as favorable conditions return. One of the best examples of this is to be seen in the regular germination of the akinetes of terrestrial species of *Cylindrospermum* immediately after a heavy rain and a thorough soaking of the dry soil. On the other hand, these resting cells may retain their viability for extremely long periods if conditions are unfavorable, and it has been shown that there was a germination of such cells in samples of dried soil that had been stored for 70 years.[1] However, akinete formation is not absolutely necessary to tide the alga over long unfavorable periods, since, in the experiments just cited, it was found that after a period of 50 years there was a growth of species that do not form akinetes.

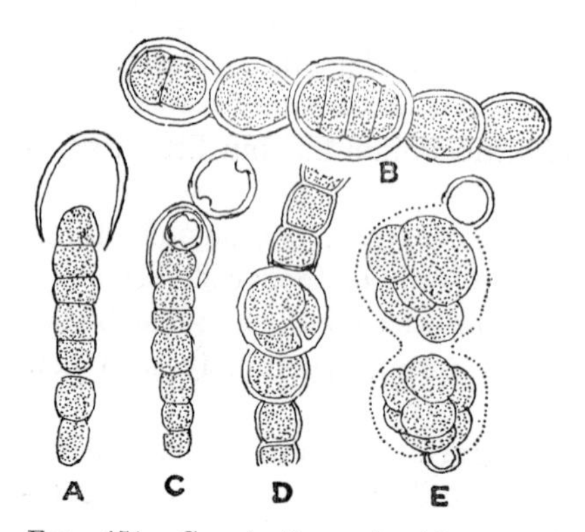

FIG. 451. Germination of akinetes of Myxophyceae. *A*, *Anabaena oscillarioides* Bory. *B–C*, *A. sphaerica* B. and F. *D–E*, *Nostoc muscorum* Kütz. (*After Bristol-Roach*, 1920.) (× 825.)

When an akinete germinates (Fig. 451), it usually grows directly into a more or less typical vegetative filament, but there may be a formation of a juvenile structure that has but little resemblance to the mature trichome.[2] Germination ordinarily begins with a transverse division of the protoplast of the akinete, and several additional transverse divisions may occur before the rupture of the spore wall. Following this, the end portion of the old wall becomes softened, or ruptures as a cap-like lid, and the young trichome grows through the opening thus formed. These germling stages are frequently motile and, while few-celled, they may advance from, and retract into, the old akinete wall.[3] Infrequently, as in *Nostoc commune* Vauch., germination begins with a rejuvenescence which results in a gelatinization of the inner spore-wall layer and a bursting of the outer layer. After these changes in the wall, the undivided contents of the akinete are extruded or there may be a single transverse division before extrusion occurs.

Endospores, which are regularly formed by all genera of the Chamaesiphonales and occasionally found elsewhere [as in certain individuals of *Gomphosphaeria aponina* Kütz.[4] and of *Phormidium autumnale* (Ag.) Gom.[5]], are an entirely different type of spore. Endospores, also called *gonidia* or *conidia*, are formed by a repeated division of the protoplast

[1] BRISTOL-ROACH, 1919, 1920. [2] BRISTOL-ROACH, 1920.
[3] HARDER, 1918. [4] SCHMIDLE, 1901. [5] BRAND, 1903.

within a cell wall. The numerous endospores within the old cell wall, the sporangium wall, are usually spherical in shape, but, as in the case of certain marine Chamaesiphonales, they may be angular by mutual compression. A distinction is sometimes made between this type of internal division, which involves the entire protoplast, and that found in *Chamaesiphon* where the endospores are successively cut off at the distal end of the protoplast (Fig. 484, page 572). Spores produced in this fashion have been called *exospores*,[1] but such a distinction is needless since the exospore is only a specialized type of endospore.

Sometimes certain Chroococcales have successive cell divisions following one another so closely that the daughter cells are very much smaller than ordinary vegetative cells. These *nannospores*[2] look very much like endospores but are not true spores.

Heterocysts. Most filamentous Myxophyceae, except the Oscillatoriaceae, produce the special type of cell known as a *heterocyst*. Heterocysts differ from vegetative cells, and from spores, in the organization of their walls and in their transparent contents. These bodies usually occur singly, and they may be strictly terminal in position (*Rivularia*, *Cylindrospermum*, *Anabaenopsis*), or they may be intercalary (*Anabaena*, *Nostoc*, *Stigonema*). Hererocysts arise by a metamorphosis of vegetative cells and usually only from recently divided ones. The metamorphosis may involve a change into a shape different from that of the vegetative cell (*Diphlocolon*), but more commonly there is no appreciable change from the shape characteristic of the vegetative cell. The first noticeable step in the development of a heterocyst is the secretion of a new wall layer internal to that originally surrounding the cell. Depending on the parent cell's terminal or intercalary position in the trichome, there is a pore at one or both poles of the new wall. Cytoplasmic connections with adjoining vegetative cells are usually evident in the developing heterocyst, but as it approaches maturity, the cytoplasmic connections are replaced by prominent button-like thickenings of wall material, the *polar nodules*. The protoplast within the heterocyst becomes more and more transparent after the polar nodules have been formed. Preparations stained with Heidenhain's iron-alum-haemotoxylin show that the transparent appearance of the cell is not due to an emptying of its contents but to a transformation of the protoplast into some homogeneous, viscous substance.

The nature and function of the heterocyst are a topic that has been debated at length but present-day opinion is more or less unanimous that they are spore-like in nature.[3] The general agreement concerning the nature of these bodies has come through the accumulation of well-authenticated

[1] GEITLER, 1925. [2] GEITLER, 1925, 1930*A*.

[3] BRAND, 1901, 1903; CANABAEUS, 1929; GEITLER, 1921; PRINGSHEIM, N., 1855.

cases in which the protoplasts do not disappear from the heterocyst, and in which the heterocyst germinates to form a new filament (Fig. 452). The changes in the nature and structure of the cell wall show that the heterocysts are not analogous to akinetes. The nonakinete nature of the heterocyst is also seen in *Anabaena Cycadeae* Reinke[1] where it has been shown that there is a formation, and a subsequent germination, of endospores within a heterocyst. These observations indicate that heterocysts are reproductive structures, but structures which have become functionless as such, except in occasional instances.

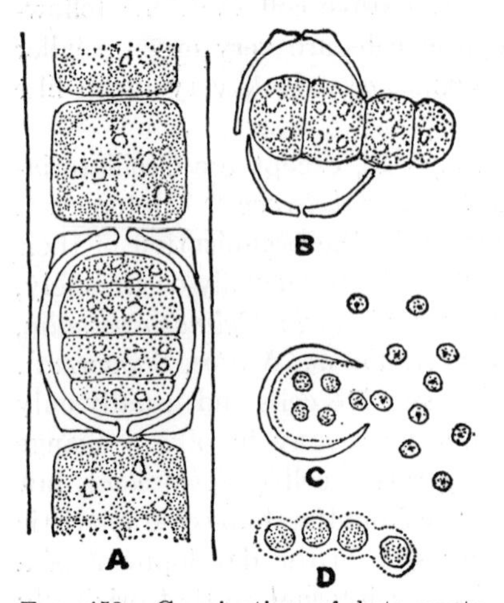

FIG. 452. Germination of heterocysts of Myxophyceae. *A*, *Anabaena hallensis* (Jancz.) B. and F. *B*, *Nostoc commune* Vauch. *C–D*, *Anabaena Cycadeae* Reinke. (*A–B*, *after Geitler*, 1921; *C–D*, *after Spratt*, 1911.) (*A–B*, × 2500; *C–D*, × 2200.)

Although ordinarily functionless as spores, or sporangia, heterocysts have, in many instances, taken on certain secondary functions. Sometimes they have a definite relationship to the place where akinetes are formed and, as was shown on page 549, certain species only develop akinetes next to a heterocyst. In other cases, as in *Anabaenopsis* and certain species of *Nostoc*, the filaments always break where two heterocysts adjoin or at the juncture of a heterocyst and a vegetative cell. In such cases, heterocysts serve as a specific device for colony reproduction. Genera with a true branching (*Stigonema*) or with a false branching (*Tolypothrix*) may have a definite correlation between position of heterocysts and the origin of true or false branches.

Classification. There is no need to enter into a discussion of the various systems proposed for a classification of the Myxophyceae since most of them are in accord concerning related groups of genera. The systems that have been proposed differ chiefly in the rank given to groups of related genera. The classification given here follows the practice of those who recognize but three orders in the class.

ORDER 1. CHROOCOCCALES

The assemblage of genera placed in this order is one characterized by negative rather than by positive characters and is the residuum left after an exclusion of all strictly filamentous genera and of all genera that regu-

[1] SPRATT, 1911.

larly form endospores. Generic differences within the order may be relative rather than absolute, and in certain cases it is extremely difficult to distinguish between juvenile thalli of many-celled genera and genera which have thalli with a few cells when fully mature.

It is also impossible to arrange the various genera in anything like a logical evolutionary sequence. A separation of the coccoid from the bacilloid genera appears logical at first glance, but the two types intergrade so closely that such a separation is more or less artificial. From the evolutionary standpoint, the unicellular members of the order are more primitive than those that are regularly colonial; but one cannot state with certainty that all known unicellular genera are really primitive.

Cell division and colony fragmentation are the only methods of reproduction. Endospores have been recorded for a couple of the genera, but in these genera they are not regularly formed. The order is divided into two families.

Family 1. Chroococcaceae

A few of the genera in this family are unicellular; the majority are colonial. When the cells are united in colonies, there is no tendency toward a pseudofilamentous organization in the colony or any part of it.

The genera found in this country differ as follows:*

1. Cells free-living 2
1. Cells endophytic in colorless algae 21
    2. Cells arranged to form a colony of distinctive shape 3
    2. Cells, if in colonies, not in a colony of distinctive shape 8
3. Cells arranged in a hollow sphere 4
3. Cells not arranged in a hollow sphere 6
    4. Interior of colony with radiating gelatinous strands .... 20. **Gomphosphaeria**
    4. Interior of colony without radiating strands 5
5. Cells spherical to ellipsoidal .... 18. **Coelosphaerium**
5. Cells pyriform .... 19. **Marssoniella**
    6. Colony cubical .... 7. **Eucapsis**
    6. Colony a monostromatic sheet 7
7. Cells in regular transverse and vertical rows .... 16. **Merismopedia**
7. Cells irregularly arranged .... 17. **Holopedium**
    8. Cells spherical 9
    8. Cells ellipsoidal, cylindrical, or spindle-shaped 14
9. Cells solitary or in colonies with less than 50 cells 10
9. Cells in colonies with hundreds of cells 12
    10. Gelatinous sheath around cells inconspicuous .... 3. **Synechocystis**
    10. Gelatinous sheath around cells clearly evident 11
11. Sheath colorless .... 1. **Chroococcus**

* *Tetrapedia* Reinsch has been found in the United States, but it is one of very doubtful validity and probably based upon an erroneous interpretation of certain Chlorococcales.

11. Sheath variously colored .... 2. **Gloeocapsa**
    12. Each cell with an evident gelatinous sheath .... 6. **Chondrocystis**
    12. Individual cells without evident sheaths 13
13. Cells densely aggregated .... 5. **Polycystis**
13. Cells remote from one another .... 4. **Aphanocapsa**
    14. Cells spindle-shaped .... 13. **Dactylococcopsis**
    14. Cells ellipsoidal or cylindrical 15
15. Cells solitary or in colonies of less than 50 cells 16
15. Cells in colonies with hundreds of cells 19
    16. Each cell with an evident gelatinous sheath 17
    16. Each cell without an evident gelatinous sheath 18
17. All cellular sheaths of uniform thickness .... 9. **Gloeothece**
17. Sheaths of certain cells prolonged into a stipe .... 10. **Chroothece**
    18. Cells solitary or 2 to 4 grouped pole to pole .... 8. **Synechococcus**
    18. Cells in colonies of a few cells .... 11. **Rhabdoderma**
19. Colony tubular, with pointed ends .... 12. **Bacillosiphon**
19. Colony not tubular 20
    20. Each cell with an evident sheath .... 15. **Anacystis**
    20. Each cell without an evident sheath .... 14. **Aphanothece**
21. Host epiphytic, with long gelatinous setae .... 22. **Gloeochaete**
21. Host free-floating, with 2 to 8 cells inside a common wall .... 21. **Glaucocystis**

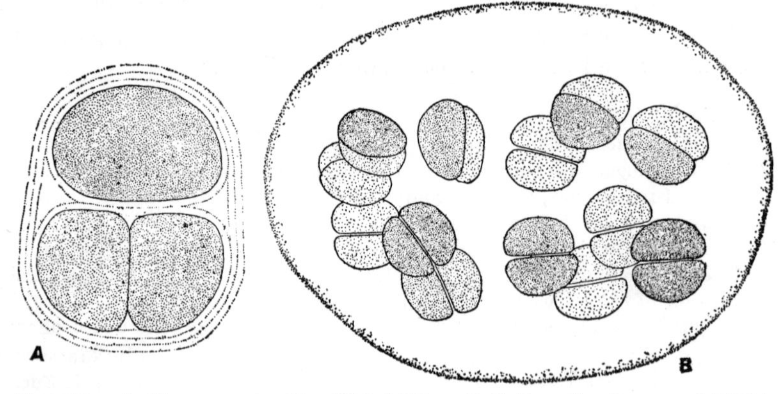

FIG. 453. *A*, *Chroococcus turgidus* (Kütz.) Näg. *B*, *C. limneticus* Lemm. (× 825.)

1. **Chroococcus** Nägeli, 1849. The spherical cells of *Chroococcus* (Fig. 453), which are usually hemispherical for some time after division, are surrounded by a hyaline, homogeneous or lamellated, gelatinous sheath. The cells may be united in colonies of two to several by persistence and distension of the sheath of the original parent cell. The cell contents are variously colored and either homogeneous or granular.

Most of the criteria established to distinguish between *Chroococcus* and *Gloeocapsa* break down at one point or another. The most consistent one seems to be that of Daily[1] who differentiates between the two on the basis of color or lack of

[1] DAILY, 1942.

color in the gelatinous sheath. When delimited in such a fashion, one can say with certainty that *C. limneticus* Lemm. (Fig. 453*B*), *C. rufescens* (Bréb.) Näg., and *C. turgidus* (Kütz.) Näg. (Fig. 453*A*) occur in this country. For descriptions of them, see Daily (1942).

2. **Gloeocapsa** Kützing, 1843. This genus resembles *Chroococcus* in having colonies composed of a few spherical cells, each surrounded by a homogeneous to lamellated sheath. It differs from *Chroococcus* in that the sheath is colored yellow, brown, red, blue, or violet. As in *Chroococcus*, reproduction is by cell division and fragmentation of colonies.

Over 20 species have been reported from the United States, but certain of them undoubtedly belong elsewhere or should be considered synonyms. Typically *Gloeocapsa* is found growing in extensive strata on damp rocks. Among the indubitable species found in this country are *G. aurata* Stiz., *G. magma* (Bréb.) Kütz. (Fig. 454), and *G. rupestris* Kütz. For descriptions of them, see Daily (1942).

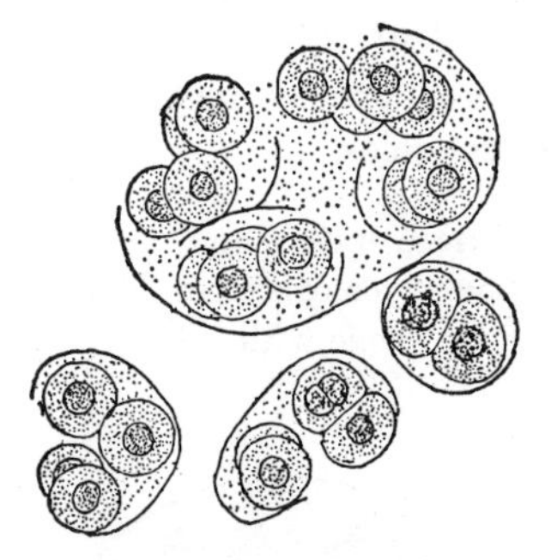

Fig. 454. *Gloeocapsa magma* (Bréb.) Kütz. (× 1665.)

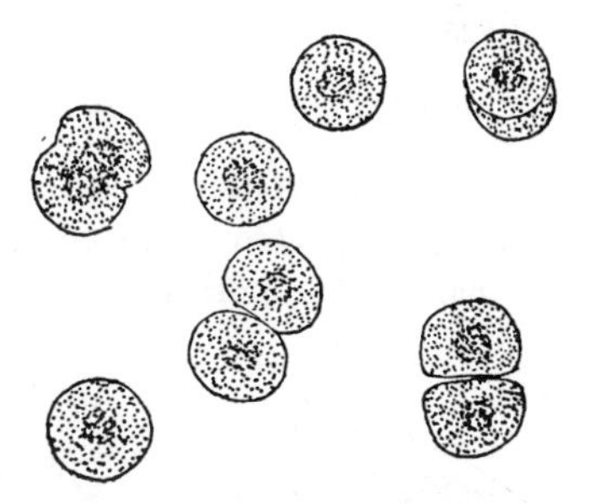

Fig. 455. *Synechocystis aquatilis* Sauv. (× 1950.)

3. **Synechocystis** Sauvageau, 1892. The cells of this alga are spherical, except immediately after division, and are without an evident gelatinous sheath. They may be solitary or aggregated in colonies of a few cells.

Synechocystis stands in much the same relationship to genera with globose cells as does *Synechococcus* with respect to genera with cylindrical cells. The species recorded for the United States are *S. aquatilis* Sauv. (Fig. 455), *S. minuscula* Woronichin, and *S. thermalis* Copeland. For a description of *S. thermalis*, see J. J. Copeland (1936); for the other two, see Geitler (1930*A*).

4. **Aphanocapsa** Nägeli, 1849. Adult colonies of *Aphanocapsa* (Fig. 456) are always many-celled and with a firm colonial matrix in which there are no evident individual sheaths around the cells. The cells are spherical, except immediately after division, and lie some distance from one another within the colonial matrix. Protoplasts of cells are usually without evident granular inclusions.

Aphanocapsa is usually aquatic in habit, and either free-floating or adhering to submerged vegetation or rocks. About a dozen species have been found in this country. For descriptions of most of the species of the genus, see Geitler (1930*A*).

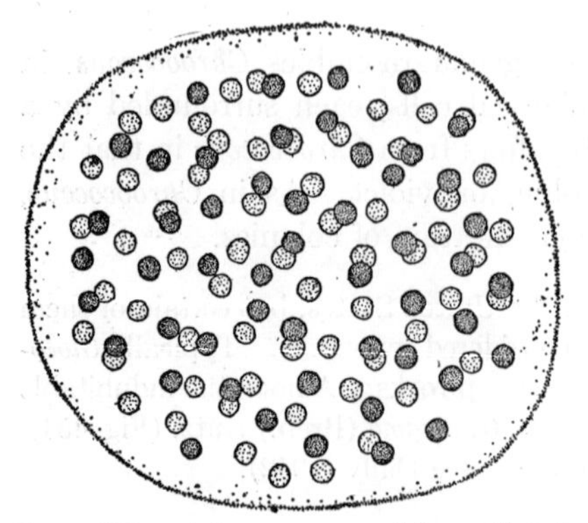

FIG. 456. *Aphanocapsa pulchra* (Kütz.) Rab. (× 825.)

5. **Polycystis** Kützing, 1849 (*Microcystis*[1] Kützing, 1833; *Clathrocystis* Henfrey, 1856). The colonies of this alga are always free-floating, and either spherical, ellipsoidal, irregularly elongate, or clathrate masses of microscopic or macroscopic size. The cells are spherical and lie close to one another with a common gelatinous matrix of so watery a consistency that the margins of the matrix are usually not evident unless demonstrated by means of special techniques. The cells frequently contain numerous pseudovacuoles.

Polycystis is essentially a planktonic genus of fresh waters and one that often causes "water blooms" in hard-water lakes. Some phycologists consider the genus one with many species; other phycologists consider it one with a few species with

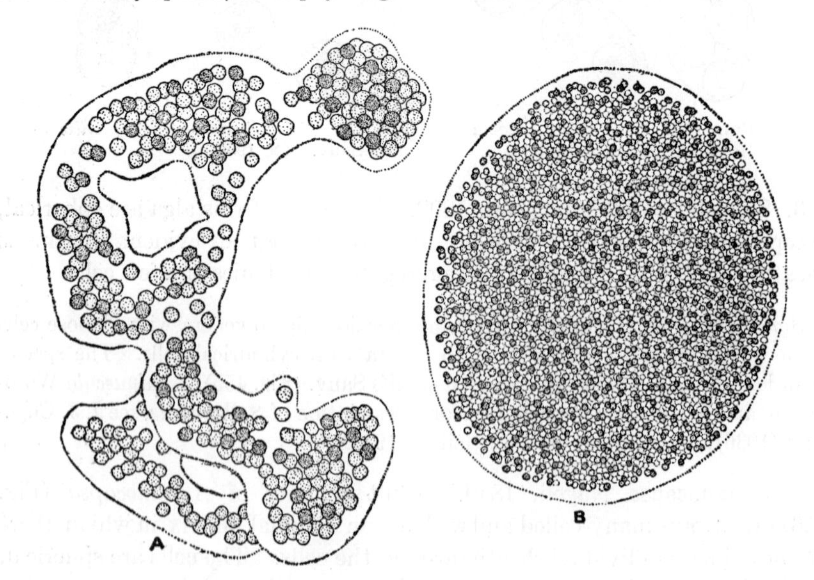

FIG. 457. *A*, *Polycystis aeruginosa* Kütz. *B*, *P. incerta* Lemm. (× 825.)

[1] Drouet and Daily, in Daily (1942), have shown why the generic name *Microcystis* is untenable.

numerous growth forms. According to the latter interpretation, the species found in this country are *P. aeruginosa* Kütz. (Fig. 457*A*), *P. glauca* (Wolle) Drouet and Daily, and *P. incerta* Lemm. (Fig. 457*B*). For a description of *P. glauca* as *Microcystis glauca* (Wolle) Drouet and Daily, see Drouet and Daily (1939); for the other two, see Daily (1942).

6. **Chondrocystis** Lemmermann, 1899. *Chondrocystis* grows in cartilaginous cushion-like masses, sometimes several feet in thickness and with the colonies in the bottom portion incrusted with lime. The individual cells are globose, and each is surrounded by a narrow but distinct sheath. External to the narrow sheath is a much wider homogeneous one, which may enclose from one to several cells. These latter sheaths are so densely aggregated that they are angular by mutual compression.

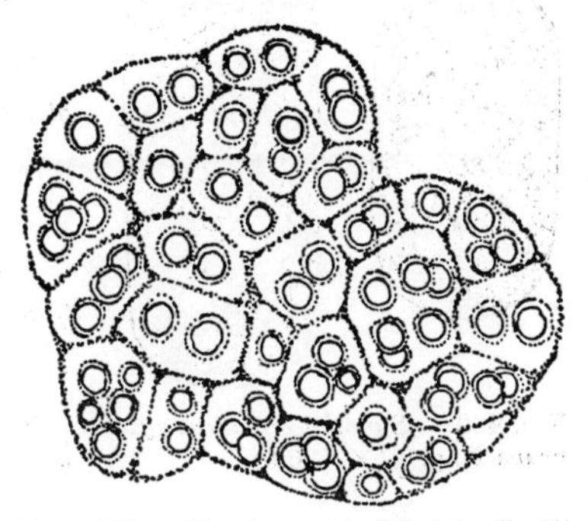

FIG. 458. *Chondrocystis Schauinslandii* Lemm. (*After Lemmermann,* 1905.) (× 750.)

This is a genus of somewhat questionable validity. There is but one species, *C. Schauinslandii* Lemm. (Fig. 458), and North Dakota[1] is the only place where it has been found in this country. For a description of it, see Lemmermann (1905).

7. **Eucapsis** Clements and Shantz, 1909. This genus differs sharply from all other Chroococcales in its regularly cubical colonies. It resembles *Merismopedia* in the regular spacing of the cells, but cell divisions are successively in three, instead of in two planes. The result is a cubical, instead of a flat, colony. The number of cells in a colony is regularly a multiple of two and may be over 500.

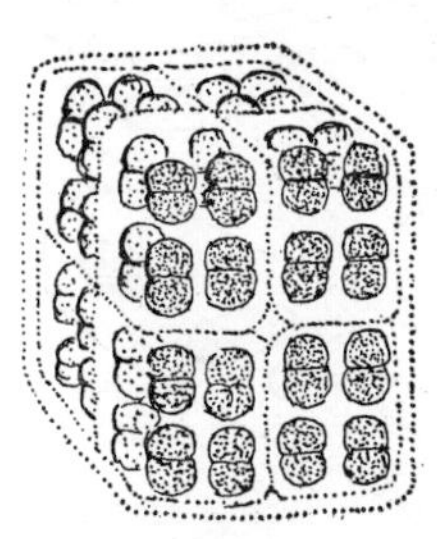

FIG. 459. *Eucapsis alpina* Clements and Shantz. (*After Clements and Shantz,* 1909.) (× 500.)

The single species, *E. alpina* Clements and Shantz (Fig. 459), has been reported from a number of localities in this country. In some collections, the number of cells in all colonies was small, and it has been held[2] that in such cases the alga identified as *Eucapsis* should be considered a species of *Chroococcus*. For a description of *E. alpina*, see Clements and Shantz (1909).

8. **Synechococcus** Nägeli, 1849. *Synechococcus* has broadly ellipsoidal to straight or curved cylindrical cells that are without or with a very inconspicuous gelatinous sheath. Cell division is transverse and usually

[1] MOORE and CARTER, 1923. [2] DROUET. 1942.

followed by a separation of the two daughter cells, but sometimes they remain united pole to pole to form colonies of from two to four cells.

Twelve species have been recorded for this country. One species, *S. aeruginosus* Näg. (Fig. 460) is widely distributed, all others are known only from hot springs of Yellowstone National Park. For a description of *S. aeruginosa*, see Daily (1942); for descriptions of the thermal species, see J. J. Copeland (1936).

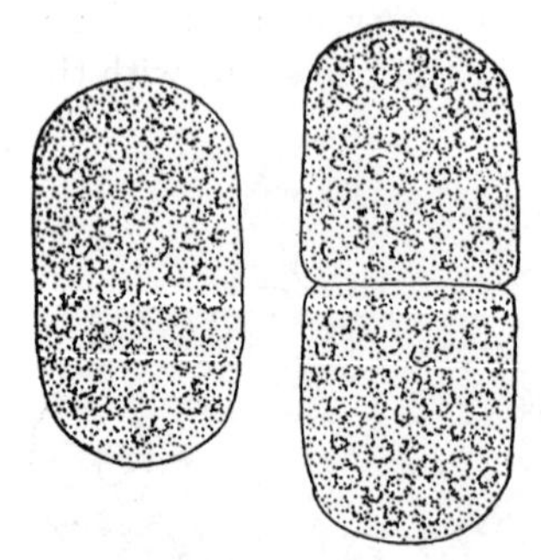

Fig. 460. *Synechococcus aeruginosus* Näg. (× 1300.)

9. **Gloeothece** Nägeli, 1849. Cells of *Gloeothece* are cylindrical, with broadly rounded poles, and surrounded by a firm envelope. Cell division is at right angles to the long axis. After division, the envelopes of daughter cells may adhere to one another, but the adhesion is so insecure that cells are constantly breaking away so that a colony never contains more than a few cells.

Gloeothece occurs free-floating among other aquatic algae or grows in moist places such as damp rocks. The species found in this country are *G. confluens* Näg., *G. distans* Stiz., *G. Goeppertiana* (Hilse) Forti, *G. linearis* Näg. (Fig. 461*A*), and *G. membranacea* (Rab.) Born. For descriptions of them, see Geitler (1930*A*).

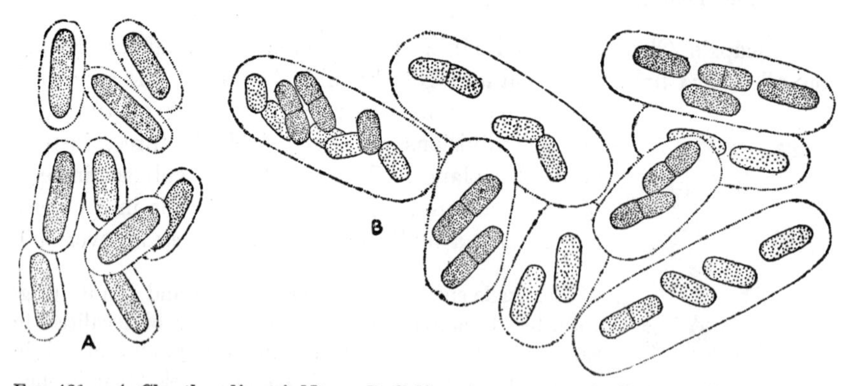

Fig. 461. *A*, *Gloeothece linearis* Näg. *B*, *G. linearis* var. *composita* G. M. Smith. (× 1000.)

10. **Chroothece** Hansgirg, 1884. The cells of *Chroothece* are broadly ellipsoidal to cylindrical, and each has a broad firm gelatinous envelope. The cells may be solitary or seriately jointed pole to pole in colonies of a few cells. Solitary cells and cells united in colonies frequently have one pole of the gelatinous sheath continued in a homogeneous or lamellated stripe-like prolongation.

The only record[1] for the occurrence of this genus in the United States is the finding of *C. monococca* (Kütz.) Hansg. (Fig. 462) in Ohio and Indiana. For a description of it, see Daily (1942).

11. **Rhabdoderma** Schmidle and Lauterborn, 1900. *Rhabdoderma* has cylindrical, straight or arcuate, cells with broadly rounded poles. The cells lie embedded within a common homogeneous colonial envelope and

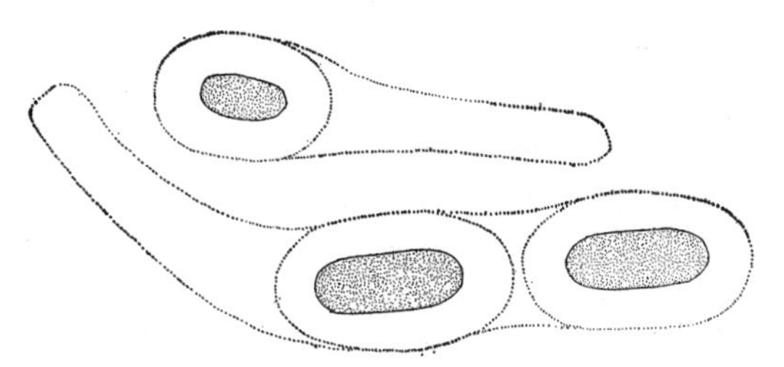

FIG. 462. *Chroothece monococca* (Kütz.) Hansg. (*After Daily*, 1942.)

lie so oriented that their long axes are parallel to one another. The colonies never contain more than a few cells.

In the United States, this genus is known only from the plankton. The two species found in this country are *R. lineare* Schmidle and Lauterborn (Fig. 463) and *R. sigmoidea* Moore and Carter. For descriptions of them, see Geitler (1930*A*).

12. **Bacillosiphon** J. J. Copeland, 1936. This genus has hundreds of cells united in a colony. The colonial matrix is elongate, spindle-shaped, and frequently with the superficial portion incrusted and impregnated with lime. The cells are cylindrical with broadly rounded poles, and all lie with their long axes parallel to the long axis of the colonial matrix. Colony reproduction is by escape of cells through the polar portion of the colonial matrix.

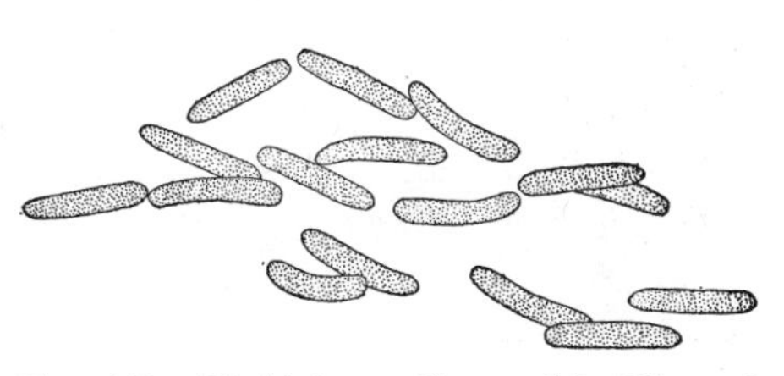

FIG. 463. *Rhabdoderma lineare* Schmidle and Lauterborn. (× 1000.)

The single species, *B. induratus* J. J. Copeland (Fig. 464) is known only from the Mammoth Hot Springs area of Yellowstone National Park. For a description of it, see J. J. Copeland (1936).

13. **Dactylococcopsis** Hansgirg, 1888. The linear, arcuate, or sigmoid cells of this genus differ markedly from those of other Chroococcaceae

[1] DAILY, 1942.

with elongate cells in that they have sharply pointed ends. The cells are united in colonies that never contain a large number of cells, and the cellular envelopes are completely fused with one another to form a homogeneous colonial matrix.

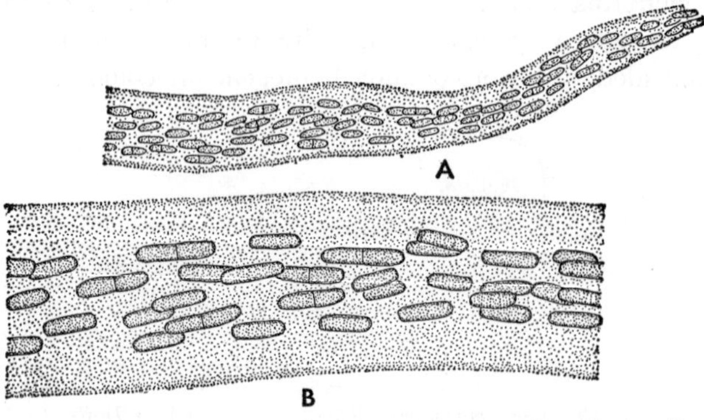

Fig. 464. *Bacillosiphon induratus* J. J. Copeland. *A*, apex of a colony. *B*, cells. (*After J. J. Copeland*, 1936.) (*A*, × 340; *B*, × 820.)

Two species, *D. antarctica* Fritsch and *D. Smithii* R. and F. Chodat (Fig. 465), are known for this country. For descriptions of them, see Geitler (1930*A*).

14. **Aphanothece** Nägeli, 1849. The cells of *Aphanothece* (Fig. 466) are cylindrical, with broadly rounded poles, and are without evident individual sheaths. They are united in colonies composed of many cells and are irregularly oriented within the homogeneous, globose to amorphous, colonial matrix.

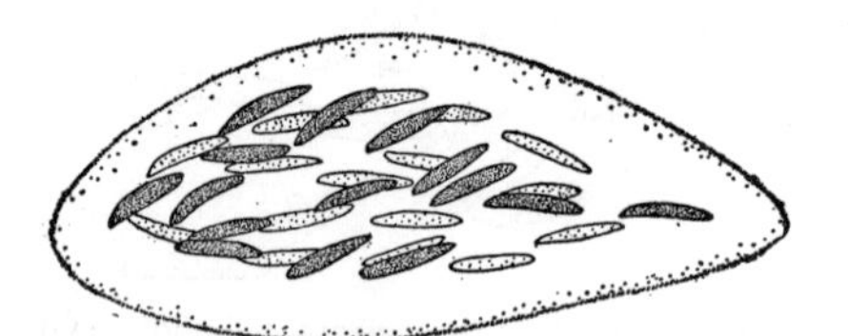

Fig. 465. *Dactylococcopsis Smithii* R. and F. Chodat. (× 825.)

This genus has been placed[1] as a synonym of *Anacystis*, but the lack of individual cellular sheaths mentioned in the description of the type species (*A. microscopica* Näg.) seems to justify considering the two distinct. *Aphanothece* is usually aquatic in habit, and certain species are known only from the plankton. About a dozen species have been reported from the United States. For descriptions of most of the species of the genus, see Geitler (1930*A*).

15. **Anacystis** Meneghini, 1837. The cells of *Anacystis* are cylindrical with broadly rounded poles, and with a length 1½ to 3 times the breadth. There is a more or less evident sheath around each cell or pair of daughter cells. The outer portions of sheaths of cells are confluent with one another to form a homogeneous colonial matrix containing many cells and thus an

[1] Daily, 1942.

amorphous colony of macroscopic size. The colonial matrix of old colonies may be yellowish or brownish.

Aphanotheca and *Aphanocapsa* have been considered[1] synonyms of *Anacystis*, but it seems better to transfer to *Anacystis* only those species of *Aphanothece* in which

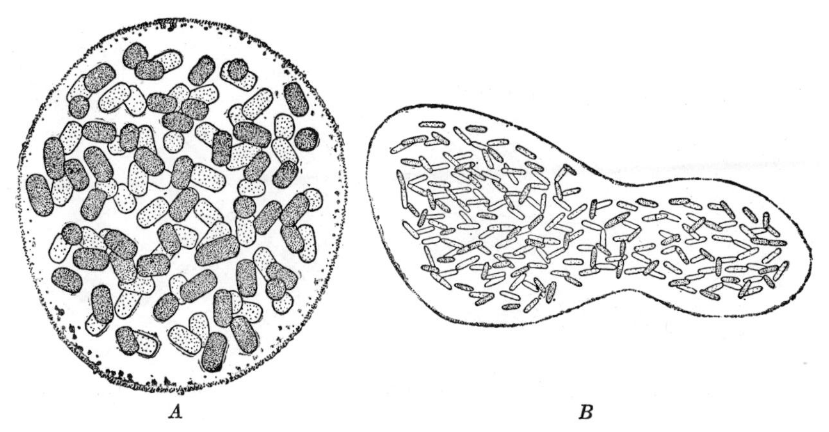

FIG. 466. *A*, *Aphanothece stagnina* (Spreng.) A. Br. *B*, *A. clathrata* W. and G. S. West. (*A*, × 825; *B*, × 1000.)

there is a clearly evident sheath around the individual cells. When restricted in this manner, the species found in this country are *A. marginata* Menegh., *A. peniocystis* (Kütz.) Drouet and Daily, and *A. rupestris* (Lyngb.) Drouet and Daily (Fig 467), all of which are usually found growing on damp rocks. For descriptions of them see Daily (1942).

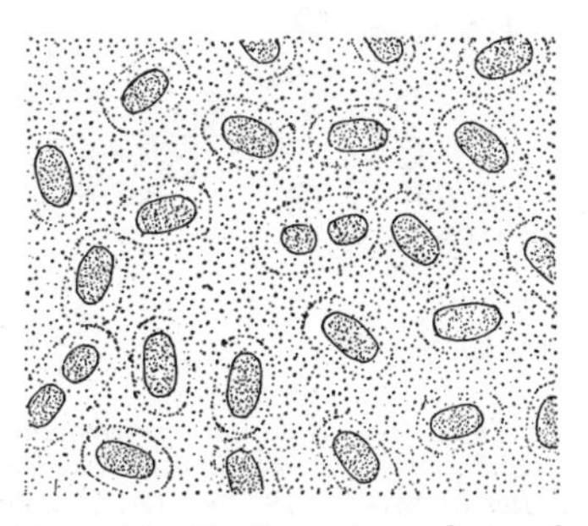

FIG. 467. Portion of a colony of *Anacystis rupestris* (Lyngb.) Drouet and Dailey.

16. **Merismopedia** Meyen, 1839. This distinctive genus has many-celled free-floating colonies, one cell in thickness, in which the cells are regularly arranged in vertical and transverse rows. The cells are broadly ellipsoidal. Individual sheaths around the cells are rarely distinct and are usually fused with one another to form a homogeneous colorless colonial matrix. The regular arrangement of cells in a colony results from the fact that successive cell divisions are at right angles to each other and in two planes. Small colonies are usually perfectly flat; large colonies, although one cell in thickness, are usually more or less bent and distorted.

[1] DAILY, 1942.

Merismopedia generally occurs sparingly intermingled with other free-floating algae of ponds and semipermanent pools. It is also a frequent, though never important, component of the fresh-water plankton. The species found in this country are *M. angularis* Thompson, *M. convoluta* Bréb., *M. elegans* A. Br. (Fig. 468*A*), *M. glauca* (Ehr.) Näg., *M. major* (G. M. Smith) Geitler, *M. punctata* Meyen (Fig. 468*B*), and *M. tenuissima* Lemm. For a description of *M. angularis*, see Thompson (1938); for the others, see Geitler (1930*A*).

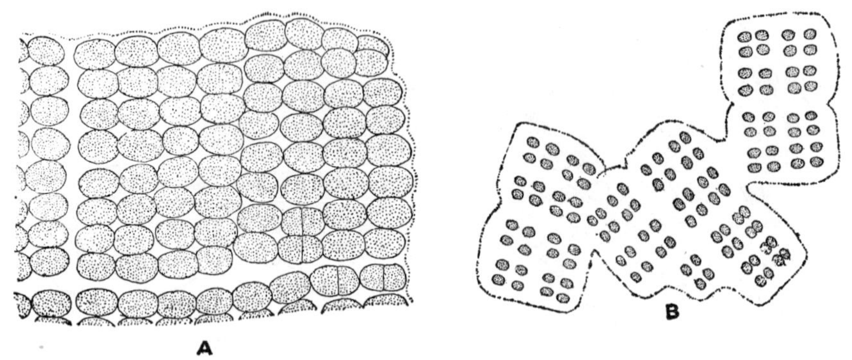

FIG. 468. *A*, *Merismopedia elegans* A. Br. *B*, *M. punctata* Meyen. (× 600.)

17. **Holopedium** Lagerheim, 1883. In its monostromatic colonies *Holopedium* resembles *Merismopedia*, but it differs in that the cells are irregularly arranged instead of in rectilinear series. The irregular cellular arrangement is due to the fact that successive cell divisions are not at right angles to each other. *Holopedium* also differs in that the long axes of the ellipsoidal cells lie perpendicular to the flattened face of a colony.

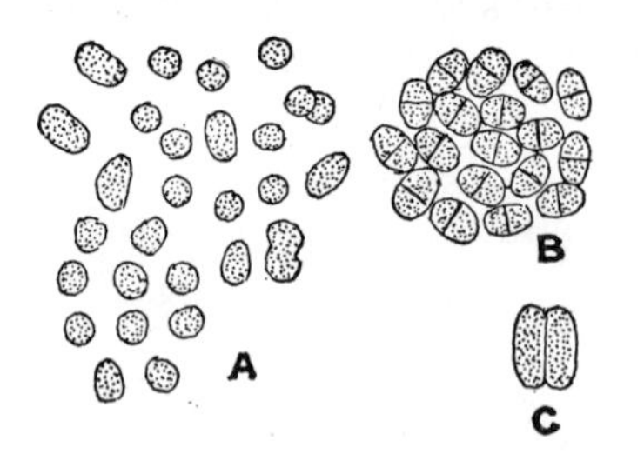

FIG. 469. *A*, *Holopedium irregulare* Lagerh. *B–C*, *H. geminatum* Lagerh. (*After Lagerheim*, 1883.) (*A*, × 960; *B–C*, × 480.)

All four species known for this country are rare algae. These are *H. geminatum* Lagerh. (Fig. 469*B–C*), *H. irregulare* Lagerh. (Fig. 469*A*), *H. obvolutum* Tiffany, and *H. pulchellum* Buell. For a description of *H. obvolutum*, see Tiffany (1934); for *H. pulchellum*, see Buell (1938); for the other two, see Geitler (1930*A*).

18. **Coelosphaerium** Nägeli, 1849. The spherical, ellipsoidal, reniform, or irregularly shaped colonies of this alga have a hyaline, homogeneous or radially fibrillar, colonial matrix in which the cells lie in a single layer a short distance inward from the surface. The individual cells may lie

equidistant from one another or be unevenly spaced. In one species, the cells are ovoid to subpyriform and lie with their long axes radial to the center of a colony. The other species have spherical to subspherical cells. According to the species, the protoplasts are homogeneous and of a pale to a bright blue-green color, or densely filled with pseudovacuoles.

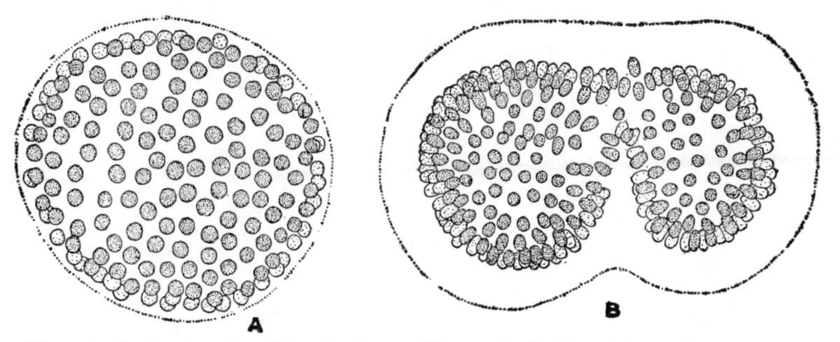

FIG. 470. *A, Coelosphaerium Kuetzingianum* Näg. *B, C. Naegelianum* Unger. (× 600.)

Certain species of *Coelosphaerium* are widely distributed plankton organisms and at times an important component of "water blooms." The species found in this country are *C. confertum* W. and G. S. West, *C. dubium* Grun., *C. Kuetzingianum* Näg. (Fig. 470*A*), *C. minutissimum* Lemm., and *C. Naegelianum* Unger (Fig. 470*B*). For descriptions of them, see Geitler (1930*A*).

19. **Marssoniella** Lemmermann, 1900. The chief distinctions between this genus and *Coelosphaerium* are the smaller number of cells in a colony, the pyriform shape of the cells, and their radial arrangement with their broader ends toward the center of a colony. The colonial matrix is exceedingly delicate and usually visible only when demonstrated by the india-ink method.

FIG. 471. *Marssoniella elegans* Lemm. (× 800.)

M. elegans Lemm. (Fig. 471) has been found in certain of the North Central states. For a description of it, see Geitler (1930*A*).

20. **Gomphosphaeria** Kützing, 1836. As in *Coelosphaerium*, the cells lie in a single layer a short distance in from the surface of the colonial matrix. Here, the cells show a much stronger tendency to remain closely appressed in twos or fours for some time after division and to be more unevenly spaced than in *Coelosphaerium*. The shape of individual cells may be spherical or pyriform; their protoplasts may be homogeneous or finely granular in structure and of a pale gray to a bright blue-green color.

Gomphosphaeria has a colonial matrix quite different from that of *Coelosphaerium*. In *Gomphosphaeria* the sheath surrounding a cell is stalked, and cell division is accompanied by a division of the distal portion of the stalk. The ultimate result of repeated cell division is a radiately branched mass of dense gelatinous material at the center of a colony, and one in which each ultimate branch terminates in a cell.

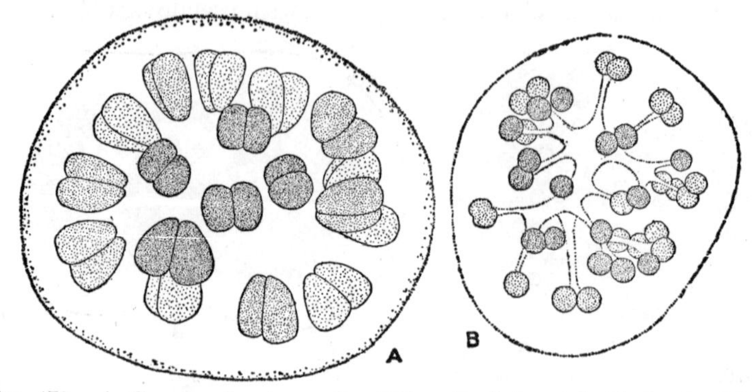

Fig. 472. *A*, *Gomphosphaeria aponina* Kütz. *B*, *G. lacustris* Chod. (× 825.)

Both of the two species found in this country, *G. aponina* Kütz. (Fig. 472*A*) and *G. lacustris* Chod. (Fig. 472*B*), are widely distributed. For descriptions of them, see G. M. Smith (1920).

21. **Glaucocystis** Itzigsohn, 1854. For a long time, *Glaucocystis* was described as an *Oöcystis*-like organism with two to eight ovoid to spherical

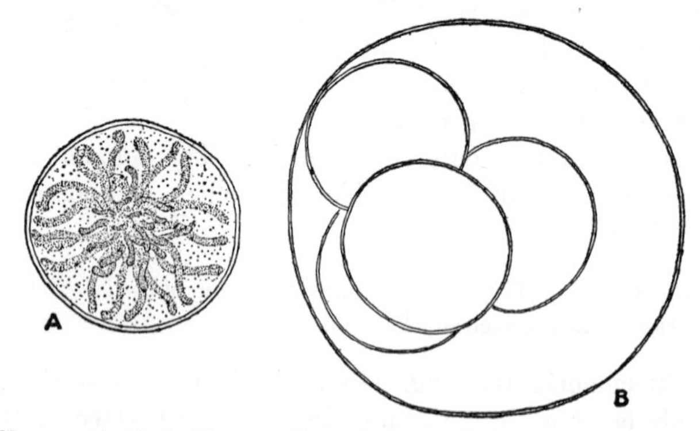

Fig. 473. *Glaucocystis Nostochinearum* Itz. *A*, host cell containing several cells of *Glaucocystis*. *B*, four-celled colony of host. (× 600.)

autospores, with polar and equatorial nodules, inside an expanded parent-cell wall. The protoplast was described as having an axial chromatophore of a bright blue-green color, with numerous vermiform projections toward the plasma membrane. Then the view was expressed[1] that this

[1] Geitler, 1924.

"organism" is in reality an association of two algae: a rod-shaped member of the Chroococcales, and a colorless member of the Chlorophyceae closely allied to *Oöcystis*. When interpreted in this manner, the generic name has been restricted[1] to the myxophycean component of the association. There has, as yet, been no demonstration that the chloroplast-like organisms can live independently of the host, but the fact that the blue-green components of another "alga" of this sort (*Cyanoptyche* Pascher) can live and divide independently of the host[1] gives a justification for accepting the view that the chloroplast-like structures in *Glaucocystis* are blue-green algae.

Glaucocystis with its host is usually found sparingly intermingled with other free-floating algae in soft-water regions. The three species found in this country are

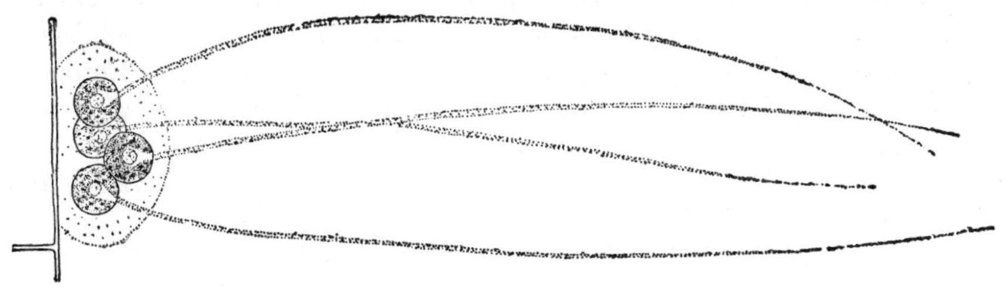

FIG. 474. *Gloeochaete Wittrockiana* Lagerh. (× 400.)

G. duplex Prescott, *G. Nostochinearum* Itz. (Fig. 473), and *G. oöcystiforme* Prescott. For a description of *G. Nostochinearum*, see Brunnthaler (1915); for the other two, see Prescott (1944).

22. Gloeochaete Lagerheim, 1883. *Gloeochaete* is another instance of endophytism of a unicellular myxophycean within the cells of a colorless member of the Chlorophyceae. The host is a colorless member of the Tetrasporaceae, which grows epiphytic upon filamentous algae, and has one, two, four, or eight cells embedded within a rather broad hyaline matrix. Each cell is furnished with one or two long gelatinous bristles. The host reproduced by means of flagellated zoospores which have a single nucleus and two contractile vacuoles at the base of the flagella.[2] The cells of the myxophycean endophyte are ovoid to sausage-shaped, and they lie close together in a cup-shaped region corresponding in shape and position to the cup-shaped chloroplasts in cells of Volvocales. Starch is often present in such abundance in the host that it is quite difficult to make out individual cells of the endophyte.

Until the dualism of the association called *Gloeochaete* was pointed out,[3] this "alga" appeared quite out of harmony with other Myxophyceae because it had long

[1] PASCHER, 1929. [2] DANGEARD, 1889; KORSHIKOV, 1917.
[3] GEITLER, 1925; PASCHER, 1929.

gelatinous bristles. It could not be placed in the Chlorophyceae because of its brilliant blue-green "chloroplasts." With the recognition of its dual nature, the specific name *Gloeochaete* has been restricted[1] to the myxophycean component of the association. The single species, *G. Witrockiana* Lagerh. (Fig. 474) has been found at several widely separated stations in this country. For a description of it, see Lagerheim (1883).

FAMILY 2. ENTOPHYSALIDACEAE

Members of this family differ from the Chroococcaceae in the pseudofilamentous arrangement of their cells. All cells in a colony may be in pseudofilamentous arrangement, or the pseudofilamentous tendency may be restricted to cells near the surface of a colony. Certain Chamaesiphonales, as *Pleurocapsa*, sometimes have a pseudofilamentous tendency, but such Chamaesiphonales may always be distinguished from Entophysalidaceae by their formation of endospores. Cells of genera belonging to Entophysalidaceae may have their sheaths distinct from, or confluent with, one another.

The only method of reproduction in the family is by cell division and fragmentation of colonies.

The two genera found in this country differ as follows:

1. Entire colony pseudofilamentous........................ 2. **Heterohormogonium**
1. Only the upper portion of colony pseudofilamentous............ 1. **Entophysalis**

1. **Entophysalis** Kützing, 1843. The spherical cells of this genus are united in *Gloeocapsa*-like groups of two or four, with a tough homogeneous sheath around each group. The entire mass of *Gloeosapsa*-like groups may be arranged in psuedofilamentous vertical series, or the groups may be laterally united to form an extended stratum from which arise numerous pseudofilamentous vertical outgrowths. In either case, the plant mass is an expanded layer of leathery texture.

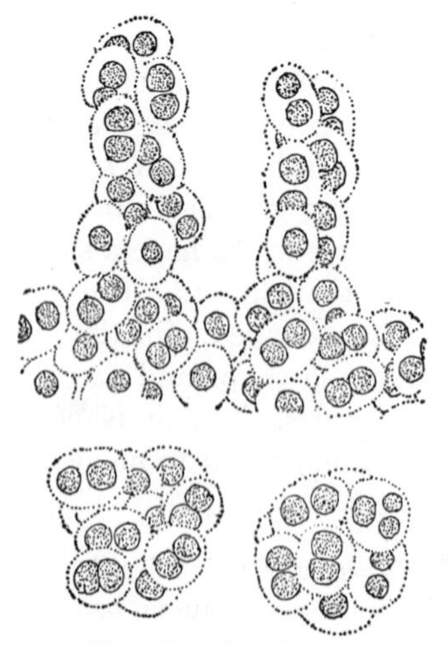

FIG. 475. *Entophysalis magnoliae* Farlow. (× 975.)

Entophysalis is a genus more frequently encountered in marine than in fresh-water habitats. The three species found in fresh waters in this country are *E. Cornuana* Sauv. (*Radiasia Cornuana* Sauv.), *E. magnoliae* Farlow (Fig. 475), and *E. rivularis* (Kütz.) Drouet (*Oncobrysa Cesatini* Rab.). For descriptions of them, see Geitler (1931).

[1] PASCHER, 1929.

2. **Heterohormogonium** J. J. Copeland, 1936. This distinctive alga has broadly ellipsoidal cells uniseriately arranged a slight distance from one another within a tubular homogeneous gelatinous sheath. The cells are so oriented that their long axes are perpendicular to the long axis of the sheath. Cell division is usually in the longitudinal axis of a cell. Now and then two or three adjacent cells divide transversely. A lateral separation of the two series of daughter cells and a subsequent splitting of the sheath next to the uniseriate portion result in a branched colony.

Reproduction is by breaking of a colony and by liberation of single cells at the tips of a colony.

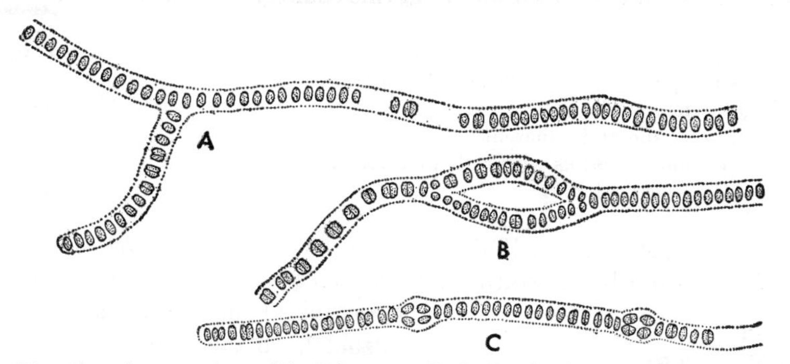

FIG. 476. *Heterohormogonium schizodichotomum* J. J. Copeland. *A*, portion of a colony. *B–C*, stages in formation of branches. (*After Copeland*, 1936.) (× 1235.)

The systematic position of this genus is uncertain. Its lack of a truly filamentous organization excludes it from the Oscillatoriaceae, and its lack of endospores excludes it from the Chamaesiphonales. The pseudofilamentous organization suggests that it belongs to the Entophysalidaceae rather than the Chroococcaceae. The single species, *H. schizodichotomum* J. J. Copeland (Fig. 476), is known only from the Mammoth Hot Springs of Yellowstone National Park. For a description of it, see J. J. Copeland (1936).

ORDER 2. CHAMAESIPHONALES

Members of this order regularly have a formation of endospores. In this respect, they are quite different from other Myxophyceae, and the regular production of endospores by any genus is a sufficient justification for referring it to the order. Some genera have unicellular thalli; other genera are multicellular and amorphous, crustose, or pseudofilamentous.

Chamaesiphonales are of much commoner occurrence in the ocean than in fresh waters. Certain of the fresh-water species are apt to be overlooked because of the lack of a distinctive vegetative organization. According to the structure of the plant body and the manner in which endospores are formed, the order is divided into families, three of which have representatives in the fresh-water flora of this country.

Family 1. Pleurocapsaceae

Thalli of members of this family are multicellular as a result of vegetative cell division. The thalli may have the cells organized into a continuous stratum, or they may be differentiated into a chroococcoid basal portion and an erect filamentous portion. Endospores are the only type of spore, and they may be formed in all cells of a thallus or only in certain cells. In the latter case, the cells functioning as sporangia are usually larger than other cells and definitely localized, either in the basal, median, or upper portion of a thallus.

The genera found in fresh waters in this country differ as follows:

1. Thallus perforating a calcareous substratum .... 4. **Hyella**
1. Thallus not perforating a calcareous substratum 2
    2. Thallus not sessile .... 3. **Myxosarcina**
    2. Thallus sessile and adherent to substratum 3
3. Upper portion of thallus with cells in vertical rows .... 1. **Pleurocapsa**
3. Cells not in vertical rows .... 2. **Xenococcus**

1. **Pleurocapsa** Thuret, 1885. The crustose thalli of *Pleurocapsa* are differentiated into a basal stratum and an erect portion. The basal portion is organized into a much-branched, rhizoidal, filamentous system; the erect portion is also filamentous, either uniseriate or multiseriate; and the sheaths enclosing the branches are laterally fused with one another. The distal portions of erect filaments are usually branched. Sporangia may be of the same size as vegetative cells or considerably larger. They may be terminal or intercalary in position.

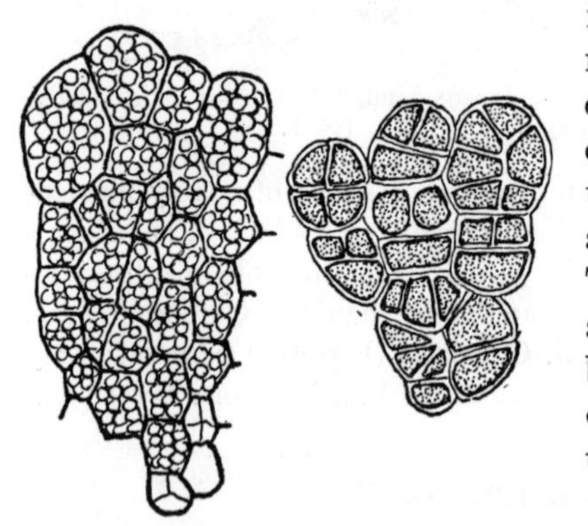

Fig. 477. *Pleurocapsa minor* Hansg., em. Geitler. (*After Geitler,* 1925*B*.) (× 800.)

The species found in the United States are *P. fluviatilis* Lagerh., *P. minor* Hansg. em. Geitler (Fig. 477) and *P. entophysaloides*, Setchell and Gardner, the latter a marine species which is also found in Great Salt Lake. For descriptions of them, see Geitler (1931).

2. **Xenococcus** Thuret, 1880. Cells of this sessile alga are usually arranged in a stratum one cell in thickness. Many of the species have cells that are angular by mutual compression, but sometimes the cells are rather remote from one another and rounded. In the latter case, a gelatinous envelope is frequently evident around each cell. The cell contents are a pale blue-green to a dark violet and usually homogeneous in structure.

Growth of a colony is by cell division in a plane vertical to the substratum. Any cell of a colony may also have a division of its contents into many endospores.

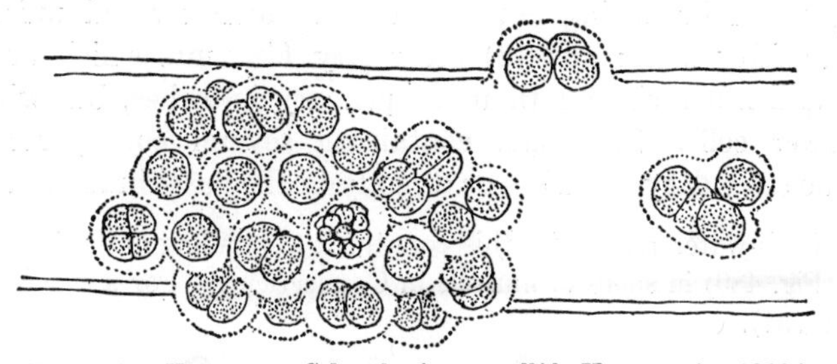

FIG. 478. *Xenococcus Schousboei* var. *pallida* Hansg. (× 1300.)

The two species recorded from fresh waters in this country are *X. Kerneri* Hansg. and *H. Schousboei* Thur. (Fig. 478). For descriptions of them, see Geitler (1931).

3. **Myxosarcina** Printz, 1921. This genus has the cells united in cuboidal, globose, or ovoid free-living colonies with 50 to 60 cells surrounded by a firm gelatinous envelope. The cells lie close to one another, and

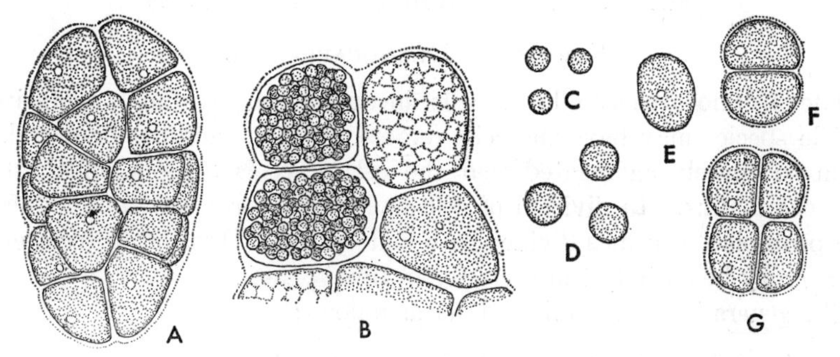

FIG. 479. *Myxosarcina amethystina* J. J. Copeland. *A*, adult colony. *B*, portion of a colony producing endospores. *C–E*, germination of endospores. *F–G*, juvenile colonies. (*After Copeland*, 1936.)

portions of cells in contact with other cells are cubical or polygonal. Reproduction is by repeated division of the contents of a cell into a large number of endospores any one of which, after escape from the parent cell, is capable of developing into a new colony.

The only species thus far found in this country is *M. amethystina* J. J. Copeland (Fig. 479), and it is known only from a single hot spring in Yellowstone National Park. For a description of it, see J. J. Copeland (1936).

4. **Hyella** Bornet and Flahault, 1888. The thalli of this alga are tangled filamentous masses growing within the shells of mollusks or, in the case of

some marine species, within thalli of other algae. Two kinds of filaments are recognizable within the host: (1) primary or basal filaments which grow mainly in a horizontal direction through the substratum and by repeated branching and rebranching develop into a much tangled uniseriate or multiseriate felty mass; and (2) secondary filaments which are always uniseriate and much shorter than the primary branches, but with longer and narrower cells. Sporangia containing several endospores develop either in the secondary branches or directly in the basal branches.

The sole fresh-water record for this country is the finding of *H. fontana* Huber and Jadin (Fig. 480) in shells of mussels in Connecticut.[1] For a description of it, see Geitler (1931).

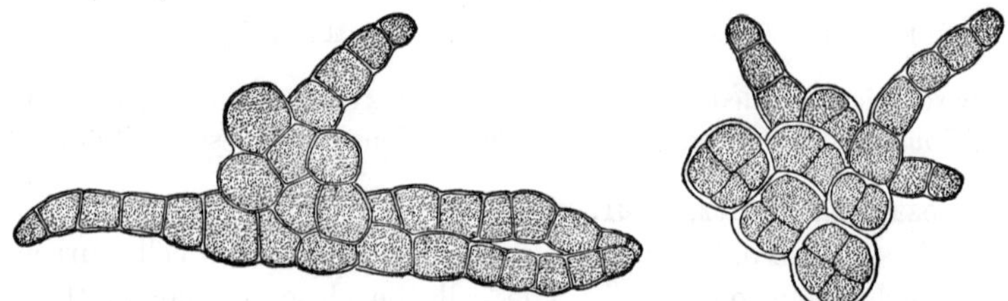

Fig. 480. *Hyella fontana* Huber and Jadin. (× 650.)

Family 2. Dermocarpaceae

The Dermocarpaceae have cells incapable of dividing vegetatively. Sessile species may have the cells remote from one another or may have them so densely aggregated that the group seems to be a multicellular crustose thallus. In division of cell contents to form endospores, the entire protoplast becomes divided into endospores. Division to form endospores may be entirely transverse or in three planes.

The genera in the local flora differ as follows:

1. Cells sessile 2
1. Cells not sessile .... 2. **Pluto**
   2. Divisions forming endospores all transverse .... 3. **Stichosiphon**
   2. Divisions forming endospores in three planes .... 1. **Dermocarpa**

1. **Dermocarpa** Crouan, 1858. *Dermocarpa* (Fig. 481) is unicellular and with spherical, ovoid, or pyriform cells. The cells are always sessile and, when one end is broader than the other, the narrower end is usually affixed to the substratum. Frequently the sessile cells lie so close to one another that the alga seems to be multicellular. Endospores are formed by repeated division of the protoplast in three

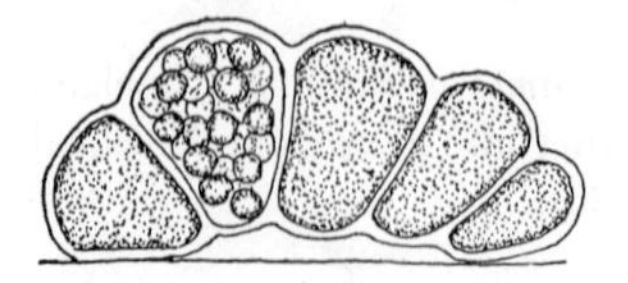

Fig. 481. *Dermocarpa pacifica* Setchell and Gardner, a marine species.

[1] Collins, 1897.

planes. The number of endospores may be only four, but in most species a cell forms a large number of endospores.

The species found in this country are *D. Gardneriana* Drouet, *D. Hollenbergii* Drouet, *D. minuta* Drouet, *D. rostrata* J. J. Copeland, *D. Setchellii* Drouet, and *D. Solheimii* Drouet. For a description of *D. rostrata*, see J. J. Copeland (1934); for the others, see Drouet (1942, 1943).

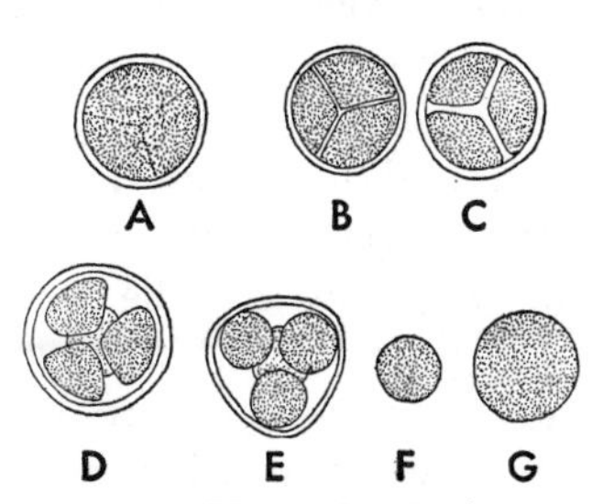

FIG. 482. *Pluto caldarius* (Tilden) J. J. Copeland. *A–E*, stages in development of endospores. *F*, endospore. *G*, nearly mature vegetative cell. (*After Copeland*, 1936.) (× 1500.)

2. **Pluto** J. J. Copeland, 1936. *Pluto* has free-living solitary spherical cells without an evident gelatinous sheath. The cells do not divide vegetatively. Reproduction is by a pyramidate division of the cell contents to form four endospores which, after liberation from the parent-cell wall, enlarge directly into vegetative cells.

It has been held[1] that this genus should not be considered generically distinct from *Dermocarpa*, but the free-living habit and the distinctive division forming the endospores seem to warrant its recognition. The single species, *P. caldarius* (Tilden) J. J. Copeland (Fig. 482) is widespread in hot springs in the western part of this country. For a description of it, see J. J. Copeland (1936).

3. **Stichosiphon** Geitler, 1931. This epiphytic unicellular alga has ellipsoidal to pyriform cells surrounded by a broad gelatinous sheath. Sooner or later there is a repeated transverse division of the entire cell contents to form a single row of endospores which are liberated through an opening at the free end of the cellular sheath.

Members of this rare genus were first described as species of *Chamaesiphon*, but later segregated as a separate genus because endospore formation is not restricted to the apex of a protoplast. The only record for the occurrence of *Stichosiphon* in this country is the finding of *S. regularis* Geitler (Fig. 483) in Florida.[2] For a description of it, see Geitler (1931).

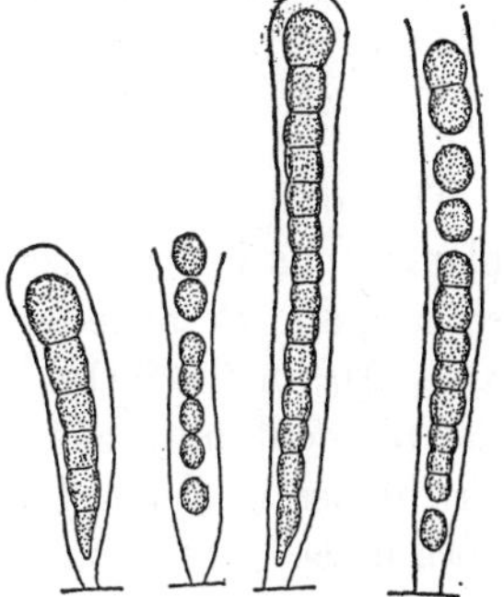

FIG. 483. *Stichosiphon regularis* Geitler. (*After Whelden*, 1941.)

FAMILY 3. CHAMAESIPHONACEAE

The single genus of this family, *Chamaesiphon*, has sessile cells incapable of dividing vegetatively. The formation of endospores (frequently called *exospores*) begins at the distal end of a cell, is basipetalous, and never continues to the base of the protoplast.

[1] DROUET, 1943*A*. [2] WHELDEN, 1941.

1. **Chamaesiphon** Braun and Grunow, 1865. The cells of *Chamaesiphon* may be ovoid, pyriform, or cylindrical. Proportions between length and breadth of cylindrical cells range from a length 2 to 3 up to 20 to 25 times the diameter. *Chamaesiphon* usually grows epiphytically upon filamentous Chlorophyceae. These epiphytic cells may be remote from one another, but more frequently they are gregarious and in a single layer; or, as a result of germination of endospores *in situ*, they may lie in a stratum more than one cell in thickness. The vegetative cells are enclosed by a sheath, which is usually thin and hyaline but which may be thick and colored. At the basal end of a cell, the sheath is sometimes drawn out into a distinct stipe.

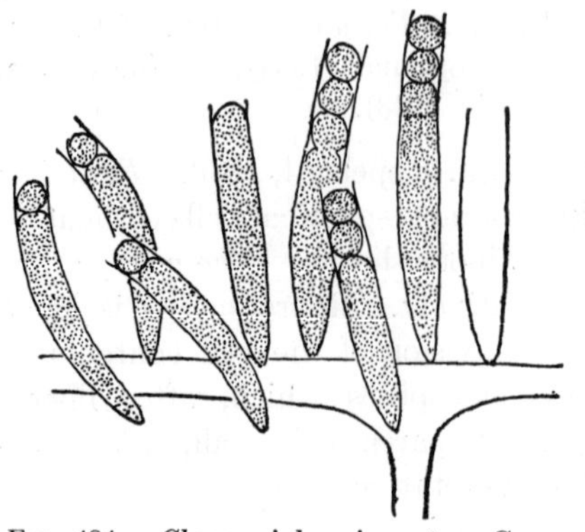

FIG. 484. *Chamaesiphon incrustans* Grun. (× 1200.)

Endospores are formed in basipetalous succession, and the major part of a protoplast may be devoted to spore formation, or spore formation may be restricted to the apex of a cell.

Species of *Chamaesiphon* are most frequently encountered on old filaments of *Oedogonium*, and of Cladophoraceae. They are rarely present on actively growing filamentous algae. The species found in this country are *C. confervicola* A. Br., *C. curvatus* Nordst., *C. cylindricus* Petersen, *C. gracilis* Rab., *C. incrustans* Grun. (Fig. 484), *C. minimus* Schmidle, *C. minutus* (Rostaf.) Lemm, and *C. polonicus* (Rostaf.) Hansg. For descriptions of them, see Geitler (1931).

ORDER 3. OSCILLATORIALES

This order includes all genera in which the thallus is multicellular and filamentous in organization. The trichomes may be uniseriate or multiseriate, and branched or unbranched. They may be naked or enclosed in a sheath; in the latter case, a single trichome may lie within a sheath, or the filament may consist of several trichomes within a common sheath. Trichomes may be of the same diameter throughout, attenuated toward both ends, or attenuated from base to apex; and straight, arcuate, or spirally or irregularly twisted. Sheaths surrounding trichomes are, according to the species, homogeneous or lamellated, firm or mucous, and hyaline or variously colored. The cell shape is quite constant for any species and may be spherical, ellipsoidal, discoid, or cylindrical. All cells of a filament may be vegetative in nature; or there may be a formation of heterocysts, akinetes, or both.

Reproduction is by breaking of a trichome into few-celled segments (hormogonia) and by a formation of akinetes.

SUBORDER 1. OSCILLATORINEAE

Members of this suborder reproduce by means of hormogonia only, and never form either heterocysts or akinetes. The trichomes are always uniseriate, unbranched and, except for the apical and immediately adjacent cells, are of the same diameter throughout. A majority of the genera have firm or gelatinous, homogeneous or lamellated, hyaline or colored sheaths about the trichomes. Some genera have several trichomes within a common sheath; others have only one trichome within a sheath.

There is but one family, the Oscillatoriaceae.

FAMILY 1. OSCILLATORIACEAE

Generic differences within the family are based upon number of trichomes within a sheath and upon structure of the sheath.

The genera found in the United States differ as follows:*

1. Trichomes without a sheath 2
1. Trichomes with a sheath 5
  2. Trichomes straight, curved, or in irregular spirals 3
  2. Trichomes in regular spirals 4
3. Trichomes with less than 20 cells .... 4. **Borzia**
3. Trichomes with hundreds of cells .... 3. **Oscillatoria**
  4. Dissepiments distinct .... 2. **Arthrospira**
  4. Dissepiments lacking .... 1. **Spirulina**
5. One trichome within a sheath 6
5. More than one trichome within a sheath 11
  6. Sheaths watery laterally confluent with one another 7
  6. Sheaths firm, not confluent with one another 9
7. Filaments growing in erect tufts .... 10. **Symploca**
7. Filaments not growing in erect tufts 8
  8. Filaments interwoven in an extended stratum .... 6. **Phormidium**
  8. Filaments parallel, in a free-floating scale-like mass .... 7. **Trichodesmium**
9. Trichomes with less than 20 cells .... 5. **Romeria**
9. Trichomes with hundreds of cells 10
  10. Sheaths colorless or brownish .... 8. **Lyngbya**
  10. Sheaths purplish .... 9. **Porphyrosiphon**
11. Sheaths firm .... 13. **Schizothrix**
11. Sheaths watery 12
  12. Many trichomes within a common sheath .... 11. **Microcoleus**
  12. Few trichomes within a common sheath .... 12. **Hydrocoleum**

1. **Spirulina** Turpin, 1827. Filaments which do not show a transverse septation and which are regularly twisted into narrow spirals, with successive turns close together or remote from one another, are usually re-

* *Dasygloea* Thwaites is to be excluded from the algal flora of this country because the material identified as *D. amorpha* Berk. has been shown (Drouet, 1939) to be *Schizothrix Muelleri* Näg.

ferred to *Spirulina*. Critical study[1] has shown that several species which appear to be unseptate are, in reality, septate and should be transferred to *Arthrospira*. The species which have no septa also differ in that the central body is not transversely interrupted throughout the length of a cell.[2]

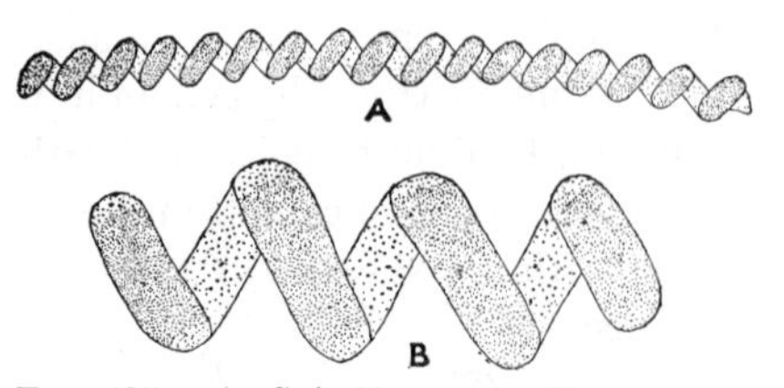

FIG. 485. *A*, *Spirulina major* Kütz. *B*, *S. princeps* (W. and G. S. West) G. S. West. (× 1000.)

Fresh-water species of *Spirulina* do not grow in such extensive strata as do marine and brackish-water species. Over a dozen species have been recorded for this country, but for certain of them it is uncertain whether they should be placed in *Spirulina* or *Arthrospira*. Among the species found in this country which seem to belong to the genus are *S. labyrinthiformis* Gom., *S. major* Kütz. (Fig. 485*A*), *S. princeps* (W. and G. S. West) G. S. West (Fig. 485*B*), and *S. subtilissima* Kütz. For descriptions of them, see Geitler (1932).

2. **Arthrospira** Stizenberger, 1852. The trichomes of *Arthrospira* are twisted to form regular spirals and are not enclosed by a sheath. In some cases, the trichome is evidently multicellular; in other cases, as noted in connection with *Spirulina*, the transverse walls of trichomes are very obscure and demonstrable only by staining.

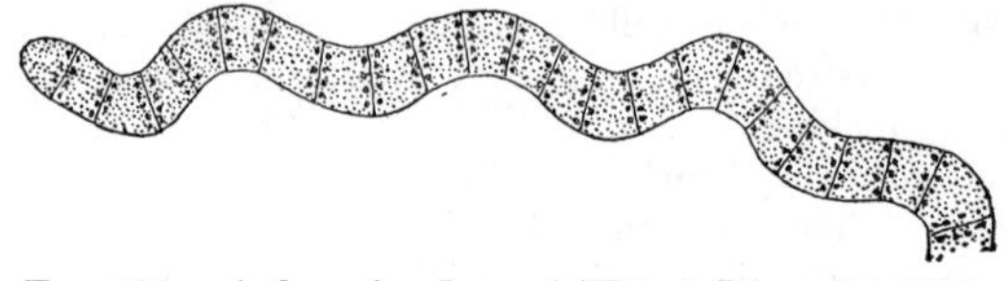

FIG. 486. *Arthrospira Jenneri* (Kütz.) Stiz. (× 800.)

The two species found in this country which unquestionably belong to *Arthrospira* are *A. Gomontiana* Setch. and *A. Jenneri* (Kütz.) Stiz (Fig. 486). For descriptions of them, see Geitler (1932).

3. **Oscillatoria** Vaucher, 1803. The filaments of *Oscillatoria* (Fig. 487) may occur singly or interwoven with one another to form a stratum of indefinite extent. The individual trichomes are unbranched, cylindrical, and entirely without, or with barely perceptible, evanescent sheaths. Species with narrow trichomes have cylindrical cells in which the length may be equal to or greater than the breadth; those with broad trichomes have cells in which the diameter is always greater than the length. Free ends of trichomes may be rounded and of approximately the same diameter as the rest of a trichome; or the terminal cells may taper to a subacute

[1] CROW, 1927; DOBELL, 1912; SCHMID, 1920, 1921*A*. [2] DOBELL, 1912.

point. Apical cells of trichomes may have walls like those of other cells, or the free face may be thickened into a *calyptra*. According to the species, the protoplasts are homogeneous, granulate (with the granules definitely or indefinitely distributed), or with numerous pseudovacuoles which completely obscure the protoplast's structure.

Oscillatoria is one of the most ubiquitous of algae. The determination of species is extremely difficult, and Gomont's monograph[1] is indispensable for exact determination of members of this genus. Over 40 species are definitely known to occur in the United States.

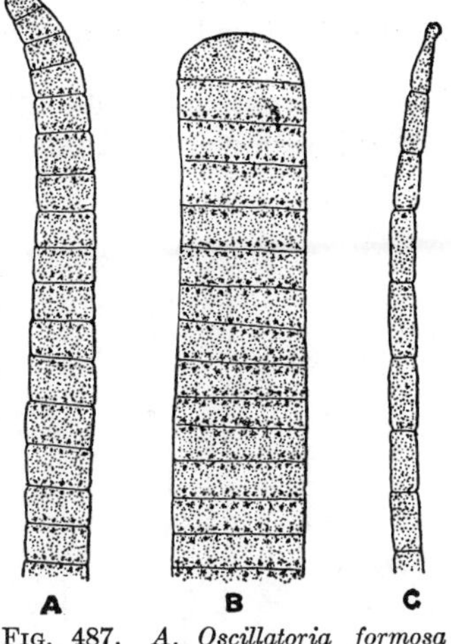

FIG. 487. *A*, *Oscillatoria formosa* Bory. *B*, *O. limosa* Ag. *C*, *O. splendida* Grev. (× 825.)

4. **Borzia** Cohn, 1883. Trichomes of *Borzia* are without a gelatinous sheath and are usually composed of 3 to 8 cells. Terminal cells of a trichome are hemispherical, and other cells are barrel-shaped.

This genus may be looked upon as one that is in a permanent hormogonial condition. As far as the United States is concerned, it is known only from Indiana where *B. trilocularis* Cohn (Fig. 488) has been collected[2] at two different stations. For a description of it, see Gomont (1892*A*).

5. **Romeria** Koczwara, 1932. The trichomes of *Romeria* are short, curved to spirally twisted, and rarely with more than a dozen uniseriately

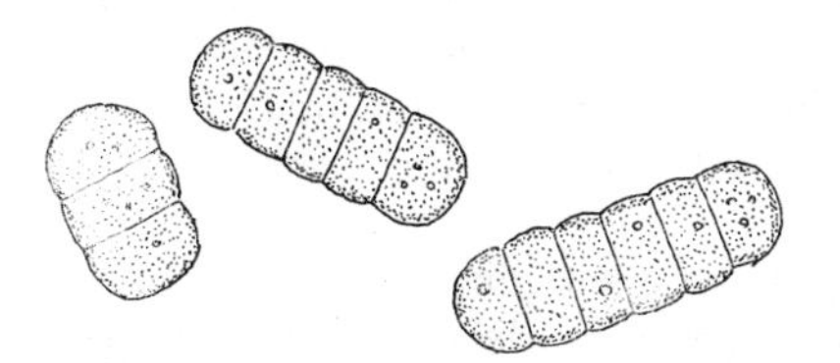

FIG. 488. *Borzia trilocularis* Cohn. (*After Gomont*, 1892*A*.) (× 1200.)

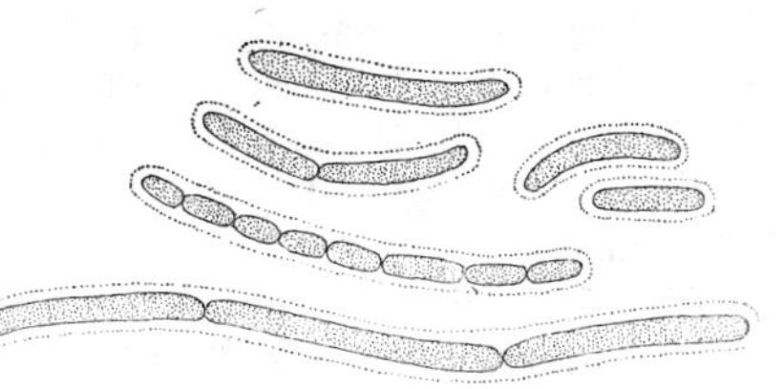

FIG. 489. *Romeria elegans* var. *nivicola* Kol. (*After Kol.* 1941.) (× 1000.)

arranged cells. There may or may not be an evident gelatinous sheath external to a trichome. The cells are cylindrical, with more or less broadly rounded poles, and with a length twice to several times the breadth. Heterocysts and akinetes are never formed, and reproduction is by fragmentation of trichomes.

[1] GOMONT, 1892, 1892*A*. [2] DAILY, 1943.

It is uncertain whether this genus should be considered one of the Chroococcaceae and allied to *Gloeothece*, or be considered one of the Oscillatoriaceae. The homogeneous sheath extending the entire length of a chain of cells indicates that it should be placed among the Oscillatoriaceae. In this country, the genus is known[1] only from snow fields in Yellowstone National Park. For a description of the alga found there, *R. elegans* var. *nivicola* Kol (Fig. 489), see Kol (1941).

6. **Phormidium** Kützing, 1843. The trichomes of *Phormidium* (Fig. 490) show much the same range of form as those of *Oscillatoria* and, as far as their shape is concerned, no morphological distinctions can be made between the two, except that cells of many *Phormidium* species are barrel-shaped rather than discoid or cylindrical. Trichomes of *Phormidium* are always enclosed by a watery gelatinous sheath, and very commonly sheaths of filaments are confluent with one another. The plant mass resulting from this coalescence of sheaths may have the trichomes approximately parallel to, or densely interwoven with, one another.

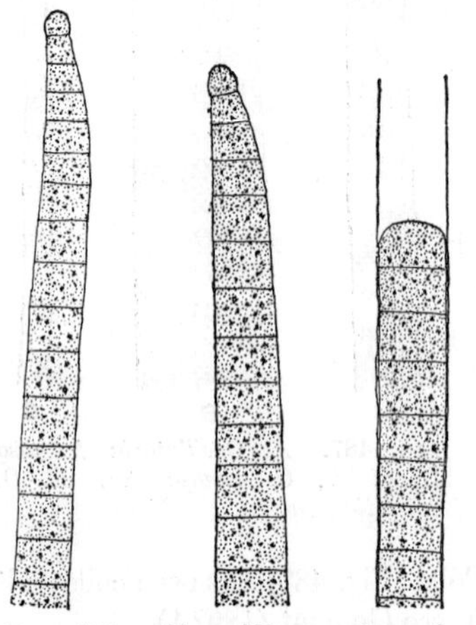

FIG. 490. *Phormidium autumnale* (Ag.) Gom. (× 800.)

In structure of its sheath, *Phormidium* is a genus which stands intermediate between *Oscillatoria*, which lacks a sheath, and *Lyngbya*, which has a clearly defined one. Although it is sometimes difficult to demonstrate the individual sheaths surrounding trichomes of *Phormidium*, there is little difficulty in distinguishing between the confluence of sheaths characteristic of it, and the lack of sheaths characteristic of *Oscillatoria*. The various species of the two genera were indiscriminately transferred from one genus to the other until this feature was recognized by Gomont,[2] the first to clear up the taxonomy of the two genera.

Phormidium is primarily subaerial in habit and often grows in extensive patches on moist rocks or damp soil. More than 25 species are definitely known to occur in this country. For monographic treatments of the genus, see Gomont (1892, 1892*A*), and Geitler (1932).

7. **Trichodesmium** Ehrenberg, 1830. The filaments of this alga are laterally joined to one another to form free-floating, spindle- to scale-shaped colonies. The trichomes are cylindrical, straight or spirally twisted, and of the same diameter throughout or with the apices slightly attenuated. Sheaths surrounding the trichomes are so delicate as to be almost imperceptible. According to the species, the cells are cylindrical or barrel-

[1] KOL, 1941. [2] GOMONT, 1892, 1892*A*.

shaped, and with homogeneous to granulose protoplasts, or with the protoplast filled with numerous pseudovacuoles.

T. lacustre Klebahn (Fig. 491), the only fresh-water species of the genus, is a widely distributed plankton organism in this country. For a description of it, see G. M. Smith (1920).

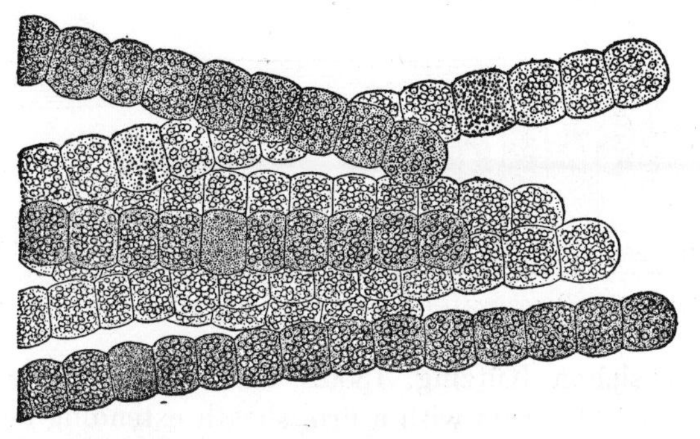

FIG. 491. *Trichodesmium lacustre* Klebahn. (× 1000.)

8. **Lyngbya** Agardh, 1824. Filaments of *Lyngbya* (Fig. 492) may occur singly, or interwoven into a free-floating mass or an extended stratum. The genus is sharply differentiated from preceding members of the family by the firm, relatively thin, hyaline to yellowish-brown, homogeneous or

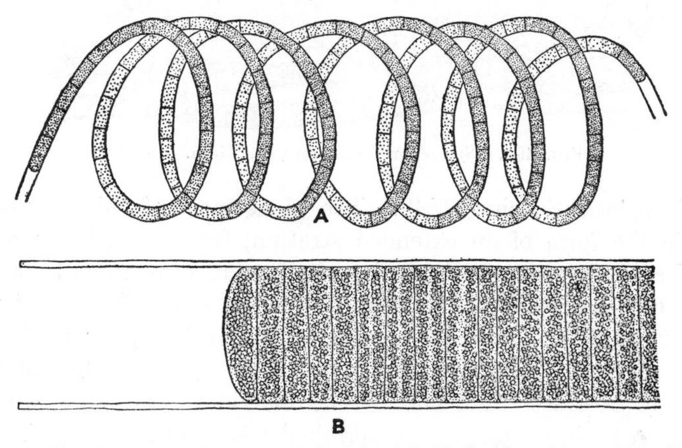

FIG. 492. *A*, *Lyngbya contorta* Lemm. *B*, *L. Birgei* G. M. Smith. (*A*, × 1000; *B*, × 825.)

lamellated sheaths which enclose but a single trichome and which generally project for some distance beyond it. The trichomes are cylindrical, more commonly with rounded than with slightly attenuated apices, and either straight, flexed, or twisted into regular spirals. Cell contents may be homogeneous, granulose, or with numerous pseudovacuoles; and gray, pale to bright blue-green, or variously colored.

Fresh-water species of this genus are more often aquatic than aerial in habit, and certain of the aquatic species are found only in the plankton. Over 35 species have been reported for this country. For monographic treatments of the genus, see Gomont (1892*A*), and Geitler (1932).

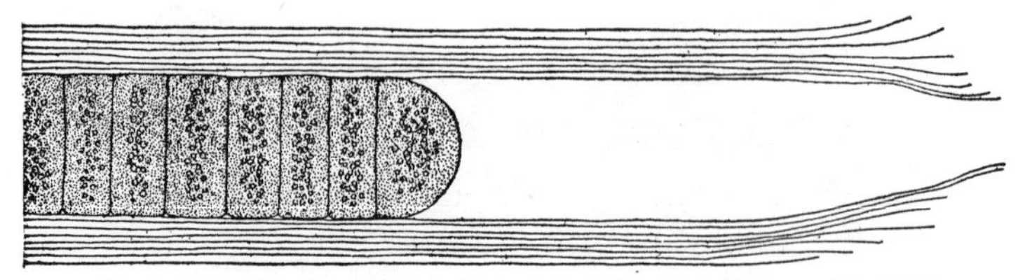

Fig. 493. *Porphyrosiphon Notarisii* (Menegh.) Kütz. (× 600.)

9. **Phorphyrosiphon** Kützing, 1850. This alga is quite like *Lyngbya* in its unbranched filaments with a firm sheath extending beyond the ends of the single trichome. The chief feature distinguishing it from *Lyngbya* is the purplish-red color of the many-layered sheath.

The only species, *P. Notarisii* (Menegh.) Kütz. (Fig. 493) is widely distributed in the United States. For a description of it, see Gomont (1892).

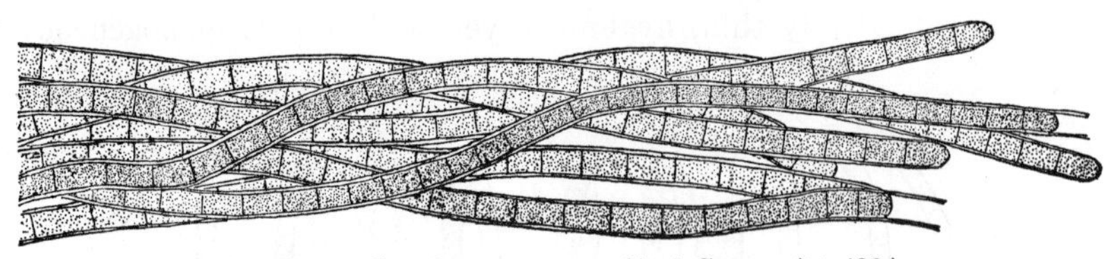

Fig. 494. *Symploca muscorum* (Ag.) Gom. (× 480.)

10. **Symploca** Kützing, 1843. The plant mass of this subaerial alga grows in the form of an extended stratum, from which arise numerous vertical conical tufts. Individual trichomes are surrounded by a definite, firm or mucous, sheath (Fig. 494). In many cases, sheaths of the filaments are confluent with one another in the median portion of filaments but not at their extremities. Under such conditions, the plant mass appears to be falsely branched. The individual trichomes are not attenuated at their apices, but the terminal cells may be rounded, and with or without a calyptra.

Symploca is usually found growing on moist cliffs or on damp soil. Eleven species have been found in the algal flora of this country. For descriptions of most of the species of the genus, see Geitler (1932).

11. **Microcoleus** Desmazières, 1823. The filaments of this alga have a wide, cylindrical, unbranched, homogeneous sheath of an extremely

gelatinous nature. Within a sheath is a central core of many parallel trichomes which are spirally and tightly interwoven. The individual trichomes have acute or obtuse ends and conical or capitate apical cells.

Marine and fresh-water species of *Microcoleus* (Fig. 495) are usually found on damp soil or on mud. About a dozen fresh-water species are known for the United States. For descriptions of most of the species of the genus, see Geitler (1932).

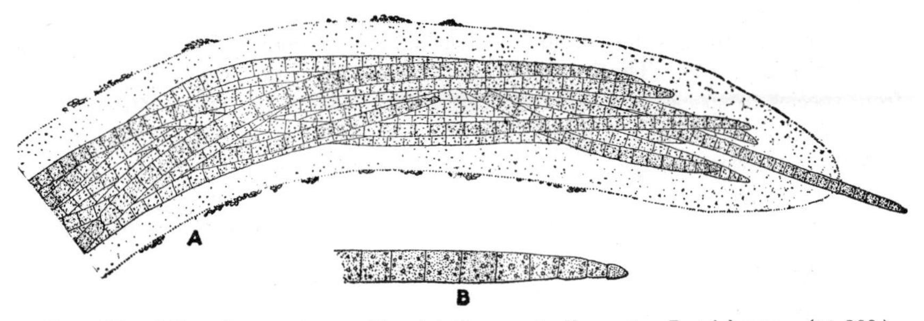

FIG. 495. *Microcoleus vaginatus* (Vauch.) Gom. *A*, filament. *B*, trichome. (× 300.)

12. **Hydrocoleum** Kützing, 1843. *Hydrocoleum* has much the same type of wide diffluent sheath as *Microcoleus* but contains only a few loosely aggregated trichomes within a common sheath. Its sheaths may at times

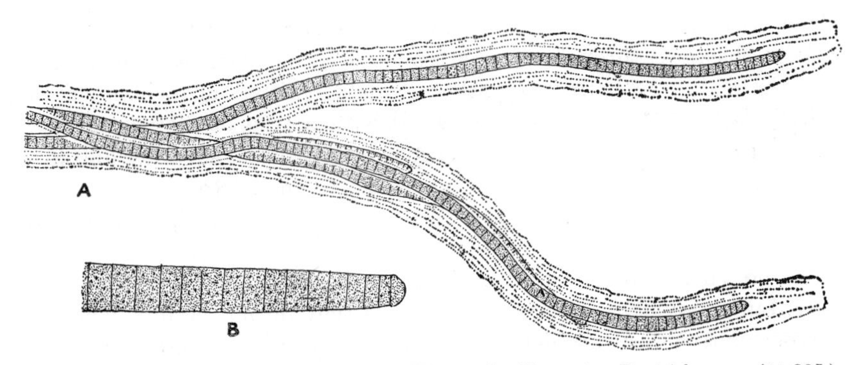

FIG. 496. *Hydrocoleum homeotrichum* Kütz. *A*, filament. *B*, trichome. (× 325.)

have a certain amount of branching and show some evidence of lamellation. The individual trichomes differ from those of *Microcoleus* in that they have more or less attenuated ends, capitate apical cells, and interior cells that are always broader than they are long.

This genus grows in both aerial and aquatic habitats. The species found in this country are *H. glutinosum* (Ag.) Gon., *H. Groesbeckianum* Drouet, and *H. homoeotrichum* Kütz. (Fig. 496). For a description of *H. Groesbeckianum*, see Drouet (1943); for the other two, see Gomont (1892).

13. **Schizothrix** Kützing, 1843. In contrast with the two foregoing genera, the filaments of *Schizothrix* (Fig. 497) are enclosed by a firm wide sheath that is usually lamellated and in certain species is always colored when old. In the median portion of a filament, there are two or more trichomes twisted around one another. The sheaths branch freely toward their ends, and the ultimate branchlets usually contain but a single trichome.

Although *Schizothrix* may occur free-floating, it is more often found on damp soil or on dripping rocks. The general appearance of the plant mass is quite variable from species to species and has been utilized in dividing the genus into sections. About 30 species occur in this country. For descriptions of most species of the genus, see Geitler (1932).

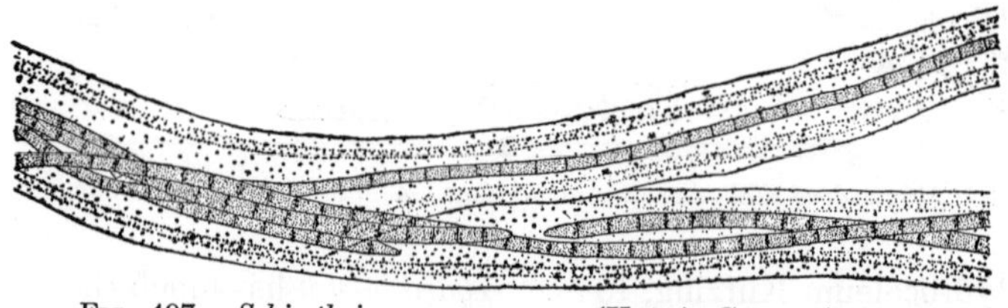

FIG. 497. *Schizothrix purpurascens* (Kütz.) Gom. (× 300.)

SUBORDER 2. NOSTOCHINEAE

The trichomes of this suborder may be uniseriate or multiseriate; of uniform thickness throughout or attenuated either from base to apex or from the middle to both extremities; straight, twisted in regular spirals, or irregularly contorted; unbranched or with a true or false branching. Sheaths are usually present in the various genera of the suborder and may enclose one or many trichomes. According to the species or genus, the sheaths are hyaline or colored, homogeneous or stratified, and firm or gelatinous.

The presence of heterocystis is the most distinctive feature of the suborder, but there are a few genera in which they do not occur. The first heterocysts to appear are developed when a trichome is quite young, and additional ones may be formed throughout the further growth of the trichome. Heterocysts may be formed at definite places in a trichome, or at no definite place; and singly or in very short series. Many of the genera that develop heterocysts also produce akinetes; other genera with heterocysts do not form akinetes.

The basic features on which the suborder is divided into four families are the organization and structure of the trichome.

FAMILY 1. NOSTOCACEAE

Trichomes of Nostocaceae are always unbranched, uniseriate, and without any appreciable attenuation at their apices. They are always surrounded by a sheath but may be straight, in regular spirals, or irregularly twisted. Sheaths surrounding trichomes are homogeneous. In rare cases, they are firm and narrow; usually they are copious, gelatinous, hyaline or colored, and distinct or confluent with one another to form colonies containing many trichomes. The cells are spherical or cylindrical and with or without constrictions at the cross walls. Protoplasts of vegetative cells may have a homogeneous or granulose structure, and their color may be blue-green or otherwise.

Heterocysts are regularly formed by all genera, and they may be terminal or intercalary, and solitary or catenate. Akinetes are also of frequent occurrence in all genera, and they may be formed adjacent to, or remote from, the heterocysts. They are usually larger than vegetative cells and often of different shape.

The genera found in this country differ as follows:

1. Heterocysts always terminal 2
1. Heterocysts intercalary 3
   2. Heterocysts at only one end of a trichome .... 7. **Cylindrospermum**
   2. Heterocysts at both ends of a trichome .... 3. **Anabaenopsis**
3. Length of cells less than breadth .... 8. **Nodularia**
3. Length of cells equal to or greater than breadth 4
   4. In colonies with trichomes parallel 5
   4. If in colonies, not with trichomes parallel 6
5. Colonies small, plate- or scale-like .... 6. **Aphanizomenon**
5. Colonies large, tubular, hollow .... 5. **Wollea**
   6. Trichomes solitary or intertwined in an amorphous mass 7
   6. Trichomes much twisted into a mass of definite form with a firm gelatinous envelope .... 4. **Nostoc**
7. Sheaths of trichomes firm and narrow .... 2. **Aulosira**
7. Sheaths of trichomes watery and broad .... 1. **Anabaena**

1. **Anabaena** Bory, 1822. Filaments of *Anabaena* (Fig. 498) occur either singly or in floccose colonies, and free-floating or in a delicate mucous stratum. The trichomes are of the same thickness throughout or slightly attenuated at their apices; straight, circinate, or irregularly contorted; and occur singly within a sheath. Sheaths surrounding trichomes are always hyaline and generally of so watery a nature that they cannot be seen unless demonstrated by special methods. They may be broad or narrow, and many plankton species have sheaths several times broader than the vegetative cells. The cells are usually spherical or barrel-shaped, rarely cylindrical, and never discoid. Protoplasts of vegetative cells are either homogeneous, granulose, or filled with numerous pseudovacuoles; and gray, blue-green, or variously colored. Heterocysts are usually of the same shape as vegetative cells, though slightly larger, are always intercalary in origin, are generally solitary, and several are usually present in any trichome. Akinetes may develop only next to heterocysts, only remote from them, or

in both positions; and singly or in very short catenate series. They are always larger than vegetative cells, and generally cylindrical and with rounded ends.

Descriptions of *Anabaena* and of *Nostoc* often leave the impression that it is difficult to distinguish between the two. In actual practice one has no such difficulty, since *Nostoc* always has a firm gelatinous envelope in which the trichomes are always much contorted; whereas *Anabaena* always has an extremely watery gelatinous sheath, never forms colonies of definite form, and, except for certain plankton species, never has contorted trichomes.

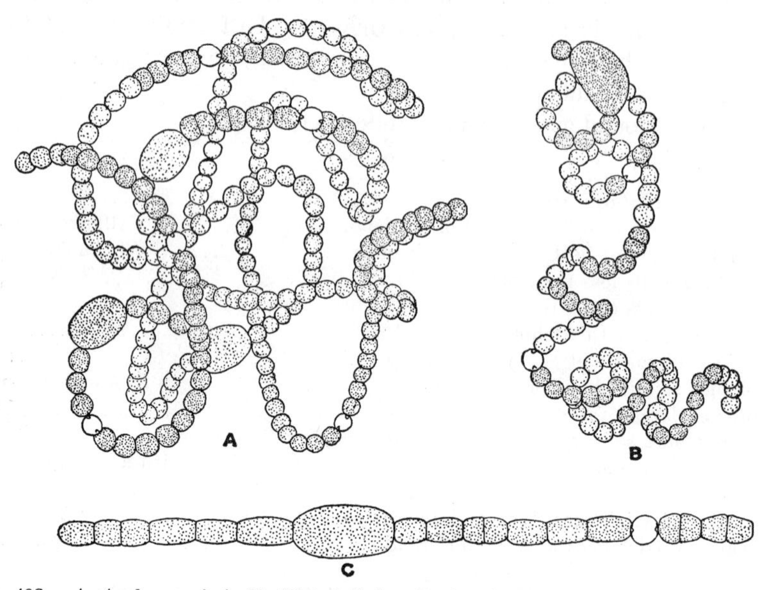

FIG. 498. *A, Anabaena circinalis* (Kütz.) Rab. *B, A. spiroides* Klebahn. *C, A. Levanderi* Lemm. (× 600.)

The shape and size of vegetative cells, heterocysts, and akinetes, as well as the relative position of heterocysts and akinetes, are all essential characters in determining species of *Anabaena*. For this reason, attempts to make specific determinations from immature trichomes are futile. *Anabaena* is primarily an aquatic alga and one not usually found in temporary pools. It often occurs in abundance in permanent and semipermanent pools and practically pure collections of a single species are not uncommon. Several species are known only from the plankton of ponds and lakes, and these may occur in such profusion as to cause a "water bloom." About 30 species have been found in the United States. For descriptions of most species of the genus, see Geitler (1932).

2. **Aulosira** Kirchner, 1878. Filaments of *Aulosira* are straight or curved, of uniform diameter throughout, and contain but one trichome.

The trichome is composed of cylindrical cells with flattened ends, and the cell length is usually equal to or greater than the breadth. The sheaths are narrow, homogeneous, and of a firm texture. Heterocysts are always intercalary and more or less cylindrical. Akinetes are usually formed in catenate series which may be adjacent to or remote from heterocysts.

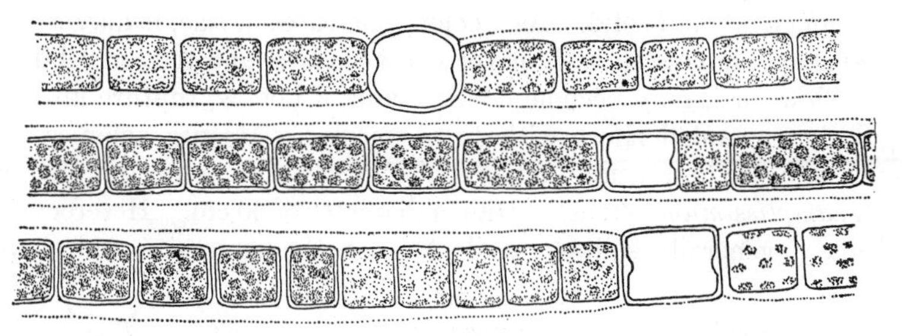

499. *Aulosira implexa* B. and F. (*After Bornet and Flahault,* 1885*A*.) (× 725.)

Aulosira resembles *Anabaena* but differs in shape of the cells and in the firmer sheath. *A. implexa* Bornet and Flahault (Fig. 499) has been found at several widely separated localities in the United States. For a description of it, see Daily (1943).

3. **Anabaenopsis** Woloszyńska, 1912; emend., Miller, 1923. This genus, first considered a section of *Anabaena* but later given generic rank, differs in having the heterocysts strictly terminal and in having these at both ends of a trichome. The cell structure is similar to that of *Anabaena*; but unlike other genera with strictly terminal heterocysts, the akinetes may be formed remote from the heterocysts. The terminal position of heterocysts is due to both cells from an intercalary cell division developing into heterocysts (Fig. 500*B* and *D*). The trichome breaks either while the heterocysts are maturing or just after they have matured. Thus it frequently happens that the heterocyst at one end of a trichome is mature and that at the other end is immature (Fig. 500*A*).

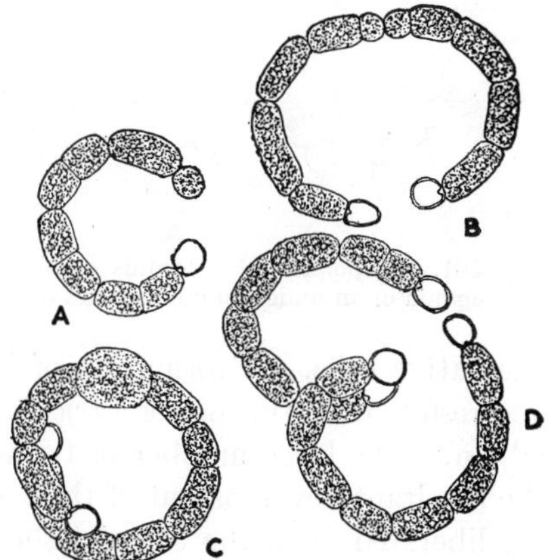

FIG. 500. *Anabaenopsis Elenkinii* Miller *A*, filament with a ripe and an immature heterocyst. *B*, a filament with both heterocysts mature. *C*, filament with an akinete. *D*, fragmentation of a filament. (× 650.)

Anabaenopsis is primarily a plankton alga. The species found in this country are *A. Arnoldii* Apetk., *A. circularis* (W. and G. S. West) Miller, *A. Elenkinii*

Miller (Fig. 500), and *A. Raciborskii* Wolosz. For descriptions of them, see Geitler (1932).

4. **Nostoc** Vaucher, 1803. The trichomes of *Nostoc* (Fig. 501) are always much contorted and embedded within a gelatinous sheath of an exceedingly firm consistency. Young colonies of *Nostoc* are of microscopic size, approximately spherical, and solid. As a colony increases in size, it may retain its original shape, have a smooth or verrucose surface, and remain solid or become hollow; or it may become lobulate and, if hollow, rupture into a firm irregularly expanded sheet with torn margins. Mature colonies are of macroscopic size and usually a few centimeters in diameter, but they may, as in *N. amplissimum* Setchell, attain a diameter of 50 cm. Heterocysts are intercalary, generally solitary, and of much the same size and shape as

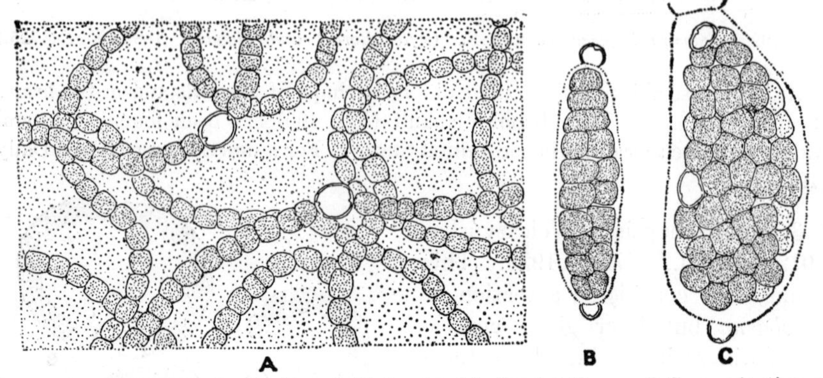

FIG. 501. *A*, portion of a thallus of *Nostoc Linckia* (Roth.) Born. *B-C*, germination of hormogonia of an undetermined species of *Nostoc*. (× 860.)

vegetative cells. A formation of hormogonia is very common in *Nostoc* and results from rupture of a trichome where heterocyst and vegetative cell adjoin. The large number of trichomes within an adult colony is largely the resultant development of these hormogonia into trichomes without being liberated from the colonial sheath. Akinetes do not usually develop until a colony is mature. When akinete formation does take place, many more vegetative cells are metamorphosed into akinetes than is the case in *Anabaena*, and it is not unusual to find that all vegetative cells between two successive heterocysts have developed into akinetes.

Nostoc is a common alga of both terrestrial and aquatic habitats. Terrestrial species grow on bare soil or intermingled with leafy plants, especially mosses. It may also be a subterranean alga and has been found at a depth of about a meter below the surface of the soil.[1] Strictly aquatic species occur either free-floating in pools, lying on the bottom of pools, or attached to submerged vegetation. Some

[1] MOORE and CARTER, 1926.

species favor swiftly running water, and it is not unusual to find them clothing the rocks in small mountain streams. Aside from their symbiotic association with fungi to form lichens, the chief association with other plants is the endophytism within thalli of bryophytes. Approximately 20 species occur in the United States. For descriptions of most species of *Nostoc*, see Geitler (1932).

5. **Wollea** Bornet and Flahault, 1888. Mature colonies of *Wollea* are vertically elongated, somewhat membranaceous, unbranched tubes 5 to 10 cm. long, that are closed at the distal end. The trichomes within the colonial matrix are numerous, erect, and mostly parallel with one another. The cells are cylindrical; the heterocysts are barrel-shaped; and the akinetes, which are formed in short catenate series adjacent to heterocysts, are cylindrical and considerably longer than vegetative cells.

There is but one species, *W. saccata* (Wolle) Bornet and Flahault (Fig. 502). For a description of it, see Geitler (1932).

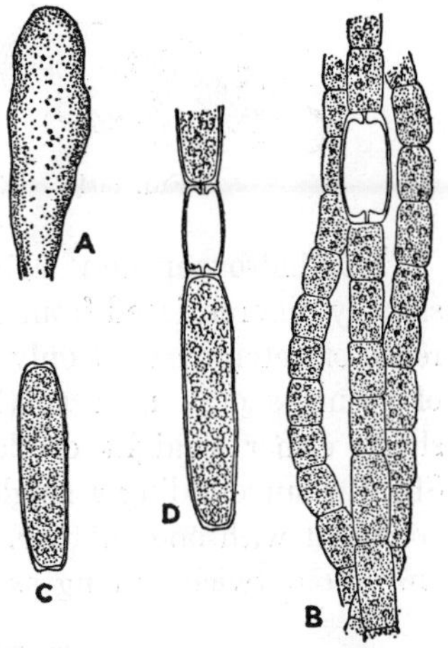

FIG. 502. *Wollea saccata* (Wolle) B. and F. *A*, habit sketch of a colony. *B*, trichomes. *C-D*, akinetes. (*A*, ×½; *B-D*, × 925.)

6. **Aphanizomenon** Morren, 1838. Trichomes of this alga are either straight, flexed, or curved, and laterally joined to one another in small, macroscopic, free-floating, feathery, or scale-like colonies. The sheaths surrounding individual trichomes are exceedingly delicate and confluent

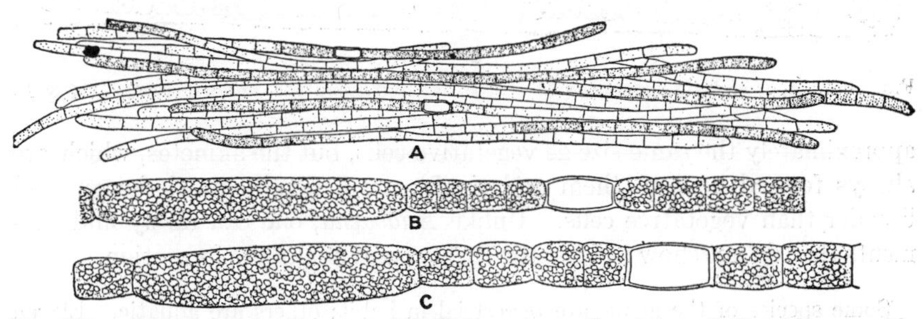

FIG. 503. *Aphanizomenon flos-aquae* (L.) Ralfs. *A*, small colony. *B-C*, trichomes with akinetes. (*A*, ×400; *B-C*, × 1000.)

with one another. The cells are cylindrical or barrel-shaped and longer than they are broad. Heterocysts are intercalary and cylindrical; akinetes are cylindrical, with a length 5 to 12 times the breadth, and not formed adjacent to heterocysts.

All members of the genus are known only from the fresh-water plankton. The single species found in this country, *A. flos-aquae* (L.) Ralfs (Fig. 503), is widely distributed but rarely found in abundance. For a description of it, see G. M. Smith (1920).

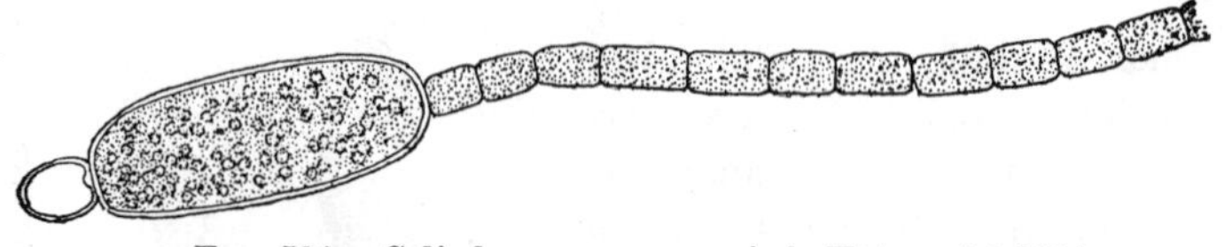

FIG. 504. *Cylindrospermum musciocla* Kütz. (× 900.)

7. **Cylindrospermum** Kützing, 1843. *Cylindrospermum* (Fig. 504) is sharply differentiated from other genera of the family by the regular occurrence of heterocysts at only one end of the trichomes and by the formation of akinetes only next to the heterocysts. The trichomes are generally short, either straight or flexed, and of the same diameter throughout. Sheaths surrounding a single trichome are extremely mucous and may be confluent with one another. The cells are cylindrical, with rounded ends, and about twice as long as broad. The strictly terminal heterocysts are

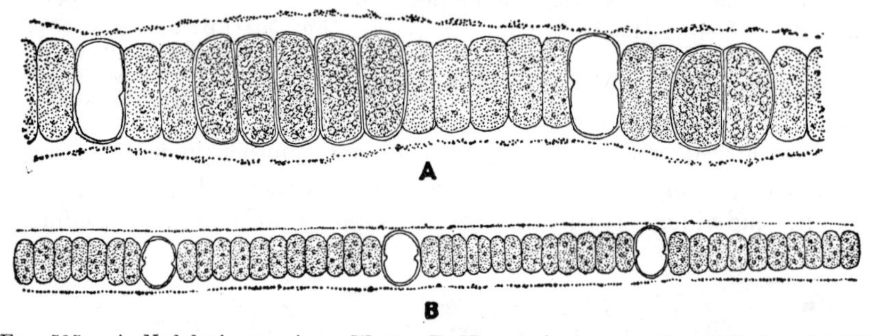

FIG. 505. *A, Nodularia spumigena* Mert. *B, N, spumigena* var. *minor* Fritsch. (× 800.)

approximately the same size as vegetative cells, but the akinetes, which are always formed next to them and singly, are generally much longer and broader than vegetative cells. Unlike *Anabaena*, one can rarely find filaments that do not show at least the early stages of akinete formation.

Some species of the genus are terrestrial in habit; others are aquatic. Eleven species have been reported for the United States. For descriptions of most of the species of the genus, see Geitler (1932).

8. **Nodularia** Mertens, 1822. This genus differs from all others of the family in that vegetative cells, heterocysts, and sometimes the akinetes, are broader than they are long. The trichomes, which are borne singly within

a sheath, are of the same diameter throughout or slightly attenuated at their apices, and straight or slightly flexed. There is usually a distinct fairly firm sheath, and the filaments are more commonly free from one another than with their sheaths confluent. Heterocysts are always intercalary, discoid, and generally somewhat broader than vegetative cells. Akinetes arise from cells remote from heterocysts and are usually formed in catenate series of two to a dozen. Sometimes all cells between two successive heterocysts develop into akinetes.

Nodularia is more commonly aquatic than terrestrial and usually grows sparingly intermingled with other free-floating algae of pools and ditches. The species found in this country are *N. Harveyana* (Thw.) Thuret, *N. sphaerocarpa* Bornet and Flahault, and *N. spumigena* Mert. (Fig. 505). For descriptions of them, see Geitler (1932).

FAMILY 2. SCYTONEMATACEAE

Genera referred to this family are characterized by uniseriate falsely branched trichomes that are of the same diameter throughout or somewhat attenuated toward their apices. The filaments always have a firm, sharply defined, sheath which may be hyaline or colored, and homogeneous or lamellated. In the majority of genera, there is but a single trichome within a sheath. False branching results from segmentation of a trichome into hormogonia, followed by a development of the hormogonia into trichomes without their liberation from the sheath. Ends of trichomes developing from hormogonia grow through the old sheath of the parent filament, either singly or in pairs, and then secrete a sheath of their own. More rarely, a common sheath encloses several parallel or much contorted trichomes.

Heterocysts are generally present at all stages of development, akinetes are frequently lacking. In many genera, the region of outgrowth of false branches is more or less definitely correlated with the position of heterocysts. Heterocysts are never found in *Plectonema*, but the false branching makes it more logical to put this genus in the Scytonemataceae than in the Oscillatoriaceae.

The genera found in this country differ as follows:

1. Filaments with heterocysts 2
1. Filaments without heterocysts .... 3. **Plectonema**
    2. One trichome within a sheath 3
    2. More than one trichome within a sheath 5
3. False branching abundant 4
3. False branching sparse, heterocysts basal, or basal and intercalary .. 6. **Fremyella**
    4. False branches usually arising singly .... 2. **Tolypothrix**
    4. False branches usually arising in pairs .... 1. **Scytonema**
5. Trichomes much contorted .... 4. **Diplocolon**
5. Trichomes parallel .... 5. **Desmonema**

1. **Scytonema** Agardh, 1824. The primary feature distinguishing this alga from *Tolypothrix* is the lateral origin of false branches in pairs, and at a point approximately midway between two heterocysts. Here and there in a filament there may be *Tolypothrix*-like false branches, but these do not lie next to heterocysts in the manner so characteristic of *Tolypothrix*. Trichomes of *Scytonema* (Fig. 506) are usually of the same diameter throughout and with cylindrical cells. Sheaths surrounding trichomes are always of an exceedingly firm texture, hyaline or colored and if colored, the color is generally yellowish or brownish. Sheaths may be homogeneous or lamellated, and with the lamellae parallel or oblique. Species with oblique lamellae have them running in an arc from the interior to periphery of a sheath where they end in a wing-like expansion. The heterocysts are intercalary and are borne singly or in twos or threes. They are of approximately the same size as vegetative cells. Akinetes are of rare occurrence and are but little larger than vegetative cells.

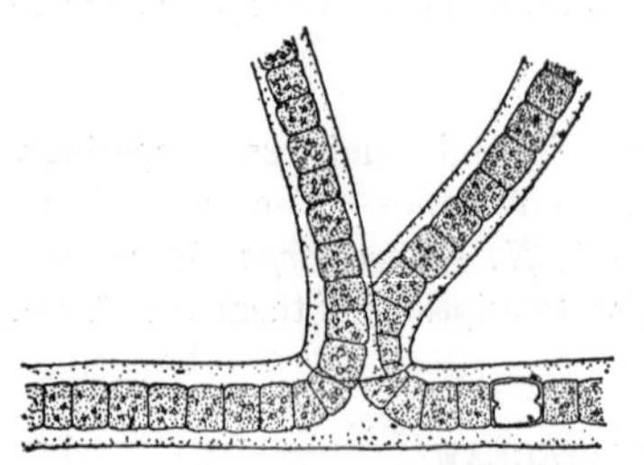

FIG. 506. *Scytonema arcangelii* B. and F. (× 375.)

Scytonema (Fig. 506) is usually found in subaerial habitats and with the filaments interwoven into a felty mass of considerable extent. Some species seem to grow best on damp soil, others on the dripping faces of rocky cliffs. Approximately 20 species occur in this country. For descriptions of most of the species of the genus, see Geitler (1932).

2. **Tolypothrix** Kützing, 1843. False branches of this alga arise singly and immediately adjacent to heterocysts. Pairs of false branches are occasionally present in a filament, but these *Scytonema*-like branches are of much rarer occurrence in *Tolypothrix* than are *Tolypothrix*-like branches in *Scytonema*. The general appearance of the filament of *Tolypothrix* is quite like that of *Scytonema*, but the sheaths are generally narrower and never have the oblique lamellation found in that genus. The heterocysts are always intercalary and may be solitary or in short series of two to six.

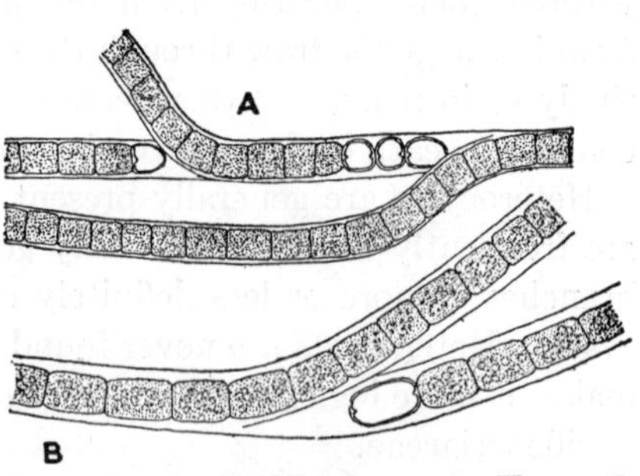

FIG. 507. *A*, *Tolypothrix tenuis* Kütz. *B*, *T. lanata* (Desv.) Wartm. (× 375.)

Tolypothrix is more frequently aquatic than subaerial. When aquatic, it usually grows in small clumps or tufts, and either intermingled with other free-floating algae or attached to submerged stones or wood. The species found in this country are *T. byssoidea* (Berk.) Kirchn., *T. distorta* (Fl. Dan.) Wartm., *T. lanata* (Desv.) Wartm. (Fig. 507*B*), *T. limbata* Thur., *T. penicillata* (Ag.) Thur., *T. Setchellii* Collins, and *T. tenuis* Kütz. (Fig. 507*A*). For descriptions of them, see Geitler (1932).

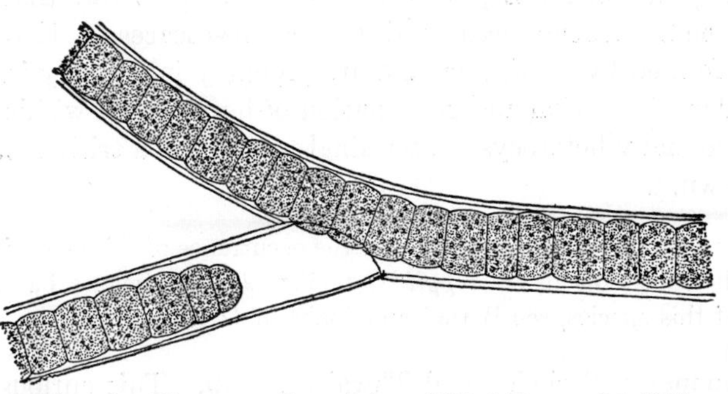

Fig. 508. *Plectonema Tomasiniana* (Kütz.) Born. (× 485.)

3. **Plectonema** Thuret, 1875. Filaments of this alga have the same firm sheath as *Tolypothrix* and the same false branching, but they never have heterocysts. The trichomes have discoidal or cylindrical cells, which are usually somewhat constricted at the cross walls. Sheaths surrounding tri-

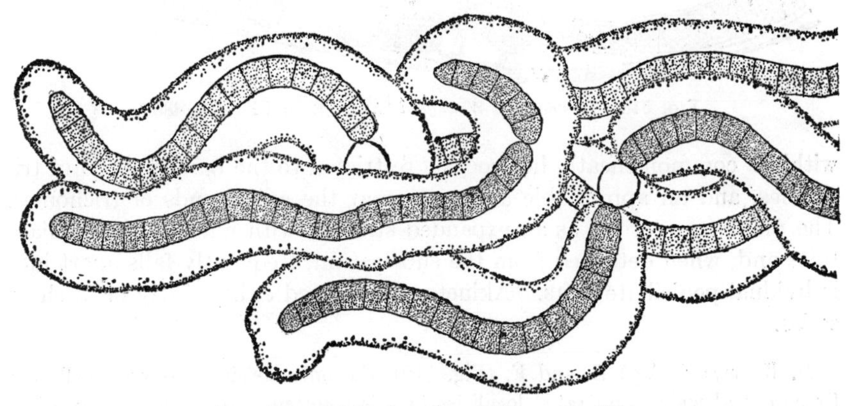

Fig. 509. *Diplocolon Heppii* Näg. (× 650.)

chomes are relatively narrow, firm, homogeneous or lamellated, and hyaline or brownish yellow. The sole method of reproduction is by hormogonia; akinetes and heterocysts are never formed.

Plectonema (Fig. 509) is found in aquatic and subaerial habitats, particularly among mosses and liverworts growing on damp rocks. Approximately a dozen species have been recorded for the United States. For descriptions of most of the species of the genus, see Geitler (1932).

4. **Diplocolon** Nägeli, 1857. This genus differs from others of the family in having several contorted trichomes within a common sheath. The trichomes may be somewhat attenuated at their apices, and each is surrounded by a sheath of its own. The general appearance of the filaments is like that of certain species of *Nostoc*, but the false branching shows that *Diplocolon* is one of the Scytonemataceae and not the Nostocaceae. Heterocysts are regularly formed by *Diplocolon* and are probably intercalary in origin, but the abundant formation and germination of hormogonia within a common sheath give many heterocysts a terminal position in a trichome. Akinetes are unknown.

The only well-authenticated record for the occurrence of this genus in the United States is the collection[1] of *D. Heppii* Näg. (Fig. 509) at Niagara Falls. For a description of this species, see Bornet and Flahault (1887).

5. **Desmonema** Berkeley and Thwaites, 1849. This curious genus differs from other genera of the family in having several parallel trichomes

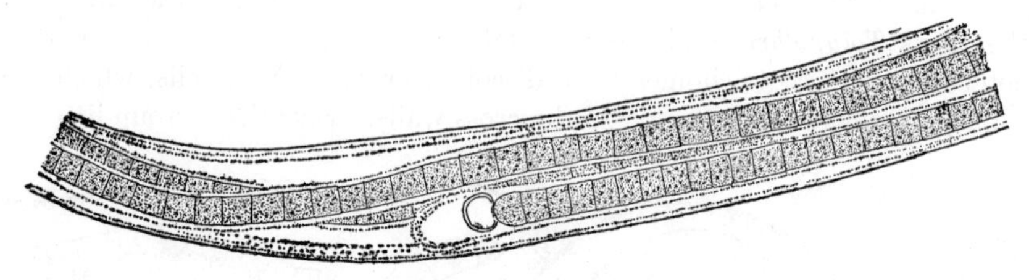

FIG. 510. *Desmonema Wrangelii* (Ag.) B. and F. (× 485.)

within a common sheath, heterocysts restricted to the basal portion of trichomes, and an appreciable attenuation at the apical ends of trichomes. The plant mass grows as an expanded stratum with numerous penicillate tufts and, when detached from the substratum, frequently falls apart into individual penicillate tufts. Akinetes are formed either singly or in short series.

D. Wrangelii (Ag.) B. and F. (Fig. 510), the only species, has been collected from several widely separated localities in this country. For a description of it, see Geitler (1932).

6. **Fremyella** J. De Toni, 1936 (*Microchaete* Thuret, 1875). The filaments of *Fremyella* always have only one trichome within a firm, narrow, homogeneous or lamellated sheath. The trichomes are sometimes somewhat attenuated at their free ends and always with heterocysts at their

[1] WOLLE, 1877.

basal ends. Sometimes there are also intercalary heterocysts. A formation of akinetes is of common occurrence and may take place adjacent to, or remote from, heterocysts, and singly or in short series. The filaments grow irregularly intertwined with one another in stellate or cushion-shaped tufts. False branching, which is the justification for including this genus in the Scytonemataceae, is found only here and there in filaments but seems to be present with sufficient regularity to warrant placing it in the Scytonemataceae instead of in a special family closely allied to the Nostocaceae.

Fremyella may grow in standing water and generally on water weeds, or it may grow on damp rocks. The species found in this country are *F. diplosiphon* (B. and F.) Drouet, *F. robusta* (Setchell), and *F. tenera* (Thur.) J. De Toni (Fig. 511). For descriptions of them as species of *Microchaete*, see Geitler (1932).

FAMILY 3. STIGONEMATACEAE

The character immediately separating the Stigonemataceae from other families of the suborder is the presence of true branches in the trichomes of the various genera. The main axis from which branches arise is usually conspicuous and may be uniseriate, biseriate, or multiseriate. Branches may have narrower and longer cells than the main axis. Both branches and axis may be uniseriate, both may be multiseriate, or the axis may be multiseriate and the branches uniseriate. Branches usually arise near the growing apex of a trichome and result from cell division at right angles to the previous plane of division. However, they may also be formed in older parts of the main axis. Branches may grow from all sides of a main axis, or they may be strictly unilateral; they may be without secondary branches or repeatedly branched. According to the genus, the sheath surrounding a trichome is narrow or copious, homogeneous or lamellated, and hyaline or variously colored. If colored, the color is generally yellowish, brownish, or blackish.

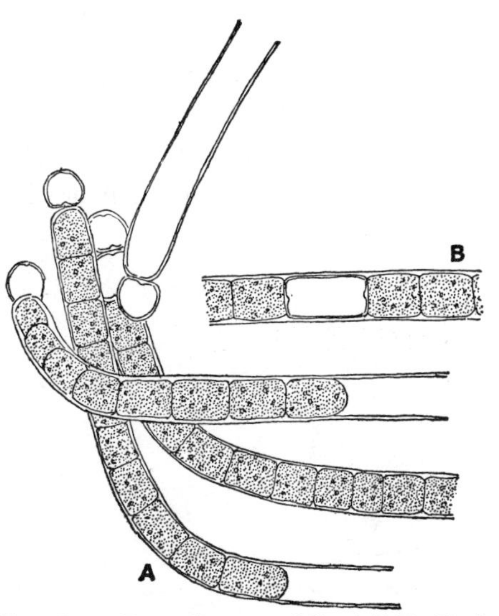

FIG. 511. *Fremyella tenera* (Thur.) J. De Toni. (× 800.)

Heterocysts are formed by most genera of the family and may be terminal or intercalary. Genera with intercalary heterocysts generally produce them abundantly in the main axis or the branches and, if the trichome is multiseriate, develop heterocysts from superficial instead of interior cells. In a few genera, heterocysts are formed terminally on short lateral branches. Akinetes have been recorded for most of the genera, but they are of rare occurrence.

The genera found in this country differ as follows:

1. Trichomes with heterocysts.. 2
1. Trichomes without heterocysts.. 6
   2. Sheaths distinct, free from one another or partly confluent................ 3
   2. Sheaths confluent with one another........................ 4. **Nostochopsis**
3. Filaments free from one another.. 4
3. Filaments united in a cushion-like mass........................... 3. **Capsosira**
   4. Trichomes wholly or partly multiseriate......................... 1. **Stigonema**
   4. Trichomes wholly uniseriate... 5
5. Branches mostly parallel to axis bearing them..................... 6. **Thalpophila**
5. Branches not parallel to axis bearing them....................... 2. **Hapalosiphon**
   6. Branching predominately dichotomous........................ 7. **Colteronema**
   6. Branching not predominately dichotomous.................... 5. **Albrightia**

1. **Stigonema** Agardh, 1824. The branching trichomes of *Stigonema* have main axes that are partly or wholly multiseriate. Branches may arise

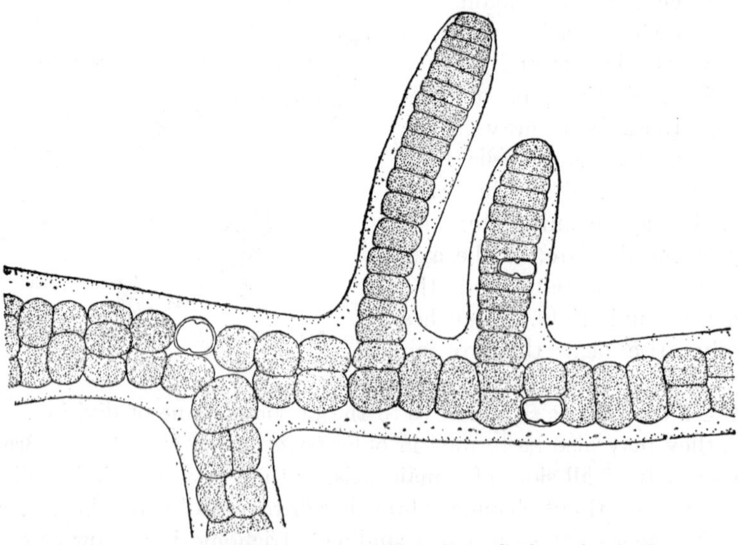

FIG. 512. *Stigonema turfaceum* (Engl. Bot.) Cooke. (× 500.)

from all sides of an axis and be repeatedly rebranched, or may be borne unilaterally and unbranched. The sheath surrounding a trichome is always of a firm texture, with a smooth or rough surface, homogeneous or lamellated, and hyaline or colored a yellowish brown, brown, or black.

Cells of *Stigonema* are usually spherical, but in young branches and axes they may be flattened by mutual pressure. In old filaments, the cells frequently lie some distance from one another, and each is surrounded by a sheath of its own. Such old filaments are not unlike elongate colonies of *Chroococcus* or *Aphanocapsa*. Heterocysts may be formed in the main axis or in the branches. If formed in a multiseriate portion of a trichome, they develop from superficial rather than deep-seated cells. A formation of hormogonia is confined largely to the ends of young branches, and they may develop with or without relationship to position of the heterocysts.

The specific limits between *Stigonema* and *Hapalosiphon* are not clear-cut. Certain *Stigonema* species show a close approach to *Hapalosiphon* in their practically uniseriate axes; and certain *Hapalosiphon* species resemble *Stigonema* in having occasional secondary branches and in having all cells in a trichome of the same shape. *Stigonema* grows on wet rocks, on moist earth, and in free-floating clumps intermingled with other algae. It is more frequently found on wet rocks than the two other types of habitat. The species found in this country are *S. hormoides* (Kütz.) B. and F., *S. informe* Kütz., *S. mamillosum* (Lyngb.) Ag., *S. minutum* (Ag.) Hass., *S. ocellatum* (Dillw.) Thur., *S. panniforme* (Ag.) B. and F., *S. thermale* (Schwabe) Borzi, and *S. turfaceum* (Engl. Bot.) Cooke (Fig. 512). For descriptions of them, see Geitler (1931).

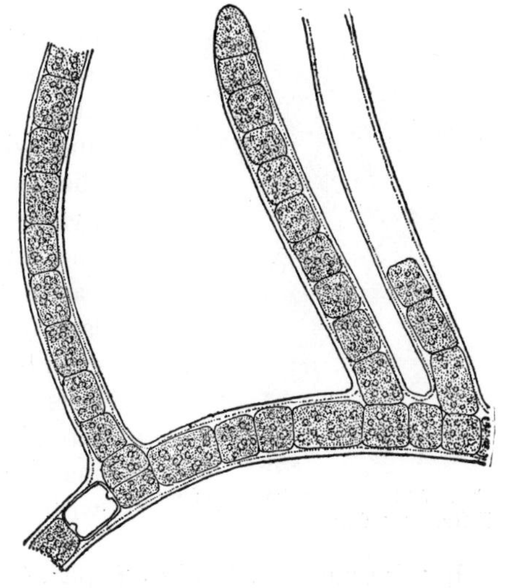

Fig. 513. *Hapalosiphon pumilus* (Kütz.) Hansg. (× 650.)

2. **Hapalosiphon** Nägeli, 1849. Filaments of *Hapalosiphon* are mostly uniseriate, though occasionally biseriate in part, and freely branched, but with the branching largely unilateral, and only here and there are branches secondarily branched. Sheaths surrounding a trichome are narrow but firm, usually colorless, and with a smooth surface. Vegetative cells are generally cylindrical, though in some cases they are subspherical. Cells of axes are usually but little longer than broad; those of branches, though of the same diameter, are frequently longer than they are broad. Heterocysts are intercalary and developed almost exclusively from cells of the main axis. Akinetes, when present, are formed in abundance and often with a development of all cells in the main axis into them. As

is the case with *Stigonema*, the formation of hormogonia is restricted to the branches. Hormogonia of *Hapalosiphon* are not strictly terminal and usually contain many cells.

Hapalosiphon is found more frequently in aquatic than in terrestrial habitats, and is more abundant in soft-water than in hard-water areas. The species found in the United States are *H. aureus* W. and G. S. West, *H. flexuosus* Borzi, *H. fontinalis* (Ag.) Born, *H. hibernicus* W. and G. S. West, *H. laminosus* (Kütz.) Hansg., *H. pumilus* (Kütz.) Hansg. (Fig. 513), and *H. Welwitschii* W., and G. S. West. For a description of *H. laminosus* as *Mastigocladus laminosus* (Kütz.) Cohn, see Geitler (1931); for descriptions of the others as species of *Hapalosiphon*, see the same.

3. **Capsosira** Kützing, 1849. The thalli of *Capsosira* are small hemispheres (1 to 2 mm. in diameter) and always with their lower surface affixed to some firm substratum. When viewed in vertical section, they often have concentric greenish and yellowish zones. The filaments are uniseriate, repeatedly branched, and without a definite main axis. The sheaths surrounding trichomes are thick, hyaline or yellow, and not confluent with one another. The cells are subspherical or barrel-shaped. Formation of hormogonia is largely restricted to the ends of branches, and each hormogonium contains about a dozen cells. Heterocysts may be intercalary or lateral and are formed in a *Stigonema*-like manner. The akinetes are subspherical and with a thick brownish wall.

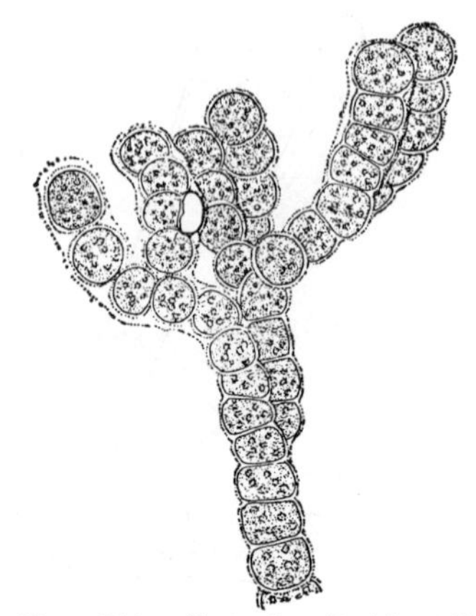

FIG. 514. *Capsosira Brebissonii* Kütz. (× 650.)

In this country *C. Brebissonii* Kütz. (Fig. 514), the only species, is known only from New England. For a description of it, see Geitler (1931).

4. **Nostochopsis** Wood, 1869. The thallus of *Nostochopsis* is of a firm gelatinous texture, of definite macroscopic shape, and either solid or hollow. When young, it is always sessile, and it may remain sessile throughout its entire development; or it may become detached and free-floating. The trichomes within a thallus are uniseriate, freely branched, and with the branches often elongate and torulose. Sheaths surrounding trichomes are wholly confluent with one another. The cells are always longer than broad and either cylindrical or barrel-shaped. Heterocysts may be intercalary, but more commonly they are formed from the terminal cells of short lateral branches.

Nostochopsis differs markedly from other Stigonemataceae found in this country in that most heterocysts are borne terminally on short lateral branchlets. *N. lobatus* Wood (Fig. 515), the only species in the local flora, has been found in several of the eastern states. For a description of it, see Geitler (1931).

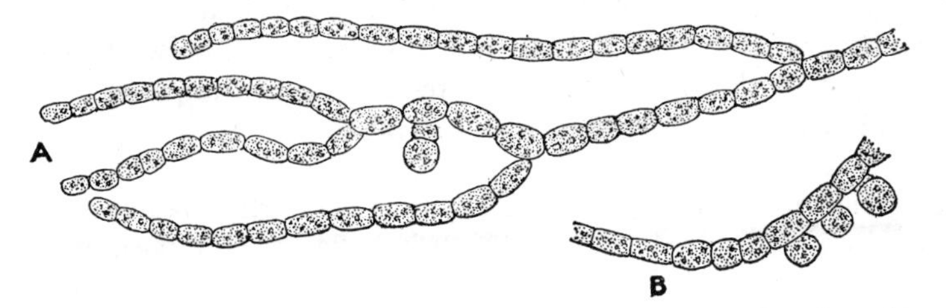

FIG. 515. *Nostochopsis lobatus* Wood, with immature heterocysts on short lateral branches (× 860.)

5. **Albrightia** J. J. Copeland, 1936. *Albrightia* has uniseriate, sparsely branched, relatively long trichomes with major axes and branches of the same size. A filament has a conspicuous sheath of firm texture that is usually homogeneous in structure. The cells are ellipsoidal, and cell division is restricted largely to the tips of branches. Reproduction is by means

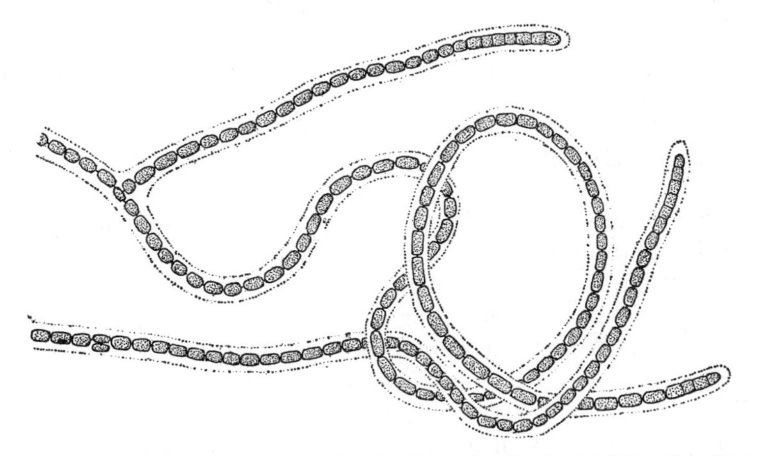

FIG. 516. *Albrightia tortuosa* J. J. Copeland. (*After Copeland* 1936.) (× 650.)

of hormogonia composed of a few cells. There is no formation of heterocysts or of akinetes.

The single species, *A. tortuosa* J. J. Copeland (Fig. 516) is known only from hot springs in the Lower Geyser Basin of Yellowstone National Park. For a description of it, see J. J. Copeland (1936).

6. **Thalpophila** Borzi, 1906. The trichomes of *Thalpophila* are uniseriate, branched, with the branching predominately unilateral, and with the branches lying parallel to the axis bearing them. There is but one trichome within a filament. Branching of a trichome may be so profuse that there may be hundreds of branches organized into a cord-like strand. Heterocysts are intercalary, but there is no correlation between their position and that of the branches. Akinetes are formed adjacent to heterocysts and in catenate series of as many as 50.

In this country *Thalpophila* is known only from the swiftly flowing portion of the overflow from a continuously erupting geyser in Yellowstone National Park.[1] For

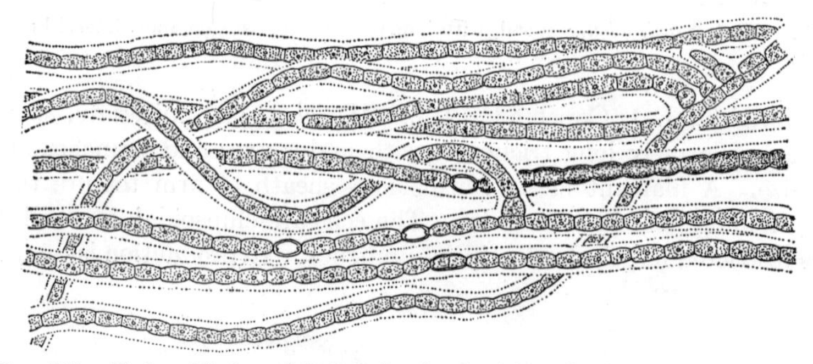

FIG. 517. *Thalpophila imperialis* J. J. Copeland. (*After Copeland*, 1636.) (× 700.)

a description of the species, *T. imperialis* J. J. Copeland (Fig. 517), growing in this rill, see J. J. Copeland (1936).

7. **Colteronema** J. J. Copeland, 1936. The filaments of this genus are branched and with the branching predominantly dichotomous through the terminal cell of a branch functioning as an apical cell that divided vertically. A filament is differentiated into a prostrate branched portion and a series of erect branches. A trichome is uniseriate and composed of ellipsoidal cells joined pole to pole. The sheath surrounding a trichome is divergently lamellated and also transversely lamellated near the ends of branches. There is no formation of heterocysts or of akinetes. Reproduction is by means of hormogonia which are formed near the ends of branches.

The single species, *C. funebre* J. J. Copeland (Fig. 518), is known[2] only from a single hot spring in Yellowstone National Park. For a description of it, see J. J. Copeland (1936).

[1] COPELAND, J. J., 1936. [2] COPELAND, J. J., 1936.

Family 4. Rivulariaceae

Genera belonging to this family have uniseriate trichomes that are conspicuously attenuated from base to apex, or from the middle toward both extremities. There may be a single trichome within an unbranched sheath, or the sheath may be falsely branched and contain several trichomes. Sheaths surrounding trichomes are of a firm texture, homogeneous or lamellated, and hyaline or colored. Frequently they are more gelatinized at their distal ends and broader, or the gelatinization may be so extensive that they are wholly confluent with one another to form a homogeneous colonial matrix.

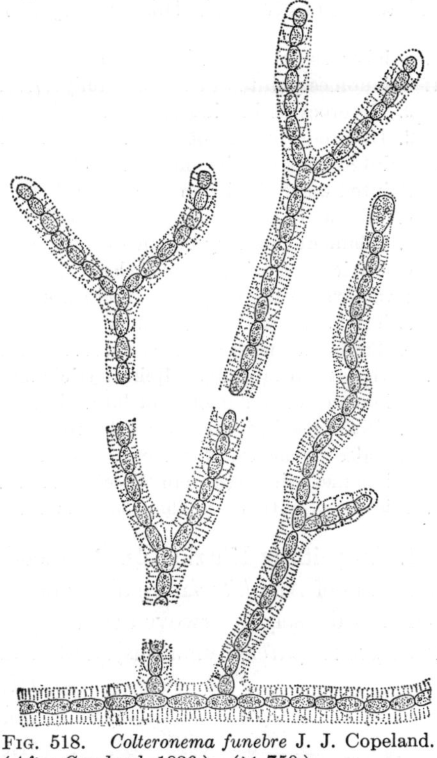

Fig. 518. *Colteronema funebre* J. J. Copeland. (*After Copeland,* 1936.) (× 750.)

Heterocysts are regularly formed by most genera of the family, but some genera never form them. If the genus is one with heterocysts, certain of them are always basal and borne singly or in short series of two, three, or more. There may also be intercalary heterocysts. The false branching so characteristic of the family may result from breaking of a trichome just below an intercalary heterocyst, the upper portion of the basal half then growing through the original sheath and secreting a sheath of its own. Indefinite repetition of this results in repeatedly and falsely branched filaments which are united with one another into spherical, hemispherical, penicillate, or caespitose colonies. False branching may also result from germination of hormogonia within the sheath of the parent filament. Hormogonia are usually formed toward the attenuated end of a trichome and, as they germinate, one end becomes attenuated to a fine hair-like point and the other develops a heterocyst. After differentiation of the two extremities, further cell divisions are restricted to the lower portion of a trichome and are most numerous in the portion adjoining the heterocyst. Sometimes both ends of a hormogonium become attenuated, and the young trichome breaks transversely into two parts in the plane where two adjoining heterocysts have been formed in its median portion.

Some of the genera which regularly form heterocysts also form akinetes; others lack akinetes. Akinetes are generally formed singly and next to basal heterocysts. They are much longer and somewhat broader than vegetative cells.

The genera found in this country differ as follows:

1. Trichomes pointed at both ends 9
1. Trichomes pointed at one end only 2
 2. Heterocysts lacking 3
 2. Heterocysts present 4
3. Pointed ends of trichomes parallel .... 1. **Amphithrix**
3. Pointed ends of trichomes not parallel .... 2. **Calothrix**
 4. Filaments united in spherical or hemispherical colonies 5
 4. Filaments solitary or not in rounded colonies 7
5. One trichome within a sheath 6
5. Two to several trichomes within a sheath .... 6. **Sacconema**
 6. Trichomes without akinetes .... 4. **Rivularia**
 6. Trichomes with akinetes .... 5. **Gloeotrichia**
7. With several laterally adjoined trichomes in a sheath .... 3. **Dichothrix**
7. Trichomes of a filament not laterally adjoined 8
 8. With false branches at regular intervals .... 7. **Scytonemopsis**
 8. False branches, if present, not at regular intervals .... 2. **Calothrix**
9. Trichomes with less than 20 cells .... 9. **Rhaphidiopsis**
9. Trichomes with more than 50 cells .... 8. **Hammatoidea**

1. **Amphithrix** Kützing, 1843; emend., Bornet and Flahault, 1886. The filaments of *Amphithrix* are distromatic and consist of a lower portion composed of densely interwoven trichomes (so closely packed that they appear to be pseudoparenchymatous), and of an upper portion with numerous erect trichomes attenuated to hair-like points at their distal ends. The erect trichomes are parallel to one another. Heterocysts and akinetes are never formed. Reproduction is by means of hormogonia, which may be formed singly or in series.

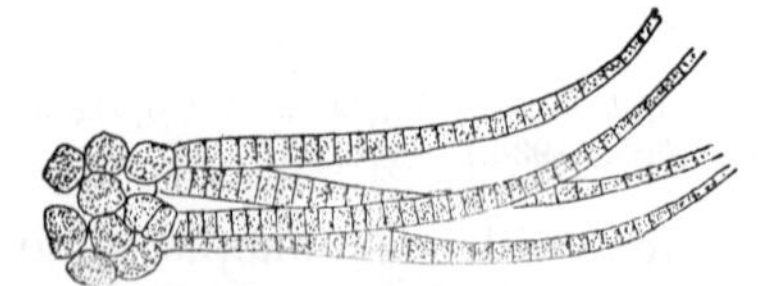

Fig. 519. *Amphithrix janthina* (Mont.) B. and F. (× 1300.)

A. janthina (Mont.) B. and F. (Fig. 519) is known from states bordering both the eastern and western seaboards of this country. For a description of it, see Geitler (1931).

2. **Calothrix** Agardh, 1824. Trichomes of *Calothrix* (Fig. 520) may taper from base to apex and terminate in a fine hair-like point, or the basal portion may be cylindrical and the attenuation restricted to the upper portion

of a trichome. In a few species, attenuation at the distal end is quite abrupt. Sheaths surrounding trichomes are of the same thickness throughout, homogeneous or distinctly stratified, and hyaline or colored. There is but a single trichome within a sheath, but the filaments may be simple or with false branches here and there. Vegetative cells in the lower portion of a trichome are discoid and with or without constrictions at the transverse walls; cells toward the apex are often cylindrical. Heterocysts may be intercalary, but a typical *Calothrix* trichome always has a basal heterocyst, which, not infrequently, lies external to the sheath of the filament. A

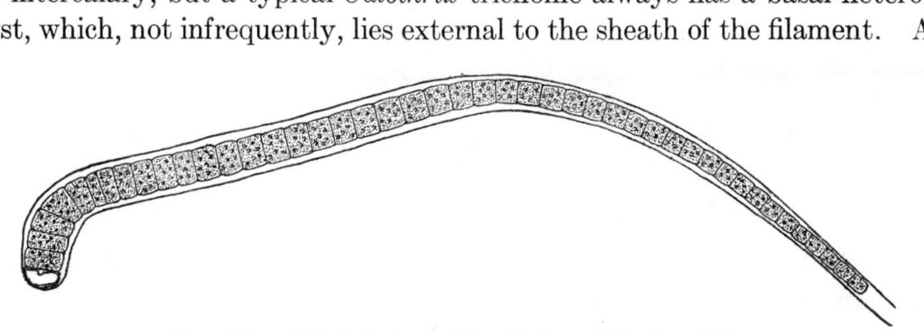

FIG. 520. *Calothrix fusca* (Kütz.) B. and F. (× 975).

few species never form heterocysts. Akinetes are known for a few species only.

The filaments may occur singly or united with one another to form strata of macroscopic or microscopic size. Sometimes the stratum is penicilliform, pulvinate, or stellate.

Calothrix (Fig. 520) grows attached to submerged rocks or to woodwork, and in flowing or standing water. The plant mass may be incrusted with, or free from, lime. Certain species grow epiphytic upon other algae. More than a dozen species have been found in this country. For descriptions of most species of the genus, see Geitler (1931).

3. **Dichothrix** Zanardini, 1858. *Dichothrix* is closely related to *Calothrix* but differs from it in having several trichomes, each enclosed by its own sheath, that lie more or less parallel to one another within a common sheath. The filaments of *Dichothrix* are freely and falsely branched, but the ultimate branchlets usually contain one trichome only. Trichomes of *Dichothrix* may have the same attenuation from base to apex as is found in *Calothrix*, or they may be attenuated in the distal portion only. Sheaths surrounding the trichomes may be hyaline, yellowish, or deep orange-brown; homogeneous or stratified. If stratified, the lamellae may be parallel or divergent. Heterocysts are usually solitary and basal, but there may be additional intercalary ones.

Species of *Dichothrix* are not uncommon upon submerged rocks in streams and ponds. and on moist cliffs. Submerged plant masses may be smooth and plushlike, or distinctly tufted. The species found in this country are *D. Baueriana* (Grun.) B. and F., *D. calcarea* Tilden, *D. compacta* (Ag.) B. and F., *D. gypsophila* (Kütz) B. and F., *D. Hosfordii* (Wolle) Born., *D. inyoensis* Drouet, *D. Nordstedtii* B. and F., and *D. Orsiniana* (Kütz.) B. and F. (Fig. 521). For a description of *D. inyoensis*, see Drouet (1943); for the others, see Geitler (1931).

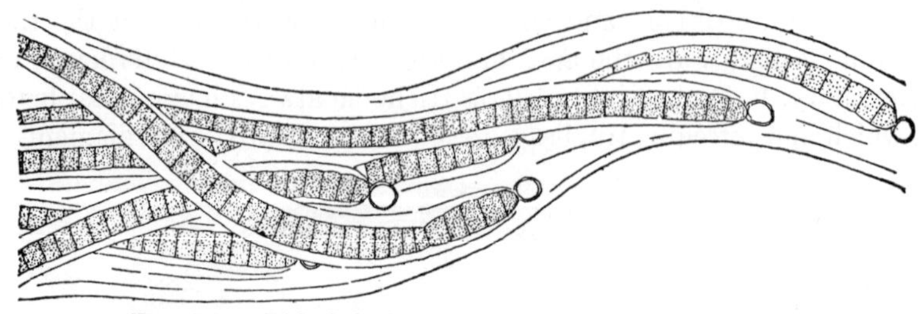

FIG. 521. *Dichothrix Orsiniana* (Kütz.) B. and F. (× 400.)

4. **Rivularia** Roth, 1797; emend., Agardh, 1812. *Rivularia* differs from preceding members of the family in having the sheaths surrounding the individual trichomes partially or wholly confluent with one another, and in having the trichomes radiately arranged within a hemispherical, globose, or irregularly expanded plant mass of macroscopic size. The trichomes are usually attenuated from base to apex and have basal heterocysts. The sheaths surrounding them may be distinct toward the lower portion and either homogeneous or lamellated, but they are always more or less con-

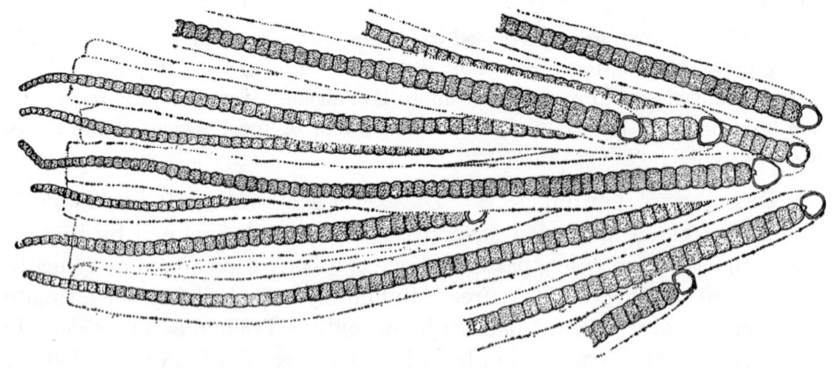

FIG. 522. *Rivularia dura* Roth. (× 485.)

fluent with one another at their distal ends. The radiate arrangement of trichomes within a thallus is the result of repeated false branching in the basal portion of trichomes, but there is usually so much displacement of the branches that the false branching can be demonstrated only in juvenile colonies. Akinetes are not formed by species of *Rivularia*.

Species of *Rivularia* grow upon submerged stones, woodwork, and water plants. They are also of frequent occurrence on wet rocks of cliffs. The thalli are of an exceedingly firm consistency and often so tough that they can be crushed only with difficulty. Sometimes thalli are heavily incrusted with lime. The species found in this country are *R. Biasolettiana* Menegh., *R. compacta* Collins, *R. dura* Roth (Fig. 522), *R. globiceps* G. S. West, *R. haematites* (DC) Ag., *R. minutula* (Kütz.) F., and *R. planktonica* Elenkin. For a description of *R. compacta*, see Collins (1901); for the others, see Geitler (1931).

5. **Gloeotrichia** J. G. Agardh, 1842. *Gloeotrichia* differs from *Rivularia* only in its regular formation of akinetes and in the gelatinous texture of its thalli. Trichomes of *Gloeotrichia* have the same regular attenuation from base to apex, but they are enclosed by more gelatinous sheaths which are

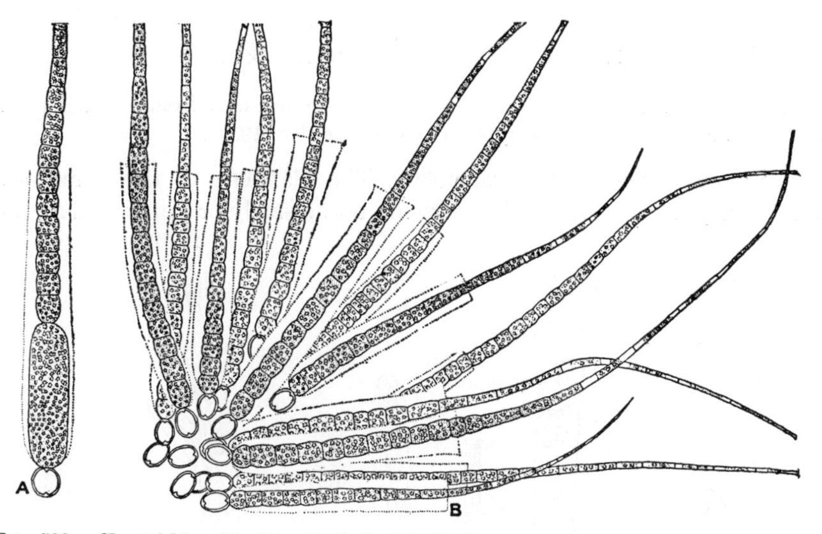

Fig. 523. *Gloeotrichia echinulata* (J. E. Smith) Richter. *A*, filament with an akinete. *B*, portion of a sterile colony. (× 400.)

often wholly confluent with one another. The genus always has basal heterocysts and sometimes intercalary ones in addition. The akinetes are always elongate and at the base of trichomes. There may be but a single akinete, in which case it lies next to the heterocyst. If more than one are present, they may be formed in short catenate series or separated from one another by two or three intervening vegetative cells.

Gloeotrichia is always aquatic. It may be either free-floating or sessile at all stages of its development; or it may be sessile at first and free-floating later on. The species found in this country are *G. echinulata* (J. E. Smith) Richt. (Fig. 523), *G. natans* (Hedw.) Rab., *G. Pilgeri* Schmidle, and *G. Pisum* (Ag.) Thur. For descriptions of them, see Geitler (1931).

6. **Sacconema** Borzi, 1882. This genus has a gelatinous thallus much like that of *Rivularia* and *Gloeotrichia*, but there are usually two or more

trichomes within a common sheath. The individual trichomes are attenuated from base to apex, and the sheaths surrounding them are lamellated and with expanded funnel-like apices. The heterocysts are basal and solitary. Akinetes are formed at the base of the trichomes.

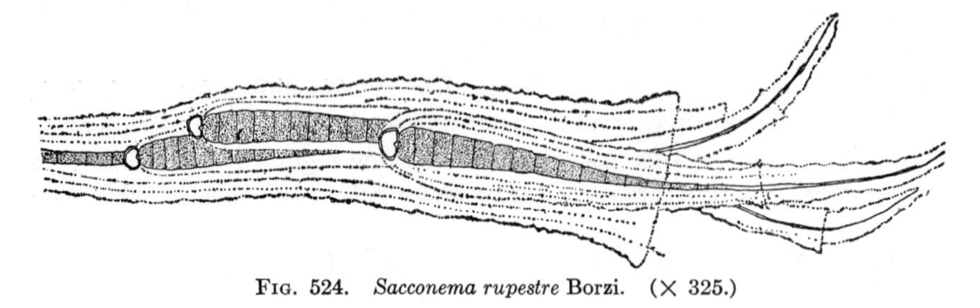

FIG. 524. *Sacconema rupestre* Borzi. (× 325.)

In this country, *S. rupestre* Borzi (Fig. 524) is known only from Massachusetts and Connecticut. For a description of it, see Geitler (1931).

7. **Scytonemopsis** Kisselawa, 1930. The filaments of this genus are falsely branched and with false branches borne singly or in pairs along the

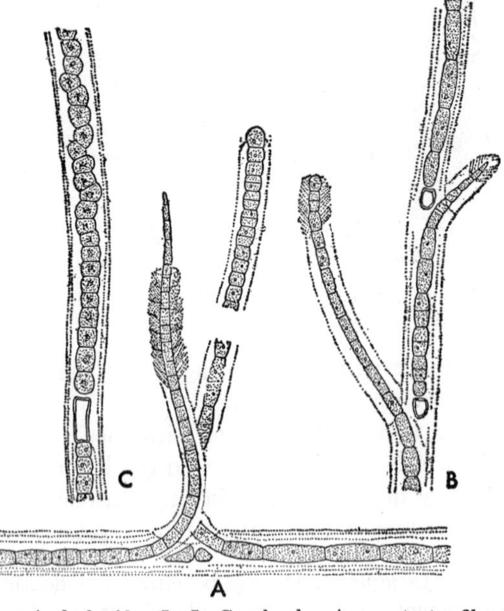

FIG. 525. *Scytonemopsis hydnoides* J. J. Copeland. *A*, prostrate filament with paired branches. *B*, portion of an erect filament. *C*, old filament with akinetes. (*After Copeland*, 1936.) (× 350.)

length of a filament. The trichomes are very gradually attenuated at one end. Sheaths surrounding trichomes may be homogeneous or lamellated, and with the layers parallel or divergent. Heterocysts are usually borne terminally at the base of a trichome, but there are frequently also intercalary ones. Akinetes are formed in catenate series remote from heterocysts.

The only record for the occurrence of the genus in this country is in hot springs of the Upper Geyser Basin of Yellowstone National Park. For a description of the species, *S. hydnoides* J. J. Copeland (Fig. 525), found in these hot springs, see J. J. Copeland (1936).

8. **Hammatoidea** W. and G. S. West, 1897. *Hammatoidea* has unbranched uniseriate trichomes that are conspicuously attenuated at both ends. The sheaths are firm, stratified, and may be yellowish brown in old filaments. The filaments are falsely branched, and the false branches are frequently U-shaped. There is no formation of heterocysts or of akinetes. Reproduction is by means of hormogonia.

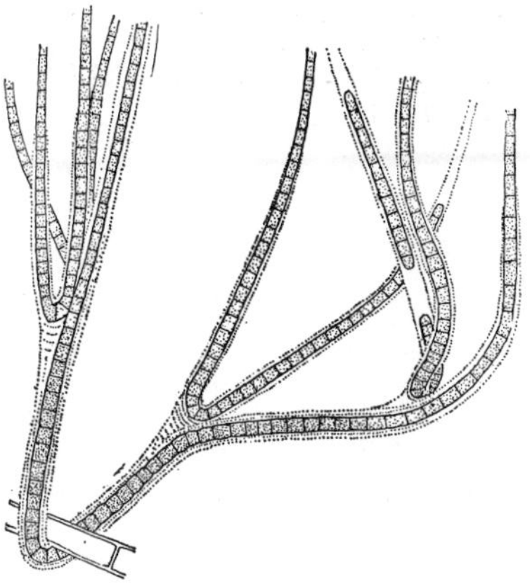

FIG. 526. *Hammatoidea Normanii* W. and G. S. West. (*After W. and G. S. West*, 1897.) (× 260.)

This very rare genus is usually found epiphytic upon gelatinous envelopes of other algae. Both *H. Normanii* W. and G. S. West (Fig. 526) and *H. yellowstonensis* J. J. Copeland have been recorded from but one locality in this country. For a description of *H. Normanii*, see Geitler (1931); for *H. yellowstonensis*, see J. J. Copeland (1936).

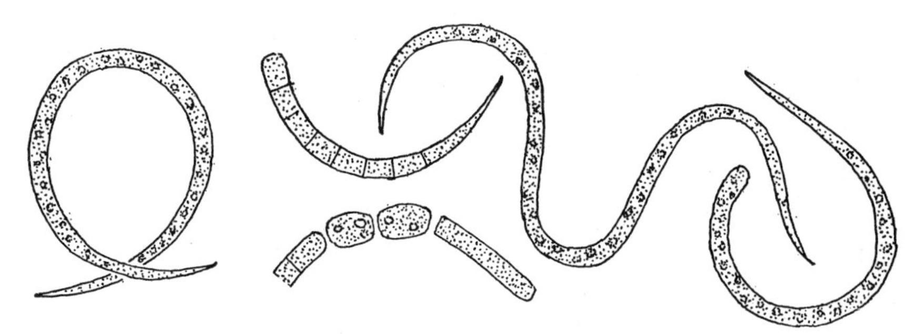

FIG. 527. *Raphidiopsis curvata* Fritsch and Rich. (*After Fritsch and Rich*, 1929.) (× 600.)

9. **Raphidiopsis** Fritsch and Rich, 1929. The solitary, more or less curved trichomes of this alga are short, usually with less than 20 cells, and without a sheath. Both poles of a trichome may taper to a fine point, or one pole may be tapered and the other be broadly rounded. The cells are cylindrical and may or may not contain numerous pseudovacuoles. There

is no formation of heterocysts, but akinetes may be produced singly or in pairs midway between the ends of a trichome. Reproduction is usually by a transverse breaking of a trichome into two equal halves.

There is but one species, *R. curvata* Fritsch and Rich (Fig. 527), and in this country it is known only from Ohio.[1] For a description of it, see Fritsch and Rich (1929).

[1] Daily, 1945.

CHAPTER 10

DIVISION RHODOPHYTA

The Rhodophyta, or red algae, have multicellular thalli of microscopic or macroscopic size and often of distinctive shape. Red algae differ from all other algae in structure of their sexual organs, in mode of fertilization, and in having fertilization followed by formation of a spore-producing structure, the so-called *cystocarp*. Pigments of Rhodophyta are localized in chromatophores. In addition to chlorophylls, a carotene, and a xanthophyll, the chromatophores contain r-phycoerythrin and r-phycocyanin (see Table I, page 3). Many Rhodophyta have the r-phycoerythrin present in such abundance as to mask the other pigments and thus give the thalli a distinctive red color. However, color is not a certain criterion for recognizing "red" algae because in many marine species growing high in the intertidal zone, and in a majority of fresh-water species, the color may be greenish, olive-green, or brownish.

The division contains but one class, the Rhodophyceae.

Occurrence. Fresh-water Rhodophyceae constitute but an insignificant portion of a class that has many representatives in the ocean. Such genera as do occur in fresh waters are usually without representatives in the ocean. With the exception of *Porphyridium*, a genus of somewhat doubtful affinities, all fresh-water Rhodophyceae are aquatic in habit. Furthermore, the great majority of these are rather closely restricted to the well-aerated waters of rapids, falls, and dams in cold rapidly flowing streams. Well-aerated and cold waters are not essential for growth of all fresh-water Rhodophyta, and *Asterocytis* and *Compsopogon* may be cited as genera which are found in quiet and relatively warm waters.

The Cell Wall. The wall surrounding a protoplast is differentiated into a relatively firm, thin, inner portion and an outer gelatinous portion of variable breadth. The inner portion has been shown[1] to contain considerable amounts of cellulose. The nature of the gelatinous materials has been studied most extensively in marine algae yielding gel-forming substances including agar and carrageenin. The gel substances are not identical in all cases, but all appear to be galactan etheral sulphates.[2]

Members of the Bangioideae have cell walls without evident perforations. Cells of Florideae, on the other hand, regularly have a pore-like opening of

[1] Kylin, 1942. [2] Tseng, 1945.

variable size in the wall between two sister cells, and an evident cytoplasmic connection between protoplasts of sister cells.

Structure of the Protoplast. Protoplasts of Bangioideae are usually without a vacuole; those of Florideae usually have a large central vacuole and the protoplasm restricted to a peripheral layer next to the cell wall.

The protoplasts have their pigments localized in definite chromatophores. Many Bangioideae and certain primitive Florideae have but a single axial stellate chromatophore within a cell. At the center of these chromatophores is a dense colorless proteinaceous body, the pyrenoid. These pyrenoids are of the "naked" type and lack the encircling sheath of starch granules found around pyrenoids of Chlorophyceae. Cells of a few Bangiales and of many Florideae have more than one chromatophore, and in some cases the number runs into the thousands. When there is more than one chromatophore, these are usually disk-shaped and parietal.

Chromatophores of Rhodophyceae contain green (chlorophylls), yellow-orange (carotene), yellow-brown (xanthophyll), red (r-phycoerythrin), and blue (r-phycocyanin) pigments. Variations in proportions of these pigments account for the diversity of shades and colors of the chromatophores and consequently of the thallus. Intense illumination seems to favor the formation of r-phycocyanin and retard the formation of r-phycoerythrin in both fresh-water and marine Rhodophyceae. Since most fresh-water members of the class grow but a few centimeters below the surface of the water and where there is little screening out of sunlight, they rarely have the red color typical of a majority of marine species. On the other hand, when fresh-water species are found growing 15 to 30 m. below the surface of a lake, they are usually of a deep red color.[1] These deep-water species are growing under conditions of greatly diminished illumination, and where the intensity of violet rays is more than 100 times that of the red rays (see Table III, page 15).

Food reserves of Rhodophyceae are generally stored in the cytoplasm, outside the chromatophores, and in the form of small granules. The granules consist of floridean starch, a carbohydrate allied to the starch of green plants and to glycogen but not identical with either.[2] Rhodophyceae may also form the soluble carbohydrate floridoside, a compound composed of one molecule of galactose and one of glycerin.[3]

The great majority of red algae have cells that are uninucleate at all times. A nucleus has a distinct nuclear membrane, a nucleolus, and a certain amount of chromatic material. Nuclear division is similar to that of higher plants and may be mitotic or meiotic according to the place where it occurs.

[1] OBERDORFER, 1927; ZIMMERMANN, 1927. [2] KYLIN, 1943.
[3] COLIN and AUGIER, 1933.

Asexual Reproduction. Vegetative multiplication by fragmentation of a thallus is of much less frequent occurrence among Rhodophyceae than among other multicellular algae.

Many Rhodophyceae reproduce asexually by means of spores. Perhaps the simplest instance of spore formation is to be found in *Asterocytis* where the protoplasts of certain cells in a filament are ejected from the enveloping gelatinous sheath (Fig. 528*A*) and drift away and develop into new filaments if they lodge on a suitable substratum.[1] Certain other Rhodophyceae have a division and redivision of vegetative cells into spores (Fig. 528*B*). These *neutral spores* are naked and usually amoeboid when first liberated, but within a day or two they round up, secrete a wall, and by cell division develop into a new thallus (Fig. 528*C–G*).

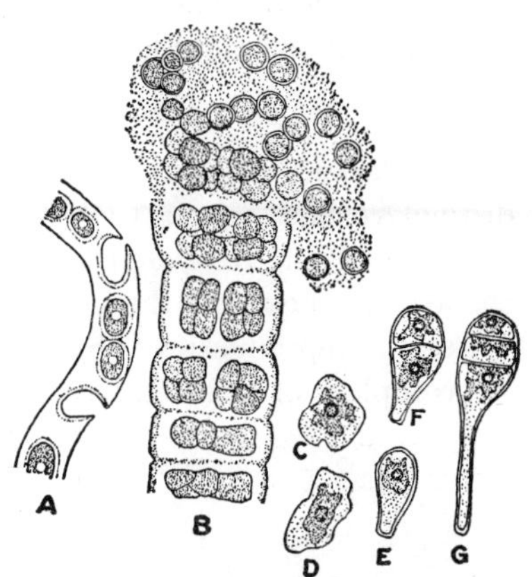

FIG. 528. Liberation and germination of neutral spores of Bangiales. *A*, *Asterocytis ramosa* (Thw.) Gobi. *B-G*, *Bangia fuscopurpurea* (Dillw.) Lyngb. *B*, liberation of spores. *C-G*, germination of spores. (*A*, *after Wille*, 1906; *B*, *after Darbishire*, 1898; *C-G*, *after Kylin*, 1922.) (*A*, × 240; *B*, × 210; *C-G*, × 260.)

Still other Rhodophyceae form definite sporangia containing either one, four, or more spores. If a sporangium contains one spore, it is a *monosporangium* and the spore a *monospore*; if there are four spores, the sporangium is a *tetrasporangium* and the spores are *tetraspores*; if there are more than four spores, the sporangium is a *polysporangium* and the spores are *polyspores*. None of these spores develop flagella either before or after liberation from a sporangium, and all dispersal of spores is effected by water currents or wave action. Liberation of spores is by a rupture of the sporangial wall and, when first liberated, the spores are naked. Liberated spores may form a wall before they become lodged on some firm substratum, or walls may not be formed until after lodgment of spores.

Sexual Reproduction. Sexual reproduction is of widespread occurrence among Rhodophyceae. Gametic union is effected by water currents carrying a nonflagellated male gamete (*spermatium*) to the prolonged distal end (*trichogyne*) of the one-celled female sex organ (*carpogonium*). After a spermatium becomes lodged against a trichogyne, the spermatial nucleus migrates into the trichogyne, down it, and fuses with the single nucleus in the protoplast of the carpogonium. In some Rhodophyceae, the resultant zygote divides to form a number of *carpospores*. In other Rhodophyceae,

[1] WILLE, 1900.

carpospores are formed on a filamentous structure. This filamentous structure may grow directly from the carpogonium, or from a thallus cell into which there has been a migration of the zygote nucleus or one of its daughter nuclei. The mass of spores and structures associated with them constitutes the *cystocarp*.

Classification. The Rhodophyceae are divided into two subclasses: the Bangioideae and the Florideae.

SUBCLASS 1. BANGIOIDEAE

The Bangioideae, sometimes called the Protoflorideae, have a number of distinctive characters. Cell division may take place anywhere in a thallus, instead of being restricted to apical cells, and there are never cytoplasmic connections between the cells.

Asexual reproduction is by means of naked nonflagellated neutral spores. A protoplast of a vegetative cell may be liberated and become a neutral spore (Fig. 528*A*); or a vegetative cell may divide to form a number of neutral spores (Fig. 528*B–G*); or a vegetative cell may divide into two daughter cells, one remaining vegetative, the other producing one or more neutral spores.

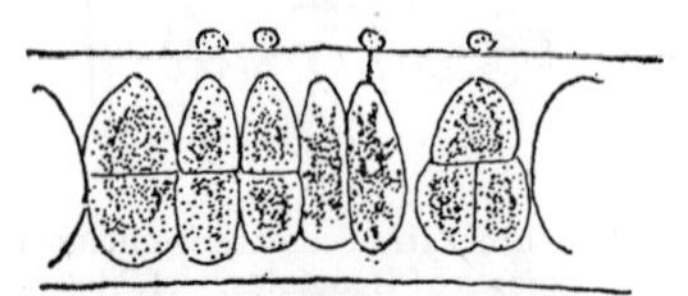

Fig. 529. Fertilization and development of carpospores of *Porphyra leucosticta* Thur. (*After Berthold*, 1882.) (× 450.)

Sexual reproduction, so far as it has been recorded in the subclass, is by division of vegetative cells to form a number of spermatia which are carried to, and fuse with, vegetative cells that have become metamorphosed into carpogonia. Gametic union is followed by a repeated bipartition of the zygote into a number of carpospores (Fig. 529).

There is but one order, the Bangiales.

ORDER 1. BANGIALES

Thalli of Bangiales may be unicellular; with simple or branched filaments that are uniseriate or multiseriate; or with expanded blades either one or two cells in thickness.

Since sexual reproduction is known for but few members of the order, the segregation into families is based chiefly upon the manner in which neutral spores are formed.

Family 1. Goniotrichaceae

Members of this family have the cells united in branched uniseriate filaments. Asexual reproduction is by escape of protoplasts from vegetative cells and a functioning of the liberated protoplasts as neutral spores. Sexual reproduction has not been found in any member of the family.

Asterocystis is the only genus found in fresh waters of this country.

1. **Asterocytis** Gobi, 1878. The cells of *Asterocytis* are spherical, or cylindrical and with broadly rounded poles. Within them is a single bright blue-green stellate chromatophore with a single large pyrenoid at its center. Each cell is surrounded by a broad gelatinous sheath, that is quite distinct from the colonial sheath. The plane of division is always at right angles to the long axis of a cell, and repeated cell division results in a uniseriate filament in which the cells are held together by a common gelatinous envelope. In rapidly growing filaments, the cells lie close together; in slowly growing ones, they lie some distance from one another. Branching of filaments is "false," and the false branches arise by a cell in a filament changing its orientation and then dividing repeatedly.[1] *Chroococcus*-like or *Stigonema*-like (Fig. 530*B*) stages may result from a division of cells in all planes.

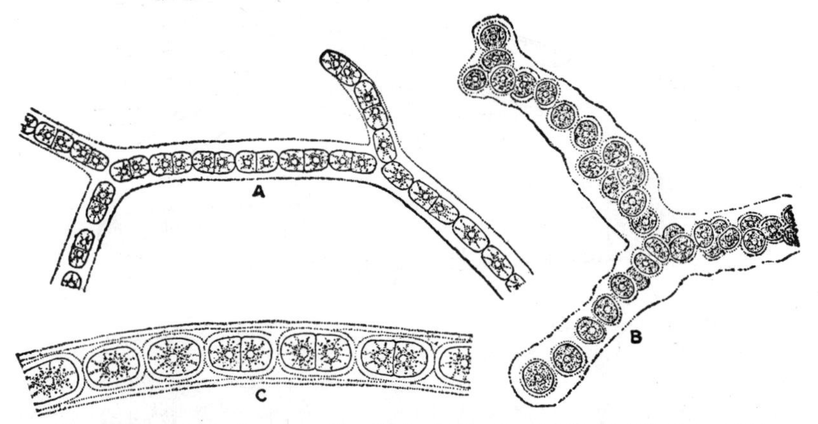

FIG. 530. *Asterocytis smaragdina* (Reinsch) Forti. (*A-B*, × 325; *C*, × 650.)

Vegetative cells of a filament may function directly as neutral spores and be ejected from the colonial sheath as naked protoplasts. These spores secrete a wall and grow into new filaments when they lodge on a suitable substratum.[2]

A. smaragdina (Reinsch) Forti (Fig. 530) has been collected at several widely separated localities in this country and has usually been found epiphytic upon *Cladophora*. For a description of it, see Pascher and Schiller (1925).

FAMILY 2. ERYTHROTRICHIACEAE

Genera of this family may have the cells seriately united in branched or unbranched filaments, or organized into laminate or cylindrical thalli. Cells of most genera have a single axial stellate chloroplast, but those of certain genera have numerous discoid chromatophores.

[1] GEITLER, 1924A. [2] GEITLER, 1924*A*; WILLE, 1900.

Neutral spores are formed singly within the smaller of two daughter cells formed by diagonal division of a vegetative cell.

All the genera but *Compsopogon* are marine.

1. **Compsopogon** Montagne, 1850. The bluish to violet-green thalli of *Compsopogon* are filamentous, more than one cell broad in older portions, and freely branched. Near their extremities, the branches consist of a uniseriate row of discoid cells any one of which may divide transversely. A short distance back from the apex, the cells divide vertically, and repeated vertical divisions result in differentiation of a peripheral layer of cells around a central axial cell. The axial cell enlarges greatly and does not divide farther; the peripheral cells divide repeatedly and may produce a corticating

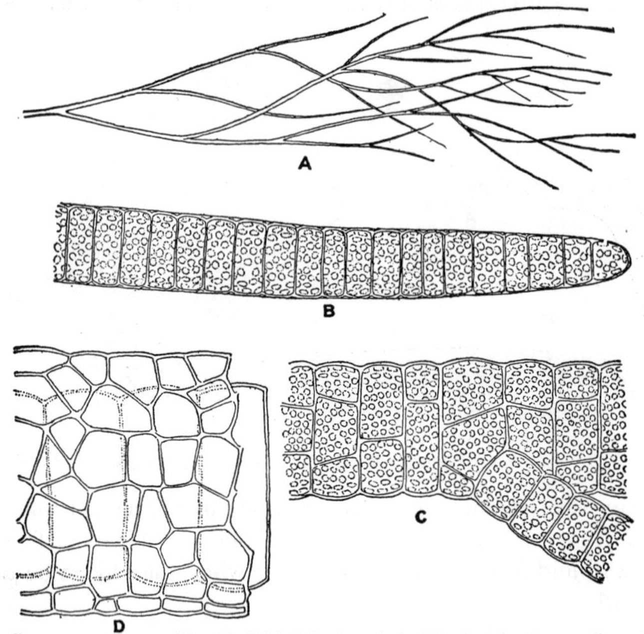

Fig. 531. *Compsopogon coeruleus* (Balbis) Mont. *A*, habit sketch of a portion of a thallus. *B*, apex of a branch. *C*, surface view of a young branch in which the central axis is corticated. *D*, older portion of the same branch. (*A*, natural size; *B-D* $\times$ 325.)

tissue that may be more than one cell in thickness. Each of the original cells in the distal portion of a filament thus becomes converted into a corticated segment, externally evident by slight constrictions in older parts of a filament. Filaments of *Compsopogon* may be free-floating or sessile. If sessile, they are attached by rhizoidal outgrowths, analagous to those of *Bangia*, from corticating cells in lower segments of a filament.

The cells are uninucleate and contain many spherical chromatophores. Some of the chromatophores lie immediately about the nucleus; but the majority are toward the periphery of the protoplast.

Any superficial corticating cell of a filament, as well as cells in the uniseriate portion, may divide vertically into two daughter cells, one of which develops a papillate outgrowth and becomes densely filled with chromatophores. The protoplast of this cell is then discharged as a naked neutral spore and the mechanism causing its discharge is the increased turgidity of the sterile sister cell. Instead of immediately becoming a neutral spore, the fertile cell may divide and redivide into a number of smaller cells each of which is discharged as a naked neutral spore.[1]

C. coeruleus (Balbis) Mont. (Fig. 531) is widely distributed in the southern part of this country, and in Arizona it may grow in such profusion in irrigation ditches as to be a nuisance. It has also been collected in Massachusetts[2] and Ohio[3] where it is thought to have been introduced with plants imported from warmer regions by nurserymen. When introduced into these northern areas, it has not become a permanent member of the flora because it is unable to survive the cold winter months. For a description of *C. coeruleus*, see Wolle (1887).

Family 3. Bangiaceae

The Bangiaceae differ from other Bangiales in that there is a repeated division of a vegetative cell to form many neutral spores. Members of this family are also known to reproduce sexually and to have a direct division of the zygote into carpospores.

Bangia is the only genus with species found in fresh water.

1. **Bangia** Lyngbye, 1819. The filaments of *Bangia* are unbranched and grow attached, generally to woodwork, by means of rhizoidal outgrowths from the lower cells. Young filaments are uniseriate; older ones are frequently more than one cell in diameter. Each cell in the filament contains a massive stellate chromatophore, with a single pyrenoid at its center. The color of the chromatophore is usually a purplish or a brownish red.

Asexual reproduction is by means of neutral spores, formed by the repeated division of vegetative cells. Sexual reproduction is by a union of spermatia, which arise by the repeated division of certain cells of male plants with metamorphosed vegetative cells (carpogonia) of female plants. After the union of the gametes the zygote divides to form four to eight carpospores.[4]

The length of the filaments, the number of cells in diameter, and the color are so dependent upon the age of the plant that all marine individuals are placed in a single species, *B. fuscopurpurea* (Dillw.) Lyngb. (Fig. 532*A*). This is considered distinct from the *Bangia* found in fresh waters[5] [*B. atropurpurea* (Roth) Ag. (Fig.

[1] Brühl and Biswas, 1923; Thaxter, 1900. [2] Collins, 1916.

[3] Masters, 1940.

[4] Berthold, 1882; Darbishire, 1898: Kylin, 1922; Rosenvinge. 1909.

[5] Hamel, 1924–1926.

532*B–D*)]. *B. atropurpurea* has been recorded[1] from the United States, but the statement that it was found "attached to wood and stones in streams more or less subject to tides" suggests that the alga found was in reality *B. fuscopurpurea*. *B. atropurpurea* is a widely distributed fresh-water alga in Europe, especially in hard waters. For a description of *B. atropurpurea*, see Pascher and Schiller (1925).

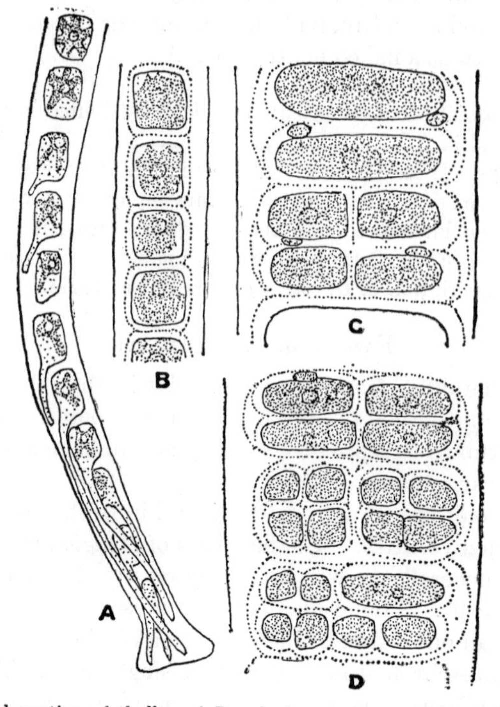

FIG. 532. *A*, basal portion of thallus of *Bangia fuscopurpurea* (Dillw.) Lyngb. *B-D*, portions of thallus of *B. atropurpurea* (Roth) Ag. (*A, after Kylin*, 1922; *B-D, after Pascher and Schiller*, 1925.) (*A*, ×380; *B-D*, × 1100.)

BANGIOIDEAE OF UNCERTAIN POSITION

1. **Porphyridium** Nägeli, 1849. This is the only one of the fresh-water Rhodophyceae that is strictly terrestrial. It grows on damp soil or on moist woodwork as a thin, gelatinous, blood-red layer of indefinite extent. The cells are approximately spherical, and each is surrounded by a gelatinous sheath.[2] In cell division there may be a bipartition of the cell contents, or the cell contents may be simultaneously divided into four or more parts.[3] After cell division, a portion of the sheath becomes drawn out into

[1] WOLLE, 1887.

[2] BRAND, 1908, 1908*A*, 1917; GEITLER, 1924*A*; LEWIS and ZIRKLE, 1920.

[3] GEITLER, 1944.

a strand or stalk, which sooner or later becomes confluent with the gelatinous matrix in which the cells are embedded (Fig. 533*C*). Each cell contains a massive, dark-red, stellate chromatophore with a central pyrenoid, but rapidly dividing cells may have chromatophores that are irregular in shape.

The only method of reproduction is cell division. Under unfavorable conditions, the cells become metamorphosed into aplanospore-like stages closely resembling vegetative cells.

P. cruentum (Smith and Sowerby) Näg. is widely distributed in this country. For a description of it, see Pascher and Schiller (1925).

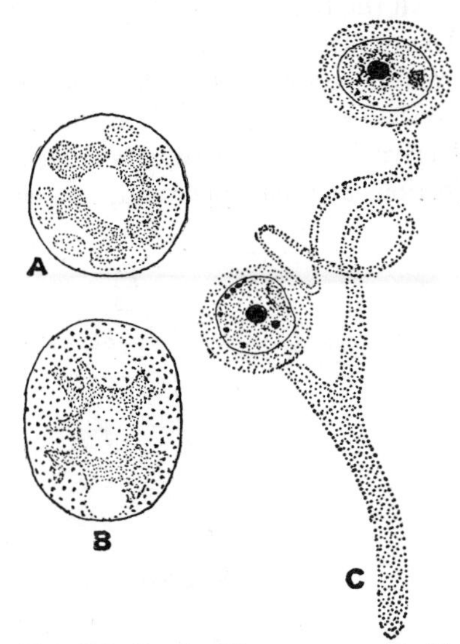

Fig. 533. *Porphyridium cruentum* (Smith and Soerly) Näg. *A*, surface view of a living cell. *B*, optical section of a living cell. *C*, from stained preparations showing cell structure and the gelatinous sheaths. (*A-B*, *after Geitler*, 1924; *C*, *after Lewis and Zirkle*, 1920.)

SUBCLASS 2. FLORIDEAE

Although thalli of Florideae exhibit the greatest diversity of form from genus to genus, they are all fundamentally alike in their restriction of cell division to apical cells, and in the cytoplasmic connection between sister cells. The male gametes (*spermatia*) are formed singly within a male sex organ (*spermatangium*). The female sex organ (*carpogonium*) is always developed from the terminal cell of a special filament, the carpogonial filament.

Gametic union is followed by a development of *gonimoblast filaments*, either from the base of the carpogonium or from another thallus cell into which has migrated a derivative from the zygote nucleus. All or certain of the cells of a gonimoblast filament develop into *carposporangia*, each containing a single *carpospore*. The carposporangia, gonimoblast filaments, and structures associated with them jointly constitute the *cystocarp*. Carpospores are liberated by rupture of the carposporangial wall and, after liberation, develop into thalli identical in vegetative structure with those producing them. If division of the zygote nucleus is meiotic, the carpospores develop into haploid thalli producing spermatangia and carpogonia. If division of the zygote nucleus is mitotic, the carpospores develop into diploid thalli that bear tetrasporangia or polysporangia in which nuclear division is meiotic. Tetraspores and polyspores develop into haploid thalli producing spermatangia and carpogonia.

Structure of the Thallus. Although the fresh-water Florideae do not exhibit the variation in size and form found among marine species, they include those that are of macroscopic size and with some internal differentiation of tissues.

All the fresh-water Florideae found in this country, with the exception of *Thorea*, are of the central filament or monoaxial type. In the simplest case, exemplified by the *Chantransia* stage of *Lemanea* (Fig. 534*B*), the successive division of the apical cell in the same plane results in a uniseriate main axis. Derivatives once or twice removed from the apical cell send out tubular projections at their upper end, and each of these soon becomes separated from

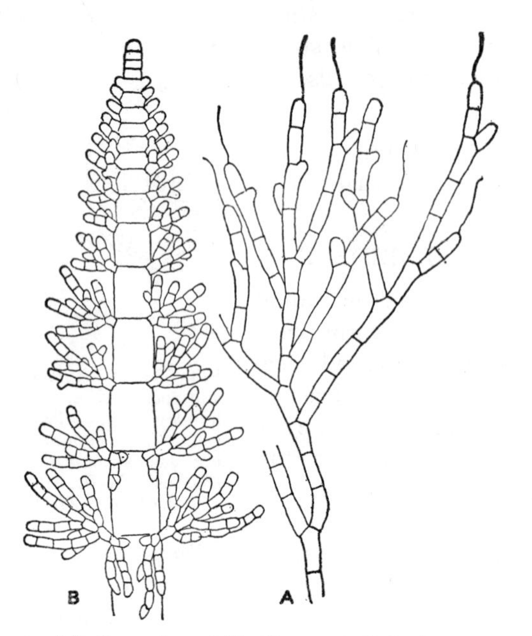

FIG. 534. Diagrams of thallus apices of Florideae. *A*, *Batrachospermum*. *B*, *Chantransia* stage of *Lemanea*.

the rest of the cell by a transverse wall. The cell thus cut off is the apical cell of a lateral branch which may grow as rapidly as the main axis.

Batrachospermum (Fig. 534*A*) has a more complicated thallus than one of the *Chantransia* type. There is a single apical cell which forms derivatives by transverse division only. Each derivative cut off by an apical cell increases in length and breadth as it comes to lie farther and farther back from the apical cell, so that the axial portion in older parts of a thallus is composed of very large cylindrical cells. Axial cells, a few cells back from the apical cell, send out four lateral projections which soon become cut off by transverse walls. Each of the cells formed in this manner is the initial of a lateral branch which forks repeatedly but never becomes very long.

Near the upper end of a thallus, the lateral branches borne by each axial cell touch one another but, as the axial cells elongate, the whorls of lateral branches become separated from one another. After the lateral branches have become well differentiated, their basal cells give rise to rhizoid-like multicellular outgrowths which grow downward and completely ensheath the large axial cell. Because of these ensheathing cells, the central axis of a thallus appears to be a multicellular structure.

At first glance, a mature thallus of *Lemanea* appears to have but little in common with the filamentous types just described, but a study of the sequence in which derivates are formed from the apical cell at the distal end shows that it has essentially the same organization. The apical cell divides only in a transverse plane (Fig. 535*A*). A short distance back from the thallus apex, each derivative cut off by the apical cell gives rise, by periclinal divisions, to an axial cell surrounded by four pericentral cells[1] (Fig. 535*B*). The axial cell, which does not divide again, eventually elongates to many times its original length. The four pericentral cells, by periclinal and anticlinal divisions, give rise to a cylindrical mass of parenchymatous tissue that ensheaths the axial cell and is in contact with it by means of four "tie cells" (Fig. 535*C–E*).

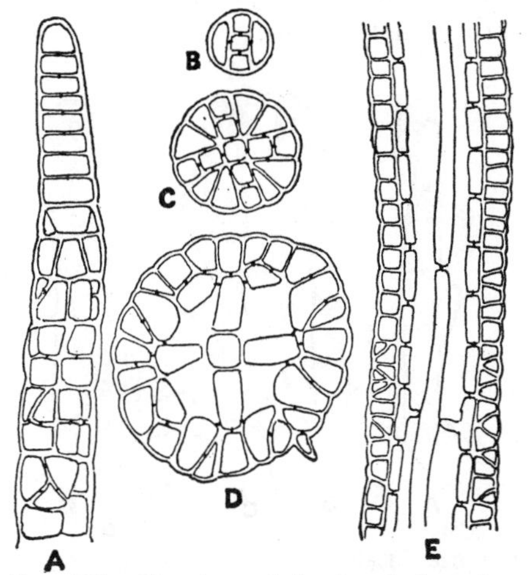

FIG. 535. Structure of the terminal portion of thallus of *Lemanea fluviatilis* Ag. *A*, surface view of apical portion. *B-D*, transverse sections at successive levels. *E*, longitudinal section some distance back from the apex. (*After Kylin*, 1923.)

The multiaxial type of thallus, represented only by *Thorea* in the fresh-water flora of this country, has a central core of axial filaments, each terminating in an apical cell. Lateral branches on axial filaments are developed only on sides not in contact with other axial filaments.

Asexual Reproduction. Among Florideae asexual reproduction of the sexual generation may be by means of either monospores or tetraspores. Reproduction by means of monospores may be restricted to juvenile (*Chantransia*) stages, as in *Thorea* and certain species of *Batrachospermum*, or there may be a formation of them by adult plants.

Sexual Reproduction. Spermatangia, each containing a single spermatium, are formed terminally at the ends of filaments. These filaments

[1] ATKINSON, 1890; KYLIN, 1923.

may stand free from one another, or they may be laterally adjoined to form a continuous stratum bearing spermatangia. The spermatia are liberated from the spermatangia by a rupture of the spermatangial wall.

Carpogonia of Florideae are borne terminally on a branched or unbranched carpogonial filament consisting of a definite or an indefinite number of cells. The carpogonium regularly has a conspicuous trichogyne. Spermatia are carried by water currents to the carpogonium. They lodge against the trichogyne, and the trichogyne wall breaks down where it is in contact with a spermatium. The spermatium nucleus then migrates into the trichogyne and down it to the nucleus at the base of the carpogonium. During this migration, it may or may not divide to form two daughter nuclei. In any case, there is a fusion of but one male nucleus with the single carpogonial nucleus.

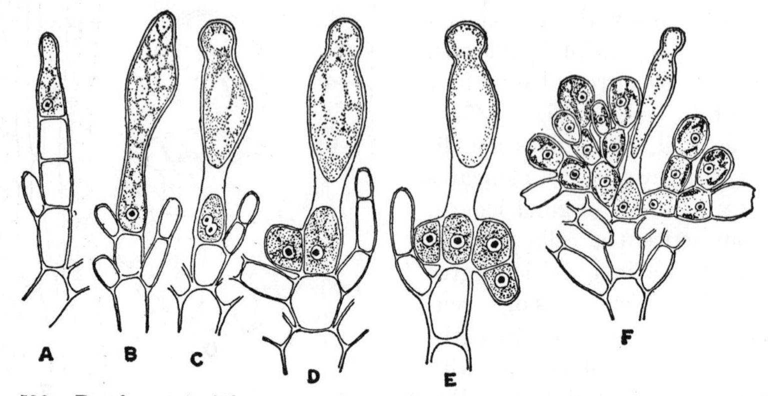

FIG. 536. Development of the carpogonium and cystocarp of *Batrachospermum moniliforme* Roth. *A*, young carpogonium. *B-C*, mature carpogonium before and after fertilization. *D-E*, early stages in development of gonimoblast filaments. *F*, mature cystocarp. (*After Kylin*, 1917.) (*A-E*, × 900; *F*, × 750.)

All fresh-water Florideae of this country are similar to *Batrachospermum* in that descendants from the zygote nucleus each migrate into a protrusion from the carpogonial base, and the protrusion becomes separated from the carpogonial base by a cross wall. Although not definitely established for *Batrachospermum*, it is very probable that division of the zygote nucleus is meiotic, a fact definitely established for certain rather closely related marine Florideae. The cells cut off at the carpogonial base are initial cells, each of which develops into a branched gonimoblast filament with carposporangia at the tips of the branches.[1] Each carposporangium contains a single carpospore that is liberated by a terminal rupture of the carposporangial wall (Fig. 536). The gonimoblast filaments, plus the carposporangia and

[1] DAVIS, B. M., 1896; KYLIN, 1917; OUSTERHOUT, 1900; SCHMIDLE, 1899.

the remains of the old carpogonium, constitute the cystocarp. Carpospores that have lodged upon a suitable substratum develop into new thalli that bear spermatangia and carpogonia when mature.

Classification. The various orders recognized among the Florideae[1] differ from one another in the place where the gonimoblast filaments develop and and in the place at which meiosis occurs in the life cycle. Only one (Nemalionales) of the six orders into which the Florideae are divided is represented in the fresh-water flora of the United States.

ORDER 1. NEMALIONALES

Thalli of Nemalionales may be of the monoaxial or multiaxial type; and microscopic, or macroscopic and of definite form. The gonimoblast filaments grow directly from the base of the carpogonium, and there is never a formation of true auxiliary cells. Most members of the order have a meiotic division of the zygote nucleus. A few genera whose nuclear cycle is unknown produce tetraspores.

Family 1. Chantransiaceae

The Chantransiaceae have microscopic, freely branched, uniseriate, filamentous thalli in which the central axis is obscure, and lateral branches are the same length as it is.

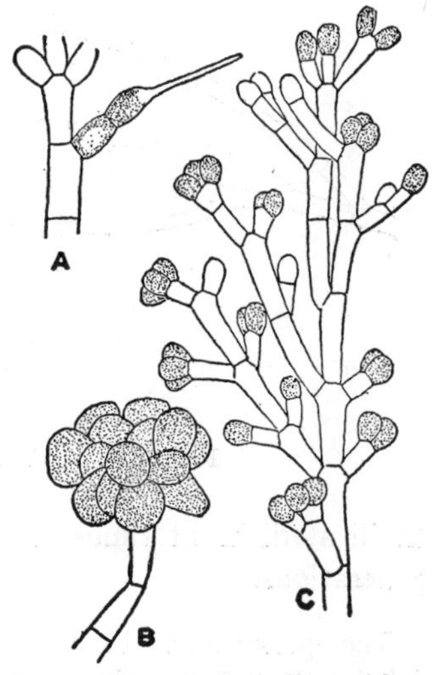

Fig. 537. *Audouinella violacea* (Kütz.) Hamel. *A*, carpogonium. *B*, cystocarp. *C*, portion of a branch with monospores. (*After Murray and Barton*, 1891.) (*A-B*, × 600; *C*, × 333.)

The only known method of reproduction in a majority of the species is by means of either monospores or tetraspores. Those species whose sexual reproduction is known have a development of gonimoblast filaments from the base of the carpogonium. Certain members of the family have both sexual and tetrasporic thalli. This suggests that they should not be placed among the Nemalionales but, until the life histories are more fully known, it is better to follow the conventional practice of including them among the Nemalionales.

There is diversity of opinion concerning the fundamental characteristics of genera within the family. The segregation according to structure of the chromatophore[2] seems the most logical. *Audouinella* is the only genus in the fresh-water flora of this country.

[1] Kylin, 1928. [2] Papenfuss. 1945.

1. **Audouinella** Bory, 1823; emend., Papenfuss, 1945. This is one of the Florideae with microscopic, freely branched, filamentous thalli with the branches equal in diameter to the poorly defined main axis and often extending beyond it. It is to be distinguished from genera with thalli of similar construction by the elongate spiral chromatophores within the cells.[1]

The thalli may be homo- or heterothallic. The spermatangia are borne in clusters on short lateral branchlets. The carpogonia (Fig. 537*A*) are borne terminally on one- or two-celled carpogonial filaments. Branched gonimoblast filaments develop from the base of a carpogonium and have solitary or seriately arranged carposporangia at their tips.[2]

Instead of producing sexual organs, a thallus may produce monosporangia (Fig. 537*C*) or tetrasporangia. The formation of sexual and tetrasporic

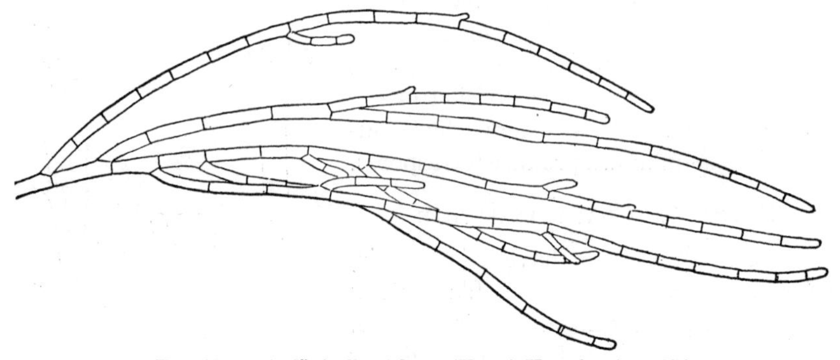

Fig. 538. *Audouinella violacea* (Kütz.) Hamel. (× 165.)

thalli at different times of the year[3] suggests that the two are alternate generations.

The species found in this country are *A. Hermanni* (Roth) Duby, *A. tenella* (Skuja) Papenfuss, and *A. violacea* (Kütz.) Hamel (Fig. 538). For a description of *A. tenella* as *Chantransia tenella* Skuja, see Skuja (1934); for the other two as species of *Audouinella*, see Hamel (1924–1926).

Family 2. Batrachospermaceae

The Batrachospermaceae have a thallus with a conspicuous central axis bearing at regular intervals transverse whorls of short lateral branches.

Monospores are usually formed only on juvenile (*Chantransia*) stages, and carpogonia and spermatangia only on adult thalli. The carposporangia are restricted to the tips of gonimoblast filaments.

[1] Papenfuss, 1945.
[2] Drew, 1935; Murray and Barton, 1891.
[3] Drew, 1935.

The two genera found in this country differ as follows:

1. Gonimoblast filaments in a compact mass .... 1. **Bartrachospermum**
1. Gonimoblast filaments widely divergent .... 2. **Sirodotia**

1. **Batrachospermum** Roth, 1797. The thallus of *Batrachospermum* (Fig. 539) is macroscopically moniliform and with an evident central axis. Its texture is gelatinous, and its color usually bluish green, olive, or violet. Older portions of a thallus consist of a single axial row of large cells, densely clothed with simple or forked vertical branches, and bearing globose clusters of lateral branches that may lie close to, or somewhat remote from, one another. The method by which derivatives are formed from an apical cell has already been noted (see page 614). Cells of lateral branches contain several discoid or elongate chromatophores, each of which contains a pyrenoid.

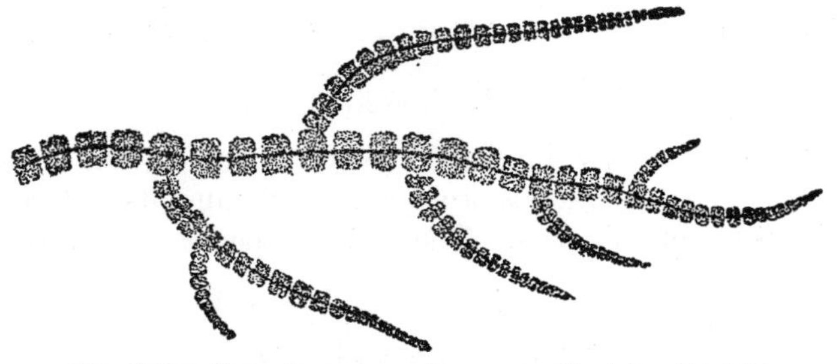

FIG. 539. *Batrachospermum Boryanum* Sirodot. (× 2.5.)

Asexual reproduction by means of monospores is usually restricted to the *Chantransia* stage. A few species, as *B. vagum* (Roth) Ag., develop monospores at the ends of short lateral branches of adult thalli.

Carpogonia and spermatangia may be formed upon the same thallus or upon separate thalli. For details concerning fertilization and development of the cystocarp, see page 616. After liberation, a carpospore germinates to form a simple filamentous plant, the *Chantransia* stage, in which apical cells of certain branches give rise to characteristic adult thalli.

Adult thalli are usually found only during the spring and summer months, and generally only in clear cold water where there is more or less shade. For this reason *Batrachospermum* is frequently found in springs or in the outflow from them. About a dozen species have been found in the United States. For a monographic treatment of the genus, see Sirodot (1884).

2. **Sirodotia** Kylin, 1912. The vegetative structure of thalli of *Sirodotia* is similar to that of *Batrachospermum*. The most conspicuous difference between the two is in the gonimoblast filaments. In *Sirodotia* they are

many cells in length, and with tufts of a few carposporangia at regular intervals.[1] Because the gonimoblast filaments do not grow in a compact mass, the cystocarp of *Sirodotia* is a diffuse structure instead of the globose one characteristic of *Batrachospermum.*

The three species known for this country have been found[2] at widely separated localities east of the Mississippi River. These species are *S. polygama* Skuja, *S. suecica* Kylin (Fig. 540), and *S. tenuissima* (Holden) Skuja. For descriptions of them, see Flint (1948).

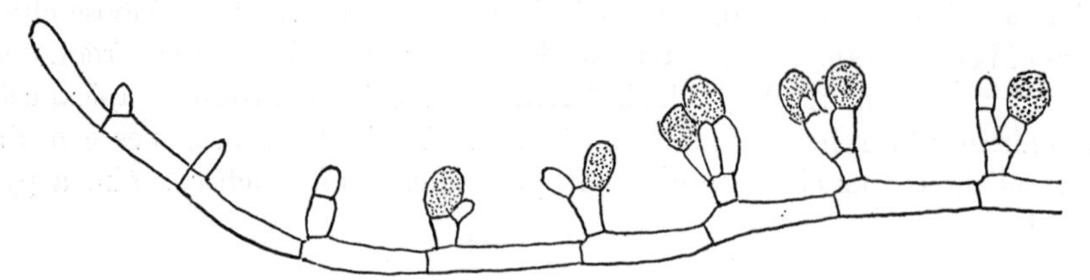

Fig. 540. Gonimoblast filament of *Sirodotia suecica* Kylin. (*After Kylin,* 1912*A*.)

Family 3. Thoreaceae

This family has been established to receive a single genus (*Thorea*), and one whose systematic position is problematical because its method of sexual reproduction is as yet unknown. The growing apex of *Thorea* is of the same

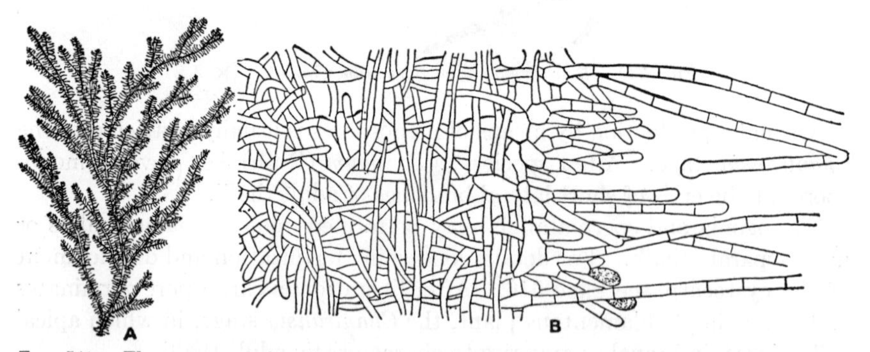

Fig. 541. *Thorea ramosissima* Bory. *A*, habit sketch. *B*, portion of thallus. (*A*, *after Wolle*, 1887; *B*, *after Hedgcock and Hunter*, 1899.) (*A*, × 2; *B*, × 225.)

multiaxial type as is found in the Nemalionaceae. If it were not for this, the genus might well be assigned to the Batrachospermaceae.

1. **Thorea** Bory, 1808. The profusely branched thallus of this alga is 2 to 3 mm. broad and may be 50 cm. or more long. Its color is either olive green, dark brown, or black. At the apex of a branch are a number of parallel axial filaments, each terminating in an apical cell.[3] Older portions of branches consist of an axial core of interlacing filaments which run in all

[1] Kylin, 1912*A*. [2] Flint, 1948. [3] Möbius, 1892.

directions, and external to them is a zone of interlacing filaments most of which run longitudinally. These, in turn, are surrounded by a layer of spherical or irregularly shaped cells from which arise short erect filaments that extend to the gelatinous envelope of a branch,[1] or even project beyond the envelope. Cells of the erect filaments and the outermost cells of the felted portion contain numerous irregularly shaped chromatophores.

Asexual reproduction is by means of monospores developed both by the *Chantransia* stage and by terminal cells of erect filaments of the adult thallus. The monospores germinate into a *Chantransia* stage, certain branches of which develop into characteristic adult thalli.

T. ramosissima Bory (Fig. 541), the only species, has been found in four of the states of this country. For a description of it, see Pascher and Schiller (1925).

Family 4. Lemaneaceae

The Lemaneaceae differ from other families in the fresh-water flora of this country both in structure of their thalli and in the manner in which carposporangia are formed by the gonimoblast filaments. The thallus is a solid or hollow parenchymatous cylinder. It has, however, a monoaxial organization (see page 615). Carposporangia are formed by all cells of a gonimoblast filament instead of only the terminal cells.

The two genera found in this country differ as follows:

1. Thalli externally differentiated into nodes and internodes.......... 1. **Lemanea**
1. Thalli not externally differentiated into nodes and internodes...... 2. **Tuomeya**

1. **Lemanea** Bory, 1808; emend., Agardh, 1828. Mature thalli of *Lemanea* are stiff, cartilaginous, thread-like, simple or branched, and tapering to a fine point at the distal ends. Older portions of a thallus are with alternate swellings and contractions at regular intervals. The color of thalli ranges from a light olive green to a very dark or blackish olive green. The structure of the thallus apex has already been described (see page 615). Mature portions of a thallus have an axial filament of very long cells that may be naked or clothed with hyphal filaments which lie parallel to it (Fig. 542*C*). Each cell of the axial filament has four cells (ray cells) perpendicular to it, and each ray cell is connected at its distal end with vertically elongate branches. The vertical branches, in turn, are repeatedly branched toward their outer face into secondary branches so close to one another that they form a parenchymatous tissue.[2]

Asexual spores have not been found in either juvenile or mature plants.

Spermatangia are borne on the thallus surface at the nodes and either in circular patches that are not in lateral contact with one another or in a con-

[1] Hedgcock and Hunter, 1899; Schmidle, 1896.
[2] Atkinson, 1890; Kylin, 1923.

tinuous transverse belt. The carpogonial filaments are borne either in the antheridial zone or midway between two antheridial zones. The gonimoblast filaments grow from the base of the carpogonium and extend inward toward the thallus center. As they approach maturity, all their cells develop into carposporangia.[1] The carpospores are liberated during summer or autumn and, when they germinate, they grow into a prostrate "protonema" stage whose cells are often so densely crowded that they are angular

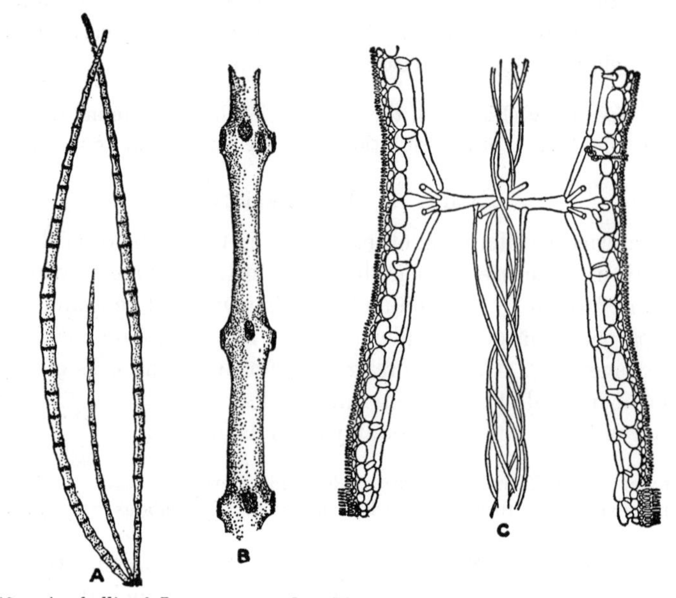

Fig. 542. *A*, thalli of *Lemanea annulata* Kütz. *B*, portion of thallus of *L. fucina* Bory *C*, diagrammatic longitudinal section of *L. australis* Atk. (*C*, *modified from Atkinson*, 1890.) (*A*, × 2; *B*, × 25.)

by mutual compression. Erect filaments growing from the protonema constitute a true *Chantransia* stage, and from certain branches of it are developed the typical adult plants.

Lemanea is an alga that is rather closely restricted to turbulent waters and is usually found only at rapids, falls, mill dams, and the like. The thalli are generally attached to the sides of stones where the currents of water press strongly. The species found in this country are *L. annulata* Kütz. (Fig. 542*A*), *L. australis* Atk., *L. fluviatilis* Ag., *L. fucina* Bory (Fig. 542*B*), *L. grandis* (Wolle) Atk., *L. nodosa* Kütz., *L. pleocarpa* Atk., and *L. torulosa* Sirodot. For a description of *L. pleocarpa* see Atkinson (1931), for the others, see Atkinson (1890).

2. **Tuomeya** Harvey, 1858. The thallus of this alga superficially resembles *Batrachospermum* in macroscopic appearance, but is of a firm ge-

[1] Kylin, 1923.

latinous texture and does not collapse into an amorphous gelatinous mass when lifted from the water. It is also more irregularly branched. At the apex of each branch is a single apical cell which divides only in a transverse plane. Each derivative a few cells back from an apical cell produces several lateral initials by vertical division, and each initial ultimately grows into a repeatedly forked lateral filament. As in *Batrachospermum*, the lateral filaments give rise to corticating threads that surround the axial filament. The erect portions of lateral filaments are so densely crowded at their dis-

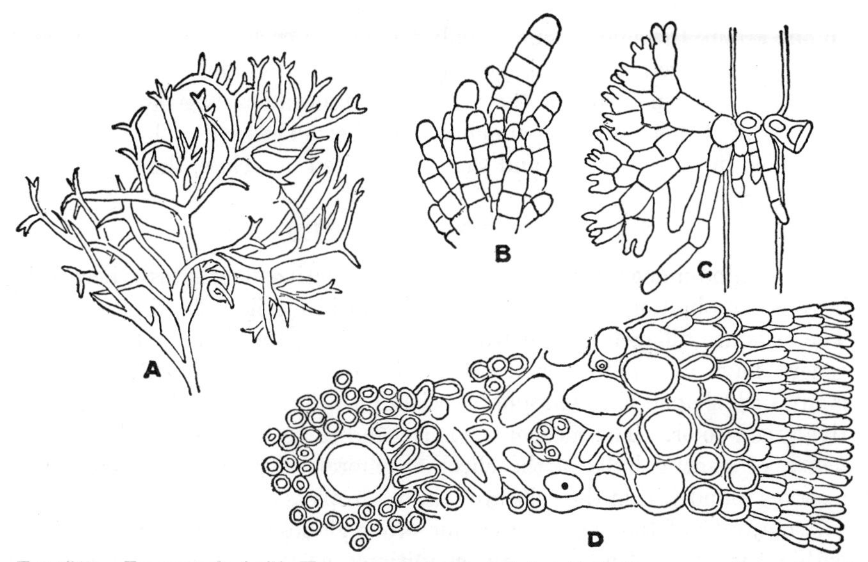

FIG. 543. *Tuomeya fluviatilis* Harvey. *A*, habit sketch. *B*, thallus apex. *C*, portion of a node. *D*, transverse section through mature portion of a thallus. (*After Setchell*, 1890.) (*B*, × 500; *C*, × 375; *D*, × 312.)

tal ends that the peripheral portion of a thallus seems to be parenchymatous. The rigidity of a thallus is due to this close apposition of lateral filaments and to the greatly thickened walls of the corticating filaments around the axial filament.

Spermatangia and carpogonia are borne on the same thallus but usually growing on separate portions of it. Carpogonial filaments are formed near the growing apex and axillary to lateral filaments. As far as is known, fertilization and development of carposporangia are similar to that of other Nemalionales.[1]

There is but one species *T. fluviatilis* Harvey (Fig. 543), and it is widely distributed east of the Mississippi River.[2] For a description of it, see Wolle (1887).

[1] SETCHELL, 1890. [2] FLINT, 1948.

CHAPTER 11

GROUPS OF UNCERTAIN SYSTEMATIC POSITION

There remain for consideration two groups, each distinctive, but neither of which can be placed in any of the divisions described on previous pages. In one group, the chloromonads, only flagellated organisms are known and these are placed in a single order, the Chloromonadales. In the other group, a majority of the genera are unicellular flagellates (cryptomonads), but there are also palmelloid and coccoid genera. Thus the group is divided into three orders and the three united in a class, the Cryptophyceae.

CHLOROMONADALES

The Chloromonadales include a few distinctive flagellates whose cells have a combination of characters not found in motile cells of other algae. The cells are biflagellate and with a reservoir at the anterior end. Some genera have chromatophores; others do not. When present, there are many disk-shaped chromatophores within a cell, and they have a distinctive green color. The meager evidence[1] indicates that there is a predominance of xanthophylls in the chromatophores. Thus far, the only food reserves found within the cells are minute droplets of oil.

A majority of the genera have numerous globose or acicular trichocysts within the cells. Under certain conditions, as when a cell is pressed or when it is treated with certain reagents, the trichocysts shoot out or elongate into threads of slime projecting from a cell.[2]

Of the two flagella, one, the swimming flagellum, projects forward and the other, the trailing flagellum, projects backward. The trailing flagellum is exceedingly delicate and easily overlooked because it lies so close to the cell. Many early accounts of Chloromonadales describe them as uniflagellate because of a failure to note the trailing flagellum. The detailed structure of flagella is known only for *Vacuolaria viridis* (Dang.) Senn, and here it has been shown[3] that each flagellum has a blepharoplast-rhizoplast neuromotor apparatus extending deep into a cell.

The cells have one or two contractile vacuoles adjacent to the reservoir,

[1] PASCHER, 1913*C*.

[2] CONRAD, 1920*A*; DROUET and COHEN, 1935; PALMER, 1942; PENARD, 1921.

[3] FOTT, 1935.

and these may be adjoined by accessory contractile vacuoles. It is uncertain whether the vacuoles discharge into the reservoir or open directly onto the surface of a cell.[1] The nucleus is conspicuous and lies near the center of a cell. It has a nuclear membrane, a chromatic network, and several nucleoli.[1] Division of a nucleus is mitotic.

Reproduction is by longitudinal division. This may take place while a cell is actively motile or after it has become immobile. The cells may also lose their flagella, become spherical, and secrete a gelatinous wall.[2] Such cells are usually called *cysts*. Cell division may take place while in this "encysted" condition and the daughter cells redivide. This results in *Gloeocystis*-like palmelloid colonies.[3]

The Chloromonadales have been placed among the Xanthophyceae[4] and among the Cryptophyceae.[5] If the Chloromonadales are to be referred to either of these classes, the relationship is exceedingly tenuous. Therefore it is better to follow those[6] who are unwilling to commit themselves and call the Chloromonadales an isolated group whose systematic position is uncertain.

The differences between the various genera are so slight that they are placed in a single family, the Chloromonadaceae. The two genera found in this country differ as follows:

1. Trichocysts not in acicular clusters .... 1. **Gonyostomum**
1. Trichocysts mostly in an acicular cluster .... 2. **Merotrichia**

1. **Gonyostomum** Diesing, 1865. Cells of this chloromonad are markedly compressed, oval to circular in front view, and with a more or less plastic periplast. They are biflagellate and, according to the species, the swimming flagellum is shorter than or longer than the trailing flagellum. There is a well-defined reservoir at the anterior end, adjoined by a contractile vacuole. There are many disk-shaped chromatophores of a distinctive bright green color. Within a cell are many acicular trichocysts, perpendicular to and just within the plasma membrane. They may be evenly distributed or more densely aggregated at the anterior end of a cell. Under certain conditions, the trichocysts become slime threads projecting a considerable distance beyond the surface of a cell.

Reproduction is by longitudinal division and takes place while the cells are motile.[7] The cells may also develop into spherical cysts with a firm gelatinous membrane.[7]

[1] Drouet and Cohen, 1935; Fott, 1935.
[2] Dangeard, 1920; Drouet and Cohen, 1935; Fott, 1935. [3] Fott, 1935.
[4] Luther, 1899. [5] Oltmanns, 1922.
[6] Fritsch, 1935; Pascher, 1913; Senn, 1900. [7] Drouet and Cohen, 1935.

G. semen (Ehr.) Diesing (Fig. 544) has been collected[1] in a pond in Massachusetts. It was found at the water's surface during the hours following sunrise, but not during midday hours. For a description of *G. semen*, see Drouet and Cohen (1935).

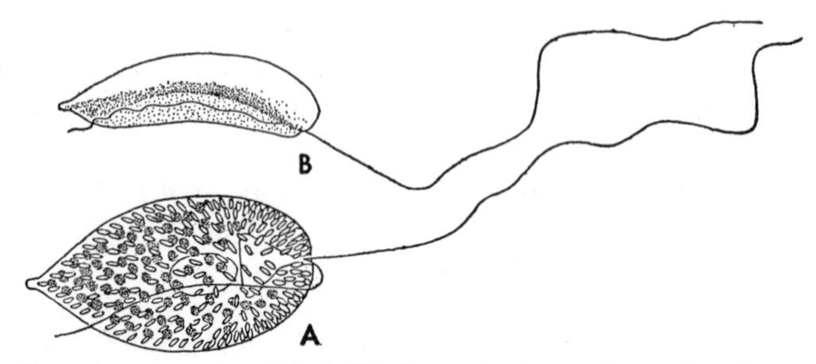

FIG. 544. *Gonyostomum semen* (Ehr.) Diesing. *A*, front view. *B*, side view. (*After Drouet and Cohen*, 1935.) (× 475.)

2. **Merotrichia** Mereschkowski, 1877. The cells of *Merotrichia* are elliptical in front view, circular in cross section, and with a nearly rigid periplast. The two flagella are inserted laterally and a short distance below the anterior end of a cell.[2] A cylindrical reservoir lies beneath the point of insertion of the flagella, and adjacent to it is a conspicuous vacuole. There is a radiately disposed cluster of acicular trichocysts at the anterior end of a cell, and there may or may not be a few isolated trichocysts elsewhere in the cell. As in *Gonyostomum*, the trichocysts may burst and become pro-

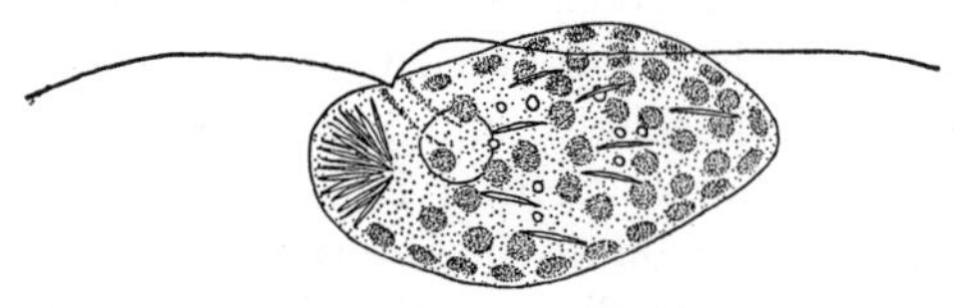

FIG. 545. *Merotrichia capitata* Skuja. (*After Skuja*, 1934*A*.) (× 550.)

jecting slime threads. The cells contain numerous disk-shaped green to yellow-green chromatophores. Minute spherical or irregularly shaped granules of food reserves lie irregularly distributed within the cytoplasm.

M. capitata Skuja (Fig. 545) has been found in a pond in Massachusetts.[3] For a description of it, see Skuja (1932).

CRYPTOPHYCEAE

Protoplasts of Cryptophyceae usually contain two chromatophores that are brownish in most cases but may be red, blue, blue-green, or grass-green. Reserve foods are usually stored as starch or starch-like compounds. Most

[1] DROUET and COHEN, 1935. [2] PALMER, 1942; SKUJA, 1932. [3] PALMER, 1942.

of the Cryptophyceae are unicellular flagellates (cryptomonads) with asymmetrical compressed cells, surrounded by a firm periplast, and with the two flagella slightly different in length. Insertion of the flagella may be terminal or lateral.

Occurrence. Some fresh-water cryptomonads grow in waters rich in organic and in nitrogenous materials; others grow in waters with small amounts of these materials. The yellowish or brownish bodies (zooxanthellae) found within various marine animals, as radiolarians and sea anemones, are considered symbiotic unicellular Cryptophyceae.

Cell Structure. Motile genera and zoospores of immobile genera have somewhat compressed cells with a superficial curved furrow extending backward from the insertion of the flagella. Many genera have, in addition to this furrow, a gullet extending inward from the point of insertion of flagella. The gullet may or may not be adjoined by trichocysts. The function of the gullet is uncertain, and it is not definitely established that cryptomonads ingest solid food.[1] There are also one or two contractile vacuoles at the anterior end of a cell, and these may discharge their contents into the gullet or to the surface of the cell.

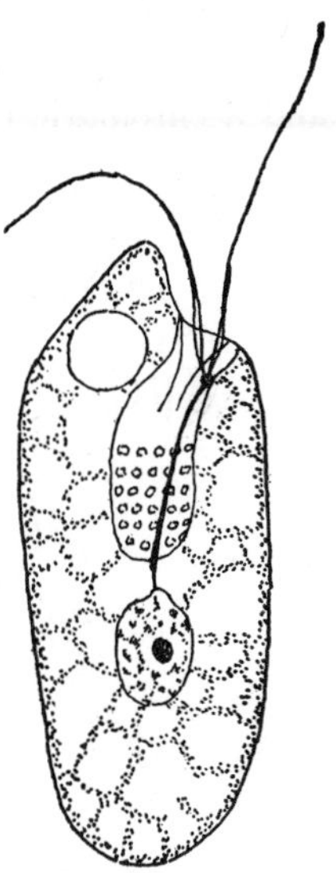

FIG. 546. *Chilomonas Paramaecium* Ehr. (*After Úlehla,* 1911.)

The flagella may or may not be inserted in a gullet. *Chilomonas* has been shown[1] to have a definite neuromotor apparatus with a blepharoplast connected to the nucleus by a conspicuous rhizoplast (Fig. 546). Flagella of cryptomonads are usually slightly different in length and somewhat flattened. Both flagella may project forward, or one may project forward and the other trail as a cell swims through the water.

Most genera have two laminate chromatophores, but certain genera have several disk-shaped chromatophores within each cell. The color of chromatophores is usually olive green to golden brown, but it may be grass-green, blue-green, blue, or red. The only study of pigments of chromatophores has been on those of a brownish zooxanthella growing in a sea anemone. This has been found[2] to have the same chlorophylls and xanthophylls as are found in Dinophyceae. Pyrenoids of some species are embedded in the chromatophores; those of other species lie in the cytoplasm. In either case, the starch or starch-like granules within a cell may

[1] ÚLEHLA, 1911.

[2] STRAIN and MANNING, 1943; STRAIN, MANNING, and HARDEN, 1944.

encircle the pyrenoids or lie along the inner face of chromatophores. Accumulation of starch and other food reserves may be due either to photosynthetic activity or to a saprophytic mode of nutrition.

In addition to the genera with flagellated motile vegetative cells, there are two genera with immobile amorphous palmelloid colonies composed of an indefinite number of cells (Fig. 547), and one genus with coccoid immobile solitary cells (Fig. 559).

Motile Cryptophyceae are uninucleate and with the nucleus toward the posterior end of a cell. The nucleus has a definite membrane, a nucleolus, and a chromatic network. Its division is mitotic.[1]

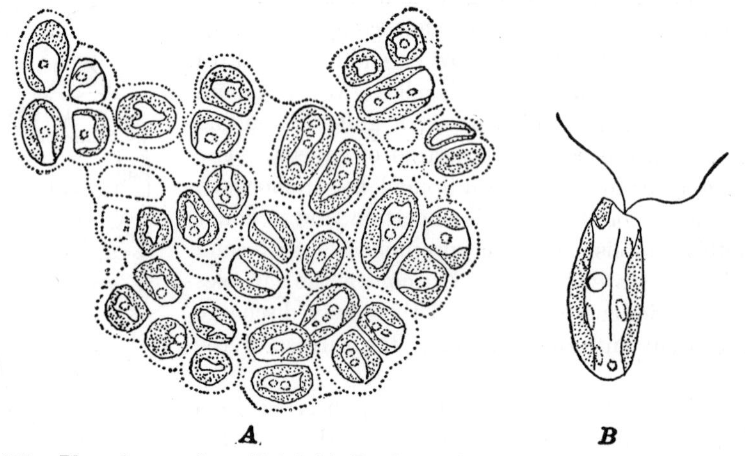

FIG. 547. *Phaeoplax marinus* (Reinisch) Pascher. *A*, vegetative colony. *B*, zoospore. (*After Reinisch*, 1911.) (*A*, × 1000; *B*, × 1500.)

Reproduction. Reproduction of Cryptophyceae with flagellated vegetative cells is by longitudinal division and usually takes place while a cell is in a temporary palmelloid condition. Cell division is preceded by a division of the nucleus and a bipartition of the gullet.[2] Palmelloid and coccoid genera reproduce by means of zoospores whose structure resembles that of a motile vegetative cell.

Relationships. The affinities of the cryptomonads were thought to be with the chrysomonads until Pascher[3] suggested that they are more closely related to the dinoflagellates. Shortly afterward he proposed[4] that the group be considered a distinct class (Cryptophyceae) coordinate in rank with the dinoflagellate series (Dinophyceae), and made these members of a division to which he gave the name Pyrrophyta. Present-day opinion[5]

[1] BELAR, 1916; REICHARDT, 1927; ÚLEHLA, 1911.

[2] REICHARDT, 1927; ÚLEHLA, 1911. [3] PASCHER, 1911*A*.

[4] PASCHER, 1914.

[5] FRITSCH, 1935; GRAHAM (in press); PRINGSHEIM, E. G., 1944.

tends to hold that the Cryptophyceae differ too markedly from the Dinophyceae to be included among the Pyrrophyta.

Classification. In spite of the paucity of known immobile genera, it has been held[1] that, just as in Xanthophyceae, Chrysophyceae, and Dinophyceae, evolution within the class has paralleled that found in Chlorophyceae. Because of this, the various types found among Cryptophyceae have been classified[1] in the same manner as analagous types in Xanthophyceae, Chrysophyceae, and Dinophyceae. According to such a classification, the genera with motile vegetative cells are placed in one order (Cryptomonadales), those with palmelloid vegetatives in a second order (Cryptocapsales), and the genus with coccoid cells in a third order (Cryptococcales).

ORDER 1. CRYPTOMONADALES

The Cryptomonadales include all genera in which the vegetative cells are flagellated and motile. Unlike the homologous orders in Chlorophyceae and Chrysophyceae, there are no genera with the cells united in motile colonies. The order has been divided[2] into three families, all closely related to one another.

FAMILY 1. CRYPTOCHRYSIDACEAE

Members of this family are unicellular free-swimming flagellates with the flagella borne near the anterior end of a cell, and with a longitudinal furrow extending backward from the point of insertion of the flagella. Genera referred to this family do not have a gullet, but they may have trichocysts.

The genera found in this country differ as follows:

1. Cells naked 2
1. Each cell surrounded by a lorica .... 6. **Cyanomastix**
  2. Cells with one or two laminate chromatophores 3
  2. Cells with several disk-shaped chromatophores .... 4. **Cyanomonas**
3. Trichocyst granules at base of cell .... 5. **Monomastix**
3. Trichocyst granules in anterior part of cell 4
  4. Chromatophores a bright blue-green .... 2. **Chroomonas**
  4. Chromatophores not a bright blue-green 5
5. Pyrenoids lacking .... 1. **Cryptochrysis**
5. Pyrenoids present .... 3. **Rhodomonas**

1. **Chryptochrysis** Pascher, 1911. *Cryptochrysis* has compressed ellipsoidal cells that are truncate at the anterior end. The two flagella are slightly different in length. A curved longitudinal furrow runs backward from the point of insertion of the flagella, and lateral to either side of it is

[1] PASCHER, 1914. [2] PASCHER, 1932.

a vertical row of trichocyst granules. There is no gullet. A cell contains two brownish to olive-green laminate chromatophores and two small vacuoles. Pyrenoids have not been reported for this genus, but small platelets of starch have been found within the cells.

Reproduction is by longitudinal division and takes place while the cells are motile.[1]

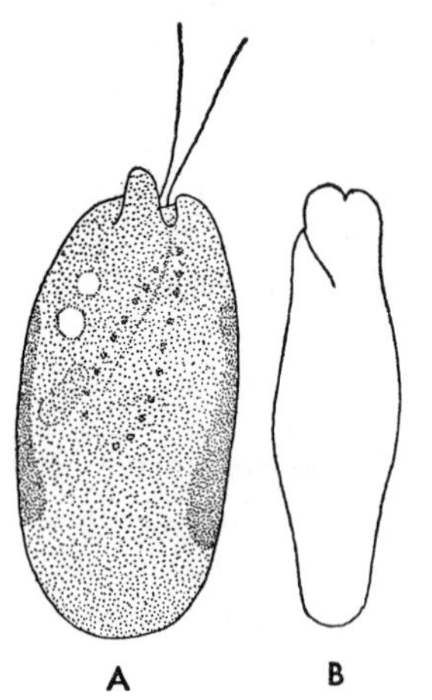

FIG. 548. *Cryptochrysis commutata* Pascher. *A*, front view. *B*, side view. (*After Prescott and Croasdale*, 1942.)

The type species, *C. commutata* Pascher (Fig. 548) has been found in Massachusetts.[2] For a description of it, see Pascher (1913).

2. **Chroomonas** Hansgirg, 1892. *Chroomonas* has compressed cells with the anterior end truncate and the posterior end rounded. There is a longitudinal surface furrow at the anterior end, and there may or may not be trichocyst granules lateral to it.[3] The flagella, which are not inserted in a gullet, are borne at the anterior end of a cell and are unequal in length. There is a single laminate chromatophore of a bright blue-green color. An eyespot mat be present or lacking. There is usually a single pyrenoid which lies free from the chromatophore and is surrounded by a sheath of starch granules.

Reproduction is by longitudinal division and may take place while a cell is in a motile or a nonmotile state.[4] Under certain conditions, there may be a development of a many-celled palmella stage.[4]

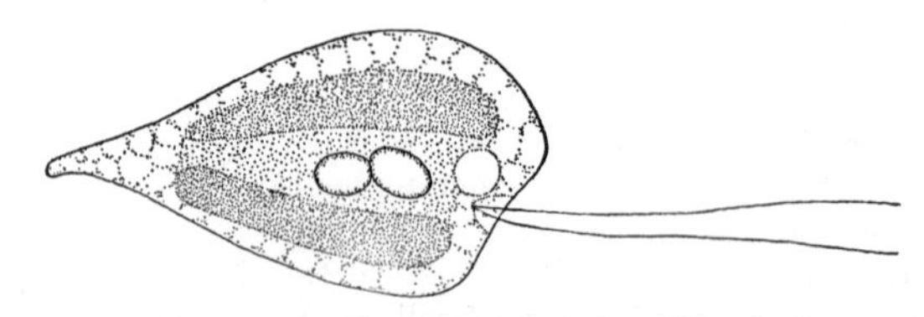

FIG. 549. *Chroomonas Nordstedtii* Hansg. (*After Lackey*, 1941.)

C. cyaneus Lackey, *C. Nordstedtii* Hansg. (Fig. 549), and *C. pulex* Pascher have been found at several widely separated stations east of the Mississippi River. It is very doubtful if *C. setonensis* Lackey[5] should be referred to *Chroomonas* because it has a conspicuous gullet. For a description of *C. cyaneus*, see Lackey, 1939; for the other two, see Pascher (1913).

[1] PASCHER, 1913*C*. [2] PRESCOTT and CROASDALE, 1942.
[3] PASCHER, 1914; ROSENBERG, 1944. [4] ROSENBERG, 1944.
[5] LACKEY, 1939

3. **Rhodomonas** Karsten, 1898. The bright red color of the chromatophores was considered the most distinctive character when this genus was first described,[1] but as found in this country[2] the color may range from pale brown to olive green. The cells are somewhat compressed, narrow at the posterior end, and with a longitudinal furrow that is flanked on either side by a vertical row of trichocyst granules. Two flagella of unequal length are borne at the anterior end. There is no gullet. Each cell usually contains a single laminate chromatophore and a single pyrenoid which lies free from the chromatophore.

R. lacustris Pascher and Ruttner (Fig. 550) has been found repeatedly in rivers of Ohio.[2] For a description of it, see Pascher, 1913.

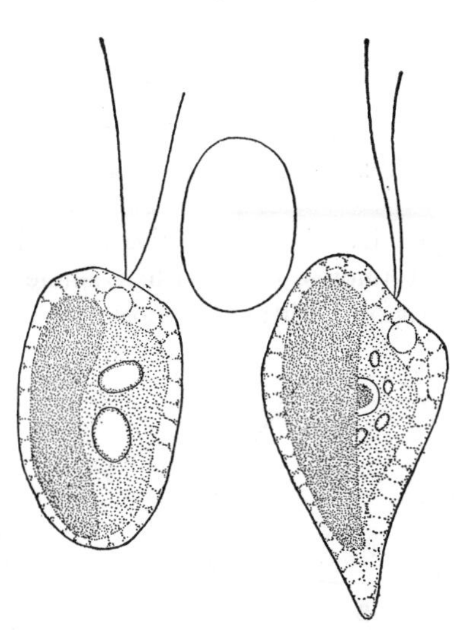

FIG. 550. *Rhodomonas lacustris* Pascher and Ruttner. *A-B*, front views. *C*, polar view. (*After Lackey*, 1941.)

4. **Cyanomonas** Oltmanns, 1904. This cryptomonad is readily distinguished from others by the presence of several blue-green chromatophores within each cell. The cells are compressed, with a longitudinal furrow, and with two flagella of unequal length at the anterior end. One species has been recorded[3] as producing immobile palmelloid colonies in which any cell may become motile and swim away.

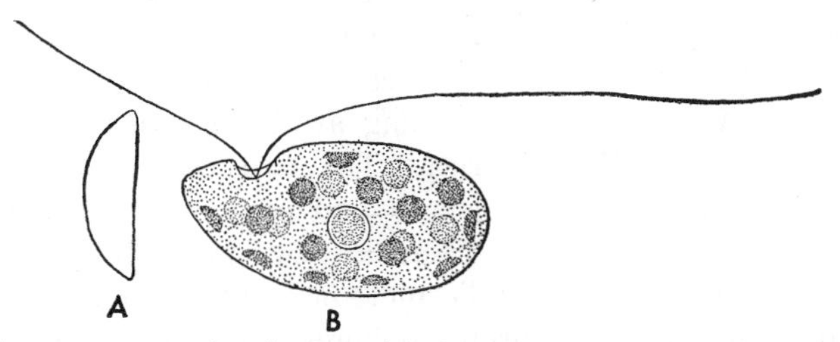

FIG. 551. *Cyanomonas coeruleus* Lackey. *A*, front view. *B*, polar view. (*After Lackey*, 1936.)

C. coeruleus Lackey (Fig. 551) has been found at two stations, one in Tennessee, the other in Alabama. For a description of it, see Lackey (1936).

[1] KARSTEN, 1898. [2] LACKEY, 1939, 1941.
[3] DAVIS, B. M., 1894*A*.

5. **Monomastix** Scherffel, 1912. *Monomastix* has elongate, cylindrical to somewhat pyriform, cells with a length threefold the breadth. The cells are naked and with a slightly plastic periplast. There is a single flagellum at the anterior end, and just beneath it is a single contractile

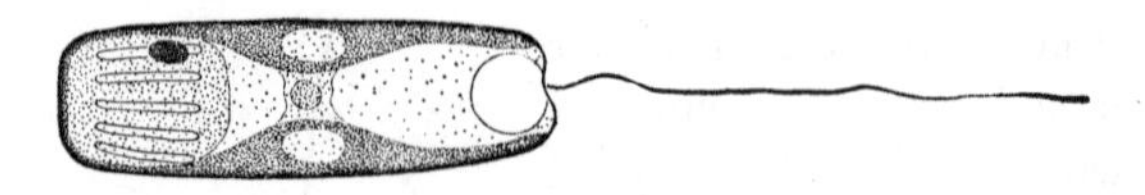

FIG. 552. *Monomastix opisthostigma* Scherffel. (*Drawn by R. H. Thompson.*) (× 1620.)

vacuole. A cell contains two lateral, parietal, laminate, grass-green chromatophores, each with a single pyrenoid. There is a single bright red eyespot. Within the posterior end of a cell and in the region between the two chromatophores are several rod-like cellular inclusions of a trichocyst-like nature.

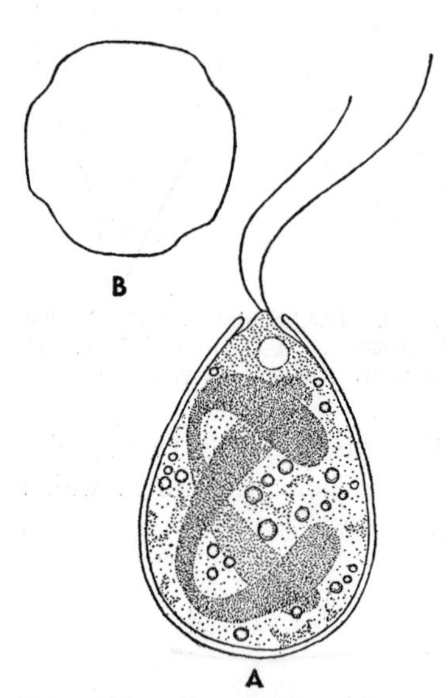

FIG. 553. *Cyanomastix Morgani* Lackey. *A*, front view. *B*, polar view. (*After Lackey*, 1936.)

Reproduction is by longitudinal division. There may also be a formation of globose cysts with stellately ridged walls.[1]

Prof. R. H. Thompson writes that he has found the only known species, *M. opisthostigma* Scherffel (Fig. 552), in Maryland. For a description of it, see Scherffel (1912).

6. **Cyanomastix** Lackey, 1936. *Cyanomastix* is the only cryptomonad with a lorica surrounding the protoplast of a cell. The protoplast completely fills a lorica and has a small papillate protuberance projecting slightly beyond the opening at the anterior end of the lorica. The two flagella are of unequal length and are not inserted in a gullet. There are two elongate band-like blue-green chromatophores, and they lie more or less twisted about each other at the center of a cell. The chromatophores contain pyrenoid-like bodies. Minute spherical bodies scattered through the cytoplasm are thought to be droplets of oil.

The single species, *C. Morgani* Lackey (Fig. 553), is known only from the type locality in Alabama. For a description of it, see Lackey (1936).

[1] SCHERFFEL, 1912.

Family 2. Cryptomonadaceae

Genera belonging to this family are unicellular free-swimming cryptomonads with the two flagella borne near the anterior end and inserted in an evident gullet.

The genera found in this country differ as follows:

1. Cells with chromatophores .... 1. **Cryptomonas**
1. Cells without chromatophores 2
   2. Cells not markedly flattened .... 2. **Chilomonas**
   2. Cells markedly flattened .... 3. **Cyathomonas**

1. **Cryptomonas** Ehrenberg, 1831. Of the motile cryptomonads with one or two large laminate chromatophores, *Cryptomonas* is the only genus with the two flagella inserted in an evident gullet at the anterior end. The cells are compressed, more or less elliptical in outline, and with the anterior end broadly rounded to truncate. The flagella are inserted in a gullet which may or may not be lined with granular trichoblasts. A single contractive vacuole lies adjacent to and empties into the gullet.[1] The color of chromatophores is usually yellowish to olive-green, but this may be red during winter months.[2] In some cases, there are numerous rod- or disk-shaped granules of starch scattered through the cytoplasm; in other cases, the starch granules are associated with pyrenoids. The nucleus is relatively large and lies in the lower half of a cell.

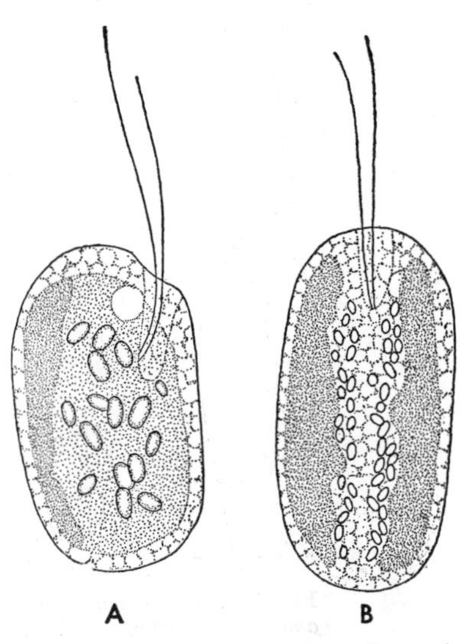

Fig. 554. *Cyrptomonas erosa* Ehr. *A*, front view. *B*, side view. (*After Lackey*, 1941.)

Cells about to divide become immobile and surrounded by mucilage. Cell division is longitudinal and in a plane that halves the gullet.[3]

Specimens of this genus have been reported from numerous localities, but the taxonomy of the genus is in so unsatisfactory a condition that most attempts to identify species are more or less in the nature of guesses. Identification has usually been based on the descriptions given by Pascher[4] and the specimens identified either as *C. erosa* Ehr. (Fig. 554) or as *C. ovata* Ehr.

[1] Lund, 1942. [2] Lackey, 1939. [3] Reichardt, 1927.
[4] Pascher, 1913.

2. **Chilomonas** Ehrenberg, 1831. In cell shape, structure of gullet, position of contractile vacuoles, and insertion of flagella *Chilomonas* resembles *Cryptomonas*. It is immediately distinguishable because of its lack of chromatophores. The nutrition is saprophytic, and numerous starch granules frequently accumulate in the cytoplasm. A conspicuous nucleus lies in the lower third of a cell.

Reproduction is by longitudinal division.

Chilomonas is generally found in waters rich in organic materials. The type species, *C. Paramaecium* Ehr. (Fig. 555), is widely distributed in the United States. For a description of it, see Pascher (1913).

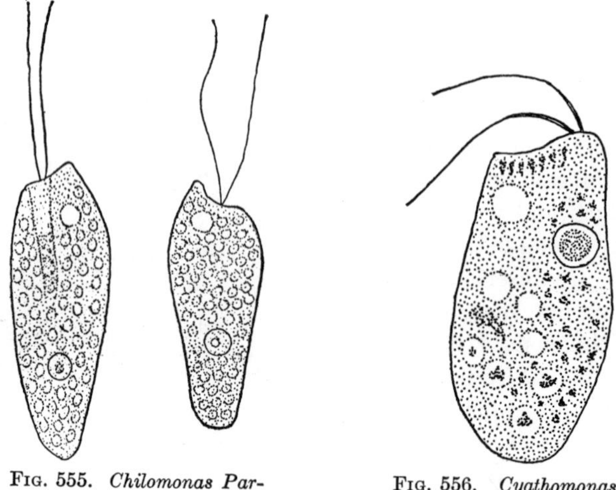

FIG. 555. *Chilomonas Paramaecium* Ehr. (*After Stein*, 1878.) (× 860.)

FIG. 556. *Cyathomonas truncata* From. (*After Bütschli*, 1883–1887.)

3. **Cyathomonas** Fromentel, 1874. *Cyathomonas* is an unpigmented cryptomonad with markedly truncate cells that are truncate at the anterior end. The two flagella are of unequal length and are inserted in a gullet[1] encircled by a transverse ring of trichoblast granules. A single contractile vacuole lies adjacent to the gullet, and the nucleus lies midway between the poles of a cell. Bacteria and other minute organisms are ingested through the gullet and come to lie in nutritive vacuoles irregularly distributed through the cell. The cells contain minute droplets of oil instead of the usual starch granules.

There is but one species, *C. truncata* From. (Fig. 556), and the only published records of its occurrence in this country[2] are from Ohio and Alabama. Dr. Lackey

[1] Úlehla, 1911. [2] Lackey, 1939, 1940.

writes that he has found it in Massachusetts, New Jersey, and Tennessee. For a description of *C. truncata*, see Pascher (1913).

FAMILY 3. NEPHROSELMIDIACEAE

Members of this family are unicellular free-swimming cryptomonads with two widely divergent laterally inserted flagella. The swimming flagellum, which projects forward, is shorter than the trailing flagellum. When swimming through the water, movement is in the direction of the major axis of a cell.

The two genera of the family differ as follows:

1. With a gullet and without an eyespot........................ 1. **Nephroselmis**
1. Without a gullet and with an eyespot.......................... 2. **Protochrysis**

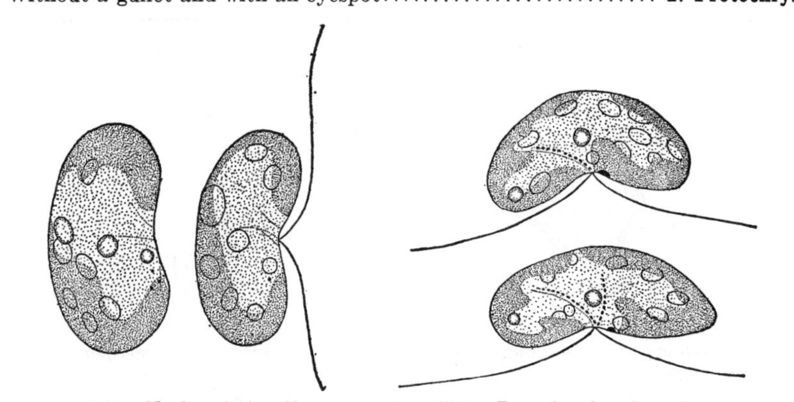

FIG. 557. *Nephroselmis olivacea* Stein. (*After Pascher*, 1912*B*.)

FIG. 558. *Protochrysis phaeophycearum* Pascher. (*After Pascher*, 1912*B*.)

1. **Nephroselmis** Stein, 1878. The cells of *Nephroselmis* are reniform and markedly flattened. The two flagella are inserted laterally and on the concave side of a cell. The swimming flagellum extends forward from the trailing flagellum backward from the point of insertion. As in other cryptomonads, there is a superficial furrow, but unlike most other cryptomonads it runs transversely across a cell from the base of the flagella. An evident gullet lies parallel to the furrow. Two contractile vacuoles lie near the base of the flagella. There are one or two brownish to blue-green, more or less cup-shaped, chromatophores. The cytoplasm internal to the chromatophores contains a pyrenoid surrounded by platelets of a starch-like nature. The cytoplasm may also contain small disk-shaped masses of starch-like compounds.[1] The nucleus lies toward the concave side of a cell.

N. olivacea Stein (Fig. 557), the only species, has been recorded from Iowa[2] and Ohio.[3] For a description of it, see Pascher (1913).

[1] PASCHER, 1912*B*. [2] PRESCOTT, 1927. [3] LACKEY, 1939.

2. **Protochrysis** Pascher, 1911. *Protochrysis* has reniform cells that are not markedly compressed. It resembles *Nephroselmis* in flagellation, transverse position of furrow, contractile vacuoles, pyrenoids, and color and shape of chromatophores. It differs in that it lacks a gullet and has an eyespot near the base of the flagella.[1]

At the time of reproduction, a cell becomes immobile and divides to form four or eight cells that remain united in a palmelloid colony for a time.[1]

P. phaeophycearum Pascher (Fig. 558) has been found[2] in Ohio and Alabama. The published record lists it as *P. viridis*, but Dr. Lackey writes that this is a clerical error. For a description of *P. phaeophycearum*, see Pascher (1913).

ORDER 2. CRYPTOCOCCALES

The Cryptococcales are immobile nonflagellate unicellular Cryptophyceae in which the protoplast is surrounded by a cellulose wall. Reproduction is by a formation of zoospores. There is but one family, the Cryptococcaceae, which contains a single genus, *Tetragonidium*.

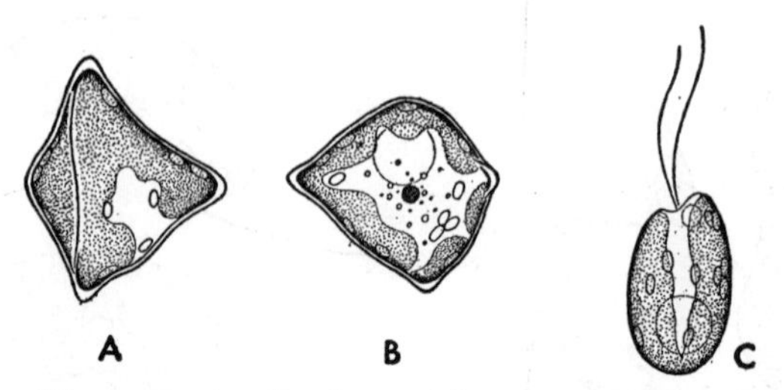

FIG. 559. *Tetragonidium verrucatum* Pascher. *A-B*, vegetative cells. *C*, zoospore. (*Drawn by R. H. Thompson.*) (× 810.)

1. **Tetragonidium** Pascher, 1914. This genus has solitary, free-floating, irregularly tetrahedral cells with a distinct wall slightly thickened at the angles. The wall is composed of cellulose. There is a single, expanded, parietal, sheet-like, brown chromatophore with lobed margins. The chromatophore contains a single pyrenoid. Granules of starch are present both around the pyrenoid and against the inner face of the chromatophore. There is a single large nucleus near one side of a cell.[3]

Reproduction is by a formation of biflagellate zoospores resembling in appearance the cells of *Cryptochrysis*.[3] After swarming for a time, the zoospores become immobile, lose their flagella, secrete a wall, and become tetrahedral vegetative cells.

Prof. R. H. Thompson writes that he has found the only known species, *T. verrucatum* Pascher (Fig. 559), in Maryland. For a description of it, see Pascher (1914).

[1] PASCHER, 1911, 1912*B*. [2] LACKEY, 1938. [3] PASCHER, 1914.

KEY TO THE FRESH-WATER ALGAE OF THE UNITED STATES

Most keys for the identification of fresh-water algae take into consideration the method of reproduction and the structure of reproductive bodies and organs. Since one often encounters algae in a sterile condition, the following key has been so constructed that vegetative and ecological characters are utilized as far as possible, and reproductive characters are included only where it is impossible to distinguish between genera except when they are in a fruiting condition.

1. Vegetative cells flagellated and, with a few exceptions, motile............ 423
1. Vegetative cells nonflagellated and immobile............................... 2
    2. Growing within or upon vertebrates.................................. 346
    2. Not growing within or upon vertebrates 3
3. Parasitic in leaves and stems of nonaquatic angiosperms................. 350
3. Free-living, epiphytic, or epizoic.. 4
    4. Protoplasts of cells unpigmented and colorless........................ 352
    4. Protoplasts pigmented and pigments usually localized in chromatophores... 5
5. Pigments not localized in chromatophores*............................... 353
5. Pigments localized in chromatophores.. 6
    6. Chromatophores grass-green or yellowish-green.......................... 7
    6. Chromatophores not grass-green or yellowish-green.................... 248
7. Chromatophores grass-green, cells containing starch........................ 8
7. Chromatophores yellowish-green, cells without starch.................... 213
    8. Cells uniseriately joined in simple or branched filaments................ 9
    8. Cells solitary or not uniseriately united in filaments.................... 69
9. Filaments unbranched.. 10
9. Filaments more or less branched... 44
    10. Cells with a more or less evident median constriction................. 11
    10. Cells without a median constriction....................................... 21
11. Length of cell several times the breadth............ **Pleurotaenium** (page 319)
11. Length of cell not over 3 to 4 times the breadth.......................... 12
    12. Apices of cells incised......................... **Micrasterias** (page 326)
    12. Apices of cells not incised.. 13
13. Cells connected by short apical processes.................................. 14
13. Cells not connected by short apical processes............................. 15
    14. Processes overlapping apices of adjoining cells.. **Onychonema** (page 330)
    14. Processes not overlapping apices of adjoining cells........................ **Sphaerozosma** (page 330)
15. End view of cells elliptical or circular. 16

* Old cells of certain Chlorophyceae seem to have the coloring matter distributed throughout the entire protoplast. These may be distinguished from Myxophyceae by the presence of starch.

15. End view of cells triangular or quadrangular 19
    16. End view elliptical 17
    16. End view circular 18
17. Cells strongly compressed .... **Spondylosium** (page 331)
17. Cells not strongly compressed .... **Desmidium** (page 334)
    18. Lateral walls longitudinally striated .... **Gymnozyga** (page 335)
    18. Lateral walls not longitudinally striated .... **Hyalotheca** (page 332)
19. Recently divided cells with apices infolded .... **Desmidium** (page 334)
19. Recently divided cells without infolded apices 20
    20. End view quadrangular .... **Phymatodocis** (page 333)
    20. End view triangular .... **Spondylosium** (page 331)
21. Length of cells 30 to 60 times breadth .... **Sphaeroplea** (page 184)
21. Length of cells less than 20 times breadth 22
    22. Chloroplasts parietal 23
    22. Chloroplasts axial 37
23. Chloroplasts spiral bands 24
23. Chloroplasts not spiral bands 25
    24. Each chloroplast usually making several spiral turns .... **Spirogyra** (page 299)
    24. Each chloroplast not making a complete spiral turn .... **Sirogonium** (page 302)
25. Filaments enclosed by a gelatinous sheath 26
25. Filaments not enclosed by a gelatinous sheath 29
    26. Poles of cells flattened .... **Ulothrix** (page 142)
    26. Poles of cells rounded 27
27. Sheaths narrow .... **Cylindrocapsa** (page 150)
27. Sheaths broad 28
    28. Cells cylindrical, often remote from one another .... **Geminella** (page 145)
    28. Cells spherical to subspherical, always touching one another .... **Radiofilum** (page 148)
29. Mature filaments with less than 20 cells 30
29. Mature filaments with more than 20 cells 31
    30. Filaments not pointed at ends .... **Stichococcus** (page 145)
    30. Filaments pointed at one or both ends .... **Raphidonema** (page 144)
31. Cells in pairs .... **Binuclearia** (page 147)
31. Cells not in pairs 32
    32. Walls of certain cells with transverse striae (apical caps) at one end .... **Oedogonium** (page 207)
    32. Walls without transverse striae 33
33. Cell wall composed of interlocked pieces that are H-shaped in optical section. **Microspora** (page 148)
33. Cell wall not composed of interlocked pieces 34
    34. Chloroplast with many pyrenoids .... **Rhizoclonium** (page 216)
    34. Chloroplast a transverse band with one or a few pyrenoids 35
35. Filaments without a basal holdfast .... **Hormidium** (page 146)
35. Filaments with a basal holdfast 36
    36. Free end of filament pointed .... **Uronema** (page 144)
    36. Free end of filament rounded .... **Ulothrix** (page 142)
37. With one axial band-shaped chloroplast 38

37. With stellate, disk-shaped, or cushion-shaped chloroplasts................ 40
    38. Chloroplast without pyrenoids................ **Mougeotiopsis** (page 295)
    38. Chloroplast with pyrenoids.. 39
39. Fertile cells filled with gelatinous material after zygote is formed............ **Debarya** (page 294)
39. Fertile cells not filled with gelatinous material after zygote is formed. ..... **Mougeotia** (page 292)
    40. With two disk-shaped chloroplasts.............. **Pleurodiscus** (page 299)
    40. With two (rarely several) stellate or cushion-shaped chloroplasts...... 41
41. Chloroplasts stellate.. 42
41. Chloroplasts cushion-shaped......................... **Zygogonium** (page 298)
    42. One chloroplast per cell........................ **Schizogonium** (page 195)
    42. Two or more chloroplasts per cell.................................. 43
43. Fertile cells filled with gelatinous material after zygote is formed............ **Zygnemopsis** (page 297)
43. Fertile cells not filled with gelatinous material after zygote is formed........ **Zygnema** (page 296)
    44. Filaments with setae or with unicellular spines....................... 45
    44. Filaments without setae or spines.................................... 50
45. Each seta with a basal sheath.................... **Coleochaete** (page 168)
45. With one-celled spines... 46
    46. Base of spines large and bulbous..................................... 47
    46. Base of spines not markedly bulbous.................................. 48
47. Chloroplast entire, with one or a few pyrenoids.......... **Fridaea** (page 178)
47. Chloroplast reticulate, with many pyrenoids....... **Bulbochaete** (page 209)
    48. Entire filament prostrate..................... **Aphanochaete** (page 163)
    48. Filament wholly or partly erect...................................... 49
49. Cells with more than one spine................... **Thamniochaete** (page 162)
49. Cells never with more than one spine.............. **Chaetonema** (page 165)
    50. Endophytic in walls of other algae............. **Entocladia** (page 160)
    50. Not endophytic in walls of other algae............................... 51
51. Perforating shells of mollusks, limestone rocks, or wood.. **Gomontia** (page 175)
51. Not growing as a perforating alga.. 52
    52. Regularly with akinetes alternating with vegetative cells............. 53
    52. Akinetes usually not present... 55
53. Thallus incrusted with lime...................... **Chlorotylium** (page 159)
53. Thallus not incrusted with lime.. 54
    54. Akinetes spherical........................... **Ctenocladus** (page 177)
    54. Akinetes cylindrical, barrel-shaped, or conical... **Pithophora** (page 218)
55. Branches closely appressed into a pseudoparenchymatous mass............ 56
55. Branches not forming a pseudoparenchymatous mass......................... 57
    56. Thallus one cell in thickness.................... **Protoderma** (page 161)
    56. Thallus more than one cell in thickness.......... **Gongrosira** (page 178)
57. Thallus with rhizoidal branches.. 58
57. Thallus without rhizoidal branches....................................... 60
    58. Branching verticillate, cells of macroscopic size..................... 59
    58. Branching not verticillate, cells of microscopic size.................. **Oedocladium** (page 210)
59. Branches of a verticil branched .... **Nitella** (page 345)

59. Branches of a verticil unbranched........................ **Tolypella** (page 346)
    60. Ends of branches acutely pointed... 61
    60. Ends of branches rounded... 65
61. With a broad-celled main axis and short lateral branches.................. 62
61. With all branches approximately the same breadth.......................... 63
    62. Main axis with alternate long and short cells.. **Draparnaldiopsis** (page 157)
    62. Main axis with all cells nearly the same length.. **Draparnaldia** (page 156)
63. Thallus of definite shape and macroscopic size...... **Chaetophora** (page 155)
63. Thallus of indefinite shape and microscopic size........................... 64
    64. Ends of branches attenuated into multicellular hairs........................ **Stigeoclonium** (page 153)
    64. Ends of branches not attenuated into multicellular hairs.................. **Leptosira** (page 174)
65. Filaments with two- to five-celled branches........ **Rhizoclonium** (page 216)
65. Filaments with many-celled branches.. 66
    66. Filaments with lemon-shaped cells.............. **Physolinum** (page 181)
    66. Filaments with cylindrical cells.. 67
67. Each cell with many pyrenoids...................... **Cladophora** (page 213)
67. Each cell with less than five pyrenoids.................................... 68
    68. Branching tending to be unilateral........... **Microthamnion** (page 159)
    68. Branching not tending to be unilateral.......... **Gongrosira** (page 178)
69. Thallus multicellular... 70
69. Thallus unicellular.. 143
    70. Thallus parenchymatous... 71
    70. Thallus not parenchymatous... 79
71. Thallus wholly prostrate and of microscopic size........................... 72
71. Thallus not prostrate and or micro- or macroscopic size.................... 75
    72. Thallus one cell in thickness.. 73
    72. Thallus more than one cell in thickness........ **Pseudoulvella** (page 162)
73. Cells with setae.. 74
73. Cells without setae................................. **Protoderma** (page 161)
    74. Base of each seta with a sheath.................. **Coleochaete** (page 168)
    74. Bases of setae without sheaths.................. **Chaetopeltis** (page 126)
75. Thallus a branched or unbranched cylinder.................................. 76
75. Thallus a monostromatic sheet.. 78
    76. Thallus an erect unbranched cylinder.. 77
    76. Thallus an erect cylinder with verticillate branching.... **Chara** (page 346)
77. Cylinder hollow, wall one cell in thickness.......... **Enteromorpha** (page 188)
77. Cylinder solid, several cells in diameter in lower portion.................. **Schizomeris** (page 192)
    78. With parietal laminate chloroplasts............ **Monostroma** (page 190)
    78. With axial stellate choroplasts...................... **Prasiola** (page 195)
79. Colonies simple or branched gelatinous tubes............................... 80
79. Colonies not gelatinous tubes.. 86
    80. Cells uniseriate or terminal at tips of tubes.............................. 81
    80. Cells irregularly arranged... 85
81. Gelatinous tube with many transverse lamellae...... **Hormotila** (page 118)
81. Gelatinous tube not transversely lamellated................................ 82
    82. Cells as broad as branches of gelatinous tubes............................ 83
    82. Cells broader than branches of gelatinous tubes............................ 84
83. Cells transversely constricted...................... **Oöcardium** (page 325)

83. Cells not transversely constricted.................. **Prasinocladus** (page 130)
    84. Cells containing starch............................ **Chlorangium** (page 129)
    84. Cells containing paramylum.................... **Colacium** (page 368)
85. Cells tending to lie in fours.......................... **Tetraspora** (page 122)
85. Cells not tending to lie in fours.................. **Palmodictyon** (page 118)
    86. Colonies always sessile (if aquatic).. 87
    86. Colonies always free-floating (if aquatic)............................ 92
87. Cells remote from one another within a common wide gelatinous envelope.....
Apiocystis (page 124)
87. Cells close together, with or without a common gelatinous envelope........ 88
    88. Cells without setae.............................. **Malleochloris** (page 128)
    88. Cells with setae.. 89
89. Setae unbranched.. 90
89. Setae branched.................................. **Dicranochaete** (page 171)
    90. Setae without sheaths at base................ **Oligochaetophora** (page 172)
    90. Base of each seta with a sheath... 91
91. Each cell with a single seta....................**Chaetosphaeridium** (page 170)
91. Certain cells with more than one seta.............. **Conochaete** (page 172)
    92. Colony with a wide gelatinous sheath.................................. 93
    92. Colony without or with narrow gelatinous sheath..................... 113
93. Cells spherical to subspherical.. 94
93. Cells not spherical to subspherical.. 102
    94. Gelatinous envelope containing remains of old parent-cell walls.. 95
    94. Gelatinous envelope not containing remains of old parent-cell walls.... 97
95. Remains of old walls forming a dichotomously branched system..............
Dictyosphaerium (page 236)
95. Remains of old walls not forming a branching system..................... 96
    96. Remains of old walls at center of colony.......... **Radiococcus** (page 253)
    96. Remains of walls scattered throughout colony.. **Schizochlamys** (page 125)
97. Envelope concentrically stratified around each cell.. **Gloeocystis** (page 116)
97. Envelope not stratified around each cell.................................. 98
    98. Chloroplast axial and stellate.................... **Asterococcus** (page 120)
    98. Chloroplast not axial and stellate...................................... 99
99. Colony spherical and of microscopic size.................................. 100
99. Colony amorphous and of indefinite extent................................. 101
    100. Chloroplasts of mature cells cup-shaped...... **Sphaerocystis** (page 116)
    100. Chloroplasts of mature cells disk-shaped, several in a cell...............
Planktosphaeria (page 255)
101. Cells in groups of four, pseudocilia often evident...... **Tetraspora** (page 122)
101. Cells not in fours, always without pseudocilia........ **Palmella** (page 115)
    102. Cells pyriform, radiately disposed.......... **Askenasyella** (page 119)
    102. Cells not pyriform.. 103
103. Cell wall smooth.. 104
103. Cell wall with spines....................................**Bohlinia** (page 263)
    104. Cells constricted in the middle.................**Cosmocladium** (page 324)
    104. Cells not constricted in the middle.................................... 105
105. Cells longer than broad and straight. 106
105. Cells longer than broad and curved.. 111
    106. With one or both poles pointed.. 107
    106. With both poles broadly rounded.. 108
107. One pole pointed, the other rounded............. **Elaktothrix** (page 134)

107. Both poles pointed .... **Quadrigula** (page 268)
    108. Long axes of cells radiating from a common center ....
    Gloeoactinium (page 268)
    108. Long axes not radiating from a common center 109
109. All cells with their long axes parallel .... **Quadrigula** (page 268)
109. Cells not having their long axes parallel 110
    110. Each cell surrounded by a distinct envelope... **Dactylothece** (page 133)
    110. Cells without individual envelopes .... **Coccomyxa** (page 132)
111. Poles of cells connected to one another by tough gelatinous strands ....
    Tomaculum (page 275)
111. Poles of cells not connected by strands 112
    112. Cells lunate, sheaths around groups of four or eight cells indistinct ......
    Kirchneriella (page 267)
    112. Cells not strongly curved, sheaths around groups of four or eight cells distinct .... **Nephrocytium** (page 261)
113. Cells separated from one another by a dark-colored substance ....
    Gloeotaenium (page 260)
113. Cells not separated by a dark-colored substance 114
    114. Cells spherical 115
    114. Cells not spherical 124
115. Cells with long setae 116
115. Cells without setae 117
    116. Cells generally in same plane, each with several setae ....
    Micractinium (page 234)
    116. Cells pyramidately arranged, each with one seta... **Errerella** (page 235)
117. With remains of old walls at center of colony 118
117. Without remains of old walls at center of colony 119
    118. Old walls forming a branching system...... **Dictyosphaerium** (page 236)
    118. Old walls not forming a branching system........ **Westella** (page 253)
119. Colony an irregular packet 120
119. Colony of definite shape 121
    120. Reproducing by means of zoospores .... **Chlorosarcina** (page 135)
    120. Not reproducing by means of zoospores .... **Protococcus** (page 166)
121. Colony a hollow sphere .... **Coelastrum** (page 248)
121. Colony cubical or plate-like 122
    122. Colony cubical, cells connected by gelatinous strands ....
    Pectodictyon (page 277).
    122. Colony plate-like 123
123. Cells connected by strands .... **Coronastrum** (page 277)
123. Cells not connected by strands .... **Dispora** (page 134)
    124. All cells in the same plane 125
    124. All cells not in the same plane 129
125. Cells elongate and with long axes parallel .... **Scenedesmus** (page 272)
125. Cells not elongate 126
    126. Cells quadrately arranged 127
    126. Cells not quadrately arranged 128
127. Cells without spines .... **Crucigenia** (page 273)
127. Free face of cells with spines .... **Tetrastrum** (page 274)
    128. Colony always two-celled and with cells angular ....
    Euastropsis (page 245)
    128. Colony with 2 to 128 angular cells, marginal and interior cells of different shape .... **Pediastrum** (page 243)

129. All or certain cells of a colony curved 130
129. None of the cells curved.. 133
130. All cells in a colony curved 131
130. Some cells curved, others not curved...................................... 132
131. End of cells pointed.................................. **Selenastrum** (page 266)
131. Ends of cells broadly rounded......................... **Nephrocytium** (page 261)
132. Center of colony with branching remains of old cell walls................ **Dimorphococcus** (page 237)
132. Center of colony without remains of old cell walls......................... **Tetrallantos** (page 274)
133. Colony a reticulate sac of cylindrical cells........... **Hydrodictyon** (page 246)
133. Colony not a reticulate sac.. 134
134. Cells elongate and in radiate clusters.. 135
134. Cells, if elongate, not in radiate clusters 136
135. Clusters of cells borne on stalks.................... **Actidesmium** (page 240)
135. Clusters of cells not borne on stalks................ **Actinastrum** (page 275)
136. Colony a hollow sphere.. 137
136. Colony not a hollow sphere... 138
137. Free face of cells with spines........................... **Sorastrum** (page 245)
137. Free face of cells without spines...................... **Coelastrum** (page 248)
138. Cell wall covered with setae......................... **Franceia** (page 263)
138. Cell wall without setae.. 139
139. Colony surrounded by greatly expanded old parent-cell wall.................. **Oöcystis** (page 259)
139. Colony not surrounded by old parent-cell wall................................. 140
140. Cells fusiform.. 141
140. Cells polygonal by mutual compression.. 142
141. Cells joined end to end in a branching series........ **Dactylococcus** (page 265)
141. Colony four-celled, cells in two parallel series....... **Tetradesmus** (page 273)
142. Reproducing by means of zoospores........... **Chlorosarcina** (page 135)
142. Not reproducing by means of zoospores........ **Protococcus** (page 166)
143. With a median constriction dividing the cell into two symmetrical halves. 144
143. Without a median constriction.. 154
144. With a length several times the breadth.. 145
144. Length not over three times the breadth.. 148
145. Apices of cells incised or depressed... 146
145. Apices of cells not incised or depressed... 147
146. With transverse rings of spines or verrucae...... **Triploceras** (page 321)
146. Without spines or verrucae......................... **Tetmemorus** (page 321)
147. Cell wall vertically plicate next to median constriction.. **Docidium** (page 320)
147. Cell wall not plicate next to median constriction... **Pleurotaenium** (page 319)
148. Cells compressed.. 149
148. Cells not compressed; end view 3 to 12 radiate.. **Staurastrum** (page 328)
149. Apices of cells incised.. 150
149. Apices of cells not incised... 151
150. Apical and lateral incisions shallow............. **Euastrum** (page 322)
150. Apical and lateral incisions deep.............. **Micrasterias** (page 326)
151. Apices of cells continued in two divergent processes.. **Staurastrum** (page 328)
151. Apices of cells not continued in processes...................................... 152
152. Cell wall without long spines.................... **Cosmarium** (page 323)
152. Cell wall with long spines.. 153
153. Central portion of cell wall with a protuberance..... **Xanthidium** (page 327)

153. Central portion of cell wall without a protuberance.. **Arthrodesmus** (page 329)
    154. Cells siphonaceous.......... 155
    154. Cells not siphonaceous.......... 157
155. Free-living.......... 156
155. Endophytic in *Sphagnum*.......... **Phyllobium** (page 231)
    156. With constrictions at regular intervals..... **Dichotomosiphon** (page 281)
    156. Not constricted, with one end green and inflated.......... **Protosiphon** (page 241)
157. Within tissues of aquatic animals or plants.......... 158
157. Not within aquatic animals or plants.......... 159
    158. Within invertebrates.......... **Chlorella** (page 251)
    158. Within leaves of plants.......... **Chlorochytrium** (page 227)
159. Cells sessile.......... 160
159. Cells not sessile.......... 166
    160. Cells with setae.......... 161
    160. Cells without setae.......... 164
161. Each cell with one seta.......... 162
161. Certain cells with more than one seta.......... 163
    162. Seta branched.......... **Dicranochaete** (page 171)
    162. Seta unbranched.......... **Chaetosphaeridium** (page 170)
163. Base of each seta with a sheath.......... **Conochaete** (page 172)
163. Setae without basal sheaths.......... **Oligochaetophora** (page 172)
    164. Affixed by a stalk as broad as cell.......... **Malleochloris** (page 128)
    164. Affixed by a stalk narrower than the cell.......... 165
165. Cells obovoid, length of stalk 2 to 4 times length of cell.......... **Stylosphaeridium** (page 128)
165. Cells elongate, length of stalk less than that of cell.... **Characium** (page 238)
    166. Cells angular.......... 167
    166. Cells spherical, ovoid, lunate, or elongate.......... 172
167. With long spines or setae at angles.......... 168
167. Without spines or setae at angles.......... 171
    168. Each angle with several setae.......... **Polyedriopsis** (page 271)
    168. Each angle with a single spine.......... 169
169. Spines hyaline, acutely pointed.......... 170
169. Spines brownish, with blunt apices.......... **Pachycladon** (page 258)
    170. Spines narrow.......... **Polyedriopsis** (page 271)
    170. Spines relatively broad.......... **Treubaria** (page 257)
171. Cells with a central portion distinct from projecting angles .... **Tetraëdron** (page 269)
171. Cells without a central portion distinct from projecting angles.......... **Cerasterias** (page 270)
    172. Cells spherical.......... 173
    172. Cells not spherical.......... 188
173. Cell wall smooth.......... 174
173. Cell wall ornamented or with spines.......... 184
    174. Cells with a broad gelatinous sheath.......... 175
    174. Cells without a gelatinous sheath.......... 177
175. Sheath unstratified, cell with several chloroplasts.. **Planktosphaeria** (page 255)
175. Sheath stratified.......... 176
    176. Chloroplast central and stellate.......... **Asterococcus** (page 120)
    176. Chloroplast parietal and cup-shaped.......... **Gloeocystis** (page 116)
177. With a central stellate chloroplast.......... 178

177. With parietal chloroplast or chloroplasts 179
  178. Cell wall of uniform thickness .... **Trebouxia** (page 224)
  178. Cell wall unevenly thickened .... **Myrmecia** (page 225)
179. Cells varying greatly in size .... **Chlorococcum** (page 222)
179. Cells not conspicuously different in size 180
  180. Cells over 50 μ in diameter, with many disk-shaped chloroplasts .... **Eremosphaera** (page 255)
  180. Cells less than 25 μ in diameter 181
181. Solitary cells intermingled with cells united in packets 182
181. Cells always solitary 183
  182. Reproducing by means of zoospores .... **Chlorosarcina** (page 135)
  182. Never reproducing by means of zoospores .... **Protococcus** (page 166)
183. With one cup-shaped chloroplast .... **Chlorella** (page 251)
183. With more than one chloroplast .... **Palmellococcus** (page 251)
  184. Wall with spines longer than diameter of cell 185
  184. Wall alveolar, reticulate, areolate, or with very short spines 187
185. Spines broad, gradually tapering to a sharp point .. **Echinosphaerella** (page 257)
185. Spines narrow, not conspicuously tapered 186
  186. Spines thickened in lower third .... **Acanthosphaera** (page 233)
  186. Spines not thickened in lower third .... **Golenkinia** (page 232)
187. Wall alveolar .... **Keriochlamys** (page 252)
187. Wall reticulate, areolate, or with very short spines .... **Trochiscia** (page 254)
  188. Cells more or less ellipsoidal 189
  188. Cells elongate, straight, or lunate 199
189. Wall without setae or spines 190
189. Wall with setae or spines 197
  190. Wall with spiral longitudinal ridges .... **Scotiella** (page 260)
  190. Wall without spiral longitudinal ridges 191
191. Wall heavy and unevenly thickened 192
191. Wall relatively thin 193
  192. With numerous parietal conical chloroplasts .. **Excentrosphaera** (page 256)
  192. With an axial chloroplast with many palisade-like processes .... **Kentrosphaera** (page 229)
193. Two stellate chloroplasts within a cell .... **Cylindrocystis** (page 307)
193. Chloroplasts not stellate 194
  194. Chloroplasts axial 195
  194. Chloroplasts parietal 196
195. With a single plate-like chloroplast extending the length of a cell .... **Mesotaenium** (page 305)
195. With two chloroplasts, both with longitudinal ribs .... **Penium** (page 318)
  196. Cells reproducing by transverse division .... **Nannochloris** (page 133)
  196. Cells reproducing by means of autospores .... **Oöcystis** (page 259)
197. Setae covering entire wall 198
197. Setae only at poles, or at poles and equator .... **Chodatella** (page 262)
  198. Autospores liberated by swelling of parent-cell wall .. **Franceia** (page 263)
  198. Autospores liberated by lateral rupture of parent-cell wall .... **Bohlinia** (page 263)
199. Ends of cells pointed 200
199. Ends of cells broadly rounded or truncate 209
  200. Poles of cells with spines 201
  200. Poles of cells without spines 203
201. Spines delicate, nearly as long as cell .... **Schroederia** (page 239)

201. Spines stout, much shorter than cell.. 202
    202. With one chloroplast completely filling cell.... **Closteridium** (page 266)
    202. With two axial chloroplasts................ **Spinoclosterium** (page 318)
203. Chloroplast or chloroplasts axial.. 204
203. Chloroplast or chloroplasts parietal.. 205
    204. With one chloroplast................................ **Roya** (page 308)
    204. With two chloroplasts........................ **Closterium** (page 316)
205. Chloroplast with an axial row of 8 to 12 pyrenoids... **Closteriopsis** (page 265)
205. Chloroplast rarely with more than one pyrenoid........................ 206
    206. Cell wall with longitudinal ribs.. 207
    206. Cell wall smooth.. 208
207. Wall with an equatorial transverse rib............ **Desmatractum** (page 225)
207. Wall without a transverse rib........................ **Scotiella** (page 260)
    208. Cells dividing transversely.................... **Ourococcus** (page 132)
    208. Cells not dividing transversely............. **Ankistrodesmus** (page 264)
209. Ends of cells truncate............................ **Gonatozygon** (page 306)
209. Ends of cells broadly rounded.. 210
    210. Chloroplasts axial.. 211
    210. Chloroplasts parietal spiral bands............. **Spirotaenia** (page 309)
211. With one chloroplast.. **Roya** (page 308)
211. With two chloroplasts.. 212
    212. Cells straight.. **Netrium** (page 308)
    212. Cells curved.................................... **Closterium** (page 316)
213. Cells uniseriately joined in simple or branched filaments................ 214
213. Cells solitary or not united in filaments.. 216
    214. Filaments unbranched.. 215
    214. Filaments branched............................ **Monocilia** (page 399)
215. Cell wall thick, H-pieces clearly evident.............. **Tribonema** (page 396)
215. Cell wall thin, H-pieces not clearly evident......... **Bumilleria** (page 398)
    216. Cells united in colonies.. 217
    216. Cells solitary.. 224
217. Colonies sessile or epiphytic.. 218
217. Colonies free-floating.. 221
    218. Colonies dendroid.. 219
    218. Colonies not dendroid.......................... **Gloeochloris** (page 377)
219. Without gelatinous stalks........................ **Ophiocytium** (page 395)
219. Cells at tips of branched gelatinous stalks.. 220
    220. Cells spherical.............................. **Mischococcus** (page 388)
    220. Cells obpyriform.......................... **Malleodendron** (page 378)
221. Colonial matrix tough and its surface wrinkled..... **Botryococcus** (page 404)
221. Colonial matrix watery and its surface smooth.. 222
    222. Matrix stratified.......................... **Chlorobotrys** (page 387)
    222. Matrix unstratified.. 223
223. Cells dividing vegetatively...................... **Gloeochloris** (page 377)
223. Cells not dividing vegetatively................... **Gloeobotrys** (page 387)
    224. Cells, if aquatic, free-floating.. 225
    224. Cells sessile and epiphytic.. 241
225. Cells siphonaceous.. 226
225. Cells not siphonaceous.. 227
    226. Tubular and of uniform diameter............... **Vaucheria** (page 402)

226. Partly globose and with colorless rhizoidal branches .... **Botrydium** (page 400)
227. Cells tetrahedral or triangular 240
227. Cells not tetrahedral or triangular 228
228. Cells hemispherical .... **Chlorogibba** (page 385)
228. Cells not hemispherical 229
229. Cells spherical to ovoid 230
229. Cells cylindrical or spindle-shaped 233
230. Cell wall smooth 231
230. Cell wall sculptured .... **Arachnochloris** (page 383)
231. Mature cells ovoid .... **Leuvenia** (page 381)
231. Mature cells spherical to subspherical 232
232. Cells varying greatly in size .... **Botrydiopsis** (page 380)
232. Cells not varying greatly in size .... **Diachros** (page 380)
233. Cells cylindrical 234
233. Cells spindle-shaped 239
234. Cell wall smooth 235
234. Cell wall ornamented 238
235. Length not over double the breadth .... **Monallantus** (page 382)
235. Length three or more times the breadth 236
236. Both poles with a spine longer than cell .... **Centritractus** (page 393)
236. Spines, if present, shorter than cell 237
237. Halves of cell wall similar in structure .... **Bumilleriopsis** (page 394)
237. Halves of cell wall unlike in structure .... **Ophiocytium** (page 395)
238. Wall with small areolae .... **Trachychloron** (page 384)
238. Wall with large areolae .... **Chlorallanthus** (page 384)
239. Poles of cells not thickened and without spines .... **Chlorocloster** (page 383)
239. Poles of cells thickened or with delicate spines ..... **Pleurogaster** (page 383)
240. Cells tetrahedral .... **Tetraedriella** (page 385)
240. Cells compressed and triangular .... **Goniochloris** (page 386)
241. Cells with evident stipes 244
241. Cells without evident stipes 242
242. Cells cylindrical .... **Characiopsis** (page 390)
242. Cells not cylindrical 243
243. Mature cells biscuit-shaped .... **Perone** (page 382)
243. Mature cells spherical .... **Lutherella** (page 392)
244. Protoplast completely surrounded by a wall 245
244. Upper end of protoplast naked and with a hair-like pseudopodium .... **Stipitococcus** (page 376)
245. Cells triangular .... **Dioxys** (page 390)
245. Cells not triangular 246
246. Stipe longer than cell .... **Peroniella** (page 391)
246. Stipe shorter than cell 247
247. Cell wall homogeneous .... **Characiopsis** (page 390)
247. Cell wall with two overlapping halves .... **Chlorothecium** (page 395)
248. Chromatophores some shade of brown 249
248. Chromatophores blue green, olive, pink, red, or purple 333
249. Cells with projecting delicate or stout pseudopodia 250
249. Cells without projecting pseudopodia 257
250. Cells solitary or in temporary colonies 252
250. Cells permanently united in colonies 251

251. Cells naked .... **Chrysidiastrum** (page 428)
251. Each cell surrounded by a lorica .... **Heliapsis** (page 428)
    252. Cells naked 253
    252. Cell surrounded by a lorica 254
253. With short pseudopodia .... **Chrysamoeba** (page 427)
253. With long pseudopodia .... **Rhizochrysis** (page 422)
    254. Cells free-floating, with stout spines .... **Diceras** (page 432)
    254. Cells sessile and epiphytic 255
255. Base of lorica extended into two long prongs encircling host alga .... **Chrysopyxis** (page 430)
255. Base of lorica without prong-like extensions 256
    256. Base of lorica flattened .... **Lagynion** (page 430)
    256. Base of lorica rounded .... **Kybotion** (page 431)
257. Cell wall smooth 258
257. Cell wall ornamented with punctae or striae 276
    258. Cells united in colonies 259
    258. Cells solitary 269
259. Cells uniseriately united in branched filaments .... **Phaeothamnion** (page 438)
259. Cells not united in filaments 260
    260. Colonies with setae 261
    260. Colonies without setae 262
261. Colonies sessile .... **Naegeliella** (page 435)
261. Colonies free-floating .... **Chrysostephanosphaeria** (page 436)
    262. Colony without a gelatinous sheath 263
    262. Colony with a gelatinous sheath 265
263. Colony with less than a half-dozen cells .... **Epichrysis** (page 440)
263. Colony with a dozen to hundreds of cells 264
    264. Colony crustose, monostromatic .... **Phaeoplaca** (page 434)
    264. Colony crustose, several cells in thickness .... **Heribaudiella** (page 512)
265. Gelatinous envelope stratified 266
265. Gelatinous envelope homogeneous 267
    266. Stratification concentric .... **Gloeodinium** (page 532)
    266. Stratification asymmetrical .... **Urococcus** (page 532)
267. Colony globose .... **Chrysocapsa** (page 433)
267. Colony not globose 268
    268. Colony a brush-like tuft .... **Hydrurus** (page 437)
    268. Colony not a brush-like tuft .... **Phaeosphaera** (page 434)
269. Cells free-floating 270
269. Cells epiphytic 273
    270. Cells tetrahedral .... **Tetragonidium** (page 636)
    270. Cells not tetrahedral 271
271. Cell with an asymmetrically stratified gelatinous envelope .... **Urococcus** (page 532)
271. Cell without a gelatinous envelope 272
    272. Cells spherical .... **Hypnodinium** (page 535)
    272. Cells lunate .... **Cystodinium** (page 534)
273. Cells globose and without a stipe .... **Epichrysis** (page 440)
273. Cells stipitate 274
    274. Cells without spines .... **Stylodinium** (page 535)
    274. Cells with spines 275
275. Cells compressed .... **Raciborskia** (page 537)

275. Cells not compressed .... **Tetradinium** (page 536)
  276. Cells solitary 295
  276. Cells not solitary 277
277. Cells sessile and borne at ends of simple or branched gelatinous stalks 278
277. Cells free-floating, or if sessile not borne at ends of gelatinous stalks 283
  278. One valve with a pseudoraphe, the other with a raphe 279
  278. Both valves with a raphe 280
279. Cuneate in girdle view .... **Rhoicosphenia** (page 485)
279. Not cuneate in girdle view .... **Achnanthes** (page 484)
  280. Ornamentation of valve longitudinally symmetrical 281
  280. Ornamentation of valve longitudinally assymmetrical. **Cymbella** (page 500)
281. Ornamentation of valve transversely symmetrical... **Brebissonia** (page 494)
281. Ornamentation of valve transversely asymmetrical 282
  282. With longitudinal lines adjoining valve margin... **Gomphoneis** (page 499)
  282. Without longitudinal lines adjoining valve margin ....
  Gomphonema (page 498)
283. Joined end to end in stellate clusters 284
283. Not joined end to end in stellate clusters 285
  284. Median portion of valve inflated .... **Tabellaria** (page 474)
  284. Median portion of valve not inflated .... **Asterionella** (page 480)
285. Joined end to end in zigzag series 286
285. Not joined end to end in zigzag series 290
  286. Ornamentation of valve radially symmetrical..... **Biddulphia** (page 470)
  286. Ornamentation of valve bilaterally symmetrical 287
287. With longitudinal internal septa 288
287. With transverse internal septa .... **Diatoma** (page 477)
  288. Septa perforate at poles and at center .... **Diatomella** (page 475)
  288. Septa with one perforation 289
289. Septa straight .... **Tabellaria** (page 474)
289. Septa curved .... **Tetracyclus** (page 475)
  290. Cells joined pole to pole in a straight filament..... **Melosira** (page 462)
  290. Cells joined valve to valve 291
291. Poles continued in long horns .... **Chaetoceros** (page 468)
291. Poles not continued in long horns 292
  292. Ornamentation of valve radially symmetrical..... **Biddulphia** (page 470)
  292. Ornamentation of valve bilaterally symmetrical 293
293. Valves longitudinally curved or arcuate .... **Eunotia** (page 482)
293. Valves not longitudinally curved or arcuate 294
  294. Wedge-shaped in girdle view .... **Meridion** (page 476)
  294. Not wedge-shaped in girdle view .... **Fragilaria** (page 479)
295. Sessile and with valve face against the substratum 296
295. Free-floating, or if sessile not with valve face against the substratum..... 297
  296. With transverse internal septa .... **Epithemia** (page 502)
  296. Without internal septa .... **Cocconeis** (page 486)
297. Wall with 20 or more intercalary bands 298
297. Wall with 2 to a few intercalary bands 299
  298. Each valve with a single long spine.......... **Rhizosolenia** (page 466)
  298. Each valve with two or three long spines .... **Attheya** (page 469)
299. Ornamentation of valve radial about a central point 300
299. Ornamentation of valve bilaterally disposed with respect to an axial or excentric line 304

300. Sides of valves undulate .... **Terpsinoë** (page 471)
300. Sides of valves not undulate 301
301. Valves with mamillate protuberances .... **Actinocyclus** (page 465)
301. Valves without mamillate protuberances 302
302. Ornamentation of valve in two concentric parts of unlike pattern .... **Cyclotella** (page 462)
302. Ornamentation of valve not in two concentric parts 303
303. Valve with radiating unornamented sectors ...... **Stephanodiscus** (page 463)
303. Entire surface of valve ornamented .... **Coscinodiscus** (page 464)
304. Both valves with a pseudoraphe 305
304. Both valves with a raphe 311
305. Valves longitudinally and transversely symmetrical 306
305. Valves not symmetrical in both axes 308
306. Frustules with internal septa 307
306. Frustules without internal septa .... **Synedra** (page 479)
307. Girdle view with parallel sides .... **Diatoma** (page 477)
307. Girdle view cuneate .... **Opephora** (page 478)
308. Valves alike at both poles 309
308. Valves inflated at one pole .... **Actinella** (page 483)
309. Valves dentate on both margins .... **Amphicampa** (page 482)
309. Valves with concave margin smoother than convex margin 310
310. Concave margin tumid at center .... **Ceratoneis** (page 483)
310. Concave margin not tumid at center .... **Eunotia** (page 482)
311. Raphe median or approximately median 312
311. Raphe marginal and in a keel 328
312. Valves symmetrical in both axes 313
312. Valves not symmetrical in both axes 326
313. Frustules without longitudinal septa 314
313. Frustules with loculiferous longitudinal septa .... **Mastogloia** (page 497)
314. Valve without a keel 315
314. Valve with a sigmoid sagittal keel .... **Amphiprora** (page 496)
315. Ornamentation of valve interrupted by longitudinal lines or by blank spaces 316
315. Ornamentation of valve not longitudinally interrupted 320
316. Central nodule not connected with longitudinal lines 317
316. Central nodule connected with longitudinal lines 319
317. Longitudinal lines parallel to sides of valve 318
317. Valve with longitudinal zigzag blank spaces .... **Anomoeoneis** (page 491)
318. With transverse striae .... **Caloneis** (page 490)
318. With transverse rows of punctae .... **Neidium** (page 490)
319. Raphe straight .... **Diploneis** (page 491)
319. Raphe sigmoid .... **Scoliopleura** (page 495)
320. Central or polar nodules elongate 321
320. Central or polar nodules not elongate 322
321. Length of central nodule at least half that of valve.. **Amphipleura** (page 492)
321. Length of central nodule less than half that of valve.. **Frustulia** (page 493)
322. Raphe sigmoid 323
322. Raphe straight 324
323. Punctae in oblique and transverse rows .... **Pleurosigma** (page 495)
323. Punctae in transverse and longitudinal rows .... **Gyrosigma** (page 494)
324. Central nodule laterally expanded into a stauros.. **Stauroneis** (page 492)
324. Central nodule not expanded into a stauros 325

325. Valve with smooth transverse costae............... **Pinnularia** (page 489)
325. Valve with transverse striae or rows of punctae...... **Navicula** (page 488)
 326. Valves with transverse costae................. **Rhopalodia** (page 503)
 326. Valves without transverse costae.................................. 327
327. Valves flat.. **Cymbella** (page 500)
327. Valves markedly convex................................ **Amphora** (page 501)
 328. With a single excentric keel next to one lateral margin of valve...... 329
 328. With a keel next to both margins of valve.......................... 331
329. Frustules without transverse septa..................................... 330
329. Frustules with transverse septa..................... **Denticula** (page 506)
 330. Valves with raphes diagonally opposite............. **Nitzschia** (page 505)
 330. Valves with raphes opposite each other........... **Hantzschia** (page 506)
331. Face of valve transversely undulate................ **Cymatopleura** (page 507)
331. Face of valve not transversely undulate................................ 332
 332. Valve face flat or spirally twisted................ **Surirella** (page 508)
 332. Valve face bent and saddle-shaped........... **Campylodiscus** (page 509)
333. Chromatophores, or what seem to be chromatophores, a bright blue-green... 334
333. Chromatophores olive-green, dark green, olive, reddish or purplish....... 336
 334. Chromatophore stellate... 335
 334. Chromatophore cup-shaped..................... **Gloeochaete** (page 565)
335. Cells uniseriate within a tubular envelope.......... **Asterocytis** (page 609)
335. Two to eight cells within old parent-cell wall......... **Glaucocystis** (page 564)
 336. Aerial or terrestrial... 337
 336. Aquatic.. 338
337. Unicellular, chromatophore reddish................. **Porphyridium** (page 612)
337. Filamentous, chromatophore orange-red and with orange-red droplets of oil..... **Trentepohlia** (page 179)
 338. Chromatophores stellate.. 339
 338. Chromatophores not stellate.. 340
339. A branched filament with a wide sheath............. **Asterocytis** (page 609)
339. An unbranched filament without a sheath............ **Bangia** (page 611)
 340. Thallus surface pseudoparenchymatous............................... 341
 340. Thallus surface filamentous.. 343
341. Externally differentiated into nodes and internodes... **Lemanea** (page 621)
341. Not differentiated into nodes and internodes........................... 342
 342. Axial row uniseriate, with very large cells....... **Compsopogon** (page 610)
 342. Axial row multiseriate, with small cells........... **Tuomeya** (page 622)
343. Cells of central axis much larger than those of branches................. 344
343. Cells of central axis same size as those of branches..................... 345
 344. Gonimoblast filaments in a compact mass... **Batrachospermum** (page 619)
 344. Gonimoblast filaments widely divergent.......... **Sirodotia** (page 619)
345. Central axis uniseriate, obscure.................... **Audouinella** (page 618)
345. Central axis multiseriate, clothed with filaments...... **Thorea** (page 620)
 346. Growing on fishes.......................... **Oödinium** (page 538)
 346. Not growing on fishes.. 347
347. Growing within amphibia.. 348
347. Growing upon turtles... 349
 348. Within eggs of *Amblystoma*....................... **Oöphila** (page 226)
 348. Within intestinal tract of frogs.............. **Euglenamorpha** (page 357)
349. Thallus filamentous.................................... **Basicladia** (page 217)
349. Thallus a pseudoparenchymatous crust........... **Dermatophyton** (page 161)
 350. Thallus unicellular.. **351**

350. Thallus multicellular .... **Cephaleuros** (page 182)

351. Thallus a branched tube of uniform diameter .... **Phyllosiphon** (page 280)

351. Thallus globose, with branched rhizoids .... **Rhodochytrium** (page 230)

352. Thallus a free-living globose cell .... **Mycanthococcus** (page 252)

352. Thallus epiphytic, cylindrical .... **Harpochytrium** (page 392)

353. Cells united in filaments (trichomes) 384

353. Cells not united in filaments 354

354. Without a formation of endospores 355

354. Regularly with a formation of endospores 377

355. Unicellular, or not pseudoparenchymatous if colonial 356

355. Colonial and pseudoparenchymatous .... **Entophysalis** (page 566)

356. Cells free-living 357

356. Cells endophytic in colorless algae 376

357. Cells united in colonies of distinctive shape 358

357. Cells, if in colonies, not in colonies of distinctive shape 363

358. Cells arranged in a hollow sphere 359

358. Cells not arranged in a hollow sphere 361

359. Interior of colony with radiating gelatinous strands. **Gomphosphaeria** (page 563)

359. Interior of colony without radiating strands 360

360. Cells spherical to ellipsoidal .... **Coelosphaerium** (page 562)

360. Cells pyriform .... **Marssoniella** (page 563)

361. Colonies cubical .... **Eucapsis** (page 557)

361. Colony a free-floating monostromatic plate 362

362. Cells in regular transverse and vertical rows .... **Merismopedia** (page 561)

362. Cells irregularly arranged .... **Holopedium** (page 562)

363. Cells spherical, except just after dividing 364

363. Cells ellipsoidal, cylindrical, or spindle-shaped 369

364. Cells solitary or in colonies of less than 50 cells 365

364. Cells in colonies with hundreds of cells 367

365. Gelatinous sheath around cells inconspicuous .... **Synechocystis** (page 555)

365. Gelatinous sheath around cells conspicuous 366

366. Sheath colorless .... **Chroococcus** (page 554)

366. Sheath variously colored .... **Gloeocapsa** (page 555)

367. Each cell with an evident sheath .... **Chondrocystis** (page 557)

367. Individual cells without evident sheaths 368

368. Cells close to one another .... **Polycystis** (page 556)

368. Cells remote from one another .... **Aphanocapsa** (page 555)

369. Cells ellipsoidal or cylindrical 370

369. Cells spindle-shaped .... **Dactylococcopsis** (page 559)

370. Cells solitary or in colonies of less than 50 cells 371

370. Cells in colonies with hundreds of cells 374

371. Each cell with an evident gelatinous sheath 372

371. Cells without evident individual sheaths 373

372. All cellular sheaths of uniform thickness .... **Gloeothece** (page 558)

372. Sheaths of certain cells prolonged into stipes .... **Chroothece** (page 558)

373. Cells solitary, or 2 to 4 grouped pole to pole .... **Synechococcus** (page 557)

373. Cells regularly in colonies of a few cells .... **Rhabdoderma** (page 559)

374. Colony tubular, with pointed ends .... **Bacillosiphon** (page 559)

374. Colony not tubular 375

375. Each cell with an evident gelatinous sheath .... **Anacystis** (page 560)

375. Individual cells without evident sheaths .... **Aphanothece** (page 560)

376. Host epiphytic, with long gelatinous setae..... **Gloeochaete** (page 565)
376. Host free-floating, 2 to 8 cells within an old parent-cell wall.............. **Glaucocystis** (page 564)
377. Unicellular, though sometimes gregarious.................................. 378
377. Cells united in colonies.. 381
378. Cells sessile.. 379
378. Cells not sessile.................................. **Pluto** (page 571)
379. Divisions forming endospores all transverse................................ 380
379. Divisions forming endospores in three planes........ **Dermocarpa** (page 570)
380. Entire protoplast dividing into endospores.... **Stichosiphon** (page 571)
380. Base of protoplast not dividing into endospores. **Chamaesiphon** (page 572)
381. Thallus perforating a calcareous substratum.............. **Hyella** (page 569)
381. Thallus not perforating a calcareous substratum........................... 382
382. Thallus not sessile........................... **Myxosarcina** (page 569)
382. Thallus sessile and adherent to substratum 383
383. Upper portion of thallus with cells in vertical rows.. **Pleurocapsa** (page 568)
383. Thallus without cells in vertical rows.............. **Xenococcus** (page 568)
384. Trichomes pointed at one or both ends 419
384. Trichomes approximately the same diameter throughout 385
385. Trichomes unbranched... 386
385. Trichomes with true or false branches..................................... 407
386. Trichomes with heterocysts... 387
386. Trichomes without heterocysts.. 394
387. Heterocysts always terminal.. 388
387. Heterocysts intercalary 389
388. Heterocysts only at one end of trichome..... **Cylindrospermum** (page 586)
388. Heterocysts at both ends of a trichome...... **Anabaenopsis** (page 583)
389. Length of cells less than breadth.................. **Nodularia** (page 586)
389. Length of cells equal to or more than breadth 390
390. Thallus with all trichomes parallel...................................... 391
390. Thallus, if with more than one trichome, not with trichomes parallel. 392
391. Thallus small, plate- or scale-like................ **Aphanizomenon** (page 585)
391. Thallus large, tubular, hollow........................... **Wollea** (page 585)
392. Trichomes solitary or intertwined in a watery amorphous mass...... 393
392. With many greatly twisted trichomes united in a firm mass of definite shape.. **Nostoc** (page 584)
393. Sheaths of trichomes narrow and firm.................. **Aulosira** (page 582)
393. Sheaths of trichomes broad and watery............. **Anabaena** (page 581)
394. Trichomes without a sheath... 395
394. Trichomes with a sheath.. 398
395. Trichomes straight, curved, or in irregular spirals....................... 396
395. Trichomes in regular spirals... 397
396. Trichomes with less than 20 cells................. **Borzia** (page 575)
396. Trichomes with hundreds of cells.............. **Oscillatoria** (page 574)
397. Cross walls distinct............................ **Arthrospira** (page 574)
397. Cross walls lacking.............................. **Spirulina** (page 573)
398. One trichome within a sheath... 399
398. More than one trichome within a sheath................................... 405
399. Cells touching one another... 400
399. Cells not touching one another.............. **Heterohormogonium** (page 567)
400. Sheaths watery, laterally confluent with one another................ 401

400. Sheaths firm, not laterally confluent 403
401. Filaments growing in erect tufts .... **Symploca** (page 578)
401. Filaments not growing in erect tufts 402
402. Filaments interwoven in an extended stratum... **Phormidium** (page 576)
402. Filaments parallel, in a free-floating scale-like mass .... **Trichodesmium** (page 576)
403. Trichomes with less than 20 cells .... **Romeria** (page 575)
403. Trichomes with hundreds of cells 404
404. Sheaths colorless or brownish .... **Lyngbya** (page 577)
404. Sheaths purplish .... **Porphyrosiphon** (page 578)
405. Sheaths firm .... **Schizothrix** (page 580)
405. Sheaths watery 406
406. Many trichomes within a common sheath..... **Microcoleus** (page 578)
406. Few trichomes within a common sheath....... **Hydrocoleum** (page 579)
407. With true branches 408
407. With false branches 414
408. Trichomes with heterocysts 409
408. Trichomes without heterocysts 413
409. Sheaths distinct, free from one another or partly confluent 410
409. Sheaths confluent with one another .... **Nostochopsis** (page 594)
410. Filaments free from one another 411
410. Filaments united in a cushion-like mass .... **Capsosira** (page 594)
411. Trichomes wholly or partly multiseriate .... **Stigonema** (page 592)
411. Trichomes wholly uniseriate 412
412. Branches parallel to axis bearing them .... **Thalpophila** (page 596)
412. Branches not parallel to axis bearing them .... **Hapalosiphon** (page 593)
413. Branching predominately dichotomous .... **Colteronema** (page 596)
413. Branching not predominately dichotomous .... **Albrightia** (page 595)
414. Trichomes with heterocysts 415
414. Trichomes without heterocysts .... **Plectonema** (page 589)
415. One trichome within a sheath 416
415. More than one trichome within a sheath 418
416. False branching abundant 417
416. False branching sparse, heterocysts basal, or basal and intercalary .... **Fremyella** (page 590)
417. False branches mostly arising singly .... **Tolypothrix** (page 588)
417. False branches mostly arising in pairs .... **Scytonema** (page 588)
418. Trichomes much contorted .... **Diplocolon** (page 590)
418. Trichomes parallel .... **Desmonema** (page 590)
419. Trichomes pointed at both ends 427
419. Trichomes pointed at one end 420
420. Heterocysts lacking 421
420. Heterocysts present 422
421. Pointed ends of trichomes parallel .... **Amphithrix** (page 598)
421. Pointed ends of trichomes not parallel .... **Calothrix** (page 598)
422. Filaments united in spherical or hemispherical thalli 423
422. Filaments solitary or not in rounded thalli 425
423. One trichome within a sheath 424
423. Two to several trichomes within a sheath .... **Sacconema** (page 601)
424. Trichomes without akinetes .... **Rivularia** (page 600)
424. Trichomes with akinetes .... **Gloeotrichia** (page 601)

425. Several laterally adjoined trichomes within a sheath... **Dichothrix** (page 599)
425. Trichomes of a filament not laterally adjoined.................. 426
426. With false branches at regular intervals........ **Scytonemopsis** (page 602)
426. False branches, if present, not at regular intervals.. **Calothrix** (page 598)
427. Trichomes with less than 20 cells................ **Raphidiopsis** (page 603)
427. Trichomes with more than 50 cells.............. **Hammatoidea** (page 603)
428. Cells with colored chromatophores.................. 429
428. Cells colorless.................. 517
429. Chromatophores grass-green.................. 430
429. Chromatophores brownish, olive, blue-green, or reddish.................. 477
430. Cells with starch and chromatophores usually with pyrenoids........ 431
430. Cells lacking starch, usually with paramylum.................. 470
431. Cells solitary.................. 432
431. Cells in colonies and all cells with flagella.................. 459
432. Cells without a wall or a lorica.................. 433
432. Cells with a wall or a lorica.................. 439
433. With one flagellum.................. **Pedinomonas** (page 68)
433. With more than one flagellum.................. 434
434. With two or four flagella.................. 435
434. With six to eight flagella.................. **Polyblepharides** (page 73)
435. Cells biflagellate.................. 436
435. Cells quadriflagellate.................. 438
436. Cells compressed.................. **Heteromastix** (page 70)
436. Cells not compressed.................. 437
437. Anterior portion of cell with longitudinal ridges.. **Stephanoptera** (page 68)
437. Cell without longitudinal ridges.................. **Dunaliella** (page 69)
438. Cells spindle-shaped.................. **Spermatozoopsis** (page 72)
438. Cells not spindle-shaped.................. **Pyramimonas** (page 71)
439. Protoplast entirely or partly in contact with wall.................. 440
439. Protoplast not touching wall or lorica.................. 452
440. Cells biflagellate.................. 441
440. Cells quadriflagellate.................. 449
441. Protoplast with numerous cytoplasmic strands extending to wall.............
Haematococcus (page 110)
441. Protoplast without cytoplasmic strands.................. 442
442. Vertical view circular.................. 443
442. Vertical view not circular.................. 446
443. Flagella inserted close together.................. 444
443. Flagella remote from each other.................. **Gloeomonas** (page 82)
444. Cells fusiform.................. **Chlorogonium** (page 81)
444. Cells not fusiform.................. 445
445. Protoplast of same shape as cell.................. **Chlamydomonas** (page 75)
445. Protoplast not of same shape as cell.................. **Sphaerellopsis** (page 77)
446. Cells compressed.................. 447
446. Cells not compressed.................. 448
447. Front view oval.................. **Platychloris** (page 79)
447. Front view circular.................. **Mesostigma** (page 79)
448. Posterior end with several conical projections.. **Brachiomonas** (page 83)
448. Posterior end without conspicuous projections.. **Lobomonas** (page 80)
449. Cells compressed.................. 450
449. Cells not compressed.................. **Carteria** (page 84)

450. Posterior end pointed 451
450. Posterior end broadly rounded .... **Platymonas** (page 85)
451. Lateral margins of cell wall wing-like .... **Scherffelia** (page 85)
451. Lateral margins of wall not wing-like .... **Chlorobrachis** (page 86)
452. Cells biflagellate 453
452. Cells quadriflagellate .... **Pedinopera** (page 93)
453. Lorica compressed 454
453. Lorica not compressed 458
454. Compressed face with projections .... **Wislouchiella** (page 89)
454. Compressed face without projections 455
455. Halves of lorica evident in motile cells .... **Phacotus** (page 87)
455. Halves of lorica not evident in motile cells 456
456. Lorica smooth 457
456. Lorica verrucose .... **Thoracomonas** (page 88)
457. Lorica much narrower at posterior pole .... **Cephalomonas** (page 90)
457. Lorica not narrower at posterior pole .... **Pteromonas** (page 89)
458. Lorica without pores .... **Coccomonas** (page 91)
458. Lorica with minute pores .... **Dysmorphococcus** (page 92)
459. Colony with a gelatinous envelope 460
459. Colony without a gelatinous envelope 468
460. Colonial envelope flattened 461
460. Colonial envelope globose 462
461. Envelope with an anterior-posterior differentiation .. **Platydorina** (page 101)
461. Envelope without anterior-posterior differentiation .... **Gonium** (page 95)
462. Cells forming a hollow sphere 463
462. Cells not forming a hollow sphere 467
463. Not over 256 cells in a colony 464
463. Over 500 cells in a colony .... **Volvox** (page 104)
464. All cells of a colony of the same size 465
464. With cells of two different sizes .... **Pleodorina** (page 102)
465. Cells close together .... **Pandorina** (page 96)
465. Cells not close together 466
466. Cells hemispherical .... **Volvulina** (page 98)
466. Cells spherical .... **Eudorina** (page 99)
467. Cells fusiform .... **Stephanosphaeria** (page 111)
467. Cells spherical .... **Stephanoon** (page 99)
468. Colonies with two or four cells .... **Pascheriella** (page 107)
468. Colonies with more than four cells 469
469. Cells biflagellate .... **Pyrobotrys** (page 108)
469. Cells quadriflagellate .... **Spondylomorum** (page 108)
470. With one flagellum 471
470. With two flagella 476
471. Protoplast without a lorica 472
471. Protoplast surrounded by a lorica 475
472. Cells plastic .... **Euglena** (page 353)
472. Cells rigid 473
473. With two elongate chromatophores .... **Cryptoglena** (page 355)
473. With numerous chromatophores 474
474. Cells compressed .... **Phacus** (page 355)
474. Cells not compressed .... **Lepocinclis** (page 354)
475. Cells free-swimming .... **Trachelomonas** (page 356)
475. Cells sessile .... **Ascoglena** (page 357)

476. Base of cell with acicular trichocysts.......... **Merotrichia** (page 626)
476. Cells without trichocysts.......................... **Eutreptia** (page 357)
477. Chromatophores brownish to olive-green.............................. 478
477. Chromatophores reddish or blue-green.................................. 510
478. Cells encircled by a transverse groove............................... 479
478. Cells not encircled by a transverse groove.......................... 487
479. Cells naked... 480
479. Cells with a wall composed of a definite number of plates............. 483
480. Transverse furrow approximately horizontal........................... 481
480. Transverse furrow descending diagonally....... **Gyrodinium** (page 525)
481. Portions above and below furrow approximately equal in size...............
Gymnodinium (page 525)
481. Portions above and below furrow unequal in size........................ 482
482. Upper portion longer and broader................ **Massartia** (page 525)
482. Lower portion longer and broader............. **Amphidinium** (page 526)
483. Plates of wall thin and smooth.. 484
483. Plates of wall thick and usually ornamented............................ 485
484. Transverse furrow completely encircling cell.. **Glenodinium** (page 527)
484. Transverse furrow incompletely encircling cell.. **Hemidinium** (page 527)
485. With long anterior and posterior horns.............. **Ceratium** (page 530)
485. Without long horns.. 486
486. With one antapical plate........................... **Gonyaulax** (page 528)
486. With two antapical plates.......................... **Peridinium** (page 529)
487. Cells uniflagellate or biflagellate....................................... 488
487. Cells triflagellate............................ **Chrysochromulina** (page 425)
488. Cells uniflagellate.. 489
488. Cells biflagellate.. 495
489. Cells without a lorica... 490
489. Cells with a lorica... 494
490. Cells with spines... 491
490. Cells without spines... 492
491. Cells solitary....................................... **Mallomonas** (page 413)
491. Cells in spherical colonies..................... **Chrysosphaerella** (page 415)
492. With one or with two chromatophores.................................... 493
492. With four chromatophores...................... **Amphichrysis** (page 411)
493. Chromatophores reticulate........................ **Chrysapsis** (page 412)
493. Chromatophores not reticulate.................... **Chromulina** (page 410)
494. Lorica globose.................................. **Chrysococcus** (page 412)
494. Lorica vase-shaped................................ **Kephyrion** (page 413)
495. Length of flagella equal or slightly different.......................... 496
495. Length of one flagellum double or more than that of the other........... 504
496. Cells solitary.. 497
496. Cells united in motile colonies... 502
497. Cells with a lorica................................. **Derepyxis** (page 416)
497. Cells without a lorica... 498
498. Flagella inserted laterally.. 499
498. Flagella inserted terminally.. 500
499. With a gullet and without an eyespot.............. **Nephroselmis** (page 635)
499. Without a gullet and with an eyespot................ **Protochrysis** (page 636)
500. With calcareous rings embedded in surface of cytoplasm.................
Hymenomonas (page 419)
500. Cytoplasmic surface without calcareous rings........................... 501

501. With pyrenoids .... **Rhodomonas** (page 631)
501. Without pyrenoids .... **Cryptochrysis** (page 629)
502. Colony with a gelatinous envelope .... **Syncrypta** (page 416)
502. Colony without a gelatinous envelope 503
503. Each cell with two parietal chromatophores .... **Synura** (page 417)
503. Each cell with two axial chromatophores .... **Skadovskiella** (page 418)
504. Cells without a lorica 505
504. Cells with a lorica 508
505. Cells solitary .... **Ochromonas** (page 420)
505. Cells united in colonies 506
506. Colony globose 507
506. Colony a flat disk .... **Cyclonexis** (page 423)
507. Center of colony with dichotomously branched gelatinous strands .... **Uroglena** (page 422)
507. Center or colony without gelatinous strands .... **Uroglenopsis** (page 421)
508. Surface of lorica smooth 509
508. Surface of lorica denticulate in optical section ... **Hyalobryon** (page 425)
509. Cells solitary and sessile .... **Epipyxis** (page 425)
509. Cells in free-swimming dendroid colonies .... **Dinobryon** (page 423)
510. Chromatophores or entire protoplast reddish 511
510. Chromatophores blue-green 514
511. Cells naked 512
511. Cells with a wall 513
512. With a gullet .... **Cryptomonas** (page 633)
512. Without a gullet .... **Rhodomonas** (page 631)
513. Protoplast connected to wall by numerous cytoplasmic strands .... **Haematococcus** (page 110)
513. Protoplast without cytoplasmic strands .... **Dunaliella** (page 69)
514. Cells naked 515
514. Cells surrounded by a lorica .... **Cyanomastix** (page 632)
515. With many disk-shaped chromatophores 516
515. Chromatophores laminate .... **Chroomonas** (page 630)
516. With numerous trichocysts .... **Gonyostomum** (page 625)
516. Without trichocysts .... **Cyanomonas** (page 631)
517. Cells uniflagellate 518
517. Cells biflagellate or quadriflagellate 526
518. With an eyespot .... **Euglena** (page 353)
518. Without an eyespot 519
519. Cells rigid 520
519. Cells more or less plastic 522
520. Cell surface with longitudinal ridges .... **Petalomonas** (page 368)
520. Cell surface without ridges 521
521. Cells radially symmetrical .... **Rhabdomonas** (page 359)
521. Cells not radially symmetrical .... **Menoidium** (page 359)
522. Cells flask-shaped .... **Urceolus** (page 367)
522. Cells more or less spindle-shaped 523
523. With pharyngeal rods 524
523. Without pharyngeal rods 525
524. Feeding exclusively upon diatoms .... **Jenningsia** (page 362)
524. Not feeding exclusively upon diatoms .... **Peranema** (page 362)
525. Periplast with spiral striae .... **Euglenopsis** (page 366)

525. Periplast without striae .... **Astasia** (page 358)
    526. Cells biflagellate 527
    526. Cells quadriflagellate .... **Polytomella** (page 72)
527. Cells containing starch 528
527. Cells not containing starch 529
    528. With a gullet at anterior end .... **Chilomonas** (page 634)
    528. Without a gullet .... **Polytoma** (page 78)
529. Cells rigid 530
529. Cells more or less plastic 536
    530. Both flagella projecting forward .... **Cyathomonas** (page 634)
    530. One flagellum projecting forward, the other trailing 531
531. Trailing flagellum the shorter 532
531. Trailing flagellum the longer 534
    532. Periplast longitudinally ridged 533
    532. Periplast not longitudinally ridged .... **Notosolenus** (page 366)
533. Ridges straight .... **Sphenomonas** (page 365)
533. Ridges curved .... **Tropidoscyphus** (page 365)
    534. Pharyngeal rods as long as cells 535
    534. Pharyngeal rods not so long as cells .... **Dinema** (page 363)
535. Cells compressed .... **Anisonema** (page 364)
535. Cells not compressed .... **Entosiphon** (page 364)
    536. Trailing flagellum the shorter 537
    536. Trailing flagellum the longer .... **Anisonema** (page 364)
537. With pharyngeal rods .... **Heteronema** (page 363)
537. Without pharyngeal rods .... **Distigma** (page 360)

BIBLIOGRAPHY

Acton, E. **1909.** Coccomyxa subellipsoidea, a new member of the Palmellaceae. *Ann. Bot.* **23:** 573–577. 1 *pl.* **1914.** Observations on the cytology of the Chroococcaceae. *Ibid.* **28:** 434–454. 2 *pl.* **1916.** On a new penetrating alga. *New Phytol.* **15:** 97–102. 2 *figs.* 1 *pl.* **1916A.** On the structure and origin of "Caldophora balls." *Ibid.* **15:** 1–10. 5 *figs.* **1916B.** Studies on nuclear division in desmids. I. Hyalotheca dissiliens (Sm.) Bréb. *Ann. Bot.* **30:** 379–382. 4 *figs.* 1 *pl.*

Agardh, C. A. **1824.** Systema algarum. Vol. 1. 312 *pp.* Lund.

Ahlstrom, E. H. **1937.** Studies on variability in the genus Dinobryon (Mastigophora). *Trans. Amer. Microsc. Soc.* **56:** 139–159. 4 *pl.*

Ahlstrom, E. H., and L. H. Tiffany. **1934.** The algal genus Tetrastrum. *Amer. Jour. Bot.* **21:** 499–507. 36 *figs.*

Allegre, C. F., and T. L. Jahn. **1943.** A survey of the genus Phacus Dujardin (Protozoa; Euglenoidina). *Trans. Amer. Microsc. Soc.* **62:** 233–244. 3 *pl.*

Allen, C. E. **1905.** Die Keimung der Zygote bei Coleochaete. *Ber. deutsch. Bot. Ges.* **23:** 285–292. 1 *pl.*

Allen, E. J., and E. W. Nelson. **1910.** On the artificial culture of marine plankton organisms. *Jour. Marine Biol. Assoc. of the United Kingdom.* N. S. **8:** 421–474.

Allen, T. F. **1883.** Notes on the American species of Tolypella. *Bull. Torrey Bot. Club.* **10:** 107–117. 6 *pl.*

Andrews, F. M. **1911.** Conjugation of two different species of Spirogyra. *Bull. Torrey Bot. Club.* **38:** 299. 1 *fig.*

Aragão, H. de B. **1910.** Untersuchungen ueber Polytomella agilis n.g., n. sp. *Memorias do Inst. Oswaldo Cruz (Rio de Janeiro).* **2:** 42–57. 1 *pl.*

Archer, W. **1864.** An endeavour to identify Palmogloes macrococca (Kütz.) with description of the plant believed to be meant, and of a new species, both, however, referable rather to the genus Mesotaenium (Näg.). *Quart. Jour. Microsc. Sci.* N.S. **4:** 109–132. 1 *pl.* **1867.** On the conjugation of Spirotaenia condensata (Bréb.) and of Spirotaenia truncata (Archer). *Ibid.* **7:** 186–193. 1 *pl.* **1874.** Double-spored or twin-spored form of Cylindrocystis Brebissonii. *Ibid.* **14:** 423. **1879.** New Closterium from New Jersey. *Ibid.* **19.** 120.

Artari, A. **1890.** Zur Entwicklungsgeschichte des Wassernetzes (Hydrodictyon utriculatum, Roth) *Bull. Soc. Imp. Nat. Moscou* N.S. **4:** 269–287. 1 *pl.* **1892.** Untersuchungen über Entwicklung und Systematik einiger Protococcoideen. *Ibid.* **6:** 222–262. 3 *pl.* **1913.** Zur Physiologie der Chlamydomonaden. Versuche und Beobachtungen an Chlamydomonas Ehrenbergii Gorosch. und verwandten Formen. *Jahrb. Wiss. Bot.* **52:** 410–466. 3 *figs.* 1 *pl.*

Askenasy, E. **1888.** Ueber die Entwicklung von Pediastrum. *Ber. deutsch. Bot. Ges.* **6:** 127–138. 1 *pl.*

Atkinson, G. F. **1890.** Monograph of the Lemaneaceae of the United States. *Ann. Bot.* **4:** 177–229. 3 *pl.* **1903.** The genus Harpochytrium in the United States. *Ann. Mycol.* **1:** 479–502. 1 *pl.* **1908.** A parasitic alga, Rhodochytrium spilanthidis Lagerheim, in North America. *Bot. Gaz.* **46:** 299–301. **1931.** Notes on the genus Lemanea in North America. *Ibid.* **92:** 225–242.

Atwell, C. B. **1889.** A phase of conjugation in Spirogyra. *Bot. Gaz.* **14:** 154.

Bachmann, H. **1907.** Vergleichende Studien über das Phytoplankton von Seen Schottlands und der Schweiz. *Arch. Hydrobiol. u. Planktonk.* **3:** 1–91. 23 *figs.*

Baker, W. B. **1926.** Studies in the life history of Euglena. I. Euglena agilis Carter. *Biol. Bull.* **51:** 321–362. 2 *figs.* 2 *pl.*

Baldwin, H. B., and G. C. Whipple. **1906.** Observed relations between dissolved oxygen, carbonic acid and algal growths in Weequahic Lake, N. J. *Rept. Am. Public Health Assoc.* **32:** 167–182.

Baumgärtel, O. **1920.** Das Problem der Cyanophyzeenzelle. *Arch. Protistenk.* **41:** 50–148. 1 *pl.*

Beger, H. **1927.** Beiträge zur Ökologie und Soziologie der luftlebigen (atmophytischen) Kieselalgen. *Ber. deutsch. Bot. Ges.* **45:** 385–407.

Beijerinck, M. W. **1890.** Culturversuche mit Zoochlorellen, Lichengonidien und anderen niederen Algen. *Bot. Zeitg.* **48:** 725–739, 741–754, 757–768, 781–785. 1 *pl.*

Belajeff, W. **1894.** Ueber Bau und Entwicklung der Spermatozoiden der Pflanzen. *Flora.* **79:** 1–48. 1 *pl.*

Belar, K. **1915.** Protozoenstudien I. *Arch. Protistenk.* **36:** 13–51. 3 *pl.* **1916.** Protozoenstudien II. *Ibid.* **36:** 241–302. 5 *figs.* 9 *pl.*

Bellido, E. C. **1927.** The technique of mounting diatom and other type slides. *Jour. Roy. Microsc. Soc. London* **1927:** 9–28. 19 *figs.* 4 *pl.*

Benecke, W. **1898:** Mechanismus und Biologie des Zerfalles der Conjugatenfäden in die einzelnen Zellen. *Jahrb. Wiss. Bot.* **32:** 453–476. 1 *fig.*

Bergon, P. **1907.** Biologie des Diatomées. Les processus de division, de rajeunissement de la cellule et de sporulation chez le Biddulphia mobiliensis Bailey. *Bull. Soc. Bot. France* **54:** 327–358. 4 *pl.*

Berliner, E. **1909.** Flagellaten-Studien. *Arch. Protistenk.* **15:** 297–325. 2 *pl.*

Bernard, C. **1908.** Protococcacées et desmidiées d'eau douce, récoltées à Java. Batavia. 230 *pp.* 16 *pl.*

Berthold, G. **1882.** Die Bangiaceen des Golfes von Neapal und der angrenzenden Meersabschnitte. *Fauna und Flora d. Golfes von Neapel.* **8.** Monographie. 1–28. 1 *pl.*

Bessey, C. E. **1884.** Hybridism in Spirogyra. *Amer. Nat.* **18:** 67–68.

Bharadwaja, Y. **1933.** A new species of Draparnaldiopsis (Draparnaldiopsis indica sp. nov.). *New Phytol.* **32:** 165–174. 2 *figs.* 1 *pl.*

Birckner, V. **1912.** Die Beobachtung von Zoosporenbildung bei Vaucheria aversa Hass. *Flora.* **104:** 167–171. 20 *figs.*

Birge, E. A., and C. Juday. **1911.** The inland lakes of Wisconsin. The dissolved gases of the water and their biological significance. *Bull. Wisconsin Geol. and Nat. Hist. Surv.* **22:** 1–259. 142 *figs.*

Blackburn, K. B. **1936.** A reinvestigation of the alga Botryococcus Braunii Kützing. *Trans. Roy. Soc. Edinburgh.* **58:** 841–854. 4 *figs.* 1 *pl.*

Blackman, F. F. **1900.** The primitive algae and the flagellata. An account of modern work bearing on the evolution of the algae. *Ann. Bot.* **14:** 647–688. 2 *figs.*

Blackman, F. F., and A. G. Tansley. **1902.** A revision of the classification of the green algae. *New Phytol.* **1:** 17–24, 47–48, 67–72, 89–96, 114–120, 133–144, 163–168, 189–192, 213–220, 238–244.

Bliding, C. **1933.** Über Sexualität und Entwicklung bei der Gattung Enteromorpha. *Svensk. Bot. Tidskr.* **27:** 233–256. 18 *figs.* **1938.** Studien über Entwicklung und Systematik in der Gattung Enteromorpha. *Bot. Notiser.* **1938:** 83–90. 6 *figs.*

BOCK, F. **1926.** Experimentelle Untersuchungen an koloniebildenden Volvocaceen. *Arch. Protistenk.* **56:** 321–356. 12 *figs.* 1 *pl.*

BOHLIN, K. **1890.** Myxochaete ett nytt slägte bland Sötvattensalgerna. *Bih. t. Kgl. Svensk. Vetensk.-Ak. Handl.* **15,** Afd. 3, No. 4: 1–7. 1 *pl.* **1897.** Studier öfver nägra slägten af Alggruppen Confervales. *Ibid.* **23,** Afd. 3, No. 3: 1–56. 2 *pl.* **1897A.** Zur Morphologie und Biologie einzelliger Algen. *Öfvers. Kgl. Svensk. Vetensk.-Ak. Förh.* **1897:** 507–529. 10 *figs.* **1897B.** Die Algen der ersten Regnellschen Expedition. I. Protococcoideen. *Bih. t. Kgl. Svensk. Vetensk.-Ak. Handl.* **23,** Afd. 3, No. 7: 1–47. 2 *pl.* **1901.** Étude sur la flore algologique d'eau douce des Açores. *Ibid.* **27,** Afd. 3, No. 4: 1–85. 1 *pl.*

BOLD, H. C. **1931.** Life history and cell structure of Chlorococcum infusionum. *Bull. Torrey Bot. Club.* **57:** 577–604. 5 *figs.* 5 *pl.* **1933.** The life history and cytology of Protosiphon botryoides. *Ibid.* **60:** 241–299. 7 *figs.* 10 *pl.* **1938.** Notes on Maryland algae. *Ibid.* **65:** 293–301. 2 *pl.* **1942.** The cultivation of algae. *Bot. Rev.* **8:** 69–138.

BORESCH, K. **1919.** Über die Einwirkung farbigen Lichtes auf die Färbung von Cyanophyceen. *Ber. deutsch. Bot. Ges.* **37:** 25–39. **1921.** Die wasserlöslichen Farbstoffe der Schizophyceen. *Biochem. Zeitschr.* **119:** 167–214. 34 *figs.* **1921A.** Die kloplementäre chromatische Adaptation. *Arch. Protistenk.* **44:** 1–70. 7 *figs.* 3 *pl.*

BORGE, O. **1894.** Über die Rhizoidenbildung bei einigen fadenförmigen Chlorophyceen. Upsala. 61 *pp.* 2 *pl.* **1897.** Zur Kenntnis der Verbreitungsweise der Algen. *Bot. Notiser.* **1897:** 210–211. **1918.** Die von Dr. A. Löfgren in São Paulo gesammelten Süsswasseralgen. *Ark. Bot.* **15,** No. 13: 1–108. 8 *pl.*

BORGERT, A. **1910.** Kern- und Zellteilung bei marinen Ceratium-Arten. *Arch. Protistenk.* **20:** 1–46. 3 *pl.*

BORNET, E., and C. FLAHAULT. **1885.** Tableau synoptique des Nostocacées filamenteuses. Hétérocystées. *Mem. Soc. Nat. Sci. Nat. et Math. Cherbourg.* **25:** 195–223; **26:** 137–152. **1885A.** Note sur le genre Aulosira. *Bul. Soc. Bot. France.* **32:** 119–122. 1 *pl.* **1886, 1886A, 1887, 1888.** Revision des Nostocacées hétérocystées contenues dans les principaux herbiers de France. *Ann. Sci. Nat. Bot. VII.* **3:** 323–381; **4:** 343–373; **5:** 51–129; **7:** 177–262. **1888A.** Deux nouveaux genres d'algues perforantes. *Jour. de Bot.* **2:** 161–165. **1889.** Sur quelques plantes vivant dans le test calcaire des mollusques. *Bull. Soc. Bot. France.* **36:** CXLVII–CLXXVI. 7 *pl.*

BORZI, A. **1883.** Studi algologici. Messina. Fasc. 1. 112 *pp.* 9 *pl.* **1895.** Studi algologici. Palermo. Fasc. 2. 257 *pp.* 22 *pl.* **1914.** Studi sulle Mixoficee. *Nuovo Gior. Bot. Ital.* **21:** 307–360.

BOYER, C. S. **1914.** A new diatom. *Proc. Acad. Nat. Sci. Phila.* **66:** 219–221. **1916.** The Diatomaceae of Philadelphia and vicinity. Philadelphia. 143 *pp.* 40 *pl.* **1927.** Synopsis of the North American Diatomaceae. Part 1. *Proc. Acad. Nat. Sci. Phila.* **78,** Suppl.: 1–228. **1927A.** Synopsis of the North American Diatomaceae. Part 2. *Ibid.* **79,** Suppl.: 229–583.

BRAND, F. **1898.** Culturversuche mit zwei Rhizoclonium-Arten. *Bot. Centralbl.* **74:** 193–202, 225–236. 1 *pl.* **1901.** Bemerkungen über Grenzzellen und über spontan rothe Inhaltskörper der Cyanophyceen. *Ber. deutsch. Bot. Ges.* **19:** 152–159. 4 *figs.* **1901A.** Ueber einigen Verhältnisse des Baues und Wachsthums von Cladophora. *Beih. Bot. Centralbl.* **10:** 481–521. 10 *figs.* **1902.** Zur näheren Kenntnis der Algengattung Trentepohlia Mart. *Ibid.* **12:** 200–225. 1 *pl.* **1902A.** Die Cladophora-Aegagropilen des Süsswassers. *Hedwigia.* **41:** 34–71. 1 *pl.* **1903.** Morphologische-physiologische Betrachtungen über Cyanophyceen.

Beih. Bot. Centralbl. **15:** 31–64. 1 *pl.* **1908.** Ueber das Chromatophor und die systematische Stellung der Blutalge (Porphyridium cruentum). *Ber. deutsch. Bot. Ges.* **26A:** 413–419. 5 *figs.* **1908A.** Weitere Bermerkungen über Porphyridium cruentum (Ag.) Naeg. *Ibid.* **26A:** 540–546. **1909.** Über die morphologischen Verhältnisse der Cladophora Basis. *Ibid.* **27:** 292–300. 5 *figs.* **1914.** Über die Beziehung der Algengattung Schizogonium Kütz. zu Prasiola Ag. *Hedwigia.* **54:** 295–310. 1 *fig.* **1917.** Ueber Beurteilung des Zellbaues kleiner Algen mit besonderem Hinweise auf Porphyridium cruentum Naeg. *Ber. deutsch. Bot. Ges.* **35:** 454–459. 3 *figs.*

Braun, A. **1851.** Betrachtungen über die Erscheinung der Verjüngung in der Natur. Leipzig. 363 *pp.* 3 *pl.* **1855.** Algarum unicellularium genera nova et minus cognita. Leipzig. 111 *pp.* 6 *pl.*

Bretschneider, L. H. **1925.** Pyramimonas utrajectina spec. nov., eine neue Polyblepharididae. *Arch. Protistenk.* **53:** 124–130. 11 *figs.*

Brinley, F. J., and L. J. Katzin. **1942.** Distribution of stream plankton in the Ohio River system. *Amer. Midland Nat.* **27:** 177–190.

Bristol-Roach, B. M. **1917.** On the life-history and cytology of Chlorochytrium grande sp. nov. *Ann. Bot.* **31:** 107–126. 2 *figs.* 2 *pl.* **1919.** On the retention of vitality by algae from old stored soils. *New Phytol.* **18:** 92–107. 2 *figs.* **1919A.** On a Malay form of Chlorococcum humicola (Näg.) Rabenh. *Jour. Linn. Soc. Bot. London.* **44:** 473–482. 2 *pl.* **1920.** On the alga-flora of some desiccated English soils: an important factor in soil biology. *Ann. Bot.* **34:** 35–80. 12 *figs.* 1 *pl.* **1920A.** A review of the genus Chlorochytrium Cohn. *Jour. Linn. Soc. Bot. London.* **45:** 1–28. 1 *fig.* 3 *pl.*

Brown, E. M. **1934.** On Oodinium ocellatum Brown, a parasitic dinoflagellate causing epidemic disease in marine fish. *Proc. Zool. Soc. London.* **1934:** 589–607. 3 *figs.* 3 *pl.*

Brown, V. E. **1930.** The cytology and binary fission of Peranema. *Quart. Jour. Microsc. Sci.* N.S. **73:** 403–419. 1 *fig.* 3 *pl.*

Brown, W. H. **1911.** Cell division in Lyngbya. *Bot. Gaz.* **51:** 390–391.

Brühl, P., and K. Biswas. **1923.** On a species of Compsopogon growing in Bengal. *Jour. Dept. Sci. Univ. Calcutta.* **5:** 1–6. 4 *pl.* (Ref. *Bot. Abstr.* **15:** No. 8253. 1926.)

Brunnthaler, J. **1913.** Die Algengattung Radiofilum Schmidle und ihre systematische Stellung. *Oesterr. Bot. Zeitschr.* **63:** 1–8. 3 *figs.* **1915.** Protococcales. In Pascher, A.: Die Süsswasserflora Deutschlands, Osterreichs und der Schweiz. **5.** Chlorophyceae **2:** 52–205. 330 *figs.*

Buchner, P. **1921.** Tier und Pflanze in intercellularer Symbiose. Berlin. 462 *pp.* 103 *figs.* 2 *pl.*

Buell, H. F. **1938.** The taxonomy of a community of blue-green algae in a Minnesota pond. *Bull. Torrey Bot. Club.* **65:** 377–396. 12 *figs.*

Bullard, C. **1921.** A method for orienting and mounting microscopical objects in glycerine. *Trans. Amer. Microsc. Soc.* **40:** 89–93.

Burkholder, P. R. **1934.** Movements in the Cyanophyceae. *Quart. Rev. Biol.* **9:** 438–459. 9 *figs.*

Bütschli, O. **1878.** Beitrage zur Kenntnis der Flagellaten und einiger verwandten Organismen. *Zeitschr. Wiss. Zool.* **30:** 205–281. 5 *pl.* **1883–1887.** Mastigophora. In H. G. Bronn's Klassen und Ordnungen des Thierreichs **1**[2]: 617–1097. 17 *pl.* **1885.** Einige Bemerkungen über gewisse Organisationsverhältnisse der sog. Cilioflagellaten und der Noctiluca. *Morph. Jahrb.* **10:** 529–577. 4 *figs.* 3 *pl.*

CALKINS, G. N. **1982.** On Uroglena, a genus of colony-building infusoria observed in certain water supplies in Massachusetts. *Ann. Rept. Massachusetts State Bd. Health.* **23:** 647–657. 4 *pl.* **1926.** The biology of the Protozoa. New York. 623 *pp.* 238 *figs.*

CALVERT, R. **1930.** Diatomaceous earth. New York.

CAMPBELL, D. H. **1902.** A university text-book of botany. New York. 579 *pp.* 493 *figs.* 15 *pl.*

CANABAEUS, L. **1929.** Ueber die Heterocysten und Gasvakuolen der Blaualgen und ihre Beziehungen zueinander. *Pflanzenforschung.* **13:** 1–48. 16 *figs.*

CARLSON, G. W. F. **1906.** Ueber Botryodictyon elegans Lemmerm. und Botryococcus Braunii Kütz. *Botaniska Studier tillägnade F. R. Kjellman. pp.* 141–146. 1 *pl.*

CARTER, N. **1919.** On the cytology of two species of Characiopsis. *New Phytol.* **18:** 177–186. 3 *figs.* **1919A.** Studies on the chloroplasts of desmids. I. *Ann. Bot.* **33:** 215–254. 5 *pl.* **1919B.** Studies on the chloroplasts of desmids. II. *Ibid.* **33:** 295–304. 1 *fig.* 2 *pl.* **1919C.** The cytology of the Cladophoraceae. *Ibid.* **33:** 467–478. 2 *figs.* 1 *pl.* **1920.** Studies on the chloroplasts of desmids. III. *Ibid.* **34:** 265–285. 4 *pl.* **1920A.** Studies on the chloroplasts of desmids. IV. *Ibid.* **34:** 303–319. 3 *pl.* **1923.** Vol. V of W. and G. S. West, A monograph of the British Desmidiaceae. London. **1926.** An investigation into the cytology and biology of the Ulvaceae. *Ann. Bot.* **40:** 665–689. 2 *pl.*

CASTLE, E. S. **1926.** Observations on motility in certain Cyanophyceae. *Biol. Bull.* **51:** 69–72. 1 *fig.*

CHADEFAUD, N. **1932.** Observation de Thamniochaete Huberi Gay en Vendée. *Rev. Algol.* **6:** 221–224. 2 *figs.* **1938.** Nouvelles recherches sur l'anatomie comparée des Eugléniens: les Pérénemines. *Ibid.* **11:** 189–220. 6 *figs.*

CHATTON, E. **1911.** Pleodorina californica à Banyuls-sur-mer. Son cycle évolutif et sa signification phylogénique. *Bull. Sci. France et Belgique.* **44:** 309–331. 2 *figs.* 1 *pl.* **1920.** Les Péridiniens parasites. *Arch. d. Zool. Expér. et Gén.* **59:** 1–475. 159 *figs.* 18 *pl.*

CHMIELEVSKY, V. **1890.** Eine Notiz über das Verhalten der Chlorophyll bänder in den Zygoten der Spirogyra Arten. *Bot. Zeitg.* **48:** 773–780. 1 *pl.*

CHODAT, R. **1894.** Matériaux pour servir à l'histoire des Protococcoidées. *Bull. Herb. Boiss.* **2:** 585–616. 8 *pl.* **1894A.** Remarques sur le Monostroma bullosum Thuret. *Bull. Soc. Bot. France.* **41:** CXXXIV–CXLII. 1 *pl.* **1984B.** Golenkinia, genre nouveau de Protococcoidées. *Jour. de Bot.* **8:** 305–308. 1 *pl.* **1895.** Ueber die Entwickelung der Eremosphaera viridis de By. *Bot. Zeitg.* **53:** 137–142. 1 *pl.* **1896.** Sur la structure et la biologie de deux algues pélagiques. *Jour. de Bot.* **10:** 333–349, 405–408. 3 *figs.* 1 *pl.* **1896A.** Sur l'évolution des Coelastrum. *Bull. Herb. Boiss.* **4:** 273–277. 10 *figs.* **1896B.** Sur la flore des neiges du col des Écandies. *Ibid.* **4:** 879–889. 1 *pl.* **1897.** Recherches sur les algues pélagiques de quelques lacs suisses et français. *Ibid.* **5:** 289–314. 5 *figs.* 3 *pl.* **1898** Recherches sur les algues littorales. *Ibid.* **6:** 431–475. 15 *fig.* 2 *pl.* **1902.** Algues vertes de la suisse. *Matér. pour la Flore Crypt. Suisse.* **1:** 1–373. 264 *figs.* **1909.** Étude critique et expérimentale sur le polymorphisme des algues. Geneva. 165 *pp.* 23 *pl.* **1913.** Monographie d'algues en culture pure. *Matér. pour la Flore Crypt. Suisse.* **4,** part 2: 1–266. 120 *figs.* 9 *pl.* **1919.** Sur un Glaucocystis et sa position systématique. *Bull. Soc. Bot. Genève.* **11:** 42–49. 2 *figs.* (Ref. *Bot. Abstr.* **6:** No. 1192, 1920.) **1925.** Algues de la région du Grand Saint-Bernard III. *Ibid.* **17:** 202–217. 7 *figs.* **1926.** Scenedesmus. Étude de génétique, de systématique expérimentale et d'hydrobiology. *Revue d'Hydrologie.* **3:** 71–258. 162 *figs.*

CHODAT, R. and F. **1925.** Esquisse planctologique de quelques lacs français. *Festschr. Carl Schröter.* **3:** 436–459. 14 *figs.*

CHODAT, R., and J. HUBER. **1894.** Sur le développement de l'Hariotinia Dangeard. *Bull. Soc. Bot. France.* **41:** CXLII–CXLVI. 6 *figs.*

CHOLNOKY, B. VON. **1927.** Ueber die Auxosporenbildung von Rhoicosphenia curvata (Kg.) Grun. *Arch. Protistenk.* **60:** 8–33. 1 *pl.* **1927A.** Beiträge zur Kenntnis der Bacillariaceen-Kolonien. *Hedwigia.* **67:** 223–236. 2 *figs.* **1927B.** Untersuchungen über die Ökologie der Epiphyten. *Arch. Hydrobiol. u. Planktonk.* **18:** 661–705. 43 *figs.* **1928.** Ueber die Wirkung von Hyper- und Hypotonischen Lösungen auf einige Diatomeen. *Internat. Rev. gesamt. Hydrobiol. Hydrog.* **19:** 452–500. 94 *figs.* **1928A.** Über die Auxosporenbildung der Anomoeoneis sculpta E. Cl. *Arch. Protistenk.* **63:** 23–57. 4 *pl.* **1929.** Beiträge zur Kenntnis der Auxosporenbildung. *Ibid.* **68:** 471–502. 2 *figs.* 3 *pl.* **1930.** Die Dauerorgane von Cladophora glomerata. *Zeitschr. Bot.* **22:** 545–585. 42 *figs.* **1933.** Die Kernteilung von Melosira arenaria nebst einigen Bemerkungen ueber ihre Auxosporenbildung. *Zeitschr. Wiss. Biol. Abt. B.* **19:** 698–719. 24 *figs.*

CIENKOWSKI, L. **1865.** Ueber einige chlorophyllhaltige Gloeocapsen. *Bot. Zeitg.* **23.** 21–27. 1 *pl.* **1870.** Ueber Palmellaceen und einige Flagellaten. *Arch. Mikrosk. Anat.* **6:** 421–438. 2 *pl.* **1876.** Zur Morphologie der Ulotricheen. *Bull. Acad. Imp. Sci. St. Pétersbourg.* **21:** 531–572. 2 *pl.*

CLEMENTS, F. E., and H. L. SHANTZ. **1909.** A new genus of blue-green algae. *Minnesota Bot. Studies.* **4:** 133–135. 1 *pl.*

CLEVE, P. T. **1894.** Synopsis of the naviculoid diatoms. *Kgl. Svensk. Vetensk.-Ak. Handl.* **26,** No. 2: 1–194. 5 *pl.* **1895.** *Ibid.* **27,** No. 2: 1–219. 4 *pl.*

COHN, F. **1856.** Mémoire sur le développement et le mode de reproduction du Sphaeroplea annulina. *Ann. Sci. Nat. Bot.* IV. **5:** 187–208. 2 *pl.* **1872.** Ueber parasitische Algen. *Beitr. Biol. Pflanzen.* **1,** part 2: 87–106. 1 *pl.*

COKER, W. C., and L. SHANOR. **1939.** A remarkable saprophytic fungoid alga. *Jour. Elisha Mitchell Sci. Soc.* **55:** 152–165. 2 *pl.*

COLIN, H., and J. AUGIER. **1933.** Floridose, trehalose, et glycogene chez les algues rouges d'eau douce. *Compt. Rend. Acad. Sci. Paris.* **197:** 423–425.

COLLINS, F. S. **1897.** Some perforating and other algae on fresh water shells. *Erythea.* **5:** 95–97. 1 *pl.* **1901.** Notes on algae. III. *Rhodora.* **3:** 132–137. **1904.** Algae of the Flume. *Ibid.* **6:** 229–231. **1907.** Some new green algae. *Ibid.* **9:** 197–202. 1 *pl.* **1909.** The green algae of North America. *Tufts College Studies. Scientific Series.* **2:** 79–480. 18 *pl.* **1912.** The green algae of North America. Supplementary paper. *Ibid.* **3:** 69–109. 2 *pl.* **1916.** Notes from the Woods Hole Laboratory—1915. *Rhodora.* **18:** 90–92. 2 *figs.* **1918.** The green algae of North America. Second supplement. *Tufts College Studies. Scientific Series.* **4,** No. 7: 1–106. 3 *pl.*

CONRAD, W. **1913.** Observations sur Eudorina elegans Ehr. *Rec. Inst. Leo Errera.* **9:** 321–343. 13 *figs.* **1913A.** Errerella bornhemiensis nov. gen., une protococcacée nouvelle. *Bull. Soc. Roy. Bot. Belgique.* **52:** 237–242. 3 *figs.* **1914.** La morphologie et la nature des enveloppes chez Hymenomonas roseola Stein et H. coccolithophora Massart et Conrad, nov. spec., et les Coccolithophoridae. *Ann. Biol. Lacustre.* **1:** 155–164. 6 *figs.* **1920.** Contribution à l'étude des Chrysomonadines. *Bull. Acad. Roy. Belgique. Cl. d. Sci.* 5 Ser., **6:** 167–189. 11 *figs.* **1920A.** Sur un flagellé nouveau à trichocystes Reckertia sagittifera, n.g., n. sp. *Ibid.* **6:** 641–555. 4 *figs.* **1922.** Contributions à l'étude des Chrysomonadines. *Rec. Inst. Leo Errera.* **10:** 333–353. 11 *figs.* **1926.** Recherches sur les Flagellates de nos eaux saumâtres. 2. Chrysomonadines. *Arch. Protistenk.* **56:** 167–231. 28 *figs.* 3 *pl.* **1926A.** Recherches sur les Flagellates de nos eaux saumâtres.

1. Dinoflagellates. *Ibid.* **55:** 63–100. 5 *figs.* 2 *pl.* **1927.** Essai d'une monographie des genres Mallomonas Perty (1852) et Pseudomallomonas Chodat (1920). *Ibid.* **59:** 423–505. 42 *figs.* 4 *pl.* **1928.** Le genre Microglena C. G. Ehrenberg (1838). *Ibid.* **60:** 415–439. 13 *figs.* **1928A.** Sur les Coccolithophoracées d'eau douce. *Ibid.* **63:** 58–66. 9 *figs.* **1928B.** Quatre flagellates nouveaux. *Ann. Protistol.* **1:** 11–18. 8 *figs.* **1933.** Revision due genre Mallomonas Perty (1851) incl. Pseudomallomonas Chodat (1920). *Mém. Mus. Roy. d'Hist. Nat.* **56:** 1–82. 70 *figs.* **1934.** Materiaux pour une monographie due genre Lepocinclis Perty. *Arch. Protistenk.* **82:** 203–249. 67 *figs.*

COPELAND, H. F. **1937.** On the pollen of Mimosoideae and the identity of the supposed alga Phytomorula. *Madroño.* **4:** 120–125. 1 *pl.*

COPELAND, J. J. **1936.** Yellowstone thermal Myxophyceae. *Ann. New York Acad. Sci.* **36:** 1–232. 73 *figs.*

CORRENS, C. **1892.** Ueber eine neue braune Süsswasseralge, Naegeliella flagellifera, nov. gen. et spec. *Ber. deutsch. Bot. Ges.* **10:** 629–636. 1 *pl.* **1893.** Ueber Apiocystis Brauniana Naeg. *Zimmermann's Beitr. z. Pflanzenzelle.* **3:** 241–259. 2 *figs.* (Ref. Just's *Bot. Jahresb.* **21**[1]**:** 88. 1896.)

COUCH, G. C., and E. L. RICE. **1948.** Vegetative habit and reproduction of Desmidium Grevillii (Kütz.) de Bary. *Amer. Jour. Bot.* **35:** 482–486. 12 *figs.*

COUCH, J. N. **1932.** Gametogenesis in Vaucheria. *Bot. Gaz.* **94:** 272–296. 35 *figs.*

CROW, W. B. **1923.** Fresh-water plankton algae from Ceylon. *Jour. Bot.* **61:** 110–114, 138–145, 164–171. 3 *figs.* **1923A.** Dimorphococcus Fritschii, a new colonial protophyte from Ceylon. *Ann. Bot.* **37:** 141–147. 5 *figs.* **1924.** Variation and species in Cyanophyceae. *Jour. Genetics.* **14:** 397–424. 8 *figs.* **1924A.** Some features of the envelope in Coelastrum. *Ann. Bot.* **38:** 398–401. 2 *figs.* **1925.** The reproductive differentiation of colonies in Chlamydomonadales. *New Phytol.* **24:** 120–123. **1927.** The generic characters of Arthrospira and Spirulina. *Trans. Amer. Microsc. Soc.* **46:** 139–148. **1927A.** Abnormal forms of Gonium. *Ann. and Mag. Nat. Hist.* IX. **19:** 593–601. 7 *figs.*

CUNNINGHAM, B. **1917.** Sexuality of filament of Spirogyra. *Bot. Gaz.* **63:** 486–500. 3 *pl.*

CUNNINGHAM, D. D. **1877.** On Mycoidea parasitica, a new genus of parasitic algae, and the part which it plays in the formation of certain lichens. *Trans. Linn. Soc. Bot. London* II. **1:** 301–316. 2 *pl.*

CZEMPYREK, H. **1930.** Beitrag zur Kenntnis der Schwärmerbildung bei der Gattung Cladophora. *Arch. Protistenk.* **72:** 433–452. 10 *figs.*

CZURDA, V. **1925.** Zur Kenntnis der Copulationsvorgänge bei Spirogyra. *Arch. Protistenk.* **51:** 439–478. 18 *figs.* 3 *pl.* **1926.** Die Reinkultur von Conjugaten. *Ibid.* **53:** 215–242. 6 *figs.* 2 *pl.* **1926A.** Über Reinkultur von Conjugaten (Nachträge). *Ibid.* **54:** 355–358. **1928.** Morphologie und Biologie des Algenstärkekornes. *Beih. Bot. Centralbl.* **45:** 97–270.

DAILY, W. A. **1942.** The Chroococcaceae of Ohio, Kentucky, and Indiana. *Amer. Midland Nat.* **27:** 636–661. 6 *pl.* **1943.** First reports for the algae Borzia, Aulosira, and Asterocytis in Indiana. *Butler Univ. Bot. Studies.* **6:** 84–86. 3 *figs.* **1945.** Additions to the filamentous Myxophyceae of Indiana, Kentucky, and Ohio. *Ibid.* **7:** 132–139.

DANGEARD, P. A. **1888.** Recherches sur les algues inférieures. *Ann. Sci. Nat. Bot.* VII. **7:** 105–175. 2 *pl.* **1889.** Mémoire sur les algues. *Le Botaniste.* **1:** 127–174. 2 *pl.* **1898.** Mémoire sur les Chlamydomonadinées ou l'histoire d'une cellule. *Ibid.* **6:** 65–292. 19 *figs.* **1901.** Étude sur la structure de la cellule et ses fonctions, le Polytoma uvella. *Ibid.* **8:** 5–58. 4 *figs.* **1901A.**

Recherches sur les Eugléniens. *Ibid.* **8:** 97–360. 53 *figs.* **1910.** Études sur le développement et la structure des organismes inférieurs. *Ibid.* **11:** 1–311. 15 *figs.* 33 *pl.* **1912.** Recherches sur quelques algues nouvelles ou peu connues. *Ibid.* **12:** I–XIX. 2 *pl.* **1916.** Note sur des cultures de Gonium sociale. *Bull. Soc. Bot. France.* **63:** 42–46.

DARBISHIRE, O. V. **1898.** Ueber Bangia pumila Aresch., eine endemische Alge der östlichen Ostsee. *Wissensch. Meeresunters.* N.F. **3.** *Abt. Kiel. pp.* 27–31. 10 *figs.*

DAVIS, B. M. **1894.** Euglenopsis: a new alga-like organism. *Ann. Bot.* **8:** 377–380. 1 *pl.* **1894A.** Notes on the life history of a blue-green motile cell. *Bot. Gaz.* **19:** 96–102. 1 *pl.* **1896.** The fertilization of Batrachospermum. *Ann. Bot.* **10:** 49–76. 2 *pl.* **1904.** Oögenesis in Vaucheria. *Bot. Gaz.* **38:** 81–98. 2 *pl.*

DEBARY, A. **1858.** Untersuchungen über die Familie der Conjugaten. Leipzig. 91 *pp.* 8 *pl.* **1871.** Über den Befruchtungsvorgang bei den Charen. *Monatsber. Akad. Wiss. Berlin.* **1871:** 227–240. 1 *pl.* **1875.** Zur keimungsgeschichte der Charen. *Bot. Zeitg.* **33:** 377–385, 393–401, 409–420. 2 *pl.*

DEBSKI, B. **1898.** Weitere Beobachtungen an Chara fragilis Desv. *Jahrb. Wiss. Bot.* **32:** 635–670. 2 *pl.*

DEFLANDRE, G. **1926.** Monographie du genre Trachelomonas Ehr. Nemours. 162 *pp.*, 15 *pl.* **1931.** Sur la structure de la membrane chez quelques Phacus. *Ann. Protistol.* **3:** 41–43. 2 *pl.* **1934.** Sur la structure des flagelles. *Ibid.* **4:** 31–54.

DILL, O. **1895.** Die Gattung Chlamydomonas und ihre nächsten Verwandten. *Jahrb. Wiss. Bot.* **28:** 323–358. 1 *pl.*

DIWALD, K. **1938.** Die ungeschlechtliche und geschlechtliche Fortpflanzung von Glenodinium lubiniensiforme spec. nov. *Flora.* **132:** 174–192. 8 *figs.*

DOBELL, C. C. **1908.** The structure and life-history of Copromonas subtilis, nov. gen. et nov. spec.: a contribution to our knowledge of the Flagellata. *Quart. Jour. Microsc. Sci.* N.S. **52:** 75–120. 3 *figs.* 2 *pl.* **1912.** Researches on the Spirochaets and related organisms. *Arch. Protistenk.* **26:** 116–240. 3 *figs.* 5 *pl.*

DODEL, A. **1876.** Ulothrix zonata. Ihre geschlechtliche und ungeschlechtliche Fortpflanzung, ein Beitrag zur Kenntnis der unteren Grenze des pflanzlichen Sexuallebens. *Jahrb. Wiss. Bot.* **10:** 417–550. 8 *pl.*

DOFLEIN, F. **1916.** Rhizochrysis. *Zool. Anz.* **47:** 153–158. 2 *figs.* **1916A.** Polytomella agilis. *Ibid.* **47:** 273–282. 5 *figs.* **1921.** Untersuchungen über Chrysomonadinen. I, II. *Arch. Protistenk.* **44:** 149–205. 3 *figs.* 4 *pl.* **1921A.** Mitteilungen über Chrysomonadinen aus dem Schwarzwald. *Zool. Anz.* **53:** 153–173. 4 *figs.* **1923.** Untersuchungen über Chrysomonadinen. III. *Arch. Protistenk.* **46:** 267–327. 5 *figs.* 7 *pl.* **1929.** Lehrbuch der Protozoenkunde. 5 ed. Neubearbeitet von E. Reichenow. Jena.

DORAISWAMI, S. **1940.** On the morphology and cytology of Eudorina indica Iyengar. *Jour. Indian Bot. Soc.* **19:** 113–139. 62 *figs.*

DOYLE, W. L. **1943.** The nutrition of the protozoa. *Biol. Rev.* **18:** 119–136.

DREW, K. M. **1935.** The life-history of Rhodochorton violaceum (Kütz.) comb. nov. (Chantransia violacea Kütz.) *Ann. Bot.* **99:** 439–450. 18 *figs.*

DROUET, F. **1938.** The Oscillatoriaceae of southern Massachusetts. *Rhodora.* **40:** 221–273. **1939.** Francis Wolle's filamentous Myxophyceae. *Bot. Ser. Field Museum Nat. Hist.* **20:** 17–64. 1 *fig.* **1942.** Studies in Myxophyceae. I. *Ibid.* **20:** 125–141. 3 *pl.* **1943.** Myxophyceae of eastern California and western Nevada. *Ibid.* **20:** 145–176. **1943A.** New species and transfers in Myxophyceae. *Amer. Midland Nat.* **30:** 671–674.

DROUET, F., and A. COHEN. **1935.** The morphology of Gonyostomum semen from Woods Hole, Massachusetts. *Biol. Bull.* **68:** 422–439. 2 *pl.*

DROUET, F., and W. A. DAILY. **1939.** The planktonic fresh-water species of Microcystis. *Bot. Ser. Field Museum Nat. Hist.* **20:** 67–83.

DUCELLIER, F. **1915.** Note sur un nouveau Coelastrum. *Bull. Soc. Bot. Genève.* **7:** 73–74. 5 *figs.*

EARDLY-WILMOT, V. L. **1928.** Diatomite, its occurrence, preparation and uses. *Canada, Dept. of Mines, Rept. Mines Branch.* **691:** 1–182. 31 *figs.* 15 *pl.*

EDDY, S. **1930.** The fresh-water armored or thecate dinoflagellates. *Trans. Amer. Microsc. Soc.* **49:** 277–321. 8 *pl.*

EDWARDS, A. M., C. JOHNSTON, and H. L. SMITH. **1877.** Practical directions for collecting, preserving, transporting, preparing, and mounting diatoms. New York. 53 *pp.*

EHRENBERG, C. G. **1870.** Ueber mächtige Gebirgs-Schichten vorherrschend aus mikroskopischen Bacillarien unter und bei der Stadt Mexiko. *Abhandl. k. Akad. Wiss. Berlin.* 1869. Physikal. Kl. Abt. **1:** 1–66. 3 *pl.*

EICHLER, A. W. **1886.** Syllabus der Vorlesungen über specielle und medicinisch-pharmaceutische Botanik. 4 ed. Berlin. 68 *pp.*

ELENKIN, A. A. **1924.** De Euglenarum sine flagello sectione nova. *Not. Syst. Inst. Crypt. Hort. Bot. Reip. Rossic.* **3:** 124–160. (Ref. Mainx, **1925A.**)

ELLIOTT, A. M. **1934.** Morphology and life history of Haematococcus pluvialis. *Arch. Protistenk.* **82:** 250–272. 7 *figs.* 2 *pl.*

ELMORE, C. J. **1922.** The diatoms (Bacillarioideae) of Nebraska. *Univ. of Nebraska Studies.* **21:** 22–214. 23 *pl.*

ENDLICHER, S. **1836.** Genera plantarum secundum ordines naturales disposita. Vindbonae.

ENGELMANN, T. W. **1883.** Farbe und Assimilation. *Bot. Zeitg.* **4:** 1–13, 17–29. **1884.** Untersuchungen über die quantitativen Beziehungen zwischen Absorption des Lichtes und Assimilation in Pflanzenzellen. *Ibid.* **42:** 81–93, 97–105.

ENTZ, G. **1918.** Über die mitotische Teilung von Polytoma uvella. *Arch. Protistenk.* **38:** 324–354. 5 *figs.* 2 *pl.* **1921.** Über die mitotische Teilung von Ceratium hirundinella. *Ibid.* **43:** 415–430. 11 *figs.* 2 *pl.* **1925.** Über Cysten und Encystierung der Süsswasser-Ceratien. *Ibid.* **51:** 131–183. 28 *figs.*

ERNST, A. **1902.** Dichotomosiphon tuberosus (A. Br.) Ernst, eine neue oogame Süsswasser-Siphonee. *Beih. Bot. Centralbl.* **13:** 115–148. 5 *pl.* **1904.** Beiträge zur Kenntnis der Codiaceen. *Ibid.* **16:** 199–236. 3 *pl.* **1904A.** Zur Morphologie und Physiologie der Fortpflanzungszellen der Gattung Vaucheria DC. *Ibid.* **16:** 367–382. 1 *pl.* **1908.** Beiträge zur Morphologie und Physiologie von Pithophora. *Ann. Jard. Bot. Buitenzorg.* **22:** 18–55. 4 *pl.*

ESCOYEZ, E. **1907.** Le noyau et la caryocinèse chez le Zygnema. *Cellule.* **24:** 355–366. 1 *pl.*

FAMINTZIN, A. **1914.** Beitrag zur Kenntnis der Zoosporen der Lichenen. *Ber. deutsch. Bot. Ges.* **32:** 218–222.

FAMINTZIN, A., and J. BORANETZKY. **1867.** Sur le changement des gonidies des lichens en zoospores. *Ann. Sci. Nat. Bot.* V. **8:** 139–144. 1 *pl.*

FECHNER, R. **1915.** Die Chemotaxis der Oscillarien und ihre Bewegungserscheinungen überhaupt. *Zeitschr. Bot.* **7:** 289–364. 10 *figs.* 1 *pl.*

FELDMANN, J. **1939.** Observations sur une algue (Dermatophyton radians Peter) vivant dans la carapace des tortues d'eau douce. *Bull. Trav. Station d'Aquicult. et Pêche Castiglione.* **1939:** 73–89. 6 *figs.* **1946.** Sur l'hétéroplastie de certaines Siphonales et leur classification. *Compt. Rend. Acad. Sci. Paris.* **222:** 752–753.

FISCHER, A. **1834.** Ueber das Vorkommen von Gypskrystallen bei den Desmidieen. *Jahrb. Wiss. Bot.* **14:** 135–184. 2 *pl.* **1897.** Untersuchungen über den Bau der Cyanophyceen und Bakterien. Jena. **1905.** Die Zelle der Cyanophyceen. *Bot. Zeitg.* **63:** 51–130. 2 *pl.*

FLAHAULT, C. **1883.** Sur le Lithoderma fontanum, algue phéosporée d'eau douce. *Bull. Soc. Bot. France.* **30:** CII–CVI.

FLINT, L. H. **1948.** Studies on fresh-water red algae. *Amer. Jour. Bot.* **35:** 428–433. 40 *figs.*

FOTT, B. **1935.** Über den inneren Bau von Vacuolaria viridis (Dangeard) Senn. *Arch. Protistenk.* **84:** 242–250. 4 *figs.* **1942.** Die planktischen Characium-arten. *Studia Bot. Cechica.* **5:** 156–166. 4 *figs.* 1 *pl.*

FOYN, B. **1929.** Vorläufige Mitteilung über die Sexualität und den Generationswechsel von Cladophora und Ulva. *Ber. deutsch. Bot. Ges.* **47:** 495–506. 2 *figs.* **1934.** Lebenszyklus, Cytologie und Sexualität der Chlorophycee Cladophora Suhriana Kützing. *Arch. Protistenk.* **83:** 1–56. 18 *figs.* 5 *pl.* **1934A.** Lebenszyklus und Sexualität der Chlorophycee Ulva Lactuca L. *Ibid.* **83:** 154–177 13 *figs.*

FRANCIS, G. **1878.** Poisonous Australian lake. *Nature.* **18:** 11–12.

FRANZÉ, R. **1893.** Zur Morphologie und Physiologie der Stigmata der Mastigophora. *Zeitschr. Wiss. Zool.* **56:** 138–164. 1 *pl.* **1894.** Die Polytomeen, eine morphologisch-entwickelungsgeschichtliche Studie. *Jahrb. Wiss. Bot.* **26:** 295–378. 11 *figs.* 4 *pl.*

FRÉMY, P. **1925.** Essai sur l'écologie des algues saxicoles aériennes et subaériennes en Normandie. *Nuova Notarisia.* **1925:** 297–304. **1930.** Les Myxophycées de l'Afrique équatoriale française. *Arch. Bot.* **3,** Mém. 2, *pp.* 1–507. 362 *figs.*

FRITSCH, F. E. **1902.** Observations on species of Aphanochaete. *Ann. Bot.* **16:** 403–412. 7 *figs.* **1902A.** The germination of zoospores in Oedogonium. *Ibid.* **16:** 412–417. 1 *fig.* **1902B.** The structure and development of the young plants in Oedogonium. *Ibid.* **16:** 467–485. 5 *figs.* **1903.** Observations on the young plants of Stigeoclonium Kütz. *Beih. Bot. Centralbl.* **13:** 368–387. 2 *pl.* **1912.** Fresh-water algae of the South Orkneys. *Rept. on the Scientific Results of the Scottish National Antarctic Expedition.* **3:** 95–134. 1 *fig.* 2 *pl.* **1916.** The algal ancestry of the higher plants. *New Phytol.* **15:** 233–250. 2 *figs.* **1916A.** The morphology and ecology of an extreme terrestrial form of Zygnema (Zygogonium) ericetorum (Kuetz.) Hansg. *Ann. Bot.* **30:** 135–149. 3 *figs.* **1918.** A first report on the fresh-water algae mostly from the Cape peninsula in the herbarium of the South African Museum. *Ann. S. African Museum.* **9:** 483–611. 43 *figs.* **1922.** The moisture relations of terrestrial algae. I. Some general observations and experiments. *Ann. Bot.* **36:** 1–20. 2 *figs.* **1922A.** The terrestrial algae. *Jour. Ecol.* **10:** 220–236. **1927.** Some aspects of the present-day investigation of the protophyta. *British Assoc. for the Adv. of Sci. Rept. of the 95th meeting, Leeds,* 1927. *pp.* 176–190. **1929.** Evolutionary sequence and affinities among protophyta. *Biol. Rev.* **4:** 103–151. 7 *figs.* **1929A.** The encrusting algal communities of certain fast-flowing streams. *New Phytol.* **28:** 165–196. 10 *figs.* 1 *pl.* **1929B.** The genus Sphaeroplea. *Ann. Bot.* **43:** 1–26. 8 *figs.* **1935.** The structure and reproduction of the algae. Vol. 1, Cambridge. 791 *pp.* 245 *figs.* **1944.** Present-day classification of the algae. *Bot. Rev.* **10:** 233–277.

FRITSCH, F. E., and F. M. HAINES. **1923.** The moisture relations of terrestrial algae. II. The changes during exposure to drought and treatment with hypertonic solutions. *Ann. Bot.* **37:** 683–728. 8 *figs.*

Fritsch, F. E., and R. P. John. **1942.** An ecological and taxonomic study of the algae of British soils. *Ann. Bot.* N.S. **6:** 371–395. 8 *figs.*

Fritsch, F. E., and F. Rich. **1929.** Freshwater algae (exclusive of diatoms) from Griqualand West. *Trans. Roy. Soc. South Africa.* **18:** 1–92. 32 *figs.*

Gabriel, C. **1925.** Sur l'existence de kystes dans l'évolution d'une Chlamydomonadacée, Brachiomonas submarina. *Compt. Rend. Soc. Biol.* **93:** 361–362.

Gaidukov, N. **1902.** Ueber den Einfluss farbigen Lichts auf die Färbung lebender Oscillarien. *Abhandl. k. Akad. Wiss. Berlin,* 1902. Anhang. Phys.-Math. Kl. 1–36. 4 *pl.* **1903.** Weitere Untersuchungen über den Einfluss farbigen Lichtes auf die Färbung der Oscillarien. *Ber. deutsch. Bot. Ges.* **21:** 484–492. 1 *pl.* **1923.** Zur Frage nach der komplementären chromatischen Adaptation. *Ibid.* **41:** 356–361.

Gain, L. **1911.** La neige verte et la neige rouge des régions antarctiques. *Bull. Museum Hist. Nat. Paris.* **17:** 479–482.

Ganong, W. F. **1905.** On balls of vegetable matter from sandy shores. *Rhodora.* **7:** 41–47. **1909.** On balls of vegetable matter from sandy shores. Second article. *Ibid.* **11:** 149–152.

Gardner, N. L. **1906.** Cytological studies in Cyanophyceae. *Univ. Calif. Publ. Bot.* **2:** 237–296. 6 *pl.* **1910.** Leuvenia, a new genus of flagellates. *Ibid.* **4:** 97–104. 1 *pl.* **1917.** New Pacific Coast marine algae. I. *Ibid.* **6:** 377–406. 5 *pl.* **1937.** A new species of Chaetomorpha from China. *Madroño.* **4:** 28–32. 1 *pl.*

Gay, F. **1891.** Recherches sur le développement et la classification de quelques algues vertes. Paris. 116 *pp.* 15 *pl.* **1891A.** Le genre Rhizoclonium. *Jour. de Bot.* **5:** 53–58. 4 *figs.* **1893.** Sur quelques algues de la flore de Montpellier. *Bull. Soc. Bot. France.* **40:** CLXXIII–CLXXVIII. 2 *figs.*

Geitler, L. **1921.** Versuch einer Lösung des Heterocysten-Problems. *Sitzber. Akad. Wiss. Wien,* Math.-Nat. Kl. **130**[1]**:** 223–245. 1 *pl.* **1921A.** Kleine Mitteilung über Blaualgen. *Oesterr. Bot. Zeitschr.* **70:** 158–167. 7 *figs.* **1923.** Studien über das Hämatochrom und die Chromatophoren von Trentepholia. *Ibid.* **73:** 76–83. 5 *figs.* **1924.** Der Zellbau von Glaucocystis Nostochinearum und Gloeochaete Wittrockiana und die Chromatophoren-Symbiosetheorie von Mereschkowsky. *Arch. Protistenk.* **47:** 1–24. 8 *figs.* 1 *pl.* **1924A.** Ueber einige wenig bekannte Süsswasserorganismen mit roten oder blaugrünen Chromatophoren. *Rev. Algologique* **1:** 357–375. 11 *figs.* **1924B.** Die Entwicklungsgeschichte von Sorastrum spinulosum und die Phylogenie der Protococcales. *Arch. Protistenk.* **47:** 440–447. 2 *figs.* **1924C.** Ueber Acanthosphaera Zachariasi und Calyptobactron indutum nov. gen. et n. sp., zwei planktonische Protococcaceen. *Oesterr. Bot. Zeitschr.* **73:** 247–261. 10 *figs.* **1925.** Cyanophyceae. in A. Pascher, Die Süsswasserflora Deutschlands, Österreichs und der Schweiz. **12:** 1–450. 560 *figs.* **1925A.** Synoptische Darstellung der Cyanophyceen in morphologischer und systematischer Hinsicht. *Beih. Bot. Centralbl.* **41:** 163–294. 4 *pl.* **1925B.** Ueber neue oder wenig bekannte interessante Cyanophyceen aus der Gruppe Chamaesiphoneae. *Arch. Protistenk.* **51:** 321–360. 21 *figs.* 2 *pl.* **1925C.** Beiträge zur Kenntnis der Flora ostholsteinischer Seen. *Ibid.* **52:** 603–611. 4 *figs.* **1925D.** Zur Kenntnis der Gattung Pyramidomonas. *Ibid.* **52:** 356–370. 8 *figs.* 1 *pl.* **1926.** Über Chromatophoren und Pyrenoide bei Peridineen. *Ibid.* **53:** 343–346. 1 *fig.* **1926A.** Zwei neue Chrysophyceen und eine neue "Syncyanose" aus dem Lunzer Untersee. *Ibid.* **56:** 291–294. 3 *figs.* **1927.** Somatische Teilung, Reduktionsteilung, Copulation und Parthenogenese bei Cocconeis placentula. *Ibid.* **59:** 506–549. 29 *figs.* 3 *pl.* **1927A.** Die Reduktionsteilung und Copulation von Cymbella lanceolata. *Ibid.* **58:** 465–507.

14 *figs.* 2 *pl.* **1927B.** Über die Auxosporen von Meridion circulare und verwandten Diatomeen-Gattungen. *Mikrokosmos.* **21:** 79–82. **1927C.** Die Schwarmer und Kieselcysten von Phaeodermatium rivulare. *Arch. Protistenk.* **58:** 272–280. 4 *figs.* **1927D.** Häufung bei einer pennaten Diatomee. *Oesterr. Bot. Zeitschr.* **76:** 98–100. 3 *figs.* **1928.** Neue Gattungen und Arten von Dinophyceen, Heterokonten und Chrysophyceen. *Arch. Protistenk.* **63:** 67–82. 8 *figs.* 1 *pl.* **1928A.** Copulation und Geschlechtsverteilung bei einer Nitzschia-Art. *Ibid.* **61:** 419–442. 13 *figs.* **1928B.** Neue Untersuchungen über die Sexualität der pennaten Diatomeen. *Biol. Centralbl.* **48:** 648–663. 10 *figs.* **1928C.** Über die Tiefenflora an Felsen im Lunzer Untersee. *Arch. Protistenk.* **62:** 96–104. 3 *figs.* **1928D.** Autogamie bei Amphora. *Oesterr. Bot. Zeitscher.* **77:** 81–91. 3 *figs.* **1928E.** Zwei neue Dinophyceenarten. *Arch. Protistenk.* **61:** 1–8. 4 *figs.* **1929.** Ueber den Bau der Kerne zweier Diatomeen. *Ibid.* **68:** 625–636. 4 *figs.* **1930.** Ein grünes Filarplasmodium und andere neue Protisten. *Ibid.* **69:** 615–636. 15 *figs.* 1 *pl.* **1930A.** Cyanophyceae. In L. Rabenhorst, Kryptogamen-Flora von Deutschland, Österreich und der Schweiz. **14:** 1–288. 141 *figs.* **1930B.** Über das Auftreten von Karotin bei Algen und die Abgrenzung der Heterokonten. *Oesterr. Bot. Zeitschr.* **79:** 319–322. **1931.** Cyanophyceae. In L. Rabenhorst, Kryptogamen-Flora von Deutschland, Österreich und der Schweiz. **14:** 289–672. 290 *figs.* **1931A.** Untersuchungen über das sexuelle Verhalten von Tetraspora lubrica. *Biol. Centralbl.* **51:** 173–187. 5 *figs.* **1932.** Cyanophyceae. in L. Rabenhorst, Kryptogamen-Flora von Deutschland, Österreich und der Schweiz. **14:** 673–1056. 223 *figs.* **1932A.** Der Formwechsel der pennaten Diatomeen (Kieselalgen). *Arch. Protistenk.* **78:** 1–226. 125 *figs.* **1935.** Reproduction and life history in diatoms. *Bot. Rev.* **1:** 149–161. **1935A.** Über zweikernig Cysten von Dinobryon divergens. *Österr. Bot. Zeitschr.* **84:** 282–286. 2 *figs.* **1939.** Gameten- und Auxosporenbildung von Synedra ulna im Vergleich mit anderen pennaten Diatomeen. *Planta.* **30:** 551–566. 6 *figs.* **1944.** Bau und Zellpolarität der grünalge Chaetopeltis orbicularis. *Beih. Bot. Centralbl.* **62:** 221–228. 4 *figs.*

GEMEINHARDT, K. **1925.** Zur Zytologie der Gattung Achnanthidium. *Ber. deutsch. Bot. Ges.* **43:** 544–550. 1 *pl.* **1926.** Die Gattung Synedra in systematischer, zytologischer und ökologischer Beziehung. *Pflanzenforschung.* **6:** 1–88. 4 *pl.* **1926A.** Poren und Streifen in der Zellwand der Diatomeen. *Ber. deutsch. Bot. Ges.* **44:** 517–526. 1 *pl.* **1927.** Sekundäres Wachstum, Porenund Streifenbildung. *Ibid.* **45:** 570–576. 1 *pl.* **1928.** Von den Gallertporen einiger Diatomeen. *Ibid.* **46:** 285–290. 1 *pl.*

GERASSIMOFF, J. J. **1898.** Ueber die Copulation der zweikernigen bei Spirogyra. *Bull. Soc. Imp. d. Nat. de Mouscou.* **1897:** 1–20. 9 *figs.*

GERLOFF, J. **1940.** Beiträge zur Kenntnis der Variabilität und Systematik der Gattung Chlamydomonas. *Arch. Protistenk.* **94:** 311–502. 48 *figs.*

GERNECK, R. **1907.** Zur Kenntnis der niederen Chlorophyceen. *Beih. Bot. Centralbl.* **21:** 221–290. 2 *pl.*

GESSNER, F. **1931.** Volvulina (Playfair) aus dem Amazonas. *Arch. Protistenk.* **74:** 259–261. 2 *figs.*

GIESENHAGEN, K. **1896.** Untersuchungen über die Characeen. *Flora.* **82:** 381–433. 25 *figs.* 1 *pl.* **1897.** Untersuchungen über die Characeen. *Ibid.* **83:** 160–202. 17 *figs.* 1 *pl.* **1898.** Untersuchungen über die Characeen. *Ibid.* **85:** 19–64. 2 *pl.* 59 *figs.*

GIFFORD, J. W. The resolution of Amphipleura pellucida. *Jour. Roy. Microsc. Soc. London.* **1892:** 173–174. 1 *pl.*

GILBERT, E. M. **1915.** Cytology of Sphaeroplea. *Science* N.S. **41:** 183.

GILBERT, P. W. **1942.** Observations on the eggs of Amblystoma maculata with special reference to the green algae found within the egg envelopes. *Ecology.* **23:** 215–227. 2 *pl.* **1944.** The alga-egg relationship in Amblystoma maculata, a case of symbiosis. *Ibid.* **25:** 366–369. 1 *fig.*

GILLAM, W. C. **1925.** The effect on live stock, of water contaminated with fresh-water algae. *Jour. Amer. Vet. Med. Assoc.* **67:** 780–784. (Ref. *Bot. Abstr.* **15,** No. 2859, 1926.)

GÖBEL, K. **1930.** Die Deutung der Characeen-Antheridien. *Flora.* **124:** 491–498. 3 *figs.*

GOBI, C. **1871.** Algologische Studien über Chroolepus Ag. *Bull. Acad. Imp. Sci. St. Pétersbourg.* **8:** 339–362. 1 *pl.* **1887.** Peroniella Hyalothecae ein neues Süsswasseralge. *Scripta Bot. Hort. Univ. Imp. Petrop.* **2:** 233–255. 1 *pl.*

GODWARD, M. B. **1933.** The genus Naegeliella in Britain. *Jour. Bot.* **71:** 33–41. 2 *figs.* **1942.** The life cycle of Stigeoclonium amoenum Kütz. *New Phytol.* **41:** 293–300. 6 *figs.* 2 *pl.*

GOETSCH, W., and L. SCHEURING. **1926.** Parasitismus und Symbiosis der Algengattung Chlorella. *Zeitschr. Morphol. u. Ökol. Tiere.* **7:** 220–253. 15 *figs.* (Ref. *Biol. Abst.* **2,** No. 12844. 1928.)

GOETZ, G. **1899.** Ueber die Entwicklung der Eiknopse bei Characeen. *Bot. Zeitg.* **57:** 1–13. 3 *figs.* 1 *pl.*

GOLENKIN, M. **1892.** Pteromonas alata Cohn. (Ein Beitrag zur Kenntniss einzelliger Algen.) *Bull. Soc. Imp. Nat. Moscou.* **5:** 417–430. 1 *pl.* **1899.** Ueber die Befruchtung bei Sphaeroplea annulina und über die Structur der Zellkerne bei einigen grünen Algen. *Ibid.* **13:** 343–361.

GOMONT, M. **1892.** Monographie des Oscillariées. (Nostocacées homocystées) 1 part. *Ann. Sci. Nat. Bot.* VII. **15:** 263–368. 9 *pl.* **1892A.** 2 part. *Ibid.* **16:** 91–264. 7 *pl.*

GOODWIN, N. **1923.** Marketing of diatomaceous earth. *Eng. Mining Jour.* **115:** 1152–1154.

GOOR, A. C. J. VAN. **1925.** Sur les pseudo-vacuoles rouges et leur signification. *Rev. Algologique.* **2:** 19–38.

GOROSCHANKIN, J. **1890.** Beiträge zur Kenntniss der Morphologie und Systematik der Chlamydomonaden. I. Chlamydomonas Braunii. *Bull. Soc. Imp. Nat. Moscou.* **4:** 498–520. 2 *pl.* **1891.** Chlamydomonas Reinhardi (Dang.) und seine Verwändten. *Ibid.* **5:** 101–142. 3 *pl.* **1905.** Chlamydomonas coccifera (mihi). *Flora.* **94:** 420–423. 1 *pl.*

GOTZ, H. **1897.** Zur Systematik der Gattung Vaucheria DC. speciell der Arten der Umgebung Basels. *Flora.* **83:** 88–134. 55 *figs.*

GRAHAM, H. W. **195?.** Pyrrophyta, in G. M. Smith (ed.), Manual of Phycology. An introduction to the morphology and biology of the algae (in press).

GREGER, J. **1915.** Beitrag zur Kenntnis der Entwicklung und Fortpflanzung der Gattung Microthamnion Naeg. *Hedwigia.* **56:** 374–380. 1 *pl.* (Ref. *Bot. Centralbl.* **131:** 78–79, 1916.)

GRIFFITHS, B. M. **1909.** On two new members of the Volvocaceae. *New Phytol.* **8:** 130–137. 3 *figs.*

GRIGGS, R. F. **1912.** The development and cytology of Rhodochytrium. *Bot. Gaz.* **53:** 127–172. 6 *pl.*

GRINTZESCO, J. **1902.** Recherches expérimentales sur la morphologie et la physiologie de Scenedesmus acutus Meyen. *Bull. Herb. Boiss.* II. **2:** 217–264, 406–421. 6 *figs.* 5 *pl.*

GRINTZESCO, J., and S. PÉTERFI. **1932.** Sur quelques espèces appartenant au genre Stichococcus de Roumanie. *Rev. Algologique.* **6:** 159–175. 7 *figs.*

GRÖNBLAD, R. **1920.** Finnländische Desmidiaceen aus Keuru. *Acta Soc. pro Fauna et Fl. Fennica.* **47,** No. 4: 1–98. 6 *pl.* **1924.** Observations on some desmids. *Ibid.* **55,** No. 3: 1–18. 2 *pl.*

GROSS, C. **1937.** The cytology of Vaucheria. *Bull. Torrey Bot. Club.* **64:** 1–15. 31 *figs.*

GROSS, F. **1937.** The life history of some marine plankton organisms. *Phil. Trans. Roy. Soc. London B.* **228:** 1–47. 4 *pl.*

GROSSE, I. **1931.** Entwicklungsgeschichte, Phasenwechsel und Sexualität bei der Gattung Ulothrix. *Arch. Protistenk.* **73:** 206–234. 20 *figs.*

GUILLIERMOND, A. **1906.** Contribution à l'étude cytologique des Cyanophycées. *Rev. Gén. Bot.* **18:** 392–408, 447–465. 5 *pl.* **1910.** À propos des corpuscles métachromatiques ou grains de volutine. *Arch. Protistenk.* **19:** 289–309. 7 *figs.* **1925.** À propos de la structure des Cyanophycées. *Compt. Rend. Soc. Biol.* **93:** 1504–1508. 22 *figs.*

GUNTHER, F. **1928.** Über den Bau und die Lebensweise der Euglenen, besonders der Arten. E. terricola, geniculata, proxima, sanguinea und lucens nov. spec. *Arch. Protistenk.* **60:** 511–590. 5 *figs.* 3 *pl.*

GUSSEWA, K. **1930.** Über die geschlechtliche und ungeschlechtliche Fortpflanzung von Oedogonium capillare Ktz. im Lichte der sie bestimmenden Verhältnisse. *Planta.* **12:** 293–326. 54 *figs.*

GUSTAFSON, A. H. **1942.** Notes on the algal flora of Michigan. *Papers Michigan Acad. Sci., Arts and Lett.* **27:** 27–36.

GUTWINSKI, R. **1902.** De algis a Dre. M. Raciborski anno 1899 in insula Java collectis. *Bull. Int. Acad. Sci. Cracovie.* **1902:** 575–617. 5 *pl.*

HAASE, G. **1910.** Studien über Euglena sanguinea. *Arch. Protistenk.* **20:** 47–59. 3 *pl.*

HAINE, W. **1918.** Control of microscopic organisms in water supplies. *Jour. New Eng. Waterworks Assoc.* **32:** 10–20.

HALL, R. P. **1923.** Morphology and binary fission of Menoidium incurvum (Fres.) Klebs. *Univ. California Publ. Zool.* **20:** 447–476. 2 *figs.* 2 *pl.* **1925.** Mitosis in Ceratium hirundinella O.F.M., with notes on nuclear phenomena in encysted forms and the question of sexual reproduction. *Ibid.* **28:** 29–64. 5 *figs.* 5 *pl.* **1925A.** Binary fission in Oxyrrhis marina Dujardin. *Ibid.* **26:** 281–324. 7 *figs.* 5 *pl.* **1933.** The method of ingestion in Peranema trichophorum and its bearing on the pharyngeal-rod ("Staborgan") problem in Euglenida. *Arch. Protistenk.* **81:** 308–317. 15 *figs.* **1934.** A note on the flagellar apparatus of Peranema trichophorum and the status of the family Peranemidae Stein. *Trans. Amer. Microsc. Soc.* **53:** 237–243. **1939.** The trophic nature of the plant-like flagellates. *Quart. Rev. Biol.* **14:** 1–12.

HALL, R. P., and T. L. JAHN. **1929.** On the comparative cytology of certain euglenoid flagellates and the systematic position of the families Euglenidae Stein and Astasiidae Bütschli. *Trans. Amer. Microsc. Soc.* **48:** 388–405. 2 *figs.* 3 *pl.*

HALL, R. P., and W. N. POWELL. **1927.** A note on the morphology and systematic position of the flagellate Peranema trichophorum. *Trans. Amer. Microsc. Soc.* **46:** 155–165. 2 *figs.* 1 *pl.* **1928.** Morphology and binary fission of Peranema trichophorum (Ehrbg.) Stein. *Biol. Bull.* **54:** 36–64. 3 *figs.* 2 *pl.*

HAMBURGER, C. **1905.** Zur Kentnis der Dunaliella salina und einer Amöbe aus Salinenwasser von Cagliari. *Arch. Protistenk.* **6:** 111–130. 7 *figs.* 1 *pl.*

HAMEL, G. **1924–1926.** Floridées de France. *Rev. Algologique.* **1:** 278–292, 427–457. **2:** 39–67, 280–309. **3:** 99–210. 47 *figs.*

HANATSCHEK, H. **1932.** Der Phasenwechsel bei der Gattung Vaucheria. *Arch. Protistenk.* **78:** 497–513. 2 *figs.*

HANNA, G. D., and H. L. DRIVER. **1924.** The study of subsurface formation in California oil-field development. *California State Mining Bur. Summary of Operations, Oil Fields. Ann. Rept.* **10³:** 5–26. 10 *figs.*

HANSGIRG, A. **1886.** Prodromus der Algenflora von Böhmen. 1 Theil. Prag. 288 *pp.* 123 *figs.* **1892.** *Ibid.* 2 Theil. 266 *pp.* 67 *figs.*

HARDER, R. **1918.** Ueber die Bewegung der Nostocaceen. *Zeitschr. Bot.* **10:** 177–244. 8 *figs.* **1922.** Lichtintensität und "chromatische Adaptation" bei den Cyanophyceen. *Ber. deutsch. Bot. Ges.* **40:** 26–32.

HARPER, R. A. **1912.** The structure and development of the colony in Gonium. *Trans. Amer. Microsc. Soc.* **31:** 65–83. 1 *pl.* **1916.** On the nature of types in Pediastrum. *Mem. New York Bot. Garden.* **6:** 91–104. 2 *figs.* **1918.** Organization, reproduction, and inheritance in Pediastrum. *Proc. Amer. Philos. Soc.* **57:** 375–439. 2 *pl.* **1918A.** Binary fission and surface tension in the development of the colony in Volvox. *Mem. Brooklyn Bot. Garden.* **1:** 154–166. 1 *pl.*

HARTMANN, M. **1919.** Über die Kern- und Zellteilung von Chlorogonium elongatum Dang. *Arch. Protistenk.* **39:** 1–33. 2 *figs.* 3 *pl.* **1921.** Die dauernd agame Zucht von Eudorina elegans, experimentelle Beiträge zum Befruchtungs- und Todproblem. *Ibid.* **43:** 223–286. 7 *figs.* 2 *pl.* **1924:** Über die Veränderung der Koloniebildung von Eudorina elegans und Gonium pectorale unter dem Einfluss äusser Bedingungen. *Ibid.* **59:** 375–395. 4 *figs.* 4 *pl.* **1929.** Ueber die Sexualität und den Generationswechsel von Chaetomorpha und Enteromorpha. *Ber. deutsch. Bot. Ges.* **47:** 485–494. 1 *fig.*

HAUPT, A. W. **1923.** Cell structure and cell division in the Cyanophyceae. *Bot. Gaz.* **75:** 170–190. 1 *pl.*

HAYE, A. **1930.** Untersuchungen über Dinobryon divergens. *Arch. Protistenk.* **72:** 295–302. 11 *figs.*

HÄYREN, E. **1928.** Grasbälle im Brackwasser bein Nystad, regio aboënsis. *Mem. Soc. Fauna et Flora Fennica.* **4:** 182–284. (Ref. *Biol. Abstr.* **4,** No. 26675. 1930.)

HAZEN, T. E. **1899.** The life history of Sphaerella lacustris (Haematococcus pluvialis). *Mem. Torrey Bot. Club.* **6:** 211–244. 2 *pl.* **1902:** The Ulotrichaceae and Chaetophoraceae of the United States. *Ibid.* **11:** 135–250. 23 *pl.* **1922.** The phylogeny of the genus Brachiomonas. *Bull. Torrey Bot. Club.* **49:** 75–92. 2 *pl.* **1922A.** New British and American species of Lobomonas, a study in morphogenesis of motile algae. *Ibid.* **49:** 123–140. 2 *pl.*

HEDGCOCK, G. C., and A. A. HUNTER. **1899.** Notes on Thorea. *Bot. Gaz.* **28:** 425–429. 1 *pl.*

HEERING, W. **1914.** Ulotrichales, Microsporales, Oedogoniales. In A. Pascher, Die Süsswasserflora Deutschlands, Österreich und der Schweiz. **6.** Chlorophyceae. **3:** 1–250. 384 *figs.* **1921.** Siphonocladiales, Siphonales. *Ibid.* **7:** Chlorophyceae, **4:** 1–103. 94 *figs.*

HEGLER, R. **1901.** Untersuchungen über die Organisation der Phycochromaceenzelle. *Jahrb. Wiss. Bot.* **36:** 229–354. 5 *figs.* 2 *pl.*

HEGNER, R. W. **1923.** Observations and experiments on Euglenoidea in the digestive tract of frog and toad tadpoles. *Biol. Bull.* **45:** 162–180. 5 *figs.*

HEIMANS, J. **1935.** Das Genus Cosmocladium. *Pflanzenforschung.* **18:** 1–132. 8 *pl.*

HEINRICHER, E. **1883.** Zur Kentniss der Algengattung Sphaeroplea. *Ber. deutsch. Bot. Ges.* **1:** 433–450. 1 *pl.*

HEINZERLING, O. **1908.** Der Bau der Diatomeenzelle. *Bibliotheca Bot.* **15,** Heft 69: 1–98. 3 *pl.*

HEMLEBEN, H. **1922.** Ueber den Kopulationsakt und die Geschlechtsverhältnisse der Zygnemales. *Bot. Arch.* **2:** 249–259, 261–277. 25 *figs.*

Hieronymus, G. **1884.** Ueber Stephalosphaera pluvialis Cohn. *Beitr. Biol. Pflanzen.* **4:** 51–78. 2 *pl.* **1892.** Ueber Dicranochaete reniformis Hieron., eine neue Protococcacea des Süsswassers. *Ibid.* **5:** 351–372. 2 *pl.*

Higinbotham, N. **1942.** Cephalomonas, a new genus of the Volvocales. *Bull. Torrey Bot. Club.* **69:** 66–668. 43 figs.

Hill, G. A. **1916.** Origin of second spiral in Spirogyra lutetiana. *Puget Sound Marine Station Publ.* **1:** 247–248. 1 *pl.*

Hirn, K. E. **1900.** Monographie und Iconographie der Oedogoniaceen. *Acta Soc. Sci. Fennicae.* **27,** No. 1: 1–394. 27 *figs.* 64 *pl.*

Hodgetts, W. J. **1916.** Dicranochaete reniformis Hieron., a fresh-water alga new to Britain. *New Phytol.* **15:** 108–116. 12 *figs.* **1918.** Uronema elongatum, a new fresh- water member of the Ulotrichaceae. *Ibid.* **17:** 159–166. 11 *figs.* **1920.** A new species of Pyramimonas. *Ibid.* **19:** 254–258. 5 *figs.* **1920A.** A new species of Spirogyra. *Ann. Bot.* **34:** 519–524. 1 *pl.* **1920B.** On the occurrence of "false branching" in the Hormidium stage of Prasiola crispa. *New Phytol.* **19:** 260–262. 4 *figs.* **1920C.** Roya anglica G. S. West, a new desmid, with an emended description of the genus Roya. *Jour. Bot.* **58:** 65–69. 5 *figs.* **1921–1922.** A study of some of the factors controlling the periodicity of fresh-water algae in nature. *New Phytol.* **20:** 150–164, 195–227; **21:** 15–33. 11 *figs.*

Hof, T., and P. Frémy. **1932.** On Myxophyceae living in strong brines. *Rec. Trav. Bot. Néerland.* **30:** 140–162. 12 *figs.*

Hofeneder, H. **1913.** Ueber eine neue koloniebildende Chrysomonadine. *Arch. Protistenk.* **29:** 293–307. 3 *figs.* 1 *pl.* **1930.** Über die animalische Ernährung von Ceratium hirundinella O. F. Müller und über die Rolle des Kernes bei dieser Zellfunktion. *Ibid.* **71:** 1–32. 9 *figs.* 2 *pl.*

Hoffmann, W. E., and J. E. Tilden. **1930.** Basicladia, a new genus of Cladophoraceae. *Bot. Gaz.* **89:** 374–384. 22 *figs.*

Hofker, J. **1928.** Die Teilung, Mikrosporen- und Auxosporenbildung von Coscinodiscus biconicus v. Breemen. *Ann. de Protistol.* **1:** 167–194. 21 figs.

Hofler, K. **1926.** Über Eisengehalt und lokale Eisenspeicherung in der Zellwand der Desmidiaceen. *Sitzungsber. Akad. Wiss. Wien.* Math.-Nat. Kl. **135**[1]**:** 103–166. 1 *pl.*

Höll, K. **1928.** Oekologie der Peridineen. *Pflanzenforschung.* **11:** 1–105. 10 *figs.*

Hoppaugh, K. W. **1930.** A taxonomic study of species of the genus Vaucheria collected in California. *Amer. Jour. Bot.* **17:** 329–347. 4 *figs.* 4 *pl.*

Hornby, A. J. W. **1918.** A new British fresh-water alga. *New Phytol.* **17:** 41–43. 4 *figs.*

Howland, L. J. **1929.** The moisture relations of terrestrial algae. IV. Periodic observations of Trentepohlia aurea Martius. *Ann. Bot.* **43:** 173–202. 15 *figs.*

Huber, G., and F. Nipkow. **1922.** Experimentelle Untersuchungen über die Entwicklung von Ceratium hirundinella O.F.W. *Zeitschr. Bot.* **14:** 337–371. 12 *figs.*

Huber, J. **1892.** Contributions à la connaissance des Chaetophorées épiphytes et endophytes et de leur affinités. *Ann. Sci. Nat. Bot.* VII. **16:** 265–359. 11 *pl.* **1894.** Sur un état particulier du Chaetonema irregulare Nowakowski. *Bull. Herb. Boiss.* **2:** 163–166. 1 *pl.*

Huber-Pestalozzi, G. **1919.** Morphologie und Entwicklungsgeschichte von Gloeotaenium Loitlesbergerianum Hansg. *Zeitschr. Bot.* **11:** 401–472. 1 *fig.* 9 *pl* **1924:** Notiz über Gloeotaenium Loitlesbergerianum Hansgirg. *Ibid.* **16:** 624–626. 3 *figs.* **1925.** Zur Morphologie und Entwicklungsgeschichte von Asterothrix (Cerasterias) raphidioides (Reinsch) Printz. *Hedwigia.* **65:** 169–178. 5 *figs.*

Hustedt, F. **1926.** Raphe und Gallertporen der Eunotioideae. *Ber. deutsch. Bot.*

Ges. **44:** 142–150. 1 *pl.* **1926A.** Koloniebildung bei der Gattung Pinnularia. *Ibid.* **44:** 394–400. 1 *pl.* **1927.** Die Kieselalgen. In L. Rabenhorst, Kryptogamen-Flora Deutschlands, Österreichs und der Schweiz. **7:** 1–272. 114 *figs.* **1928.** Zur Morphologie und Systematik der Gattung Denticula und Epithemia. *Ber. deutsch. Bot. Ges.* **46:** 148–157. 1 *pl.* **1928A.** Ueber den Bau der Raphe bei der Gattung Hantzschia Grun. *Ibid.* **46:** 157–162. 1 *pl.* **1928B.** Die Kieselalgen. In L. Rabenhorst, Kryptogamen-Flora Deutschlands, Österreichs und der Schweiz. **7:** 273–464. 144 *figs.* **1929.** Weitere Untersuchungen über die Kanalraphe der Nitzschioideae. *Ber. deutsch. Bot. Ges.* **47:** 101–104. 1 *pl.* **1929A.** Untersuchungen über die Kanalraphe der Gattung Surirella. *Ibid.* **47:** 104–110. 1 *pl.* **1929B.** Die Kieselalgen. In L. Rabenhorst, Kryptogamen-Flora Deutschlands, Österreichs und der Schweiz **7:** 465–784. 198 *figs.*

HYMAN, L. H. **1936.** Observations on Protozoa. II. Structure and mode of food ingestion in Peranema. *Quart. Jour. Microsc. Sci. N.S.* **79:** 50–56. 1 *fig.*

IKARI, J. **1921.** On the formation of auxospores and resting spores of Chaetoceras teres Cleve. *Bot. Mag. Tokyo.* **35:** 222–227. 4 *figs.* 1 *pl.*

IMHÄUSER, L. **1889.** Entwicklungsgeschichte und Formenkreis von Prasiola. *Flora.* **72:** 233–290. 4 *pl.*

IRÉNÉE-MAIRE (Fr.). **1939.** Flore desmidiale de la region de Montreal. Laprairie (Canada). 547 *pp.* 69 *pl.*

IWANOFF, L. **1900.** Beitrag zur Kentniss der Morphologie und Systematik der Chrysomonaden. *Bull. Acad. Imp. Sci. St. Pétersbourg* V. **11:** 247–262. 2 *figs.* 1 *pl.*

IYENGAR, M. O. P. **1923.** Notes on some attached forms of Zygnemaceae. *Jour. Indian Bot. Soc.* **2:** 1–9. 4 *pl.* **1925.** Note on two new species of Botrydium from India. *Ibid.* **4:** 193–201. 5 *pl.* **1937:** Fertilization in Eudorina elegans Ehrenberg. *Ibid.* **16:** 111–118. 15 *figs.* 1 *pl.* **1939:** On the life-history of Cylindrocapsa geminella Wolle. *Current Science.* **8:** 216–217. 5 *figs.*

IYENGAR, M. O. P., and K. R. RAMANATHAN. **1940.** On sexual reproduction in a Dictyosphaerium. *Jour. Indian Bot. Soc.* **18:** 195–200. 14 figs. 1 *pl.*

IYENGAR, M. O. P., and R. SUBRAHMANYAN. **1942.** On reduction division and auxospore-formation in Cyclotella Meneghiniana Kütz. (Preliminary note.) *Jour. Indian Bot. Soc.* **21:** 231–237. 14 *figs.* 1 *pl.*

JAAG, O. **1929.** Recherches expérimentales sur les gonidies des lichens appartenant aux genres Parmelia et Cladonia. Geneva. 128 *pp.* 5 *figs.* 6 *pl.* **1933.** Coccomyxa Schmidle, Monographie einer Algengattung. *Beitr. Kryptogaenfl. d. Schweiz.* **8:** 1–132. 47 *figs.* 3 *pl.*

JACOBS, D. L. **1946.** A new parasitic dinoflagellate from fresh-water fish. *Trans. Amer. Microsc. Soc.* **65:** 1–17. 3 *pl.*

JACOBSEN, H. C. **1910.** Kulturversuche mit einigen niederen Volvocaceen. *Zeitschr. Bot.* **2:** 145–188. 1 *pl.*

JAHN, T. L. **1946.** The euglenoid flagellates. *Quart. Rev. Biol.* **21:** 246–274. 6 *figs.*

JAHN, T. L., and W. R. MCKINNEN. **1937.** A colorless euglenoid flagellate Khawkinea Halli n. gen., n. sp. *Trans. Amer. Microsc. Soc.* **56:** 48–54. 1 *pl.*

JANE, F. W. **1944.** Studies on the British Volvocales. *New Phytol.* **43:** 36–48. 41 *figs.*

JANET, C. **1912.** Le Volvox. Limoges.

JIROVEC, O. **1926.** Die Plasmaeinschlüsse bei Polytoma uvella. *Arch. Protistenk.* **56:** 280–284. 2 *figs.* 1 *pl.*

JOHANSEN, D. A. **1940.** Plant microtechnique. New York. 523 *pp.* 110 *figs.*

JOHNSON, D. F. **1934.** Morphology and life history of Colacium vesicolosum Ehrbg. *Arch. Protistenk.* **83:** 241–263. 20 *figs.*

JOHNSON, L. N. **1893.** Observations on the zoospores of Draparnaldia. *Bot. Gaz.* **18:** 294–298. 1 *pl.*

JOHNSON, L. P. **1944.** Euglenae of Iowa. *Trans. Amer. Microsc. Soc.* **63:** 97–135. 7 *pl.*

JOLLOS, V. **1910.** Dinoflagellatenstudien. *Arch. Protistenk.* **19:** 178–206. 4 *pl.*

JORDE, I. **1933.** Untersuchungen über den Lebenszyklus von Urospora Aresch. und Codiolum A. Braun. *Nyt. Mag. Naturvidenskab.* **73:** 1–19. 5 *figs.* 1 *pl.*

JÖRSTADT, I. **1919.** Undersökelser over zygoternes spiring hos Ulothriz subflaccida Wille. *Nyt. Mag. Naturvidenskab.* **51:** 61–68. 1 *pl.*

JULLER, E. **1937.** Der Generations- und Phasenwechsel bei Stigeocolonium subspinosum. *Arch. Protistenk.* **89:** 55–93. 21 *figs.*

JURÄNYI, L. **1873.** Beitrag zur Morphologie der Oedogonien. *Jahrb. Wiss. Bot.* **9:** 1–35. 3 *pl.*

JUSSIEU, A. L. DE. **1789.** Genera plantarum secundum ordines naturales disposita. Paris.

JUST, L. **1882.** Phyllosiphon Arisari. *Bot. Zeitg.* **40:** 1–8, 17–26, 33–47, 49–57. 1 *pl.*

KARLING, J. S. **1927.** Variations in the mature antheridium of the Characeae: a study in morphogenesis. *Bull. Torrey Bot. Club.* **54:** 187–230. 13 *figs.* 5 *pl.* **1935.** Tetracladium Marchalianum and its relation to Asterothrix, Phycastrum, and Cerasterias. *Mycolog.* **27:** 478–495. 5 *figs.*

KARSTEN, G. **1891.** Untersuchungen ueber die Familie der Chroolepideen. *Ann. Jard. Bot. Buitenzorg.* **10:** 1–66. 6 *pl.* **1896.** Untersuchungen über Diatomeen. *Flora.* **82:** 286–296. 1 *pl.* **1897.** Untersuchungen über Diatomeen. II. *Ibid.* **83:** 33–53. 2 *pl.* **1897A.** Untersuchungen über Diatomeen. III. *Ibid.* **83:** 203–222. 1 *pl.* **1898.** Rhodomonas baltica, n.g. et sp. *Wissensch. Meeresunters.* N.F. *Abt. Kiel.* **3:** 15–16. 1 *pl.* **1899.** Die Diatomeen der Kieler Bucht. *Ibid.* **4:** 17–205. 219 *figs.* **1900.** Die Auxosporenbildung der Gattungen Cocconeis, Surirella und Cymatopleura. *Flora.* **87:** 253–283. 3 *pl.* **1904.** Die sogenannten "Mikrosporen" der Planktondiatomeen und ihre weitere Entwicklung, beobachtet an Corethron Valdiviae n. sp. *Ber. deutsch. Bot. Ges.* **22:** 544–554. 1 *pl.* **1909.** Die Entwicklung der Zygoten von Spirogyra jugalis Ktzg. *Flora.* **99:** 1–11. 1 *pl.* **1912.** Über die Reduktionsteilung bei der Auxosporenbildung von Surirella saxonica. *Zeitschr. Bot.* **4:** 417–426. 1 *pl.* **1924.** Ueber Diatomeen, ihre Fortpflanzung und verwandtschaftlichen Beziehungen. *Internat. Rev. gesamt. Hydrobiol. Hydrograph.* **12:** 116–120. **1928.** Bacillariophyta. In A. Engler and K. Prantl, Die näturlichen Pflanzenfamilien. 2d ed. **2:** 105–303. 332 *figs.*

KASANOWSKY, V. **1913.** Die Chlorophyllbänder und Verzweigung derselben bei Spirogyra Nawaschini (sp. nov). *Ber. deutsch. Bot. Ges.* **31:** 55–59. 1 *pl.*

KATER, J. M. **1925.** Morphology and life history of Polytomella citri n. sp. *Biol. Bull.* **49:** 213–236. 3 *pl.* **1929.** Morphology and division of Chlamydomonas with reference to the phylogeny of the flagellate neuromotor system. *Univ. California Publ. Zool.* **33:** 125–168. 7 *figs.* 6 *pl.*

KAUFFMANN, H. **1914.** Über den Entwicklungsgang von Cylindrocystis. *Zeitschr. Bot.* **6:** 721–774. 4 *figs.* 1 *pl.*

KAWASAKI, Y. **1937.** On the life history of Schizomeris Leibleinii Kütz. *Bot. Mag. Tokyo.* **51:** 25–30. 1 *fig.* 1 *pl.*

KEEFE, A. M. **1926.** A preserving fluid for green plants. *Science* N.S. **64:** 331–332.

KIENER, W. **1944.** Green snow in Nebraska. *Proc. Nebraska Acad. Sci. 54th Annual Meeting. p.* 12. **1946:** A list of algae chiefly from the Alpine zone of Longs Peak, Colorado. *Madroño.* **8:** 161–173.

KILLIAN, C. **1924.** Le cycle évolutif du Gloedinium montanum (Klebs). *Arch. Protistenk.* **50:** 50–66. 2 *figs.* 2 *pl.*

KINDLE, A. M. **1915.** Limestone solution on the bottom of Lake Ontario. *Amer. Jour Sci.* IV. **39:** 651–656. 3 *figs.* **1934.** Concerning "lake balls," "Cladophora balls," and "coal balls." *Amer. Midland Nat.* **15:** 752–760. 2 *figs.*

KIRCHNER, O. **1883.** Zur Entwicklungsgeschichte von Volvox minor (Stein). *Beitr. Biol. Pflanzen.* **3:** 95–103. 1 *pl.*

KLEBAHN, H. **1891.** Die Keimung von Closterium und Cosmarium. *Jahrb. Wiss. Bot.* **22:** 415–443. 2 *pl.* **1892.** Die Befruchtung von Oedogonium Boscii. *Ibid.* **24:** 235–267. 1 *pl.* **1892A.** Chaetosphaeridium Pringsheimii, novum genus et nova species algarum chlorophycearum aquae dulcis. *Ibid.* **24:** 268–282. 1 *pl.* **1893.** Zur Kritik einiger Algengattungen. *Ibid.* **25:** 278–321. 1 *pl.* **1895.** Gasvacuolen, ein Bestandtheil der Zellen der wasserblüthebildenden Phycochromaceen. *Flora.* **80:** 240–282. 1 *pl.* **1896.** Beiträge zur Kenntnis der Auxosporenbildung. I. Rhopalodia gibba (Ehr.) O. Müll. *Jahrb. Wiss. Bot.* **29:** 595–654. 1 *pl.* **1896A.** Ueber wasserblütebildende Algen, insbesondere des Plöner Seengebietes, und über das Vorkommen von Gasvacuolen bei den Phycochromaceen. *Forschungsber. Biol. Stat. Plön.* **4:** 189–206. **1897.** Bericht über einige Versuche, betreffend die Gasvacuolen von Gloiotrichia echinulata. *Ibid.* **5:** 166–179. 2 *figs.* **1899.** Die Befruchtung von Sphaeroplea annulina Ag. *Schwendener Festschr. pp.* 81–103. 1 *pl.* **1922.** Neue Untersuchungen über die Gasvacuolen. *Jahrb. Wiss. Bot.* **61:** 535–589. 8 *figs.* **1925.** Weitere Untersuchungen über die Gasvakuolen. *Ber. deutsch. Bot. Ges.* **43:** 143–159. 2 figs.

KLEBS, G. **1881.** Beiträge zur Kenntnis niederer Algenformen. *Bot. Zeitg.* **39:** 249–257, 265–272, 281–290, 297–308, 313–319, 329–336. 2 *pl.* **1883.** Über die Organisation einiger Flagellaten-Gruppen und ihre Beziehungen zu Algen und Infusorien. *Untersuch. Bot. Inst. Tübingen.* **1:** 233–360. 2 *pl.* **1885.** Ueber Bewegung und Schleimbildung bei Desmidiaceen. *Biol. Centralbl.* **5:** 353–367. **1886.** Über die Organisation der Gallerte bei einigen Algen und Flagellaten. *Untersuch. Bot. Inst. Tübingen.* **2:** 333–417. 2 *pl.* **1891:** Ueber die Bildung der Fortpflanzungszellen bei Hydrodictyon utriculatum Roth. *Bot. Zeitg.* **49:** 789–798, 805–817, 821–835, 837–846, 853–862. 1 *pl.* **1892.** Flagellatenstudien II. *Ztschr. Wiss. Zool.* **55:** 353–445. 2 *pl.* **1896.** Die Bedingungen der Fortpflanzung bei einigen Algen und Pilzen. Jena. 543 *pp.* 15 *figs.* 3 *pl.* **1912.** Ueber flagellaten- und algenähnliche Peridineen. *Verh. Naturh.-Med. Ver. Heidelberg* N.F. **11:** 369–451. 15 *figs.* 1 *pl.*

KLYVER, F. D. **1929.** Notes on the life history of Tetraspora gelatinosa (Vauch.) Desv. *Arch. Protistenk.* **66:** 290–296. 1 *pl.*

KOFOID, C. A. **1898.** On Pleodorina illinoisensis, a new species from the plankton of the Illinois River. *Bull. Illinois State Lab. of Nat. Hist.* **5:** 273–293. 2 *pl.* **1899.** On Platydorina, a new genus of the family Volvocidae, from the plankton of the Illinois River. *Ibid.* **5:** 419–440. 1 *pl.* **1907.** The plates of Ceratium with a note on the unity of the genus. *Zool. Anz.* **32:** 177–183. 8 *figs.* **1908.** Exuviation, autotomy and regeneration in Ceratium. *Univ. California Publ. Zool.* **4:** 345–386. 33 *figs.* **1909.** On Peridinium Steini Jörgensen, with a note on the nomenclature of the skeleton of the Peridinidae. *Arch. Protistenk.* **16:** 25–47. 1 *pl.* **1910.** The plankton of the Illinois River, 1894–1899. Part 2. Constituent organisms and their seasonal distribution. *Bull. Illinois State Lab. of Nat. Hist.* **8:** 1–361. 5 *pl.* **1911.** The genus Gonyaulax with notes on its skeletal morphology and a discussion of its generic and specific characters. *Univ. California Publ. Zool.* **8:** 187–269. 5 *figs.* 9 *pl.* **1914.** Phytomorula regularis, a symmetrical protophyte related to Coelastrum. *Univ. California Publ. Bot.* **6:** 35–40. 1 *pl.*

Kofoid, C. A., and O. Swezy. **1921.** The free-living unarmored Dinoflagellata. *Mem. Univ. California.* **5:** 1–538. 47 *figs.* 12 pl.

Kohl, F. G. **1903.** Ueber die Organisation und Physiologie der Cyanophyceenzelle und die mitotische Teilung ihres Kernes. Jena.

Kol, E. **1927.** Über die Bewegung mit Schleimbildung einiger Desmidiaceen aus der Hohen Tatra. *Folia Cryptogamica Szeged* (Hungary). **1:** 435–442. 2 *pl.* (Ref. *Jour. Roy. Microsc. Soc.* **1927:** 300). **1941.** The green snow of Yellowstone National Park. *Amer. Jour. Bot.* **28:** 185–191. 14 *figs.* 1 *pl.* **1947.** Vergleich der Kryovegetation der nordlichen und sudlichen Hemisphäre. *Arch. Hydrobiol.* **40:** 835–846. 11 *figs.*

Kolkwitz, R. **1926.** Zur Ökologie und Systematik von Botrydium granulatum (L.) Grev. *Ber. deutsch. Bot. Ges.* **44:** 533–540. 2 *figs.* 1 *pl.*

Korshikov, A. A. **1913.** Spermatozoopsis exultans nov. gen. et sp. *Ber. deutsch. Bot. Ges.* **31:** 174–183. 1 *pl.* **1917.** Contribution à l'étude des algues de la Russie. *Trav. de la Stat. Biol. Borodinskaja.* **4:** 219–267. 1 *pl.* (Ref. Korshikov, 1928.) **1923.** Protochlorinae, eine neue Gruppe der grünen Flagellata. *Arch. Russ. Protistol.* **2:** 148–169. 2 *pl.* **1923A.** Zur Morphologie des geschlechtlichen Prozesses bei den Volvocales. *Ibid.* **2:** 179–194. 1 *pl.* **1924.** Zur Morphologie und Systematik der Volvocales. *Ibid.* **3:** 153–197. 1 *pl.* **1924A.** Protistologische Beobachtungen. *Ibid.* **3:** 57–74. 1 *pl.* **1924B.** Über einige wenig bekannte Organismen. *Ibid.* **3:** 113–127. 4 *figs.* 1 *pl.* **1925.** Beiträge zur Morphologie und Systematik der Volvocales. I. *Ibid.* **4:** 153–197. 3 *pl.* **1926.** On some new organisms from the groups Volvocales and Protococcales, and on the genetic relations of these groups. *Arch. Protistenk.* **55:** 439–503. 15 *figs.* 9 *pl.* **1927.** Phyllocardium complanatum, a new Polyblepharidacea. *Ibid.* **58:** 441–449. 2 *pl.* **1927A.** On the validity of the genus Schizomeris Kütz. *Arch. Russ. Protistol.* **6:** 71–82. 2 *pl.* **1927B.** Skadovskiella sphagnicola, a new colonial Chrysomonad. *Arch. Protistenk.* **58:** 450–455. 1 *pl.* **1928.** Notes on some new or little known Protococcales. *Arch. Protistenk.* **62:** 416–426. 1 *pl.* **1928A.** On two new Spondylomoraceae: Pascheriella tetras n. gen. et sp., and Chlamydobotrys squarrosa n. sp. *Ibid.* **61:** 223–238. 5 *figs.* 1 *pl.* **1929:** Studies on the Chrysomonads. I. *Ibid.* **67:** 253–290. 1 *fig.* 4 *pl.* **1930.** On the occurrence of pyrenoids in Heterocontae. *Beih. Bot. Centralbl.* **46:** 470–478. 2 *figs.* **1930A.** On the origin of diatoms. *Ibid.* **46:** 460–469. 1 *fig.* **1935.** On the taxonomical position of Chaetopeltis orbicularis. *Univ. de Charkov. Trav. Inst. Bot.* **1:** 13–19. 10 *figs.* **1937.** On the sexual reproduction (oögamy) in the Micractinieae. *Proc. Kharkov A. Gorky State Univ.* Book **10:** 109–126 5 *pl.* **1938.** Contribution to the algal flora of the Gorky district. 1. *Proc. Kharkov A. Gorky State Univ.* Book **14.** *Proc. Bot. Inst.* **3:** 1–21. 3 *pl.* **1938A.** On the occurrence of Volvoluna Steinii in Ukrania. *Bull. Soc. Nat. Moscou.* **47:** 56–63. 1 *pl.*

Kostrun, G. **1944.** Entwicklung der Keimlinge und Polaritätsverhalten bei Chlorophyceen. *Wiener Bot. Zeitschr.* **93:** 172–221. 17 *figs.*

Kraemer, H. **1901.** The position of Pleurococcus and mosses on trees. *Bot. Gaz.* **32:** 422–423.

Kraskovits, G. **1905.** Ein Beitrag zur Kenntnis des Zellteilungsvorgänge bei

Oedogonium. *Sitzungsber. Akad. Wiss. Wien.* Math.-Nat. Kl. **114**[1]: 237–274. 11 *figs.* 3 *pl.*

KRASSILSTSCHIK, J. **1882.** Zur Entwicklungsgeschichte und Systematik der Gattung Polytoma Ehr. *Zool. Anz.* **5:** 426–429. **1882A.** Zur Naturgeschichte und über die systematische Stellung von Chlorogonium euchlorum Ehr. *Ibid.* **5:** 627–634.

KRICHENBAUER, H. **1937.** Beitrag zur Kenntnis der Morphologie und Entwicklungsgeschichte der Gattungen Euglena und Phacus. *Arch. Protistenk.* **90:** 8–122. 18 *figs.*

KRIEGER, W. **1933.** Die Desmidiaceen. In L. Rabenhorst, Kryptogamen-Flora von Deutschland, Österreich und der Schweiz. **13,** Abt. **1:** 1–223. 33 *figs.* 8 *pl.* **1935.** Die Desmidiaceen. *Ibid.* **13,** Abt. **1:** 224–376. 29 *pl.* **1937.** Die Desmidiaceen. *Ibid.* **13,** Abt. **1:** 377–712. 60 *pl.*

KRIGER, W. **1930.** Untersuchungen über Plankton-Chrysomonaden. *Bot. Archiv.* **29:** 257–329. 63 *figs.*

KUCKUCK, P. **1894.** Bemerkungen zur marinen Algenvegetation von Helgoland. *Wissensch. Meersuntersuch.* N. F. **1**[1]: 225–263. 29 *figs.*

KUDO, R. R. **1946.** Protozoology. Springfield, Illinois. 778 *pp.* 334 *figs.*

KURSSANOW, L. **1911.** Über Befruchtung, Reifung und Keimung bei Zygnema. *Flora.* **104:** 65–84. 4 *pl.*

KURSSANOW, L., and N. M. SCHEMAKHANOVA. **1927.** Sur la succession des phases-nucléaire chez les algues vertes. I. Le cycle de développement du Chlorochy trium Lemnae Cohn. *Arch. Russ. Protistol.* **6:** 131–146. 2 *figs.* 2 *pl.*

KUSCHAKEWITSCH, S. **1931.** Zur Kenntnis der Entwicklungsgeschichte von Volvox. *Arch. Protistenk.* **73:** 323–330. 14 *figs.* 1 *pl.*

KYLIN, H. **1906.** Zur Kenntnis einiger schwedischen Chantransia-Arten. *Bot. Studier tillägn. F. R. Kjellman.* *pp.* 113–126. 9 *figs.* **1912.** Ueber die roten und blauen Farbstoffe der Algen. *Hoppe-Seyler's Zeitschr. Physiol. Chem.* **76:** 396–425. **1912A.** Studien über die schwedischen Arten der Gattungen Batrachospermum Roth und Sirodotia nov. gen. *Nova Acta Reg. Soc. Sci. Upsaliensis* IV. **3.** No. 3: 1–40. **1912B.** Über die Farbe der Florideen und Cyanophyceen. *Svensk. Bot. Tidskr.* **6:** 531–544. **1913.** Zur Biochemie der Meeresalgen. *Hoppe-Seyler's Zeitschr. Physiol. Chem.* **83:** 171–197. **1917.** Ueber die Entwicklungsgeschichte von Batrachospermum moniliforme. *Ber. deutsch. Bot. Ges.* **35:** 155–164. 7 *figs.* **1922.** Ueber die Entwicklungsgeschichte der Bangiaceen. *Ark. Bot.* **17,** No. 5: 1–12. 7 *figs.* **1923.** Studien über die Entwicklungsgeschichte der Florideen. *Kgl. Svensk. Vetensk.-Ak. Handl.* **63,** No. 11: 1–139. 82 *figs.* **1928.** Entwicklungsgeschichtliche Florideenstudien. *Lunds Univ. Årsskr.* N. F. **24,** No. 4: 1–127. 64 *figs.* **1930.** Über Heterogamie bei Enteromorpha intestinalis. *Ber. deutsch. Bot. Ges.* **48:** 458–464. 1 *fig.* **1930A.** Some physiological remarks on the relationship of the Bangiales. *Bot. Notiser.* **1930:** 417–420. **1931.** Einige Bemerkungen über Phykoerythrin und Phycocyan. *Hoppe-Seyler's Zeitschr. Physiol. Chem.* **197:** 2–6. 2 *figs.* **1935.** Über einige kalkbohrende Chlorophyceen. *Förh. Kgl. Fysiograf. Sallsk. i Lund.* **5,** No. 19: 1–19. 17 *figs.* **1937.** Über die Farbstoffe und die Farbe der Cyanophyceen. *Ibid.* **7,** No. 12: 1–28. **1943.** Zur Biochemie der Rhodophyceen. *Ibid.* **13,** No. 6: 1–13. **1943A.** Zur Biochemie der Cyanophyceen. *Ibid.* **13,** No. 7: 1–14.

LACKEY, J. B. **1929.** The cytology of Entosiphon sulcatum (Duj.) Stein. *Arch. Protistenk.* **66:** 175–200. 24 *figs.* **1932.** Oxygen deficiency and sewage protozoa: with descriptions of some new species. *Biol. Bull.* **63:** 287–295. 1 *pl.* **1933.** The morphology of Peranema trichophorum Ehrenberg, with special reference to its kinetic elements and the classification of the Heteronemidae. *Ibid.* **65:** 238–248. 2 *pl.* **1934.** A comparison of the structure and division of Distigma proteus Ehrenberg and Astasia Dangeardii Lemm. A study in phylogeny. *Ibid.* **67:** 145–162. 26 *figs.* **1936.** Some fresh-water protozoa with blue chromatophores. *Ibid.* **71:** 492–497. 1 *pl.* **1938.** The manipulation and counting of river plankton and changes in some organisms due to formalin preservation. *U.S. Treasury Dept., Public Health Repts.* **53:** 2080–2093. 6 *figs.* **1938A.** A study of some ecological factors affecting the distribution of protozoa. *Ecol. Monogr.* **8:** 501–527. **1938B.** Scioto River forms of Chrysococcus. *Amer. Midland Nat.* **20:** 619–623. 11 *figs.* **1938C.** Protozoan plankton as indicators of pollution in a flowing stream. *U.S. Treasury Dept., Public Health Repts.* **53:** 2037–2058. **1939.** Notes on plankton flagellates from the Scioto River. *Lloydia.* **2:** 128–143. 38 *figs.* **1941.** Two groups of flagellated algae serving as indicators of clean water. *Jour. Amer. Waterworks Assn.* **33:** 1099–1110. 32 *figs.* **1942.** The plankton algae and protozoa of two Tennessee rivers. *Amer. Midland Nat.* **17:** 191–202. 2 *figs.* **1942A.** The effects of distillery wastes and waters on the microscopic flora and fauna of a small creek. *U.S. Treasury Dept., Public Health Repts.* **57:** 253–260.

Lackey, J. B., E. Wattie, J. F. Kachmar, and O. R. Placak. **1943.** Some plankton relationships in a small upland unpolluted stream. *Amer. Midland Nat.* **30:** 403–425.

Lagerheim, G. **1882.** Bidrag till kannendomen om Stockholmstrakens Pediastreer, Protococcaceer och Palmellaceer. *Öfvers. Kgl. Svensk.-Ak. Förh.* **39,** No. 2: 47–81. 2 *pl.* **1883.** Bidrag till Sveriges algflora. *Ibid.* **40,** No. 2: 37–78. 1 *pl.* **1884.** Ueber Phaeothamnion, eine neue Gattung den Süsswasseralgen. *Bih. Kgl. Svensk. Vetensk.-Ak. Handl.* **9,** No. 19: 1–14. 1 *pl.* **1884A.** Eine Präparirmethode für trockene mikroskopische Pflanzen. *Bot. Centralbl.* **18:** 182–183. **1885.** Codiolum polyrhizum n. sp. *Öfvers. Kgl. Svensk. Vetensk.-Ak. Förh.* **42,** No. 8: 21–31. 1 *pl.* **1887.** Note sur l'Uronema, nouveau genre des algues d'eau douce de l'ordre Chlorozoosporacées. *Malphigia.* **1:** 517–523. 1 *pl.* **1888.** Zur Entwicklungsgeschichte des Hydrurus. *Ber. deutsch. Bot. Ges.* **6:** 73–85. 5 *figs.* **1889.** Studien über die Gattungen Conferva und Microspora. *Flora.* **72:** 179–210. 2 *pl.* **1892.** Ueber die Fortpflanzung von Prasiola (Ag.) Menegh. *Ber. deutsch. Bot. Ges.* **10:** 366–374. 1 *pl.* **1892A.** Die Schneeflora des Pichincha. *Ibid.* **10:** 517–534. 1 *pl.* **1893.** Rhodochytrium nov. gen., eine Uebergangsform von den Protococcaceen zu den Chytridiaceen. *Bot. Zeitg.* **51:** 43–52. 1 *pl.* **1894.** Ueber die Entwickelung von Tetraëdron Kütz. und Euastropsis Lagerh., eine neue Gattung der Hydrodictyaceen. *Tromsö Museums Aarshefter.* **17:** 1–24. 1 *pl.* **1895.** Ueber das Phycoporphyrin, einen Conjugatenfarbstoff. *Videnskabs.-Selskab. Christiana Skrifter.* Mat.-Nat. Kl **1895,** No. 5: 1–25. 2 *figs.*

Lambert, F. D. **1910.** Two new species of Characium. *Rhodora.* **11:** 65–74. 1 *pl.* **1910A.** An unattached zoosporic form of Coleochaete. *Tufts College Studies.* Scientific Series. **3:** 61–68. 1 *pl.* **1930.** On the structure and development of Prasinocladus. *Zeitschr. Bot.* **23:** 227–244. 4 *figs.*

Lander, C. E. **1929.** Oögenesis and fertilization in Volvox. *Bot. Gaz.* **87:** 431–436. 1 *pl.*

Langeron, M. **1923.** Les Oscillariées parasites du tube digestif de l'homme et des animaux. *Ann. Parasitol. humaine et comp.* **1:** 75–89, 113–123. 10 *figs.* (Ref *Rev. Algologique.* **1:** 186–188. 1924.)

Lauterborn, R. **1895.** Kern und Zelltheilung von Ceratium hirundinella O.F.M

Zeitschr. Wiss. Zool. **59:** 167–190. 2 *pl.* **1896.** Untersuchungen über Bau, Kernteilung und Bewegung der Diatomeen. Leipzig. **1899.** Flagellaten aus dem Gebiete des Oberrheins. *Zeitschr. Wiss. Zool.* **65:** 369–391. 2 *pl.* **1911.** Pseudopodien bei Chrysopyxis. *Zool. Anz.* **38:** 46–51. 1 *fig.*

LEAKE, D. V. **1938.** Preliminary note on the production of motile cells in Basicladia *Proc. Oklahoma Acad. Sci.* **19:** 109–111. 4 *figs.* **1946.** Studies of development and germination of aplanospores and zoospores of Basicladia crassa Hoffman and Tilden. *Amer. Jour. Bot.* **33:** 7*a*–8*a*.

LEE, S. **1927.** Cytological study of Stigonema mammilosum. *Bot. Gaz.* **83:** 420–424. 1 *pl.*

LEVERE, M. **1932.** Monographie des espèces d'eau douce du genre Peridinium. *Archives de Bot.* **2:** 1–210. 915 *figs.* 5 *pl.*

LEMMERMANN, E. **1898.** Beiträge zur Kenntnis der Planktonalgen. I. Golenkinia Chodat, Richteriella Lemm., Franceia nov. gen., Phythelios Frenzel. Lagerheimia Chodat, Chodatella nov. gen., Schroederia nov. gen. *Hedwigia*, **37:** 303–312. 4 *figs.* 1 *pl.* **1899.** Das Genus Ophiocytium Naegeli. *Ibid.* **38:** 20–38. 4 *figs.* 2 *pl.* **1899A.** Das Phytoplankton sächsicher Teiche. *Forschungsber. Biol. Stat. Plön.* **7:** 96–135. 2 *pl.* **1900.** Beiträge zur Kenntnis der Planktonalgen, IX. Lagerheimia Marssonii nov. spec., Centratractus (Schm.) nov. gen. et spec., Synedra limnetica nov. spec., Marssoniella elegans nov. gen. et spec. *Ber. deutsch. Bot. Ges.* **18:** 272–275. **1900A.** Die Gattung Dinobryon Ehrenb. *Ibid.* **18:** 500–524. 2 *pl.* **1900B.** Die Coloniebildung von Richteriella botryoides (Schmidle) Lemm. *Ibid.* **18:** 90–91. 1 *pl.* **1900C.** Die Arten der Gattung Pteromonas Seligo. *Ibid.* **18:** 92–94. 1 *pl.* **1900D.** Das Phytoplankton brackischer Gewässer. *Ibid.* **18:** 94–98. 1 *pl.* **1904.** Das Plankton schwedischer Gewässer. *Ark. Bot.* **2,** 1–209. 2 *pl.* **1904A.** Über Entstehung neuer Planktonformen. *Ber. deutsch. Bot. Ges.* **22:** 17–20. **1905.** Die Algenflora der Sandwich-Inseln. *Bot. Jahrb. Syst., Pflanzengeschichte und Pflanzengeog.* **34:** 607–663. 2 *pl.* **1913.** Eugleninae. In A. Pascher, Die Süsswasserflora Deutschlands, Österreichs und der Schweiz. 2. Flagellata. **2:** 115–174. 197 *figs.* **1914.** Die Gattung Characiopsis Borzi. *Abh. Nat. Ver. Bremen.* **23:** 249–261. 14 *figs.* **1915.** Tetrasporales. In A. Pascher, Die Süsswasserflora Deutschlands, Österreichs und der Schweiz. 5, Chlorophyceae. **2:** 21–51. 33 *figs.*

LERCHE W. **1937.** Untersuchungen über Entwicklung und Fortpflanzung in der Gattung Dunaliella. *Arch. Protistenk.* **88:** 236–268. 5 *figs.* 3 *pl.*

LEWIS, F. J. **1898.** The action of light on Mesocarpus. *Ann. Bot.* **12:** 418–421.

LEWIS, I. F. **1907.** Notes on the morphology of Coleochaete Nitellarum. *Johns Hopkins Univ. Circ.* **195:** 201–202 (29–30). **1913.** Chlorochromonas minuta, a new flagellate from Wisconsin. *Arch. Protistenk.* **32:** 249–256. 1 *pl.* **1924.** The flora of Penikese, fifty years after. *Rhodora.* **26:** 181–195, 211–229. **1925:** A new conjugate from Woods Hole. *Amer. Jour. Bot.* **12:** 351–357. 2 *pl.*

LEWIS, I. F., and C. ZIRKLE, **1920.** Cytology and systematic position of Porphyridium cruentum Naeg. *Amer. Jour. Bot.* **7:** 333–340. 2 *pl.*

LIEBISCH, W. **1928.** Amphitetras antediluviana Ehrbg., sowie einige Beiträge zum Bau und zur Entwicklung der Diatomeenzelle. *Zeitschr. Bot.* **20:** 225–271. 22 *figs.* 2 *pl.* **1929.** Experimentelle und kritische Untersuchungen über die Pektinmembran der Diatomeen unter besonderer Berücksichtigung der Auxosporenbildung und der Kraticularzustände. *Ibid.* **22:** 1–65. 14 *figs.* 1 *pl.*

LINDEMANN, E. **1919.** Untersuchungen über Süsswasserperidineen und ihre Variationsformen. *Arch. Protistenk.* **39:** 209–262. 144 *figs.* 1 *pl.* **1928.** Peridineae. In A. Engler and K. Prantl. Die natürlichen Pflanzenfamilien. 2ed. **2:** 1–104. 92 *figs.*

LINDENBEIN, W. **1927.** Beiträge zur Cytologie der Charales. *Planta.* **4:** 437–466. 22 *figs.*

LINNAEUS, C. **1754.** Genera plantarum. 5 ed. Holmiae.

LIST, H. **1930.** Die Entwicklungsgeschichte von Cladophora glomerata Kützing. *Arch. Protistenk.* **72:** 453–481. 7 *figs.*

LIVINGSTON, B. E. **1900.** On the nature of the stimulus which causes the change of form in polymorphic green algae. *Bot. Gaz.* **30:** 289–317. 2 *pl.* **1901.** Further notes on the physiology of polymorphism in green algae. *Ibid.* **32:** 292–302. **1905.** Notes on the physiology of Stigeoclonium. *Ibid.* **39:** 297–300. 3 *figs.* **1905A.** Chemical stimulation of a green alga. *Bull. Torrey Bot. Club.* **32:** 1–34. 17 *figs.*

LLOYD, F. E. **1926.** Cell disjunction in Spirogyra. *Papers Michigan Acad. Sci. Arts and Letters.* **6:** 275–287. 1 *fig.* 1 *pl.* **1926A.** Conjugation in Spirogyra. *Trans. Roy. Canadian Inst.* **15**[1]**:** 151–193. 4 *pl.* **1926B.** Studies on Spirogyra. I. Additional studies on conjugation. *Trans. Roy. Soc. Canda.* III. **20**[5]**:** 75–99. 1 *pl.* **1926C.** Studies on Spirogyra. II. Adhesions and geniculations. *Ibid.* III. **20**[5]**:** 101–111. 1 *pl.* **1928.** Further observations on the behavior of gametes during maturation and conjugation. *Protoplasma.* **4:** 45–66. 1 *pl.*

LOEFER, J. B. **1931.** Morphology and binary fission of Heteronema acus (Ehr.) Stein. *Arch. Protistenk.* **74:** 449–470. 3 *figs.* 3 *pl.*

LOWE, W. C., and F. E. LLOYD. **1927.** Some observations on Hydrodictyon reticulatum (L.) Lagerh. with special reference to the chloroplast and organization. *Trans. Roy Soc. Canada.* III. **21**[5]**:** 279–289. 4 *pl.*

LUND, J. W. G. **1937.** Some new British algal records. *Jour. Bot.* **75:** 305–314. 4 *figs.* **1942.** Contributions to our knowledge of British algae. VIII. *Ibid.* **80:** 57–73. 11 *figs.* 1 *pl.* **1942A.** Contributions to our knowledge of British Chrysophyceae. *New Phytol.* **41:** 274–292. 11 *figs.*

LUNDELL, P. M. **1871.** De Desmidiaceis, quae in Sueciae inventae sunt, observationes criticae. *Nova Acta Reg. Soc. Sci. Upsaliensis.* III. **8:** 1–100. 5 *pl.*

LUTHER, A. **1899.** Ueber Chlorosaccus, eine neue Gattung der Süsswasseralgen, nebst einigen Bemerkungen zur Systematik verwandter Algen. *Bih. Kgl. Svensk. Vetersk.-Ak. Handl.* **24,** Afd. 3, No. 13: 1–22. 1 *pl.*

LÜTKEMÜLLER, J. **1902.** Die Zellmembran der Desmidiaceen. *Beitr. Biol. Pflanzen.* **8:** 347–414. 3 *pl.* **1905.** Zur Kenntnis der Gattung Penium Bréb. *Verh. Zool.-Bot. Ges Wien.* **55:** 332–337. **1910.** Zur Kenntnis der Desmidiaceen Böhmens. *Ibid.* **60:** 478–503. 3 *figs.* 2 *pl.*

LUTMAN, B. F. **1910.** The cell structure of Closterium Ehrenbergii and Closterium moniliferum. *Bot. Gaz.* **49:** 241–255. 2 *pl.* **1911.** Cell and nuclear division in Closterium. *Ibid.* **51:** 401–430. 1 *fig.* 2 *pl.*

MAGNUS, W., and B. SCHINDLER. **1912.** Ueber den Einfluss der Nährsalze auf die Färbung der Oscillarien. *Ber. deutsch. Bot. Ges.* **30:** 314–320.

MAINX, F. **1927.** Untersuchungen über Ernährung und Zellteilung bei Eremosphaera viridis DeBary. *Arch. Protistenk.* **57:** 1–13. 1 *fig.* 1 *pl.* **1928.** Beiträge zur Morphologie und Physiologie der Eugleninen. I Teil. *Ibid.* **60:** 305–354. 8 *figs.* 1 *pl.* **1928A.** Beiträge zur Morphologie und Physiologie der Eugleninen. II Teil. *Ibid.* **60:** 355–414. **1929.** Über die Geschlechterverteilung bei Volvox aureus. *Ibid.* **67:** 205–214. **1931.** Physiologische und genetische Untersuchungen an Oedogonien. I. Mitteilung. *Zeitschr. Bot.* **24:** 481–527. 13 *figs.* 1 *pl.* **1931A.** Gametencopulation und Zygotenkeimung bei Hydrodictyon reticulatum. *Arch. Protistenk.* **75:** 502–516. 1 *pl.*

MAIRE, R. **1908.** Remarques sur une algue parasite (Phyllosiphon Arisari Kühne). *Bull. Soc. Bot. France.* **55:** 162–164.

MANGIN, L. **1907.** Observations sur la constitution de la membrane des Péridiniens. *Compt. Rend. Acad. Sci. Paris.* **144:** 1055–1057. **1908.** Observations sur les Diatomées. *Ann. Sci. Nat. Bot.* IX. **8:** 177–219. 14 *figs.* **1911.** Modifications de la cuirasse chez quelques Péridiniens. *Internat. Rev. gesamt. Hydrobiol. Hydrograph.* **4:** 44–54. 2 *pl.* **1912.** La sporulation chez les Diatomées. *Rev. Scientifique.* **50**[2]**:** 481–486. 7 *figs.*

MANN, A. **1922.** Suggestions for the collecting and preparing of diatoms. *Proc. U. S. Nat. Museum.* **69**[15]**:** 1–8.

MANN, H. H., and C. M. HUTCHINSON. **1907.** Cephaleuros virescens Kunze, the "red rust" of tea. *Mem. Dept. Agr. in India. Bot.* **1.** No. 6: 1–33. 8 *pl.*

MANNING, W. M., C. JUDAY, and M. WOLF. **1939.** Photosynthesis of aquatic plants at different depths in Trout Lake, Wisconsin. *Trans. Wisconsin Acad.* **31:** 377–410.

MASSEE, G. **1891.** Life history of a stipitate fresh-water alga. *Jour. Linn. Soc. London Bot.* **27:** 457–462.

MAST, S. O. **1916.** The process of orientation in the colonial organism, Gonium pectorale, and a study of the structure and function of the eyespot. *Jour. Exp. Zool.* **20:** 1–17. 6 *figs.* **1928.** Structure and function of the eyespot in unicellular and colonial organisms. *Arch. Protistenk.* **60:** 197–220. 4 *figs.* 1 *pl.*

MASTERS, C. W. **1940.** Notes on subtropical plants and animals in Ohio. *Ohio Jour. Sci.* **40:** 146–148.

MCALLISTER, F. **1913.** Nuclear division in Tetraspora lubrica. *Ann. Bot.* **27:** 681–696. 1 *pl.* **1930.** Starch formation in Oedogonium and Zygnema. Abstracts of papers before the general section, Botanical Society of America, Dec. 30, 1930–Jan. 1, 1931. **1931.** The formation of the achromatic figure in Spirogyra setiformis. *Amer. Jour. Bot.* **18:** 838–853. 2 *pl.*

MCINTEER, B. B. **1939.** A check list of the algae of Kentucky. *Castanea.* **4:** 27–37.

MEINHOLD, T. **1911.** Beiträge zur Physiologie der Diatomeen. *Beitr. Biol. Pflanzen.* **10:** 353–378. 1 *pl.*

MEISTER, F. **1912.** Die Kieselalgen der Schweiz. *Beitr. Kryptogamenfl. Schweiz.* **4:** 1–254. 48 *pl.*

MELOCHE, V., G. LEADER, L. SAFRANSKI, and C. JUDAY. **1938.** The silica and diatom content of Lake Mendota water. *Trans. Wis. Acad.* **31:** 363–376.

MERESCHKOWSKY, C. **1903.** Ueber farblose Pyrenoide und gefärbte Elaeoplasten der Diatomeen. *Flora.* **92:** 77–83. 4 *figs.*

MERRIAM, M. L. **1906.** Nucear division in Zygnema. *Bot. Gaz.* **41:** 45–53. 2 *pl.*

MERTON, H. **1908.** Über den Bau und die Fortpflanzung von Peodorina illinoisensis Kofoid. *Zeitschr. Wiss. Zool.* **90:** 445–477. 2 *figs.* 2 *pl.*

METCALF, R. W., and A. B. HOLLEMAN. **1948.** Abrasive materials. In E. W. Pehrson, Minerals Yearbook for 1946. Washington. *pp.* 94–96.

METZNER, J. **1945.** A morphological and cytological study of a new form of Volvox. *Bull. Torrey Bot. Club.* **72:** 86–113, 121–136. 120 *figs.*

MEYER, A. **1895.** Untersuchungen über Stärkekörne. Jena. **1895A.** Ueber den Bau von Volvox aureus Ehrenb. und Volvox globator Ehrenb. *Bot. Centralbl.* **63:** 225–233. 4 *figs.* **1896.** Die Plasmaverbindungen und die Membranen von Volvox globator, aureus, und tertius, mit Rücksicht auf die thierischen Zellen. *Bot. Zeitg.* **54:** 187–217. 7 *figs.* 1 *pl.* **1904.** Orientirende Untersuchungen über Verbreitung, Morphologie und Chemie des Volutins. *Ibid.* **62:** 113–152. 1 *pl.*

MEYER, K. **1906.** Die Entwicklungsgeschichte der Sphaeroplea annulina Ag. *Bull. Soc. Imp. Nat. Moscou.* N.S. **19:** 60–84. 2 *pl.* **1909.** Zur Lebensgeschichte der

Trentepohlia umbrina Mart. *Bot. Zeitg.* **67:** 25–43. 2 *figs.* 2 *pl.* **1929.** Über die Auxosporenbildung bei Gomphonema geminatum. *Arch. Protistenk.* **66:** 421–435. 2 *pl.* **1930.** Über den Befruchtungsvorgang bei Chaetonema irregulare Nowak. *Ibid.* **72:** 147–157. 15 *figs.* **1935.** Zur Kenntnis der geschlechtlichen Fortpflanzung bei Eudorina und Pandorina. *Beih. Bot. Bentralbl.* **53:** 421–426. **1936.** Sur le genre Trentepohlia Mart. I. Trentepohlia Gobii sp. nov. *Bull. Soc. Nat. Moscou.* **45:** 315–321. 20 *figs.* **1936A.** Contribution à l'étude du genre Trentepohlia. II. Trentepohlia uncinata (Gobi) Hansg. *Ibid.* **45:** 425–432. 17 *figs.* **1936B.** Germination des zoospores et des gamètes chez le Trentepohlia Mart. *Ibid.* **45:** 95–103. 43 *figs.* **1937.** Contribution à l'étude du genre Trentepohlia Mart. III. Trentepohlia aurea (L.) Mart. *Ibid.* **46:** 101–110. 33 *figs.* **1938.** Contribution à l'étude du genre Trentepohlia Mart. *Ibid.* **47:** 64–68. 11 *figs.*

MIGULA, W. **1890.** Beiträge zur Kenntnis des Gonium pectorale. *Bot. Centralbl.* **44:** 72–76, 103–107, 143–147. 1 *pl.*

MILLER, V. **1906.** Beobachtungen über Actidesmium Hookeri Reinsch. *Ber. biol. Süsswasserstation kais. Naturf.-Ges. St. Petersburg.* **2:** 9–29. 1 *pl.* **1927.** Untersuchungen über die Gattung Botrydium Wallroth. I. Allgemeiner Teil. *Ber. deutsch. Bot. Ges.* **45:** 151–161. 1 *pl.* **1927A.** Untersuchungen über die Gattung Botrydium Wallroth. II. Spezieller Teil. *Ibid.* **45:** 161–170. 1 *pl.*

MITRA, A. K. **1943.** A preliminary study of some methods of asexual reproduction in Draparnaldiopsis indica. *Proc. Nat. Acad. Sci. India.* **13:** 106–111. 2 *pl.*

MÖBIUS, M. **1888.** Beitrag zur Kenntnis der Algengattung Chaetopeltis Berthold. *Ber. deutsch. Bot. Ges.* **6:** 242–248. 1 *pl.* **1891.** Beitrag zur Kenntnis der Gattung Thorea. *Ibid.* **9:** 333–344. 1 *pl.* **1892.** Bemerkungen über die systematische Stellung von Theorea Bory. *Ibid.* **10:** 266–270.

MOEWUS, F. **1933.** Untersuchungen über die Sexualität und Entwicklung von Chlorophyceen. *Arch. Protistenk.* **80:** 469–526. 8 *figs.* **1935.** Die Vererbung des Geschlechts bei verschiedenen Rassen von Protosiphon botryoides. *Ibid.* **86:** 1–57. 5 *figs.* **1935A.** Über die Vererbung des Geschlechts bei Polytoma Pascheri und bei Polytoma uvella. *Zeitschr. Indukt. Abstamm. u. Vererb.* **69:** 374–417. 12 *figs.* **1936.** Faktorenaustausch, insbesondere der Realisatoren bei Chlamydomonas-Kreuzungen. *Ber. deutsch. Bot. Ges.* **54:** (45)–(57). 3 *figs.* 1 *pl.* **1938.** Vererbung des Geschlechts bei Chlamydomonas eugametos und verwandten Arten. *Biol. Zentralbl.* **58:** 516–536. **1938A.** Die Sexualität und der Generationswechsel der Ulvaceen und Untersuchungen über die Parthenogenese der Gameten. *Arch. Protistenk.* **91:** 357–441. 25 *figs.* **1939.** Untersuchungen über die relative Sexualität von Algen. *Biol. Zentralbl.* **59:** 40–58. **1940.** Über Sexualität von Botrydium granulatum. *Ibid.* **60:** 484–498. 2 *figs.* **1940A.** Die Analyse von 42 erblichen Eigenschaften der Chlamydomonas eugametos Gruppen. *Zeitschr. Indukt. Abstamm. u. Vererb.* **78:** 418–522. 12 *figs.*

MOLISCH, H. **1903.** Die sogenannten Gasvacuolen und das Schweben gewisser Phycochromaceen. *Bot. Zeitg.* **61:** 47–58. 4 *figs.* **1903A.** Notiz über eine blaue Diatomee. *Ber. deutsch. Bot. Ges.* **21:** 23–26. 1 *pl.* **1905.** Ueber den braunen Farbstoff der Phaeophyceen und Diatomeen. *Bot. Zeitg.* **63:** 131–144. **1913.** Mikrochemie der Pflanzen. Jena. **1925.** Über die Symbiose der beiden Lebermoose Blasia pusilla L. und Cavicularia densa St. mit Nostoc. *Sci. Repts. Tohoku Imp. Univ.* 4 Ser. Biol. **1:** 169–188. 1 *fig.* 2 *pl.*

MOORE, G. T. **1897.** Notes on Uroglena americana Calk. *Bot. Gaz.* **23:** 105–112. 1 *pl.* **1901.** Eremosphaera viridis and Excentrosphaera. *Ibid.* **32:** 309–324. 3 *pl.* **1917.** Preliminary list of algae in Devils Lake, North Dakota. *Ann. Mis-*

souri Bot. Garden. **4:** 293–303. **1917A.** Chlorochytrium gloeophilum Bohl. *Ibid.* **4:** 271–278. 1 *pl.* **1918.** A wood-penetrating alga, Gomontia lignicola n. sp. *Ibid.* **5:** 211–224. 3 *figs.*

Moore, G. T., and N. Carter. **1923.** Algae from lakes in the northeastern part of North Dakota. *Ann. Missouri Bot. Garden.* **10:** 393–422. 1 *pl.* **1926.** Further studies on the subterranean algal flora of the Missouri Botanical Garden. *Ibid.* **13:** 101–140.

Moore, G. T., and J. L. Karrer. **1919.** A subterranean algal flora. *Ann. Missouri Bot. Garden.* **6:** 281–307.

Moore, S. L. **1890.** Apiocystis, a Volvocinea, a chapter in degeneration. *Jour. Linn. Soc. Bot. London.* **25:** 362–380. 3 *pl.*

Morse, D. C. **1943.** Some details of asexual reproduction in Pandorina morum. *Trans. Amer. Microsc. Soc.* **62:** 24–26.

Mottier, D. M. **1904.** The development of the spermatozoid in Chara. *Ann. Bot.* **18:** 245–254. 1 *pl.*

Müller, O. **1881.** Ueber den anatomischen Bau der Bacillariengattung Terpsinoë. *Sitzungsber. Ges. Naturf. Freunde in Berlin* **1881:** 3, (Ref. *Jour. Roy. Microsc. London.* **1881:** 783–785.). **1889.** Durchbrechung der Zellwand in ihren Beziehungen zur Ortsbewegung der Bacillariaceen. *Ber. deutsch. Bot. Ges.* **7:** 160–180. 1 *pl.* **1889A.** Auxosporen von Terpsinoë musica Ehr. *Ibid.* **7:** 181–183. 1 *pl.* **1893.** Die Ortsbewegung der Bacillariaceen. *Ibid.* **11:** 571–576. 1 *fig.* **1894.** Die Ortsbewegung der Bacillariaceen. II. *Ibid.* **12:** 136–143. 1 *fig.* **1895.** Ueber Achsen, Orientirungs- und Symmetrieebenen bei den Bacillariaceen. *Ibid.* **13:** 222–234. 1 *pl.* **1896.** Die Ortsbewegung der Bacillariaceen. III. *Ibid.* **14:** 54–64. 2 *pl.* **1896A.** Die Ortsbewegung der Bacillariaceen. IV. **14:** 111–128. 1 *pl.* **1897.** Die Ortsbewegung der Bacillariaceen. V. *Ibid.* **15:** 70–86. **1898.** Kammern und Poren in der Zellwand der Bacillariaceen. *Ibid.* **16:** 386–402. 2 *pl.* **1899.** Kammern und Poren in der Zellwand der Bacillariaceen. II. *Ibid.* **17:** 423–452. 2 *pl.* **1900.** Kammern und Poren in der Zellwand der Bacillariaceen. III. *Ibid.* **18:** 480–497. 1 *fig.* **1901.** Kammern und Poren in der Zellwand der Bacillariaceen. IV. *Ibid.* **19:** 195–210. 3 *figs.* 1 *pl.* **1908.** Die Ortsbewegung der Bacillariaceen. VI. *Ibid.* **26A:** 676–685. **1909.** Die Ortsbewegung der Bacillariaceen. VII. *Ibid.* **27:** 27–43. 1 *fig.* 1 *pl.*

Mundie, J. R. **1929.** Cytology and life history of Vaucheria geminata. *Bot. Gaz.* **87:** 397–410. 2 *pl.*

Murray, G., and E. S. Barton. **1891.** On the structure and systematic position of Chantransia; with a description of a new species. *Jour. Linn. Soc. Bot.* **28:** 209–216. 2 *pl.*

Nageli, C. **1849.** Gattungen einzelliger Algen. 137 *pp.* 8 *pl.*

Naumann, E. **1919.** Notizen zur Systematik der Süsswasseralgen. *Ark. Bot.* **16,** No. 2: 1–19. 12 *figs.*

Nelson, N. P. B. **1903.** Observations upon some algae which cause "water bloom." *Minnesota Bot. Studies.* **3:** 51–56. 1 *pl.*

Nienburg, W. **1916.** Die Perzeption des Lichtreizes bei den Oscillarien und ihre Reaktionen auf Intensitätsschwankungen. *Zeitschr. Bot.* **8:** 161–193.

Nieuwland, J. A. **1909.** Resting spores of Cosmarium bioculatum Bréb. *Amer. Midland Nat.* **1:** 4–8. 1 *pl.*

Nigrelli, R. F. **1936.** The morphology, cytology and life-history of Oodinium ocellatum Brown, a dinoflagellate parasite of marine fishes. *Zoologica (New York).* **21:** 129–164. 5 *figs.* 9 *pl.*

Nipkow H. F. **1928.** Über das verhalten der Skelette planktischer Kieselalgen im geschichteten Tiefschlamm Zürich und Baldegeersees. *Zeitschr. Hydrol.* **4:** 71–120.

NISHIMURA, M., and R. KANNO. **1927.** On the asexual reproduction of Aegagropila Sauteri (Nees) Kütz. *Bot. Mag. Tokyo.* **41:** 432–438. 2 *pl.*

NOLAND, L. E. **1928.** A combined fixative and stain for demonstrating flagella and cilia in temporary mounts. *Science* N.S. **67:** 535.

OBERDORFER, E. **1927.** Lichtverhältnisse und Algenbesiedlung im Bodensee. *Zeitschr. Bot.* **20:** 465–568. 10 *figs.*

OEHLKERS, F. **1916.** Beitrag zur Kenntnis der Kernteilungen bei den Characeen. *Ber. deutsch. Bot. Ges.* **34:** 223–227. 1 *fig.*

OHASHI, H. **1930.** Cytological study of Oedogonium. *Bot. Gaz.* **90:** 177–197. 21 *figs.* 3 *pl.*

OLIVE, E. W. **1904.** Mitotic division of the nuclei of the Cyanophyceae. *Beih. Bot. Centralbl.* **18:** 9–44. 2 *pl.*

OLTMANNS, F. **1895.** Ueber die Entwickelung der Sexualorgane bei Vaucheria. *Flora.* **80:** 388–420. 5 *pl.* **1898.** Die Entwickelung der Sexualorgane bei Coleochaete pulvinata. *Ibid.* **85:** 1–14. 2 *pl.* **1904.** Morphologie und Biologie der Algen. Vol. 1. Jena. 733 *pp.* 467 *figs.* **1922.** Morphologie und Biologie der Algen. 2d ed. Vol. 1. Jena. 459 *pp.* 287 *figs.*

OSTERHOUT, W. J. V. **1900.** Befruchtung bei Batrachospermum. *Flora.* **87:** 109–115. 1 *pl.* **1906.** The resistance of certain marine algae to changes in osmotic pressure and temperature. (Preliminary communication.) *Univ. California Publ. Bot.* **2:** 227–228.

OTROKOV, P. **1875.** Ueber das Keimen d. Zygoten v. Eudorina elegans. *Nachtr. d. kais. Ges. d. Liebh. d. Naturw. u. Anthropol.* u.s. w. (Ref. Schreiber, 1925.)

OTT, E. **1900.** Untersuchungen über den Chromatophorenbau der Süsswasser-Diatomaceen und dessen Beziehungen zur Systematik. *Sitzungsber. Akad. Wiss. Wien* (Math.-Nat. Kl.). **109**[1]**:** 769–801. 6 *pl.*

PALIK, P. **1933.** Über die Entstehung der Polyeder bei Pediastrum Boryanum (Turpin) Meneghini. *Arch. Protistenk.* **79:** 234–238. 10 *figs.*

PALLA, E. **1894.** Ueber eine neue pyrenoidlose Art und Gattung der Conjugaten. *Ber. deutsch. Bot. Ges.* **12:** 228–236. 1 *pl.*

PALM, B. **1924.** The geographical distribution of Rhodochytrium. *Ark. Bot.* **18,** No. 15: 1–8.

PALMER, C. M. **1942.** Additional records for algae, including some of the less common forms. *Butler Univ. Bot. Studies.* **5:** 224–234. 21 *figs.*

PALMER, T. C., and F. J. KEELEY. **1900.** The structure of the diatom girdle. *Proc. Acad. Nat. Sci. Phila.* **52:** 465–479. 2 *pl.*

PAPENFUSS, G. F. **1945.** Review of the Acrochaetium-Rhodochorton complex of the red algae. *Univ. California Publ. Bot.* **18:** 299–334.

PASCHER, A. **1905.** Zur Kenntnis der geschlechtlichen Fortpflanzung bei Stigeoclonium sp. [St. fasciculatum Kutz. (?)]. *Flora.* **95:** 95–107. 2 *figs.* **1907.** Studien über die Schwärmer einiger Süsswasseralgen. *Bibliotheca Bot.* **67:** 1–116. 8 *pl.* **1907A.** Über auffallende Rhizoid- und Zweigbildung bei einer Mougeotia-Art. *Flora.* **97:** 107–115. 3 *figs.* **1909.** Über merkwürdige amoeboide Stadien bei einer höheren Grünalge. *Ber. deutsch. Bot. Ges.* **27:** 143–150. 1 *pl.* **1909A.** Der Grossteich bei Hirschberg in Nord-Böhmen. I. Chrysomonaden. *Internat. Tev. gesamt. Hydrobiol. Hydrograph. Monogr. u. Abhandl.* **1:** 1–66. 3 *pl.* **1909B.** Einige neue Chrysomonade. *Ber. deutsch. Bot. Ges.* **27:** 247–254. 1 *pl.* **1910.** Ueber einige Fälle vorübergehender Koloniebildung bei Flagellaten. *Ibid.* **28:** 339–350. 1 *pl.* **1911.** Zwei braune Flagellaten. *Ibid.* **29:** 190–192. 2

figs. **1911A.** Über die Beziehungen der Cryptomonaden zu den Algen. *Ibid.* **29:** 193–203. **1911B.** Über Nannoplanktonten des Süsswassers. *Ibid.* **29:** 523–533. 1 *pl.* **1912.** Eine farblose, rhizopodial Chrysomonade. *Ibid.* **30:** 152–158. 1 *pl.* **1912A.** Ueber Rhizopoden und Palmellastadien bei Fagellaten (Chrysomonaden). *Arch. Protistenk.* **25:** 153–200. 7 *figs.* 1 *pl.* **1912B.** Braune Flagellaten mit seitlichen Geisseln. *Zeitschr. Wiss. Zool.* **100:** 177–189. 3 *figs.* **1913.** Zur Gliederung der Heterokonten. *Hedwigia.* **53:** 6–22. 8 *figs.* **1913A.** Die Süsswasserflora Deutschlands, Österreichs und der Schweiz. 2. Flagellata **2.** Chrysomonadinae *pp.* 7–95. 150 *figs.* **1913B.** Die Heterokontengattung Pseudotetraëdron. *Hedwigia.* **53:** 1–5. 6 *figs.* **1913C.** Die Süsswasserflora Deutschlands, Österreichs und der Schweiz. 2. Flagellata **2.** *pp.* 96–114. 30 *figs.* **1914.** Ueber Flagellaten und Algen. *Ber. deutsch. Bot. Ges.* **32:** 136–160. **1915.** Animalische Ernährung bei Grünalgen. *Ibid.* **33:** 427–442. 1 *pl.* **1915A.** Ueber Halosphaera. *Ibid.* **33:** 488–492. **1915B.** Ueber einige rhizopodiale, Chromatophoren führende Organismen aus der Flagellatenreihe der Chrysomonaden. *Arch. Protistenk.* **36:** 91–117. 13 *figs.* 3 *pl.* **1915C.** Über eine neue Amöbe–Dinamoeba (varians)—mit dinoflagellatenartigen Schwärmern. *Ibid.* **36:** 118–136. 1 *pl.* **1915D.** Die Süsswasserflora Deutschlands, Österreichs und der Schweiz. 5. Chlorophyceae **2.** Einzellige Chlorophyceengattungen unsicherer Stellung. *pp.* 206–229. 34 *figs.* **1916.** Rhizopodialnetze als Fangvorrichtung bei einer plasmodialen Chrysomonade. *Arch. Protistenk.* **37:** 15–30. 6 *figs.* 1 *pl.* **1916A.** Fusionsplasmodien bei Flagellaten und ihre Bedeutung für die Ableitung der Rhizopoden von den Flagellaten. *Ibid.* **37:** 31–64. 20 *figs.* 1 *pl.* **1916B.** Über die Kreuzung einzelliger, haploider Organismen; Chlamydomonas. *Ber. deutsch. Bot. Ges.* **34:** 228–242. 5 *figs.* **1917.** Flagellaten und Rhizopoden in ihren gegenseitigen Beziehungen. *Arch. Protistenk.* **38:** 1–88. 65 *figs.* **1918.** Über amoeboide Gameten, Amoebozygoten und diploide Plasmodien bei einer Chlamydomonadine. *Ber. deutsch. Bot. Ges.* **36:** 352–359. 15 *figs.* **1918A.** Amoeboide Stadien bei einer Protococcale, nebst Bemerkungen über den primitiven Charakter nicht festsitzender Algenformen. *Ibid.* **36:** 253–260. 8 *figs.* **1921.** Ueber die Übereinstimmungen zwischen den Diatomeen, Heterokonten und Chrysomonaden. *Ibid.* **39:** 236–248. 6 *figs.* **1923.** Ueber das regionale Auftreten roter Organismen in Süsswasserseen. *Bot. Arch.* **3:** 311–314. **1924.** Zur Homologisierung der Chrysomonadencysten mit den Endosporen der Diatomeen. *Arch. Protistenk.* **48:** 196–203. 4 *figs.* **1925.** Die braune Algenreihen der Chrysophyceen. *Ibid.* **52:** 489–564. 10 *figs.* 1 *pl.* **1925A.** Die Süsswasserflora Deutschlands, Österreichs und der Schweiz. **11:** Heterokontae. *pp.* 1–118. 96 *figs.* **1925B.** Neue oder wenig bekannte Protisten. VII. *Arch. Protistenk.* **51:** 549–577. 21 *figs.* **1927.** Die braune Algenreihe aus der Verwandtschaft der Dinoflagellaten (Dinophyceen). *Ibid.* **58:** 1–54. 38 *figs.* **1927A.** Die Süsswasserflora Deutschlands, Österreichs und der Schweiz. **4:** Volvocales. *pp.* 1–506. 451 *figs.* **1927B.** Neue oder wenig bekannte Flagellaten. XVIII. *Arch. Protistenk.* **58:** 577–598. 20 *figs.* **1928.** Von einer neuen Dinococcale (Cystodinium phaseolus) mit zwei verschiedenen Schwärmertypen. *Ibid.* **63:** 241–254. 7 *figs.* **1929.** Ueber einige Endosymbiosen von Blaualgen in Einzellern. *Jahrb. Wiss. Bot.* **71:** 386–462. 31 *figs.* 1 *pl.* **1929A.** Ueber die Beziehungen zwischen Lagerform und Standortsverhältnissen bei einer Gallertalge (Chrysocapsale). *Arch. Protistenk.* **68:** 637–668. 22 *figs.* **1929B.** Doppelzellige Flagellaten und Parallelentwicklungen zwischen Flagellaten und Algenschwärmern. *Ibid.* **68:** 261–304. 21 *figs.* **1929C.** Eine neue farblose Chlorophyceae. *Beih. Bot. Centralbl.* **45:** 390–400. 3 *figs.* **1930.** Neue Volvocalen

(Polyblepharidinen-Chlamydomonadinen). *Arch. Protistenk.* **69:** 103–146. 40 *figs.* **1930A.** Ein grüner Sphagnum-Epiphyt und seine Beziehung zu freilebenden Verwandten (Desmatractum, Calyptrobactron, Bernardinella). *Ibid.* **69:** 637–658. 16 *figs.* 1 *pl.* **1930B.** Ueber einen grünen assimilationsfähigen plasmodialen Organismus in den Blättern von Sphagnum. *Ibid.* **72:** 311–358. 27 *figs.* 2 *pl.* **1930C.** Zur Kenntnis der heterokonten Algen. *Ibid.* **69:** 401–451. 45 *figs.* 1 *pl.* **1931.** Systematische Übersicht über die mit Flagellaten in Zusammenhang stehenden Algenreihen und Versuch einen Einreihung dieser Algenstämme in die Stämme des Pflanzenreiches. *Beih. Bot. Centralbl.* **48:** 317–332. **1931A.** Über einen neuen einzelligen und einkernigen Organismus mit Eibefruchtung. *Ibid.* **48:** 466–480. 10 *figs.* **1932.** Über die Verbreitung endogener bzw. endoplasmatisch gebildeter Sporen bei den Algen. *Ibid.* **49:** 293–308. 14 *figs.* **1932A.** Über eine in ihrer Jugend rhizopodial und anamalisch lebende epiphytische Alge (Perone). *Ibid.* **49:** 675–685. 7 *figs.* **1932B.** Zur Kenntnis mariner Planktonten. I. Meringosphaera und ihre Verwandten. *Arch. Protistenk.* **77:** 195–218. 27 *figs.* **1932C.** Über einige neue oder kritische Heterokonten. *Ibid.* **77:** 305–359. 37 *figs.* **1932D.** Über drei auffallend konvergente zu verschiedenen Algenreihen gehörende epiphytische Gattungen. *Beih. Bot. Centralbl.* **49:** 549–568. 13 *figs.* 1 *pl.* **1937.** Heterokonten. In L. Rabenhorst, Kryptogamen-Flora Deutschlands, Österreich und der Schweiz. **11:** 1–480. 335 *figs.* **1938.** Heterokonten. *Ibid.* **11:** 481–832. 358 *figs.* **1939.** Heterokonten. *Ibid.* **11:** 833–1092. 219 *figs.* **1939A.** Über geisselbewegliche Eier, mehrköpfig Schwärmer und vollständifen Schwärmerverlust bei Sphaeroplea. *Beih. Bot. Centralbl.* **59:** 188–213. 22 *figs.* **1940.** Rhizopodiale Chrysophyceen. *Arch. Protistenk.* **93:** 331–349. 15 *figs.* 1 *pl.* **1940A.** Filarplasmodiale Ausbildungen bei Algen. *Ibid.* **94:** 295–309. 12 *figs.* **1943.** Zur Kenntnis verschiedener Ausbildungen der planktonischen Dinobryen. *Int. Rev. ges. Hydrobiol. Hydrog.* **43:** 110–123. 5 *figs.* **1943A.** Alpine Algen. I. Neue Protococcalengattungen aus den Uralpen. *Beih. Bot. Centralbl.* **62:** 175–196. 16 *figs.*

Pascher, A., and J. Schiller. **1925.** Rhodophyta. In A. Pascher, Die Süsswasserflora Deutschlands, Österreichs und der Schweiz. **11:** 134–206. 94 *figs.*

Pavillard, J. **1914.** Observations sur les Diatomées; 3 série. *Bull. Soc. Bot. France.* **61:** 164–172. 2 *figs.*

Pearsall, W. H. **1923.** A theory of diatom periodicity. *Jour. Ecol.* **11:** 165–183. 10 *figs.*

Peck, R. E. **1946.** Fossil Charophyta. *Amer. Midland Nat.* **36:** 275–278. 6 *figs.*

Peebles, F. **1909.** The life history of Sphaerella lacustris (Haematococcus pluvialis) with especial reference to the nature and behavior of the zoospores. *Centralbl. Bakt.* Abt. 2, **24:** 511–521. 28 *figs.*

Peirce, G. J. **1906.** Anthoceros and its Nostoc colonies. *Bot. Gaz.* **42:** 55–59.

Peirce, G. J., and F. A. Randolph. **1905.** Studies of irritability in algae. *Bot. Gaz.* **40:** 321–350. 27 *figs.*

Penard, E. **1921.** Studies on some flagellata. *Proc. Acad. Nat. Sci. Phila.* **73:** 105–168. 4 *pl.*

Persidsky, B. M. **1935.** The sexual process in Melosira varians. *Beih. Bot. Centralbl.* **53A:** 122–132. 23 *figs.*

Peter, A. **1886.** Ueber eine auf Thieren schmarotzende Alge. **59.** *Versammlung deutsch. Naturf. u. Aerzte.* (Ref. *Bot. Centralbl.* **28:** 125, 1886.)

Peterschilka, F. **1924.** Über die Kernteilung und die Vielkernigkeit und über die Beziehungen zwischen Epiphytismus und Kernzahl bei Rhizoclonium hieroglyphicum Kütz. *Arch. Prostistenk.* **47:** 325–349. 5 *figs.* 1 *pl.*

PETERSEN, J. B. **1915.** Studier over Danske aërofile Alger. *Mém. Acad. Roy. Sci. et Lett. Danemark.* VII. **12:** 272–379. 22 *figs.* 4 *pl.* **1918.** Om Synura uvella Stein og nogle andre Chrysomonadiner. *Vidensk. Medd. fra Dansk. naturhist. Foren.* **69:** 345–357. 1 *pl.* **1928.** The aërial algae of Iceland. In The Botany of Iceland. **2:** 328–447. 36 *figs.* **1929.** Beiträge zue Kenntnis der Flagellatengeisseln. *Bot. Tidskr.* **40:** 373–389. 1 *pl.* **1935.** Studies on the biology and taxonomy of soil algae. *Dansk Bot. Ark.* **8,** No. 9: 1–183.

PETIT, A. **1926.** Contribution à l'étude cytologique et taxonomique des bacteries. *Compt. Rend. Acad. Sci. Paris.* **182:** 717–719. 4 *figs.*

PFITZER, E. **1871.** Untersuchungen über Bau und Entwicklung der Bacillariaceen. In Hanstein, *Bot. Abhandl. a. d. Gebiet der Morph. u. Physiol.* Heft. 2. **1882.** Die Bacillariaceen (Diatomaceen). In A. Schenk, Handbuch der Botanik. **2:** 403–445. 16 *figs.*

PHILIPOSE, M. T. **1946.** A note on Pseudoulvella americana (Snow) Wille growing in Madras. *Jour. Indian Bot. Soc.* M.O.P. Iyengar Commemoration Volume *pp.* 321–325. 26 *figs.* 1 *pl.*

PHILLIPS, O. P. **1904.** A comparative study of the cytology and movements of the Cyanophyceae. *Contrib. Bot. Lab. Univ. Pennsylvania.* **2:** 237–335. 3 *pl.*

PICKETT, F. L. **1912.** A case of changed polarity in Spirogyra elongata. *Bull. Torrey Bot. Club.* **39:** 509–510. 1 *fig.* 1 *pl.*

PIEPER, A. **1913.** Die Diaphototaxis der Oscillarien. *Ber. deutsch. Bot. Ges.* **31:** 594–599.

PITELKA, D. R. **1945.** Morphology and taxonomy of flagellates of the genus Peranema Dujardin. *Jour. Morphol.* **76:** 179–192. 1 *fig.* 2 *pl.*

PLAYFAIR, G. I. **1914.** Contributions to a knowledge of the biology of the Richmond River. *Proc. Linn. Soc. New South Wales.* **39:** 93–151. 5 *pl.* **1915.** Fresh-water algae of the Lismore district: with an appendix on the algal fungi and Schizomycetes. *Ibid.* **40:** 310–362. 11 *figs.* 6 *pl.* **1916.** Oöcystis and Eremosphaera. *Ibid.* **41:** 107–147. 28 *figs.* 3 *pl.* **1918.** New and rare fresh-water algae. *Ibid.* **43:** 497–543. 11 *figs.* 5 *pl.*

POCHMANN, A. **1942.** Synopsis der Gattung Phacus. *Arch. Protistenk.* **95:** 81–252. 192 *figs.*

POCOCK, M. A. **1933.** Volvox and associated algae from Kimberley. *Ann. South African Mus.* **14:** 473–521. 7 *figs.* 12 *pl.* **1933A.** Volvox in South Africa. *Ibid.* **14:** 523–646. 10 *figs.* 12 *pl.* **1937.** Studies in South African Volvocales. *Proc. Linn. Soc. London,* Session 149, 1936–1937. *pp.* 55–58. **1937A.** Hydrodictyon in South Africa. With notes on the known species of Hydrodictyon. *Trans. Roy. Soc. South Africa.* **24:** 263–280. 2 *figs.* 2 *pl.* **1938.** Volvox tertius Meyer. With notes on the two other British species of Volvox. *Jour. Queckett Microsc. Club.* IV. **1:** 1–25. 3 *figs.* 4 *pl.*

POLJANSKY, G., and G. PETRUSCHEWSKY. **1929.** Zur Frage über die Struktur der Cyanophyceenzelle. *Arch. Protistenk.* **67:** 11–45. 1 *pl.*

POTTER, M. C. **1888.** Notes on an alga (Dermatophyton radicans Peter) growing on the European tortoise. *Jour. Linn. Soc. Bot. London.* **24:** 251–254. 1 *pl.*

POTTHOFF, H. **1927.** Untersuchungen über die Desmidiacee Hyalotheca dissiliens Bréb. forma minor. *Planta.* **4:** 261–283. 14 *figs.* 1 *pl.* **1928.** Zur Phylogenie und Entwicklungsgeschichte der Conjugaten. *Ber. deutsch. Bot. Ges.* **46:** 667–673. 3 *figs.*

POULTON, E. M. **1925.** Étude sur les Hétérokontes. Geneva. 96 *pp.* 13 *figs.* **1930.** Further studies on the Heterokontae: some Heterokontae of New England, U.S.A. *New Phytol.* **29:** 1–26. 4 *figs.*

POWERS, J. H. **1908.** Further studies in Volvox, with descriptions of three new species. *Trans. Amer. Microsc. Soc.* **28:** 141–175. 4 *pl.*

PRELL, H. **1921.** Zur Theorie der sekretorischen Ortsbewegung. I. Die Bewegung der Cyanophyceen. *Arch. Protistenk.* **42:** 99–156. 11 *figs.*

PRESCOTT, G. W. **1931.** Iowa algae. *Univ. Iowa Studies.* **13:** 1–235. 39 *pl.* **1937.** Preliminary notes on the desmids of Isle Royale, Michigan. *Papers Michigan Acad. Sci., Arts and Lett.* **22:** 201–213. 1 *pl.* **1938.** A new species and a new variety of the algal genus Vaucheria with notes on the genus. *Trans. Amer. Microsc. Soc.* **57:** 1–10. 2 *pl.* **1940.** Desmids of Isle Royale, Michigan. *Papers Michigan Acad. Sci., Arts and Lett.* **25:** 89–100. 4 *pl.* **1942.** The algae of Louisiana, with descriptions of some new forms and notes on distribution. *Trans. Amer. Microsc. Soc.* **61:** 109–119. 1 *pl.* **1944.** New species and varieties of Wisconsin algae. *Farlowia.* **1:** 347–385. 1 *fig.* 5 *pl.*

PRESCOTT, G. W., and H. T. CROASDALE. **1937.** New or noteworthy fresh-water algae of Massachusetts. *Trans. Amer. Microsc. Soc.* **56:** 269–282. 3 *pl.* **1942.** The algae of New England. II. Additions to the fresh-water algal flora of Massachusetts. *Amer. Midland Nat.* **27:** 662–676. 5 *pl.*

PRESCOTT, G. W., and A. MAGNOTTA. **1935.** Notes on Michigan desmids, with descriptions of some species and varieties new to science. *Papers Michigan Acad. Sci., Arts and Lett.* **20:** 157–169. 3 *pl.*

PRESCOTT, G. W., and A. M. SCOTT. **1942.** The fresh-water algae of southern United States. I. Desmids from Mississippi, with descriptions of new species and varieties. *Trans. Amer. Microsc. Soc.* **61:** 1–29. 4 *pl.*

PRINGSHEIM, E. G. **1914.** Zur physiologie der Schizophyceen. *Beitr. Biol. Pflanzen.* **12:** 49–108. 1 *pl.* **1927.** Enthält Polytoma Stärke? *Arch. Protistenk.* **58:** 281–284. **1936.** Zur Kenntnis saprotropher Algen und Flagellaten. I. Mitteilung. Über Anhäufungskulturen polysaprober Flagellaten. *Ibid.* **87:** 43–96. 11 *figs.* **1942.** Contribution to our knowledge of saprophytic algae and flagellata. III. Astasia, Distigma, Menoidium and Rhabdomonas. *New Phytol.* **41:** 171–205. 20 *figs.* **1944.** Some aspects of taxonomy of the Cryptophyceae. *Ibid.* **43:** 143–150. **1946.** Pure cultures of algae, their preparation and maintenance. Cambridge. 119 *pp.* 8 *figs.* **1946A.** The biphasic or soil-water culture method for growing algae and Flagellata. *Jour. Ecol.* **33:** 193–204. 2 *figs.*

PRINGSHEIM, E. G., and F. MAINX. **1926.** Untersuchungen an Polytoma uvella Ehrbg., insbesondere über Beziehungen zwischen chemotactischer Reizwirkung und chemischer Konstitution. *Planta.* **1:** 583–623.

PRINGSHEIM, N. **1855.** Ueber die Befruchtung und Keimung der Algen. *Monatsber. Akad. Wiss. Berlin.* **1855:** 133–165. 1 *pl.* **1858.** Beiträge zur Morphologie und Systematik der Algen. I. Morphologie der Oedogonieen. *Jahrb. Wiss. Bot.* **1:** 11–81. 6 *pl.* **1860.** Beiträge zur Morphologie und Systematik der Algen. III. Die Coleochaeteen. *Ibid.* **2:** 1–38. 6 *pl.* **1861.** Über die Dauerschwärmer des Wassernetzes und einige ihnen verwandte Bildung. *Monatsber. Akad. Wiss. Berlin.* **1860:** 775–794. 1 *pl.* **1870.** Über Paarung von Schwärmsporen, die morphologische Grundform der Zeugung im Pflanzenreiche. *Ibid.* **1869:** 721–738. 1 *pl.*

PRINTZ, H. **1913.** Eine systematische Übersicht der Gattung Oöcystis Nägeli. *Nyt. Mag. Naturvidensk.* **51:** 165–203. 3 *pl.* **1914.** Kristianiatraktens Protococcoideer. *Skr. Vidensk. i Kristiana.* Mat.-Nat. Kl. 1913, No. 6: 1–123. 2 *figs.* 7 *pl.* **1915.** Die Chlorophyceen des südlichen Siberiens und des Uriankailandes. *Ibid.* 1915, No. 4: 1–52. 7 *pl.* **1921.** Subaërial algae from South

Africa. *Ibid.* 1920, No. 1: 1–41. 14 *pl.* **1927.** Chlorophyceae. In A. Engler and K. Prantl, Die natürlichen Pflanzenfamilien. 2 ed. **3:** 1–463. 366 *figs.* **1939.** Vorarbeiten zu einer Monographie der Trentepohliaceae. *Nyt. Mag. Naturbidensk.* **80:** 137–210. 32 *pl.*

PROBST, T. **1916.** Ueber die ungeschlechtliche Vermehrung von Sorastrum spinulosum Naeg. *Tätigkeitsber. Naturf. Ges. Basel-Land.* (Ref. Geitler, 1924*B*.) **1926.** Über die Vermehrung von Sorastrum Nägeli, Pediastrum Meyen und Tetraedron Kuetzing. *Ibid.* **7:** 29–36. 1 *pl.* (Ref. *Bot. Centralbl.* **151:** 249, 1927.)

PROWAZEK, S. **1903.** Die Kernteilung des Entosiphon. *Arch. Protistenk.* **2:** 325–328. 12 *figs.*

PUYMALY, A. DE. **1922.** Reproduction des Vaucheria par zoospores amiboides. *Compt. Rend. Acad. Sci. Paris.* **174:** 824–827. **1922A.** Adaptation à la vie aérienne d'une Conjugée filamenteuse (Zygnema peliosporum Wittr.). *Ibid.* **175:** 1229–1231. **1924.** Sur le vacuole des algues vertes adaptées à la vie aérienne. *Ibid.* **178:** 958–960. **1927.** Sur une Spirogyra fixé, pérennant, se multipliant par marcottage. *Ibid.* **185:** 1512–1514.

RABENHORST, L. **1863.** Kryptogamen-Flora von Sachsen, der Ober-Lausitz, Thüringen und Nordböhmen mit Berücksichtigung der benachbarten Länder. Abt. 1. Leipzig. 653 *pp.*

RALFS, J. **1848.** The British Desmidieae. London. 226 *pp.* 35 *pl.*

RAMANATHAN, K. R. **1939.** On the mechanism of spore liberation in Pithophora polymorpha Wittr. *Jour. Indian Bot. Soc.* **18:** 25–29. 8 *figs.* 1 *pl.* **1939A.** The morphology, cytology and alternation of generations in Enteromorpha compressa (L.) Grev. var. lingulata (J. Ag.) Hauck. *Ann. Bot.* N.S. **3:** 375–398. 74 *figs.* **1942.** On the oögamous sexual reproduction in a Carteria. *Jour. Indian Bot. Soc.* **21:** 129–135. 18 *figs.*

RANDHAWA, M. S. **1937.** A note on aplanospores of Oedogonium. *Proc. Indian Acad. Sci.* **6:** 230–231. 1 *fig.*

RATCLIFFE, H. L. **1927.** Mitosis and cell division in Euglena spirogyra Ehrenberg. *Biol. Bull.* **53:** 109–122. 3 *pl.*

RAUWENHOFF, N. W. P. **1887.** Recherches sur le Sphaeroplea annulina Ag. *Arch. Néerland. Sci. Exactes et Nat.* **22:** 91–144. 2 *pl.*

RAYSS, T. **1915.** Le Coelastrum proboscideum Bohl. Étude de planctologie expérimentale suivi d'une revision des Coelastrum de la Suisse. *Matér. pour la Flore Crypt. Suisse.* **5**[2]**:** 1–65. 20 *pl.* **1930.** Microthamnion Kuetzingianum Naeg. *Bull. Soc. Bot. Genève.* **21:** 143–160. 9 *figs.*

REICHARDT, A. **1927.** Beiträge zur Cytologie der Protisten. *Arch. Protistenk.* **59:** 301–338. 9 *figs.* 4 *pl.*

REINISCH, O. **1911.** Eine neue Phaeocapsacee. *Ber. deutsch. Bot. Ges.* **29:** 77–83. 1 *pl.*

REINKE, J. **1879.** Zwei parasitische Algen. *Bot. Zeitg.* **37:** 473–478. 1 *pl.*

REINSCH, P. F. **1867.** Die Algenflora des mittleren Theils von Franken. Nürenberg. 238 *pp.* 13 *pl.* **1886.** Ueber das Palmellaceen genus Acanthococcus. *Ber. deutsch. Bot. Ges.* **4:** 237–244. 2 *pl.* **1888.** Familiae Polyedriearum monographia. *Notarisia.* **3:** 493–516. 5 *pl.* **1891.** Ueber das Protococcaceen Gattung Actidesmium. *Flora.* **74:** 445–459. 2 *pl.*

RHODES, R. C. **1919.** Binary fission in Collodictyon triciliatum. *Univ. California Publ. Zool.* **19:** 201–274. 4 *figs.* 8 *pl.* **1926.** Mouth and feeding habits of Heteronema acus. (Abstract.) *Anat. Record.* **34:** 152–153.

RICHTER, O. **1909.** Zur Physiologie der Diatomeen. II. Die Biologie der Nitzschia putrida Benecke. *Denkschr. kais. Akad. Wiss. Wien.* (Math.-Nat. Kl.) **84:** 1–116. 6 *figs.* 4 *pl.*

Rieth, A. **1940.** Die Auxosporenbildung bei Melosira arenaria (Moore). *Planta.* **31:** 171–183. 26 *figs.*

Robinson, C. B. **1906.** The Chareae of North America. *Bull. New York Bot. Garden.* **4:** 244–308.

Rose, J. N. **1885.** Notes on the conjugation of Spirogyra. *Bot. Gaz.* **10:** 304–306. 1 *pl.*

Rosenberg, M. **1930.** Die geschlechtliche Fortpflanzung von Botrydium granulatum Grev. *Oesterr. Bot. Zeitschr.* **79:** 289–296. 1 *pl.* **1944.** On a blue-green cryptomonad, Chroomonas Nordstedtii Hansg. *Ann. Bot.* N.S. **8:** 315–322. 6 *figs.*

Rosenvinge, L. K. **1909.** The marine algae of Denmark. Pt. 1. *Mém. Acad. Sci. et des Lett. de Danemark.* VII. 7, No. 1.

Rostafiński, J. **1882.** L'Hydrurus et ses affinités. *Ann. Sci. Nat. Bot.* VI. **14:** 5–25. 1 *pl.*

Rostafiński, J., and M. Woronin. **1877.** Ueber Botrydium granulatum. *Bot. Zeitg.* **35:** 649–671. 5 *pl.*

Roy, J., and P. Bissett. **1893–1894.** On Scottish Desmidieae. *Ann. Scottish Natur. Hist.* **1893:** 106–111, 170–180, 237–245. 1 *pl.* **1894:** 40–46, 100–105, 167–178, 241–256. 3 *pl.*

Ruinen, J. **1933.** Life-cycle and environment of Lochmiopsis sibirica Woron. *Recueil Trav. Bot. Néerland.* **30:** 725–797. 18 *figs.* 1 *pl.*

Sachs, J. **1875.** Text-book of botany, morphological and physiological. Translated by A. W. Bennett. Oxford. 858 *pp.* 461 *figs.*

Sargent, M. C. **1934.** Causes of color changes in blue-green algae. *Proc. Nat. Acad. Sci. of U. S.* **20:** 251–254.

Saunders, H. 1931. Conjugation in Spirogyra. *Ann. Bot.* **45:** 233–256. 8 *figs.*

Schaarschmidt, J. **1883.** Adutok a Gongrosirák fejlödéséhez. *Magyar növény Lapok.* **7:** 129–138. 1 *pl.* (Ref. Just's *Bot. Jahresber.* **11**[1]: 272–273. 1885.)

Schaeffer, A. A. **1918.** A new and remarkable diatom-eating flagellate, Jenningsia diatomophaga, nov. gen., nov. spec. *Trans. Amer. Microsc. Soc.* **37:** 177–182. 1 *pl.*

Schaffner, J. H. **1927.** Extraordinary sexual phenomena in plants. *Bull. Torrey Bot. Club.* **54:** 619–629.

Scherffel, A. **1901.** Kleiner Beitrag zur Phylogenie einiger Gruppen niederer Organismen. *Bot. Zeitg.* **59:** 143–158. 1 *pl.* **1901A.** Einige Beobachtungen über Oedogonien mit halbkugeliger Fusszelle. *Ber. deutsch. Bot. Ges.* **19:** 557–562. 1 *pl.* **1904.** Notizen zur Kenntnis der Chrysomonadineen. *Ibid.* **22:** 439–444. **1907.** Mehrere Stigmen bei grünen Schwärmzellen. *Ibid.* **25:** 229–230. 3 *figs.* **1908.** Asterococcus n.g. superbus (Cienk.) Scherffel und dessen angebliche Beziehungen zu Eremosphaera. *Ibid.* **26A:** 762–771. 3 *figs.* **1908A.** Einiges zur Kenntnis von Schizochlamys gelatinosa A. Br. *Ibid.* **26A:** 783–795. 1 *pl.* **1911.** Beitrag zur Kenntnis der Chrysomonadineen. *Arch. Protistenk.* **22:** 299–344. 1 *pl.* **1912.** Zwei neue trichocystenartige Bildung führende Flagellaten. *Ibid.* **27:** 94–128. 1 *pl.* **1924.** Ueber die Cyste von Monas. *Ibid.* **48:** 187–195. 6 *figs.* **1926.** Beiträge zur Kenntnis der Chytrideen. Teil III. *Ibid.* **54:** 510–528. 1 *pl.* **1927.** Beitrag zur Kenntnis der Chrysomonadineen. II. *Ibid.* **57:** 331–361. 3 *figs.* 1 *pl.* **1928.** Einiges zur Kenntnis der Copulation der Conjugaten. *Ibid.* **62:** 167–176. 3 *figs.* 1 *pl.*

Schiller, J. **1909.** Eine neue Fall von Mikrosporenbildung bei Chaetoceras Loren zianum Grun. *Ber. deutsch. Bot. Ges.* **27:** 351–361. 1 *pl.* **1923.** Beobachtungen über die Entwicklung des roten Augenfleckes bei Ulva lactuca. *Oesterr. Bot. Zeitschr.* **72:** 236–241. 6 *figs.* **1924.** Die geschlechtliche Fortpflanzung von Characium. *Ibid.* **73:** 14–23. 1 *pl.* **1927.** Über Spondylomorum caudatum n. sp, seine Fortpflanzung und Lebensweise. *Jahrb. Wiss. Bot.* **66:** 274–284. 5 *figs.* 1 *pl.* **1933.** Dinoflagellatae (Peridineae) und Verwandte. In L. Rabenhorst, Kryptogamen-Flora Deutschlands, Österreich und der Schweiz. **10,** Abt. 3, 1 Teil: 1–617. 631 *figs.*

Schilling, A. J. **1891.** Die Süsswasser-Peridineen. *Flora.* **74:** 220–299. 3 *pl.* **1913.** Dinoflagellatae. In A. Pascher, Die Süsswasserflora Deutschlands, Österreichs und der Schweiz. **3:** 1–66. 69 *figs.*

Schindler, B. **1913.** Ueber den Farbenwechsel der Oscillarien. *Zeitschr. Bot.* **5:** 497–575. 5 *figs.*

Schmid, G. **1918.** Zur Kenntnis der Oscillarienbewegung. *Flora.* **111:** 327–379. 11 *figs.* **1920.** Ueber die vermeintliche Einzelligkeit der Spirulinen. *Ber. deutsch. Bot. Ges.* **38:** 368–371. **1921.** Ueber Organisation und Schleimbildung bei Oscillatoria Jenensis und das Bewegungsverhalten künstlicher Teilstücke. *Jahrb. Wiss. Bot.* **60:** 572–627. 26 *figs.* **1921A.** Bemerkungen zu Spirulina Turp. *Arch. Protistenk.* **43:** 463–466. **1923.** Das Reizverhalten künstlicher Teilstücke, die Kontractilität und das osmotische Verhalten der Oscillatoria Jenensis. *Jahrb. Wiss. Bot.* **62:** 328–419. 6 *figs.* **1927.** Zur Ökologie der Luftalgen. *Ber. deutsch. Bot. Ges.* **45:** 518–533. 2 *figs.* 1 *pl.*

Schmidle, W. **1894.** Aus der Chlorophyceen-Flora der Torfstiche zu Virnheim. *Flora.* **78:** 42–66. 1 *pl.* **1896.** Untersuchungen über Thorea ramosissima Bory. *Hedwigia.* **35:** 1–33. 3 *pl.* **1899.** Einiges über die Befruchtung, Keimung und Haarinsertion von Batrachospermum. *Bot. Zeitg.* **57:** 125–135. 1 *pl.* **1899A.** Algologische Notizen XIII. *Allg. Bot. Zeitschr.* 1899, No. 2 (Ref. Just's *Bot. Jahresber.* **27**[1]: 150, 1901.) **1899B.** Einige Algen aus preussischen Hochmooren. *Hedwigia.* **38:** 156–176. 2 *pl.* **1900.** Ueber die Gattung Staurogenia Ktzg. *Ber. deutsch. Bot. Ges.* **18:** 149–158. 1 *pl.* **1901.** Ueber drei Algengenera. *Ibid.* **19:** 10–24. 1 *pl.* **1902.** Algen, insbesondere solche des Plankton, aus den Nyassa-See und seiner Umgebung gesammelt von Dr. Füllborn. *Bot. Jahrb.* **32:** 56–88. 3 *pl.* **1902A.** Über die Gattung Radiococcus n. gen. *Allg. Bot. Zeitschr.* **8:** 41–42. **1905.** Algologische Notizen XVI. *Ibid.* **8:** 63–65.

Schmidt, P. **1923.** Morphologie und Biologie der Melosira varians mit einem Beitrag zur Mikrosporenfrage. *Internat. Rev. gesamt. Hydrobiol. Hydrograph.* **11:** 114–147. 5 *pl.* **1927.** Ist die scharfe Trennung zwischen zentrischen und pennaten Diatomeen haltbar *Ibid.* **17:** 247–288. 5 *figs.* **1929.** Beiträge zur Karyologie und Entwicklungsgeschichte der zentrischen Diatomeen. *Ibid.* **21:** 289–334. 4 *pl.*

Schmitz, F. **1879.** Untersuchungen über die Zellkerne der Thallophyten. *Sitzungsber. Niederrheinisch. Ges. Nat. u. Heilk. Bonn.* **1879:** 345–376. **1880.** Untersuchungen über die Struktur des Protoplasmas und der Zellkerne der Pflanzenzellen. *Ibid.* **1880:** 159–198.

Schönfeldt, H. von. **1907.** Diatomaceae Germaniae. Berlin. 263 *pp.* 19 *pl.* **1913.** Bacillariales. In A. Pascher, Die Süsswasserflora Deutschlands, Österreichs und der Schweiz. **10:** 1–187. 378 *figs.*

Schreiber, E. **1925.** Zur Kenntnis der Physiologie und Sexualität höherer Volvocales. *Zeitschr. Bot.* **17:** 336–376. 2 *figs.* 1 *pl.* **1942.** Die geschlechtliche Fortpflanzung von Monostroma Grevillei (Thur.) und Cladophora rupestris. *Planta.* **32:** 414–417. 1 *fig.*

SCHRODER, B. **1898.** Pandorina morum, ihre ungeschlechtliche Vermehrung und ihre Parasiten. *Schles. Ges. vaterl. Kultur. Sitzungsber. zool.-bot. Sektion*, Dec. 8, 1898. 4 *pp.* **1898A.** Neue Beiträge zur Kenntnis der Algen des Riesengebirges. *Forschungsber. Biol. Stat. Plön.* **6:** 9–47. 2 *pl.* **1902.** Untersuchungen über Gallertbildung der Algen. *Verh. Naturh.-Med. Ver. Heidelberg.* N.F. **7:** 139–196. 2 *pl.*

SCHULZE, B. **1927.** Zur Kenntnis einiger Volvocales. *Arch. Protistenk.* **58:** 508–576. 28 *figs.* 2 *pl.*

SCHUSSNIG, B. **1911.** Beitrag zur Kenntnis von Gonium pectorale Müll. *Oesterr. Bot. Zeitschr.* **61:** 121–126. 1 *pl.* **1923.** Die Kernteilung bei Cladophora glomerata. *Ibid.* **72:** 199–222. 1 *fig.* 1 *pl.* **1928.** Die Reduktionsteilung bei Cladophora glomerata. *Ibid.* **77:** 62–67. 4 *figs.* **1928A.** Zur Entwickungsgeschichte der Siponeen. *Ber. deutsch. Bot. Ges.* **46:** 481–490. 10 *figs.* **1930.** Der Generations- und Phasenwechsel bei den Chlorophyceen (Ein historischer Rückblick). *Oesterr. Bot. Zeitschr.* **79:** 58–77. **1930A.** Der Generations- und Phasenwechsel bei den Chlorophyceen. II. Beitrag. *Ibid.* **79:** 323–332. **1930B.** Chlorophyceae. *Abstracts of Communications, 5th Internat. Bot. Congress, Cambridge. pp.* 189–190. **1930C.** Der Chromosomencyclus von Cladophora Suhriana. *Oesterr. Bot. Zeitschr.* **79:** 273–278. 4 *figs.*

SCHÜTT, F. **1896.** Bacillariales. In A. Engler and K. Prantl, Die natürlichen Pflanzenfamilien. **1,** Abt. 1[a]: 31–150. 250 *figs.*

SELIGO, A. **1887.** Untersuchungen über Flagellaten. *Beitr. Biol. Pflanzen.* **4:** 145–180. 1 *pl.*

SENN, G. **1899.** Über einige coloniebildende einzellige Algen. *Bot. Zeitg.* **57:** 39–104. 39 *figs.* 2 *pl.* **1900.** Flagellata. In A. Engler and K. Prantl, Die natürlichen Pflanzenfamilien. **1,** Abt. 1[b]: 93–188. 84 *figs.* **1911.** Oxyrrhis, Nephroselmis und einige Euflagellaten, nebst Bemerkungen über deren System. *Zeitschr. Wiss. Zool.* **97:** 605–672. 8 *figs.* 2 *pl.*

SETCHELL, W. A. **1890.** Concerning the structure and development of Tuomeya fluviatilis Harv. *Proc. Amer. Acad. Arts Sci.* **25:** 53–68. 1 *pl.*

SETCHELL, W. A., and N. L. GARDNER. **1920.** The marine algae of the Pacific Coast of North America. Part II. Chlorophyceae. *Univ. California Publ. Bot.* **8:** 139–374. 24 *pl.* **1920A.** Phycological contributions. I. *Ibid.* **7:** 279–324. 11 *pl.*

SHAW, W. R. **1894.** Pleodorina, a new genus of the Volvocineae. *Bot. Gaz.* **19:** 279–283. 1 *pl.*

SHAWHAN, F. M., and T. L. JAHN. **1947.** A survey of the genus Petalomonas Stein (Protozoa: Euglenida). *Trans. Amer. Microsc. Soc.* **66:** 182–189. 2 *pl.*

SINGH, R. N. **1939.** An investigation into the algal flora of paddy fields of the United Provinces. *Indian Jour. Agr. Sci.* **9:** 55–77. **1942.** Reproduction in Draparnaldiopsis indica Bharaewaja. *New Phytol.* **41:** 262–273. 59 *figs.* **1945.** Nuclear phases and alternation of generations in Draparnaldiopsis indica Bharadwaja. *Ibid.* **44:** 118–129. 49 *figs.*

SIRODOT, S. **1884.** Les Batrachospermes. Organisation, fonctions, développement, classification. Paris. 299 *pp.* 50 *pl.*

SKUJA, H. **1926.** Vorarbeiten zu einer Algenflora von Lettland. I. *Acta Horti Bot. Latviensis.* **1:** 33–53. 4 *figs.* **1927.** Ueber die Gattung Furcilia Stokes und ihre systematische Stellung. *Ibid.* **2:** 117–124. 2 *figs.* **1927A.** Vorarbeiten zu einer Algenflora von Lettland. III. *Ibid.* **2:** 51–116. 2 *pl.* **1928.** Vorarbeiten zu einer Algenflora von Lettland. IV. *Ibid.* **3:** 103–218. 4 *pl.* **1934.** Beitrag zur

Algenflora Lettland. I. *Ibid.* **7:** 25–86. 119 *figs.* **1934A.** Untersuchungen über die Rhodophyceen des Süsswassers. *Beih. Bot. Centralbl.* **52:** 173–192. 4 *figs.*

Skvortzow, B. B. **1925.** Wislouchiella planctonica nov. gen. et spec. of Volvocales. *Proc. Sungari River Biol. Sta.* **1**[1]**:** 27–29. 1 *pl.* **1925A.** Die Euglenaceengattung Trachelomonas Ehrenberg. Eine systematische Übersicht. *Ibid.* **1**[2]**:** 1–101. 8 *pl.* **1928.** Die Euglenaceengattung Phacus Dujardin. *Ber. deutsch. Bot. Ges.* **46:** 105–125. 1 *pl.* **1937.** Diatoms from Lake Michigan. I. *Amer. Midland Nat.* **18:** 652–658. 1 *pl.*

Smith, B. H. **1932.** The algae of Indiana. *Proc. Indiana Acad. Sci.* **41:** 177–206.

Smith, G. M. **1913.** Tetradesmus, a new four-celled coenobic alga. *Bull. Torrey Bot. Club.* **40:** 75–87. 1 *pl.* **1914.** The organization of the colony in certain four-celled coenobic algae. *Trans. Wisconsin Acad.* **17**[2]**:** 1165–1220. 7 *pl.* **1914A.** The cell structure and colony formation in Scenedesmus. *Arch. Protistenk.* **32:** 278–297. 2 *pl.* **1916.** A preliminary list of algae found in Wisconsin lakes. *Trans. Wisconsin Acad.* **18:** 531–565. **1916A.** Zoospore formation in Characium Sieboldii A. Br. *Ann. Bot.* **30:** 459–466. 2 *figs.* 1 *pl.* **1916B.** Cell structure and zoospore formation in Pediastrum Boryanum (Turp.) Menegh. *Ibid.* **30:** 467–479. 4 *figs.* 1 *pl.* **1916C.** New or interesting algae from the lakes of Wisconsin. *Bull. Torrey Bot. Club.* **43:** 471–483. 3 *pl.* **1916D.** A monograph of the algal genus Scenedesmus based upon pure culture studies. *Trans. Wisconsin Acad.* **18**[2]**:** 422–530. 9 *pl.* **1917.** The vertical distribution of Volvox in the plankton of Lake Mendota. *Amer. Jour. Bot.* **5:** 178–185. **1918.** A second list of algae found in Wisconsin lakes. *Trans. Wisconsin Acad.* **19:** 614–654. 6 *pl.* **1918A.** Cell structure and autospore formation in Tetraëdron minimum (A. Br.) Hansg. *Ann. Bot.* **32:** 459–464. 1 *pl.* **1920.** Phytoplankton of the inland lakes of Wisconsin. *Bull. Wisconsin Geol. and Nat. Hist. Surv.* **57**[1]**:** 1–243. 51 *pl.* **1922.** The phytoplankton of the Muskoka region, Ontario, Canada. *Trans. Wiscon Acad.* **20:** 323–364. 6 *pl.* **1922A.** The phytoplankton of some artificial pools near Stockholm. *Ark. Bot.* **17,** No. 13: 1–8. 28 *figs.* **1924.** Ecology of the plankton algae in the Palisades Interstate Park, including the relation of control methods to fish culture. *Roosevelt Wild Life Bull.* **2:** 94–195. 2 *figs.* 22 *pl.* **1924A.** Phytoplankton of the inland lakes of Wisconsin. *Bull. Wisconsin Geol. and Nat. Hist. Surv.* **57**[2]**:** 1–227. 38 *pl.* **1926.** The plankton algae of the Okoboji region. *Trans. Amer. Microsc. Soc.* **45:** 156–233. 20 *pl.* **1930.** Observations on some siphonaceous green algae of the Monterey peninsula. Contributions to Marine Biology. Lectures and Symopsia given at the Hopkins Marine Station, Dec. 20–21, 1929. *pp.* 222–233. 3 *figs.* **1931.** A consideration of the species of Eudorina. *Bull. Torrey Bot. Club.* **57:** 359–364. 1 *pl.* **1931A.** The phylogeny of the higher colonial Volvocales. *Ibid.* **57:** 364–365. **1931B.** Pandorina charkowiensis Korshikov. *Ibid.* **57:** 365–366. 1 *pl.* **1931C.** Gonium formosum Pascher and Gonium sociale (Duj.) Warming. *Ibid.* **57:** 366–368. 1 *pl.* **1933.** The fresh-water algae of the United States. New York. 716 *pp.* 449 *figs.* **1944.** A comparative study of the species of Volvox. *Trans. Amer. Microsc. Soc.* **63:** 265–310. 45 *figs.* **1944A.** Microaplanospores of Vaucheria. *Farlowia.* **1:** 387–389. 1 *fig.* **1946.** On the structure and reproduction of Spongomorpha coalita (Rupr.) Collins. *Jour. Indian Bot. Soc.* M. O. P. Iyengar Commemoration Volume. *pp.* 21–208. 5 *figs.* **1947.** On the reproduction of some Pacific Coast species of Ulva. *Amer. Jour. Bot.* **34:** 80–87. 38 *figs.*

Smith, G. M., and F. D. Klyver. **1929.** Draparnaldiopsis, a new member of the algal family Chaetophoraceae. *Trans. Amer. Microsc. Soc.* **48:** 196–203. 1 *fig.* 1 *pl.*

SMITH, H. L. **1878.** Description of new species of diatoms. *Amer. Quart. Microsc. Jour.* **1:** 12–18. 1 *pl.*

SMITH, W. **1856.** A synopsis of the British Diatomaceae. Vol. 2. London. 101 *pp.* 38 *pl.*

SNOW, J. W. **1899.** Ulvella americana. *Bot. Gaz.* **27:** 309–314. 1 *pl.* **1899A** Pseudo-pleurococcus nov. gen. *Ann. Bot.* **13:** 189–195. 1 *pl.* **1903.** The plank ton algae of Lake Erie, with special reference to the Chlorophyceae. *Bull. U S. Fish Commiss.* **1902:** 369–394. 4 *pl.* **1911.** Two epiphytic algae. *Bot. Gaz* **51:** 360–368. 1 *pl.* **1912.** Two epiphytic algae, a correction. *Ibid.* **53:** 347.

SONNTAG, C. F. **1922.** Contributions to the histology of the three-toed sloth (Bradypus tridactylus). *Jour. Roy. Microsc. London.* **1922:** 37–46. 13 *figs.*

SPESSARD, E. A. **1930.** Fertilization in a living Oedogonium. *Bot. Gaz.* **89:** 385–393. 11 *figs.*

SPRATT, E. A. **1911.** Some observations on the life-history of Anabaena Cycadeae. *Ann. Bot.* **25:** 369–380. 1 *pl.*

STAHL, E. **1879.** Ueber die Ruhezustände der Vaucheria geminata. *Bot. Zeitg.* **37:** 129–137. 1 *pl.* **1891.** Oedocladium protonema, eine neue Oedogoniaceengattung. *Jahrb. Wiss. Bot.* **23:** 339–348. 2 *pl.*

STEIL, W. N. **1944.** Leptosira Mediciana Borzi. *Bull. Torrey Bot. Club.* **71:** 507–511. 3 *figs.*

STEIN, F. RITTER v. **1878.** Der Organismus der Infusionsthiere. III. Abt., I. Hälfte. Leipzig. 154 *pp.* 24 *pl.* **1883.** *Ibid.* III. Abt., II. Hälfte. Leipzig. 30 *pp.* 25 *pl.*

STEINECKE, F. **1926.** Die Zweischaligheit im Membranbau von Zygnemalen und ihre Bedeutung für die Phylogenie der Conjugaten. *Bot. Archiv.* **13:** 328–339. 36 *figs.* **1929.** Hemizellulose bei Oedogonium. *Ibid.* **24:** 391–403.

STEUER, A. **1904.** Über eine Euglenoide (Eutreptia) aus dem Canale Grande von Triest. *Arch. Protistenk.* **3:** 126–137. 13 *figs.*

STOKES, A. C. **1884.** Notes on some apparently undescribed forms of fresh-water infusoria. *Amer. Jour. Sci.* III. **28:** 38–49. 10 *figs.* **1885.** Notes on some apparently undescribed forms of fresh-water infusoria, No. 2. *Ibid.* III., **29:** 313–328. 1 *pl.* **1885A.** Some new infusoria from American fresh waters. *Ann. and Mag. Nat. Hist.* V. **15:** 437–449. 1 *pl.* **1886.** Some new infusoria from American fresh waters. *Ibid.* V. **17:** 98–112. 1 *pl.* **1886A.** Notices of new fresh-water infusoria. *Proc. Amer. Phil. Soc.* **23:** 562–568. 1 *pl.*

STRAIN, H. H. **1944.** Chloroplast pigments. *Ann. Rev. Biochem.* **13:** 591–610 **1948.** Occurrence and properties of chloroplast pigments. *Carnegie Inst Washington Year Book* **47:** 97–100. **195?.** The pigments of algae. In G. M Smith (ed.), Manual of phycology, an introduction to the algae and their biology (in press).

STRAIN, H. H., and W. M. MANNING. **1943.** Pigments of algae. In H. A. Spoehr, Annual report of the chairman of the division of biology. *Carnegie Inst. Washington Year Book.* **42:** 79–82.

STRAIN, H. H., W. M. MANNING, and G. HARDEN. **1944.** Xanthophylls and carotenes of diatoms, brown algae, dinoflagellates, and sea anemones. *Biol. Bull* **86:** 169–191.

STRASBURGER, E. **1875.** Ueber Zellbildung und Zelltheilung. Jena. 256 *pp.* 7 *pl.* **1880.** Zellbildung und Zelltheilung. 3d ed. Jena. 392 *pp.* 14 *pl.* **1892.** Schwärmsporen, Gameten, pflanzliche Spermatozoiden und das Wesen der Befruchtung. *Histol. Beitr.* **4:** 47–158. 1 *pl.*

STREHLOW, K. **1929.** Ueber die Sexualität einiger Volvocales. *Zeitschr. Bot.* **21:** 625–692. 17 *figs.*

STRÖM, K. M. **1921.** The germination of the zoogonidia of Stigeoclonium tenue. *Nyt. Mag. Naturvidensk.* **59:** 9–11. 1 *pl.* **1921A.** Resting spores of Pediastrum. *Ibid.* **59:** 11–12. 1 *pl.* **1924.** Vorarbeiten zu einer Monographie der Chroococcaceen von N. Wille. *Ibid.* **62:** 169–209. 8 *pl.* **1924A.** Studies in the ecology and geographical distribution of fresh-water algae. *Rev. Algologique.* **1:** 127–155. 2 *figs.* **1926.** Norwegian mountain algae. *Videnskabs.-selskab. Christiania Skrifter* (Mat.-Nat. Kl.). 1926[6]: 1–263. 25 *pl.*

SUBRAHMANYAN, R. **1948.** On somatic division, reduction division, auxospore formation, and sex-differentiation in Navicula halophila. *Jour. Indian Bot. Soc.*, M.O.P. Iyengar Commemoration Volume. *pp.* 239–266. 81 *figs.* 2 *pl.*

SUNDARALINGAM, V. S. **1948.** The cytology and spermatogenesis in Chara zeylanica Willd. *Jour. Indian Bot. Soc.*, M.O.P. Iyengar Commemoration Volume. *pp.* 289–303. 44 *figs.*

SUNESON, S. **1947.** Notes on the life-history of Monostroma. *Sv. Bot. Tidsskr.* **41:** 235–246. 2 *figs.*

SUSSENGUTH, K. **1926.** Über die Eiseninkrustation von Golenkinia radiata Chodat. *Kryptog. Forsch. Bayr. Bot. Ges.* **7:** 49L. (Ref. *Bot. Centralbl.* **151:** 461. 1927.)

SUSSKI, E. P. **1929.** Die komplementäre chromatische Adaptation bei Oscillatoria Engelmanniana Gaiduk. *Beitr. Biol. Pflanzen.* **17:** 45–50.

SVEDELIUS, N. **1930.** Über die sogenannten Süsswasser-Lithodermen. *Leitschr. Bot.* **23:** 892–918. 13 *figs.*

SWINGLE, W. T. **1894.** Cephaleuros mycoidea and Phyllosiphon, two species of parasitic algae new to North America. *Proc. Amdr. Assoc. Adv. Sci.* **42:** 260.

TAFT, C. E. **1937.** The life history of a new species of Mesotaenium. *Bull. Torrey Bot. Club.* **64:** 75–79. 13 *figs.* **1939.** Additions to the algae of Michigan. *Ibid.* **66:** 77–85. 12 *figs.* **1940.** Asexual and sexual reproduction in Platydorina caudata Kofoid. *Trans. Amer. Microsc. Soc.* **59:** 1–11. 2 *pl.* **1941.** Inversion of the developing coenobium in Pandorina morum Bory. *Ibid.* **60:** 327–328. **1942.** A new species of Cylindrocystis. *Ohio Jour. Sci.* **42:** 122. 6 *figs.* **1945.** The validity of the algal genus Thamniastrum. *Bull. Torrey Bot. Club.* **72:** 246–247. 1 *fig.* **1945A.** Pectodictyon, a new genus in the family Scenedesmaceae. *Trans. Amer. Microsc. Soc.* **64:** 25–28. 1 *pl.*

TAKEDA, H. **1916.** Dysmorphococcus variabilis, gen. et sp. nov. *Ann. Bot.* **30:** 151–156. 15 *figs.*

TAKESIGE, T. **1937.** Die Bedeutung der Symbiosis zwischen einigen endophytischen Blaualgen und ihren Wirtspflanzen. *Bot. Mag. Tokyo.* **51:** 514–524. 1 *pl.*

TANNREUTHER, G. W. **1923.** Nutrition and reproduction in Euglena. *Arch. Entwicklungsmech. d. Organismen.* **52:** 367–383. 52 *figs.*

TAYLOR, W. R. **1928.** The marine algae of Florida with special reference to the Dry Tortugas. *Carnegie Inst. Wash. Publ.* **379:** 1–219. 37 *pl.* **1928A.** The alpine algal vegetation of the mountains of British Columbia. *Proc. Acad. Nat. Sci. Phila.* **80:** 45–114. 3 *figs.* 5 *pl.* **1935.** The fresh-water algae of Newfoundland. Part II. *Papers Mich. Acad. Sci., Arts, Lett.* **20:** 185–230. 17 *pl.*

TEILING, E. **1916.** Tetrallantos, eine neue Gattung der Protococcoideen. *Svensk. Bot. Tidsskr.* **10:** 59–66. 16 *figs.* **1916A.** En kaledonisk fytoplanktonformation. *Ibid.* **10:** 506–519.

TEODORESCO, E. C. **1905.** Organisation et développment du Dunaliella, nouveau genre de Volvocacée-Polyblépharidée. *Beih. Bot. Centralbl.* **18**[1]**:** 215–232. **5** *figs.* 2 *pl.* **1906.** Observations morphologiques et biologiques sur le genre Dunaliella. *Rev. Gén. Botanique.* **18:** 353–371, 409–427. 25 *figs.* 2 *pl.*

THAXTER, R. **1900.** Note on the structure and reproduction of Compsopogon. *Bot. Gaz.* **29:** 259–267. 1 *pl.*

THOMAS, N. **1913.** Notes on Cephaleuros. *Ann. Bot.* **27:** 781–792. 1 *pl.*

THOMPSON, R. H. **1938.** A preliminary survey of the fresh-water algae of eastern Kansas. *Univ. Kansas Sci. Bull.* **25:** 5–83. 12 *pl.* **1938A.** Coronastrum: a new genus of algae in the family Scenedesmaceae. *Amer. Jour. Bot.* **25:** 692–694. 10 *figs.* **1947.** Fresh-water dinoflagellates of Maryland. *State of Maryland, Board of Natural Resources. Publ.* **67:** 1–28. 4 *pl.* **1949.** Immobile Dinophyceae. I. New records and a new species. *Amer. Jour. Bot.* **36:** 301–308. 34 *figs.*

TIFFANY, L. H. **1921.** Algal food of a young gizzard shad. *Ohio Jour. Sci.* **21:** 272–275. **1924.** A physiological study of growth and reproduction among certain green algae. *Ibid.* **24:** 65–98. 1 *pl.* **1924A.** Some new forms of Spirogyra and Oedogonium. *Ibid.* **24:** 180–187. 3 *pl.* **1926.** The algal collection of a single fish. *Papers Michigan Acad. Sci., Arts, Lett.* **6:** 303–306. **1926A.** The filamentous algae of northwestern Iowa with special reference to the Oedogoniaceae. *Trans. Amer. Microsc. Soc.* **45:** 69–132. 16 *pl.* **1928.** The algal genus Bulbochaete. *Ibid.* **47:** 121–177. 10 *pl.* **1930.** The Oedogoniaceae. Columbus. 188 *pp.* 64 *pl.* **1934.** The plankton algae of the west end of Lake Erie. *Ohio State Univ., Contrib. Franz Theodore Stone Lab.* **6:** 1–112. 15 *pl.* **1935.** Homothallism and other variations in Pleodorina californica Shaw. *Arch. Protistenk.* **85:** 140–144. 4 *figs.* **1936.** Wille's collection of Puerto Rican fresh-water algae. *Brittonia.* **2:** 165–176. 3 *pl.* **1937.** The filamentous algae of the west end of Lake Erie. *Amer. Midland Nat.* **18:** 911–951. 9 *pl.* **1937A.** Oedogoniales. *North American Flora.* **11,** Part 1: 1–102. 36 *pl.*

TIFFANY, L. H., and E. N. TRANSEAU. **1927.** Oedogonium periodicity in the North Central States. *Trans. Amer. Microsc. Soc.* **46:** 166–174. 3 *figs.*

TILDEN, J. E. **1896.** A contribution to the life history of Pilinia diluta Wood and Stigeoclonium flagelliferum Atz. *Minnesota Bot. Studies.* **1:** 601–635. 5 *pl.* **1935.** The algae and their life relations. Minneapolis. 550 *pp.* 257 *figs.*

TIMBERLAKE, H. G. **1901.** Starch-formation in Hydrodictyon utriculatum. *Ann. Bot.* **15:** 619–635. 1 *pl.* **1902.** Development and structure of the swarm spores of Hydrodictyon. *Trans. Wisconsin Acad.* **13:** 486–522. 2 *pl.*

TOBLER, F. **1917.** Ein neues tropisches Phyllosiphon, seine Lebensweise und Entwicklung. *Jahrb. Wiss. Bot.* **58:** 1–28. 11 *figs.* 1 *pl.*

TRANSEAU, E. N. **1913.** The periodicity of algae in Illinois. *Trans. Amer. Microsc. Soc.* **32:** 31–40. 8 *figs.* **1913A.** Annotated list of the algae of Eastern Illinois. *Trans. Illinois Acad. Sci.* **6:** 69–89. **1913B.** The life history of Gloeotaenium. *Bot. Gaz.* **55:** 66–73. 1 *pl.* **1914.** New species of green algae. *Amer. Jour. Bot.* **1:** 289–301. 5 *pl.* **1915.** Notes on the Zygnemales. *Ohio Jour. Sci.* **16:** 17–31. **1916.** The periodicity of freshwater algae. *Amer. Jour. Bot.* **3:** 121–133. 3 *figs.* **1917.** The algae of Michigan. *Ohio Jour. Sci.* **17:** 217–232. **1919.** Hybrids among species of Spirogyra. *Amer. Naturalist.* **53:** 109–119. 7 *figs.* **1925.** The genus Debarya. *Ohio Jour. Sci.* **25:** 193–199. 2 *pl.* **1926.** The genus Mougeotia. *Ibid.* **26:** 311–331. 7 *pl.* **1934.** The genera of the Zygnemataceae. *Trans. Amer. Microsc. Soc.* **53:** 201–207. **1943.** Two new Ulotrichales. *Ohio Jour. Sci.* **43:** 212–213. 1 *fig.* **195?.** The Zygnemataceae. *Graduate School Monogr., Ohio State Univ., Contrib. in Botany.* No. 1. (in press).

TREBOUX, O. **1912.** Die freilebende Alge und die Gonidie Cystococcus humicola in bezug auf die Flechtensymbiose. *Ber. deutsch. Bot. Ges.* **30:** 69–80.

TROITZKAJA, O. V. **1924.** Zur Morphologie und Entwicklungsgeschichte von Uroglenopsis americana (Calkins) Lemmerm. *Arch. Protistenk.* **49:** 260–277. 13 *figs.*

TRÖNDLE, A. **1911.** Über die Reduktionsteilung in den Zygoten von Spirogyra und über die Bedeutung der Synapsis. *Zeitschr. Bot.* **3:** 593–619. 20 *figs.* 1 *pl.*

TSENG, C. K. **1945.** The terminology of seaweed colloids. *Science.* **101:** 597–602.

T'SERCLAES, J. **1922.** Le noyau et la division nucléaire dans le Cladophora glomerata. *Cellule.* **32:** 313–326. 2 *pl.*

TURNER, C. **1922.** The life history of Staurastrum Dickiei var. parallelum (Nordst.). *Proc. Linn. Soc. London.* **1922:** 59–63. 1 *pl.* (Ref. *Jour. Roy. Microsc. London.* **1922:** 419. 1922.)

TURNER, W. B. **1892.** The fresh-water algae (principally Desmidieae) of East India. *Kgl. Svensk. Vetensk.-Ak. Handl.* **25,** No. 5: 1–187. 23 *pl.*

TUTTLE, A. H. **1910.** Mitosis in Oedogonium. *Jour. Exper. Zool.* **9:** 143–157. 18 *figs.*

ÚLEHLA, V. **1911.** Die Stellung der Gattung Cyathomonas From. im System der Flagellaten. *Ber. deutsch. Bot. Ges.* **29:** 284–292. 2 *figs.*

ULLRICH, H. **1926.** Ueber die Bewegung von Beggiatoa mirabilis und Oscillatoria Jenensis. *Planta.* **2:** 295–324. 8 *figs.* **1929.** Über die Bewegungen der Beggiatoaceen und Oscillatoriaceen. *Ibid.* **9:** 144–194. 15 *figs.*

USPENSKAJA, W. J. **1930.** Über die Physiologie der Ernährung und die Formen von Draparnaldia glomerata Agardh. *Zeitschr. Bot.* **22:** 337–393. 12 *figs.*

USPENSKI, E. E. **1927.** Eisen als Faktor für die Verbreitung niederer Wasserpflanzen. *Pflanzenforschung.* **9:** 1–104.

VAN HEURCK, H. **1880-1885.** Synopsis des diatomées de Belgique. Antwerp. 235 *pp.* 132 *pl.* **1890.** Structure of diatom valves. *Jour. Roy. Microsc. Soc. London.* **1890:** 104–106. 2 *pl.*

VIRIEUX, J. M. **1908.** Note sur le Dichotomosiphon tuberosus (A.Br.) Ernst et le Mischococcus confervicola Naeg. *Bull. de la Soc. d'Hist. Nat. du Doubs.* **1910:** 1–8. 1 *pl.* (Ref. *Just's Bot. Jahresber.* **38**[1]: 399. 1913.)

VISHCER, W. **1919.** Sur le polymorphisme de l'Ankistrodesmus Braunii (Naegeli) Collins. Étude de planctologie expérimentale. *Rev. d'Hydrologie.* **1919:** 1–48. 7 *figs.* 2 *pl.* **1927.** Zur Biologie von Coelastrum proboscideum und einigen andern Grünalgen. *Verhandl. Naturforsch. Ges. Basel.* **38:** 386–415. 10 *figs.* 1 *pl.* (Ref. *Biol. Abstr.* **4.** No. 25605. 1930.) **1933.** Ueber einige kritische Gattungen und die Systematik der Chaetophorales. *Beih. Bot. Centralbl.* **51:** 1–101. 40 *figs.*

VLK, W. **1931.** Über die Struktur der Heterokontengeisseln. *Beih. Bot. Centralbl.* **48:** 214–220. 15 *figs.* **1938.** Über den Bau der Geissel. *Arch. Protistenk.* **90:** 448–488. 12 *figs.* 1 *pl.*

WAGER, H. **1899.** On the eyespot and flagellum in Euglena viridis. *Jour. Linn. Soc. Zool. London.* **27:** 463–481. 1 *pl.*

WALLICH, C. G. **1860.** Descriptions of Desmidiaceae from Lower Bengal. *Ann. and Mag. Nat. Hist.* III. **5:** 184–197, 273–285. 4 *pl.*

WALZ, J. **1866.** Beitrag zur Morphologie und Systematik der Gattung Vaucheria DC. *Jahrb. Wiss. Bot.* **5:** 127–160. 3 *pl.*

WARD, H. B., and G. C. WHIPPLE. **1918.** Freshwater Biology. New York.

WARD, H. M. **1883.** On the structure, development, and life history of a tropical epiphyllous lichen. *Trans. Linn. Soc. London.* II. *Bot.* **2:** 87–119. 4 *pl.* **1899.** Some methods for use in the culture of algae. *Ann. Bot.* **13:** 563–566. 1 *pl.*

Watson, J. B., and J. E. Tilden. **1930.** The algal genus Schizomeris and the occurrence of Schizomeris Leibleinii Kützing in Minnesota. *Trans. Amer. Microsc. Soc.* **49:** 160–167. 1 *pl.*

Weatherwax, P. **1915.** Some peculiarities in Spirogyra dubia. *Proc. Indiana Acad. Sci.* **1914:** 203–206. 5 *figs.*

Wehrle, E. **1927.** Studien über Wasserstoffionenkonzentrationsverhältnisse und Besiedelung an Algenstandorten in der Umgebung von Freiburg im Breisgau. *Zeitschr. Bot.* **19:** 209–287. 9 *figs.*

Wenrich, D. H. **1924.** Studies on Euglenamorpha Hegneri n.g., n. sp., a euglenoid flagellate found in tadpoles. *Biol. Bull.* **47:** 149–174. 4 *pl.* **1937.** Protozoological methods. In C. E. McClung, Handbook of microscopical technique. 2 ed., *pp.* 522–551.

Werner, E. **1910.** Der Bau des Panzers von Ceratium hirundinella. *Ber. deutsch. Bot. Ges.* **28:** 103–107. 1 *pl.*

Wesenberg-Lund, C. **1904.** Studier over de danske söers Plankton. Copenhagen. 223 + 44 *pp.* 10 *pl.* **1905.** A comparative study of the lakes of Scotland and Denmark. *Proc. Roy. Soc. Edinburgh.* **25:** 401–448. 2 *pl.*

Wesley, O. C. **1928.** Asexual reproduction in Coleochaete. *Bot. Gaz.* **86:** 1–31. 41 *figs.* 2 *pl.* **1930.** Spermatogenesis in Coleochaete scutata. *Ibid.* **89:** 180–191. 2 *pl.*

West, G. S. **1902.** On some algae from hot springs. *Jour. Botany.* **40:** 241–248. 1 *pl.* **1904.** A treatise on the British fresh-water algae. Cambridge. 372 *pp.* 166 *figs.* **1904A.** West Indian fresh-water algae. *Jour. Botany.* **42:** 281–294. 1 *pl.* **1908.** Some critical green algae. *Jour. Linn. Soc. Bot. London.* **38:** 279–289. 2 *pl.* **1909.** The algae of the Yan Yean Reservoir, Victoria; a biological and oecological study. *Ibid.* **39:** 1–88. 10 *figs.* 6 *pl.* **1909A.** A biological investigation of the Peridineae of Sutton Park, Warwickshire. *New Phytol.* **8:** 191–196. 7 *figs.* **1911.** Diplochaete Collins and Polychaetophora W. and G. S. West. *Jour. Botany.* **49:** 88–89. **1912.** Resting spores of Surirella spirale Ktz. *Ibid.* **50:** 325–326. 1 *fig.* **1912A.** New and interesting British freshwater algae. *Ibid.* **50:** 328–331. 2 *figs.* **1912B.** Observations upon two species of Oedogonium, with some remarks upon the origin of the dwarf males. *Ibid.* **50:** 321–325. 1 *fig.* **1915.** The genus Tetradesmus. *Ibid.* **53:** 82–84. 1 *fig.* **1915*A*.** Observations on the structure and life-history of Mesotaenium caldariorum (Lagerh.) Hansg. *Ibid.* **53:** 78–81. 2 *figs.* **1916.** Algae. Vol. 1. Myxophyceae, Peridinieae, Bacillarieae, Chlorophyceae, together with a brief summary of the occurrence and distribution of fresh-water algae. Cambridge. 475 *pp.* 271 *figs.* **1916A.** On a new marine genus of the Volvocaceae. *Jour. Botany.* **54:** 2–4. 1 *fig.* **1916B.** On two species of Pteromonas. *Ibid.* **54:** 7–9. 2 *figs.*

West, G. S., and F. E. Fritsch. **1927.** A treatise on the British fresh-water algae. New and revised edition. Cambridge. 534 *pp.* 207 *figs.*

West, G. S., and O. E. Hood. **1911.** The structure of the cell wall and the apical growth in the genus Trentepohlia. *New Phytol.* **10:** 241–248. 6 *figs.*

West, G. S., and C. B. Starkey. **1915.** A contribution to the cytology and life-history of Zygnema ericetorum (Kütz.) Hansg., with some remarks on the "genus" Zygogonium. *New Phytol.* **14:** 194–205. 5 *figs.*

West, W. **1892.** Algae of the English Lake District. *Jour. Roy. Microsc. Soc. London* **1892:** 713–748. 2 *pl.*

West, W. and G. S. **1895.** New American algae. *Jour. Botany.* **33:** 52. **1895A.** On some North American Desmidiaceae. *Trans. Linn. Soc. Bot. London.* II. **5:** 229–274. 7 *pl.* **1896.** On some new and interesting fresh-water algae. *Jour. Roy. Microsc. Soc. London.* **1896:** 149–165. 2 *pl.* **1897.** Welwitsch's African fresh-water algae. *Jour. Botany.* **35:** 1–7, 33–42, 77–89, 113–122, 172–183. **1898.** Observations on the Conjugatae. *Ann. Bot.* **12:** 29–58. 2 *pl.* **1901.** Fresh-water Chlorophyceae. *Bot. Tidsskr.* **24:** 73–103. 3 *pl.* **1901A.** The alga-flora of Yorkshire: a complete account of the known fresh-water algae of the county. *Bot. Trans. Yorkshire Natural. Union.* **5:** 1–239. **1903.** Notes on fresh-water algae. III. *Jour. Botany.* **41:** 33–41, 74–82. 3 *pl.* **1903A.** Scottish fresh-water plankton. *Jour. Linn. Soc. Bot. London.* **35:** 519–556. 5 *pl.* **1904.** A monograph of the British Desmidiaceae. Vol. 1. London. 224 *pp.* 32 *pl.* **1905.** A further contribution to the fresh-water plankton of the Scottish lochs. *Trans. Roy. Soc. Edinburgh.* **41:** 477–518. 7 *pl.* **1905A.** A monograph of the British Desmidiaceae. Vol. 2. London. 204 *pp.* 32 *pl.* **1906.** A comparative study of the plankton of some Irish lakes. *Trans. Roy. Irish Acad.* **33,** Sec. *B*: 77–116. 6 *pl.* **1907.** Fresh-water algae from Burma, including a few from Bengal and Madras. *Ann. Roy. Bot. Garden Calcutta.* **6:** 175–260. 7 *pl.* **1908.** A monograph of the British Desmidiaceae. Vol. 3. London. 274 *pp.* 41 *pl.* **1909.** The British fresh-water phytoplankton, with special reference to the desmid-plankton and the distribution of British desmids. *Proc. Roy. Soc. London B.* **81:** 165–206. 6 *figs.* **1909A.** The phytoplankton of the English Lake District. *The Naturalist.* **1909:** 115–193, 260–331. 8 *figs.* 3 *pl.* **1912.** On the periodicity of the phytoplankton of some British lakes. *Jour. Linn. Soc. Bot. London.* **40:** 395–432. 4 *figs.* 1 *pl.* **1912A.** A monograph of the British Desmidiaceae. Vol. 4. London. 191 *pp.* 33 *pl.*

Wettstein, F. v. **1921.** Das Vorkommen von Chitin und seine Verwertung als systematisch-phylogenetisches Merkmal im Pflanzenrech. *Sitzungsber. Akad. Wiss. Wien.* (Mat.-Nat. Kl.). **130**[1]**:** 3–20.

Whelden, R. M. **1939.** Notes on New England algae. I. Cyclonexis and Actidesmium. *Rhodora.* **41:** 133–137. 7 *figs.* **1941.** Some observations on the fresh-water algae of Florida. *Jour. Elisha Mitchell Sci. Soc.* **57:** 261–272. 2 *pl.* **1943.** Notes on New England algae. III. Some interesting algae from Maine. *Farlowia.* **1:** 9–23. 1 *pl.*

Whipple, G. C. **1927.** The microscopy of drinking water. 4th ed. New York.

Whitford, L. A. **1938.** A new green alga, Oedocladium Lewisii. *Bull. Torrey Bot. Club.* **65:** 23–26. 1 *pl.* **1943.** The fresh-water algae of North Carolina. *Jour. Elisha Mitchell Sci. Soc.* **59:** 131–170. 1 *pl.* **1946.** Structure and composition of a recently described fresh-water alga; Phaeospora perforata (Chrysophyceae). *Amer. Jour. Bot.* **33:** 11*A*.

Wiedling, S. A skeleton-free diatom. *Bot. Notiser.* **1941:** 33–36. 2 *figs.*

Wille, N. **1881.** Om Hvileceller hos Conferva (L.). *Öfvers. Kgl. Svensk. Vetensk.-Ak. Förh.* **38,** No. 8: 1–26. 2 *pl.* **1883.** Ueber Akineten und Aplanosporen bei den Algen. *Bot. Centralbl.* **16:** 215–219. **1883A.** Om slagten Gongrosira Kütz. *Öfvers. Kgl. Svensk. Vetensk.-Ak. Förh.* 15 *pp.* 1 *pl.* (Ref. *Just's Bot. Jahresber.* **11**[1]: 270–271. 1885.) **1887.** Ueber die Schwärmzellen und deren Copulation bei Trenepohlia Mart. *Jahrb. Wiss. Bot.* **18:** 426–434. 1 *pl.* **1887A.** Ueber die Zelltheilung bei Oedogonium. *Ibid.* **18:** 443–454. 2 *pl.* **1887B.** Ueber die Gattung Gongrosira Kütz. *Ibid.* **18:** 484–491. 2 *pl.* **1890.** Conjugatae und

Chlorophyceae. In A. Engler and K. Prantl, Die natürlichen Pflanzenfamilien. **1**[2]: 1–161. 108 *figs.* **1897.** Om Faeröernes Ferkskvandsalger og om Ferskvandsalgernas Spredningsmaader. *Bot. Notiser.* **1897:** 1–32, 49–61. 1 *pl.* **1898.** Planktonalgen aus norwegischen Süsswasserseen. *Biol. Centralbl.* **18:** 302. **1899.** New forms of green algae. *Rhodora.* **1:** 149–150. **1900.** Asterocytis ramosa (Thw.) Gobi. *Nyt. Mag. Naturvidenskab.* **38:** 7–10. 1 *pl.* **1901.** Eine submarine Form von Prasiola crispa (Lightf.). *Vidensk. Selsk. Skr. Christiana.* (Math.-Nat. Kl.) **1900**[6]**:** 13–18. 1 *pl.* **1903.** Über Pteromonas nivalis (Shuttlw.) Chodat. *Nyt. Mag. Naturvidenskab.* **41:** 167–171. 1 *pl.* **1903A.** Über Cerasterias nivalis Bohlin. *Ibid.* **41:** 171–176. 1 *pl.* **1903B.** Über ene neue Art der Gattung Carteria Diesing. *Ibid.* **41:** 89–94. 1 *pl.* **1906.** Über die Zoosporen von Gomontia polyrrhiza (Lagerh.) Born. et Flah. *Vidensk. Selks. Skr. Christiana.* (Math.-Nat. Kl.) **1906**[3]**:** 29–33. 1 *pl.* **1906A.** Über die Entwicklung von Prasiola furfuraceae (Fl.D.) Menegh. *Ibid.* **1906**[3]**:** 1–12. 1 *pl.* **1908.** Zur Entwicklungsgeschichte der Gattung Oöcystis. *Ber. deutsch. Bot. Ges.* **26A:** 812–822. 1 *pl.* **1909.** Conjugatae und Chlorophyceae. In A. Engler and K. Prantl, Die natürlichen Pflanzenfamilien. Nachträge zum Teil 1, Abt. 2: 1–134. 70 *figs.* **1912.** Om Udviklingen af Ulothrix flaccida Kütz. *Svensk. Bot. Tidsskr.* **6:** 447–458. 1 *pl.* **1913.** Studien in Agardh's Herbarium. *Nyt. Mag. Naturvidenskab.* **51:** 1–20. 1 *pl.* **1918.** Über die Variabilität bei der Gattung Scenedesmus Meyen. *Ibid.* **56:** 1–22. 1 *pl.* **1918A.** Das Keimen der Aplanosporen bei der Gattung Coelastrum Naegl. *Ibid.* **56:** 23–27. 1 *pl.* **1922.** Phycoerythrin bei den Myxophyceen. *Ber. deutsch. Bot. Ges.* **40:** 188–192. 1 *fig.*

Williams, M. **1926.** Oögenesis and spermatogenesis in Vaucheria geminata. *Proc. Linn. Soc. New South Wales.* **51:** 283–295. 16 *figs.*

Winchell, A. N., and E. R. Miller. **1918.** The dustfall of March 9, 1918. *Amer. Jour. Sci.* IV. **46:** 599–609. 3 *figs.* **1922.** The great dustfall of March 19, 1920. *Ibid.* IV. **3:** 349–364. 1 *fig.*

Winston, J. R. **1938.** Algal fruit spot of orange. *Phytopath.* **28:** 283–286. 2 *figs.*

Wisselingh, C. van. **1908.** Über die Karyokinese bei Oedogonium. *Beih. Bot. Centralbl.* **23:** 137–156. 1 *pl.* **1908A.** Über den Ring und die Zellwand bei Oedogonium. *Ibid.* **23:** 157–190. 4 *pl.*

Wittrock, V. B. **1867.** Algologiska Studier. I. Om utvecklingen af Staurospermum punctatum nov. spec. Upsala. 67 *pp.* 2 *pl.* **1877.** On the development and systematic arrangement of the Pithophoraceae, a new order of the algae. *Nova Acta Reg. Soc. Sci. Upsaliensis.* III. Vol. extraord. 1–80 *pp.* 6 *pl.* **1886.** Om Binucleria, ett nytt Confervacé-Slägte. *Bih. Kgl. Svensk. Vetensk.-Ak. Handl.* **12,** Afd. 3, No. 1: 1–10. 1 *pl.*

Wolf, F. A. **1930.** A parasitic alga, Cephaleuros virsecens Kunze, on citrus and certain other fruits. *Jour. Elisha Mitchell Sci. Soc.* **45:** 187–205. 5 *pl.*

Wolle, F. **1884.** Fresh-water algae VIII. *Bull. Torrey Bot. Club.* **11:** 13–17. 1 *pl.* **1884A.** Desmids of the United States. Bethlehem, Pa. 168 *pp.* 53 *pl.* **1887.** Fresh-water algae of the United States. Bethlehem, Pennsylvania. 364 *pp.* 210 *pl.*

Wollenweber, W. **1909.** Untersuchungen über die Algengattung Haematoccus Berlin. 60 *pp.* 5 *pl.*

Woloszyńska, J. **1917.** Beitrag zur Kenntnis der Algenflora Litauens. *Bull. Acad. Sci. Cracovie. B.* **1917:** 123–130. 2 *figs.* 1 *pl.* **1917A.** Neue Peridineen-Arten nebst Bemerkung den Bau der Hülle bei Gymno- und Glenodinium. *Ibid. B.* **1917:** 114–122. 3 *pl.* **1919.** Die Algen der Tatraseen und -Tümpel. *Ibid.*

1918: 196–200. 1 *pl.* **1924.** Über die sogennanten "Schleimfäden" bei Gymnodinium fuscum. *Acata Soc. Bot. Poloniae.* **2,** No. 3: 1–4. 1 *pl.* **1925.** Beiträge zur Kenntnis der Süsswasser-Dinoflagellaten Polens. *Ibid.* **3,** No. 1: 1–16. 7 *figs.*

WOOD, H. C. **1872.** A contribution to the history of the fresh-water algae of North America. *Smithsonian Contributions to Knowledge.* **19,** No. 241: 1–262. 21 *pl.*

WOOD, R. D. **1947.** Characeae of the Put-in Bay Region of Lake Erie (Ohio). *Ohio Jour. Sci.* **47:** 240–258. 3 *pl.* **1948.** A review of the genus Nitella (Characeae) of North America. *Farlowia.* **3:** 331–398. 2 *pl.*

WORONIN, M. **1872.** Recherches sur les gonidies du lichen Parmelia pulverulenta Ach. *Ann. Sci. Nat. Bot.* V. **16:** 317–325, 1 *pl.*

WURDACK, M. E. **1923.** Chemical composition of the walls of certain algae. *Ohio Jour. Sci.* **23:** 181–191.

YABE, Y. **1932.** On the sexual reproduction of Prasiola japonica Yatabe. *Sci. Rept. Tokyo Bunrika Daigaku.* Sec. *B.* **1:** 39–40. 1 *pl.*

YAMADA, Y. **1932.** Notes on some Japanese algae. III. *Jour. Faculty Sci. Hokkaido Imp. Univ.* 5 Ser. **1:** 109–123. 5 *figs.* 5 *pl.*

YAMADA, Y., and T. KANDA. **1941.** On the culture experiment of Monostroma zostericola and Enteromorpha nana var. minima. *Sci. Papers Inst. Algol. Res. Hokkaido Imp. Univ.* **2:** 217–226. 8 *figs.* 4 *pl.*

YAMADA, Y., and E. SAITO. **1938.** On some culture experiments with the swarmers of certain species belonging to the Ulvaceae. *Sci. Papers Inst. Algol. Res. Hokkaido Imp. Univ.* **2:** 35–51. 12 *figs.* 1 *pl.*

ZACHARIAS, O. **1894.** Ueber den Bau der Monaden und Familiestöcke von Uroglena Volvox Ehrb. *Zool. Anz.* **17:** 353–356.

ZEDERBAUER, E. **1904.** Geschlechtliche und ungeschlechtliche Fortpflanzung von Ceratium hirundinella. *Ber. deutsch. Bot. Ges.* **22:** 1–8. 1 *pl.*

ZIMMERMANN, W. **1921.** Zur Entwicklungsgeschichte und Zytologie von Volvox. *Juhrb. Wiss. Bot.* **60:** 256–294. 2 *figs.* 1 *pl.* **1925.** Die ungeschlechtliche Entwicklung von Volvox. *Naturwissensch.* **13:** 397–402. 3 *figs.* **1925A.** Helgoländer Meeresalgen. I–VI. Beiträge zur Morphologie, Physiologie und Oekologie der Algen. *Wissencsh. Meeresuntersuch.* N.F. *Abt. Helgoland.* **16:** 1–25. 1 *pl.* **1927.** Ueber Algenbestände aus der Tiefzone des Bodensees. Zur Ökilogie und Soziologie der Tiefseepflanzen. *Zeitschr. Bot.* **20:** 1–35. 5 *figs.* 2 *pl.*

INDEX

Page references in **boldface** refer to pages on which entries are defined, described or illustrated. Generic names printed in *italics* are synonyms.

D

I

J

K

L

T